W9-CGY-021

GEOMETRY

By the permission of Johnny Hart and News America Syndicate

SECOND EDITION

GEOMETRY

HAROLD R. JACOBS

W. H. FREEMAN AND COMPANY
New York

The covers depict a polyhedron composed of five intersecting cubes. The computer-generated figure on the front cover was created by Thomas Banchoff and Paul Strasse of Brown University Graphics Group. The photograph on the back cover is of a model constructed by the author.

Library of Congress Cataloging in Publication Data

Jacobs, Harold R.
Geometry.

Includes index.
1. Geometry. I. Title.
QA453.J26 1987 516'.2 85–7034
ISBN 0–7167–1745–X

Printed in the United States of America

Ninth printing, 2002, RRD

Contents

3

RAYS AND ANGLES 83

4

CONGRUENT TRIANGLES 119

5

INEQUALITIES 167

6

PARALLEL LINES 191

7

QUADRILATERALS 235

8

TRANSFORMATIONS 269

15

GEOMETRIC SOLIDS 535

16

NON-EUCLIDEAN GEOMETRIES 579

17

COORDINATE GEOMETRY 605

A Letter to the Student

Pythagoras was a Greek geometer who lived about 2500 years ago. He wondered whether he could teach geometry even to a reluctant student. After finding such a student, Pythagoras agreed to pay him an *obel* for each theorem that he learned. Because the student was very poor, he worked diligently. After a time, however, the student realized that he had become more interested in geometry than in the money he was accumulating. In fact, he became so intrigued with his studies that he begged Pythagoras to go faster, now offering to pay him back an *obel* for each new theorem. Eventually, Pythagoras got all of his money back!

What is it that could have so fascinated the student? Perhaps it was the logic of geometry; for geometry was the first system of ideas to be developed in which a few simple statements were assumed and then used to derive a rich and attractive array of results. Such a system is called *deductive*. The beauty of geometry as a deductive system has inspired writers in other fields to organize their ideas in the same way. Sir Isaac Newton's *Principia*, in which he tried to present physics as a deductive system, and the philosopher Spinoza's *Ethics* are especially noteworthy examples.

Geometry is also fascinating and useful because of its wide range of applications. We are surrounded by objects with pleasing or useful shapes. We see arcs of circles in rainbows, hexagons in honeycombs, cubes in salt crystals, and spheres in soap bubbles. Biochemists have learned that shape is essential to function; the double helix of the DNA molecule is a well-known example. Architects have made use of a wide variety of geometric shapes—from the pyramids of Egypt and the tepees of the American Indians to the complex skyscrapers of the twentieth century. In astronomy, recognizing that we live on a spherical earth and determining its size is a geometrical problem. Eratosthenes used simple geometry to do this in 300 B.C. Since his time, astronomers have applied geometry to determine the orbits of the planets, the positions and motions of the stars, and even the size of the universe.

The word *geometry* came from the ancient Greeks, and it was one of them, Euclid, who organized the ideas that we will study. We will also consider some interesting contributions to the subject that were made in India during the Dark Ages, when European scholarship was almost nonexistent, and in Europe during the Renaissance, when the pursuit of knowledge was revived. And we will briefly survey the "non-Euclidean" geometries developed in the nineteenth century and learn how Einstein used them in his theory of the nature of space.

As you embark on your study of geometry, you may wonder how successful you will be in your efforts. Because geometry is a logical system, it is necessary to spend time on a regular basis to master the basic ideas contained within it. A positive attitude and a willingness to try to do your best are also important factors in determining how you will do. I wrote this book to help my students and others discover the beauty and excitement of geometry that Pythagoras's student found. I hope that it will enable you to find your studies an enjoyable and rewarding endeavor.

Harold Jacobs

Roy Bishop

Acknowledgments

I want to thank the countless teachers and students whose suggestions and comments over the ten years since publication of the First Edition of *Geometry* have helped shape this revision. The long span of time and the number of people involved make it impossible for me to individually acknowledge these contributions. Many teachers sent comments to me and others made helpful suggestions during conversations at mathematics meetings.

The users of the First Edition have, in effect, provided me with the results of a very large-scale class testing program, involving tens of thousands of teachers and over a million students. I have put their ideas to work in my revision, and I can only hope that users of the Second Edition will be as free with their comments, criticism, and praise as were the users of the First Edition.

In the three years when I was rewriting *Geometry,* a much smaller number of colleagues gave detailed comments on the work or class tested the various drafts of this revision. For either or both, I am grateful to: Steven Bergen, Richard Brady, Don Chakerian, Barbara Fracassa, Gary Froelich, Akiba Harris, Hector Hirigoyen, Geoffrey Hirsch, Gayle Machado, Mel Noble, Jack Patton, Linda Rasmussen, Sydney Shanks, Bill Shutters, Mel Stave, and Sharon Tello.

Old Euclid drew a circle
On a sand-beach long ago.
He bounded and enclosed it
With angles thus and so.
His set of solemn graybeards
Nodded and argued much
Of arc and of circumference,
Diameters and such.
A silent child stood by them
From morning until noon
Because they drew such charming
Round pictures of the moon.

VACHEL LINDSAY

INTRODUCTION

Euclid, the Surfer, and the Spotter

In about 300 B.C., a man named Euclid wrote what has become one of the most successful books of all time. Euclid taught at the university at Alexandria, the main seaport of Egypt, and his book, the *Elements*, contained much of the mathematics then known. Its fame was almost immediate. Since Euclid's time, the *Elements* has been translated into more languages and published in more editions than any other book except the Bible.

Although very little of the mathematics in the *Elements* was original, what made the book unique was its logical organization of the subject, beginning with a few simple principles and deriving from them everything else.

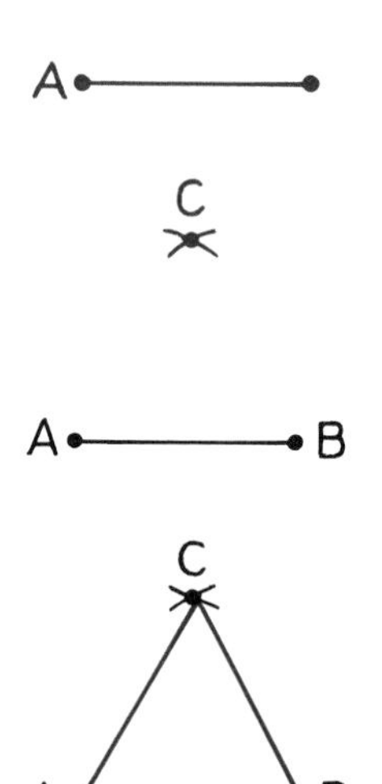

The *Elements* begins with an explanation of how to draw a triangle with three sides of equal length. Euclid's method requires the use of two tools: a straightedge for drawing straight lines and a compass for drawing circles. These tools have been used ever since in making geometric drawings called *constructions.*

A triangle having three sides of equal length is called *equilateral.* To construct an equilateral triangle, we begin by using the straightedge to draw a segment for one side. The segment is named AB in the figure at the left above. Next, we adjust the radius of the compass (the distance between pencil point and metal point) so that it is equal to the length of the segment. We draw two arcs having this radius and A and B as their centers so that they intersect as shown in the second figure. Finally, we draw two line segments from the point of intersection (labeled C in the figure) to points A and B to form the triangle.

Exercises

1. Use a straightedge and a compass to construct an equilateral triangle each of whose sides is 12 centimeters long.

Next, we will consider a couple of geometric problems about equilateral triangles. To make them easier to understand, they will be presented in the form of a story.

The Puzzles of the Surfer and the Spotter

One night a ship is wrecked in a storm at sea and only two members of the crew survive. They manage to swim to a deserted tropical island where they fall asleep exhausted. After exploring the island the next morning, one of the men decides that he would like to spend some time there surfing on the beaches. The other man, however, wants to escape and decides to use his time looking for a ship that might rescue him.

The island is overgrown with vegetation and happens to be in the shape of an equilateral triangle, each side being 12 kilometers (about 7.5 miles) long.

Wanting to be in the best possible position to spot any ship that might sail by, the man who hopes to escape (we will call him the "spotter") goes to one of the corners of the island. Because he doesn't know which corner is best, he decides to rotate from one to another, spending a day on each. He wants to build a shelter somewhere on the island and a path from it to each corner so that the sum of the lengths of the three paths is a minimum. (Digging up the vegetation to clear the paths is not an easy job.) Where should the spotter build his house?

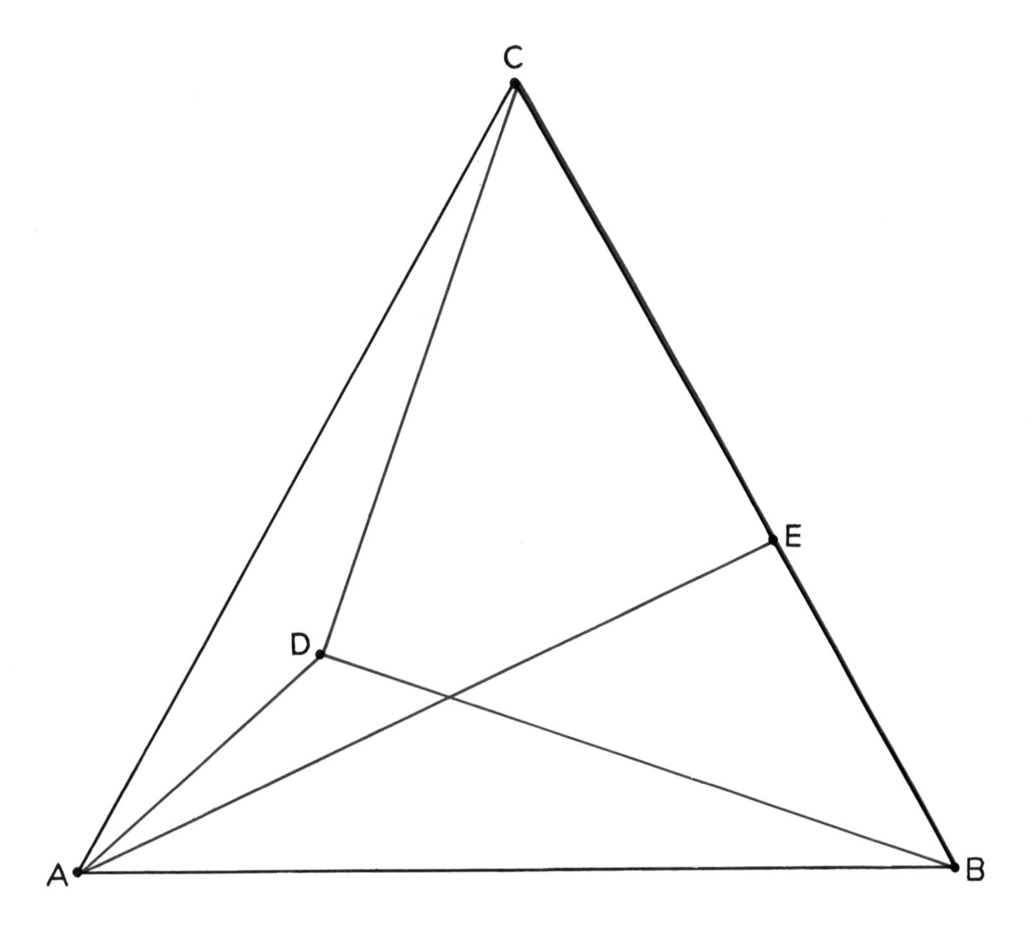

The figure above is a scale drawing of the island in which 1 centimeter (cm) represents 1 kilometer (km). Suppose that the spotter builds his house at point D. The three paths that he has to clear have the following lengths: DA = 4.4 km, DB = 9.1 km, and DC = 8.0 km. Check these measurements with your ruler, remembering that 1 cm represents 1 km. The sum of these lengths is 21.5 km.

If the spotter builds his house at point E, the path lengths are: EA = 10.5 km, EB = 5.0 km, and EC = 7.0 km, and their sum is 22.5 km. So point D is a better place for him to build than point E. But where is the best place?

2. Use the equilateral triangle that you drew in the first exercise to represent the island. Choose several different points on it; for each point, measure the distance between it and each of the corners to the nearest 0.1 cm, and find their sum as illustrated for points D and E above.

3. On the basis of your work in Exercise 2, where do you think is the best place for the spotter to build his house? Also, how many kilometers of path does he have to clear? (Remember that 1 cm on your map represents 1 km.)

4. Where do you think is the *worst* place on the island for the spotter to locate? How many kilometers of path would he have to clear from it?

Now consider the problem of where the surfer should build *his* house. He likes the beaches along all three sides of the island and decides to spend an equal amount of time on each. To make the paths from his house to each beach as short as possible, he

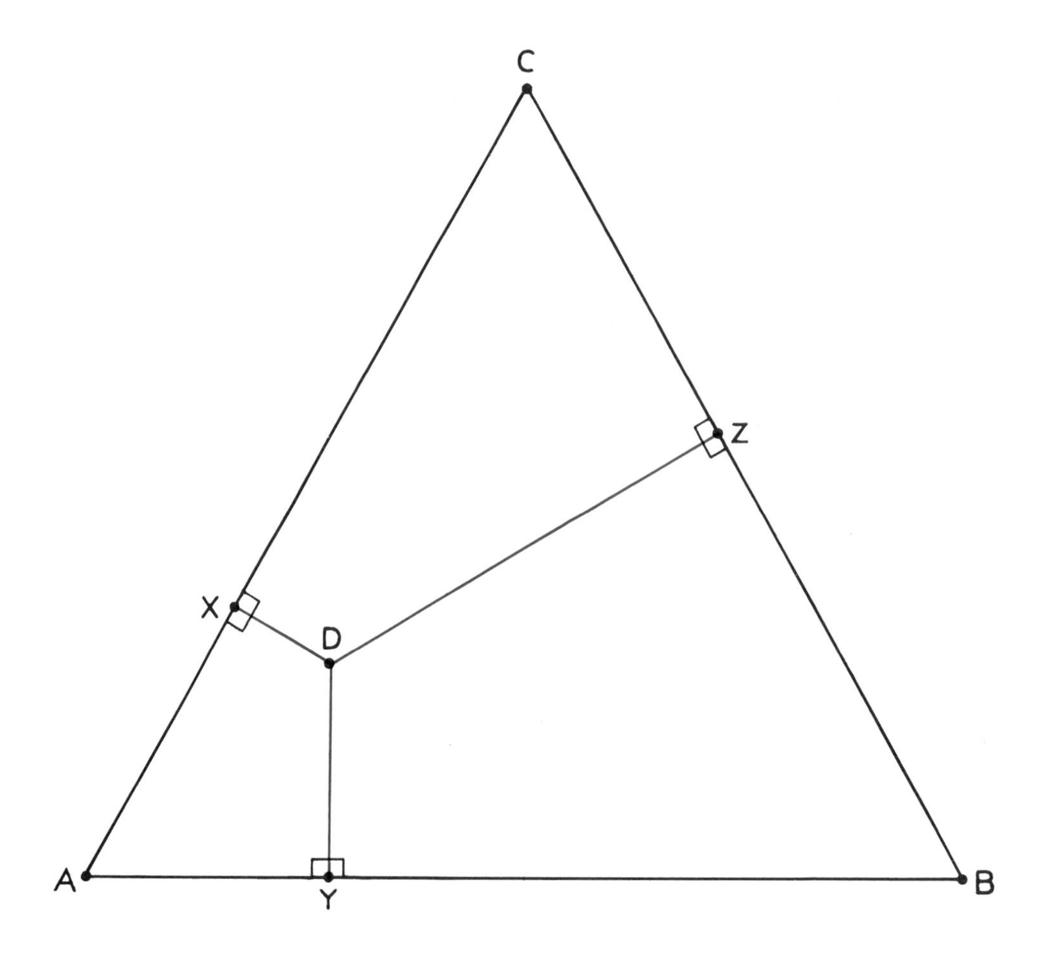

constructs them so that they are perpendicular to the lines of the beaches. For example, in the figure above, if the surfer built his house at point D, the three paths to the beaches would be as shown. Path DX is perpendicular* to beach AC, path DY is perpendicular to beach AB, and path DZ is perpendicular to beach BC. The lengths of the three paths are: DX = 1.5 km, DY = 2.8 km, and DZ = 6.1 km, so their sum is 10.4 km.

The surfer, like the spotter, wants to locate his house so that the sum of the lengths of the paths is a minimum. Where is the best place on the island for him?

5. Use a straightedge and compass to construct another equilateral triangle whose sides are 12 centimeters long. Choose several different points on it; for each point, measure the perpendicular distance from it to each of the sides to the nearest 0.1 cm, and find their sum as illustrated for point D above.
6. On the basis of your work in Exercise 5, where do you think is the best place for the surfer to build his house? Also, how many kilometers of path does he have to clear?
7. Where do you think is the worst place for the surfer to locate? How many kilometers of path would he have to clear from it?

*Two lines that are perpendicular form right angles and these angles are shown by the small squares at the points of intersection, X, Y, and Z.

Chapter 1

THE NATURE OF DEDUCTIVE REASONING

Drawing by Tom Henderson; reprinted by the permission of the American Legion Magazine

"Help yourselves to the spaghetti, folks. Peg will serve the meatballs."

Lesson 1
Drawing Conclusions

One goal in studying geometry is to develop the ability to think critically. An understanding of the methods of deductive reasoning is fundamental in the development of critical thinking, and so it is to this subject that we first turn our attention.

Misunderstandings are very common in everyday life. Like the couple in this cartoon, who have been told what is going to happen next but who are nevertheless probably in for a surprise, we often draw conclusions that are incorrect or at least questionable.

Exercises

Passages from several books are quoted below. Each is followed by a series of statements that are conclusions that might be drawn on the basis of accepting all of the information in the passage as literally true. Some of these conclusions are true, some are false, and some are questionable—that is, from the information in the passage, it cannot be definitely determined whether they are true or false.

Write the numbers of the statements on your paper and mark each "true," "false," or "not certain." Where you feel that someone might not agree with you, briefly explain the basis for your answer.

Set I

"Mother used to send a box of candy every Christmas to the people the Airedale bit. The list finally contained forty or more names. Nobody could understand why we didn't get rid of the dog. I didn't understand it very well myself, but we didn't get rid of him. I think that one or two people tried to poison Muggs—he acted poisoned once in a while—and old Major Moberly fired at him once with his service revolver near the Seneca Hotel in East Broad Street—but Muggs lived to be almost eleven years old and even when he could hardly get around he bit a

Congressman who had called to see my father on business."

JAMES THURBER, *My Life and Hard Times*

1. Muggs was an Airedale dog.
2. People who received boxes of candy from Mother at Christmas had been bitten by the dog.
3. At least one person had tried to poison Muggs.
4. The dog had bitten at least forty people.
5. Mother couldn't understand why we didn't get rid of Muggs.
6. Muggs tried to bite Major Moberly.
7. Major Moberly tried to kill Muggs.
8. Major Moberly missed Muggs when he fired at him.
9. The Seneca Hotel was on East Broad Street.
10. Muggs was eleven years old when he died.
11. Muggs died of old age.

"'In answer to your question what we got out of English so far I am answering that so far I got without a doubt nothing out of English. Teachers were sourcastic sourpuses or nervous wrecks. Half the time they were from other subjects or only subs . . ..

"'Also no place to learn. Last term we had no desks to write only wet slabs from the fawcets because our English was in the Science Lab and before that we had no chairs because of being held in Gym where we had to squatt.

"'Even the regulars Mrs. Lewis made it so boreing I wore myself out yawning, and Mr. Loomis (a Math) hated teaching and us.'"

BEL KAUFMAN, *Up the Down Staircase*

12. The student writing this essay dislikes English.
13. The student writing this essay answered the question concerning what he or she had learned from his or her English class.
14. There are several spelling mistakes in this essay.
15. The student writing this essay is a poor speller.
16. The faucets in the science room leaked.
17. The gym had no bleachers because the students had to squat on the floor.
18. The school is overcrowded.
19. Mrs. Lewis was not one of the substitute teachers.
20. Mr. Loomis would rather teach mathematics than English.

Set II

"'Excellent!' said Gandalf, as he stepped from behind a tree, and helped Bilbo to climb out of a thornbush. Then Bilbo understood. It was the wizard's voice that had kept the trolls bickering and quarrelling, until the light came and made an end of them.

"The next thing was to untie the sacks and let out the dwarves. They were nearly suffocated, and very annoyed: they had not at all enjoyed lying there listening to the trolls making plans for roasting them and squashing them and mincing them. They had to hear Bilbo's account of what had happened to him twice over, before they were satisfied."

J. R. R. TOLKIEN, *The Hobbit*

21. Dwarves are not very large because they will fit inside sacks.
22. The trolls had been arguing with each other.
23. Gandalf was a wizard.
24. We know that the trolls had planned to eat the dwarves.
25. Trolls are fond of mince pie.
26. Bilbo had climbed into a thornbush.
27. Dwarves don't carry knives or else they would have cut their way out of the sacks.
28. Gandalf and Bilbo let the dwarves out of the sacks.
29. The dwarves hadn't been able to hear anything while they were tied up in the sacks.
30. Gandalf had been hiding behind a tree.
31. Maybe the dwarves deserved to have a good fright.
32. There were seven dwarves.
33. The sunlight changed the trolls into stone.
34. Gandalf fooled the trolls.
35. Bilbo knew all along what had been happening.
36. Bilbo told the dwarves what had happened to him more than once.

Set III

Study the following photograph carefully.
Try to form some conclusions that you think seem reasonable.

Carl Nessensohn, Wide World Photos, Inc.

Courtesy of Avis Rent A Car System, Inc.

Courtesy of Columbia Broadcasting System

Lesson 2

Conditional Statements

The advertisements shown above and on the next page have something in common: each has a headline that begins with the word "if." Statements consisting of two clauses, one of which begins with the word "if" or "when" or some equivalent word, are called *conditional statements*. Such statements are often used when the purpose is to establish certain conclusions, and so they are very common in the field of advertising. They are also important in mathematics in writing deductive proofs.

A conditional statement can be represented symbolically by

"If a, then b,"

or, even more briefly, by

$$a \rightarrow b.$$

The letter a represents the "if" clause, or *hypothesis*, and the letter b represents the "then" clause, or *conclusion*. (The word "then," being

Courtesy of Qantas—Australia's round-the-world airline

understood, is usually omitted.) The symbols

$$a \rightarrow b$$

are read as "if a, then b," or as "a implies b." For example, in the first advertisement, a represents the words "Avis is out of cars" and b represents the words "we'll get you one from our competition." In the second advertisement, a stands for "life is discovered on Mars" and b for "it will come as news to you."

It is helpful in learning how to relate two or more conditional statements to each other to represent them with circle diagrams. These are often called *Euler diagrams* after an eighteenth-century Swiss mathematician, Leonhard Euler, who first used them.

To represent a conditional statement with an Euler diagram, we draw two circles, one inside the other. The interior of the smaller circle represents a, the hypothesis, and the interior of the larger circle represents b, the conclusion. Notice that, if a point is inside circle a, it is also inside circle b. Or, more briefly, "if a, then b," which is what the diagram is intended to represent. The headline of the advertisement above, represented by an Euler diagram, looks like the figure shown at the top of the next page.

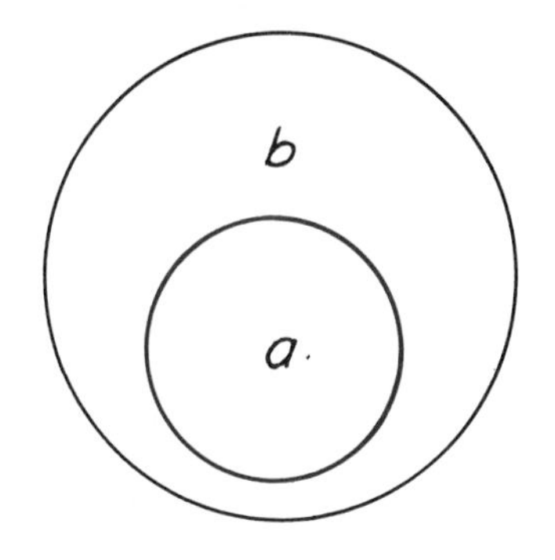

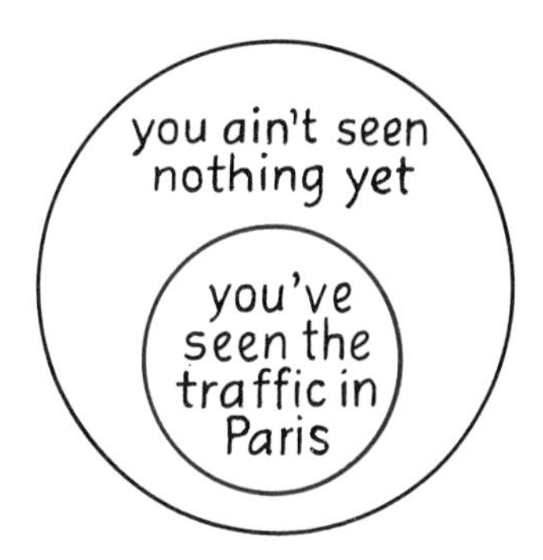

Exercises

Set I

1. What is a statement that can be represented symbolically by "If *a*, then *b*" called?

2. In the statement "If *a*, then *b*," what do the letters "*a*" and "*b*" represent?

3. What does an Euler diagram consist of?

The following exercises refer to this statement:

If you live in the Ozarks,
then you live in the United States.

4. What is the hypothesis of this statement?

5. What is its conclusion?

6. Rewrite the statement in the form "*b* if *a*."

The following exercises refer to this statement:

It is cold outside if it is snowing.

7. What is the hypothesis of this statement?

8. What is its conclusion?

9. Rewrite the statement in the form "If *a*, then *b*."

Rewrite each of the following sentences in "if-then" form. Be careful not to change the meanings of any of the sentences; for example, if you write a true statement in "if-then" form so that it turns out to be false, something is wrong.

Example: No ghost has a shadow.

Possible answers:

If a creature is a ghost,
it does not have a shadow.
If you are a ghost,
then you do not have a shadow.

10. All leap years have 366 days.

11. Koala bears eat nothing but eucalyptus leaves.

12. When the cat is in the birdcage, it isn't there to sing.

13. Licorice ice cream has a peculiar color.

14. Smokey the Bear wouldn't have to do commercials for a living if money grew on trees.

15. A heavy object stored in the attic of a jungle mansion may crash down on the occupants.

16. People who live in grass houses shouldn't stow thrones.

17. Use the stairs instead of the elevator in case of fire.

18. No genuine phone number begins with 555.

Set II

19. Write the conditional statement represented by this Euler diagram.

20. Rewrite the conditional statement you have just written so that the conclusion appears first.

An Euler diagram, such as the one shown below, separates the paper into three regions: they are numbered 1, 2, and 3 in the figure.

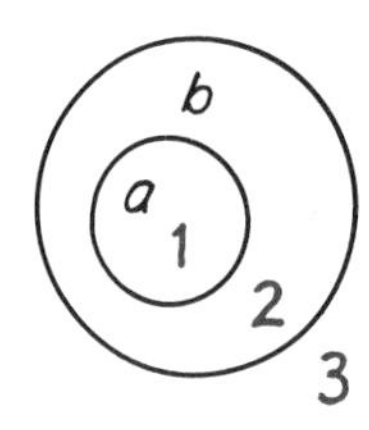

21. Which region is inside circle a?
22. Which regions are inside circle b?
23. Which regions are outside circle a?
24. Which region is outside circle b?
25. Draw an Euler diagram to represent the statement

    If x, then y

    and a second Euler diagram to represent the statement

    If y, then x.

26. Which of your diagrams, the first or the second, also illustrates the statement

    If not x, then not y?

27. What other "if-then" statement does your other diagram illustrate?
28. How many "if-then" statements does an Euler diagram represent?
29. Draw an Euler diagram to represent the following statement:

    All athletes who compete
    in the Olympics are amateurs.

30. Which of the following statements are illustrated by your diagram?
    a) If an athlete competes in the Olympics, then the athlete is an amateur.
    b) If an athlete is an amateur, then the athlete competes in the Olympics.
    c) If an athlete does not compete in the Olympics, then the athlete is not an amateur.
    d) If an athlete is not an amateur, then the athlete does not compete in the Olympics.
31. The statement "All athletes who compete in the Olympics are amateurs" is true. Which of the statements listed in Exercise 30 are true?
32. Draw an Euler diagram to represent the statement:

    If you live in Texas,
    you do not have to pay
    any state income tax.

33. Which of the following statements are illustrated by your diagram?
    a) If you do not have to pay any state income tax, then you live in Texas.
    b) If you do not live in Texas, then you have to pay state income tax.
    c) If you have to pay state income tax, then you do not live in Texas.
    d) You do not have to pay any state income tax if you live in Texas.
34. The statement "If you live in Texas, you do not have to pay any state income tax" is true. Which of the statements listed in Exercise 33 are true?

Set III

By the permission of Johnny Hart and News America Syndicate

Peter, the fellow on the wheel in this cartoon, has made a remark in the form of a conditional statement about God wanting B.C. to fly.

1. What is the statement?

A similar conditional statement about Peter riding on a wheel is implied by B.C.'s remark about the bug screen.

2. What is it?

3. What idea follows logically from this statement and the fact that God did not give Peter a bug screen?

Lesson 3

Equivalent Statements

Lewis Carroll, the author of *Alice's Adventures in Wonderland* and *Through the Looking Glass,* was a mathematics teacher who wrote stories as a hobby. His books contain many amusing examples of both good and deliberately poor logic and, as a result, have long been favorites among mathematicians. Consider the following conversation held at the Mad Hatter's Tea Party.

"Then you should say what you mean," the March Hare went on.
"I do," Alice hastily replied; "at least—at least I mean what I say—that's the same thing, you know."
"Not the same thing a bit!" said the Hatter. "Why, you might just as well say that 'I see what I eat' is the same thing as 'I eat what I see'!"
"You might just as well say," added the March Hare, "that 'I like what I get' is the same thing as 'I get what I like'!"
"You might just as well say," added the Dormouse, who seemed to be talking in his sleep, "that 'I breathe when I sleep' is the same thing as 'I sleep when I breathe'!"
"It *is* the same thing with you," said the Hatter, and here the conversation dropped, and the party sat still for a minute.

Carroll is playing here with pairs of related statements, and the Hatter, the Hare, and the Dormouse are right: the sentences in each pair do not say the same thing at all. Consider the Dormouse's example. If we change his two statements into "if-then" form, we get

If I sleep, then I breathe,

and

If I breathe, then I sleep.

Although both statements may be true of the Dormouse, the first statement is true and the second statement is false for ordinary beings.

Notice that the hypothesis, "I sleep," and the conclusion, "I breathe," of the first statement are interchanged in the second. The second statement is called the *converse* of the first.

► The *converse* of a conditional statement is formed by interchanging its hypothesis and conclusion. In symbols, the converse of $a \rightarrow b$ is $b \rightarrow a$.

The converse of a true statement may be false. It is also possible that it may be true, but in either case a statement and its converse do not have the same meaning. In other words, accepting a statement as true does not require us to accept its converse as true.

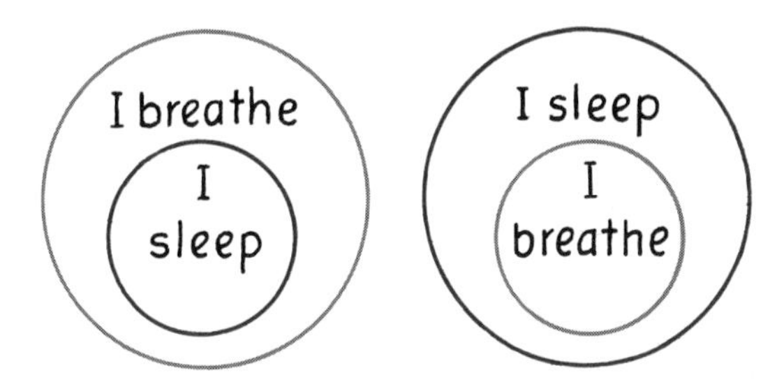

Euler diagrams for the Dormouse's sentences are shown here. Notice that, if a point is not inside the larger circle in one of these diagrams, then it is not inside the smaller circle. For this reason, each diagram illustrates *two* statements. The first one represents not only

If I sleep, then I breathe,

but also

If I do not breathe, then I do not sleep.

The second of these statements is called the *contrapositive* of the first.

► The *contrapositive* of a conditional statement is formed by interchanging its hypothesis and conclusion and denying both. In symbols, the contrapositive of $a \rightarrow b$ is not $b \rightarrow$ not a.

Because both a statement and its contrapositive are represented by the same diagram, they are said to be *logically equivalent.* One cannot be true and the other false; they are either both true or both false.

The second diagram represents not only

If I breathe, then I sleep,

but also

If I do not sleep, then I do not breathe.

Because both statements are represented by the same diagram, they are also logically equivalent. The first is the converse of the Dormouse's original statement and the second is its *inverse.*

► The *inverse* of a conditional statement is formed by denying both its hypothesis and conclusion. In symbols, the inverse of $a \rightarrow b$ is not $a \rightarrow$ not b.

The relations between statements considered in this lesson are summarized below.

logically equivalent	A conditional statement: $a \rightarrow b$ The contrapositive of the statement: not $b \rightarrow$ not a
logically equivalent	The converse of the statement: $b \rightarrow a$ The inverse of the statement: not $a \rightarrow$ not b

Exercises

Set I

How are each of the following statements formed?

1. The converse of a conditional statement.
2. The inverse of a conditional statement.
3. The contrapositive of a conditional statement.
4. Which of the following statements are illustrated by this Euler diagram?

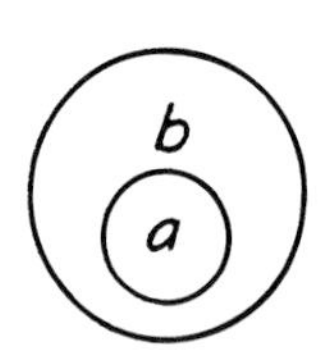

a) $a \rightarrow b$
b) $b \rightarrow a$
c) not $a \rightarrow$ not b
d) not $b \rightarrow$ not a

5. What fact does your answer to Exercise 4 illustrate about a conditional statement and its contrapositive?
6. Write in symbols the two conditional statements illustrated by this Euler diagram.

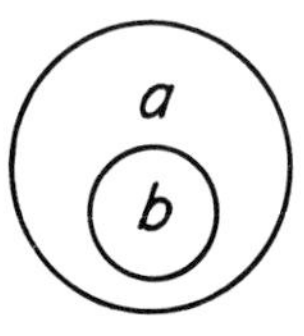

7. What fact does your answer to Exercise 6 illustrate about the converse and the inverse of a conditional statement?

Consider the statement

If your temperature is more than 102°, then you have a fever.

8. Is this statement true?
9. If the statement is represented by the symbols $a \rightarrow b$, what words do a and b represent?
10. Write the statement that is represented by not $a \rightarrow$ not b as a complete sentence.

11. Is it true?
12. What is it called with respect to the original statement?
13. Does it have the same meaning as the original statement?
14. Write the statement that is represented by $b \to a$ as a complete sentence.
15. Is it true?
16. What is it called with respect to the original statement?
17. Does it have the same meaning as the original statement?

The following statement is true:

If you are a U.S. astronaut,
you are not more than six feet tall.

18. Write the converse of this statement.
19. Is it true?
20. Write the inverse of this statement.
21. Is it true?
22. Write the contrapositive of this statement.
23. Is it true?

If a conditional statement is true, can its

24. converse be false?
25. inverse be false?
26. contrapositive be false?

Set II

Each of the lettered statements below is followed by some other statements. Identify the relation of each of them to the lettered statement if possible. Write "converse," "inverse," "contrapositive," or "original statement," as appropriate.

Statement A: If you live in Atlantis, then you need a snorkel.

Example: If you need a snorkel, then you live in Atlantis.

Answer: Converse.

27. If you do not live in Atlantis, then you do not need a snorkel.
28. If you do not need a snorkel, then you do not live in Atlantis.
29. You need a snorkel if you live in Atlantis.

Statement B: If you are older than ninety, the Chop Chop Studio will give you free karate lessons.

30. If the Chop Chop Studio won't give you free karate lessons, then you aren't older than ninety.
31. If you are ninety or younger, the Chop Chop Studio will not give you free karate lessons.

Statement C: All St. Louis Cardinals catch flies.

32. If someone is a St. Louis Cardinal, he catches flies.
33. If someone catches flies, he is a St. Louis Cardinal.
34. If someone is not a St. Louis Cardinal, he does not catch flies.

Statement D: Lady kangaroos do not need handbags.

35. If a kangaroo is not a lady, it needs a handbag.
36. If it needs a handbag, then it is not a lady kangaroo.

Write the indicated statement for each of the following sentences.

37. If the moon is full, the vampires are out. *Converse.*

38. If a giraffe has a sore throat, then gargling doesn't help much. *Contrapositive.*

39. If we have been receiving signals from Jupiter, it may not be wise to go there. *Inverse.*

40. You cannot comprehend geometry if you do not know how to reason deductively. *Converse.*

Set III

The advertising slogan

"When you're out of Schlitz, you're out of beer"

is a rather unusual one.

1. Is it necessarily true?

Write each of the following statements and tell whether it is true or false.

2. Its converse.

3. Its inverse.

4. Its contrapositive.

Broom-Hilda © 1973 The Chicago Tribune

Lesson 4
Definitions

The word "duck" has more than one meaning, as the behavior of the characters in this cartoon clearly demonstrates. Whenever a word has more than one meaning, we can usually tell from the context in which it is being used which meaning is intended. Even words that have just one definition in a dictionary may mean slightly different things to different people. These different meanings depend on each person's past experiences with the word and, if the interpretations differ enough, misunderstandings or even disagreements may result. Variations in a word's meaning seldom cause much of a problem in everyday communication, but they are very serious if we are trying to reason in a precise way. For this reason, definitions play an important role in mathematics.

When we define a word in geometry, the word and its definition are understood to have exactly the same meaning. For example, if we define the word "duck" as "an aquatic bird having a flat bill and webbed feet," we can say not only that

> If an animal is a duck, then it is an aquatic bird having a flat bill and webbed feet

but also that

> If an animal is an aquatic bird having a flat bill and webbed feet, then it is a duck.

Notice that each of these statements is the converse of the other. You know that, in general, the converse of a true statement is not necessarily true. It is a consequence of the nature of a definition, however, that its *converse is always true.*

The reason that the converse of a definition is always true is that because "a", the word being defined, and "b", its definition, have the same meaning, "a" and "b" are interchangeable. Consequently, every definition can be written as either $a \to b$ or $b \to a$.

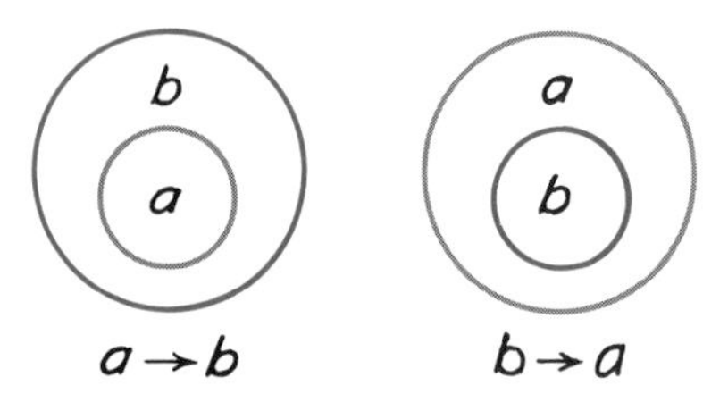

The two Euler diagrams shown here represent these two ways. Notice that the second diagram also illustrates the statement "a, if b" because in it a point is inside circle a if it is inside circle b. The first diagram also illustrates the statement "a only if b" because in it a point is inside circle a only if it is inside circle b. These observations indicate that if both a statement and its converse are true, we can write

"a if b" *and* "a only if b",

or more briefly,

"a if and only if b".

The phrase "if and only if" is represented by the abbreviation "iff" and by the symbol $\leftrightarrow$; $a \leftrightarrow b$ means both $a \to b$ and $b \to a$.

Because the converse of every definition is true, we usually write definitions in the form "a iff b." For example, the definition of duck given in this lesson can be written as

An animal is a duck if and only if it is an aquatic bird having a flat bill and webbed feet.

Exercises

Set I

1. When a word is defined in geometry, what is understood to be true about the word and its definition?
2. Is the converse of a true statement necessarily true?
3. Is the converse of a definition necessarily true?
4. What does the abbreviation "iff" stand for?
5. Write in symbols the two statements represented by $a \leftrightarrow b$.

Decide which of the following statements are good definitions of the italicized words by determining whether or not their converses are true.

Example: If it is *New Year's Day,* then it is January 1.

Answer: A good definition because, if it is January 1, it is New Year's Day.

6. If it is *New Year's Day,* then it is a holiday.
7. A *camera* is a device for taking pictures.
8. A *skunk* is an animal that has black and white fur.
9. *Dry ice* is frozen carbon dioxide.
10. A *ukulele* is a musical instrument that has four strings.
11. An object is *fragile* if it is easily broken or damaged.
12. The following statement is a definition of *extraterrestrial creature:*

    An extraterrestrial creature is a being from a place other than the earth.

    Which of the following statements must be true?
    a) If a creature is extraterrestrial, then it is a being from a place other than the earth.
    b) If a creature is a being from a place other than the earth, then it is extraterrestrial.
    c) If a creature is not extraterrestrial, then it is not a being from a place other than the earth.
    d) If a creature is not a being from a place other than the earth, then it is not extraterrestrial.

Set II

Compare the following two sentences:

If it is your birthday,
then you get some presents.
Only if it is your birthday,
do you get some presents.

13. Is the first sentence true for you?
14. Is the second sentence true for you?
15. Do both sentences say the same thing?

The following questions are about this Euler diagram. Is it necessarily true that

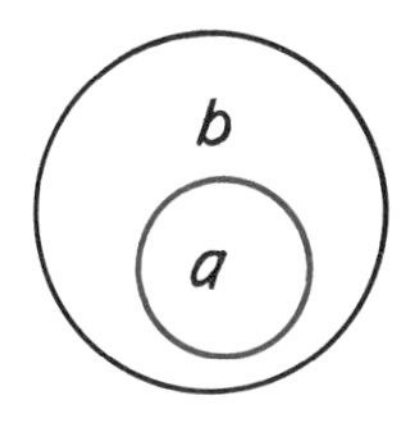

16. a point is inside circle *a* if it is inside circle *b*?
17. a point is inside circle *a* only if it is inside circle *b*?
18. a point is inside circle *b* if it is inside circle *a*?
19. a point is inside circle *b* only if it is inside circle *a*?
20. Which of the following statements does the Euler diagram represent?
    a) *a* if *b*
    b) *a* only if *b*
    c) *b* if *a*
    d) *b* only if *a*
21. The following statement is a rule of water polo:

    The ball is out of play when it completely crosses the goal line.

    Which of the following statements must be true?
    a) If the ball completely crosses the goal line, then it is out of play.
    b) The ball is out of play if it has completely crossed the goal line.
    c) The ball is out of play only if it has completely crossed the goal line.

d) If the ball is out of play, then it has completely crossed the goal line.

22. The following statement is a definition of *marzipan:*

    Marzipan is molded candy made from almonds.

    Which of the following statements must be true?
    a) If something is marzipan, it is molded candy made from almonds.
    b) Something is marzipan only if it is molded candy made from almonds.

23. Write the two conditional statements that are equivalent to the statement:

    It is a whodunit iff it is a detective story.

24. Rewrite the following two statements as a single statement. Make the statement as brief as you can.

    If a car is a convertible, then it has a removable top.
    If a car has a removable top, then it is a convertible.

Set III

According to the by-laws of the Lodge of the Chocolate Moose, you can become a member only if you are crazy about chocolate. Kermit wants to join the lodge and he is crazy about chocolate.

The members of the lodge never break their by-laws. Does this mean that they must let Kermit become a member? Explain why or why not.

"His logic certainly isn't my logic."

Lesson 5

Valid and Invalid Deductions

Suppose that during a trial it is established that whoever committed the crime is color-blind and that Mr. Black and Miss White are two of the main suspects. If the prosecutor produces proof that Mr. Black is color-blind, must the jury conclude that he is guilty? If Miss White's lawyer proves that she is not color-blind, does it follow that she must be innocent?

One way to check these conclusions is with an Euler diagram. The

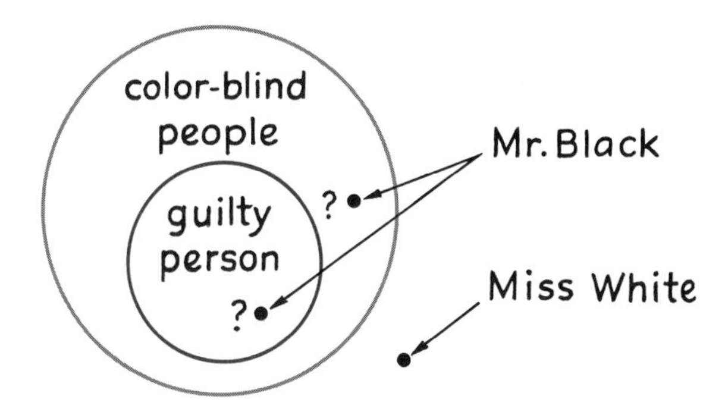

diagram shown here illustrates the conditional statement

If the person is guilty, then he or she is color-blind.

We then look at the diagram to see in which region each suspect belongs: in both circles, in just the larger circle, or in neither. We can't tell where Mr. Black belongs, so no conclusion is justified. Because Miss White is outside the larger circle, she cannot be inside the smaller one, and so she is not the guilty person.

Another way to check the two conclusions is as follows. The argument about Mr. Black is:

> If a person is guilty, then he or she is color-blind.
> Mr. Black is color-blind.
> Therefore, Mr. Black is guilty.

The first sentence in this argument is a conditional statement. The second and third sentences can be combined to form a second conditional statement:

> If a person is color-blind, then he or she is guilty.

What is the relation of this statement to the first one? The hypothesis and conclusion of the second statement and those of the first are interchanged, and so the second statement is the *converse* of the first. Because a statement and its converse are not logically equivalent, the second statement may be false even though the first statement is true. The conclusion "Mr. Black is guilty" does not necessarily follow from the other two statements, and so it is *not a valid deduction.*

How about Miss White? The argument for her is:

> If a person is guilty, then he or she is color-blind.
> Miss White is not color-blind.
> Therefore, Miss White is not guilty.

By combining the second and third sentences to form the conditional statement

> If a person is not color-blind, then he or she is not guilty,

we see that it is the *contrapositive* of the first statement of the argument. If we accept the first statement as true, we cannot help but accept the second statement as true also, because a statement and its contrapositive are logically equivalent. The deduction "Miss White is not guilty" is valid.

Exercises

Set I

1. Which of the following statements is logically equivalent to a conditional statement?
   a) Its converse.
   b) Its inverse.
   c) Its contrapositive.

Refer to the figure below to tell whether or not the third statement in each of the following deductions follows logically from the first two statements.

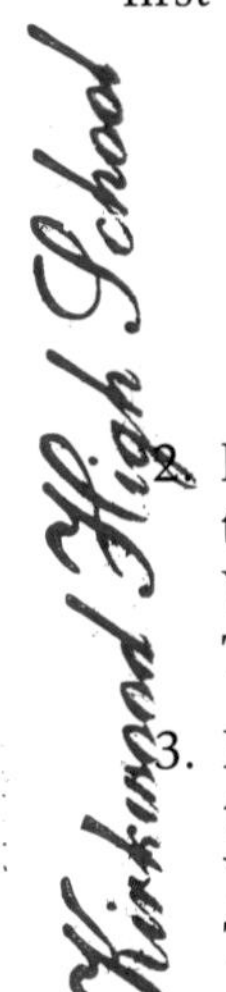

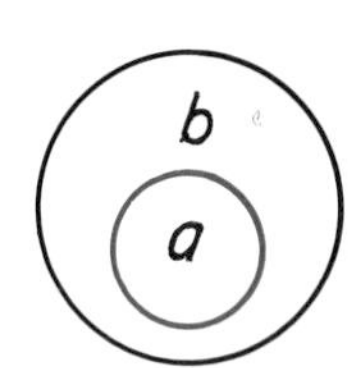

2. If you choose a point in circle a, then that point is also in circle b.
   You choose a point in circle a.
   Therefore, that point is also in circle b.
3. If you choose a point in circle a, then that point is also in circle b.
   You choose a point in circle b.
   Therefore, that point is also in circle a.
4. If you choose a point in circle a, then that point is also in circle b.
   You choose a point that is not in circle a.
   Therefore, that point is not in circle b.
5. If you choose a point in circle a, then that point is also in circle b.
   You choose a point that is not in circle b.
   Therefore, that point is not in circle a.

Express the patterns of the deductions for Exercises 2 through 5 in symbols.

Example: Exercise 2.

Answer: $a \rightarrow b$
a
Therefore, b.

6. Exercise 3.
7. Exercise 4.
8. Exercise 5.

The deduction for Exercise 3 is *invalid;* it makes the *converse* error.

9. The deduction for Exercise 4 is also invalid. What error does it make?
10. The deduction for Exercise 5 is valid. What word names its pattern?

Express the patterns of the following deductions in symbols. Refer to the patterns to tell whether the deductions are *valid* or *invalid.* If a deduction is invalid, name the error (*converse* error or *inverse* error) that has been made.

Example: If you enjoy golf, you like to beat around the bush.
Ollie doesn't enjoy golf.
Therefore, Ollie doesn't like to beat around the bush.

Answer: $a \rightarrow b$
not a
Therefore, not b.
Invalid—the inverse error.

11. If you see spots in front of your eyes, you're looking at a leopard.
    You're looking at a leopard.
    Therefore, you see spots in front of your eyes.
12. If you see strands in front of your eyes, your hair is too long.
    You don't see strands in front of your eyes.
    Therefore, your hair is not too long.
13. If you forgot your pencil, you may borrow one of mine.
    You forgot your pencil.
    Therefore, you may borrow a pencil from me.
14. If you brush your teeth with Brylcreem, you misunderstood the commercial.
    You didn't misunderstand the commercial.
    Therefore, you don't brush your teeth with Brylcreem.

15. All moths are attracted to candle flames.
    This insect is not a moth.
    Therefore, this insect is not attracted to candle flames.

16. All carbonated soft drinks contain bubbles.
    You are drinking something that contains bubbles.
    Therefore, it is a carbonated soft drink.

17. No graduate of the White Elephant Memory School ever forgets.
    Eloise is very forgetful.
    Therefore, Eloise did not graduate from the White Elephant Memory School.

18. The students will stop paying attention if the class is boring.
    The class isn't boring.
    Therefore, the students will pay attention.

Set II

In each of the following exercises, two statements are given that are to be accepted as true. If possible, write a third statement that can be deduced from these statements. Otherwise, write "no deduction possible."

Example: If your dog has a license, then you don't have to do all the driving.
You don't have to do all the driving.

Answer: No deduction possible. (To conclude that "your dog has a license" would be to make the *converse* error.)

19. If I have reached the party to whom I am speaking, then I have dialed correctly.
    I have indeed reached the party to whom I am speaking.

20. If the Jolly Green Giant started turning blue, he should put on a sweater.
    The Jolly Green Giant has not started to turn blue.

21. If there is a fly in your soup, then you shouldn't be too quick to swallow each spoonful.
    You may swallow each spoonful quickly.

22. If I had a chimp for a nephew, then I'd be a monkey's uncle.
    I'm a monkey's uncle.

Write the first statement in each of the following exercises in "if-then" form before following the directions given for Exercises 19 through 22.

23. All Polaroid cameras take self-developing pictures.
    That camera is not a Polaroid.

24. All night owls hoot it up.
    Fred never gives a hoot.

25. No flying saucer can travel faster than the speed of light.
    The object hovering overhead is not a flying saucer.

26. His name ends in "o" if he is one of the Marx brothers.
    Groucho's name ends in "o."

Set III

The first sentence of the advertisement below is a conditional statement that most people would probably accept as true. Compare it with the last sentence of the advertisement. Assuming that "you know just what to do" means that you will decide to buy a Volkswagen station wagon, is the advertisement's logic valid? Does this conclusion follow logically from what has been said before? Explain.

If the world looked like this,
and you wanted to buy a car that sticks out a little,
you probably wouldn't buy a Volkswagen Station Wagon.
But in case you haven't noticed, the world doesn't look like this.
So if you've wanted to buy a car that sticks out a little,
you know just what to do.

Courtesy of Volkswagen of America, Inc.

Lesson 6
Arguments with Two Premises

"How is it you can all talk so nicely?" Alice said. . . . "I've been in many gardens before, but none of the flowers could talk."
"Put your hand down, and feel the ground," said the Tiger-Lily. "Then you'll know why."
Alice did so. "It's very hard," she said, "but I don't see what that has to do with it."
"In most gardens," the Tiger-Lily said, "they make the beds too soft—so that the flowers are always asleep."

LEWIS CARROLL, *Through the Looking Glass*

Perhaps the mums wouldn't talk even if they were awake! The argument that the Tiger-Lily is presenting to Alice is essentially this:

If the flower bed is too soft, the flowers are always asleep.
If the flowers are asleep, they don't talk.
Therefore, if the flower bed is too soft, the flowers don't talk.

Notice that this argument consists of three conditional statements. If you accept the first two statements as being true, would you also have to accept the third statement as true?
We can illustrate the first two statements, called the *premises* of the argument, with Euler diagrams.

If we combine these two diagrams into one, we get the diagram on the left below. Omitting the second circle leaves the diagram shown at the right below, which illustrates the third statement, called the *conclusion* of the argument.

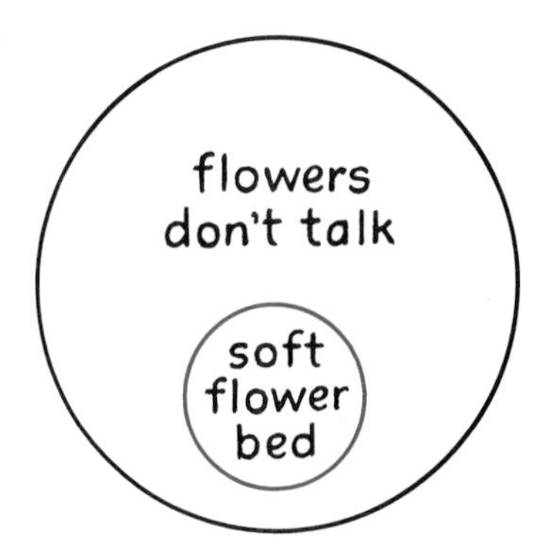

In more general terms, if we have two premises in which *the conclusion of the first is the same as the hypothesis of the second* ($a \rightarrow b$, $b \rightarrow c$) then from them we can derive a third statement, $a \rightarrow c$. Such an argument is called a *syllogism.*

► A *syllogism* is an argument of the form

$a \rightarrow b$ (first premise)
$b \rightarrow c$ (second premise)
Therefore, $a \rightarrow c$ (conclusion)

If both premises of a syllogism are true, then the conclusion must be true as well. If one or both of the premises are false, then the conclusion may be false. This does not mean, however, that the reasoning is incorrect. In other words, the truth or falsehood of the statements used has nothing to do with the validity of an argument. Consider, for example, this syllogism:

If you live in Whangamata, then you live in Auckland.
If you live in Auckland, then you live in New Zealand.
Therefore, if you live in Whangamata, you live in New Zealand.

It is not necessary to look in an atlas to know that this is a valid argument. The conclusion is a logical consequence of the premises.

Exercises

Set I

A syllogism consists of three conditional statements.

1. What are the first two statements called?
2. What is the third statement called?
3. With what part of the second statement must the conclusion of the first statement be identical?
4. If the first two statements of a syllogism are true, must the third statement also be true?
5. What could cause the third statement of a syllogism to be false?

Tell whether each of the following pairs of premises can be used to form a syllogism.

6. $a \to b$
   $a \to c$
7. $a \to b$
   $b \to c$
8. $a \to b$
   $c \to b$
9. $a \to b$
   $c \to a$

Tell whether each of the following arguments is a syllogism.

10. If a duck flies upside-down, it gets dizzy.
    If a duck gets dizzy, it quacks up.
    Therefore, if a duck flies upside-down, it quacks up.
11. If a penny has an Indian head on it, it was made before 1910.
    If a penny has an Indian head on it, it is worth more than one cent.
    Therefore, if a penny was made before 1910, it is worth more than one cent.
12. If you go to Waikiki Beach, you will see beautiful sunsets.
    If you are in Hawaii, you will see beautiful sunsets.
    Therefore, if you go to Waikiki Beach, you are in Hawaii.

Rewrite the statements in each of the following arguments in "if-then" form. Then, use symbols to represent the form of the argument and tell whether it is valid or invalid.

Example: All movies directed by Alfred Hitchcock have suspenseful plots.
The movie *North by Northwest* has a very suspenseful plot.
Therefore, *North by Northwest* was directed by Alfred Hitchcock.

Answer: If a movie was directed by Alfred Hitchcock, it has a suspenseful plot.
If the movie is *North by Northwest,* it has a very suspenseful plot.
Therefore, if the movie is *North by Northwest,* it was directed by Alfred Hitchcock.
The form of this argument is

$$a \to b$$
$$c \to b$$

Therefore, $c \to a$.

The argument is not valid.

13. All donkeys have long ears.
    All long-eared creatures are habitual eavesdroppers.
    Therefore, all donkeys are habitual eavesdroppers.
14. The second batter on a major league baseball team is always a good hit-and-run player.
    The second batter on a major league baseball team is always a good bunter.
    Therefore, if someone on a major league baseball team is a good hit-and-run player, he is also a good bunter.
15. People who shoplift are dishonest.
    No dishonest person is trustworthy.
    Therefore, people who shoplift are not trustworthy.

16. A clock marked "25% off" will not be accurate.
    A clock is not accurate if it loses 15 minutes every hour.
    Therefore, a clock marked "25% off" will lose 15 minutes every hour.
17. If the conclusion of an argument is true, does it follow that the argument is valid?
18. If the conclusion of an argument is false, does it follow that the argument is invalid?
19. Is it possible for all three statements of a valid argument to be false?
20. Is it possible for all three statements of an invalid argument to be true?

Set II

At first glance, it may not seem that any conclusion can be drawn from the following pair of premises.

> If your name is in *Who's Who*,
> then you know what's what.
> If you're not sure of where's where,
> then you don't know what's what.

21. Write the contrapositive of the second premise.

Because a statement and its contrapositive are logically equivalent, the contrapositive of the second premise can be combined with the first premise to obtain a conclusion.

22. What is the conclusion?

The following statements are pairs of premises from which it may or may not be possible to derive conclusions. You may need to consider the contrapositive of one of the statements before being able to tell whether a conclusion is justified. Write either the conclusion statement or "no conclusion" as your answer.

23. If you are faster than a speeding bullet, then you are more powerful than a locomotive.
    If you are more powerful than a locomotive, then you are able to leap tall buildings with a single bound.
24. If it is April first, then the soldiers are tired.
    If they have just had a March of thirty-one days, then the soldiers are tired.
25. The check will cost a lot
    if you eat in an expensive restaurant.
    It helps to have a mint
    if the check costs a lot.
26. If you are afraid of earthquakes,
    then you shouldn't live in California.
    If you are crazy,
    then you are not afraid of earthquakes.
27. If Captain Spaulding is in the jungle,
    he can't play cards.
    Captain Spaulding can't play cards
    if there are too many cheetahs.

Set III

If an argument is logically valid and the premises upon which it is based are true, then its conclusion is also true. Yet, on the surface, the following argument seems to contradict this.

> Breadcrumbs are better than nothing.
> Nothing is better than a really good steak.
> Therefore, breadcrumbs are better than a really good steak.

Of all the people who agree with the first two statements in this argument probably not one would accept its conclusion. Because the argument seems to be a logical one, can you explain what is wrong? (Hint: Try writing each statement in "if-then" form.)

Lesson 7
Direct Proof: Arguments with Several Premises

The American artist Rube Goldberg was so well known for his cleverly ridiculous inventions that, according to one dictionary,* his name has come to mean "having a fantastically complicated, improvised appearance" and "deviously complex and impractical." One of Goldberg's inventions, a way to keep cool, is shown in the cartoon above.

The cartoon illustrates the following chain of events:

> If the man pushes the wheelbarrow,
> the pulley turns the kicking wheel.
> If the pulley turns the kicking wheel,
> the bear is annoyed and eats the doll.
> If the bear is annoyed and eats the doll,
> the string connected to the mechanical bird is pulled.
> If the string connected to the mechanical bird is pulled,
> the bird says: "Do you love me?"
> If the mechanical bird says: "Do you love me?",
> the love-bird nods its head.
> If the love-bird nods its head,
> the fan will make a nice breeze on the man's face.

Notice how the conclusion of each statement matches the hypothesis of the following one. Using letters to represent each hypothesis and conclusion makes the underlying pattern more obvious:

$$a \to b, \quad b \to c, \quad c \to d, \quad d \to e, \quad e \to f, \quad f \to g.$$

---

* *The Random House Dictionary of the English Language*, 1966.

From this pattern, we can draw the conclusion

$$a \rightarrow g.$$

In words, the conclusion is: Therefore, if the man pushes the wheelbarrow, then the fan will make a nice breeze in his face.

This argument is a simple example of a *direct proof.* The first six statements in the argument on page 32 are called its *premises.* They lead to the statement at the top of this page. This statement, called the *conclusion* of the argument, might be considered a *theorem.*

► A *theorem* is a statement that is proved by reasoning deductively from already accepted statements.

Exercises

Set I

The Euler diagram shown here illustrates the following argument:

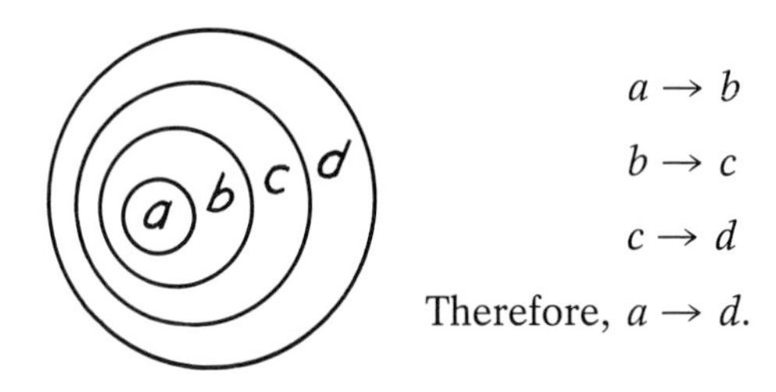

$a \rightarrow b$

$b \rightarrow c$

$c \rightarrow d$

Therefore, $a \rightarrow d$.

1. What are the first three statements in this argument called?
2. What is the last statement called?
3. If the first three statements are true, must the last statement also be true?
4. What is a statement called that is proved from already accepted statements?
5. What theorem is proved by the following statements?

   If you go to a 3-D movie,
   two images will appear on the screen.
   If two images appear on the screen,
   you must see one with each eye.
   If you see one image with each eye,
   you must wear Polaroid glasses.

Each of the following exercises consists of a "theorem" and a proof in which one or more statements have been omitted. After studying the relations of the statements given, write the missing statements.

6. *Theorem.*
   If the electricity goes off during the night, you will be late for school.

   Proof.
   If the electricity goes off during the night, your clock will be wrong.
   If your clock is wrong,
   you won't realize what time it is.
   (What statement belongs here?)

7. *Theorem.*
   If there is a total eclipse of the sun,
   the temperature can be determined without a thermometer.

   Proof.
   If there is a total eclipse of the sun,
   the sky becomes dark.
   (What statement belongs here?)
   If the crickets think that it is night, they start chirping.
   (What statement belongs here?)
   If the temperature is estimated by counting cricket chirps,
   it can be determined without a thermometer.

8. Copy the following statements, rearranging them in logical order.

   If you take a plane, you will go to the airport.
   If you go to Dallas, you will take a plane.
   If you see all the cabs lined up, you will see the yellow rows of taxis.
   If you go to the airport, you will see all the cabs lined up.

9. What "theorem" is proved by the four statements you have written for Exercise 8?

10. Copy the following statements, writing them in "if-then" form and rearranging them in logical order.

    NASA would send some mice on a lunar mission if they were eager astronauts.
    If mice were sent on a lunar mission, the eyes of the entire world would be watching them on television.
    If the moon were made of green cheese, mice would make eager astronauts.
    It would be one giant peep for mouse kind if the eyes of the entire world were watching them on television.

11. What "theorem" is proved by the four statements you have written for Exercise 10?

Set II

12. The following logic exercise was written by Lewis Carroll.

    Theorem.
    Babies cannot manage crocodiles.

    Proof.
    Babies are illogical.
    Nobody is despised who can manage a crocodile.
    Illogical persons are despised.

    Carroll deliberately made the proof difficult to follow by not stating his sentences in "if-then" form and by not stating them in logical order. Make the proof more understandable by rewriting it, doing both of these things.

13. The following proof looks incorrect but is actually valid. Show why there is nothing wrong with it by replacing one of the statements with one that is logically equivalent.

    Theorem.
    If Sherlock Holmes has to solve a case, he explains his conclusions to Watson.

    Proof.
    If Sherlock Holmes has to solve a case, he reasons deductively.
    If Sherlock Holmes does not figure out who did it, he has not reasoned deductively.
    If Sherlock Holmes figures out who did it, he explains his conclusions to Watson.

Reorganize the proofs below so that they are easier to follow. (They should be rewritten on your paper.)

14. *Theorem.*
    If there is no Great Pumpkin, Snoopy won't have pie for dinner.

    Proof.
    If Lucy plays a trick on Charlie Brown, he will be upset.
    If Linus is mistaken, Lucy is pleased.
    If Lucy becomes unruly, she plays a trick on Charlie Brown.
    If there is no Great Pumpkin, then Linus is mistaken.
    If Charlie Brown forgets to feed Snoopy, Snoopy won't have pie for dinner.

If Lucy is pleased,
she becomes unruly.
Charlie Brown forgets to feed Snoopy if
he is upset.

15. *Theorem.*
If this is the last exercise,
it is not easy.

Proof.
If this is arranged in logical order,
it is not the last exercise.
If I can't understand an exercise,
I get dizzy when I try to do it.
If an exercise is easy,
it does not give me trouble.
If I get dizzy while trying to do an exercise, it is giving me trouble.
If an exercise is not arranged in a logical order, I can't understand it.

Set III

A driver makes the following statements about automobiles:

A front-wheel drive gives a good hold on the road.
A heavy car must have good brakes.
Any powerful car is high-priced.
Light cars do not have a good hold on the road.
A low-powered car cannot have good brakes.*

Is it logical for this driver to buy a cheap front-wheel drive? Show your proof.

*From *100 Games of Logic,* by Pierre Berloquin (Scribners, 1977).

Lesson 8
Indirect Proof

In the story "The Lady or the Tiger?" by Frank Stockton, a young man is forced to choose between two rooms, one of which contains a lady and the other a tiger. If he chooses the room containing the lady, he gets to marry her. If he chooses the other room, he has to face a fierce tiger. The result of the man's choice is not revealed in the story.

Many puzzles have been invented on this theme. In one of these puzzles,* the king puts signs on the doors of the rooms, telling the young man that one of the signs is true and one is false. In which room is the lady?

A	B
IN THIS ROOM THERE IS A LADY, AND IN THE OTHER ROOM THERE IS A TIGER.	IN ONE OF THESE ROOMS THERE IS A LADY, AND IN ONE OF THESE ROOMS THERE IS A TIGER.

If we assume that sign A is true, then the lady is in room A and the tiger is in room B. If this is so, then sign B is also true. But this conclusion contradicts the fact that one sign is true and the other sign is false. This means that our assumption that sign A is true is wrong. So sign A is false. Consequently the tiger is in room A and the lady is in room B.

We have arrived at this conclusion by reasoning *indirectly*.

---

* *The Lady or the Tiger? and Other Logic Puzzles,* by Raymond Smullyan (Knopf, 1982), page 15.

► In an *indirect proof*, an assumption is made at the beginning that leads to a contradiction. The contradiction indicates that the assumption is false and the desired conclusion is true.

The pattern of an indirect proof is illustrated here with the puzzle of the lady or the tiger.

Theorem.
In the puzzle of the lady or the tiger, the lady is in room B.

Proof.

Suppose that the lady is not in room B.	⟵	We assume the *opposite of the desired conclusion.*
If she is not in room B, then she is in room A.		
If she is in room A, then both sign A and sign B are true.		
This contradicts the fact that one sign is true and one is false.	⟵	The *contradiction.*
Therefore, what we supposed is false and the lady is in room B.	⟵	We end with the *desired conclusion.*

Both the direct and indirect methods of proof require making a chain of conditional statements. In the direct method, the chain is made by beginning with a, the hypothesis of the theorem, and ending with b, its conclusion. In the indirect method, the chain is made by beginning with *not* b, the opposite of the conclusion of the theorem, and ending in a contradiction with a. The contradiction tells us that, if a is true, b cannot be false. This means that $a \rightarrow b$.

Exercises

Set I

To prove a theorem indirectly, we begin by assuming that its conclusion is false. List the assumption with which an indirect proof of each of the following statements would begin.

Example: If a tailor wants to make a coat last, he makes the pants first.

Answer: Suppose that the tailor does not make the pants first.

1. If a chicken could talk, it would speak foul language.
2. If a teacher is cross-eyed, he has no control over his pupils.
3. If a proof is indirect, then it leads to a contradiction.

4. The following statements can be rearranged to make an indirect proof of this theorem.

   Theorem.
   A bridge hand must contain more than three cards of the same suit.

   Copy the statements, rearranging them in logical order.

   If a bridge hand does not contain more than three cards of the same suit,
   it does not contain more than twelve cards.
   Therefore, what we supposed is false and a bridge hand must contain more than three cards of the same suit.
   Suppose that a bridge hand does *not* contain more than three cards of the same suit.
   This contradicts the fact that a bridge hand contains thirteen cards.

5. This unfinished tick-tack-toe game once appeared in an IBM advertisement. If it

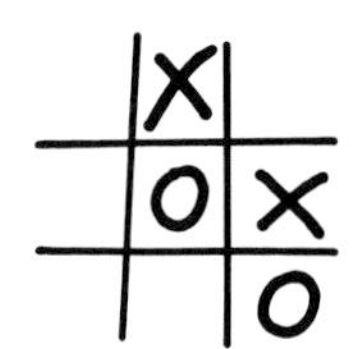

   is assumed that neither player is stupid, we can prove the following theorem.

   Theorem.
   The player using X went first.

   Copy and complete the following indirect proof of this theorem.

   Proof.
   Suppose that _______________.
   If _______________,
   the player using O went first.
   If the player using O went first,
   then it is O's turn to play now.
   If _______________,
   then X will lose the game.
   If X loses the game,
   then X is stupid.
   This contradicts the fact that _______________.
   Therefore, what we supposed is false and the player using X went first.

6. A backward geometry student named Dilcue is taking a true-false quiz of five questions. Refer to the following facts to complete the proof that the last answer on the quiz is false.

   Fact 1. If the first answer is true, the next one is false.
   Fact 2. The last answer is the same as the first answer.
   Fact 3. The second answer is true.

   Theorem.
   The last answer on the quiz is false.

   Proof.
   Suppose that _______________.
   If _______________,
   then it is true.
   If the last answer is true,
   then the first answer _________.
   If the first answer _________,
   then the second one _________.
   This contradicts the fact that _______________.
   Therefore, what we supposed is false and _______________.

7. Although blood types cannot be used to prove that a particular man is the father of a given child, they can be used to prove that he is not. The table below shows the possible blood types of the children of parents having certain blood types.

Parents	*Children*
O and O	O
O and A	O or A
O and B	O or B
O and AB	A or B

Suppose that a woman has type O blood, and her child also has type O blood. Complete the following indirect proof that a man having type AB blood cannot be the child's father.

Theorem.
A man having type AB blood cannot be the child's father.

Proof.
Suppose that __________.
If __________,
the child has either __________ blood.
This contradicts the fact that __________.
Therefore, what we supposed is false
and __________.

Set II

Lorelei's boy friend has given her a "diamond" ring but she isn't certain that it is genuine.

Consider the following argument:

If a stone is a diamond,
its index of refraction is more than 2.
The stone in Lorelei's ring has an index
of refraction of 2.4.
Therefore, the stone in Lorelei's ring
is a diamond.

8. Assuming that the first two statements in this argument are true, does it follow that the third statement is true?
9. Explain why or why not.

The following true statements can be used to prove that the stone in Lorelei's ring is a diamond.

If a stone's hardness is less than 10,
it can be scratched by corundum.
Therefore, what we supposed is false
and the stone in Lorelei's ring is a diamond.
If the stone is not a diamond,
its hardness is less than 10.
Suppose that the stone in Lorelei's ring
is not a diamond.
This contradicts the fact that the stone
in Lorelei's ring cannot be scratched by corundum.

10. Copy the five statements above, rearranging them in logical order.
11. What kind of a proof have you written?
12. While driving her 1954 Chevy to the market, Miss Piggy suddenly worries that she may have locked her keys in her apartment. Use one or more of the following statements to draw a conclusion about this. Explain your reasoning.

    Miss Piggy never leaves her apartment unlocked.
    She keeps every key she owns on one large key ring.
    She frequently gets herself into a jam.

13. The Chicago Bears are playing the Green Bay Packers, and the score at half-time is 7 to 6 with the Bears ahead. In the second half, all points are made by making either touchdowns (6 points each) or touchdowns with conversions (7 points each). The final score was 18 to 13. Which team won the game? Explain your reasoning.

Set III

Bashful, Dopey, and Sneezy are arguing about how many friends Grumpy has. Bashful claims that Grumpy has at least fifty friends. Dopey says that Grumpy certainly doesn't have that many, whereas Sneezy remarks that Grumpy must have at least one friend.

If what *only one* of the three is saying is true, how many friends does Grumpy actually have? Explain your reasoning.

Lesson 9

A Deductive System

B.C.'s predicament in trying to learn the meaning of "ecology" from Wiley's dictionary illustrates the fact that it is impossible to define everything without going around in circles. It is also impossible to *prove* everything without going around in circles.

To avoid eventually coming back to the point from which we begin, we must leave at least a few words undefined and a few statements unproved. The words left undefined, called the *undefined terms*, can be used as a basis for building definitions of other words. The statements left unproved, called *postulates*, can be used as a basis for building proofs of other statements.

► A *postulate* is a statement that is assumed to be true without proof.

From the *undefined terms*, then, we can construct definitions of other words. From the *postulates* and *definitions*, we can construct proofs, either directly or indirectly, of the statements we call *theorems*. The structure that is built by means of logic in this way is called *a deductive system.*

A deductive system is very much like a game. In learning how to play a game, you have to learn the meanings of the terms used in it (the definitions), and you have to learn the rules (the postulates.) To be a good football player, for example, a person has to be familiar with a lot of terms and rules. *The Official N.C.A.A. Football Guide* lists 197 rules, many of which contain more than one part.

Furthermore, for the game to be playable, it is important that the rules be both *sufficient* and *consistent.* By sufficient, we mean that they tell what to do in every possible situation and, by consistent, we mean that they neither contradict each other nor lead to contradictions.

The postulates of a deductive system are also like the rules of a game in that they can be changed. The game of football is somewhat different now from what it was originally like because some of its rules have been changed since the game was invented.

What a given deductive system is like, then, depends on the postulates used in it. Geometry is a good example of a deductive system in which changes in the postulates bring about interesting changes in the theorems that can be derived. Now that you are acquainted with the nature of a deductive system, you are ready to learn how to "play the game" of geometry.

Exercises

Set I

Is it possible to

1. define everything without going around in circles?
2. prove everything without going around in circles?
3. What is the difference between a postulate and a theorem?
4. What do you have to learn when you are learning how to play a game?

What does it mean to say that the rules of a game are

5. sufficient?
6. consistent?
7. Suppose that you want to learn Italian. Someone has given you an Italian dictionary that they bought in Italy. You want to learn the meaning of the word "sorpréndere" and, when you look it up, you find that the definition is "meravigliare." Because you don't know what "meravigliare" means, you then look it up. Do you think that you will be able to learn the meaning of "sorpréndere" from this dictionary? Explain why or why not.

The following dialogue is from a scene in a classroom in the Laurel and Hardy film *Pardon Us.*

Teacher: What is a comet?
A student: A star with a tail on it.
Teacher: Can anyone give us an example?
Laurel: Rin-Tin-Tin?

8. Which words in the other student's answer did Laurel misunderstand?
9. Why did he say "Rin-Tin-Tin"?
10. Explain why the following rules for the first throw in a game with two dice are not sufficient.

    Rule 1. If the sum of the numbers on the two dice is 7 or 11, the player who threw them wins.
    Rule 2. If the sum of the numbers on the two dice is 2, 3, or 12, the player who threw them loses.

11. Explain why the following rules for the first throw in a game with two dice are not consistent.

    Rule 1. If both dice turn up the same number, the player who threw them wins.
    Rule 2. If the sum of the numbers on the two dice is even, the player who threw them loses.
    Rule 3. If the sum of the numbers on the two dice is odd, the player must throw them again.

The following statements appear on the customer agreement for obtaining a credit card.

Statement 1. A transaction finance charge is a charge made if a new advance is added to your account.
Statement 2. If you go over your credit limit, you will be charged a fee.
Statement 3. A supercheck is a check designed for use with your credit card account.
Statement 4. If you are charged a fee, the fee will be added to your new balance.
Statement 5. If your card is lost or stolen, you agree to report it immediately.

12. Which of these statements are definitions?
13. What words do they define?
14. Which statements are postulates?
15. Which two statements can be combined to form a syllogism?
16. What theorem is proved by the syllogism?

Set II

Theorem.
To travel to Australia,
you must have your picture taken.

Postulate.
To travel to Australia,
you must have a tourist visa.

Postulate.
To have a passport,
you must have your picture taken.

17. What postulate is needed in addition to the two postulates above to prove the theorem above?
18. Use it, together with the other postulates, to write a direct proof of the theorem.

Mr. Boddy has been murdered.

19. What should you assume to prove indirectly that Colonel Mustard did it?
20. Use your assumption and the following facts to prove that Colonel Mustard did it.

    Fact 1. If Miss Scarlet did it, she used a revolver.
    Fact 2. No bullets were fired.
    Fact 3. If Colonel Mustard didn't do it, then Miss Scarlet did it.
    Fact 4. If a revolver was used, a bullet was fired.

Consider the following statements:

Statement 1. If someone is a felon, he cannot vote.
Statement 2. A person is a felon iff he has committed a serious crime.
Statement 3. Dillinger has committed a serious crime.

21. Which one of these statements is a definition?
22. Write the two conditional statements that are equivalent to it.
23. Use the statements to write a direct proof that Dillinger cannot vote.
24. Use them to write an indirect proof that Dillinger cannot vote.

Set III

The following puzzle is by Kobon Fujimura, Japan's leading inventor of puzzles.*

Four people, A, B, C, and D, went to a department store together. One bought a watch, one a book, one a pair of shoes, and one a camera. The first, second, third, and fourth floors carry those items but not necessarily in the order given for the purchases.

On the basis of the following clues, determine who bought what on which floor. Explain your reasoning.

Clue 1. B went to the first floor.
Clue 2. Watches are sold on the fourth floor.
Clue 3. D went to the second floor.
Clue 4. A bought a book.
Clue 5. B did not buy a camera.

---

* *The Tokyo Puzzles* (Scribners, 1978).

Chapter 1 / Summary and Review

The following list includes the most important concepts in the Introduction and Chapter 1.

Basic Ideas

Conditional statement, parts of a 9
Constructions, tools for 2
Contrapositive 15
Converse 15
Definition 19–20
Direct proof 32–33
Equivalent statements 15
Euclid 1
Euler diagram: used to
 represent a conditional statement 10, 15
 test the validity of an argument 23, 29
If and only if 20
Indirect proof 36–37
Inverse 16
Postulate 40
Premises of an argument 28–29
Syllogism 29
Theorem 33
Undefined term 40

Exercises

Set I

Write in "if-then" form:

1. All limericks have five lines.
2. I will make a fortune when I perfect my perpetual motion machine.
3. No toadstools are edible.

"In *that* direction," the Cat said, waving its right paw round, "lives a Hatter: and in *that* direction," waving the other paw, "lives a March Hare. Visit either you like: they're both mad."

"But I don't want to go among mad people," Alice remarked.

"Oh, you can't help that," said the Cat: "we're all mad here. I'm mad. You're mad."

"How do you know I'm mad?" said Alice.

"You must be," said the Cat, "or you wouldn't have come here."

Alice didn't think that proved it at all. . . .

LEWIS CARROLL, *Alice in Wonderland*

4. The Cat is claiming that

 If you are not mad,
 then you wouldn't have come here.

 Which of the following statements would the Cat have to agree with?

 a) If you are not here, then you are not mad.
 b) If you are mad, then you are here.
 c) If you are here, then you are mad.

The following pairs of statements appear in the book *Knots* by the psychiatrist R. D. Laing.* What relation (converse, inverse, or contrapositive) does the second statement have to the first one in each case?

5. I never got what I wanted.
 I always got what I did not want.
6. What I want, I can't get.
 What I get, I don't want.
7. I get what I deserve.
 I deserve what I get.

Write the indicated statement for each of the following sentences.

8. Where there's smoke, there's fire. *Converse.*
9. If it isn't an Eastman, it isn't a Kodak. *Contrapositive.*
10. If you want to drive a baby buggy, just tickle its feet. *Inverse.*

---

*Pantheon Books, 1971.

By the permission of Johnny Hart and News America Syndicate

Peter's argument in this cartoon is:

All apes have tails.
You do not have a tail.
Therefore, you are not an ape.

11. Is this argument valid?

12. Does it prove what B.C. asked?

If possible, state conclusions that can be derived from each of the following pairs of premises. Otherwise, write "no conclusion."

13. When you sneeze, you must close your eyes.
    Your eyes are closed.

14. If the ground hog does not see its shadow on February 2, spring weather is on the way.
    This year the ground hog saw its shadow on February 2.

15. If a sailor went to sleep on his watch, he would have to be very small.
    Popeye is not very small.

The following statement is a definition of *daredevil:*

You are a daredevil
iff you are recklessly bold.

16. Write the two conditional statements that are equivalent to this statement.

Set II

Euclid gathered together the
geometric knowledge of his time,
and arranged it
not just in a hodge-podge manner,
but,
he started with what he thought were
self-evident truths
and then proceeded to
PROVE all the rest by
LOGIC.
A splendid idea, as you will admit.
And his system has served
as a model
ever since.

LILLIAN LIEBER,
The Education of T. C. Mits

17. What are Euclid's "self-evident truths" called?

18. What are the rest of the truths proved by logic called?

19. What is another name for the logic used in proofs?

20. Is the following a good definition? Explain why or why not.

    A *riddle* is a question that
    requires thought to answer.

Compare the following two sentences:

If you play Monopoly,
you throw dice.
Only if you play Monopoly
do you throw dice.

21. Is the first sentence true?

22. Is the second sentence true?

23. Which sentence does this Euler diagram illustrate?

Sherlock Holmes once said to Watson:

"How often have I said to you
that when you have eliminated the
impossible, whatever remains,
however improbable, must be the truth?"*

24. What method of proof is Holmes describing?

Father: "Snooks, stop making that same noise!"
Baby Snooks: "This ain't the same noise, daddy. It's another one just like it!"

25. Explain the basis for disagreement in this conversation.

The following proof is not arranged in logical order.

If a rabbit's name is Harvey,
he is invisible.
A rabbit will not be taken seriously
if he is thought to be imaginary.
If a rabbit is over six feet tall,
his name is Harvey.
If only a rabbit's best friends can see him,
everyone else will think he is imaginary.
If a rabbit is invisible,
then only his best friends can see him.

26. Rewrite it so that it is easier to follow.

27. What type of proof is it?

28. What theorem does it prove?

A mad chemist asserts that he has created a liquid in which all substances will dissolve. He shows you a test tube which, he says, contains the liquid.

29. Is this possible? Explain your reasoning.

30. What type of proof does your reasoning illustrate?

*Sir Arthur Conan Doyle, *The Sign of the Four*.

ALGEBRA REVIEW

Operations

Operation	*Name of result*
Addition: $x + y$	Sum
Subtraction: $x - y$	Difference
Multiplication: xy	Product
Division: $\frac{x}{y}$	Quotient

Properties of Operations

Property	*Operation*	
	Addition	Multiplication
Commutative	$x + y = y + x$	$xy = yx$
Associative	$(x + y) + z = x + (y + z)$	$(xy)z = x(yz)$
Identity	$x + 0 = x$	$1 \cdot x = x$
Inverse	$x + (-x) = 0$	$x \cdot \frac{1}{x} = 1$
Distributive	$x(y + z) = xy + xz$	

Definitions

Subtraction: $x - y = x + (-y)$

Division: $\frac{x}{y} = x \cdot \frac{1}{y}$

Absolute Value

The absolute value of x is written $|x|$.
If x is positive or zero, then $|x| = x$.
If x is negative, then $|x| = -x$.

Exercises

Find the indicated answers.

Example 1: The sum of -5 and -8.
Answer: -5 + -8 = -13.

Example 2: The quotient of 12 and -4.
Answer: $\frac{12}{-4} = -3$.

1. The sum of 10 and -2.
2. The difference between 10 and -2.
3. The product of 10 and -2.
4. The quotient of 10 and -2.
5. The sum of -6 and -4.
6. The product of -6 and -4.
7. The quotient of -15 and -5.
8. The difference between -15 and -5.
9. The product of -7 and 9.
10. The difference between -7 and 9.
11. The quotient of -9 and 36.
12. The sum of -9 and 36.

Simplify.

Example: $|-3|$.
Answer: 3.

13. $|0|$
14. $|7 + 8|$
15. $|4 - 9|$
16. $|-12| + |-6|$
17. $|-10 + 2|$

Simplify.

Example 1: $x + x + x + x + x$.
Answer: $5x$.

Example 2: $x \cdot x \cdot x \cdot x \cdot x$.
Answer: x^5.

18. $x + x + y + y + y$
19. $x \cdot x + y \cdot y \cdot y$
20. $6x + x$
21. $6x \cdot x$
22. $x^2 + x^2 + x^2 + x^2$
23. $x^2 \cdot x^2 \cdot x^2 \cdot x^2$
24. $9x^3 - x^3$
25. $9x^3 \cdot x^3$
26. $4 + (7 + x)$
27. $4(7x)$
28. $4(7 + x)$
29. $8x - 5x$
30. $8x(-5x)$
31. $x + (5 - x)$
32. $x(5 - x)$
33. $(6x + y) + (3x + y)$
34. $(6x + y) - (3x + y)$
35. $x^2 + (x^3 - x^2)$
36. $x^2(x^3 - x^2)$
37. $7x(1 + 10x)$
38. $7x + (1 + 10x)$
39. $(4 + x^4) + (5 + x^5)$
40. $(4x^4)(5x^5)$

Chapter 2

POINTS, LINES, AND PLANES

Marion and Tony Morrison/South American Pictures

Lesson 1
Points, Lines, and Planes

In our study of a deductive system, we learned that it is impossible to define everything without going around in circles.* To avoid eventually returning to the point from which we began, we must begin with some words that are left undefined. Among these words, the *undefined terms,* are "point," "line," and "plane." Because we have no definitions that tell us what these words mean, we will give them some meaning by assuming some relations between them. These assumptions, our first *postulates,* are considered in this lesson.

One of the world's unsolved mysteries concerns the desert of southern Peru. Stretching for miles across the land are gigantic drawings of birds and other animals, and lines—many lines—leading in every direction. The drawings and lines were put there by the Nazca Indians more than 1,500 years ago. For what purpose, no one knows.†

The lines, some of which can be seen in this photograph, are incredibly straight. It is thought that the Nazcas stretched strings from posts

---

*See page 40.

†Further information about the Nazca lines can be found in *The World's Last Mysteries* (A Reader's Digest Book, 1978) and *Pathways to the Gods,* by Tony Morrison (Harper and Row, 1978).

to aid in their construction. The figure at the right shows that just two posts are sufficient to determine the direction of one of these strings. Thinking of the posts as points and the string as a line suggests the following postulate.

Postulate 1
If there are two points, then there is exactly one line that contains them.

This can be said more briefly in the form:

Two points determine a line.

For this postulate to be true, it would seem that a line not only must be *straight,* but also must *extend without end* in both directions. The arrowheads in the figure above indicate that the line extends endlessly.

A second postulate about points and a line is sort of a converse of the first.

Postulate 2
If there is a line, then there are at least two points on the line.

There are, in fact, infinitely many points on a given line. Such points are called *collinear.*

Definitions
Points are ***collinear*** iff there is a line that contains all of them. Points are ***noncollinear*** iff there is no line that contains all of them.

Although a line cannot be drawn so that it contains three noncollinear points, it is always possible to draw a *plane* that contains them.

Postulate 3
If there are three noncollinear points, then there is exactly one plane that contains them.

This postulate can be stated more briefly in the form:

Three noncollinear points determine a plane.

The figure at the right shows three noncollinear points, A, B, and C, and part of the plane, P, that contains them. For the postulate to be true, a plane must be *flat* and have *no boundaries.* Although the part of the plane shown here is bounded by edges, the complete plane extends beyond them.

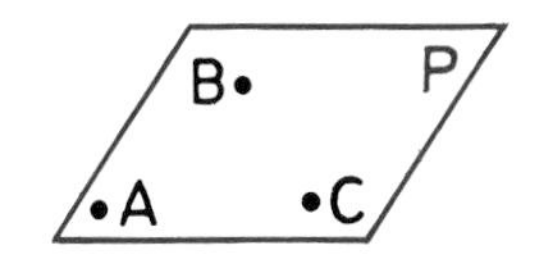

The three points are also said to be *coplanar.*

Definition
Points are ***coplanar*** iff there is a plane that contains all of them.

This figure shows a plane, P, that contains the two points A and B.

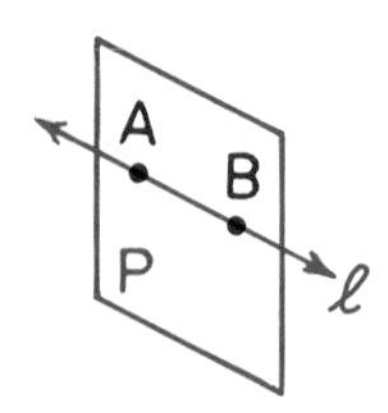

These points, in turn, determine line ℓ. Because the line is straight, the plane is flat, and both go on endlessly, it seems reasonable to make one more assumption:

► **Postulate 4**
If two points lie in a plane, then the line that contains them lies in the plane.

In doing some of the exercises of this lesson, you will see how we can use these four postulates as the basis for beginning to build a deductive system. They can be used to prove some *theorems* about points, lines, and planes.

Exercises

Set I

What belongs in the indicated space in each of the following postulates and definitions?

1. If there are two points, then there is exactly |||||||| that contains them.
2. If there is a line, then there are |||||||| on the line.
3. Points are |||||||| iff there is a line that contains all of them.
4. If there are three noncollinear points, then there is exactly |||||||| that contains them.
5. Points are |||||||| iff there is a plane that contains all of them.
6. If two points lie in a plane, then the line that contains them ||||||||.

Consider the following definition:

Points are noncollinear iff there is no line that contains all of them.

7. Write the two "if-then" statements that follow from this definition.
8. What relation do your two statements have to each other?

Tell whether each of the following statements is true or false.

9. Any two points determine a line.
10. Any two points are collinear.
11. Any two points are coplanar.
12. Any two points determine a plane.
13. Any three points are coplanar.
14. Any three points determine a plane.
15. Any four points are coplanar.

Tell which of the following figures illustrates each of the following point-line-plane relations.

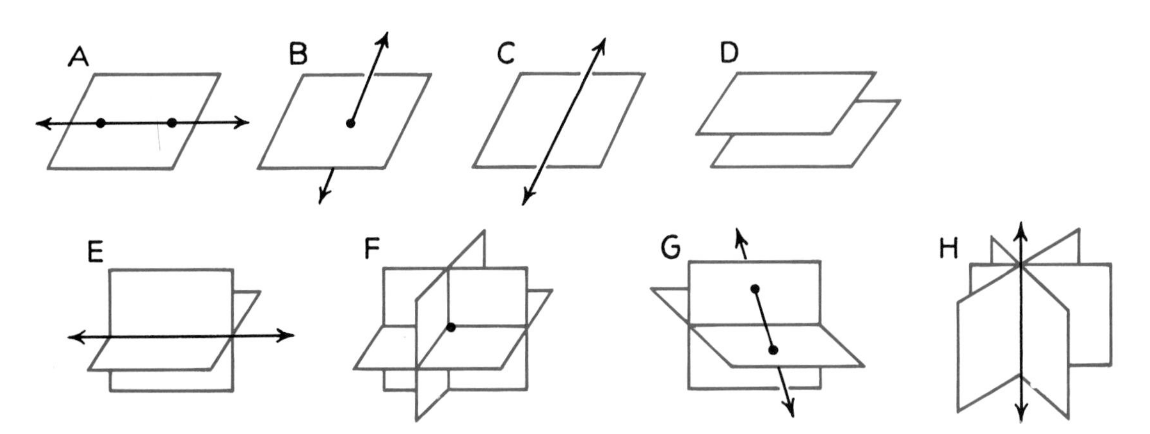

16. Two planes that do not intersect.
17. A line and a plane that intersect in exactly one point.
18. A line and a plane that do not intersect.
19. A line that is contained in a plane.
20. Two planes that intersect in a line.
21. Three planes that intersect in a point.
22. Three planes that intersect in a line.
23. A line that intersects two planes in different points.

Set II

The following theorem can be proved by using the direct method.

Theorem.
If there is a line and a point not on the line, then there is exactly one plane that contains them.

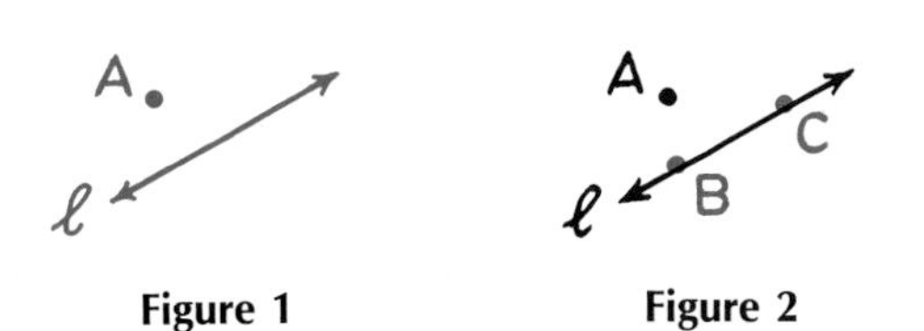

Figure 1 **Figure 2**

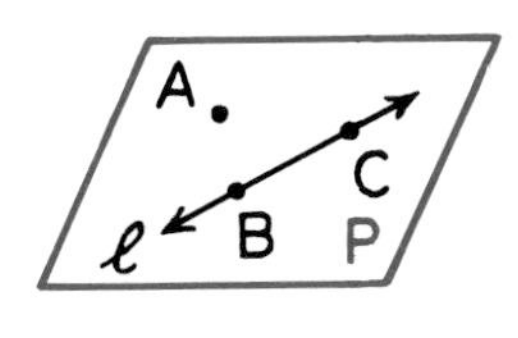

Figure 3

State, as a complete sentence, the *postulate* or *definition* that justifies each numbered statement.

Proof.
Figure 1 shows a line, ℓ, and a point, A, not on the line.

24. We can choose two more points, B and C, on line ℓ. (See Figure 2.)
25. Points A, B, and C are noncollinear.
26. We can draw exactly one plane, P, that contains points A, B, and C. (See Figure 3.)
27. Plane P contains line ℓ.

The following theorem can be proved by using the indirect method.

Theorem.
If two lines intersect, then they intersect in no more than one point.

28. What is the conclusion of this theorem?

29. With what assumption should the proof begin?

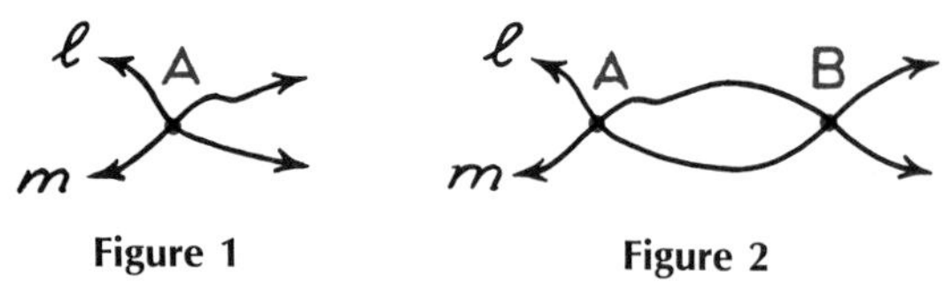

Figure 1 **Figure 2**

(The lines in these figures have been deliberately drawn so that they do not look straight.)

If lines ℓ and m intersect in a second point B, one of our *postulates* is contradicted.

30. Which postulate is it? (State it as a complete sentence.)

31. What does this contradiction indicate about what we assumed?

32. What conclusion follows?

A proof of the following theorem is suggested by the figures below it.

Theorem.
If two lines intersect, then there is exactly one plane that contains them.

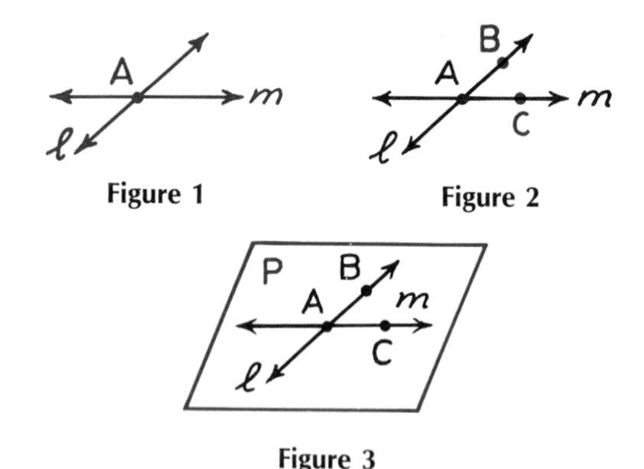

Figure 1 **Figure 2**

Figure 3

Use these figures as a guide in copying and completing the following direct proof of this theorem.

Proof.

33. Figure 1 shows ||||||||.

34. We can choose a second point B on line ℓ and ||||||||. (See Figure 2.)

35. Points A, B, and C are ||||||||.

36. We can draw exactly one ||||||||. (See Figure 3.)

37. Plane P contains ||||||||.

A proof of the following theorem is suggested by the figures below it.

Theorem.
If a line intersects a plane that does not contain it, then they intersect in no more than one point.

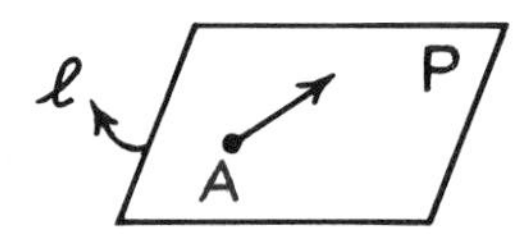

Figure 1

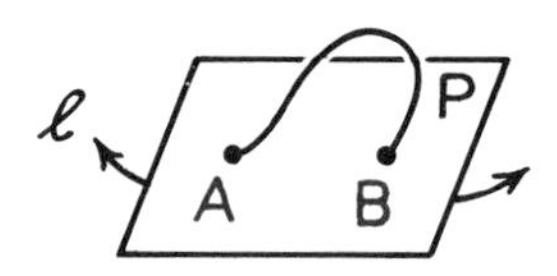

Figure 2

Use these figures as a guide in copying and completing the following indirect proof of this theorem.

Proof.

38. Suppose that line ℓ intersects ||||||||.

39. If it does, then points A and B lie in plane P, but line ℓ ||||||||. (See Figure 2.)

40. This contradicts the postulate that says ||||||||.

41. Therefore, what we supposed is false and ||||||||.

Set III

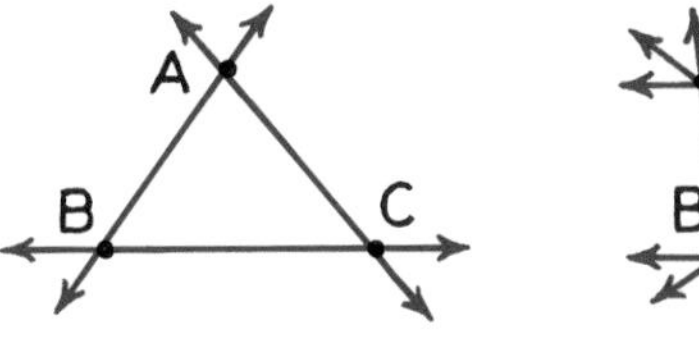

Figure 1

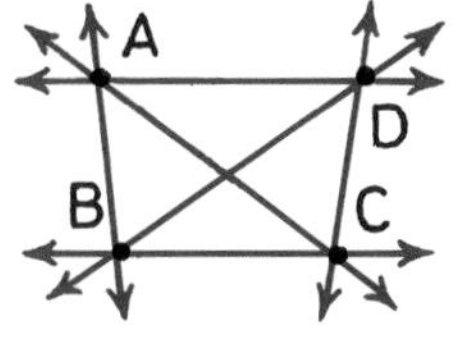

Figure 2

Figure 1 shows three noncollinear points and all of the lines (three) that are determined by these points. Figure 2 shows four points, no three of which are collinear, and all of the lines (six) that they determine.

1. Make a drawing to illustrate five points, no three of which are collinear, and all of the lines that they determine. How many lines are there in all?
2. Do the same for six points, no three of which are collinear.
3. Can you guess how many lines are determined by *ten* points, no three of which are collinear, without making a drawing? If you can, explain the basis for your answer.

Lesson 2

The Ruler Postulate

The concept of distance is a very basic one in geometry. In fact, it was used in the puzzles of the surfer and the spotter with which we began our study of the subject. In this lesson, we will consider in more detail what distance is and how it is measured.

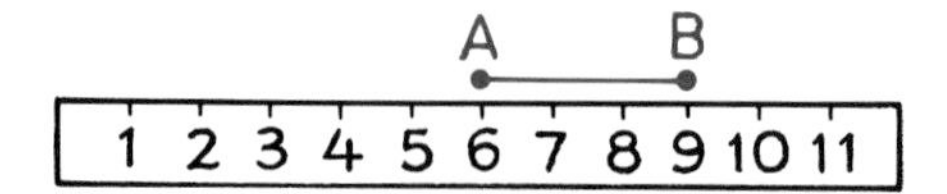

Sally, in measuring Snoopy's mouth, has noticed that two points (A and B in the figure above) correspond to the numbers 6 and 9, respectively. From this, she has concluded that the distance between the points is

$$9 - 6 = 3.$$

Because the numbers on the ruler are 1 inch apart, Snoopy's mouth measures 3 inches across.

If the ruler is moved, the coordinates of A and B will change. In the figure shown here, they have become 4 and 7, respectively.

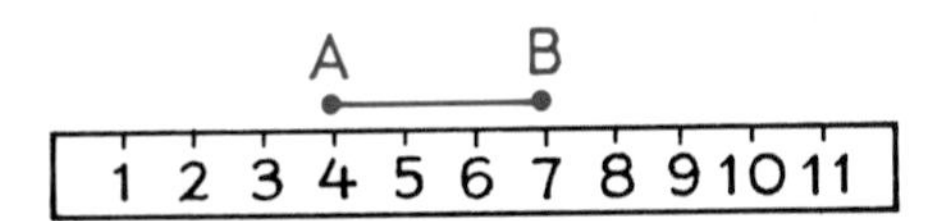

The distance between the two points, however, does not change because

$$7 - 4 = 3.$$

To find the distance in each example, we subtracted the smaller coordinate from the larger one. This was done to make the distance *positive.* In geometry, it is convenient to treat *all distances as positive.*

Another way to make sure that the distance between two points is positive is to subtract either coordinate from the other and take the *absolute value* of the result. For example, to find the distance illustrated above, we can write either

$$|7 - 4| = |3| = 3$$

or

$$|4 - 7| = |-3| = 3.$$

The figure at the right represents two points, A and B, whose coordinates are a and b, respectively. The distance between points A and B is represented by the symbol AB and is equal to $|a - b|$, or $|b - a|$. More briefly,

$$AB = |a - b| = |b - a|.$$

The properties of a ruler described in this lesson can be summarized in the following postulate.

► **Postulate 5** (The Ruler Postulate)
The points on a line can be numbered so that
a) to every point there corresponds exactly one real number called its coordinate,
b) to every real number there corresponds exactly one point,
c) to every pair of points there corresponds exactly one real number called the distance between the points,
d) and the distance between two points is the absolute value of the difference between their coordinates.

We will use this postulate frequently, and it would be inconvenient to write it out every time. After you are certain that you know what it means, you may refer to it as simply the "Ruler Postulate."

Exercises

Set I

Refer to the figure below to answer the following questions.

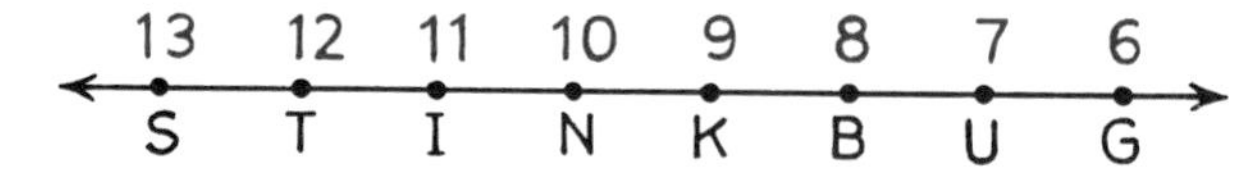

1. What is the coordinate of point N?
2. Which point corresponds to 7?
3. What is the distance between S and B?

Find each of the following distances in the figure below.

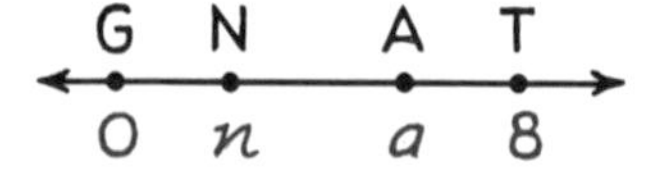

M A N T I S
-20 -7 0 7 11 32

4. NI.
5. IS.
6. MN.
7. AT.
8. MA.
9. MS.

Refer to the same figure to tell whether each of the following equations is true or false.

Example: TI + IS = TS.
Answer: True (because 4 + 21 = 25).

10. AN = NT.
11. MA + AN = MN.
12. MA = AT.
13. AI + IS = AS.

Find a number or expression for the distance between points having the following coordinates.

14. 1 and 100.
15. 15 and -15.
16. 0.7 and 0.07.
17. $\frac{1}{2}$ and $\frac{1}{5}$.
18. x and 0.
19. y and z.

Refer to the figure below to answer the following questions.

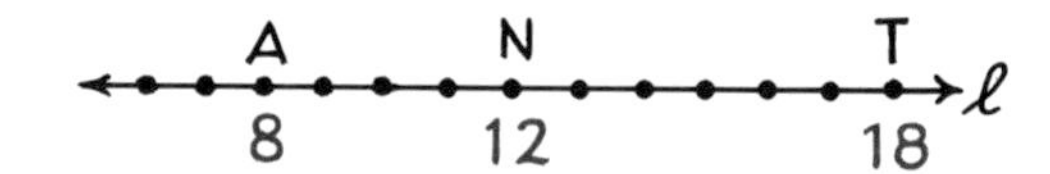

20. How many points on line ℓ are 5 units from point N?
21. What are their coordinates?
22. How many points on line ℓ are the same distance from A that they are from T?
23. What are their coordinates?

Set II

G N A T
O n a 8

In this figure, the coordinate of each point is shown below it. Write an expression for each of the following distances.

Example: AT.
Answer: $8 - a$.

24. GN.
25. NT.
26. NA.

27. Copy the figure below and, using the following clues, name as many of the points as you can.

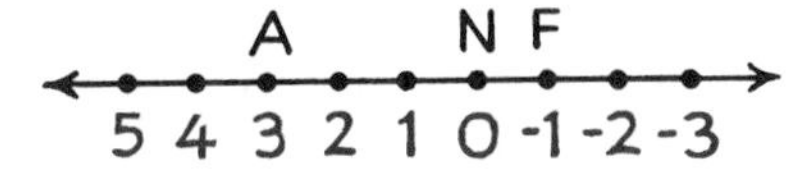

Clue 1. The coordinate of L is -2.
Clue 2. AN = NY. (Assume that A and Y are different points.)
Clue 3. The point R corresponds to 4.
Clue 4. The coordinate of O is the opposite of the coordinate of F.
Clue 5. The coordinate of D is one larger than the coordinate of R.
Clue 6. GF = 3.

The points in each of the following figures are evenly spaced along the lines. Find the missing numbers.

Example:

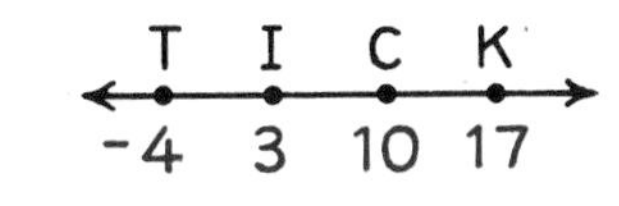

Answer:

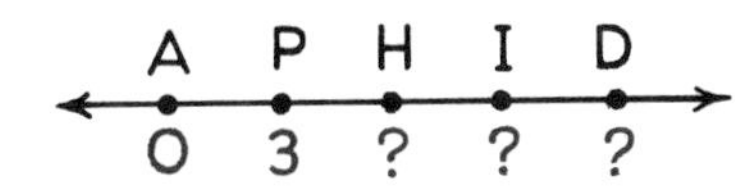

28.

A P H I D
0 3 ? ? ?

29.

L O U S E
5 ? ? ? 15

30.

M I D G E
? 8 ? 2 ?

31.

R O A C H
? 1 -3 ? ?

Copy the figure below and mark it as necessary to find each of the following.

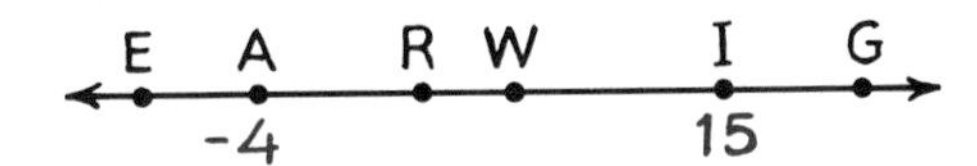

32. Find AI.

33. Find the coordinate of R, given that AR = 7.

34. Find the coordinate of W, given that WI = 9.

35. Find RW.

36. Find the coordinate of E, given that EI = 24.

37. Find the coordinate of G, given that AG = 25.

Set III

Variations of the following puzzle have appeared in many places.

> A bookworm eats its way from page 1 of volume 1 to the last page of volume 20 of an encyclopedia. The books are arranged in order on a bookshelf in the normal way. If the inside of each volume is 4 cm thick and each cover is 0.2 cm thick, through what distance did the bookworm chew?

Most people who try to solve this puzzle get the wrong answer and you probably will too if you don't watch out.

Lesson 3

Properties of Equality

Mathematics is a very precise subject. Answers that are "mostly a matter of opinion" are avoided by starting with statements that have been accepted as true. As you already know, these statements are called *postulates,* and statements that are proved by reasoning deductively with them are called *theorems.*

Because equations are used extensively in geometry, we need some algebraic postulates in addition to the geometric postulates introduced in Lessons 1 and 2 of this chapter. You will probably recognize most of these postulates from your study of algebra.

If the symbols a and b represent real numbers, what do we mean by the statement $a = b$? Simply that a and b *represent the same number.* The statements in the list below are direct consequences of this idea. Because some of them can be proved by using the others, we will refer to them as *properties* of equality.

In the interest of simplicity, the properties are stated symbolically, the letters representing real numbers. The properties have been given names with which you may identify them when you use them.

► The Reflexive Property
$a = a$. (Any number is equal to itself.)

► The Symmetric Property
If $a = b$, then $b = a$.

► The Transitive Property
If $a = b$ and $b = c$, then $a = c$.

► The Substitution Property
If $a = b$, then a can be substituted for b in any equation.

► The Addition and Subtraction Properties
If $a = b$, then $a \pm c = b \pm c$.
If $a = b$ and $c = d$, then $a \pm c = b \pm d$.

► The Multiplication Properties
If $a = b$, then $ac = bc$.
If $a = b$ and $c = d$, then $ac = bd$.

► The Division Properties
If $a = b$ and $c \neq 0$, then $\frac{a}{c} = \frac{b}{c}$.

If $a = b$ and $c = d \neq 0$, then $\frac{a}{c} = \frac{b}{d}$.

► The Square Roots Property*
If $a = b \geq 0$, then $\sqrt{a} = \sqrt{b}$.

Here is an example of how some of these properties, together with the properties of real numbers, are used to solve an algebraic equation.

Equation:	$7 + x = 4(x - 2)$	
Solution:	$4(x - 2) = 4x - 8$	Distributive property.
	$7 + x = 4x - 8$	Transitive property.
	$7 = 3x - 8$	Subtraction property (x was subtracted from each side).
	$15 = 3x$	Addition property (8 was added to each side).
	$5 = x$	Division property (each side was divided by 3).
	$x = 5$	Symmetric property.

*For this property to be meaningful, it is important to understand that the symbol $\sqrt{n}$ represents the *positive* square root of n. For example, 9 has *two* square roots: 3 and -3. The symbol $\sqrt{9}$, however, represents only one of them: 3.

Exercises

Set I

Match the following statements with the names of the properties they illustrate.

1. If $x = y$ and $y = z$, then $x = z$.	(A) Addition.
2. If $x = y$, then $xz = yz$.	(B) Transitive.
3. If $x = y \geq 0$, then $\sqrt{x} = \sqrt{y}$.	(C) Symmetric.
4. If $x = y$, then $y = x$.	(D) Multiplication.
5. $x = x$.	(E) Square roots.
6. If $w = x$ and $y = z$, then $w + y = x + z$.	(F) Reflexive.

Name the property illustrated by each of the following statements.

7. $ab = ab$.
8. If $a - b = c$, then $a = b + c$.
9. If $a = b - 1$ and $b - 1 = c$, then $a = c$.
10. If $a + b = c + d$ and $a = c$, then $b = d$.
11. If $a = b - c$, then $b - c = a$.
12. If $a + b = c$, then $\frac{a + b}{c} = 1$.

Use the property named to complete each of the following statements.

13. The symmetric property: If $x + 7 = y$, then |||||||||.
14. The multiplication property: If $x = 2$ and $y = 5$, then |||||||||.
15. The square roots property: If $x^2 = y \geq 0$, then |||||||||.
16. The reflexive property: $x - y =$ |||||||||.
17. The subtraction property: If $x + 2 = y$, then $x =$ |||||||||.
18. The transitive property: If $x + 7 = 2y$ and $2y = 5x$, then |||||||||.
19. The division property: If $xy = 7$, then $x =$ |||||||||.

Use the substitution property to complete each of the following statements.

Example: If $x - y = 10$ and $y = 3$, then |||||||||.
Answer: $x - 3 = 10$.

20. If $12 + x = 5y$ and $x = 2y$, then |||||||||.
21. If $x^2 - 1 = y^2$ and $y = 4$, then |||||||||.
22. If $\frac{x}{3} = \frac{y + 2}{x}$ and $y + 2 = 7$, then |||||||||.
23. If $3x = 8 - y$ and $x = y + 1$, then |||||||||.

Set II

In each of the following exercises, a figure is shown together with some given facts about it. Tell which fact (or facts) and what property of equality can be used to prove the conclusion stated in each exercise.

S O D A

Fact 1. SO = OD.
Fact 2. OD + DA = OA.

Example 1: 2SO = 2OD.
Answer: Fact 1 and the multiplication property.

Example 2: SO + DA = OA.
Answer: Facts 1 and 2 and the substitution property.

24. OD = SO.

25. OD = OA − DA.

26. SO + OD + DA = OD + OA.

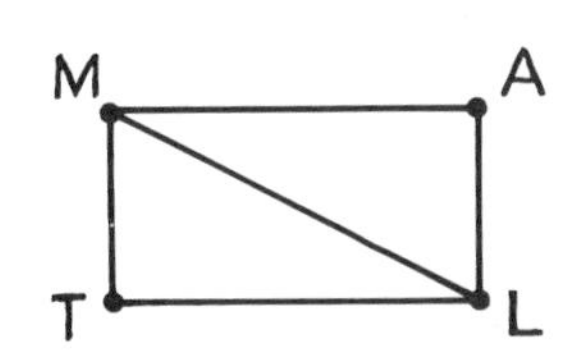

Fact 1. MA = 2MT.
Fact 2. $MT^2 = AL^2$.
Fact 3. MA = TL.

27. $\frac{1}{2}$MA = MT.

28. MT = AL.

29. MA + ML = TL + ML.

30. TL = 2MT.

31. $AL^2 = MT^2$.

32. ML = LM. (Name only the property of equality.)

33. MA − TL = 0.

34. $\frac{MA}{TL} = 1$.

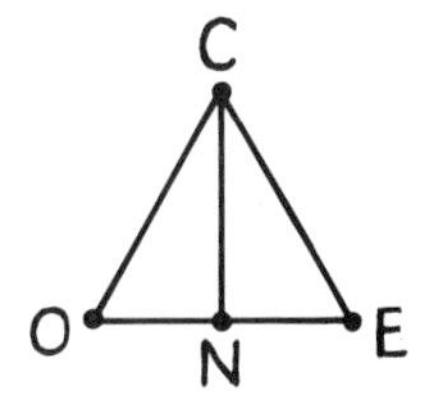

Fact 1. CO = OE.
Fact 2. OE = CE.
Fact 3. NE = OE − ON.

35. CO = CE.

36. OE − ON = NE.

37. NE = CE − ON.

38. CO + OE = 2OE.

39. CN = CN.

40. $\frac{CO}{OE} = \frac{OE}{CE}$.

Set III

Consider the following three statements:

Snail A is as slow as itself.
If snail A is as slow as snail B, then snail B is as slow as snail A.
If snail A is as slow as snail B and snail B is as slow as snail C, then snail A is as slow as snail C.

These three statements sound like the reflexive, symmetric, and transitive properties of equality. A relation that is reflexive, symmetric, and transitive is called an *equivalence relation.*

Which of the following do you think are equivalence relations? Explain.

1. Is older than.
2. Is the same color as.
3. Is a friend of.

Lesson 4

Betweenness of Points

The map of the United States shown here appears to be "upside down." This is because we are used to thinking of the North Pole as the "top" of the world. If the South Pole were chosen as the top instead, the directions of north and south as customarily shown on maps would probably be reversed. If the development of civilization in the southern hemisphere had preceded that in the northern, it is likely that our maps would look like the one shown above.

Look at the points on the map representing the cities of Santa Fe, Las Vegas, and San Francisco. Which city is between the other two? The answer seems fairly obvious: Las Vegas is. Now look at the points representing St. Louis, Minneapolis, and Denver. Which one of these cities is between the other two? In this case, the question doesn't even seem to make sense.

The difference in the two situations is that the points representing the cities of Santa Fe, Las Vegas, and San Francisco seem to be *collinear,* whereas the points representing St. Louis, Minneapolis, and Denver are not. Because of this, the idea of "betweenness" of points is limited in geometry to sets of points that are collinear.

In the figure at the left, points A, B, and C are collinear because there is a line that contains all of them. If we place a ruler in line with the points, we see that the fact that B is between A and C is related to the fact that the coordinate of B is between the coordinates of A and C; that is, $4 < 5 < 7$.*

*Recall that the symbol $<$ means "is less than"; the symbol $>$ means "is greater than."

This relation suggests the following definition.

► **Definition** (Betweenness of Points)
Suppose that points A, B, and C are collinear with coordinates a, b, and c, respectively. Point B is between points A and C (written A-B-C) iff either $a < b < c$ or $a > b > c$.

A direct consequence of this definition is the fact that if one point is between two others, the three distances that they determine are related in a very simple way. We will prove this as a theorem.

► **Theorem 1** (Betweenness of Points Theorem)
If A-B-C, then AB + BC = AC.

To prove this theorem, we have to consider two possibilities. If A-B-C, then either $a < b < c$ or $a > b > c$. The first possibility, $a < b < c$, will be considered here and the second possibility considered in one of the exercises.

To make the proof easier to read, the statements in it are listed at the left and the reasons for them at the right.

Proof.

Statements	*Reasons*
1. If A-B-C, then either $a < b < c$ or $a > b > c$.	Definition of betweenness of points.
2. If $a < b < c$, then AB $= \lvert a - b \rvert = b - a$ and BC $= \lvert b - c \rvert = c - b$.	Ruler postulate.
3. AB + BC $= (b - a) + (c - b)$ $= b - a + c - b$ $= c - a$.	Addition property of equality.
4. But AC $= \lvert a - c \rvert = c - a$.	Ruler postulate.
5. So AB + BC = AC.	Substitution property of equality.

Exercises

Set I

Points Y, P, and A are collinear with coordinates 1, 7, and 3, respectively.

1. Draw a figure to scale that illustrates this.

2. Which point is between the other two?

3. Copy and complete the following inequality relating the coordinates of the three points: $1 < ▒ < ▒$.

4. Find each of the following distances: YA, AP, and YP.

5. Write the equation relating YA, AP, and YP that follows from the Betweenness of Points Theorem.

The following exercises refer to this figure.

N E I G H

2 3 4 5 6 7 8 9

6. Is it true that E-G-H?
7. Use the definition of betweenness of points to show why or why not.
8. Is it true that EG + GH = EH?
9. Is it true that H-E-N?
10. Is it true that HE = EN?

Points M, E, and W have coordinates m, e, and w, respectively; M-E-W. Tell whether each of the following statements *must be true, may be true,* or *is false.*

11. Points M, E, and W are collinear.
12. Point E is between M and W.
13. $m < e < w$.
14. $m > w > e$.
15. ME + EW = MW.
16. ME = EW.

In the figure below, point A is between points B and Y.

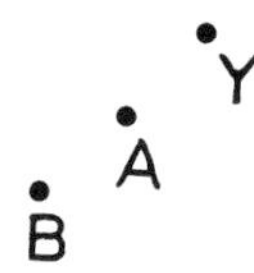

17. Write a three-letter symbol to represent this fact.
18. Write an equation that follows from this fact.
19. Does it follow that BA = AY?

In the figure below, points C, R, O, A, and K are collinear in the order shown.

Refer to the figure to copy and complete the following equations.

Example: CR + ______ = CA.
Answer: CR + RA = CA.

20. CA + AK = ______.
21. KR = KA + ______.
22. CR + RO + OA = ______.
23. RA − RO = ______.
24. AC − ______ = OC.

Set II

Points B, A, R, and K are four different points on a line, but not necessarily in that order. Tell what betweenness relation must be true if each of the following equations is true.

Example: AK + KB = AB.
Answer: A-K-B.

25. AK = AR + RK.
26. BR = BK.
27. RB − RA = AB.
28. KB = RB − RK.

29. In the figure below, we can prove that if YE = LP, Y-E-L, and E-L-P, then YL = EP. For convenience, we will refer to the parts of the hypothesis as "given" and the conclusion as "prove."

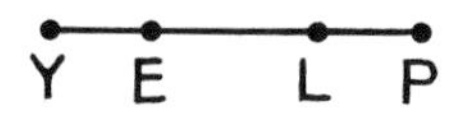

Given: YE = LP, Y-E-L, and E-L-P.
Prove: YL = EP.

Copy and complete the proof.

Statements	Reasons
1. YE = LP.	Given.
2. YE + EL = EL + LP.	________ (Why does this equation follow from the previous one?)
3. Y-E-L and E-L-P.	Given.
4. YE + EL = ________ and EL + LP = ________.	Betweenness of Points Theorem.
5. ________ = ________.	Substitution (steps 2 and 4).

30. To prove the Betweenness of Points Theorem, it is necessary to consider two possibilities. If A-B-C, then either $a < b < c$ or $a > b > c$. Use the proof of the first possibility (see page 65) as a guide in copying and completing the following proof of the second possibility.

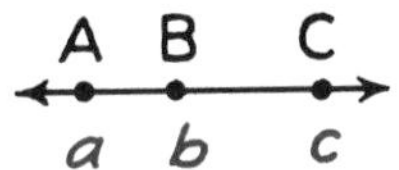

Statements	Reasons
1. If A-B-C, then either $a < b < c$ or $a > b > c$.	Definition of betweenness of points.
2. If $a > b > c$, then AB = ________ = ________ and BC = ________ = ________.	Ruler postulate.
3. AB + BC = (________) + (________) = ________ + ________ = ________.	________
4. But AC = ________ = ________.	________
5. So AB + BC = AC.	________

Set III

Five geometry students are sitting in a row.

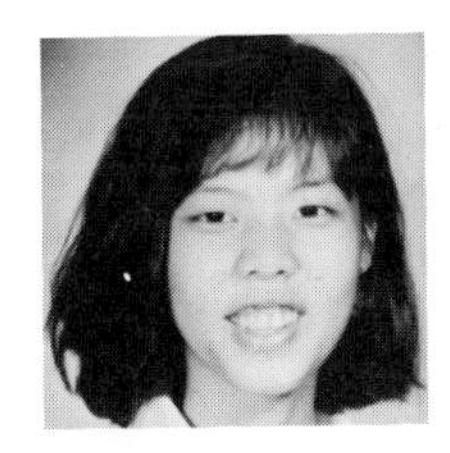

Linh

Brad

Tammy

Lee

Tiina

Can you figure out the order in which they are seated from the following clues?

Clue 1. Lee is the same distance from Linh that Linh is from Brad.
Clue 2. Tiina is seated between Tammy and Linh.
Clue 3. Brad is sitting next to Tiina.
Clue 4. Tiina is not seated between Brad and Tammy.

If you can, show your method.

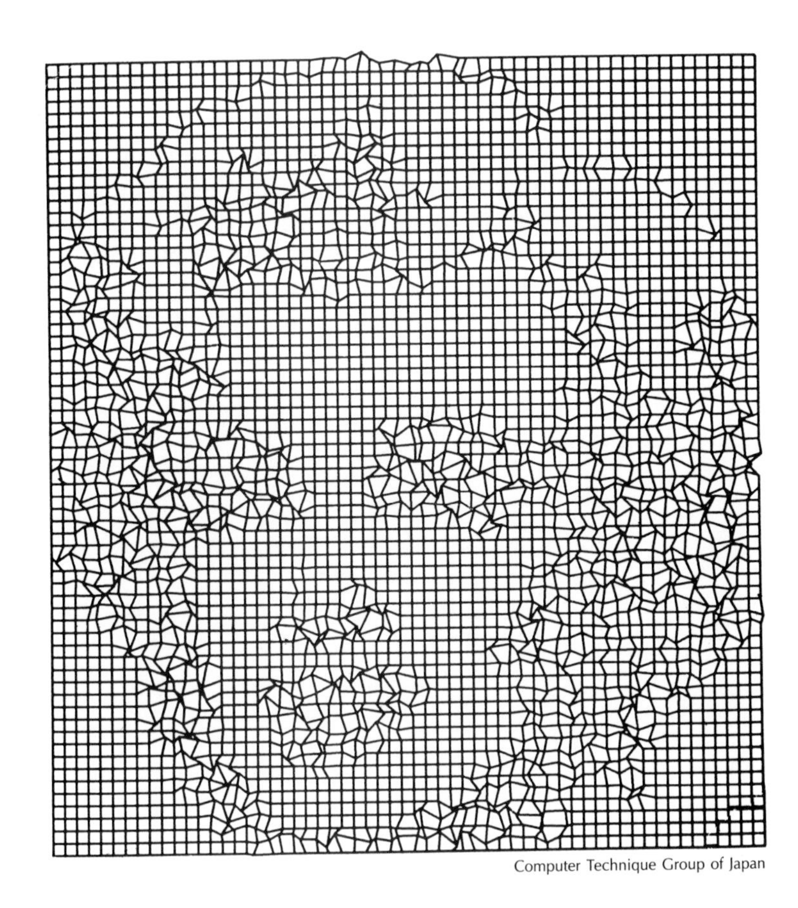
Computer Technique Group of Japan

Lesson 5
Line Segments

This picture of Marilyn Monroe was produced by a computer programmed to transform a photograph into a net pattern. The net consists entirely of line segments.

▶ **Definition**
A ***line segment*** is the set of two points and all the points between them.

To distinguish between the *line* through points A and B and the *line segment* AB, we will use different symbols. The symbol $\overleftrightarrow{AB}$ represents the *line*, and the symbol $\overline{AB}$ represents the *line segment*. The figure at the right illustrates $\overline{AB}$; the points A and B are called its *endpoints*.

A line segment, then, is a set of points; its length is a number.

▶ **Definition**
The ***length of a line segment*** is the distance between its endpoints.

For simplicity, we will refer to line segments that have equal lengths as "equal segments."

▶ **Definition**
A ***midpoint of a line segment*** is a point between its endpoints that divides it into two equal segments.

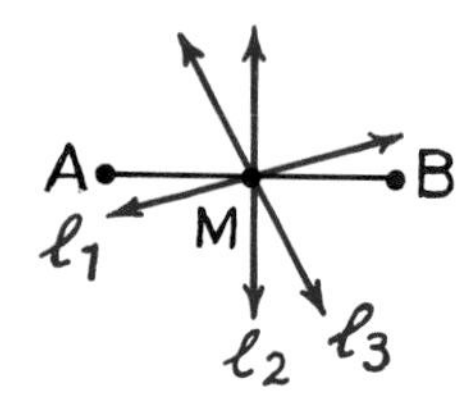

To *bisect* means to divide into two equal parts. A midpoint of a line segment, therefore, bisects it, as does any line, ray, or segment that intersects it in a midpoint. In the figure at the left, if M is a midpoint of $\overline{AB}$, then ℓ_1, ℓ_2, and ℓ_3 are three *bisectors* of $\overline{AB}$.

How many midpoints does a line segment have? We will make the following assumption.

► **Postulate 6** (The Midpoint Postulate)
A line segment has exactly one midpoint.

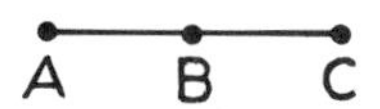

If B is the midpoint of $\overline{AC}$ in this figure, then AB = BC. It is also true that $AB = \frac{1}{2}AC$ and $BC = \frac{1}{2}AC$.

► **Theorem 2**
The midpoint of a line segment divides it into segments half as long as the line segment.

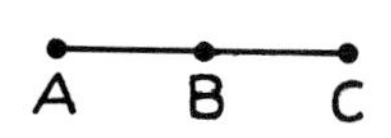

The figure at the left illustrates the theorem. We will refer to the hypothesis of the theorem as "given" and the conclusion as "prove." In terms of the figure, the "given" and "prove" are:

Given: B is the midpoint of $\overline{AC}$.
Prove: $AB = \frac{1}{2}AC$ and $BC = \frac{1}{2}AC$.

Proof.

Statements	***Reasons***
1. B is the midpoint of $\overline{AC}$.	Given.
2. A-B-C and AB = BC.	The midpoint of a line segment is the point between its endpoints that divides it into two equal segments.
3. AB + BC = AC.	The Betweenness of Points Theorem.
4. AB + AB = AC, so 2AB = AC.	Substitution (steps 2 and 3).
5. $AB = \frac{1}{2}AC$.	Multiplication $\left(\text{by } \frac{1}{2}\right)$.
6. $BC = \frac{1}{2}AC$.	Substitution (steps 2 and 5).

Exercises

Set I

Match the following symbols with the words they represent.

1. JO. (A) A pair of points.
2. $\overline{JO}$. (B) Betweenness of points.
3. $\overleftrightarrow{JO}$. (C) A number.
4. J and O. (D) Equal lengths.
5. J-O-Y. (E) A line.
6. JO = OY. (F) A line segment.
7. Write the definition of midpoint of a line segment using the words "if and only if."
8. How many midpoints does a line segment have?
9. How many bisectors does a line segment have?
10. Write Theorem 2 in "if-then" form.
11. Write the converse of Theorem 2. Do you think it is true?

In the figure below, point M is the midpoint of $\overline{AY}$. Use the coordinates given to find

A M Y
7 15

12. AY.
13. AM.
14. the coordinate of point M.
15. How do $\overline{AM}$ and $\overline{MY}$ compare in length?
16. How do $\overline{AM}$ and $\overline{AY}$ compare in length?

In the figure below, S-U-E.

S U E

17. Write the equation that follows from this fact.

In the same figure, U is the midpoint of $\overline{SE}$.

18. Write an equation relating the lengths SU and UE that follows from this fact.
19. Write an equation relating the lengths UE and SE that follows from this fact.

Points I and S trisect $\overline{LA}$.

L I S A
30 6

20. What does "trisect" mean?
21. Write an equation relating the lengths IS and LA.
22. Write an equation relating the lengths LS and SA.

Use the coordinates given to find

23. LA.
24. LI.
25. the coordinates of points I and S.

Set II

In the figure below, points I, N, and A are collinear. Tell what is wrong with each of the following statements.

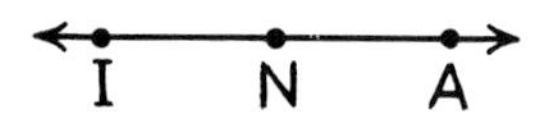

26. If IN = NA, then N is the midpoint of $\overleftrightarrow{IA}$.
27. If I-N-A, then $\overline{IN} + \overline{NA} = \overline{IA}$.
28. If IN = NA, then I = A by dividing both sides by N.

In the figure below, B is the midpoint of $\overline{AC}$; the coordinate of each point is given below it and $a < b < c$.

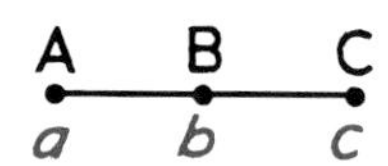

Write an expression for each of the following in terms of the given coordinates.

Example: AC in terms of a and c.
Answer: $c - a$.

29. AB in terms of a and b.
30. AB in terms of a and c.
31. Set the expressions you have written for Exercises 29 and 30 equal and solve the resulting equation for b in terms of a and c.

In the figure below, A is the midpoint of $\overline{MY}$. If m is the coordinate of M and MA = r,

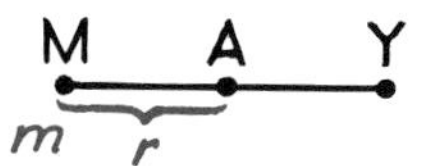

32. what could be the coordinate of A?
33. what could be the coordinate of Y?

34. Copy and complete the lettered statements in the following proof.

Given: I is the midpoint of both $\overline{AE}$ and $\overline{LC}$; AE = LC.
Prove: AI = LI.

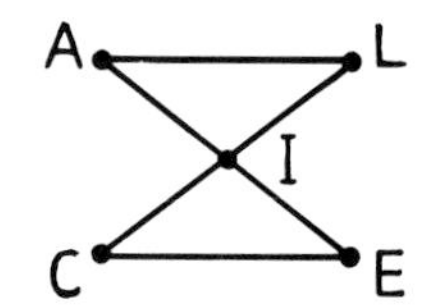

Statements	***Reasons***
1. I is the midpoint of $\overline{AE}$.	Given.
2. a) AI = $\frac{1}{2}$____.	The midpoint of a line segment divides it into segments half as long as the line segment.
3. b) AE = ____.	Given.
4. c) ____	Substitution (steps 2 and 3).
5. I is the midpoint of $\overline{LC}$.	Given.
6. d) LI = ____.	Same as reason for step 2.
7. e) ____	Substitution (steps 4 and 6).

35. Copy and complete the lettered statements in the following proof.

Given: E is the midpoint of $\overline{BT}$ and T is the midpoint of $\overline{EH}$.
Prove: BT = EH.

B E T H

Statements	***Reasons***
1. a) |||||||||||	Given.
2. b) BE = ||||||||||| and ET = |||||||||||.	The midpoint of a line segment divides it into two equal segments.
3. c) |||||||||||	Transitive.
4. d) BE = ||||||||||| and TH = |||||||||||.	The midpoint of a line segment divides it into segments half as long as the line segment.
5. $\frac{1}{2}BT = \frac{1}{2}EH$.	Substitution (steps 3 and 4).
6. e) |||||||||||	Multiplication.

Set III

An Experiment with an Unexpected Result

Cut out a strip of paper about 10 inches long and 1 inch wide. Make it into a loop, turn one end over, and tape the two ends together as shown in the first photograph. Now cut the loop along its center as shown in the second photograph.

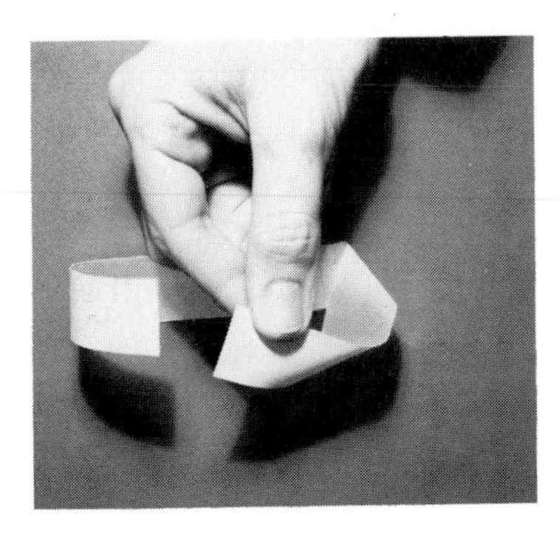

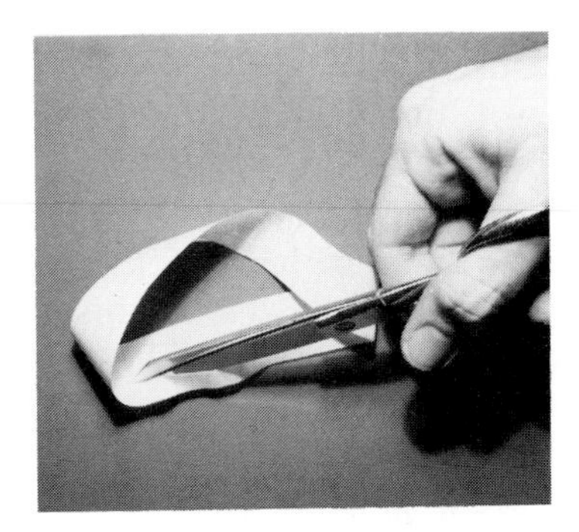

You know that to bisect is to cut or divide into two equal parts. Does cutting all the way around the loop bisect it? Explain your answer.

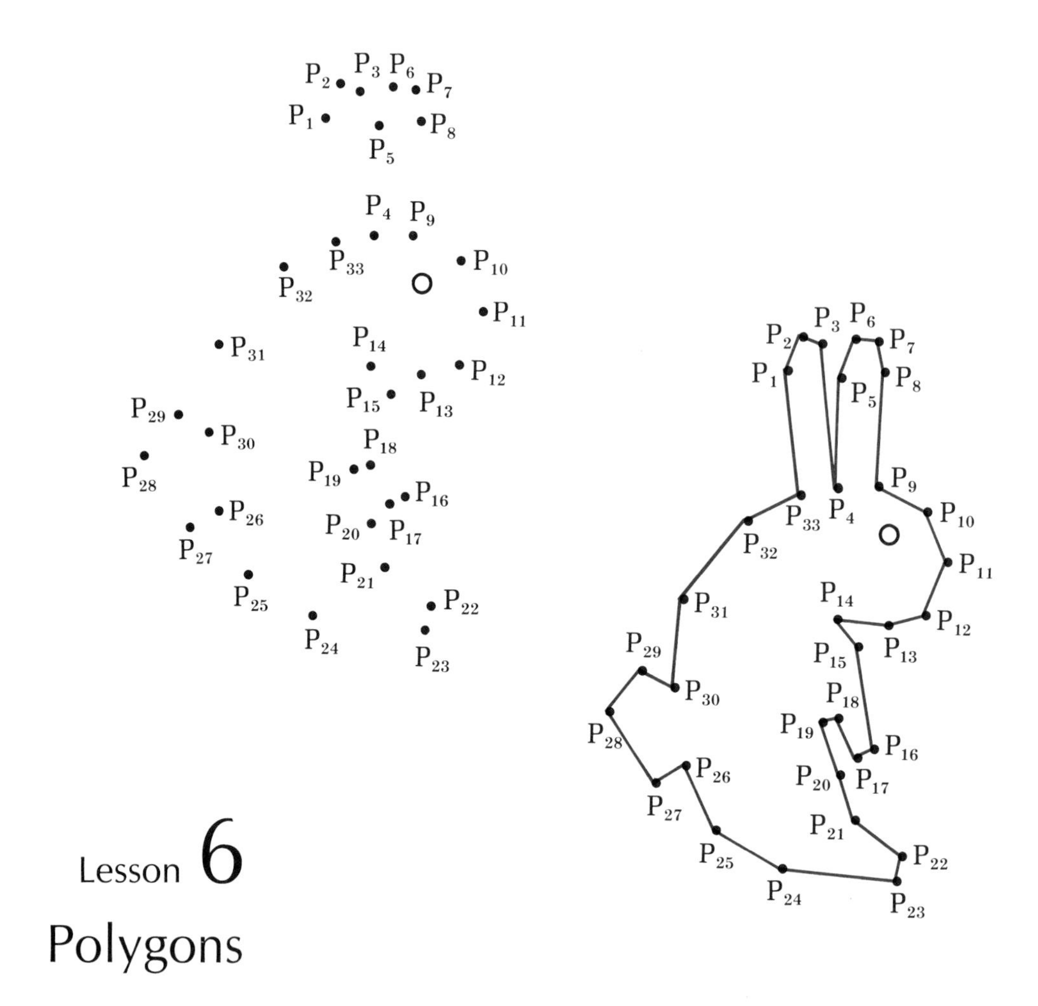

Lesson 6
Polygons

A popular children's puzzle consists of a set of numbered dots that are to be joined in sequence to make a picture. An example of such a puzzle, together with its solution, is shown above. This puzzle is somewhat unusual in that it looks like two different things, depending on how you view it. Turn the page 90° counterclockwise and you may see something entirely different.

In solving such a puzzle, a set of points is joined by a series of line segments in a certain order. In the example shown above, the last point has also been joined to the first to make a closed figure. The result is an example of a *polygon.* The term "polygon" is derived from a Greek word meaning "many angled." We will base our definition on the way in which the figure is drawn.

► **Definition**

Let $P_1, P_2, \ldots, P_n$ be a set of at least three points in a plane such that no three consecutive points are collinear. If the segments $\overline{P_1P_2}, \overline{P_2P_3}, \ldots, \overline{P_nP_1}$ intersect only at their endpoints, they form a ***polygon.***

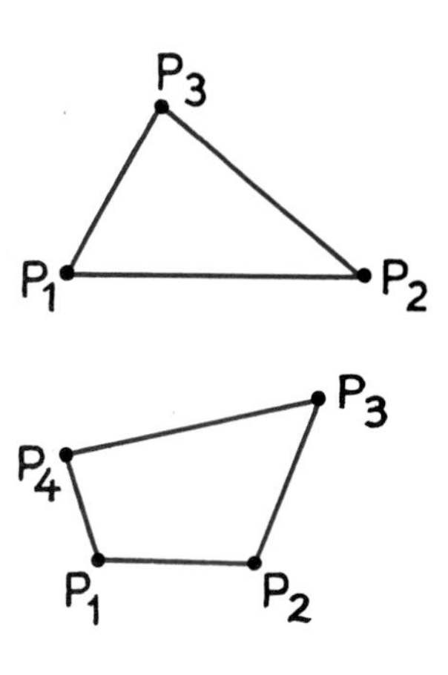

The simplest polygon is the one for which $n = 3$. In this case, the three points P_1, P_2, P_3 determine the three line segments $\overline{P_1P_2}$, $\overline{P_2P_3}$, $\overline{P_3P_1}$. The polygon is a *triangle.*

If $n = 4$, the four points P_1, P_2, P_3, P_4 determine the four line segments $\overline{P_1P_2}$, $\overline{P_2P_3}$, $\overline{P_3P_4}$, $\overline{P_4P_1}$ and the polygon is a *quadrilateral.*

The points that determine a polygon are called its *vertices* and the line segments are called its *sides.* For example, polygon ABCDEF shown at the left below has six vertices and six sides. The second figure shows that the polygon also has nine *diagonals.*

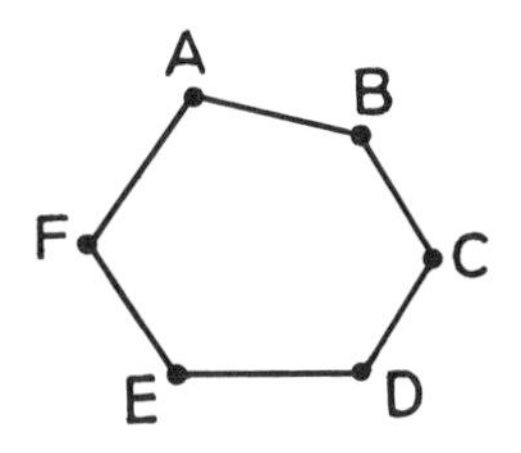

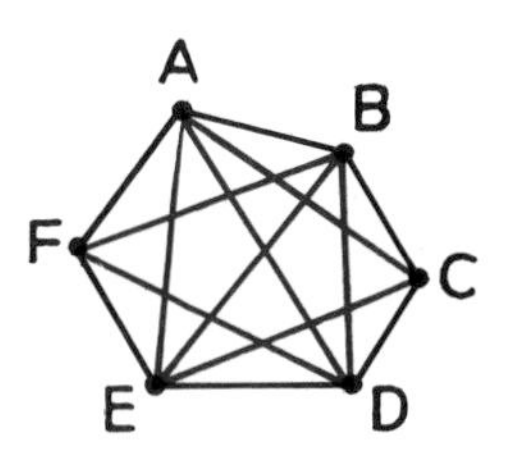

► **Definition**

A ***diagonal*** of a polygon is a line segment that joins any two nonconsecutive vertices.

Polygons are named according to the number of sides that they have. Those to which we will refer the most frequently are listed in the table at the right.

Number of sides	*Name of polygon*
3	Triangle
4	Quadrilateral
5	Pentagon
6	Hexagon
7	Heptagon
8	Octagon
9	Nonagon
10	Decagon
12	Dodecagon

In general, a polygon having n sides is called an *n-gon.* The "rabbit-duck" polygon, for example, is a 33-gon.

The distance traveled in drawing a polygon is called its *perimeter.* We will represent the word "perimeter" by ρ (rho, a letter of the Greek alphabet).

► **Definition**

The ***perimeter*** of a polygon is the sum of the lengths of its sides.

Polygons are classified as being either *convex* or *concave.* The two figures below illustrate the difference.

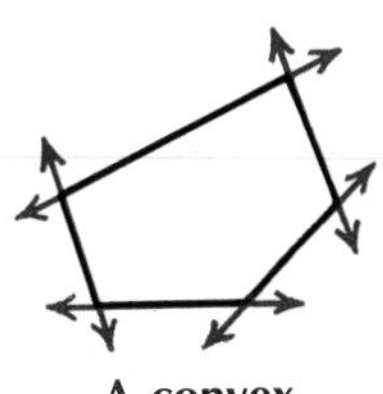
A convex polygon

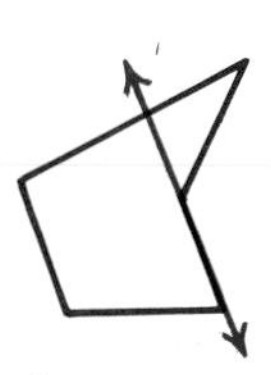
A concave polygon

► **Definitions**

A polygon is ***convex*** iff, for each line that contains a side of the polygon, the rest of the polygon lies on one side of the line. A polygon that is not convex is ***concave.***

Exercises

Set I

The following exercises refer to polygon HOAX shown here.

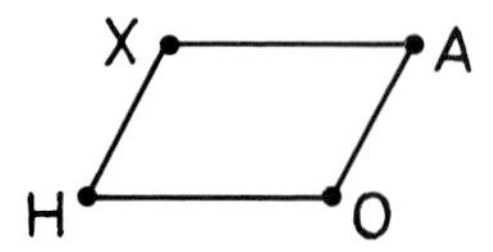

1. What are H, O, A, and X called with respect to the polygon?
2. What are $\overline{HO}$, $\overline{OA}$, $\overline{AX}$, and $\overline{XH}$ called?
3. What is HO + OA + AX + XH called?
4. If $\overline{HA}$ and $\overline{OX}$ were drawn, what would they be called?
5. What type of polygon is HOAX: *convex* or *concave?*
6. What is HOAX called with respect to the number of its sides?
7. Name each of the polygons below according to the number of its sides.

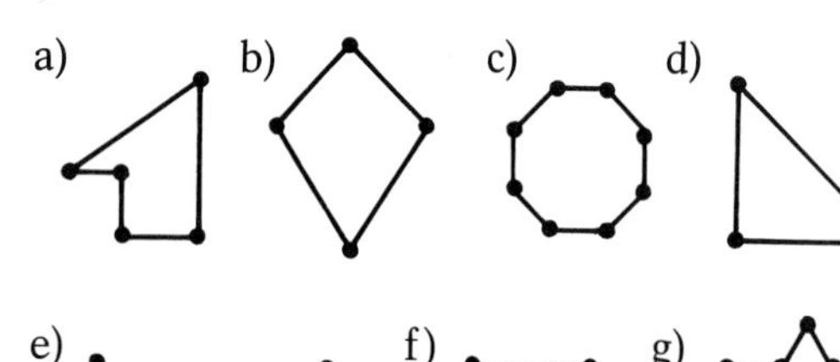

8. Which polygons in Exercise 7 are convex?
9. Is the first figure below a polygon?

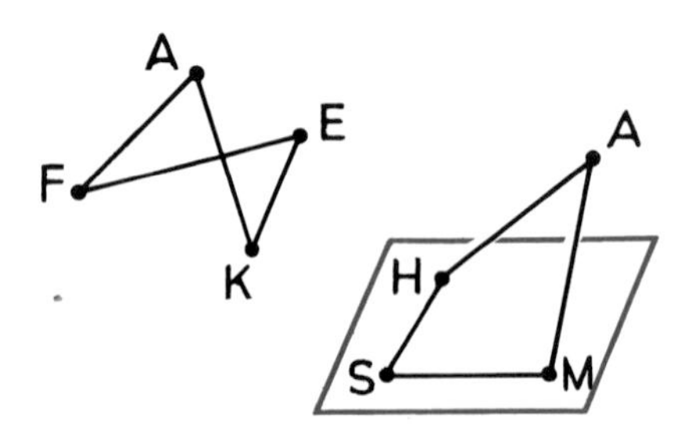

10. Use the definition of a polygon to explain why or why not.
11. Is the second figure below Exercise 9 a polygon?
12. Use the definition of a polygon to explain why or why not.

The four points below can be connected with four line segments in three different ways to form the three different polygons shown.

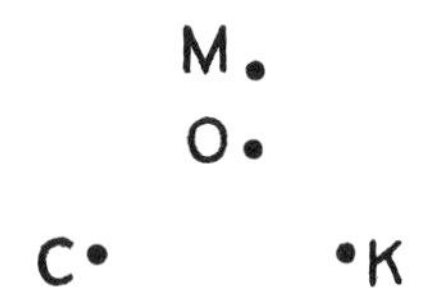

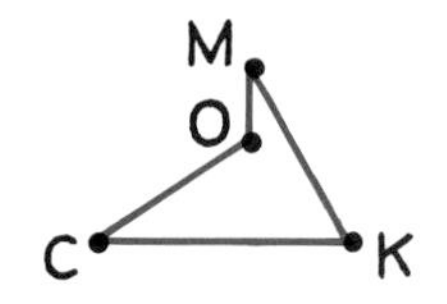

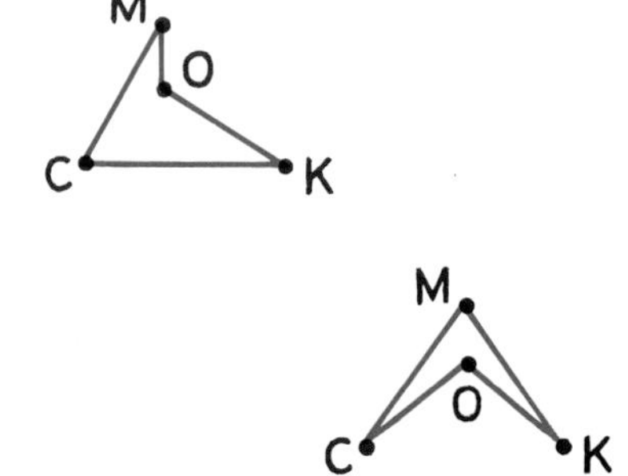

13. Trace the five points below and connect them with five line segments to form a polygon. Repeat this procedure to form as many different polygons from the five points as possible.

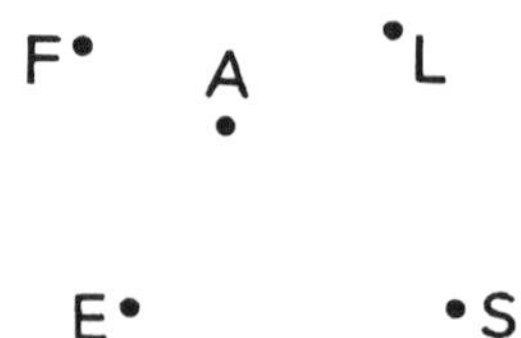

14. Trace the six points below and connect them with six line segments to form a polygon.

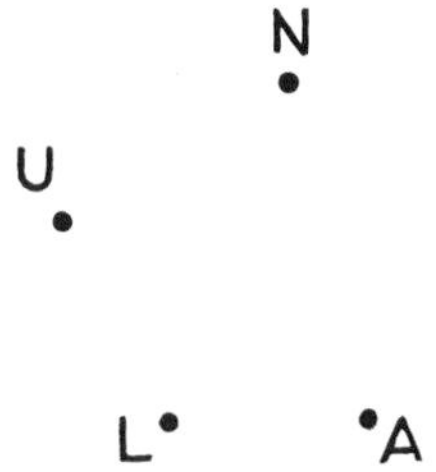

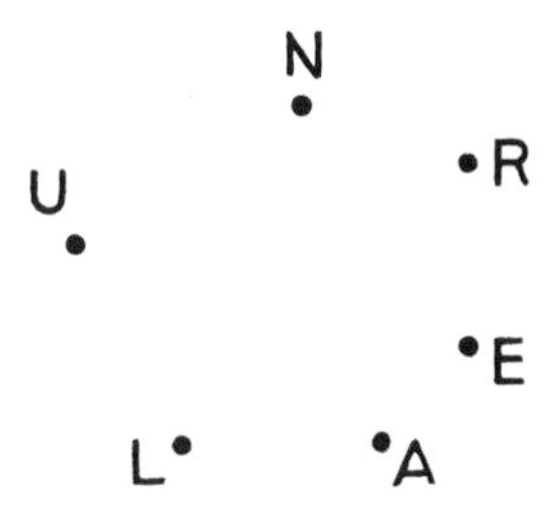

15. How many different polygons do you think these six points determine?

16. Define *perimeter of a polygon.*

Write an expression for the perimeter of each of the following polygons.

Example:

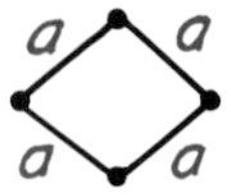

Answer: $4a$.

17.

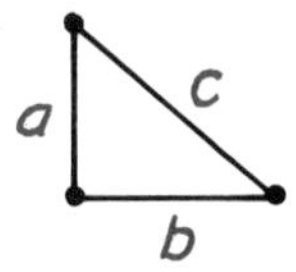

18.

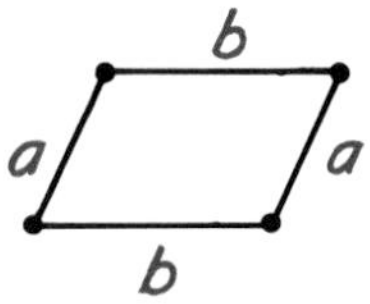

19.

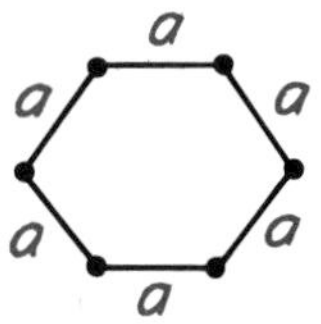

20.

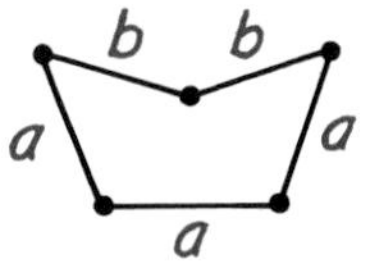

Set II

This figure shows a convex pentagon in which diagonals have been drawn from vertex B.

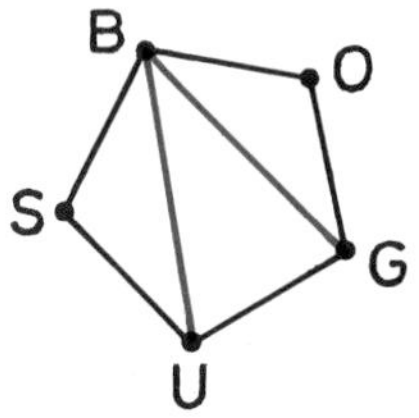

21. How many diagonals were drawn?

22. How many triangles are formed?

23. Draw a convex hexagon and all of the diagonals from one vertex.

24. How many diagonals did you draw?

25. How many triangles were formed?

26. If all of the diagonals from one vertex of a convex octagon are drawn, how many diagonals do you think there would be?

27. How many triangles do you think would be formed?

28. If all of the diagonals from one vertex of a convex n-gon are drawn, how many diagonals do you think there would be?

29. How many triangles do you think would be formed?

Copy and complete the following proofs.

30. *Given:* Y is the midpoint of $\overline{PO}$;
PO = NO.
Prove: NO = 2YO.

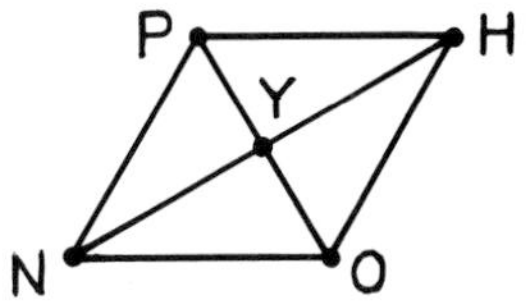

Proof.

Statements	***Reasons***
1. Y is the midpoint of $\overline{PO}$.	Given.
2. YO $= \frac{1}{2}$______.	______
3. PO = NO.	______
4. YO $= \frac{1}{2}$______.	Substitution (steps 2 and 3).
5. 2YO = ______.	______
6. ______	______

31.

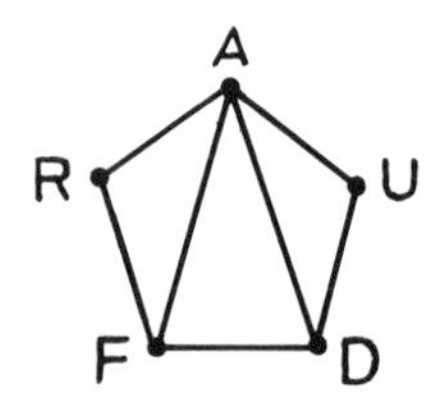

Given: FR = DU, RA = UA, and AD = AF.

Prove: ρ FRAD = ρ DUAF.

Proof.

<table>
<tr><th>Statements</th><th>Reasons</th></tr>
<tr><td>1. FR = DU, RA = UA, and AD = AF.</td><td>||||||||</td></tr>
<tr><td>2. FR + RA + AD = DU + UA + AF.</td><td>||||||||</td></tr>
<tr><td>3. FR + RA + AD + |||||||| = DU + UA + AF + ||||||||.</td><td>Same as preceding reason.</td></tr>
<tr><td>4. ρ FRAD = FR + RA + AD + ||||||||; ρ DUAF = DU + UA + AF + ||||||||.</td><td>||||||||</td></tr>
<tr><td>5. ||||||||</td><td>||||||||</td></tr>
</table>

Set III

The 33-gon at the beginning of this lesson has an interesting shape.

1. Use your imagination to invent a different example of a polygon that has an interesting shape.
2. How many vertices and sides does your polygon have?
3. Is it concave or convex?

By the permission of Johnny Hart and News America Syndicate

Chapter 2 / Summary and Review

You should be familiar with the following concepts introduced in Chapter 2.

Basic Ideas

Addition and subtraction properties 61
Betweenness of points 65
Collinear points 51
Convex and concave polygons 75
Coordinate of a point 57
Coplanar points 51
Diagonal 75
Distance between two points 57
Division properties 61
Length of a line segment 69
Line 51
Line segment 69
Midpoint of a line segment 69
Multiplication properties 61
Perimeter 75
Plane 51
Polygon 74
Polygons, names of 75
Reflexive property 61
Square roots property 61
Substitution property 61
Symmetric property 61
Transitive property 61

Postulates

1. If there are two points, then there is exactly one line that contains them. (Two points determine a line.) 51
2. If there is a line, then there are at least two points on the line. 51
3. If there are three noncollinear points, then there is exactly one plane that contains them. (Three noncollinear points determine a plane.) 51
4. If two points lie in a plane, then the line that contains them lies in the plane. 52
5. *The Ruler Postulate.* The points on a line can be numbered so that
   a) to every point there corresponds exactly one real number called its coordinate,
   b) to every real number there corresponds exactly one point,
   c) to every pair of points there corresponds exactly one real number called the distance between the points,
   d) and the distance between two points is the absolute value of the difference between their coordinates. 57
6. *The Midpoint Postulate.* A line segment has exactly one midpoint. 70

Theorems

1. *The Betweenness of Points Theorem.* If A-B-C, then AB + BC = AC. 65
2. The midpoint of a line segment divides it into segments half as long as the line segment. 70

Exercises

Set I

The following exercises refer to the figure below, which represents three planes.

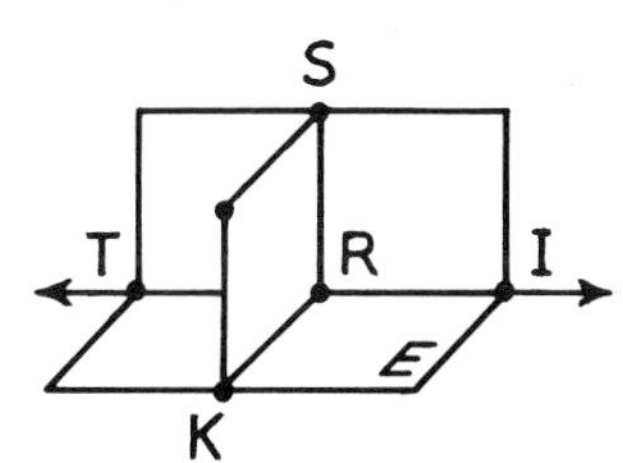

1. What relation do points T, R, and I have?
2. How many planes in the figure contain point R?
3. How many planes in the figure contain the two points K and R?

4. How many planes in the figure contain the three points K, R, and S?
5. State the postulate illustrated by your answer to Exercise 4.
6. What relation do points S, T, and I have?
7. If points T and I lie in plane E, can you conclude that $\overleftrightarrow{TI}$ must also lie in plane E?
8. State the postulate that is the basis for your answer to Exercise 7.

In the figure below, point E is between points B and L.

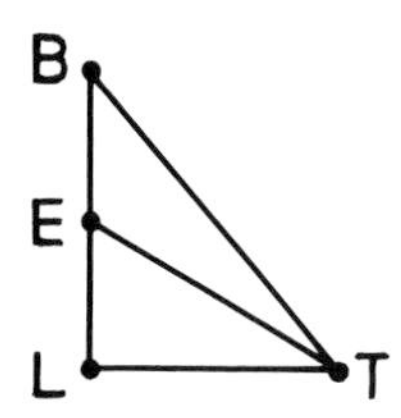

9. Write a three-letter symbol to represent this fact.
10. Write the equation that follows from this fact.

Point E is the midpoint of $\overline{BL}$.

11. Write an equation relating BE and EL that follows from this fact.
12. Write an equation relating EL and BL that follows from this fact.

In the figure below, H is the midpoint of $\overline{WP}$ and O is the midpoint of $\overline{HP}$.

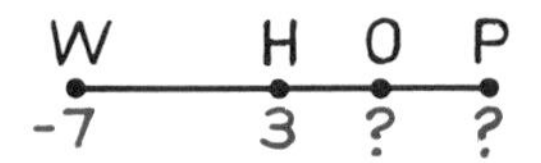

13. Find WH.
14. Find the coordinate of P.
15. Find HO.
16. Find the coordinate of O.

Set II

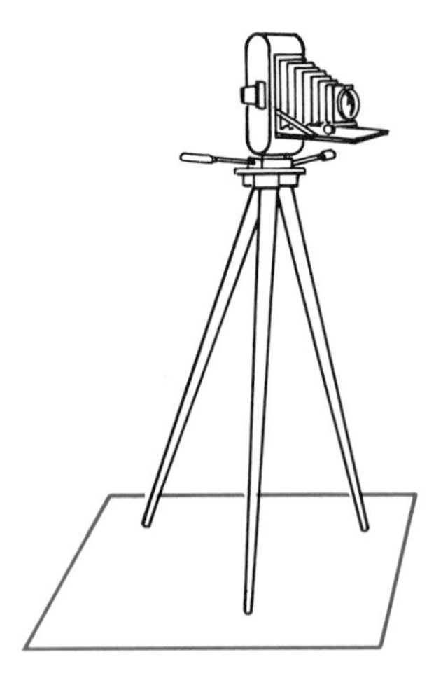

A tripod is a device used to provide steady support to a camera.

17. State the postulate that is the basis for the fact that a tripod has three legs.
18. What might happen if a tripod had four legs instead of three?

The points in the figure below are evenly spaced along the line.

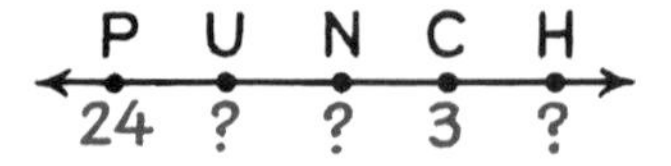

19. Find the missing numbers.

Copy the figure below.

W H A C K

−17 31

Use the coordinates shown and the following information to find the numbers in Exercises 20 through 25.

HA = 11, AC = 24, H is the midpoint of $\overline{WC}$.

20. HK.
21. The coordinate of A.
22. The coordinate of C.
23. HC.
24. WH.
25. The coordinate of W.

The following exercises refer to this figure.

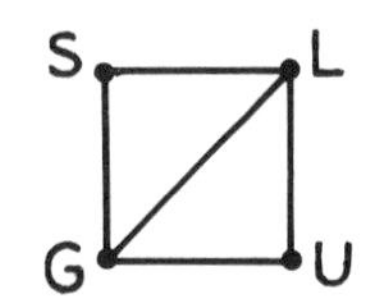

Tell which fact (or facts) and what property of equality can be used to prove the conclusion stated in each exercise.

Fact 1. SL = LU.
Fact 2. $LG^2 = LU^2 + GU^2$.
Fact 3. LU = GU.

26. SL = GU.
27. $LG^2 - GU^2 = LU^2$.
28. $LG^2 = SL^2 + GU^2$.
29. GU = LU.
30. $SL \cdot LU = LU \cdot GU$.

Points C, L, O, U, and T are five different points on a line, but not necessarily in that order. Tell what betweenness relation must be true if each of the following equations is true.

31. LT + TU = LU.
32. CO = CU.
33. OL = OT − TL.

Copy and complete the following proofs.

34.

Given: P-K-E; K is the midpoint of $\overline{OE}$.
Prove: PK = PE − OK.

Proof.

Statements	***Reasons***
1. P-K-E.	———
2. ———	Betweenness of Points Theorem.
3. K is the midpoint of $\overline{OE}$.	Given.
4. ———	The midpoint of a line segment divides it into two equal segments.
5. ———	Substitution (steps 2 and 4).
6. PK = PE − OK.	———

35.

Given: A is the midpoint of $\overline{WT}$; $2WA^2 = SW^2$.
Prove: $WT = \sqrt{2}SW$.

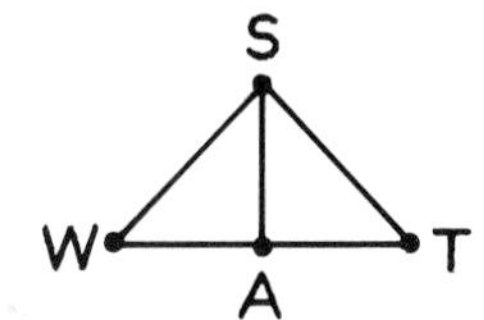

Proof.

Statements	***Reasons***
1. A is the midpoint of $\overline{WT}$.	———
2. $WA = \frac{1}{2}WT$.	———
3. ———	Given.
4. $2\left(\frac{1}{2}WT\right)^2 = SW^2$, $2\left(\frac{1}{4}WT^2\right) = SW^2$, $\frac{1}{2}WT^2 = SW^2$.	———
5. $WT^2 = 2SW^2$.	———
6. ———	———

Algebra Review

Solving Linear Equations

Some useful steps in solving linear equations are:

Step 1. Use the distributive property to eliminate parentheses.

Step 2. Simplify both sides of the equation as much as possible.

Step 3. Add and/or subtract to get all of the terms containing the variable on one side of the equation and all of the other terms on the other side.

Step 4. Multiply or divide to get the variable alone on one side.

Exercises

Simplify the following expressions.

Example 1: $7x - 3x$
Answer: $4x$

Example 2: $7(3x)$
Answer: $21x$

1. $8x + 5x$
2. $10x - x$
3. $10x(x)$
4. $4x - 12x$
5. $4(12x)$
6. $2x + 2 + 9x$
7. $6x - 6 - x$
8. $15x - 3 - 5x - 2$

Use the distributive property to eliminate the parentheses.

Example: $4(2x - 7)$
Answer: $8x - 28$

9. $5(x + 3)$
10. $11(4 - x)$
11. $6(3x + 1)$
12. $8(5 - 7x)$
13. $(x + 1)x$
14. $(5 - x)12$
15. $9x(x - 2)$
16. $2x(10 + 3x)$

Solve the following equations.

Example: $4(2x - 7) = 3x$
Solution: $8x - 28 = 3x$
(Using the distributive property.)
$5x - 28 = 0$
(Subtracting $3x$ from each side.)
$5x = 28$
(Adding 28 to each side.)
$x = \frac{28}{5} = 5.6.$
(Dividing each side by 5.)

17. $5x - 3 = 47$
18. $9 + 2x = 25$
19. $4x + 7x = 33$
20. $10x = x + 54$
21. $6x - 1 = 5x + 12$
22. $2x + 9 = 7x - 36$
23. $8 - x = x + 22$
24. $3x - 5 = 10x + 30$
25. $4(x - 11) = 3x + 16$
26. $x(x + 7) = x^2$
27. $5(x + 2) = 2(x - 13)$
28. $6(4x - 1) = 7(15 + 3x)$
29. $8 + 2(x + 3) = 10$
30. $x + 7(x - 5) = 2(5 - x)$
31. $4(x + 9) + x(x - 1) = x(6 + x)$

Chapter 3

RAYS AND ANGLES

NASA

Lesson 1
Rays and Angles

This photograph is the first one made that shows both the earth and the moon. It was taken in 1977 from the Voyager 1 spacecraft at a distance of more than seven million miles.

The distance between the earth and the moon, about 240,000 miles, has been known for a long time. In fact, the Greek astronomer Hipparchus used geometry to figure it out back in the second century B.C. Since then, the accuracy to which the distance is known has been improved to the extent that we can now tell the distance at any moment (it is actually continually changing) to within a *few inches.* This is possible because of a lunar laser reflector left on the moon during the first landing in 1969.

The distance is measured by bouncing a laser beam from the earth off the moon and back. Because the path of the beam from the earth to the moon appears to be part of a line bounded by two endpoints, it is a good physical model of a *line segment.* If the laser were beamed away from the moon and if it went infinitely far off into space, it would illustrate a *ray.* A ray, in contrast to a line segment, has only one endpoint.

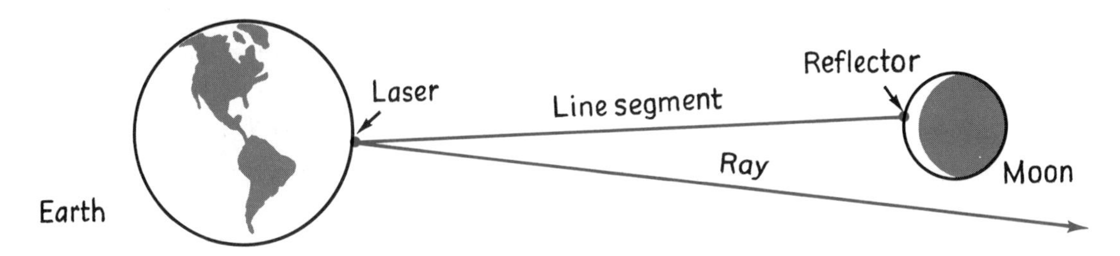

To get an idea of how a ray may be precisely defined, imagine the laser beam slowly being turned away from the moon until it just misses the moon and goes off into space. The ray shown starts at point A and passes the moon at point B. It includes all of the points on the line $\overleftrightarrow{AB}$ that are between A and B and all of those points on the line that are on the other side of B with respect to A.

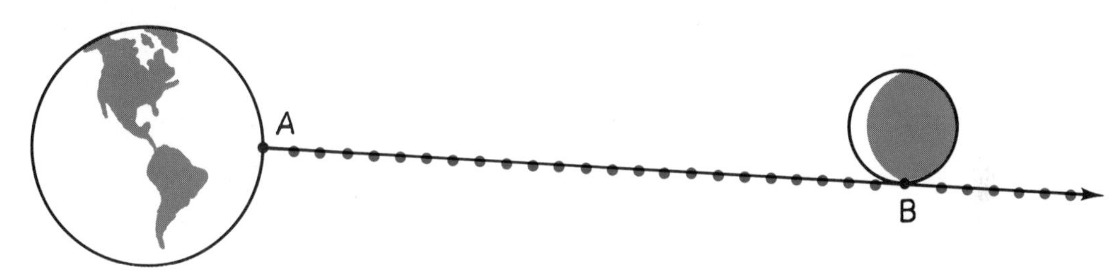

These ideas are included in the following definition.

► **Definition**
A ***ray*** $\overrightarrow{AB}$ is the set of points A, B, and all points X such that either A-X-B or A-B-X.

Notice the symbol used to represent a ray in this definition. The endpoint of a ray is always named first, followed by the name of any other point on it.

Two rays that have the same endpoint form an angle.

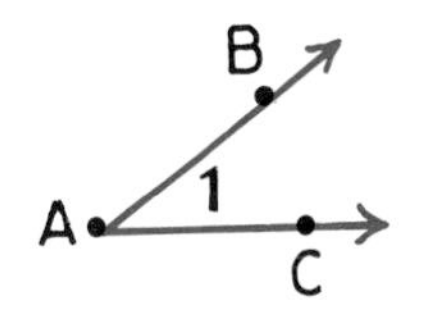

► **Definition**
An ***angle*** is a pair of rays that have the same endpoint.

The rays are called the *sides* of the angle, and their common endpoint is called the *vertex* of the angle.

Angles can be named in several different ways. Using the symbol ∡ to mean "angle," the angle shown above can be named ∡BAC or ∡CAB (where the vertex is named in the middle). Because no other angle is shown that has the same vertex, this angle can also be simply called ∡A. It can also be named with the number written inside: ∡1.

If two rays with a common endpoint point in opposite directions, they are called *opposite rays.*

► **Definition**
Rays $\overrightarrow{AB}$ and $\overrightarrow{AC}$ are ***opposite rays*** iff B-A-C.

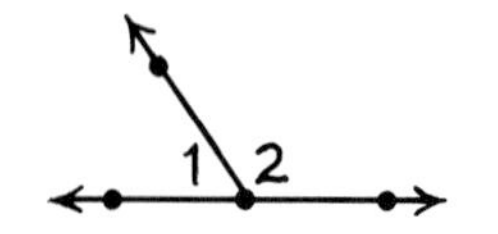

∡1 and ∡2 are a linear pair.

The idea of opposite rays can be used to define two special angle relations. The first relation—two angles that are a *linear pair*—is illustrated by this figure.

► **Definition**
Two angles are a ***linear pair*** iff they have a common side and their other sides are opposite rays.

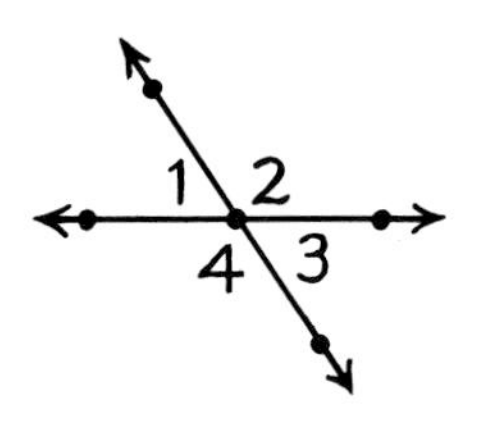

∡1 and ∡3 are vertical angles;
∡2 and ∡4 are vertical angles.

The second relation—two angles that are *vertical angles*—is illustrated by this figure.

► **Definition**
Two angles are ***vertical angles*** iff the sides of one angle are opposite rays to the sides of the other.

Exercises

Set I

The following exercises refer to the figure below.

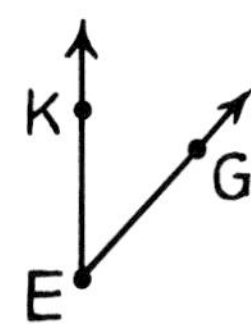

1. Name the vertex of this angle.
2. Name its sides.
3. Name the angle in three different ways.

It is possible to answer the following questions about ∡TUB without looking at a figure.

4. What point is its vertex?
5. What rays are its sides?

In the figure below, U-R-N.

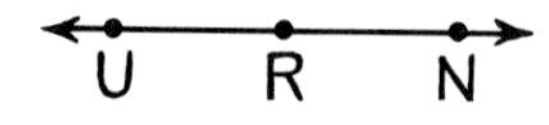

6. What are $\overrightarrow{RU}$ and $\overrightarrow{RN}$ called?
7. Do $\overrightarrow{RU}$ and $\overrightarrow{RN}$ form an angle?
8. Use the definition of "angle" to explain why or why not.

The following exercises refer to the figure below.

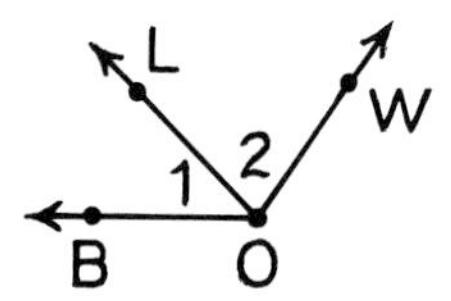

9. Name ∡1 with three letters.
10. Name ∡2 with three letters.
11. Name the common side of ∡1 and ∡2.
12. Are ∡1 and ∡2 a linear pair?
13. Use the definition of linear pair to explain why or why not.
14. How many angles are illustrated in the figure?

In the figure below, $\overrightarrow{AC}$ and $\overrightarrow{AE}$ are opposite rays and $\overrightarrow{AR}$ and $\overrightarrow{AT}$ are opposite rays.

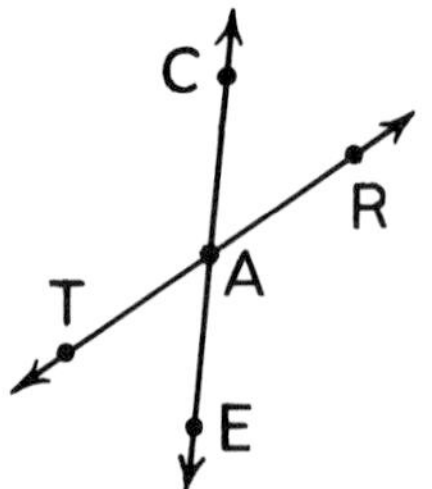

15. Name an angle in the figure that forms a linear pair with ∡CAT.

16. Are ∡CAT and ∡EAR vertical angles?

17. Use the definition of vertical angles to explain why or why not.

18. How many angles are illustrated in the figure?

Match the following symbols with the lettered phrases.

19. $\overline{JR}$.

20. $\overrightarrow{JR}$.

21. $\overleftrightarrow{JR}$.

22. ∡JAR.

A. The pair of rays $\overrightarrow{AJ}$ and $\overrightarrow{AR}$.

B. The set of points J and R and all points A such that J, A, and R are collinear.

C. The set of points J and R and all points A such that J-A-R.

D. The set of points J and R and all points A such that either J-A-R or J-R-A.

Set II

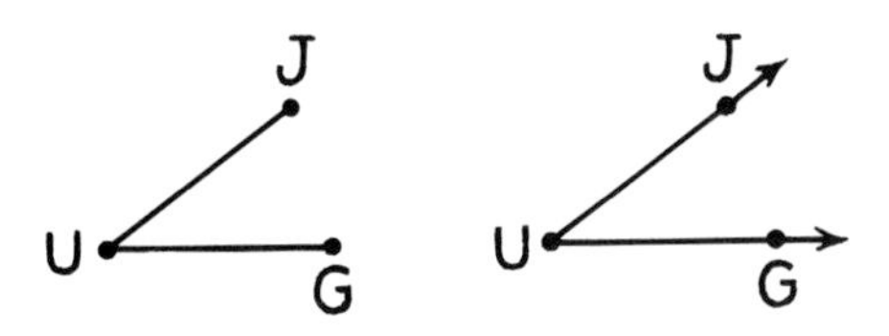

Even though the first figure at the left above is not an angle because its sides are line segments, it *determines* the angle shown in the second figure.

The following exercises refer to the figure below, in which A-S-E.

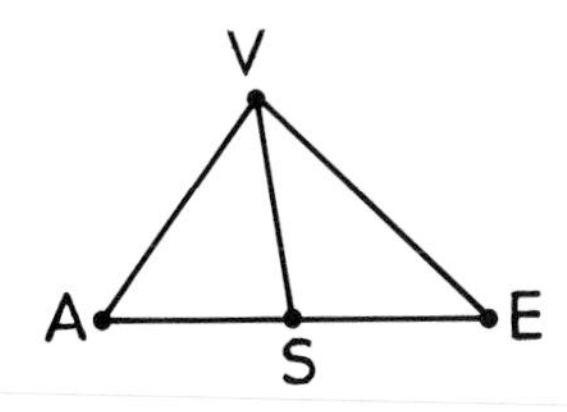

23. The figure contains six line segments, of which three are $\overline{AV}$, $\overline{VS}$, and $\overline{AS}$. Name the other three.

24. The figure determines two angles that can each be named with a single letter. Name them.

25. Name the three angles whose vertex is V.

26. How many angles altogether does the figure determine?

In the figure below, points H, T, and S are collinear. Tell whether each of the following statements is correct or incorrect. If a statement is incorrect, tell how it could be made correct.

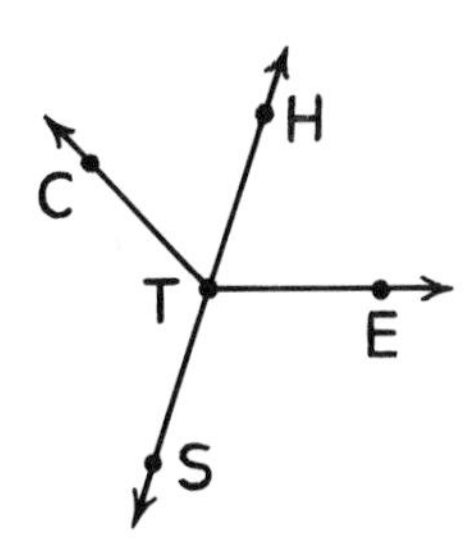

27. $\overrightarrow{TS}$ and $\overrightarrow{ST}$ are opposite rays.

28. The sides of ∡T are $\overrightarrow{TC}$ and $\overrightarrow{TE}$.

29. $\overrightarrow{SH}$ and $\overrightarrow{ST}$ are the same ray.

30. $\overrightarrow{HT} + \overrightarrow{TS} = \overrightarrow{HS}$.

31. $\overrightarrow{TH}$ and $\overrightarrow{TS}$ are a linear pair.

32. Point T is the vertex of six angles in the figure.

33. Copy and complete the following proof.

Given: F-A-S and L-A-K.

Prove: ∡FAL and ∡KAS are vertical angles.

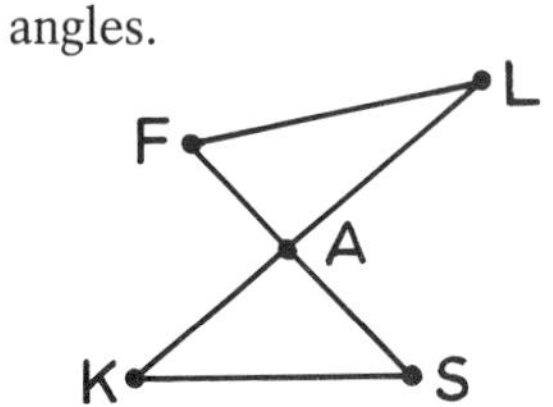

Proof.

Statements	***Reasons***
1. F-A-S and L-A-K.	________
2. ________ and ________.	If B-A-C, then $\overrightarrow{AB}$ and $\overrightarrow{AC}$ are ________.
3. ________	Two angles are vertical angles if ________.

34. Copy and complete the following proof.

Given: ∡TAK and ∡KAN are a linear pair.

Prove: T-A-N.

T A K N

Proof.

Statements	***Reasons***
1. ________	________
2. $\overrightarrow{AT}$ and $\overrightarrow{AN}$ are ________.	If two angles are a linear pair, then ________.
3. ________	If $\overrightarrow{AB}$ and $\overrightarrow{AC}$ are opposite rays, then ________.

Set III

This figure shows ten frogs arranged so that one row contains four frogs and three rows contain three frogs each.

Can you figure out a way to move exactly three frogs so that there are five rows of frogs with four frogs in each row? If you can, make a drawing to illustrate your answer.

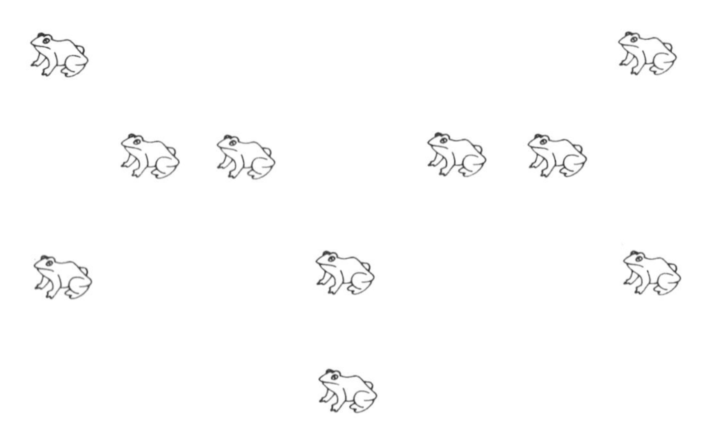

By the permission of Johnny Hart and News America Syndicate

Lesson 2

The Protractor Postulate

The ancient Babylonians chose 360 as their number for measuring angles more than four thousand years ago. The unit based on this number, the *degree*, has been used to measure angles in geometry ever since.

To measure an angle, we need a *protractor*. The protractor allows us to assign numbers to the sides of the angle and from these numbers we can figure out another number called the measure of the angle.

The first figure below shows a circular protractor placed on a set of rays that have the same endpoint. We will refer to a set of all the rays

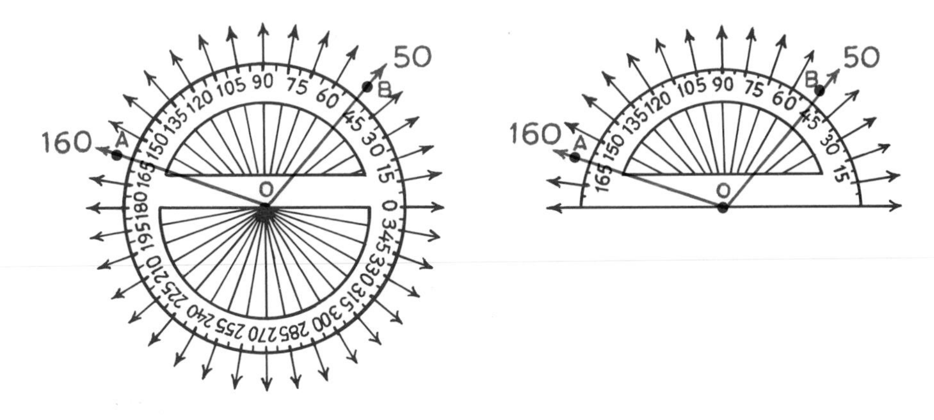

that lie in the same plane and that have the same endpoint as a *rotation of rays*. If the rays are evenly spaced and if there are 360 of them, then the angle formed by two consecutive rays is a *degree*. There are 36 evenly spaced rays in the figure, and so the rays in the figure form angles of 10 degrees.

To measure $\measuredangle AOB$, we observe that its sides have the numbers 160 and 50. These numbers are called the *coordinates* of the rays and the *measure of the angle,* which we will represent by the symbol $\angle AOB$, is found by subtracting the coordinates:

$$\angle AOB = 160 - 50 = 110°.$$

The second figure on page 89 shows a semicircular protractor being used to measure the same angle. We will call the rays that correspond to it a *half-rotation of rays.* A semicircular protractor is sufficient to measure any angle, and it is the one that is most frequently used.

The figures below show how a semicircular protractor can be used to measure several different angles. Notice that the *center* of the protractor is always placed at the *vertex* of the angle.

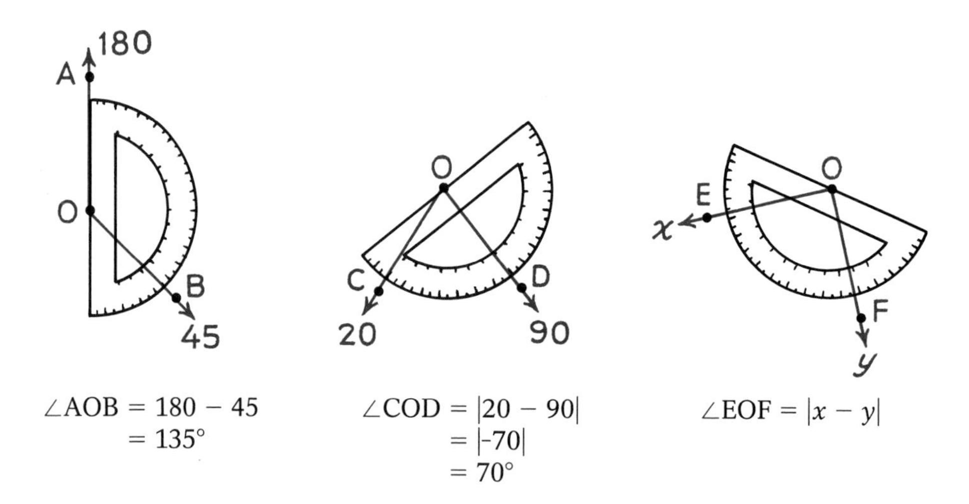

$\angle AOB = 180 - 45$
$= 135°$

$\angle COD = |20 - 90|$
$= |-70|$
$= 70°$

$\angle EOF = |x - y|$

The properties of a protractor illustrated in this lesson can be summarized in the following postulate. Notice that it is similar in many ways to the Ruler Postulate.

► **Postulate 7** (The Protractor Postulate)
The rays in a half-rotation can be numbered so that
a) to every ray there corresponds exactly one real number called its coordinate,
b) to every real number from 0 to 180 inclusive there corresponds exactly one ray,
c) to every pair of rays there corresponds exactly one real number called the measure of the angle that they determine,
d) and the measure of an angle is the absolute value of the difference between the coordinates of its rays.

As in the case of the Ruler Postulate, after you are certain that you know what the Protractor Postulate means, you may simply refer to it by name.

Angles are classified according to their measures as follows.

► **Definitions**
An angle is ***acute*** iff its measure is less than 90°.
An angle is a ***right*** angle iff its measure is 90°.
An angle is ***obtuse*** iff its measure is more than 90° but less than 180°.
An angle is a ***straight*** angle iff its measure is 180°.

If an angle of one degree is divided into 60 equal angles, each angle has a measure of one *minute,* and if an angle of one minute is divided into 60 equal angles, each angle has a measure of one *second.* This table summarizes these relations:

In words	*In symbols*
1 degree = 60 minutes	$1° = 60'$
1 minute = 60 seconds	$1' = 60''$

Exercises

Set I

If a protractor is placed on ∡ABC so that 30 corresponds to $\overrightarrow{BA}$, then 75 corresponds to $\overrightarrow{BC}$.

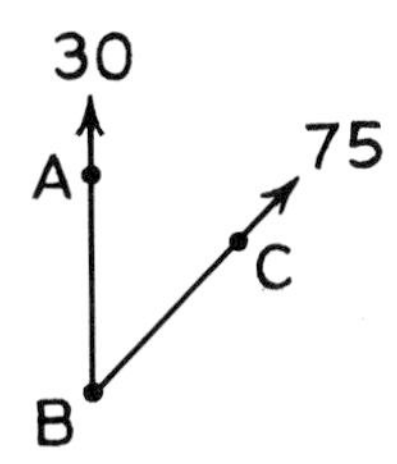

1. What are the numbers 30 and 75 called?
2. What number corresponds to ∡ABC?
3. How is it found?
4. What is it called?

The symbols ∡ABC and ∠ABC mean different things.

5. What does ∡ABC represent?
6. What does ∠ABC represent?

Obtuse Ollie thinks that ∡X is larger than ∡Y.

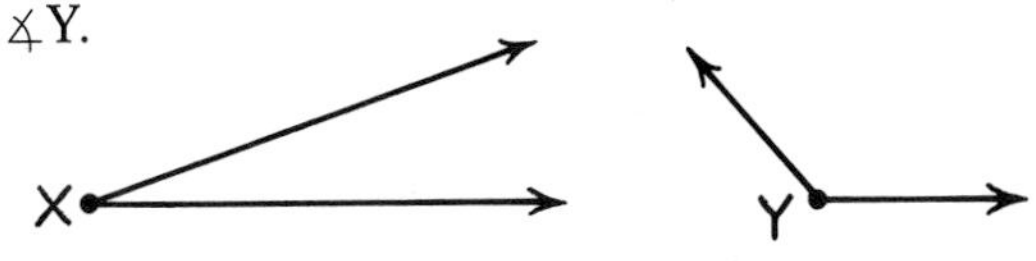

7. Is he right?
8. Why does he think this?
9. How could Acute Alice set him straight?

The following exercises refer to the figure below. Tell who is correct in each case.

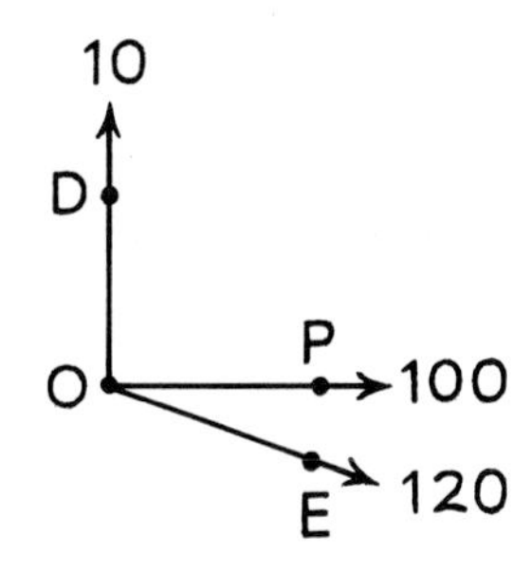

10. When asked to find the measure of the acute angle, Ollie said "10°" and Alice said "20°."
11. When asked to name the coordinate of $\overrightarrow{OP}$, Ollie said "100°" and Alice said "100."
12. When asked to name the right angle, Ollie wrote "∠DOP" and Alice wrote "∡DOP."

13. When asked to write an equation for the measure of the obtuse angle, Ollie wrote "$\angle$DOE = 110°" and Alice wrote "$\measuredangle$DOE = 110°."

Find the measure of each of the following angles in the figure below.

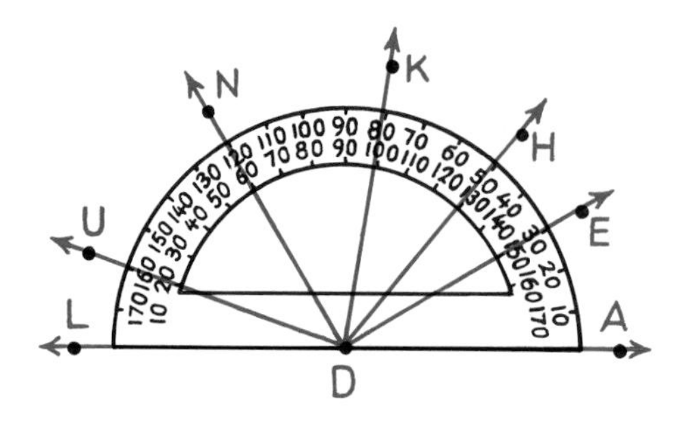

14. $\angle$LDK.

15. $\angle$KDH.

16. $\angle$EDN.

17. $\angle$ADU.

18. $\angle$NDH.

19. $\angle$LDA.

The following exercises refer to the angles in Exercises 14 through 19.

20. Name the angles that are acute.

21. Name the right angle.

22. Name the angles that are obtuse.

23. Name the straight angle.

Set II

When asked to find half of 19°, Obtuse Ollie used his calculator and got 9°5′.

24. Is this correct?

Find each of the following in degrees and minutes.

Example: 20.7°.

Answer: 0.7° = 0.7 × 60 = 42′, and so 20.7° = 20°42′.

25. 5.2°.

26. 11.35°.

27. 0.05°.

28. One third of 100°.

29. One fourth of 75°.

Copy the figure below and mark it as necessary to find each of the following numbers.

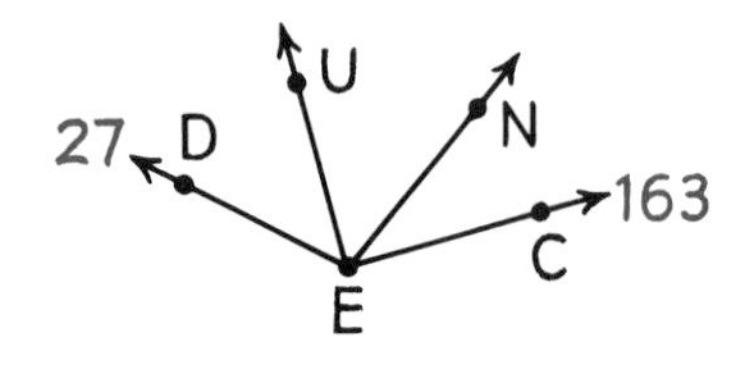

30. Find $\angle$DEC.

31. Find the coordinate of $\overrightarrow{EU}$, given that $\angle$DEU = 48°.

32. Find the coordinate of $\overrightarrow{EN}$, given that $\angle$CEN = 35°.

33. Find $\angle$UEN.

The rays in each of the following figures are evenly spaced about point O. Find the missing numbers.

34.

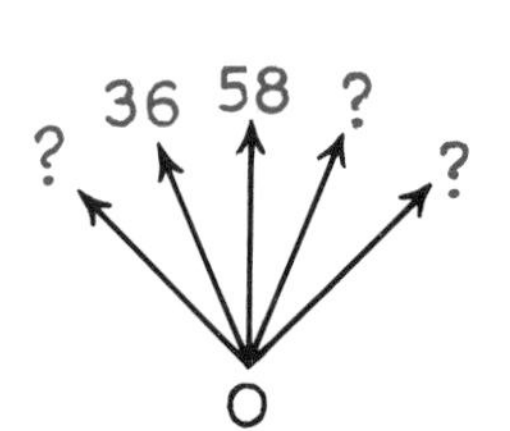

35.

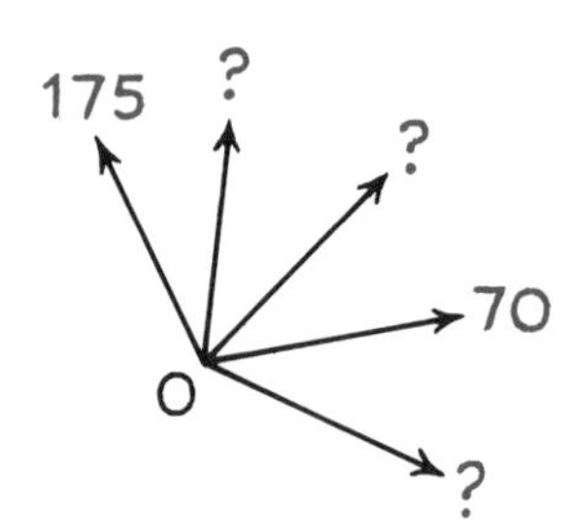

Set III

The Federal Aviation Agency requires that, as some airplanes come in for a landing, the angle of descent must be 3° for the last four miles.

1. Use your ruler and protractor to make a scale drawing illustrating this. Let 1 inch represent 1 mile.

2. Use your drawing to estimate the approximate altitude in feet at which such an airplane should be four miles before landing (1 mile = 5,280 feet).

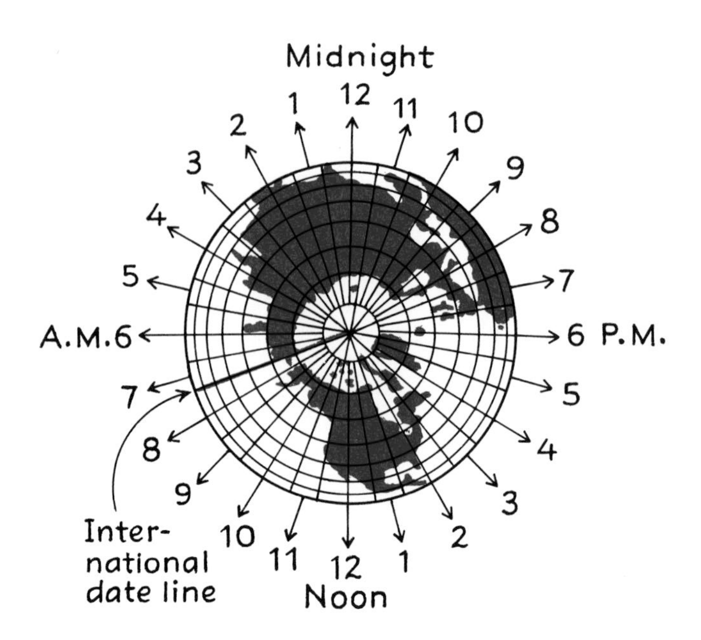

Lesson 3

Betweenness of Rays

The world is divided into 24 time zones that correspond to the 24 hours in a day. These zones are shown in this figure, in which the earth is viewed from the North Pole.

As the earth rotates, the ray marking the International Date Line rotates with it but the rays marking the time zones remain fixed. It is easy to see that, for the position of the earth shown, the ray marking the International Date Line is between the rays marking 5 A.M. and 12 noon. If we compare the ray marking the International Date Line to the rays marking 3 P.M. and 10 P.M., however, it is not at all obvious that one ray is between the other two.

This observation suggests that the idea of betweenness of rays, like that of betweenness of points, must be carefully defined.

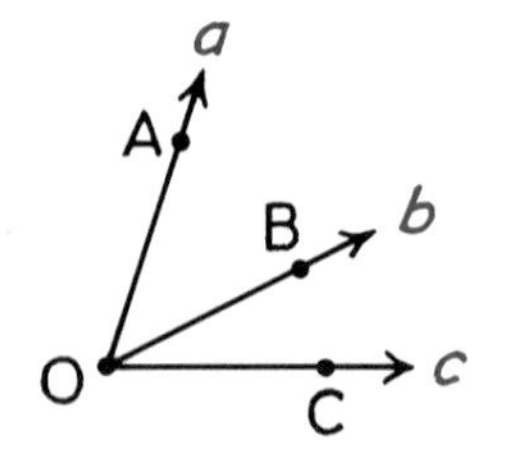

► **Definition** (Betweenness of Rays)
Suppose that $\overrightarrow{OA}$, $\overrightarrow{OB}$, and $\overrightarrow{OC}$ are in a half-rotation with coordinates a, b, and c respectively. $\overrightarrow{OB}$ is between $\overrightarrow{OA}$ and $\overrightarrow{OC}$ (written $\overrightarrow{OA}$-$\overrightarrow{OB}$-$\overrightarrow{OC}$) iff either $a < b < c$ or $a > b > c$.

Using this definition, we can prove a theorem relating the measures of the three angles determined by the rays.

► **Theorem 3** (Betweenness of Rays Theorem)
If $\overrightarrow{OA}$-$\overrightarrow{OB}$-$\overrightarrow{OC}$, then $\angle AOB + \angle BOC = \angle AOC$.

Proof.
If $\overrightarrow{OA}$-$\overrightarrow{OB}$-$\overrightarrow{OC}$, then either $a < b < c$ or $a > b > c$. (Definition of betweenness of rays.)

Case 1. Suppose that $a < b < c$.

If $a < b < c$, then $\angle AOB = b - a$, $\angle BOC = c - b$, and $\angle AOC = c - a$. (Protractor postulate.)

$$\begin{aligned}\angle AOB + \angle BOC &= (b - a) + (c - b) \text{ (Addition)}\\ &= b - a + c - b\\ &= c - a.\end{aligned}$$

But

$$\angle AOC = c - a.$$

So

$$\angle AOB + \angle BOC = \angle AOC \text{ (Substitution).}$$

Case 2, in which $a > b > c$, is proved in a similar fashion.

If $\angle AOB = \angle BOC$, then $\overrightarrow{OB}$ *bisects* $\measuredangle AOC$.

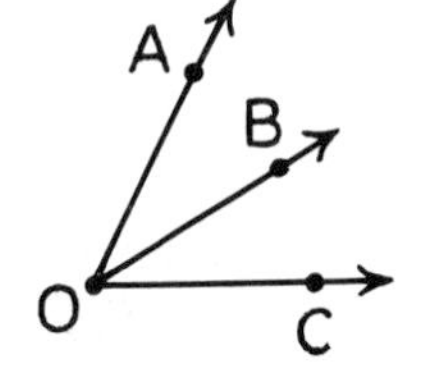

► **Definition**
A ***ray bisects an angle*** iff it is between the sides of the angle and divides it into two equal angles.

This definition leads to a theorem about a ray that bisects an angle.

► **Theorem 4** (Angle Bisector Theorem)
A ray that bisects an angle divides it into angles half as large as the angle.

This theorem is comparable to the theorem about the point that is the midpoint of a line segment. Because the proof is nearly identical to it (Theorem 2 proved on page 70), it is not given here.

Exercises

Set I

The following exercises refer to the figure below.

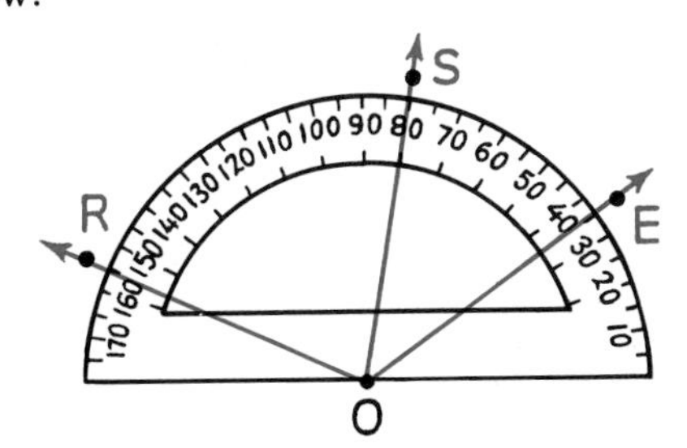

1. What are the coordinates of $\overrightarrow{OR}$, $\overrightarrow{OS}$, and $\overrightarrow{OE}$?
2. Which ray is between the other two?
3. Use the definition of betweenness of rays to show why.
4. Find the following angle measures: $\angle ROS$, $\angle SOE$, and $\angle ROE$.
5. Write the equation relating $\angle ROS$, $\angle SOE$, and $\angle ROE$ that follows from the Betweenness of Rays Theorem.

$\overrightarrow{BU}$ is between $\overrightarrow{BL}$ and $\overrightarrow{BE}$.

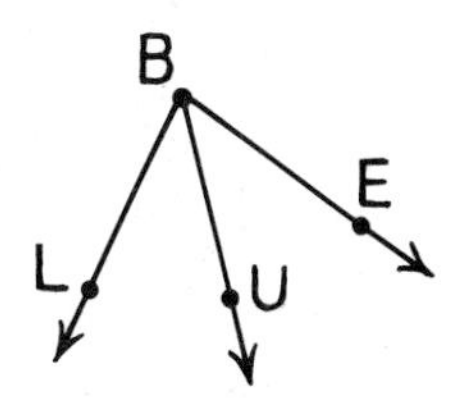

6. Write the symbol to represent this fact.

7. Write an equation that follows from this fact.

8. Does it follow that $\angle$LBU = $\angle$UBE?

In the figure below, $\overrightarrow{KI}$ bisects $\measuredangle$PKN. Use the coordinates to find

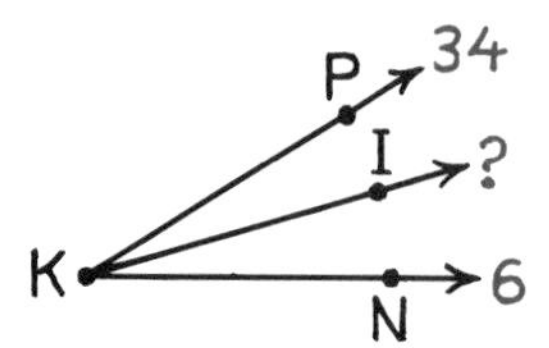

9. $\angle$PKN.

10. $\angle$IKN.

11. the coordinate of $\overrightarrow{KI}$.

12. How do $\measuredangle$PKI and $\measuredangle$IKN compare in measure?

13. How do $\measuredangle$PKI and $\measuredangle$PKN compare in measure?

The following exercises refer to the figure below. Tell which fact (or facts) and what property of equality can be used to prove the conclusion stated in each exercise.

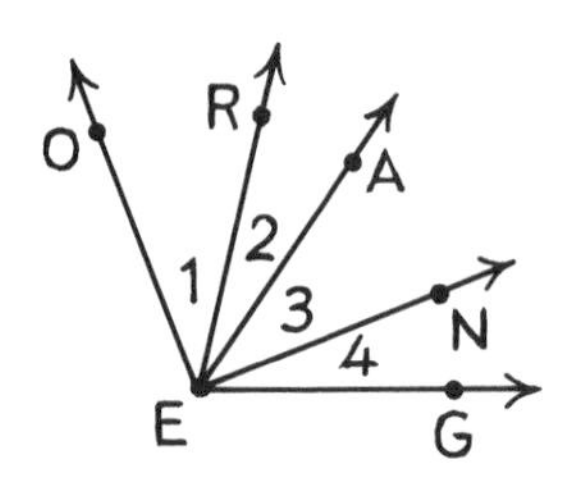

Fact 1. $\angle$OEA = $\angle$REN.
Fact 2. $\angle 2 = \angle 4$.
Fact 3. $\angle$REN = $\angle 2 + \angle 3$.

14. $\angle$REN $- \angle 3 = \angle 2$.

15. $\angle$OEA = $\angle 2 + \angle 3$.

16. $\angle$REN = $\angle$OEA.

17. $\angle 2 + \angle 3 = \angle 3 + \angle 4$.

18. $\angle$REN = $\angle 4 + \angle 3$.

Set II

Match each of the following statements about this figure with a theorem or definition below.

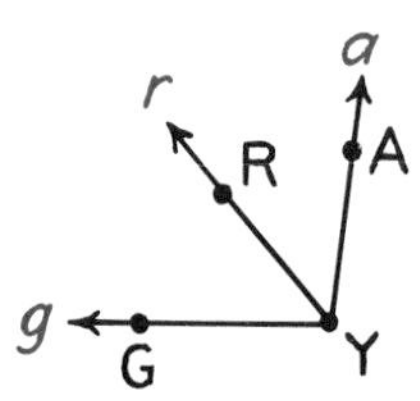

19. If $\overrightarrow{YG}$-$\overrightarrow{YR}$-$\overrightarrow{YA}$, then $\angle$GYR + $\angle$RYA = $\angle$GYA.

20. If $\overrightarrow{YR}$ bisects $\measuredangle$GYA, then $\angle$GYR = $\angle$RYA.

21. If $g > r > a$, then $\overrightarrow{YG}$-$\overrightarrow{YR}$-$\overrightarrow{YA}$.

22. If $\overrightarrow{YR}$ bisects $\measuredangle$GYA, then $\angle \text{GYR} = \frac{1}{2}\angle \text{GYA}$.

A. The Angle Bisector Theorem.

B. Definition of betweenness of rays.

C. Definition of angle bisector.

D. The Betweenness of Rays Theorem.

In the figure below, the coordinates of $\overrightarrow{ER}$ and $\overrightarrow{ED}$ are 20 and 90 respectively. Suppose that the coordinate of $\overrightarrow{EX}$ is x. What can you conclude about x if

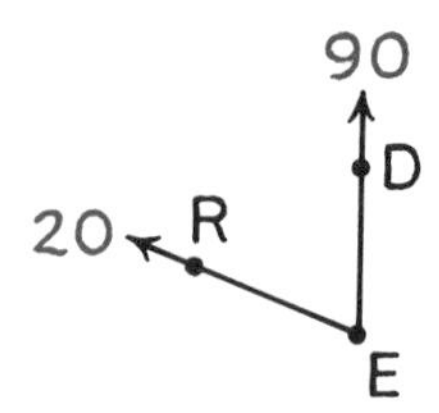

23. $\overrightarrow{ER}$-$\overrightarrow{EX}$-$\overrightarrow{ED}$?

24. $\overrightarrow{EX}$ bisects $\measuredangle$RED?

25. $\overrightarrow{ER}$-$\overrightarrow{ED}$-$\overrightarrow{EX}$?

26. $\overrightarrow{ED}$ bisects $\measuredangle$REX?

In the figure below, $\overrightarrow{SN}$ bisects $\measuredangle ASD$, $\angle ASN = (5x)^\circ$, and $\angle NSD = (3x + 28)^\circ$.

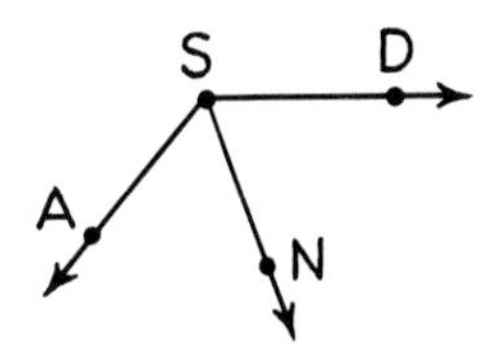

27. Find x.

28. Find $\angle ASN$.

29. Find $\angle NSD$.

30. Find $\angle ASD$.

In the figure below, $\overrightarrow{JA}$-$\overrightarrow{JD}$-$\overrightarrow{JE}$, $\angle AJD = (4x + 5)^\circ$, $\angle DJE = (10x - 3)^\circ$, and $\angle AJE = (15x - 4)^\circ$.

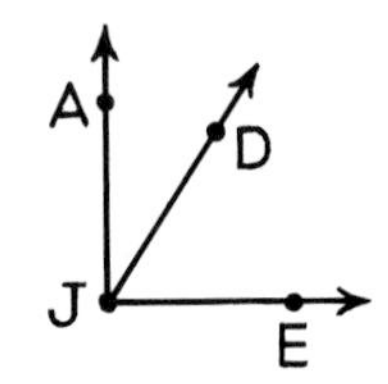

31. Find x.

32. Find $\angle AJD$.

33. Find $\angle DJE$.

34. Find $\angle AJE$.

Copy and complete the following proofs.

35.

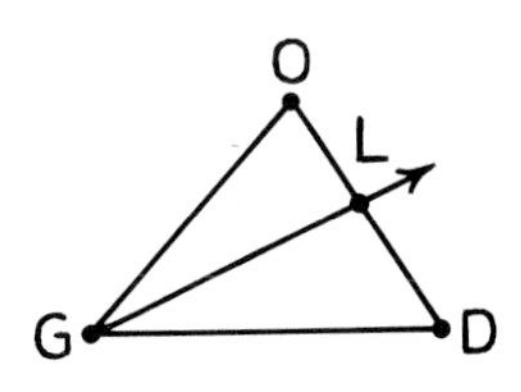

Given: $\overrightarrow{GL}$ bisects $\measuredangle OGD$.

Prove: $\angle OGD = 2\angle OGL$.

Proof.

Statements	***Reasons***
1. ____	____
2. $\angle OGL = \frac{1}{2}\angle OGD$.	____
3. $2\angle OGL = \angle OGD$.	____
4. ____	____

36.

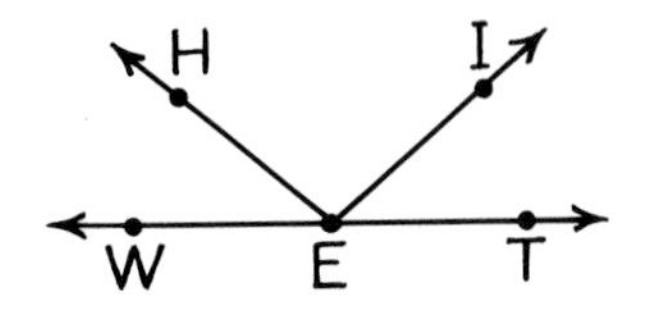

Given: $\overrightarrow{EW}$-$\overrightarrow{EH}$-$\overrightarrow{EI}$; $\angle WEI = \angle HET$.

Prove: $\angle WEH + \angle HEI = \angle HET$.

Proof.

Statements	***Reasons***
1. $\overrightarrow{EW}$-$\overrightarrow{EH}$-$\overrightarrow{EI}$.	____
2. ____	The Betweenness of Rays Theorem.
3. $\angle WEI = \angle HET$.	____
4. ____	____

Set III

The first mechanical clocks were made in the fourteenth century and had only hour hands. Clocks did not have minute hands until the seventeenth century, and second hands did not appear until the eighteenth century.

You know that minutes and seconds are units of angle measure as well as time.

1. Through how many minutes of angle does the minute hand of a clock move during one minute of time? Show your method.

2. Through how many seconds of angle does the second hand of a clock move during one second of time? Show your method.

From *Fifty Years of Popular Mechanics 1902–1952*, edited by Edward L. Throm (Simon & Schuster, 1952)

Lesson 4

Complementary and Supplementary Angles

In the early days of automobiles, motorists frequently had to get out of their cars and push them when they came to a slight grade. This photograph from a 1902 issue of *Popular Mechanics* shows a car being driven up a ramp to the top of a house. The purpose of the photograph was to demonstrate the power and grade-climbing ability of the car, both of which were unusual for the time.

The drawing below represents the ramp ($\overline{BC}$), the side of the house ($\overline{CD}$), and the ground ($\overleftrightarrow{AD}$). The ramp rose at a 23° angle with the ground and made a 67° angle with the side of the house. Notice that the sum of the measures of these angles is 90°, a special number with regard to angle measurement. The two angles are said to be *complementary*.

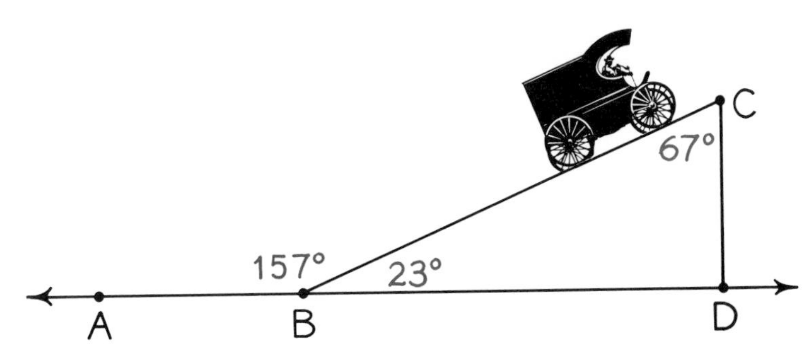

► **Definition**
Two angles are ***complementary*** iff the sum of their measures is 90°.

Each angle is called the *complement* of the other.

Because the sum of the measures of two complementary angles is 90°, the measure of the complement of an angle is found by subtracting the measure of the angle from 90°.

The sum of the measures of the two angles made by the ramp with the ground (∡ABC and ∡CBD) is 180°, another special number in angle measurement. Such angles are called *supplementary.*

► **Definition**
Two angles are ***supplementary*** iff the sum of their measures is 180°.

Each angle is called the *supplement* of the other, and the measure of the supplement of an angle is found by subtracting the measure of the angle from 180°.

The three angles in the figure below have been drawn so that ∡A is complementary to ∡B and ∡B is complementary to ∡C. In what way are ∡A and ∡C related?

If "complementariness" of angles were transitive like equality of numbers, we could say that ∡A and ∡C are complementary. This, however, is not true of the angles in the figure. If ∠A = 42°, then ∠B = 48°.

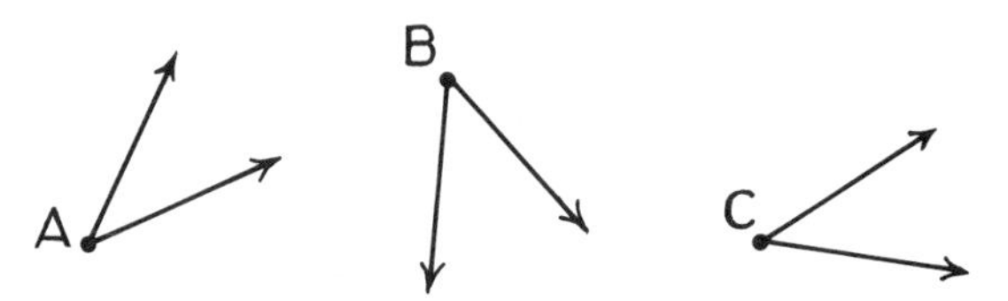

If ∠B = 48°, then ∠C = 42°. The fact that ∡A and ∡C are not complementary but equal suggests the following theorem.

► **Theorem 5**
Complements of the same angle (or equal angles) are equal.

The proof for complements of the same angle is given here.

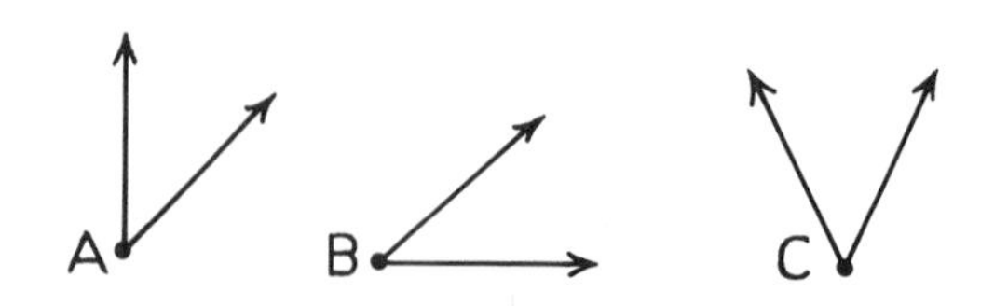

Given: ∡A and ∡B are complements of ∡C.
Prove: ∠A = ∠B.

Proof.

Statements	Reasons
1. $\measuredangle A$ and $\measuredangle B$ are complements of $\measuredangle C$.	Given.
2. $\angle A + \angle C = 90°$, $\angle B + \angle C = 90°$.	If two angles are complementary, the sum of their measures is 90°.
3. $\angle A = 90° - \angle C$, $\angle B = 90° - \angle C$.	Subtraction.
4. $\angle A = \angle B$.	Substitution.

A similar theorem is true for supplementary angles.

► **Theorem 6**
Supplements of the same angle (or equal angles) are equal.

The proof for supplements of equal angles is given here.

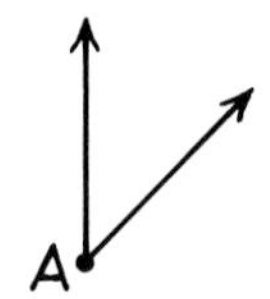

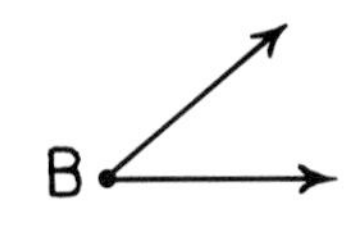

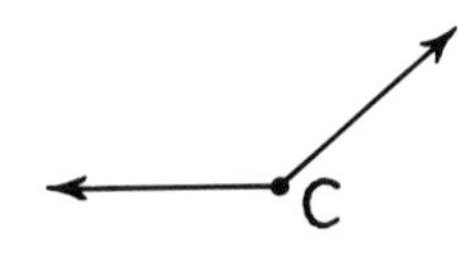

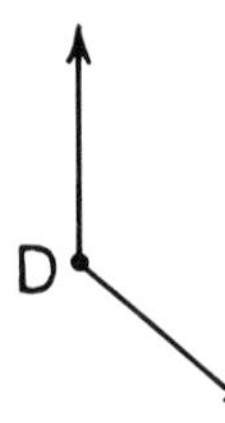

Given: $\measuredangle A$ is the supplement of $\measuredangle C$ and $\measuredangle B$ is the supplement of $\measuredangle D$; $\angle C = \angle D$.

Prove: $\angle A = \angle B$.

Proof.

Statements	Reasons
1. $\measuredangle A$ is the supplement of $\measuredangle C$ and $\measuredangle B$ is the supplement of $\measuredangle D$.	Given.
2. $\angle A + \angle C = 180°$, $\angle B + \angle D = 180°$.	If two angles are supplementary, the sum of their measures is 180°.
3. $\angle A = 180° - \angle C$, $\angle B = 180° - \angle D$.	Subtraction.
4. $\angle C = \angle D$.	Given.
5. $\angle A = 180° - \angle D$.	Substitution (steps 3 and 4).
6. $\angle A = \angle B$.	Substitution (steps 3 and 5).

Exercises

Set I

Use the definitions and theorems of this lesson to complete each of the following statements.

1. If $\measuredangle A$ and $\measuredangle B$ are supplementary, then ||||||||.
2. If $\measuredangle A$ and $\measuredangle B$ are supplements of $\measuredangle C$, then ||||||||.
3. If $\measuredangle D$ and $\measuredangle E$ are complementary, then ||||||||.
4. If $\measuredangle D$ and $\measuredangle E$ are complements of equal angles, then ||||||||.

The following exercises refer to the figure below.

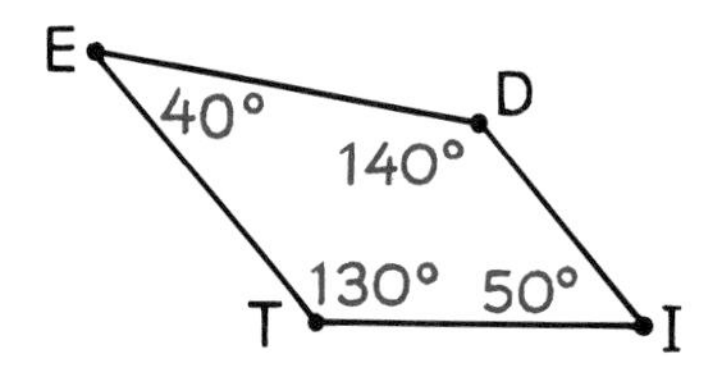

5. Which angles in the figure are acute?
6. Which angles are obtuse?
7. Which angles in the figure are complementary?
8. Which angles are supplementary? (There are two pairs.)

To answer each of the following questions, write "acute," "right," "obtuse," or "straight."

9. What kind of angle is the supplement of an acute angle?
10. What kind of angle is the complement of an acute angle?
11. What kind of angle does not have a supplement?
12. What kind of angles are both supplementary and equal?

Obtuse Ollie says: "$\angle F = 30°$, $\angle A = 90°$, and $\angle B = 60°$. Therefore, they are supplementary."

13. Why does he think this?
14. Why is he wrong?
15. Would it be correct to say that $\measuredangle A$ is complementary?

Angle B is supplementary to angle I, and angle I is supplementary to angle Z.

16. What can you conclude about angle B and angle Z?
17. Why?
18. Is "supplementariness" of angles transitive?

Set II

Copy the figure below and mark it as necessary to find each of the following numbers.

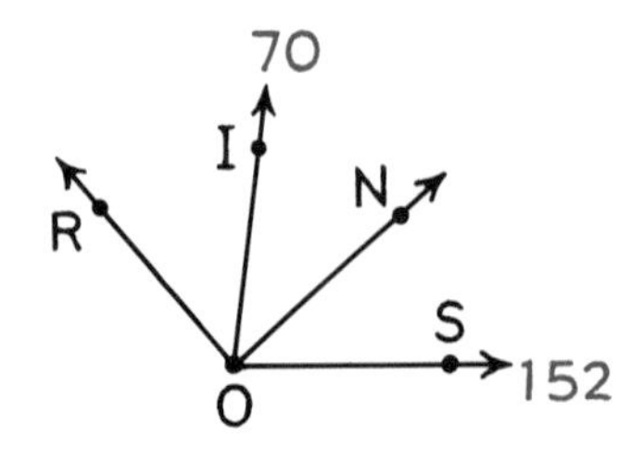

19. Find $\angle IOS$.
20. Find $\angle ION$, given that $\overrightarrow{ON}$ bisects $\measuredangle IOS$.
21. Find the coordinate of $\overrightarrow{ON}$.
22. Find $\angle ROI$, given that $\measuredangle ROI$ and $\measuredangle ION$ are complementary.
23. Find the coordinate of $\overrightarrow{OR}$.

In the figure below, $\measuredangle DHS$ and $\measuredangle AHS$ are complementary; $\angle DHS = (2x + 5)^\circ$ and $\angle AHS = (x - 5)^\circ$.

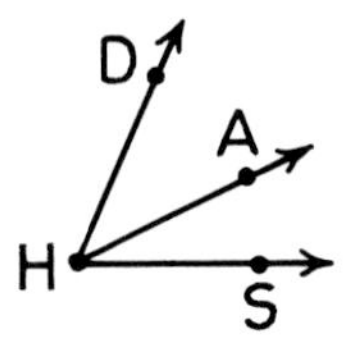

24. Find x.
25. Find $\angle DHS$.
26. Find $\angle AHS$.
27. Find $\angle DHA$.

In the figure below, $\measuredangle BDL$ and $\measuredangle BDO$ are supplementary; $\angle BDL = (150 - x)^\circ$ and $\angle BDO = (5x - 18)^\circ$.

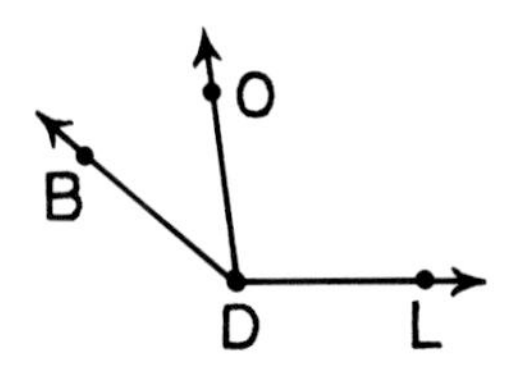

28. Find x.
29. Find $\angle BDL$.
30. Find $\angle BDO$.
31. Find $\angle ODL$.

Copy and complete the following proofs.

32.

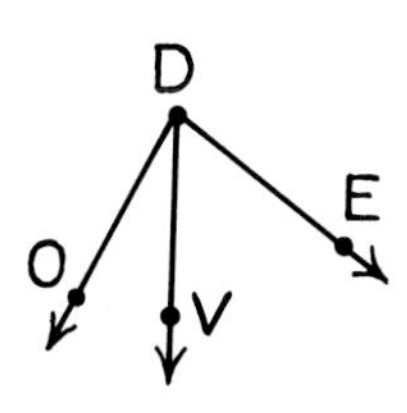

Given: $\overrightarrow{DO}$-$\overrightarrow{DV}$-$\overrightarrow{DE}$; $\measuredangle ODE$ is a right angle.

Prove: $\measuredangle ODV$ and $\measuredangle VDE$ are complementary.

Proof.

Statements	***Reasons***
1. $\overrightarrow{DO}$-$\overrightarrow{DV}$-$\overrightarrow{DE}$.	|||||||||
2. |||||||||	The Betweenness of Rays Theorem.
3. |||||||||	Given.
4. $\angle ODE = 90^\circ$.	|||||||||
5. |||||||||	Transitive (steps 2 and 4).
6. |||||||||	|||||||||

33.

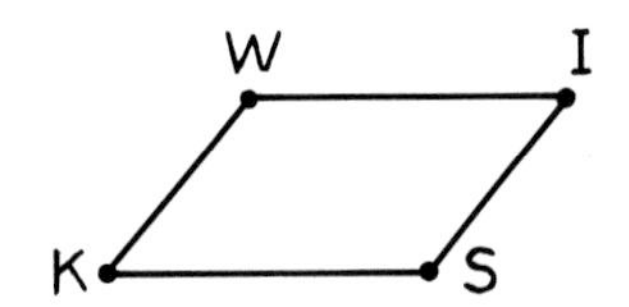

Given: $\measuredangle K$ and $\measuredangle S$ are supplementary; $\angle K = \angle I$.

Prove: $\angle S = 180^\circ - \angle I$.

Proof.

Statements	***Reasons***
1. $\measuredangle K$ and $\measuredangle S$ are supplementary.	|||||||||
2. |||||||||	If two angles are supplementary, the sum of their measures is 180°.
3. $\angle K = \angle I$.	|||||||||
4. |||||||||	Substitution (steps 2 and 3).
5. |||||||||	|||||||||

Set III

An amateur mathematician named Wrigley decided to invent a new angle relation: "doublemintary angles." Wrigley defined two angles to be doublemintary iff the measure of one angle is twice that of the other. He called the larger angle the "doublemint" of the smaller.

1. What is the measure of an angle whose complement and doublemint are equal? Show your method.
2. What is the measure of an angle whose doublemint and supplement are equal? Show your method.
3. Can you find an angle whose supplement is the doublemint of its complement?

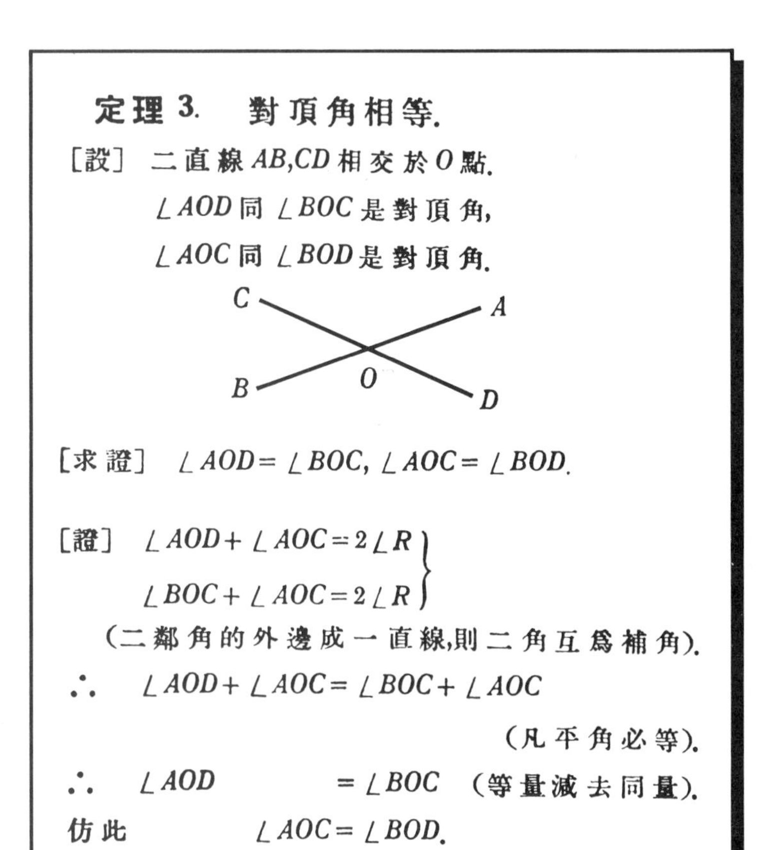

定理 3.　對頂角相等.

〔設〕 二直線 AB, CD 相交於 O 點.

$\angle AOD$ 同 $\angle BOC$ 是對頂角,

$\angle AOC$ 同 $\angle BOD$ 是對頂角.

〔求證〕 $\angle AOD = \angle BOC$, $\angle AOC = \angle BOD$.

〔證〕 $\left.\begin{array}{l}\angle AOD + \angle AOC = 2\angle R \\ \angle BOC + \angle AOC = 2\angle R\end{array}\right\}$

(二鄰角的外邊成一直線,則二角互爲補角).

$\therefore$ $\angle AOD + \angle AOC = \angle BOC + \angle AOC$

(凡平角必等).

$\therefore$ $\angle AOD = \angle BOC$ (等量減去同量).

仿此 $\angle AOC = \angle BOD$.

Lesson 5

Linear Pairs and Vertical Angles

If you looked at a geometry book used by students in a foreign country, you would probably be able to guess what many of the proofs were about even if you didn't understand the language. The reason is that most of the symbols used in mathematics are the same throughout the world. For example, the passage from a Chinese geometry book shown here is a proof of a useful theorem about vertical angles. Before considering this theorem, we will establish a couple of theorems about the angles in a linear pair.

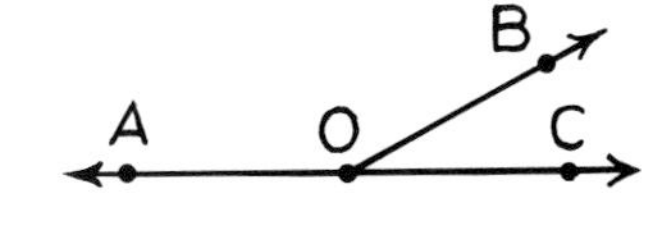

The figure at the right illustrates a linear pair: ∡AOB and ∡BOC have a common side ($\overrightarrow{OB}$) and their other sides ($\overrightarrow{OA}$ and $\overrightarrow{OC}$) are opposite rays. The three rays are part of a half-rotation, and so the Protractor Postulate can be used to assign coordinates to them. If $\overrightarrow{OA}$ is given the coordinate 0, then $\overrightarrow{OC}$, its opposite ray, will have the coordinate 180, and $\overrightarrow{OB}$ will have some coordinate between 0 and 180. Because $\overrightarrow{OA}$-$\overrightarrow{OB}$-$\overrightarrow{OC}$, $\angle AOB + \angle BOC = \angle AOC$. But $\angle AOC = |180 - 0| = 180°$, and so $\angle AOB + \angle BOC = 180°$. This proves that ∡AOB and ∡BOC are supplementary and we state the result as the following theorem.

▶ **Theorem 7**
If two angles are a linear pair, then they are supplementary.

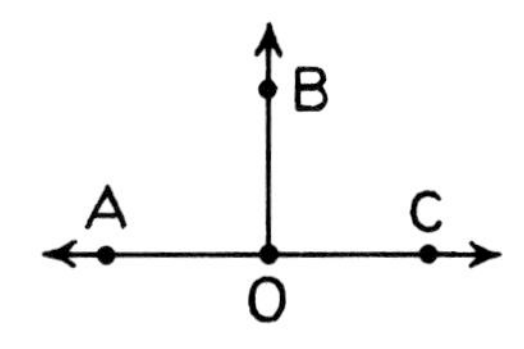

The figure at the left illustrates a linear pair in which $\angle AOB = \angle BOC$. Because $\measuredangle AOB$ and $\measuredangle BOC$ are also supplementary (Theorem 7), $\angle AOB + \angle BOC = 180°$. Substituting $\angle AOB$ for $\angle BOC$ in this equation, we get $\angle AOB + \angle AOB = 180°$, or $2\angle AOB = 180°$. Dividing by 2, we get $\angle AOB = 90°$, and so $\measuredangle AOB$ is a right angle. It immediately follows that $\measuredangle BOC$ must also be a right angle. These results are stated as the next theorem.

▶ **Theorem 8**
If the two angles in a linear pair are equal, then each is a right angle.

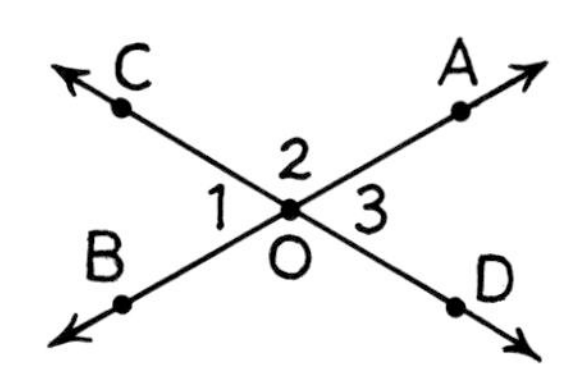

Now look at the figure from the Chinese geometry book shown here with the segments drawn as rays. The rays are opposite rays: $\overrightarrow{OB}$ is opposite $\overrightarrow{OA}$ and $\overrightarrow{OC}$ is opposite $\overrightarrow{OD}$. It follows that $\measuredangle 1$ and $\measuredangle 2$ are a linear pair, that $\measuredangle 2$ and $\measuredangle 3$ are a linear pair, and that $\measuredangle 1$ and $\measuredangle 3$ are vertical angles. Because the angles in a linear pair are supplementary (Theorem 7), $\measuredangle 1$ and $\measuredangle 3$ are supplements of $\measuredangle 2$. Because supplements of the same angle are equal (Theorem 6), $\angle 1 = \angle 3$. This result is a useful fact about vertical angles.

▶ **Theorem 9**
If two angles are vertical angles, then they are equal.

Exercises

Set I

Complete the following definitions.

1. Two angles are a linear pair iff ||||||||||.
2. Two angles are vertical angles iff ||||||||||.

Complete the following statements of the theorems of this lesson.

3. If two angles are a linear pair, then ||||||||||.
4. If two angles are vertical angles, then ||||||||||.
5. If the two angles in a linear pair are equal, then ||||||||||.

According to Theorem 7, if two angles are a linear pair, then they are supplementary.

6. State the converse of this theorem.
7. Is it true?
8. State the contrapositive of this theorem.
9. Is it true?

According to Theorem 9, if two angles are vertical angles, then they are equal.

10. State the inverse of this theorem.
11. Is it true?
12. State the contrapositive of this theorem.
13. Is it true?

In the figure below, S-O-Y, R-O-T, and $\angle SOT = x°$.

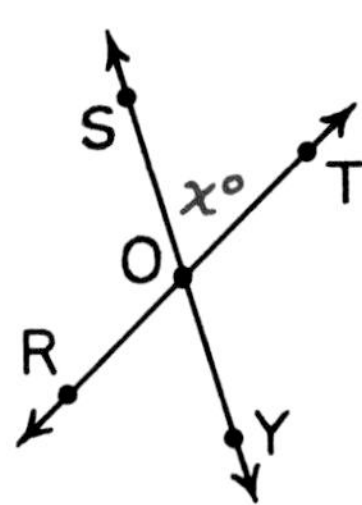

14. What can you conclude about $\overrightarrow{OS}$ and $\overrightarrow{OY}$ and about $\overrightarrow{OR}$ and $\overrightarrow{OT}$?

15. What conclusions can you draw about $\measuredangle$SOT and $\measuredangle$ROY?

16. Express $\angle$ROY in terms of x.

17. What conclusions can you draw about $\measuredangle$SOT and $\measuredangle$TOY?

18. Express $\angle$TOY in terms of x.

The figure in the column at the right represents two vertical angles that are also complementary.

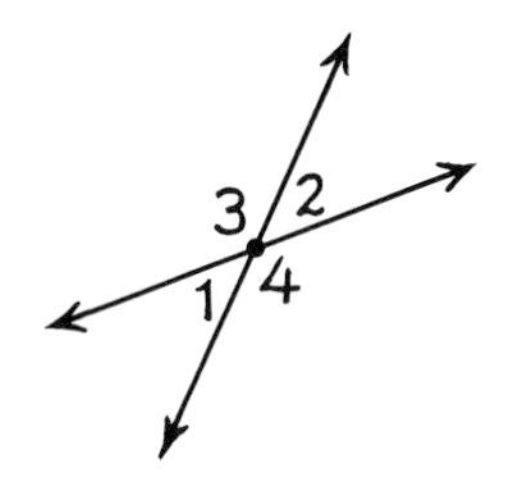

19. What can you conclude about $\measuredangle$1 and $\measuredangle$2 given that they are vertical angles?

20. What can you conclude about $\measuredangle$1 and $\measuredangle$2 given that they are complementary?

21. Use both of these facts to find the measures of $\measuredangle$1 and $\measuredangle$2.

22. Find the measures of $\measuredangle$3 and $\measuredangle$4.

Set II

In the figure below, $\overrightarrow{EF}$ and $\overrightarrow{EB}$ are opposite rays, $\overrightarrow{EA}$ and $\overrightarrow{EL}$ are opposite rays, $\angle FEA = (x + 120)°$, and $\angle BEL = (24 - 5x)°$.

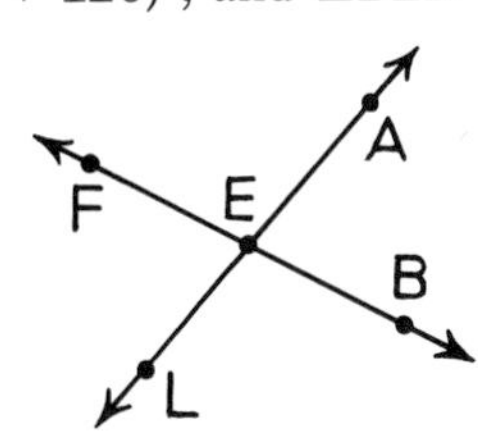

23. Find x.

24. Find $\angle$FEA.

25. Find $\angle$BEL.

In the figure below, $\overrightarrow{LA}$ and $\overrightarrow{LE}$ are opposite rays, $\angle TLA = (6x - 1)°$, and $\angle TLE = (20x - 1)°$.

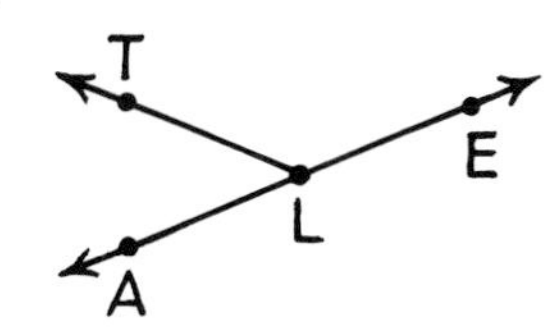

26. Find x.

27. Find $\angle$TLA.

28. Find $\angle$TLE.

Copy and complete the following proofs.

29.

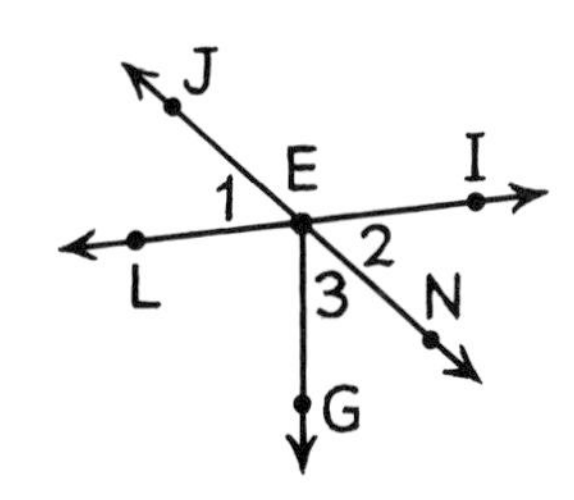

Given: $\measuredangle$1 and $\measuredangle$2 are vertical angles; $\overrightarrow{EN}$ bisects $\measuredangle$IEG.

Prove: $\angle 1 = \angle 3$.

Proof.

Statements	***Reasons***
1. $\measuredangle$1 and $\measuredangle$2 are vertical angles.	||||||||
2. ||||||||	If two angles are vertical angles, they are equal.
3. ||||||||	Given.
4. $\angle 2 = \angle 3$.	||||||||
5. ||||||||	||||||||

30.

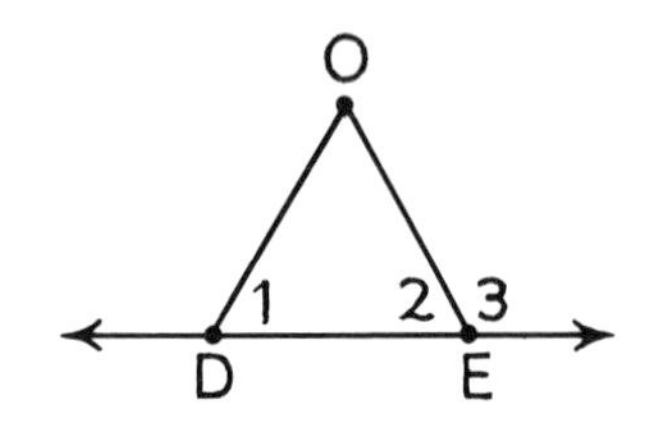

Given: ∡2 and ∡3 are a linear pair; ∡1 and ∡3 are supplementary.

Prove: ∠1 = ∠2.

Proof.

Statements	Reasons
1. ▥	▥
2. ▥	If two angles are a linear pair, then they are supplementary.
3. ∡1 and ∡3 are supplementary.	▥
4. ▥	▥

Write your own proofs for each of the following exercises. Copy the figure, "given," and "prove" before writing your statements and reasons.

31.

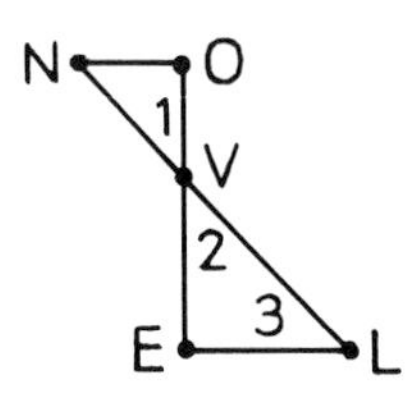

Given: ∡1 and ∡2 are vertical angles; ∠2 = ∠3.

Prove: ∠1 = ∠3.

32.

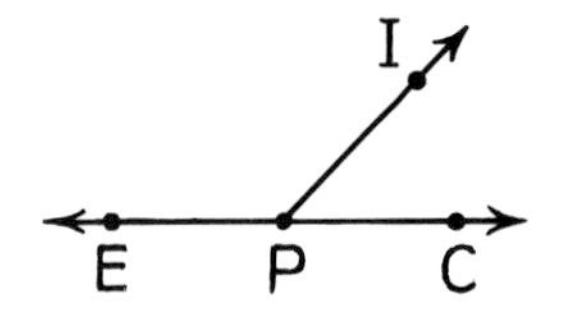

Given: ∡EPI and ∡IPC are a linear pair.

Prove: ∠IPC = 180° − ∠EPI.

33.

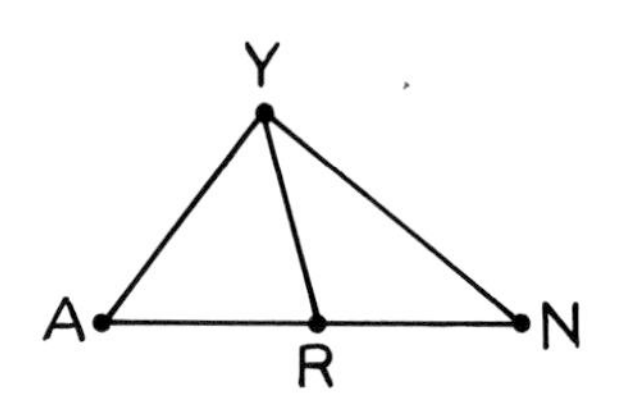

Given: ∡AYR and ∡RYN are complementary; $\overrightarrow{YA}$-$\overrightarrow{YR}$-$\overrightarrow{YN}$.

Prove: ∠AYN = 90°.

34.

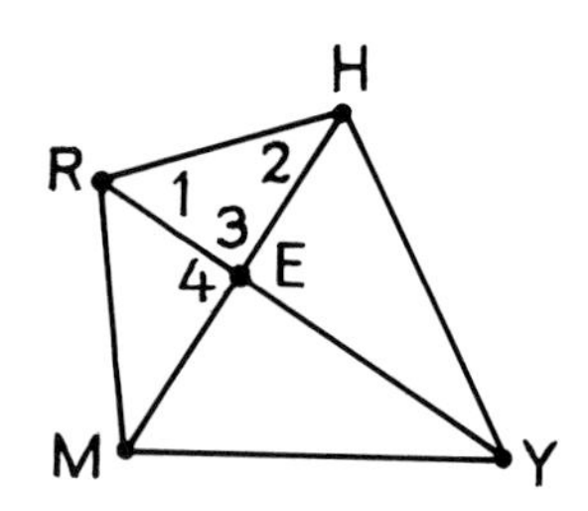

Given: ∡3 and ∡4 are a linear pair and ∠3 = ∠4; ∠1 + ∠2 = ∠3.

Prove: ∡1 and ∡2 are complementary.

Set III

The proof of the vertical angle theorem from the Chinese geometry book is shown again here.

A → 定理 3. 對頂角相等.

B → [設] 二直線 AB, CD 相交於 O 點. ← C

$\angle AOD$ 同 $\angle BOC$ 是對頂角,
$\angle AOC$ 同 $\angle BOD$ 是對頂角. ← D

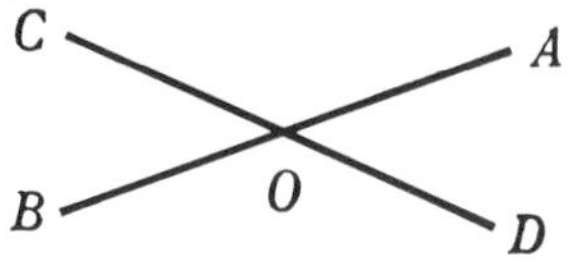

C A

B O D

E → [求證] $\angle AOD = \angle BOC,\ \angle AOC = \angle BOD.$

F → [證] $\left.\begin{array}{l}\angle AOD + \angle AOC = 2\angle R \\ \angle BOC + \angle AOC = 2\angle R\end{array}\right\}$ ← G

(二鄰角的外邊成一直線,則二角互爲補角). ← H

$\therefore\ \angle AOD + \angle AOC = \angle BOC + \angle AOC$

(凡平角必等). ← I

$\therefore\ \angle AOD = \angle BOC$ (等量減去同量). ← J

仿此 $\angle AOC = \angle BOD.$

The symbols in the box labeled A say:
"Theorem 3. If two angles are vertical angles, then they are equal."

The symbol in the box labeled B says: "Given."

1. What do you think the symbols in box C say?
2. Box D?
3. Box E?
4. Box F?
5. What do you think the symbols $2\angle R$ mean in box G given that there is no angle R in the figure?
6. The symbols in box H are the reason for the equations in box G. State the theorem in this lesson that you think they represent.
7. What do you think the symbols in box I say?
8. Box J?

Lesson 6

Parallel and Perpendicular Lines

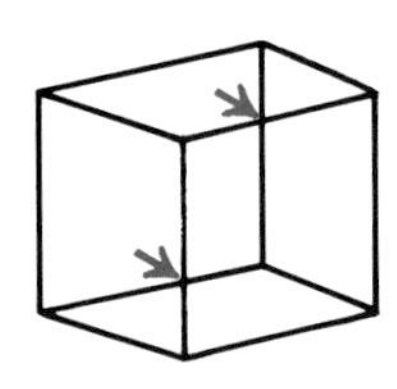

This picture is part of a lithograph made by the Dutch artist Maurits Escher titled *Belvedere.* The young man holds a strange cubelike structure in his hands that is not like any ordinary cube. It is based on the drawing shown at the left. The arrows indicate the two places in which the edges appear to cross each other. In a real cube, the edges would have one of the first two relations shown below. Escher has chosen the third, an impossibility.

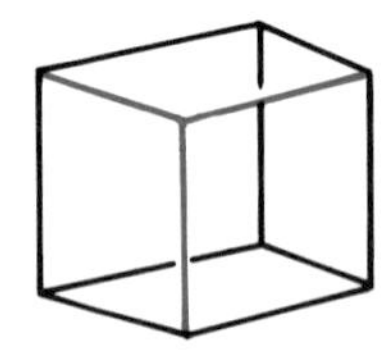

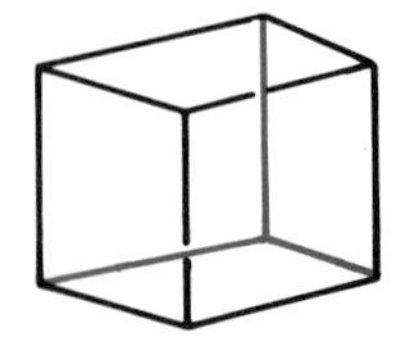

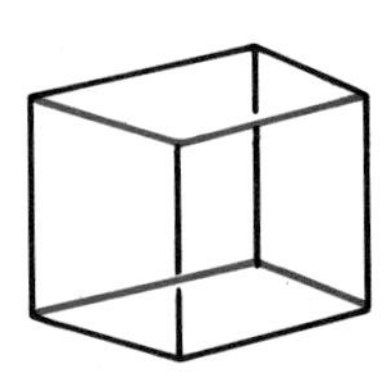

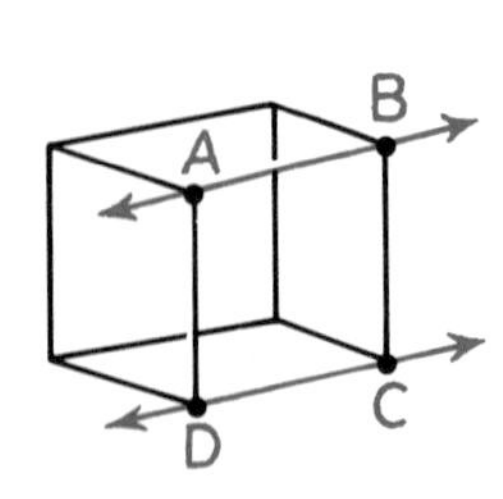

Some of the edges of a cube, such as the pair shown in the figure at the left, are said to be *parallel* because they lie in parallel lines.

► **Definition**
Two lines are ***parallel*** iff they lie in the same plane and do not intersect.

The symbol for "parallel" is $\|$; to indicate that $\overleftrightarrow{AB}$ and $\overleftrightarrow{DC}$ are parallel, we write $\overleftrightarrow{AB} \parallel \overleftrightarrow{DC}$. Rays and line segments are parallel iff the lines that contain them are parallel.

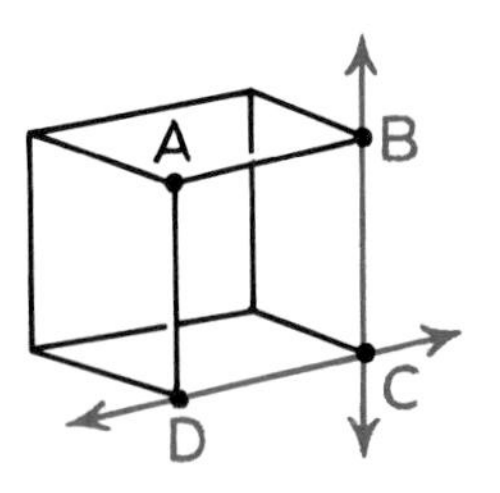

Other edges of a cube, such as the ones shown in this figure, are said to be *perpendicular* because they lie in perpendicular lines.

► **Definition**
Two lines are ***perpendicular*** iff they form a right angle.

The symbol for "perpendicular" is $\perp$; to indicate that $\overleftrightarrow{BC}$ and $\overleftrightarrow{CD}$ are perpendicular, we write $\overleftrightarrow{BC} \perp \overleftrightarrow{CD}$. Rays and line segments are perpendicular iff the lines that contain them are perpendicular.

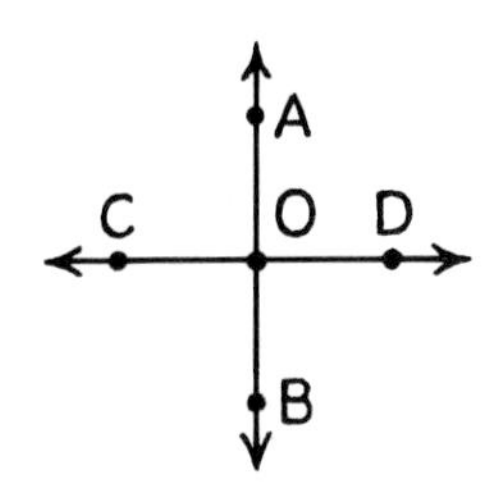

This figure shows two lines, $\overleftrightarrow{AB}$ and $\overleftrightarrow{CD}$, intersecting at point O. According to our definition of perpendicular lines, if $\measuredangle AOC$ is a right angle, then it follows that $\overleftrightarrow{AB} \perp \overleftrightarrow{CD}$. It also follows that $\measuredangle AOD$, $\measuredangle DOB$, and $\measuredangle BOC$ must also be right angles. This is easily proved because of the linear pairs and vertical angles in the figure.

Notice that $\measuredangle AOD$ forms a linear pair with $\measuredangle AOC$ as does $\measuredangle BOC$. The two angles in a linear pair are supplementary. If $\measuredangle AOC$ is a right angle, it has a measure of 90°, and so $\measuredangle AOD$ and $\measuredangle BOC$, its supplements, must also have measures of 90°. Also, $\measuredangle DOB$ and $\measuredangle AOC$ are vertical angles. If two angles are vertical angles, they are equal. Therefore $\measuredangle DOB$ must also have a measure of 90° and be a right angle.

These results are stated as the following theorem.

► **Theorem 10**
If two lines are perpendicular, they form four right angles.

The fact that any angle that is a right angle has a measure of 90° establishes one more useful theorem.

► **Theorem 11**
Any two right angles are equal.

Exercises

Set I

Define each of the following terms.

1. Parallel lines.
2. Perpendicular lines.

Complete the statements of the following theorems.

3. If two lines are perpendicular, ______.
4. Any two right angles ______.

The following exercises refer to the figure below.

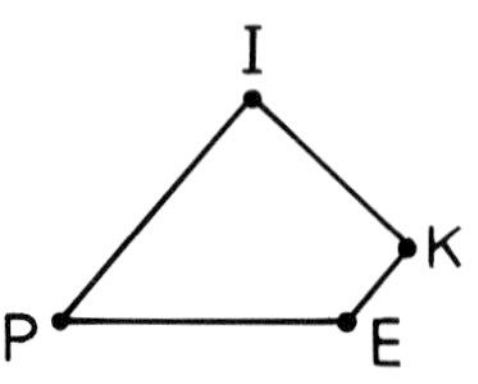

5. What is a polygon having this number of sides called?
6. Is it convex or concave?
7. Name a pair of sides that appear to be parallel.
8. Name two pairs of sides that appear to be perpendicular.
9. Which sides of PIKE appear to be equal?
10. Which angles of PIKE appear to be equal?

Refer to the figures below to decide whether the following statements about a set of lines in a plane seem to be true or false.

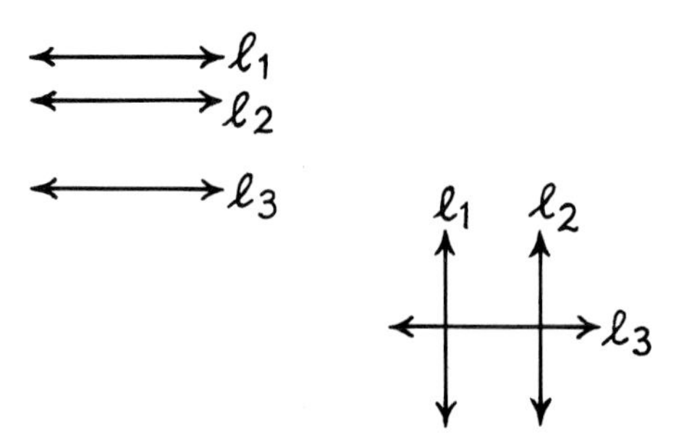

11. If two lines are parallel to a third line, then they are parallel to each other.
12. If two lines are perpendicular to a third line, then they are perpendicular to each other.
13. If a line is perpendicular to one of two parallel lines, then it is also perpendicular to the other.
14. If a line is parallel to one of two perpendicular lines, then it is also parallel to the other.

Theorem 10 states that "if two lines are perpendicular, they form four right angles." Remember that the hypothesis of a theorem is "given"; the "given" for Theorem 10 for this figure is:

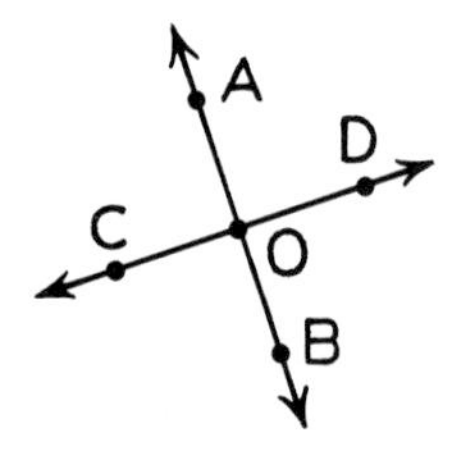

Given: $\overleftrightarrow{AB} \perp \overleftrightarrow{CD}$.

15. What is the "prove" for Theorem 10 for this figure?

Theorem 11 states that "any two right angles are equal."

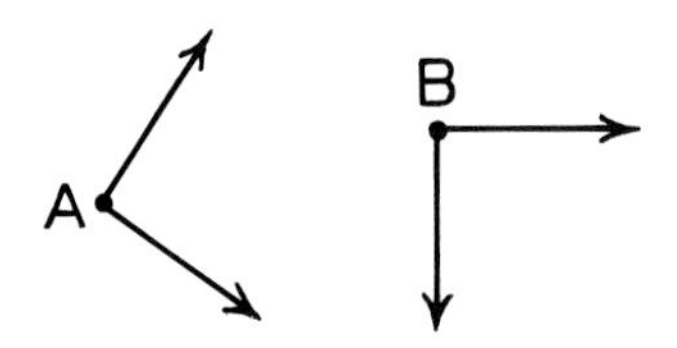

16. What is the "given" for Theorem 11 for the figure above?
17. What is the "prove"?

The figure below represents a cube.

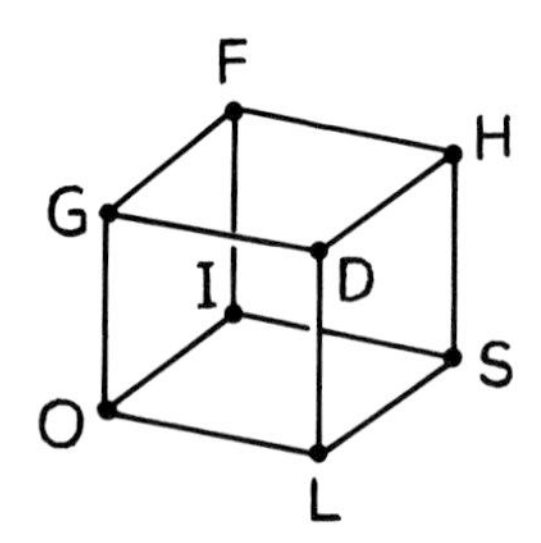

18. Name the edges that appear to be perpendicular to $\overline{DL}$.
19. Name the edges that appear to be parallel to $\overline{DL}$.
20. Name an edge that appears to be neither perpendicular nor parallel to $\overline{DL}$.

In the figure below, the four rays form angles having the measures as indicated. Tell who is correct in each case: Obtuse Ollie, Acute Alice, both of them, or neither one.

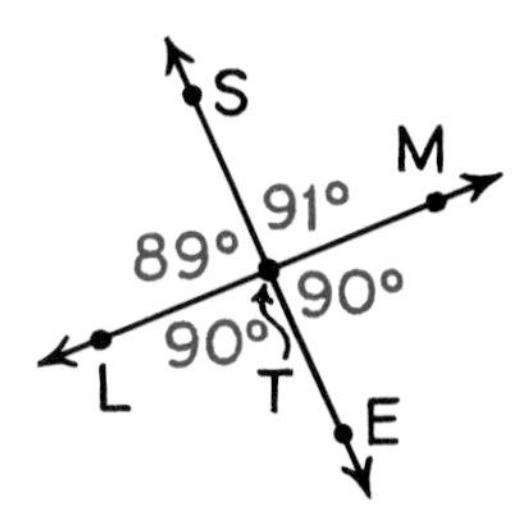

21. When asked if the figure contains a pair of opposite rays, Ollie said "$\overrightarrow{TS}$ and $\overrightarrow{TE}$" and Alice said "$\overrightarrow{TL}$ and $\overrightarrow{TM}$."

22. When asked if any rays in the figure are perpendicular, Ollie said "$\overrightarrow{TL}$ and $\overrightarrow{TE}$" and Alice said "$\overrightarrow{TE}$ and $\overrightarrow{TM}$."

23. When asked if the figure contains any vertical angles, Ollie said "$\measuredangle$STL and $\measuredangle$MTE" and Alice said "$\measuredangle$STM and $\measuredangle$LTE."

24. When asked if the figure contains any linear pairs, Ollie said "$\measuredangle$LTS and $\measuredangle$STM" and Alice said "$\measuredangle$STM and $\measuredangle$MTE."

Set II

Copy and complete the following proofs.

25.

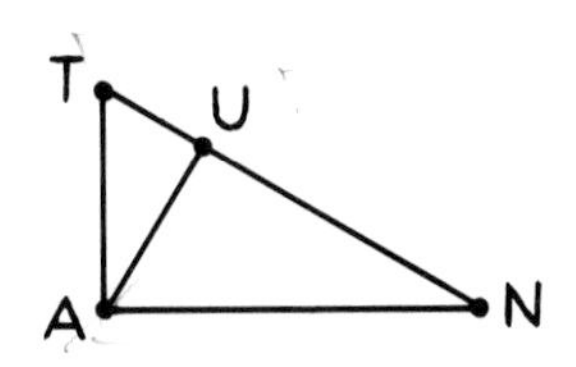

Given: $\overline{AU} \perp \overline{TN}$ and $\overline{TA} \perp \overline{AN}$.

Prove: $\measuredangle$TUA and $\measuredangle$TAN are supplementary.

Proof.

Statements	Reasons
1. ________	________
2. $\measuredangle$TUA and $\measuredangle$TAN are right angles.	________
3. $\angle TUA = 90°$ and $\angle TAN = 90°$.	________
4. ________	Addition.
5. ________	________

26.

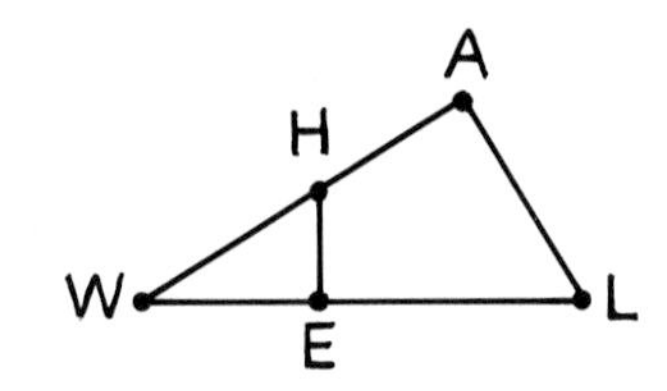

Given: $\overline{HE} \perp \overline{WL}$; $\angle HEW = \angle A$.

Prove: $\angle A = 90°$.

Proof.

Statements	Reasons
1. $\overline{HE} \perp \overline{WL}$.	________
2. $\measuredangle$HEW is a right angle.	________
3. ________	A right angle has a measure of 90°.
4. ________	Given.
5. ________	________

Write your own proofs for each of the following exercises. Copy the figure, "given," and "prove" before writing your statements and reasons.

27.

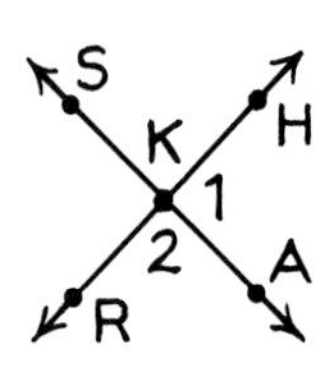

Given: $\measuredangle 1$ and $\measuredangle 2$ are a linear pair and $\angle 1 = \angle 2$.

Prove: $\overleftrightarrow{SA} \perp \overleftrightarrow{HR}$.

28.

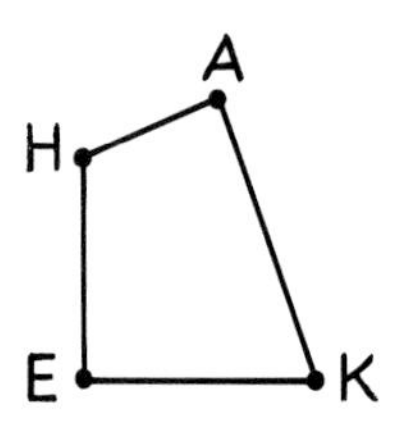

Given: $\overline{HE} \perp \overline{EK}$ and $\overline{HA} \perp \overline{AK}$.

Prove: $\angle E = \angle A$.

29.

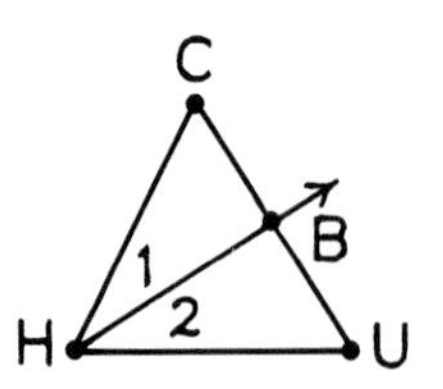

Given: $\overrightarrow{HB}$ bisects $\measuredangle CHU$; $\angle 1 + \angle U = 90°$.

Prove: $\measuredangle 2$ and $\measuredangle U$ are complementary.

30.

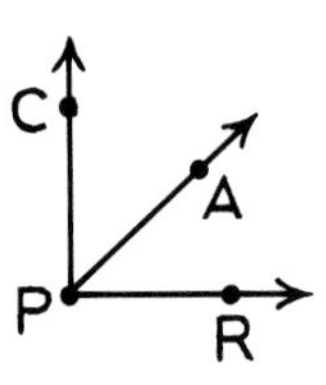

Given: $\overrightarrow{PC} \perp \overrightarrow{PR}$; $\overrightarrow{PA}$ bisects $\measuredangle CPR$.

Prove: $\angle CPA = 45°$.

Set III

Photograph by C. F. Cochran

This is a photograph of a wooden crate similar in nature to the "cube" in Escher's picture. The crate actually exists, but not everything in it is exactly what it appears to be in the photograph.

Can you figure out how such a crate could be built and photographed? If so, make a drawing to show how.

Chapter 3 / Summary and Review

Basic Ideas

Acute angle 91
Angle 85
  measure of 90
Betweenness of rays 94
Complementary angles 100
Coordinate of a ray 90
Degree 89
Linear pair 86
Minute 91
Obtuse angle 91
Opposite rays 85
Parallel lines 111
Perpendicular lines 111
Ray 85
  that bisects an angle 95
Right angle 91
Second 91
Straight angle 91
Supplementary angles 100
Vertical angles 86

Postulate

7. *The Protractor Postulate.* The rays in a half-rotation can be numbered so that
   a) to every ray there corresponds exactly one real number called its coordinate,
   b) to every real number from 0 to 180 inclusive there corresponds exactly one ray,
   c) to every pair of rays there corresponds exactly one real number called the measure of the angle that they determine,
   d) and the measure of an angle is the absolute value of the difference between the coordinates of its rays. 90

Theorems

3. *The Betweenness of Rays Theorem.* If $\overrightarrow{OA}$-$\overrightarrow{OB}$-$\overrightarrow{OC}$, then $\angle AOB + \angle BOC = \angle AOC$. 94

4. *The Angle Bisector Theorem.* A ray that bisects an angle divides it into angles half as large as the angle. 95

5. Complements of the same angle (or equal angles) are equal. 100

6. Supplements of the same angle (or equal angles) are equal. 101

7. If two angles are a linear pair, then they are supplementary. 106

8. If the two angles in a linear pair are equal, then each is a right angle. 106

9. If two angles are vertical angles, then they are equal. 106

10. If two lines are perpendicular, they form four right angles. 111

11. Any two right angles are equal. 111

Exercises

Set I

Complete the following definition statements by using as few words or symbols as you can.

1. ∡A and ∡B are supplementary iff ||||||||.
2. ∡C is acute iff ||||||||.
3. $\overrightarrow{DE}$ bisects ∡FDG iff |||||||| and ||||||||.
4. $\overrightarrow{HI}$ is the set of points H and I and all points J such that either |||||||| or ||||||||.
5. $\overrightarrow{KL}$ and $\overrightarrow{KM}$ are opposite rays iff ||||||||.

Complete the statements of the following theorems.

6. If two angles are a linear pair, ||||||||.
7. If a ray bisects an angle, ||||||||.
8. Complements ||||||||.
9. If two angles are vertical angles, ||||||||.
10. If two lines are perpendicular, ||||||||.
11. Any two right angles ||||||||.
12. If the two angles in a linear pair are equal, ||||||||.

The following exercises refer to the figure below.

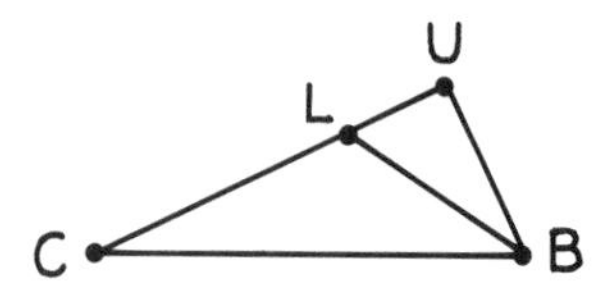

13. Name an angle that appears to be obtuse.
14. Name an angle that appears to be right.
15. How many angles of the figure appear to be acute?

Set II

The following definition is incorrectly stated and, as a result, it is false.

> Two angles are complementary iff their measures are 90°.

16. How should it be changed to make it true? (Hint: Look at the cartoon at the beginning of this chapter review.)

In the figure below, $\overrightarrow{OJ}$ and $\overrightarrow{OE}$ are opposite rays and ∠JOR = ∠ROE. Use this information to tell whether each of the following statements *must be true, may be true,* or *appears to be false.*

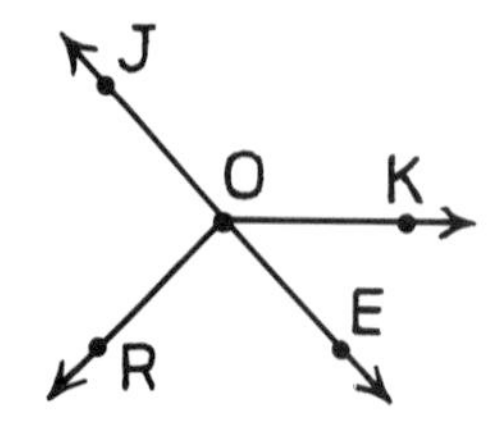

17. ∡JOR and ∡ROE are a linear pair.
18. ∡JOR and ∡ROE are right angles.
19. $\overrightarrow{OR} \perp \overrightarrow{OE}$.
20. ∡JOR and ∡KOE are vertical angles.
21. $\overrightarrow{OR}$-$\overrightarrow{OJ}$-$\overrightarrow{OK}$.
22. ∠JOK = ∠ROK.
23. ∡JOK and ∡KOE are supplementary.
24. $\overrightarrow{OE}$ bisects ∡ROK.

In the figure below, ∡A is the complement of ∡C and ∡C is the complement of ∡E. Also, ∠A = $(x + 30)°$ and ∠E = $(9 - 6x)°$.

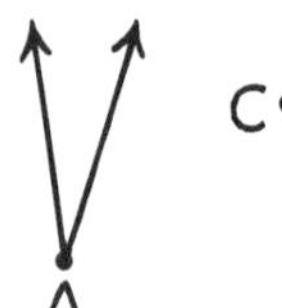

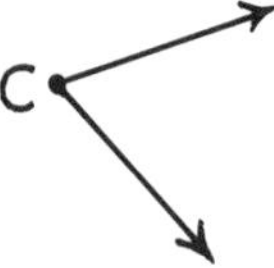

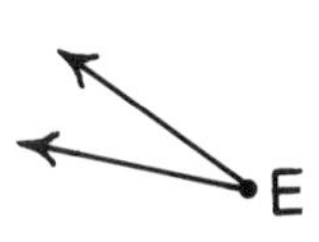

25. Find x.

26. Find $\angle A$.

27. Find $\angle C$.

28. Find $\angle E$.

Copy and complete the following proof.

29.

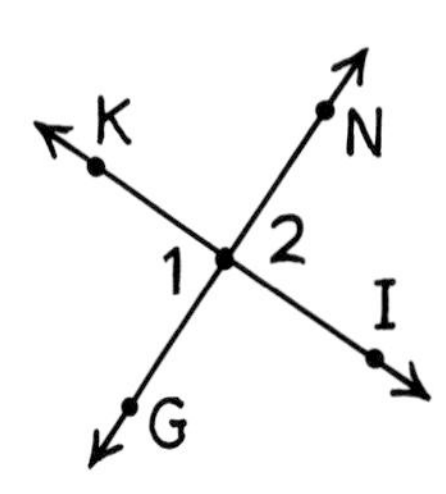

Given: $\measuredangle 1$ and $\measuredangle 2$ are vertical angles; $\measuredangle 1$ and $\measuredangle 2$ are supplementary.

Prove: $\overleftrightarrow{KI} \perp \overleftrightarrow{NG}$.

Proof.

Statements	***Reasons***
1. $\measuredangle 1$ and $\measuredangle 2$ are vertical angles.	________
2. ________	Vertical angles are equal.
3. $\measuredangle 1$ and $\measuredangle 2$ are supplementary.	________
4. ________	If two angles are supplementary, the sum of their measures is 180°.
5. $\angle 1 + \angle 1 = 180°$; so $2\angle 1 = 180°$.	________
6. ________	Division.
7. $\measuredangle 1$ is a right angle.	________
8. ________	________

Write your own proofs for each of the following exercises. Copy the figure, "given," and "prove" before writing your statements and reasons.

30.

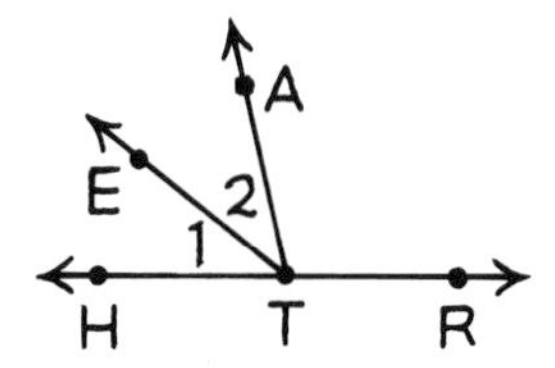

Given: $\measuredangle 1$ and $\measuredangle ETR$ are a linear pair; $\angle 1 = \angle 2$.

Prove: $\angle 2 + \angle ETR = 180°$.

31.

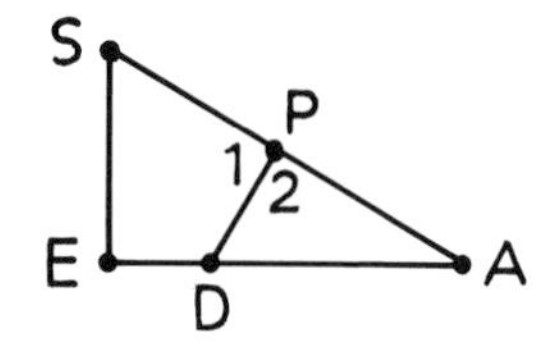

Given: $\measuredangle 1$ and $\measuredangle 2$ are a linear pair; $\angle 1 = \angle E$ and $\angle E = \angle 2$.

Prove: $\overline{SA} \perp \overline{PD}$.

32.

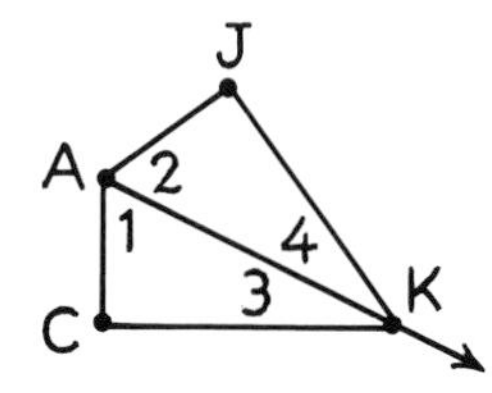

Given: $\overrightarrow{AK}$ bisects $\measuredangle JAC$; $\measuredangle 1$ and $\measuredangle 3$ are complementary.

Prove: $\angle 2 = 90° - \angle 3$.

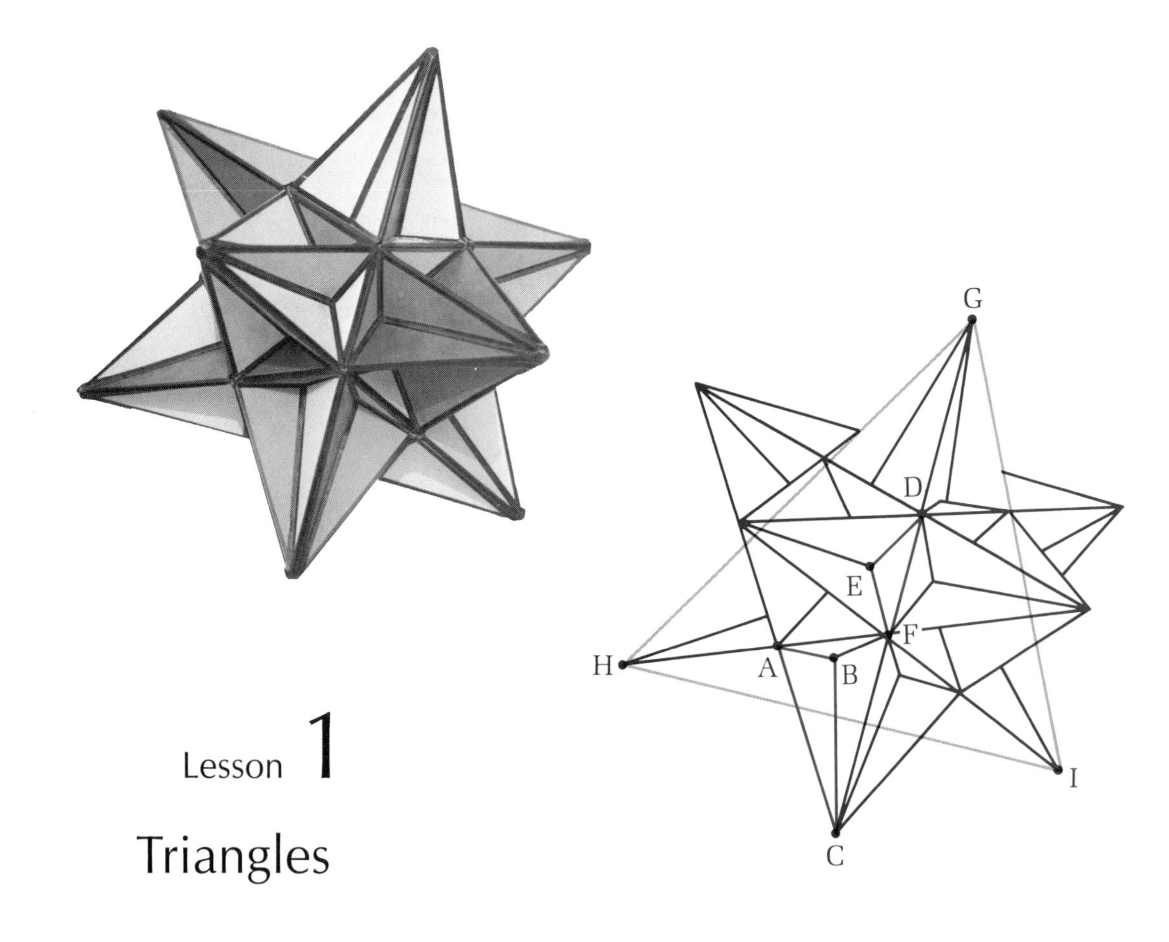

Lesson 1
Triangles

The beautiful geometric solid shown in the photograph above is one that Euclid never got to see. It was discovered in 1809 by a French mathematician, Louis Poinsot, and is called a "great icosahedron."

The edges of the great icosahedron form a large number of triangles having several different shapes. Although the arrangement of these triangles in the solid is very complex, the triangles themselves are very simple figures. As you already know, a triangle is a polygon.

► **Definition**
A ***triangle*** is a polygon that has three sides.

The symbol for triangle is $\triangle$; to refer to the triangle shown at the left, we write $\triangle ABC$. The *vertices* of $\triangle ABC$ are the points A, B, and C; the *sides* of the triangle are the segments $\overline{AB}$, $\overline{BC}$, and $\overline{AC}$; and the *angles* of the triangle are $\angle A$, $\angle B$, and $\angle C$.

The words "included" and "opposite" are used to describe the relation of the positions of the sides and angles of a given triangle. In $\triangle ABC$, side $\overline{AB}$ is included by $\angle A$ and $\angle B$ and is opposite $\angle C$; $\angle B$ is included by sides $\overline{BA}$ and $\overline{BC}$ and is opposite side $\overline{AC}$. In general, a *side* of a triangle is *included* by the two angles whose vertices are its endpoints and is *opposite* the third angle. An *angle* of a triangle is *included* by the two sides of the triangle that lie on the sides of the angle and is *opposite* the third side. Every triangle is said to contain *six parts:* its sides and its angles.

Triangles are named with respect to the relative lengths of their sides. Most of the triangles in the great icosahedron have the shape of △ABC in the figure at the beginning of this lesson. The triangle has no equal sides and is called *scalene.* The solid also contains some triangles that have two equal sides, such as △DEF, called *isosceles.* And there are some triangles that have three equal sides, such as △GHI, called *equilateral.*

► **Definitions**

A triangle is:

scalene iff it has no equal sides;
isosceles iff it has at least two equal sides;
equilateral iff all of its sides are equal.

Triangles are also classified according to the measures of their angles.

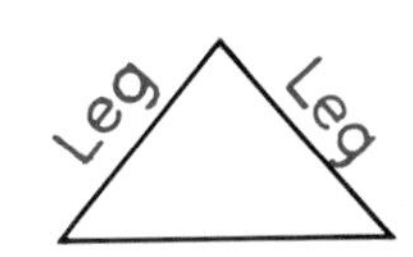

An isosceles triangle

► **Definitions**

A triangle is:

acute iff all three of its angles are acute;
right iff it has a right angle;
obtuse iff it has an obtuse angle;
equiangular iff all of its angles are equal.

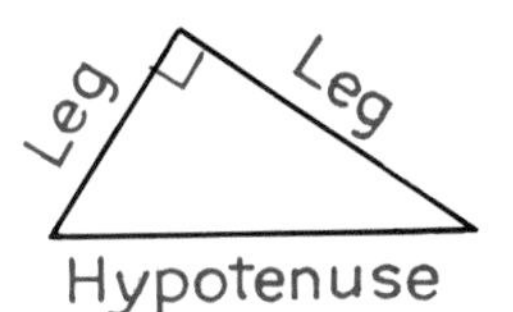

A right triangle

Some of the parts of two types of triangles have special names. In an isosceles triangle that has exactly two equal sides, the equal sides are called the *legs.* In a right triangle, the sides that include the right angle are called the *legs* and the side opposite the right angle is called the *hypotenuse.*

Exercises

Set I

The following exercises refer to △JAN, in which JA = JN.

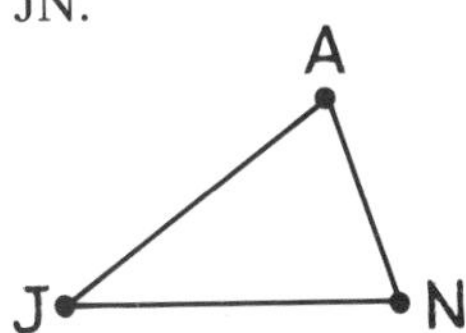

1. What kind of triangle is △JAN with respect to its sides?
2. What kind of triangle is △JAN with respect to its angles?
3. Name its vertices.
4. Name its legs.

The following exercises refer to △FEB, which has no equal sides.

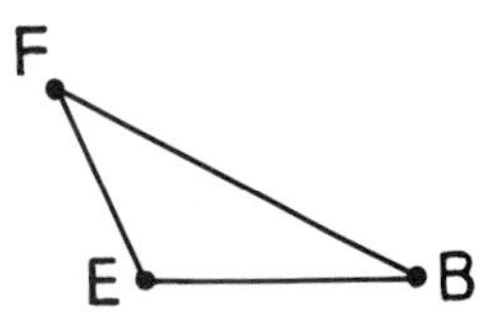

5. What kind of triangle is △FEB with respect to its sides?
6. What kind of triangle is △FEB with respect to its angles?
7. Name the sides of the triangle that include ∡F.

8. Name the side opposite ∡E.
9. Name the angle opposite side $\overline{EF}$.
10. Name the angles that include side $\overline{FB}$.

The following exercises refer to △MAY, in which ∡Y is a right angle.

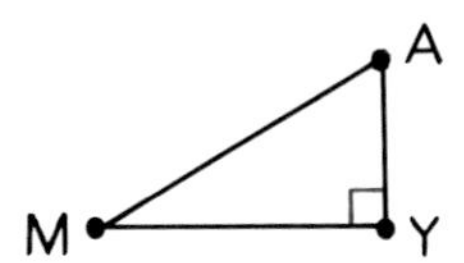

11. What kind of triangle is △MAY with respect to its sides?
12. What kind of triangle is △MAY with respect to its angles?
13. Name its legs.
14. Name its hypotenuse.

Review the definitions of isosceles and equilateral triangles before answering the following questions.

15. If a triangle is equilateral, is it also isosceles?
16. Why or why not?
17. If a triangle is isosceles, is it also equilateral?

In the figure at the top of the next column, M-H-R and C-H-A. Name all of the triangles in the figure that have each of the following as one of their parts.

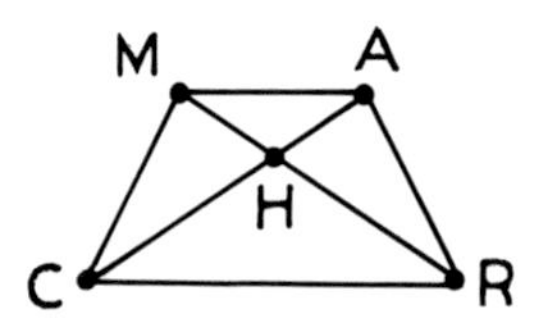

Example: $\overline{CH}$.
Answer: △CHM and △CHR.

18. $\overline{MR}$.
19. $\overline{AR}$.
20. ∡CHR.
21. ∡AMR.

In the figure below, $\overline{PI} \perp \overline{LR}$.

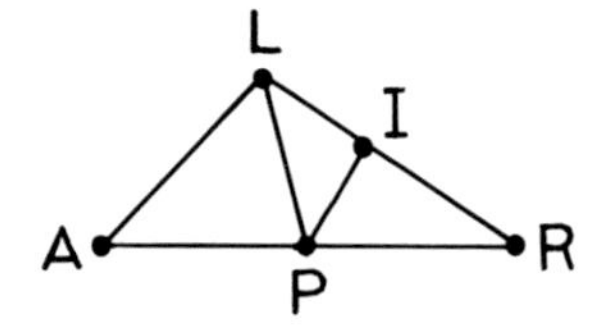

22. Which two triangles in the figure are right triangles?
23. Which triangle is an acute triangle?
24. Which two triangles are obtuse triangles?
25. How many angles in the figure have their vertex at L?
26. How many angles in the figure have their vertex at P?
27. How many angles in the figure can be named with just one letter?

Set II

The table below lists some different types of triangles. For example, "*g*" represents an obtuse scalene triangle.

	Scalene	***Isosceles***	***Equilateral***
Acute	*a*	*b*	*c*
Right	*d*	*e*	*f*
Obtuse	*g*	*h*	*i*

28. Use a straightedge to draw an example of each type of triangle in the table that you think can exist. If you think the figure cannot exist, write "impossible."

Use your ruler and protractor to draw triangles having the following parts.

29. △AUG, in which AG = 5 cm, ∠A = 60°, and AU = 3 cm.

30. $\triangle$OCT, in which $\angle O = 60°$, OT = 3 cm, and $\angle T = 95°$.

31. $\triangle$NOV, in which NV = 3 cm, $\angle V = 60°$, and VO = 3 cm.

32. $\triangle$DEC, in which $\angle D = 95°$, DC = 3 cm, and $\angle C = 60°$.

33. How many parts of each triangle did you need to be able to draw it?

Refer to your drawings to answer the following questions. (You may need to make some measurements.)

34. Tell, using two words, what type of triangle each of your figures appears to be.

35. Name the two triangles that seem to have the same size and shape.

Set III

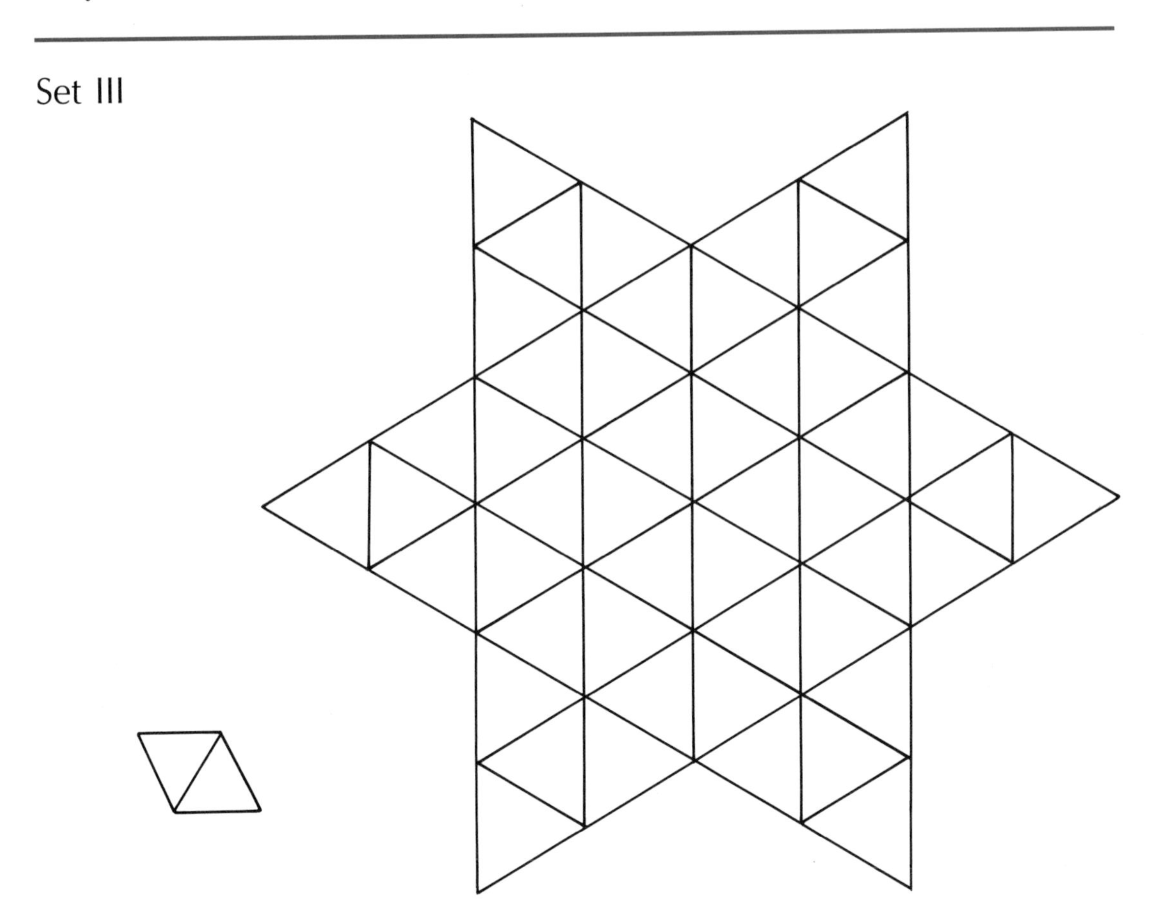

"Polyiamonds" are figures made by fitting together two or more equilateral triangles. The simplest polyiamond, a "diamond," consists of two equilateral triangles.

There are twelve different polyiamonds that contain six equilateral triangles each. They are shown on the next page. Trace and cut them out and see if you can figure out how to fit *eight* of them together to form the star shown above.* A good way to begin is to

---

*From "Mathematical Games" by Martin Gardner. Copyright © 1964 by Scientific American, Inc. All rights reserved.

figure out which polyiamonds can be used to fill in the six triangles at the corners of the star. In solving the puzzle, it may be necessary to turn some of the pieces over.

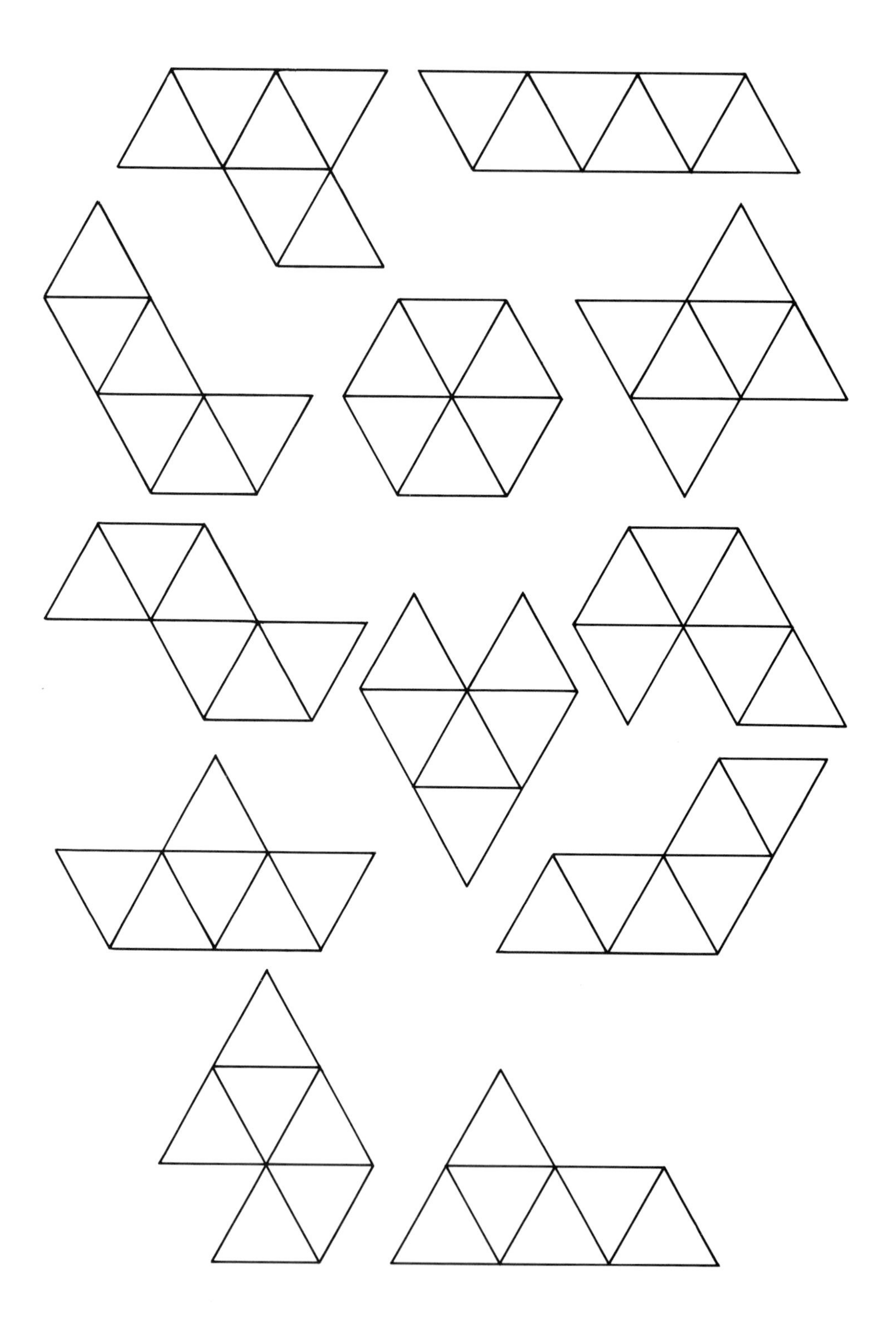

Lesson 2

Congruent Polygons

In this remarkable drawing by Escher, the artist has succeeded in dividing a plane into reptiles, all of which are identical in size and shape. Although only a part of the plane is shown, the repeating reptile pattern can be extended indefinitely.

The reptiles are in the shape of polygons having 50 vertices each. If a tracing were made of one of these polygons, it could be placed so as to coincide exactly with any one of the others. In doing this, a correspondence is established between the vertices of the polygons. For example, the vertex labeled A in the first polygon below would correspond to the vertex labeled A′ in the second polygon. The vertex labeled B would correspond to the vertex labeled B′, and so on.

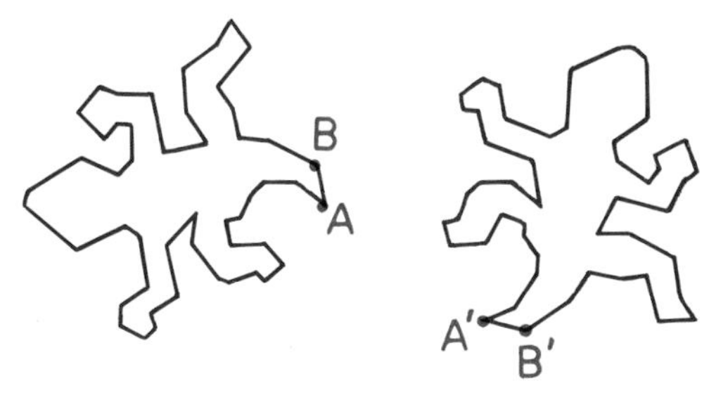

In order for the polygons to fit exactly together, all of their corresponding angles and sides would have to be equal: $\angle A = \angle A'$, $AB = A'B'$, $\angle B = \angle B'$, and so on. Such polygons are said to be *congruent.*

Definition

Two polygons are ***congruent*** iff there is a correspondence between their vertices such that all of their corresponding sides and angles are equal.

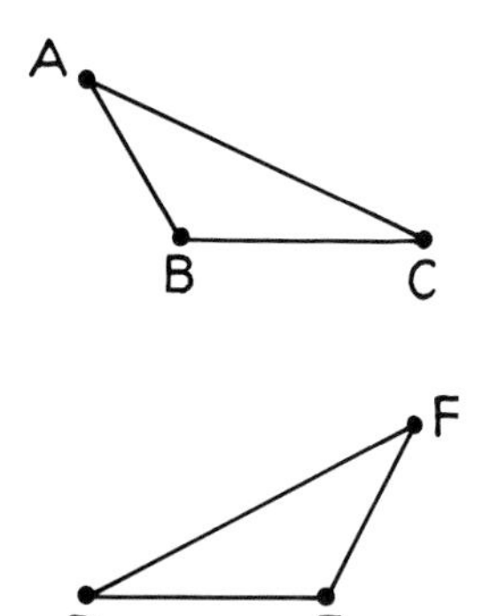

Look at the two triangles at the left. If a tracing of $\triangle ABC$ were picked up and turned over, it could be placed on $\triangle DEF$. Vertex A would fall on F, B would fall on E, and C would fall on D. To indicate this, we write

$$A \leftrightarrow F, \quad B \leftrightarrow E, \quad \text{and} \quad C \leftrightarrow D,$$

or, more briefly,

$$ABC \leftrightarrow FED.$$

If all of the corresponding sides and angles of the two triangles are equal, the triangles are congruent. The symbol for congruence is $\cong$; to indicate that the two triangles are congruent, we write

$$\triangle ABC \cong \triangle FED.$$

The equations for the corresponding parts of the triangles can be written from this correspondence:

$$AB = FE, \quad AC = FD, \quad BC = ED,$$

$$\angle A = \angle F, \quad \angle B = \angle E, \quad \text{and} \quad \angle C = \angle D.$$

Although a triangle has six parts—three sides and three angles—not all of them are needed to draw it. The first figure below suggests that two sides and the included angle are sufficient to determine a triangle's vertices, and hence its shape. The second figure suggests that two angles and the included side are also sufficient to determine a triangle's shape.

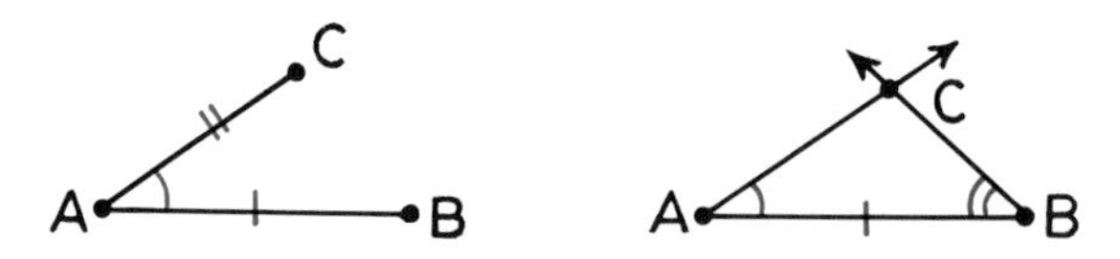

We state these observations as a couple of postulates about congruent triangles. The tick marks and arcs in the figures identify equal sides and angles.

Postulate 8 (The S.A.S. Congruence Postulate)

If two sides and the included angle of one triangle are equal to two sides and the included angle of another triangle, the triangles are congruent.

▶ **Postulate 9** (The A.S.A. Congruence Postulate)
If two angles and the included side of one triangle are equal to two angles and the included side of another triangle, the triangles are congruent.

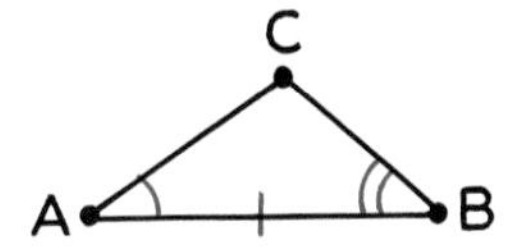

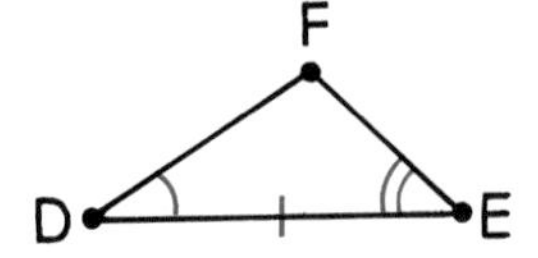

Exercises

Set I

The two quadrilaterals below are congruent.

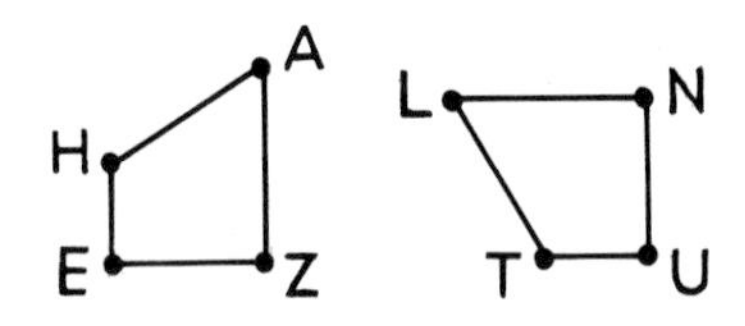

1. If a tracing of HAZE were placed on LNUT, which of its vertices would fit on L?
2. Which vertex would fit on N?
3. Copy and complete the following congruence correspondence between the vertices of the quadrilaterals: HAZE ↔ ▒▒▒▒.

The triangles below are congruent.

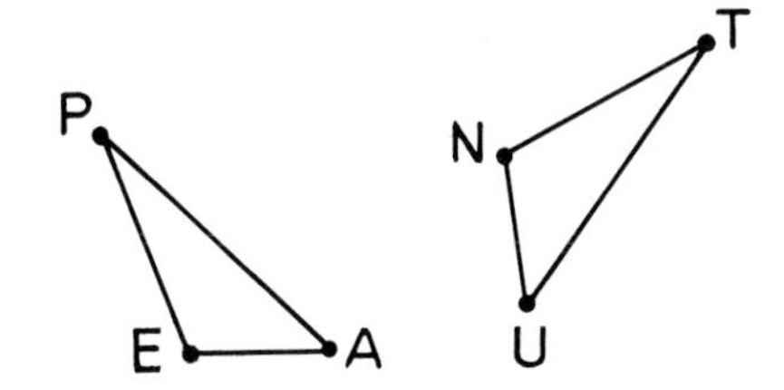

4. Copy and complete the following correspondence to show this:

PEA ↔ ▒▒▒▒.

Copy and complete the following equations for some of the equal angles and sides of △PEA and △NUT.

5. ∠P = ∠▒▒▒▒.
6. ∠U = ∠▒▒▒▒.
7. PA = ▒▒▒▒.
8. NU = ▒▒▒▒.

The equal sides and angles of two congruent triangles can be read from a congruence correspondence between them. Use the fact that

$$\triangle CAS \cong \triangle HEW$$

to copy and complete the following equations.

9. ∠S = ▒▒▒▒.
10. ∠E = ▒▒▒▒.
11. AS = ▒▒▒▒.
12. HW = ▒▒▒▒.

The following questions refer to △NUT.

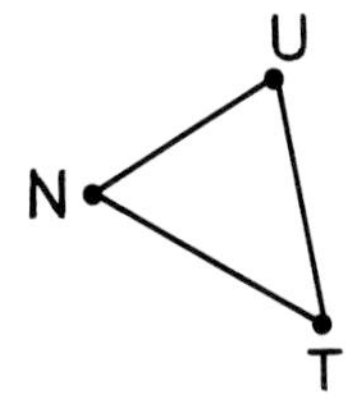

13. Which side of △NUT is included by ∡N and ∡T?
14. Which angle is included by sides $\overline{NT}$ and $\overline{UT}$?
15. Which angles include $\overline{UT}$?
16. Which sides include ∡U?

In each of the following figures, the tick marks and arcs identify equal parts. If it is possible to conclude that the triangles are congruent on the basis of one of the congruence postulates, name it. Otherwise, write "no conclusion possible."

Example:

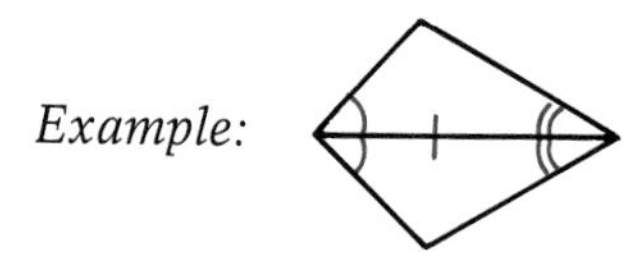

Answer: A.S.A.

17.

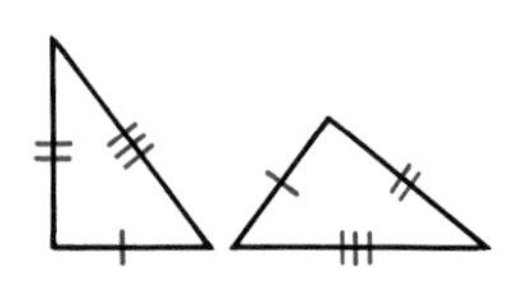

18.

19.

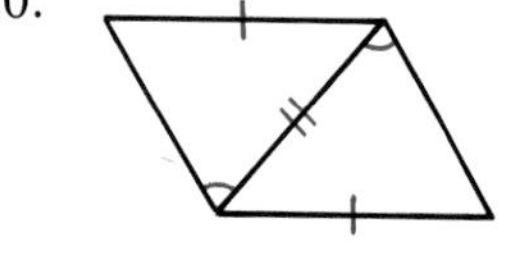

20.

21.

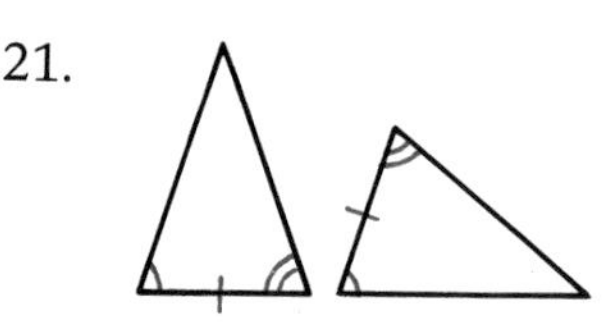

22.

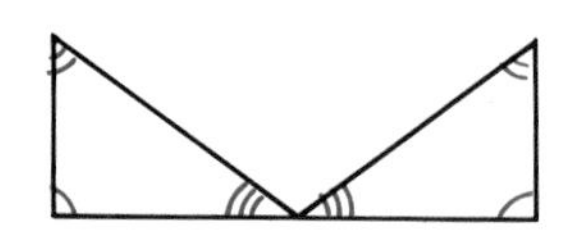

23.

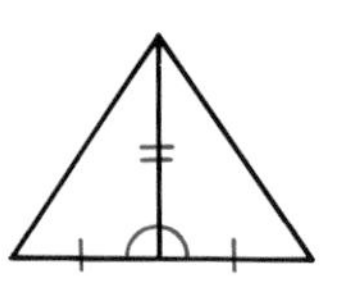

24.

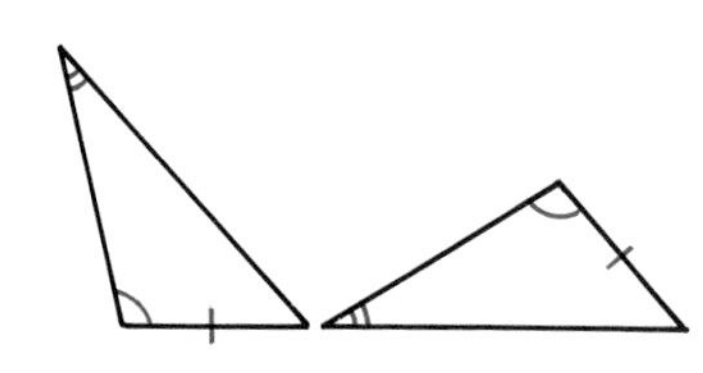

25.

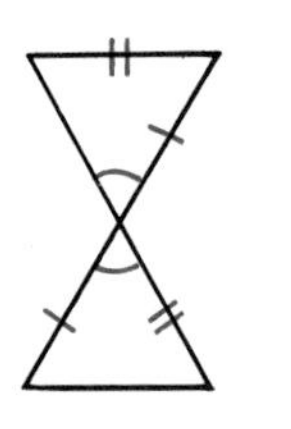

Set II

26. According to the definition of congruent polygons, how many pairs of parts must be equal if two triangles are congruent?

27. According to the S.A.S. and A.S.A. congruence postulates, how many pairs of equal parts are sufficient to show that two triangles are congruent?

In the triangles shown below, ∠A = ∠O and ∠L = ∠D.

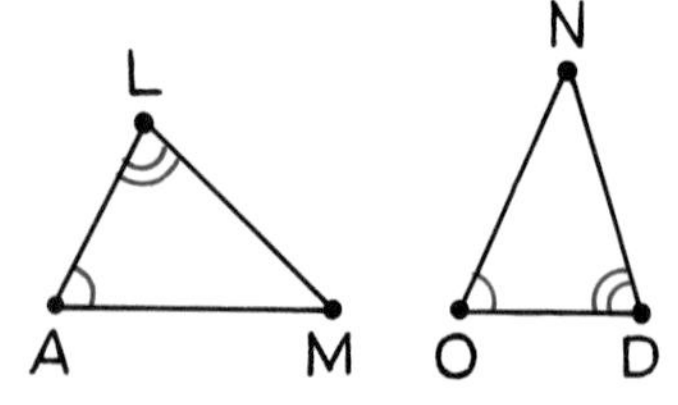

28. Name the other pair of equal parts needed to prove that the triangles are congruent.

29. State the postulate that would be the basis for showing that the triangles are congruent.

In the triangles shown below, AW = UT and ∠A = ∠T.

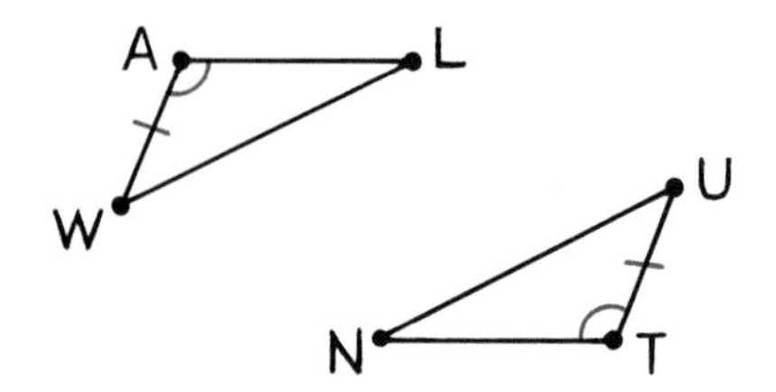

30. Name the other pair of equal parts needed to show that the triangles are congruent by the S.A.S. postulate.

31. Name the other pair of equal parts needed to show that the triangles are congruent by the A.S.A. postulate.

In the figure below, PI = PE and $\overrightarrow{PN}$ bisects ∡IPE.

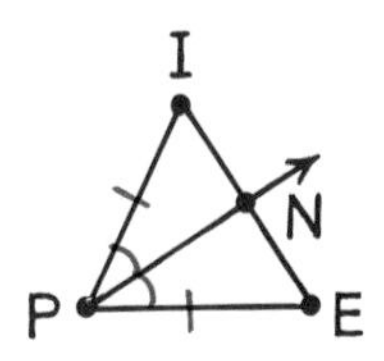

32. Why is ∠IPN = ∠NPE?

33. Why is PN = PN?

34. Why is △PIN ≅ △PEN?

In the figure below, C is the midpoint of $\overline{EA}$, ∡E and ∡A are right angles, and ∡PCE and ∡ACN are vertical angles.

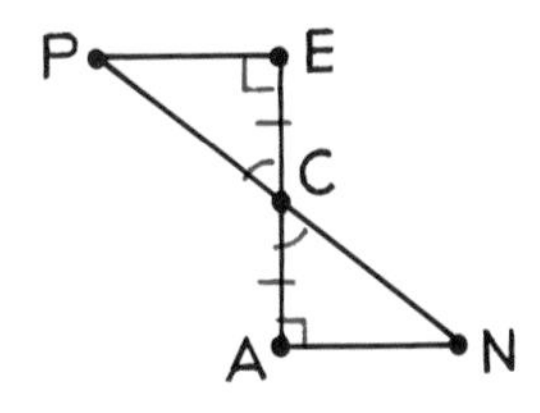

35. Why is EC = CA?

36. Why is ∠E = ∠A?

37. Why is ∠PCE = ∠ACN?

38. Why is △PEC ≅ △NAC?

Set III

The seven reptiles shown here are congruent, yet one is different from the others in a very basic way. Which reptile is it and why?

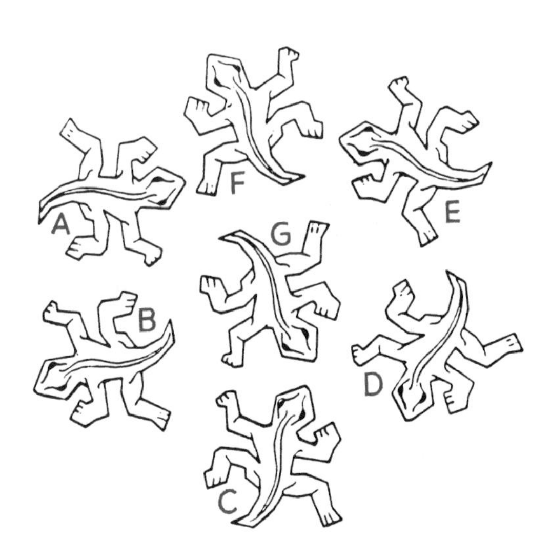

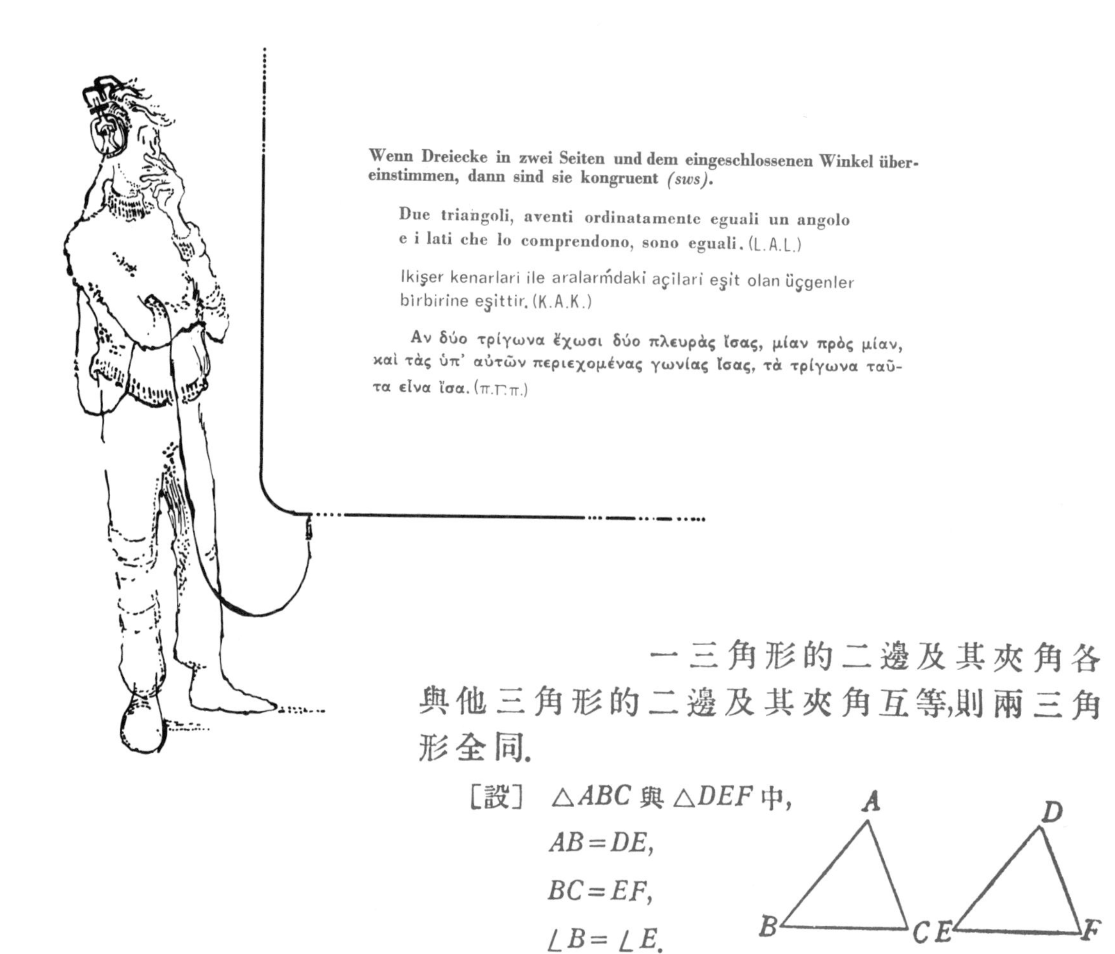

Lesson 3

Proving Triangles Congruent

The postulate for proving triangles congruent that we call "S.A.S." is known by different names in other countries. In Germany, it is the "S.W.S." postulate because the German words for side and angle are "seite" and "winkel." In Italy it is "L.A.L.," in Turkey it is "K.A.K.," and in Greece the postulate is called "Π.Γ.Π." (pi-gamma-pi). It is stated above in each of these languages and in Chinese as well. In the Chinese version, the hypothesis and conclusion of the postulate are restated in terms of the figure.

Although there are other ways by which triangles can be proved congruent, we will use at first just the two postulates introduced in Lesson 2 in this chapter: S.A.S. and A.S.A. You should not only know the English names for them, but also be able to state each one as a complete sentence. Here are examples of proofs based on them.

Example 1.

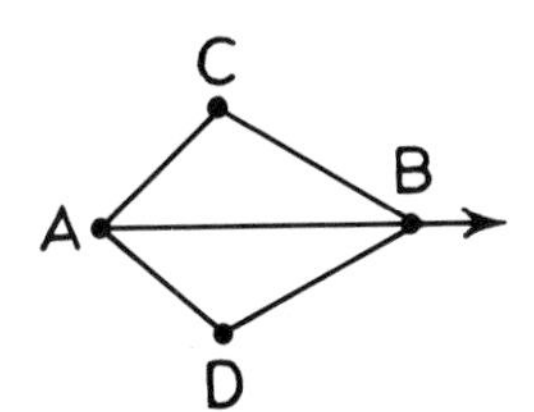

Given: AC = AD; $\overrightarrow{AB}$ bisects ∡CAD.

Prove: △ABC ≅ △ABD.

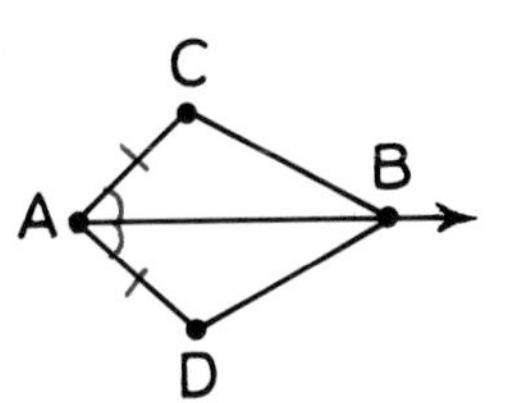

After marking the given information on the figure at the left above, it looks like the figure at the right. Two pairs of parts are not enough to prove that triangles are congruent. We need a third pair. In this case, it is the common side, $\overline{AB}$; because AB = AB, the triangles are congruent by the S.A.S. postulate.

Proof.

Statements	Reasons
1. AC = AD.	Given.
2. $\overrightarrow{AB}$ bisects ∡CAD.	Given.
3. ∠CAB = ∠BAD.	If a ray bisects an angle, it divides it into two equal angles.
4. AB = AB.	Reflexive.
5. △ABC ≅ △ABD.	S.A.S. postulate.

Example 2.

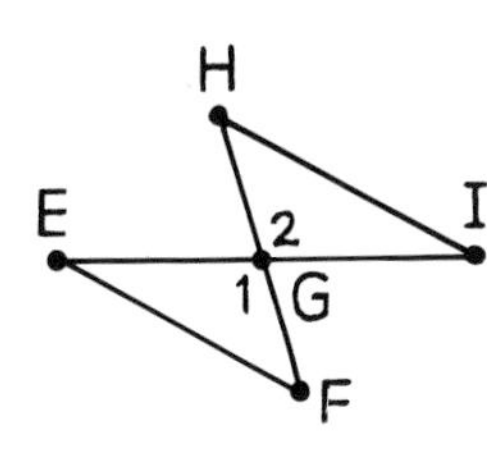

Given: ∡1 and ∡2 are vertical angles; G is the midpoint of $\overline{HF}$; ∠H = ∠F.

Prove: △GHI ≅ △GFE.

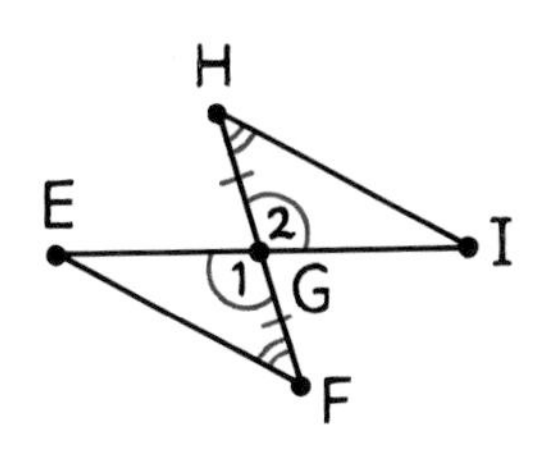

After marking the given information on the figure at the left above, it looks like the figure at the right. We can prove the triangles congruent by the A.S.A. postulate.

Proof.

Statements	Reasons
1. ∡1 and ∡2 are vertical angles.	Given.
2. ∠1 = ∠2.	Vertical angles are equal.
3. G is the midpoint of $\overline{HF}$.	Given.
4. HG = GF.	The midpoint of a line segment divides it into two equal segments.
5. ∠H = ∠F.	Given.
6. △GHI ≅ △GFE.	A.S.A. postulate.

Exercises

Set I

In France, one of the congruence postulates is stated in these words:

> "Si deux triangles ont deux *côtés* égaux chacun à chacun comprenant un *angle* égal, ils sont égaux."

1. Refer to the figure below to tell our name for this postulate.

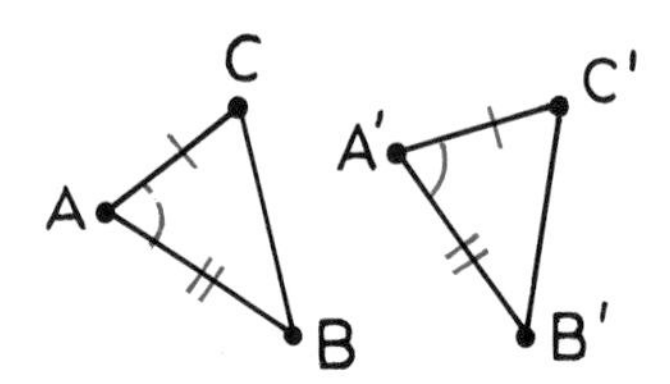

2. State the postulate in English as a complete sentence.
3. What do you think the name of this congruence postulate is in French?
4. What do you think the name of the other congruence postulate is in French?
5. State it in English as a complete sentence.

In the figure below, △CHI ≅ △MES, CI = $2x + 3$, and MS = $30 - x$. Also, ∠H = $(36 - 5y)°$ and ∠E = $(3y + 148)°$.

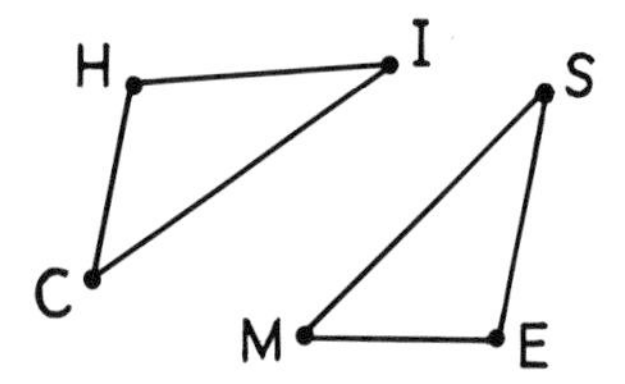

6. Find x.
7. Find CI.
8. Find MS.
9. Find y.
10. Find ∠H.
11. Find ∠E.

In the figure below, B-O-N, A-O-J, ∠1 = ∠2, and AO = OJ. Use this information to tell whether each of the following statements *must be true, may be true,* or *appears to be false.*

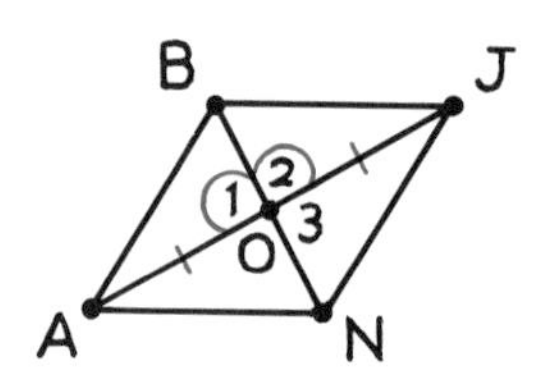

12. △AOB ≅ △JOB.
13. AB = BJ.
14. △ABJ is isosceles.
15. △ABN is equilateral.
16. $\overline{BN} \perp \overline{AJ}$.
17. $\overline{BN}$ and $\overline{AJ}$ bisect each other.
18. $\overline{AB} \parallel \overline{NJ}$.
19. ∡ABO and ∡JBO are a linear pair.
20. ∠1 = ∠3.
21. $\overrightarrow{BN}$ bisects ∡ABJ.

Copy each of the following figures and then use the given information to mark the parts that could be used to prove the triangles congruent.

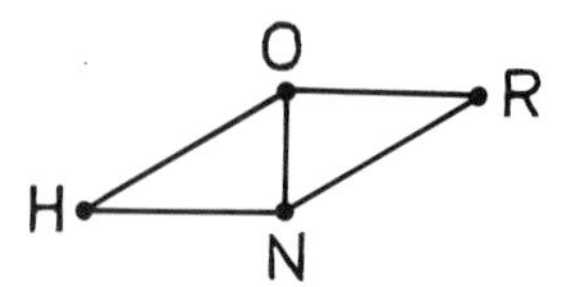

Example:

Given: $\overline{HN} \perp \overline{ON}$ and $\overline{ON} \perp \overline{OR}$; HN = OR.
Prove: △HON ≅ △RNO.

Answer:

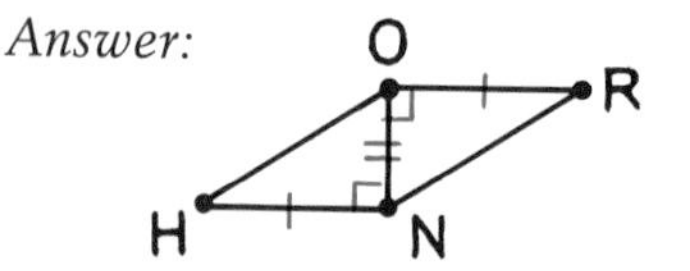

22. 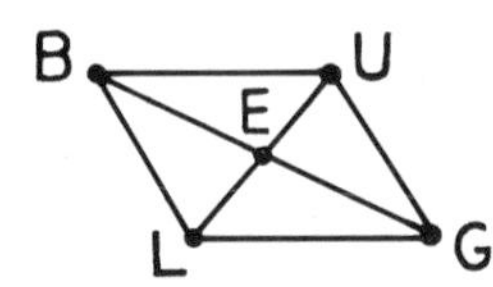

Given: $\angle LBG = \angle BGU$;
E is the midpoint of $\overline{BG}$;
BL = UG.
Prove: $\triangle BLE \cong \triangle GUE$.

23. 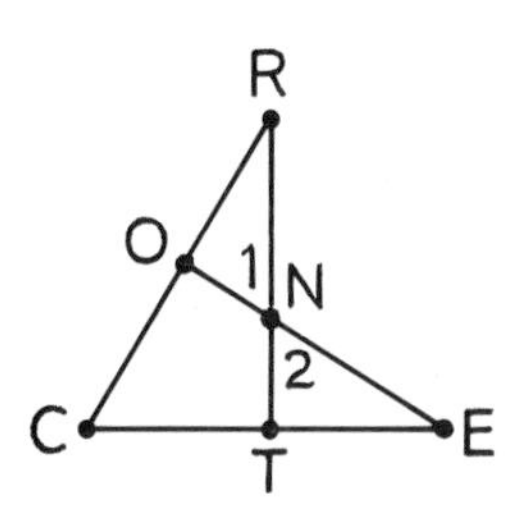

Given: $\angle RON = \angle NTE$;
ON = NT;
$\angle 1 = \angle C$ and $\angle 2 = \angle C$.
Prove: $\triangle RON \cong \triangle ETN$.

24. 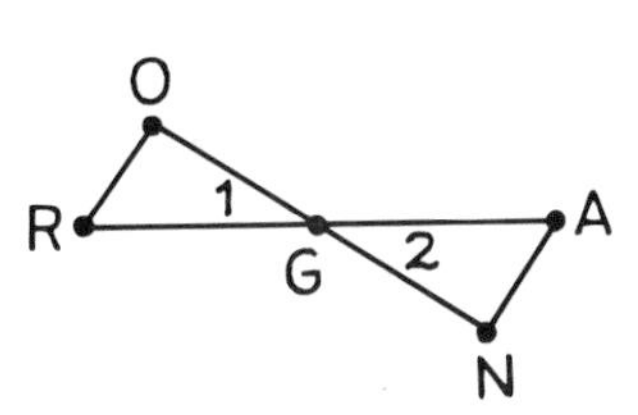

Given: $\measuredangle O$ and $\measuredangle N$ are right angles;
$\overline{RA}$ bisects $\overline{ON}$;
$\measuredangle 1$ and $\measuredangle 2$ are vertical angles.
Prove: $\triangle ROG \cong \triangle ANG$.

25. 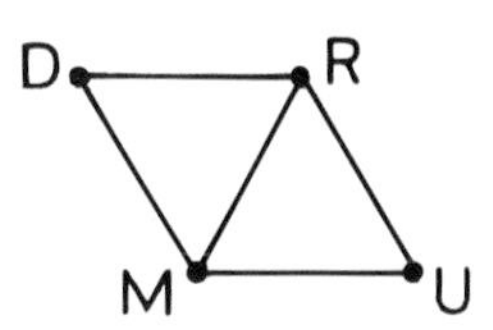

Given: $\triangle DRM$ and $\triangle URM$ are equilateral;
$\overrightarrow{RM}$ bisects $\measuredangle DRU$.
Prove: $\triangle DRM \cong \triangle URM$.

26. Copy and complete the following proof.

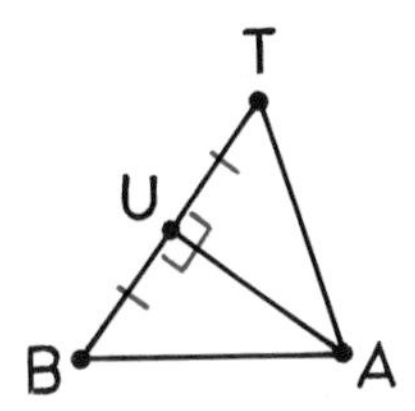

Given: U is the midpoint of $\overline{TB}$;
$\overline{AU} \perp \overline{TB}$.
Prove: $\triangle TUA \cong \triangle BUA$.

Proof.

Statements	***Reasons***
1. U is the midpoint of $\overline{TB}$.	_____
2. _____	The midpoint of a line segment divides it into two equal segments.
3. $\overline{AU} \perp \overline{TB}$.	_____
4. _____	Perpendicular lines form right angles.
5. $\angle TUA = \angle BUA$.	_____
6. UA = UA.	_____
7. _____	_____

Set II

Write complete proofs for each of the following exercises. Copy the figure, "given," and "prove" before writing your statements and reasons.

27. 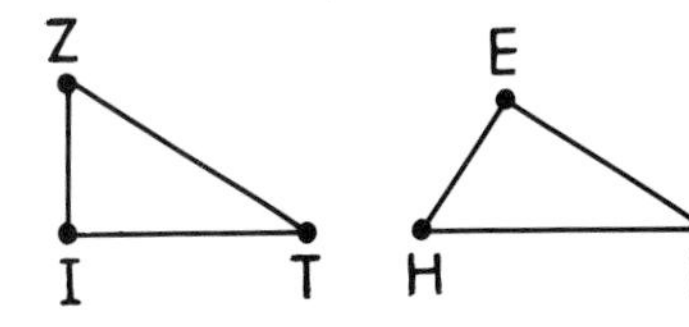

Given: ZI = HE;
$\measuredangle I$ and $\measuredangle E$ are right angles;
$\angle Z = \angle H$.
Prove: $\triangle ZIT \cong \triangle HER$.

28.

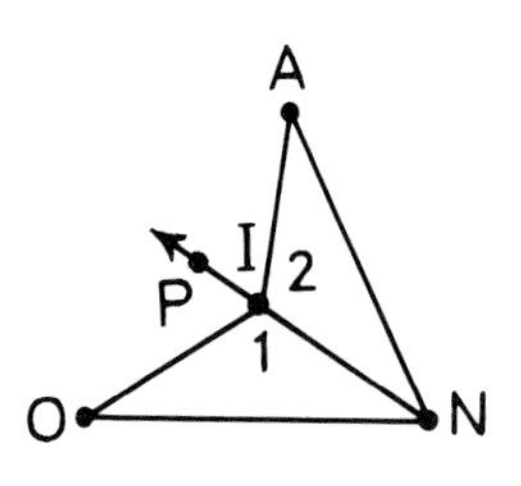

Given: $\overrightarrow{NP}$ bisects ∡ANO; AN = ON.
Prove: △INA ≅ △INO.

29.

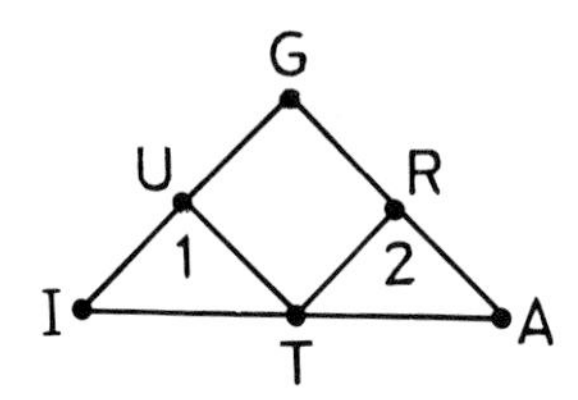

Given: $\overline{UT} \perp \overline{GI}$ and $\overline{RT} \perp \overline{GA}$; UI = RT and UT = RA.
Prove: △IUT ≅ △TRA.

30.

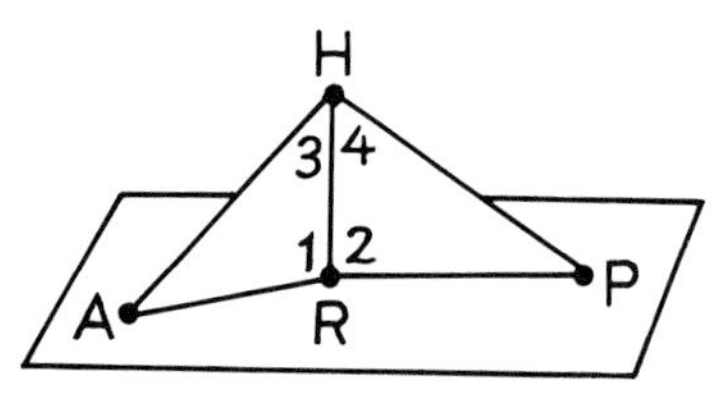

Given: ∠1 = ∠2; ∡3 and ∡4 are complements of ∡P.
Prove: △HRA ≅ △HRP.

31.

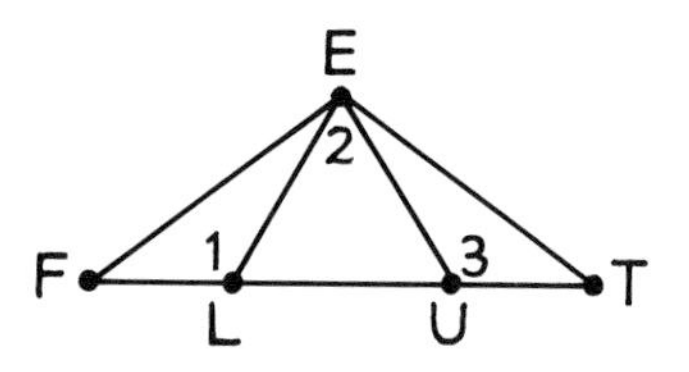

Given: △LUE is equilateral; ∡1 and ∡3 are supplements of ∡2; FL = UT.
Prove: △FLE ≅ △TUE.

32.

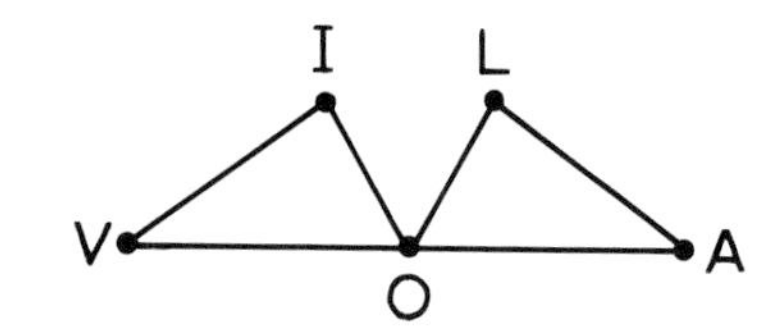

Given: O is the midpoint of $\overline{VA}$; ∠V = ∠A; $\overrightarrow{OI}$ bisects ∡VOL and $\overrightarrow{OL}$ bisects ∡IOA.
Prove: △VIO ≅ △ALO.

Set III

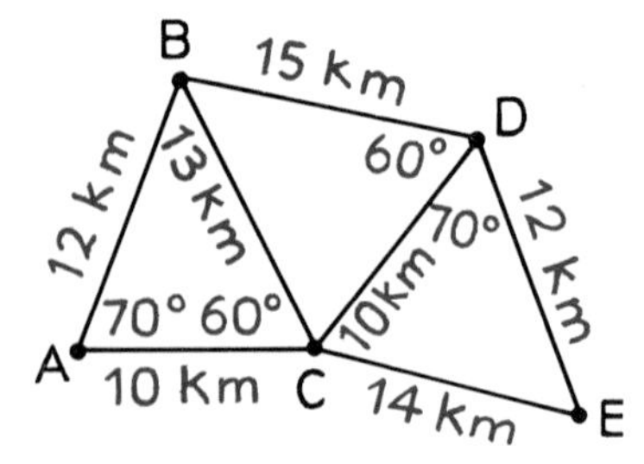

This road map is Obtuse Ollie's first attempt at map-making. It shows five towns and the roads joining them.

After looking at Ollie's map for awhile, Acute Alice noticed that he had not drawn it to scale. Then she noticed that at least one of the numbers on Ollie's map must be incorrect. How could she tell this?

"Very, very exclusive."

Courtesy of Virgil Partch

Lesson 4

Proving Corresponding Parts Equal

There does not seem to be any easy way for the two cowboys in this cartoon to get over to that drive-in. Yet there *is* an easy way for them to find out how far away it is. In fact, if they know the approximate length of their horses' strides, they can measure the distance without any special instruments or even getting off their horses.

The method of measurement uses congruent triangles and the following fact, which is a restatement of the definition of congruent polygons.

► **Definition**
If two triangles are *congruent,* their corresponding parts are equal.

We will assume that the two men are directly across from the drive-in, so that in the figure at the right point A represents their position, point D represents the drive-in, and line ℓ represents the edge of the cliff they are standing beside. The problem is to find the distance AD. Notice that $\overline{AD}$ is perpendicular to line ℓ.

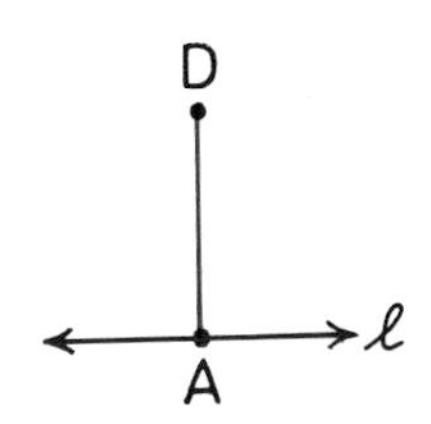

First, the two men can ride a short distance along the edge of the cliff to a point that we will call B, counting strides as they go along. One man can then stop while the other rides on for an equal distance to a point we will call C (see the second figure).

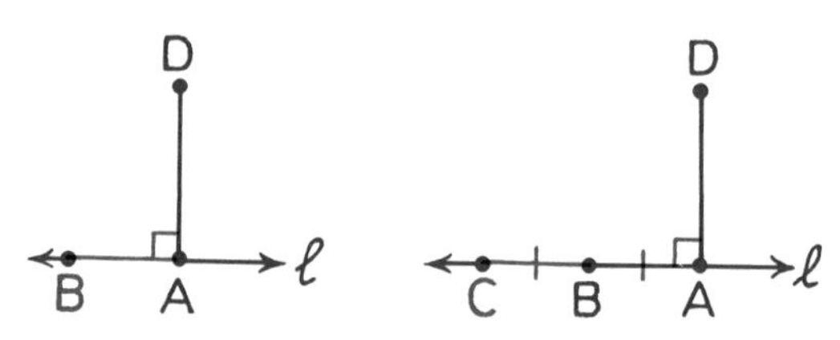

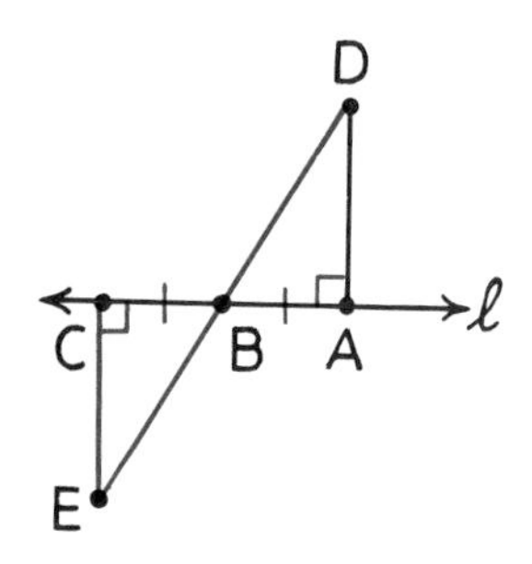

Finally, the man at C can turn directly away from the cliff and ride until he comes to the point at which the man who stayed behind him is directly between him and the drive-in. If we call this point E, the distance from C to E is equal to the distance from A to D. Do you see why?

It is because the two triangles in the figure are congruent, and because $\overline{CE}$ and $\overline{AD}$ are corresponding sides of these triangles, they must be equal. The parts that have been marked in the figure below show why the triangles are congruent. The proof below is based upon this information.

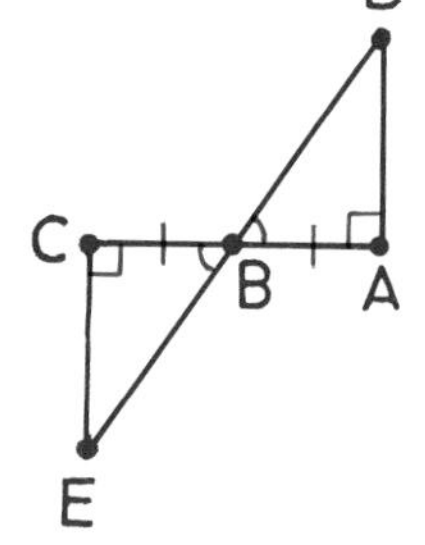

Given: $\overline{AD} \perp \overline{CA}$ and $\overline{CE} \perp \overline{CA}$; CB = BA; ∡CBE and ∡DBA are vertical angles.

Prove: CE = AD.

Proof.

Statements	Reasons
1. $\overline{AD} \perp \overline{CA}$ and $\overline{CE} \perp \overline{CA}$.	Given.
2. ∡C and ∡A are right angles.	Perpendicular lines form right angles.
3. ∠C = ∠A.	Any two right angles are equal.
4. CB = BA.	Given.
5. ∡CBE and ∡DBA are vertical angles.	Given.
6. ∠CBE = ∠DBA.	Vertical angles are equal.
7. △CBE ≅ △ABD.	A.S.A. postulate.
8. CE = AD.	If two triangles are congruent, their corresponding sides are equal.

The fact that corresponding parts of congruent triangles are equal can also be used to prove the following theorem.

▶ **Theorem 12**
Two triangles congruent to a third triangle are congruent to each other.

Exercises

Set I

1. Complete the following definition of congruent triangles.

   If two triangles are congruent, their corresponding ||||||||.

Tell whether each of the following statements is a definition or a theorem and then tell whether its converse is true or false.

Example: If a triangle is obtuse, then one of its angles is obtuse.
Answer: Definition. Its converse is true.

2. If two lines are perpendicular, then they form a right angle.
3. If two angles are a linear pair, then they are supplementary.
4. If two angles are complementary, then the sum of their measures is 90°.
5. If two angles are vertical angles, then they are equal.

In the figure below, △KOA ≅ △AEK. Copy and complete the following equations for their equal parts.

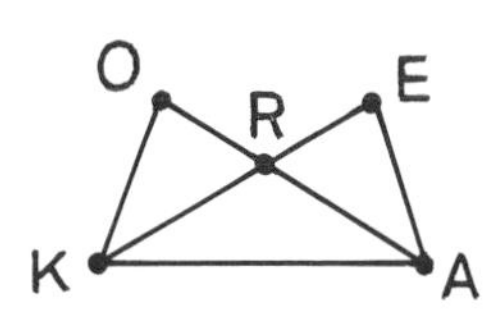

6. KO = ||||||||.
7. EK = ||||||||.
8. KA = ||||||||.
9. ∠O = ||||||||.
10. ∠OKA = ||||||||.
11. ∠AKE = ||||||||.

In the figure below, $\overrightarrow{IA}$ bisects ∡RIN, ∡IAR and ∡IAN are right angles, and R-A-N. When asked to write some conclusions that can be drawn from this information, Obtuse Ollie wrote the following. Rewrite each conclusion using the correct symbols.

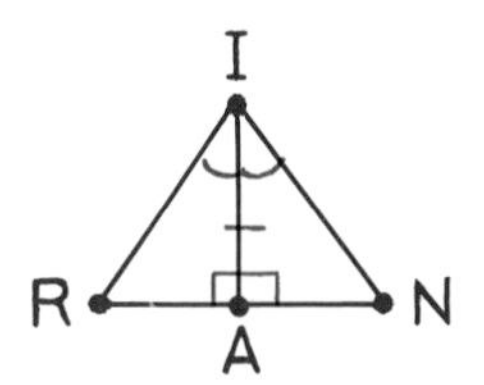

Example: ∡IAR = 90°.
Answer: ∠IAR = 90°.

12. ∠RAI and ∠IAN are a linear pair.
13. ∠I = ∠I.
14. △IAR = △IAN.
15. $\overline{RA}$ = $\overline{AN}$.
16. A is the midpoint of RN.
17. IA ⊥ RN.

In the figure below, ∡E and ∡T are supplements of ∡GYT, EY = YT, ∡1 and ∡2 are vertical angles, and G-Y-P.

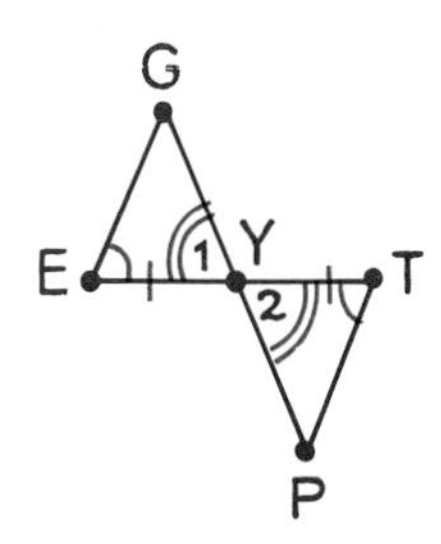

18. Why is ∠E = ∠T?
19. Why is ∠1 = ∠2?
20. Why is △EGY ≅ △TPY?

21. Why is GY = PY?

22. Why is Y the midpoint of $\overline{GP}$?

In the figure below, CH = CD, $\overrightarrow{CA}$ bisects ∡HCD, and ∡1 and ∡2 are a linear pair.

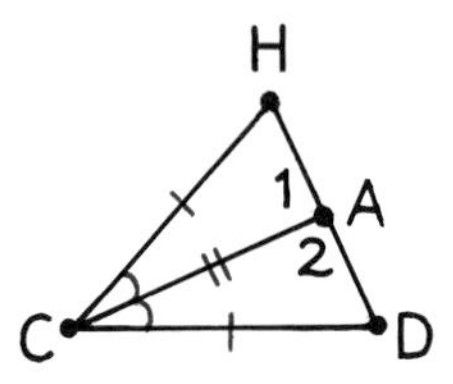

23. Why is ∠HCA = ∠ACD?

24. Why is △CHA ≅ △CDA?

25. Why is ∠1 = ∠2?

26. Why are ∡1 and ∡2 right angles?

27. Why is $\overline{CA} \perp \overline{HD}$?

28. Copy and complete the following proof of Theorem 12.

Two triangles congruent to a third triangle are congruent to each other.

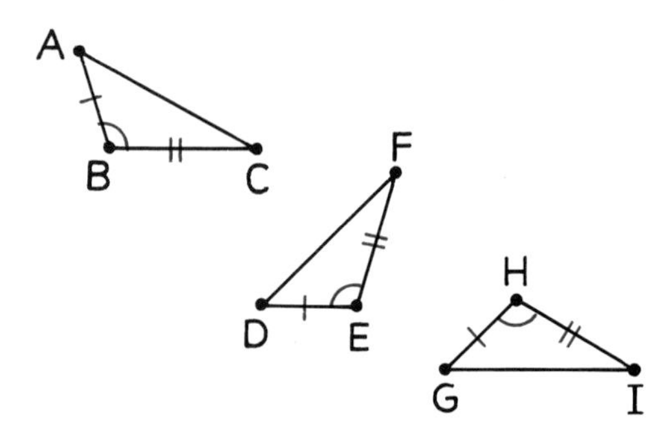

Given: △ABC ≅ △DEF and △DEF ≅ △GHI.
Prove: △ABC ≅ △GHI.

Proof.

Statements	***Reasons***
1. ____	____
2. AB = DE and DE = GH, ∠B = ∠E and ∠E = ∠H, BC = EF and EF = HI.	____
3. AB = GH, ∠B = ∠H, and BC = HI.	____
4. ____	____

Set II

29. Give the missing statements and reasons in the following proof.

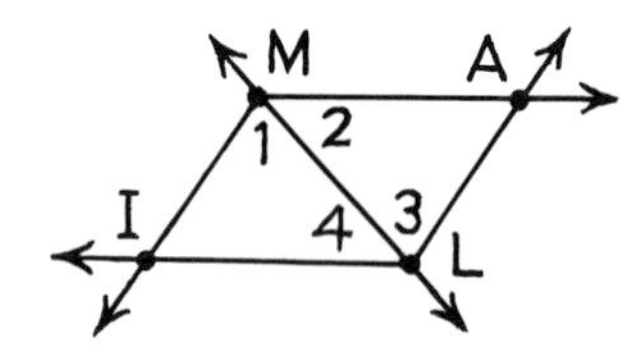

Given: △MAL ≅ △LIM; $\overrightarrow{MI}$-$\overrightarrow{ML}$-$\overrightarrow{MA}$ and $\overrightarrow{LI}$-$\overrightarrow{LM}$-$\overrightarrow{LA}$.
Prove: ∠IMA = ∠ALI.

Proof.

Statements	***Reasons***
1. △MAL ≅ △LIM.	a) ____
2. ∠1 = ∠3 and ∠2 = ∠4.	b) ____
3. ∠1 + ∠2 = ∠3 + ∠4.	c) ____
4. d) ____	Given.
5. ∠1 + ∠2 = ∠IMA and ∠3 + ∠4 = ∠ALI.	Betweenness of Rays Theorem.
6. e) ____	f) ____

Write complete proofs for each of the following exercises. Copy the figure, "given," and "prove" before writing your statements and reasons.

30.

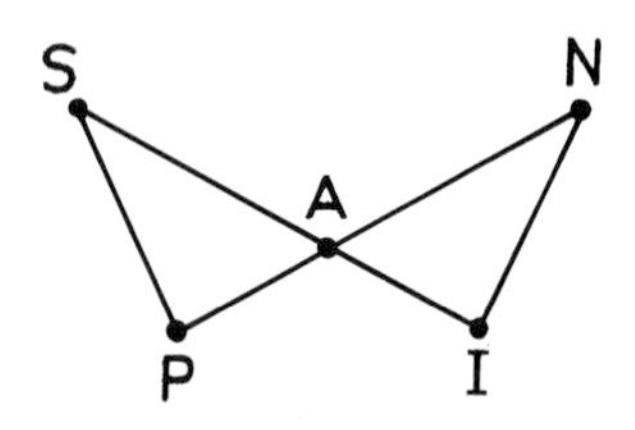

Given: ∡SAP and ∡NAI are vertical angles; SA = NA and AP = AI.
Prove: SP = NI.

31.

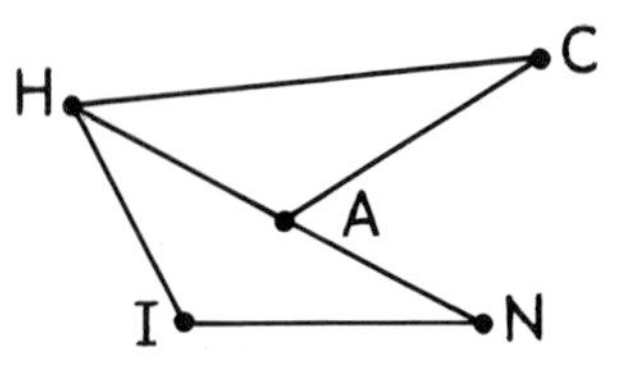

Given: $\overrightarrow{HN}$ bisects ∡CHI; HC = HN; ∠C = ∠N.
Prove: ∠HAC = ∠I.

32.

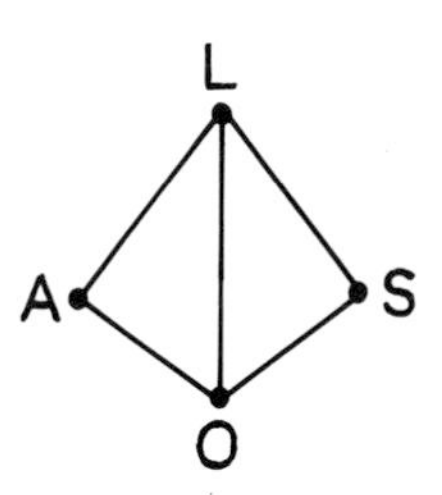

Given: △LAO ≅ △LSO; $\overline{LA} \perp \overline{AO}$.
Prove: ∠S = 90°.

33.

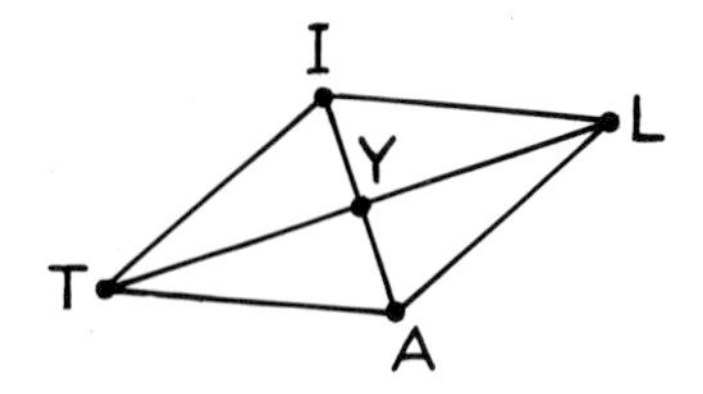

Given: △ITY ≅ △ATY and △ILY ≅ △ATY; T-Y-L.
Prove: Y is the midpoint of $\overline{TL}$.

34.

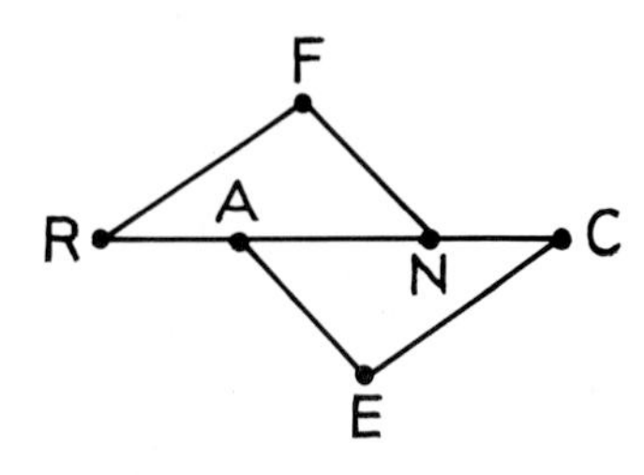

Given: △FRN ≅ △ECA; R-A-N and A-N-C.
Prove: RA = NC.

35.

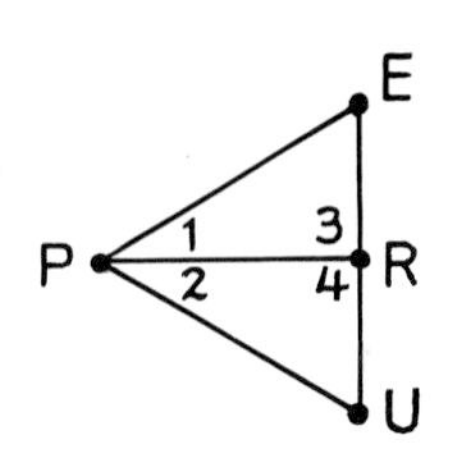

Given: PE = PU; ∠1 = ∠2; ∡3 and ∡4 are a linear pair.
Prove: $\overline{PR} \perp \overline{EU}$.

Set III

While swimming across Lake Minnehaha, Obtuse Ollie wondered how long the lake was. Acute Alice suggested that they start from opposite ends of the lake and walk as shown on the figure at the right until they met at the point labeled P.

What do you think they might do next in order to determine the length of the lake? Explain your reasoning.

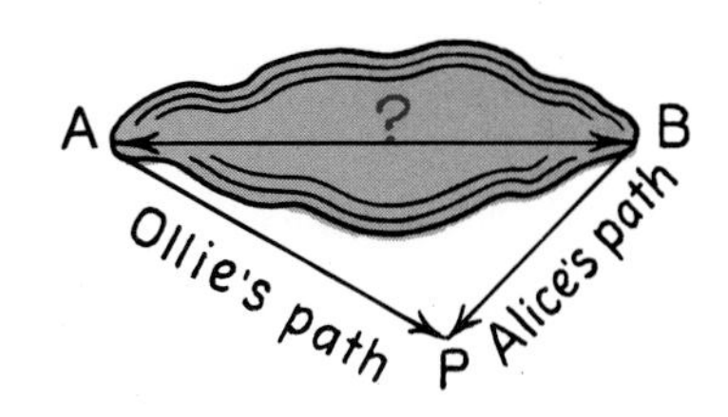

Lesson 5

The Isosceles Triangle Theorem

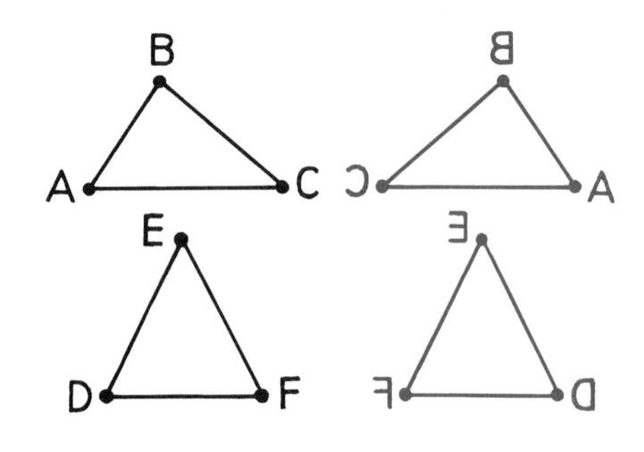

In Lewis Carroll's *Through The Looking Glass,* Alice wonders what it is like on the other side of the mirror over her fireplace. She climbs up on the mantel and passes through the glass into a room in which everything is reversed.

The figures at the left show what a couple of triangles would look like to Alice on the other side of the looking glass. The first triangle is scalene, and its mirror image looks different from the triangle itself. The reflection of the second triangle, however, does not seem to be reversed (even though the letters naming its vertices are). Because of this, the following correspondence between the vertices of the triangle and its reflection seems to be a congruence:

$$\text{DEF} \leftrightarrow \text{FED}.$$

It is easy to prove that it is and, because $\measuredangle$D corresponds to $\measuredangle$F in this congruence, these angles must be equal. In other words, the angles opposite the equal sides in an isosceles triangle are equal.

► **Theorem 13**
If two sides of a triangle are equal, the angles opposite them are equal.

Given: In $\triangle$DEF, ED = EF.
Prove: $\angle$D = $\angle$F.

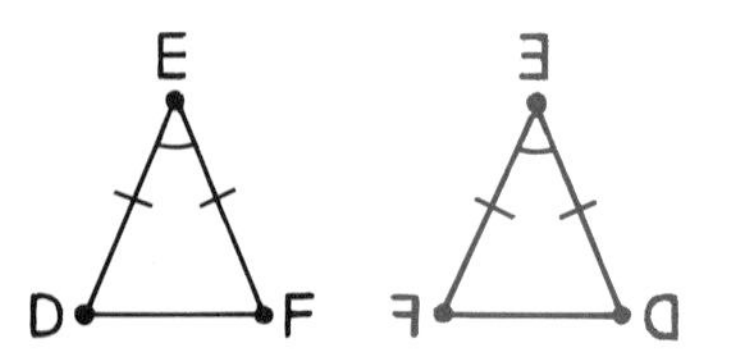

The figures at the left show two views of the same triangle: one from "in front of" and one from "behind the looking glass."

Proof.

Statements	Reasons
1. In $\triangle DEF$, $ED = EF$.	Given.
2. $\angle E = \angle E$.	Reflexive.
3. $EF = ED$.	Symmetric.
4. $\triangle DEF \cong \triangle FED$.	S.A.S. postulate.
5. $\angle D = \angle F$.	If two triangles are congruent, the corresponding angles are equal.

Theorem 13 can be used to prove a related theorem, called a *corollary,* about equilateral triangles.

► **Definition**
A ***corollary*** is a theorem that can be easily proved as a consequence of another theorem.

► **Corollary to Theorem 13**
If a triangle is equilateral, it is also equiangular.

Given: $\triangle ABC$ is equilateral.
Prove: $\triangle ABC$ is equiangular.

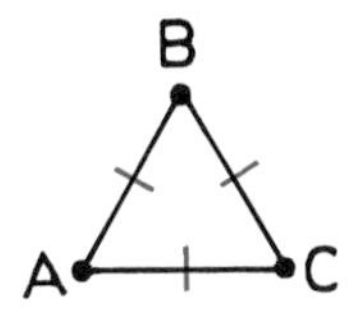

Proof.

Statements	Reasons
1. $\triangle ABC$ is equilateral.	Given.
2. $BC = AC$ and $AC = AB$.	If a triangle is equilateral, all of its sides are equal.
3. $\angle A = \angle B$ and $\angle B = \angle C$.	If two sides of a triangle are equal, the angles opposite them are equal.
4. $\angle A = \angle C$.	Transitive.
5. $\triangle ABC$ is equiangular.	If all of its angles are equal, a triangle is equiangular.

The converses of Theorem 13 and its corollary are also true. They are stated below and their proofs are considered in the exercises.

► **Theorem 14**
If two angles of a triangle are equal, the sides opposite them are equal.

► **Corollary to Theorem 14**
If a triangle is equiangular, it is also equilateral.

Exercises

Set I

Copy and complete the following definitions.

1. A triangle is isosceles iff ||||||||.
2. A triangle is equilateral iff ||||||||.
3. A triangle is equiangular iff ||||||||.

Copy and complete the following statements of the theorems of this lesson.

4. If two sides of a triangle are equal, ||||||||.
5. If two angles of a triangle are equal, ||||||||.
6. If a triangle is equilateral, ||||||||.
7. If a triangle is equiangular, ||||||||.

Tell whether each of the following statements is true or false.

8. If a triangle is isosceles, at least two of its angles must be equal.
9. If a triangle is equiangular, all three of its sides must be equal.
10. If a triangle is scalene, none of its angles can be equal.
11. If a triangle is equilateral, it is also isosceles.

In the figure below, PI = PE and $\angle$N = $\angle$IEN.

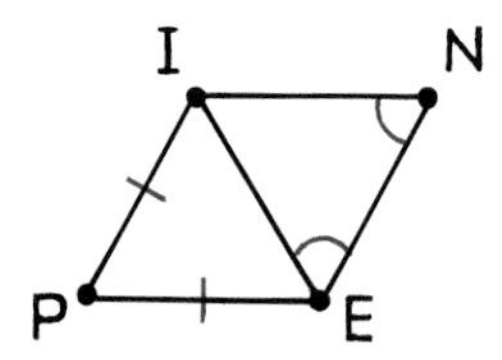

12. Which angle of $\triangle$PIE is opposite $\overline{\text{PI}}$?
13. Which angles of $\triangle$PIE are equal?
14. Which side of $\triangle$INE is opposite $\measuredangle$N?
15. Which sides of $\triangle$INE are equal?

In $\triangle$PEA, PE = EA.

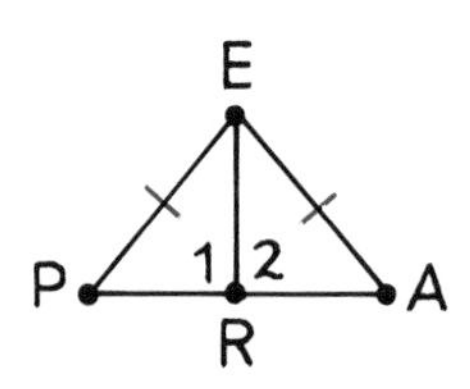

16. Obtuse Ollie says that $\angle 1 = \angle 2$ because "if two sides of a triangle are equal, the angles opposite them are equal." Explain why this is not correct.
17. Could you conclude that $\angle 1 = \angle 2$ if PE = EA *and* $\angle$PER = $\angle$REA?

In $\triangle$LAH, $\angle$A = $\angle$H.

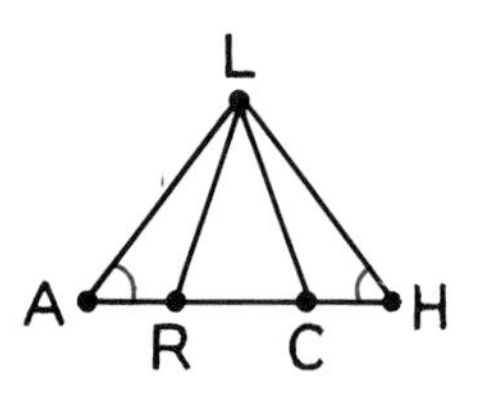

18. Can you conclude that LR = LC?
19. Can you conclude that LH = LA?
20. Why or why not?

In the figure below, $\triangle$PLM is equilateral and L-U-M. Use this information to tell whether each of the following statements *must be true, may be true,* or *appears to be false.*

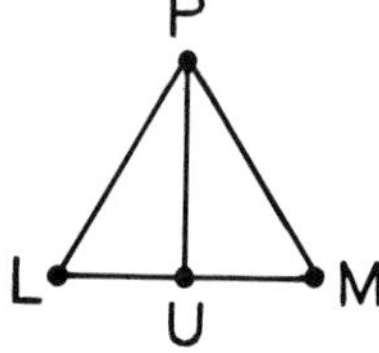

21. $\triangle$PLM is isosceles.
22. LM = PM.
23. $\angle$PUL = $\angle$PUM.
24. LU = UM.
25. $\angle$LPM = $\angle$M.
26. $\triangle$PUL $\cong$ $\triangle$PUM.

Set II

27. Copy and complete the following proof of Theorem 14:

    If two angles of a triangle are equal, the sides opposite them are equal.

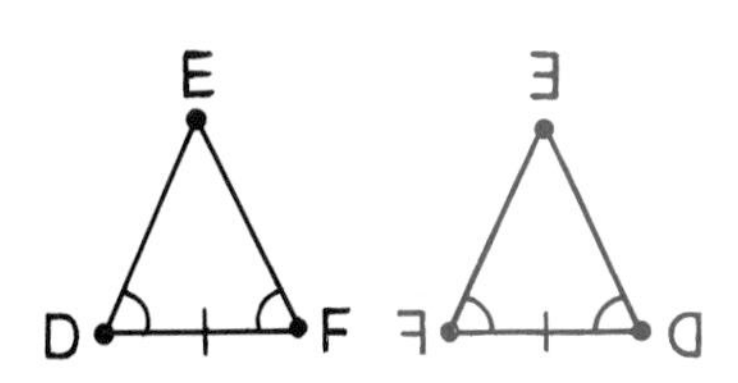

Given: In △DEF, ∠D = ∠F.
Prove: ED = EF.

Proof.

Statements — **Reasons**

1. |||||||| — ||||||||
2. DF = FD. — ||||||||
3. ∠F = ∠D. — ||||||||
4. |||||||| — A.S.A. postulate.
5. |||||||| — ||||||||

28. Copy and complete the following proof of the corollary to Theorem 14:

    If a triangle is equiangular, then it is equilateral.

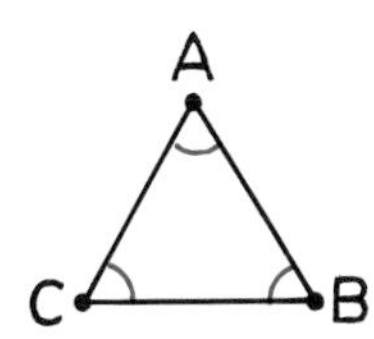

Given: △ABC is equiangular.
Prove: △ABC is equilateral.

Proof.

Statements — **Reasons**

1. |||||||| — ||||||||
2. ∠C = ∠A and ∠A = ∠B. — ||||||||
3. AB = BC and BC = AC. — ||||||||
4. |||||||| — Transitive.
5. |||||||| — ||||||||

29. Give the missing statements and reasons in the following proof.

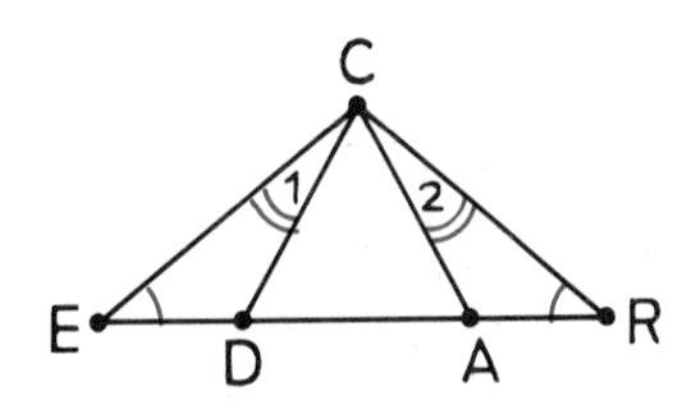

Given: ∠E = ∠R; ∠1 = ∠2.
Prove: △CDA is isosceles.

Proof.

Statements — **Reasons**

1. ∠E = ∠R. — a) ||||||||
2. b) |||||||| — If two angles of a triangle are equal, the sides opposite them are equal.
3. c) |||||||| — Given.
4. △CED ≅ △CRA. — d) ||||||||
5. e) |||||||| — If two triangles are congruent, their corresponding sides are equal.
6. f) |||||||| — g) ||||||||

Write complete proofs for each of the following exercises. Copy the figure, "given," and "prove" before writing your statements and reasons.

30.

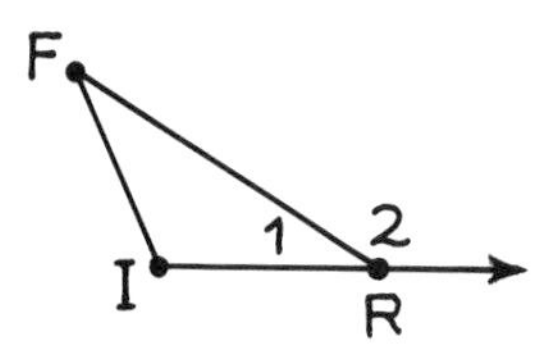

Given: ∡1 and ∡2 are a linear pair; ∡F and ∡2 are supplementary.
Prove: $\triangle$FIR is isosceles.

31.

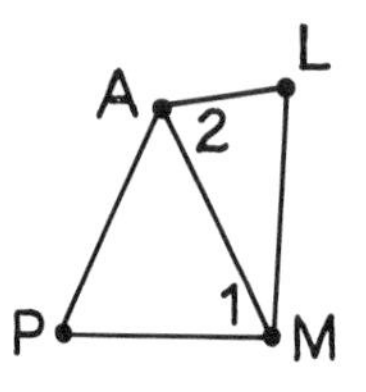

Given: $\angle 1 = \angle P$; $\angle L = \angle 2$.
Prove: AP = LM.

32.

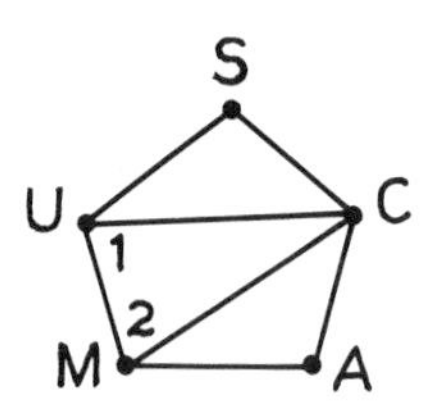

Given: SU = AM, $\angle S = \angle A$, and CS = CA.
Prove: $\angle 1 = \angle 2$.

33.

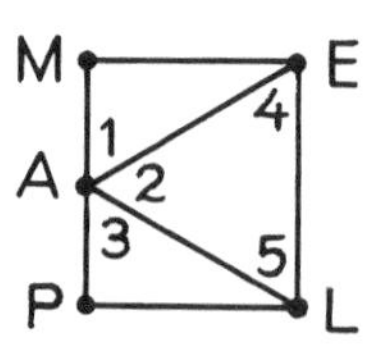

Given: $\triangle$AEL is equiangular; A is the midpoint of $\overline{MP}$; $\angle 1 = \angle 3$.
Prove: $\triangle MAE \cong \triangle PAL$.

34.

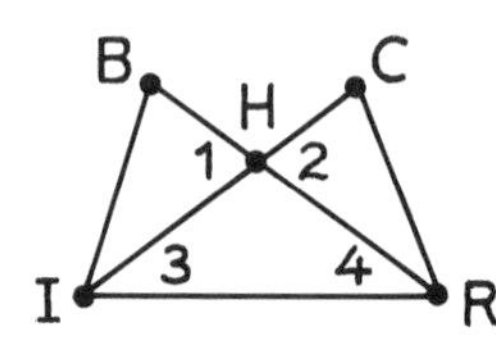

Given: ∡1 and ∡2 are vertical angles; $\angle 3 = \angle 4$; BH = HC.
Prove: $\triangle BHI \cong \triangle CHR$.

35.

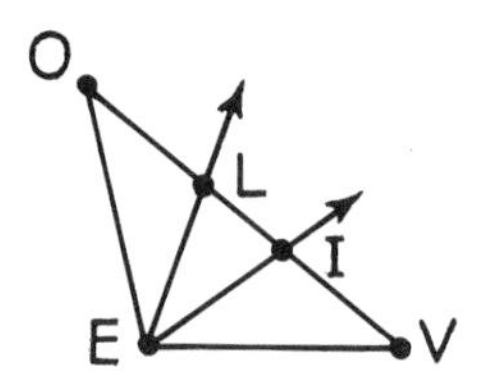

Given: In $\triangle$OEV, $\overrightarrow{EL}$ bisects ∡OEI and $\overrightarrow{EI}$ bisects ∡LEV; OE = EV.
Prove: OL = IV.

Set III

A Looking Glass Proof.
Alice found a book in the room behind the mirror written in a language that she didn't recognize. It started out like this:

JABBERWOCKY

'Twas brillig, and the slithy toves
Did gyre and gimble in the wabe:
All mimsy were the borogoves,
And the mome raths outgrabe.

If the book had been about geometry rather than poetry, it might have had problems like the following one. Can you write a "looking glass" proof of it?

Given: $\angle A = \angle C$; $\angle 1 = \angle 2$.
Prove: AD = DC.

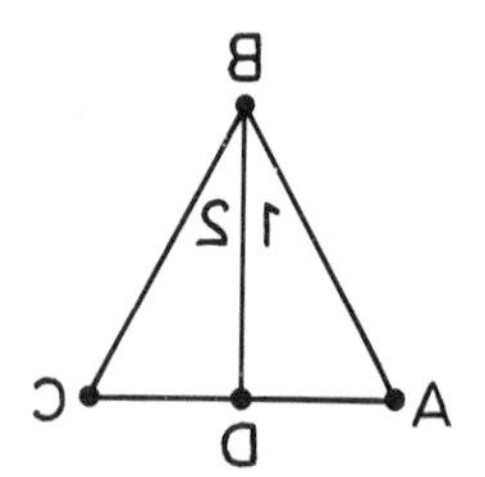

Prints and Photographs Division, Library of Congress. Courtesy of the Library of Congress

Lesson 6

The S.S.S. Congruence Theorem

Alexander Graham Bell, the inventor of the telephone, was also interested in flight. Several years before the Wright brothers built their first airplane, Bell designed and flew a number of giant man-carrying kites. One of the kites is pictured in this photograph, in which Bell and his wife are shown sharing a kiss.

Bell's kites were made from three-dimensional networks of triangles. The triangle was used because it is a rigid form. If models of a triangle and a quadrilateral are made by threading drinking straws together, the triangle will hold its shape but the quadrilateral will collapse. The shape of the triangle, then, is completely determined by the lengths of its sides.

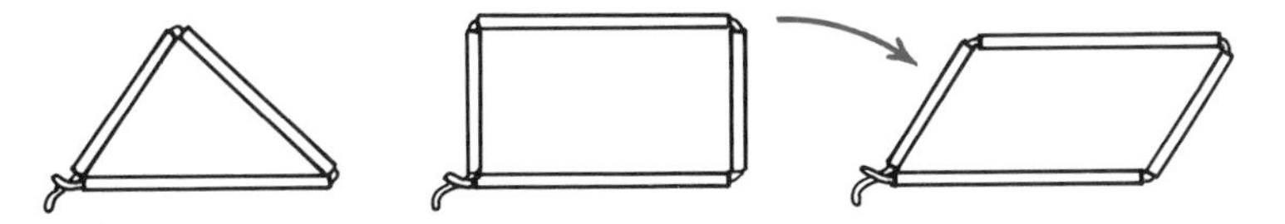

This means that if the sides of one triangle are equal to the sides of another, the triangles must be congruent. This can be proved, and we will refer to it as the S.S.S. Congruence Theorem.

► **Theorem 15** (The S.S.S. Congruence Theorem)
If the three sides of one triangle are equal to the three sides of another triangle, the triangles are congruent.

Given: AB = DE, BC = EF, and AC = DF.
Prove: △ABC ≅ △DEF.

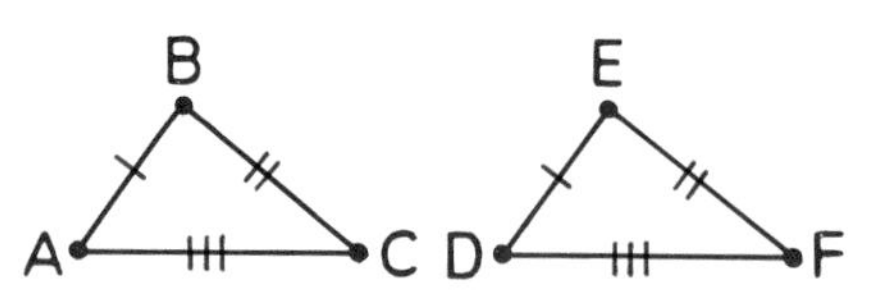

To prove the triangles congruent, we will draw a third triangle and show that △ABC and △DEF are both congruent to it.

Proof.

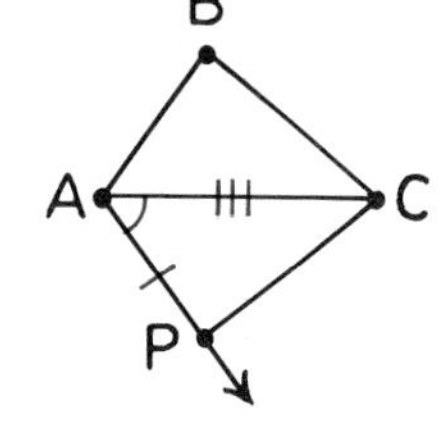

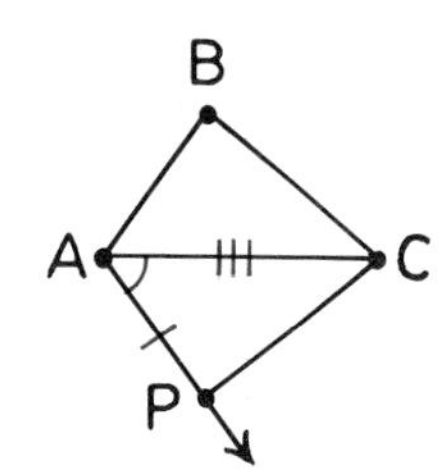

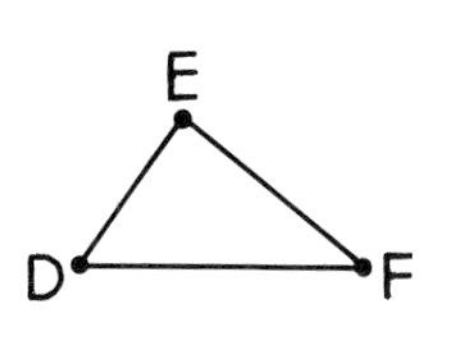

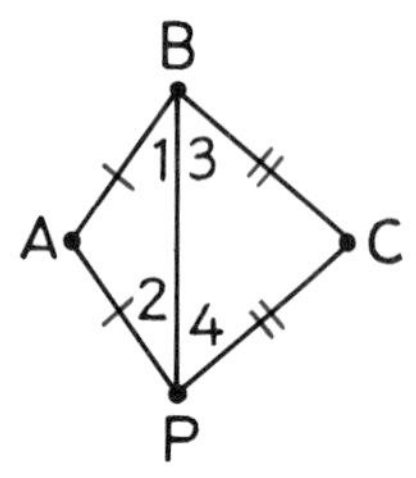

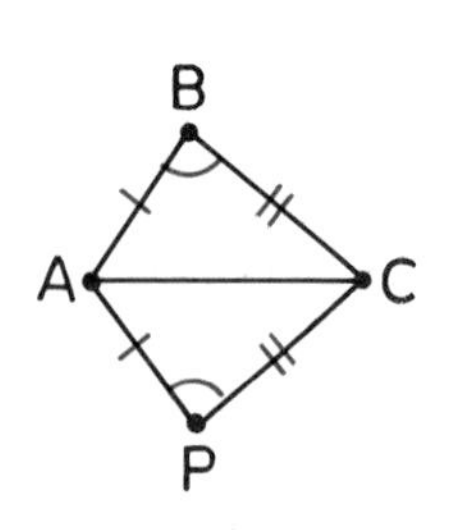

Statements	*Reasons*
1. AB = DE, BC = EF, and AC = DF.	Given.
2. Draw $\overrightarrow{AP}$ so that ∠CAP = ∠D.	Protractor postulate.
3. Choose point P so that AP = DE.	Ruler postulate.
4. Draw $\overline{CP}$.	Two points determine a line.
5. △APC ≅ △DEF.	S.A.S. postulate.
6. Draw $\overline{BP}$.	Same as step 4.
7. AB = AP.	Substitution (steps 1 and 3).
8. ∠1 = ∠2.	If two sides of a triangle are equal, the angles opposite them are equal.
9. EF = PC.	If two triangles are congruent, the corresponding sides are equal.
10. BC = PC.	Transitive (steps 1 and 9).
11. ∠3 = ∠4.	Same as step 8.
12. ∠1 + ∠3 = ∠2 + ∠4.	Addition (steps 8 and 11).
13. ∠1 + ∠3 = ∠ABC and ∠2 + ∠4 = ∠APC.	Betweenness of Rays Theorem.
14. ∠ABC = ∠APC.	Substitution (steps 12 and 13).
15. △ABC ≅ △APC.	S.A.S. postulate (steps 7, 14, and 10).
16. △ABC ≅ △DEF.	Two triangles congruent to a third triangle are congruent to each other.

Exercises

Set I

The following exercises refer to the figures below.

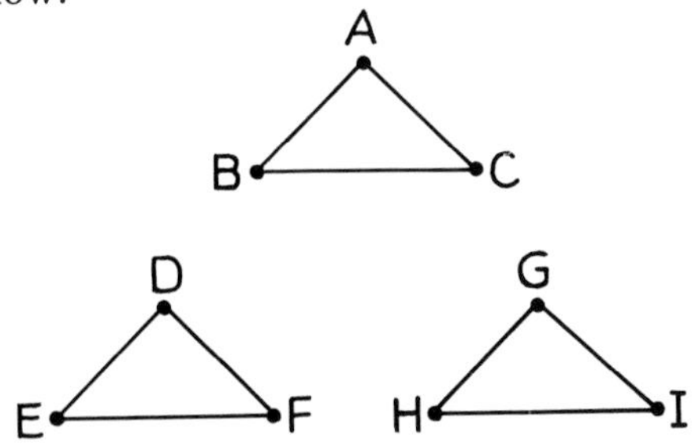

1. If $\angle ABC = \angle DEF$ and $\angle DEF = \angle GHI$, does it necessarily follow that $\angle ABC = \angle GHI$?
2. Explain why or why not.
3. If $\measuredangle ABC$ is complementary to $\measuredangle DEF$ and $\measuredangle DEF$ is complementary to $\measuredangle GHI$, does it necessarily follow that $\measuredangle ABC$ is complementary to $\measuredangle GHI$?
4. Explain why or why not.
5. If $\triangle ABC \cong \triangle DEF$ and $\triangle DEF \cong \triangle GHI$, does it necessarily follow that $\triangle ABC \cong \triangle GHI$?
6. Explain why or why not.

In $\triangle THR$ and $\triangle ONG$, TH = ON, HR = NG, and TR = OG.

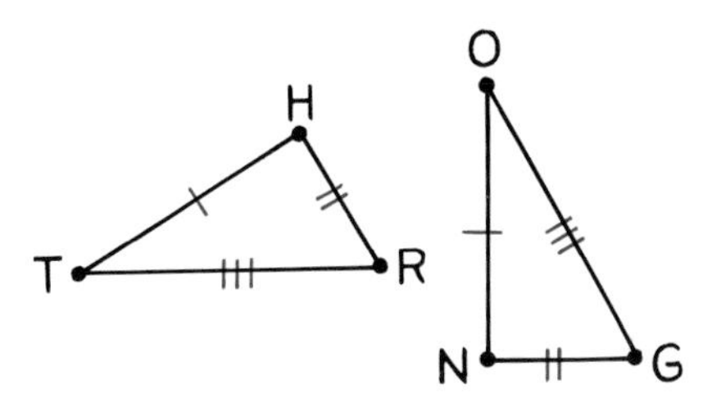

7. Is $\triangle THR \cong \triangle ONG$?
8. State the postulate or theorem that is the basis for your answer.

In $\triangle LEG$ and $\triangle ION$, $\angle L = \angle I$, $\angle E = \angle O$, and $\angle G = \angle N$.

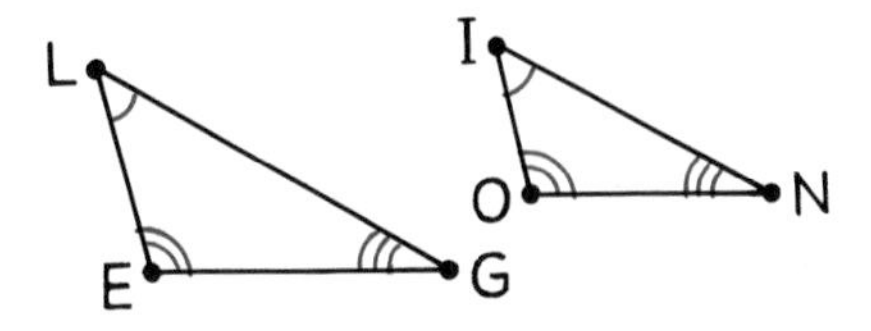

9. Is $\triangle LEG \cong \triangle ION$?
10. If the three angles of one triangle are equal to the three angles of another triangle, does it follow that the triangles are congruent?

Is it possible to prove that $\triangle PAK \cong \triangle CKA$ from each of the following sets of facts? If so, name the congruence postulate or theorem that could be used.

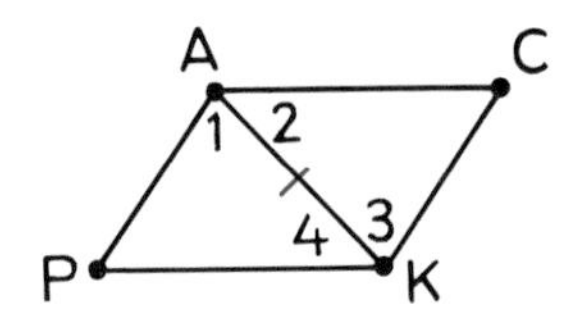

Example: $\angle 1 = \angle 3$ and $\angle 2 = \angle 4$.
Answer: Yes; A.S.A. postulate.

11. AP = KC and PK = CA.
12. AC = KP and $\angle C = \angle P$.
13. $\angle 1 = \angle 3$ and AP = KC.

In the figure below, SR = WA and SA = WR. Use this information to tell whether each of the following statements *must be true, may be true,* or *appears to be false.*

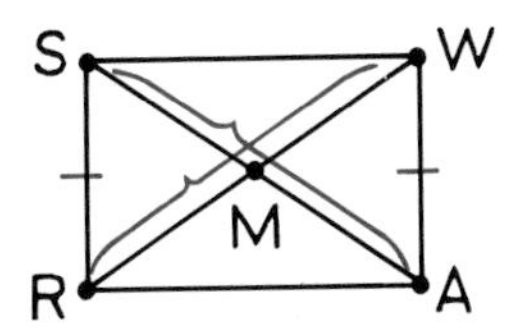

14. $\triangle SRA \cong \triangle WAR$.

15. ∠SAR = ∠WRA.

16. MR = MA.

17. SW = RA.

18. M is the midpoint of $\overline{SA}$ and $\overline{RW}$.

19. $\overline{SA} \perp \overline{RW}$.

20. $\overline{SR} \parallel \overline{WA}$.

Set II

21. Give the missing statements and reasons in the following proof.

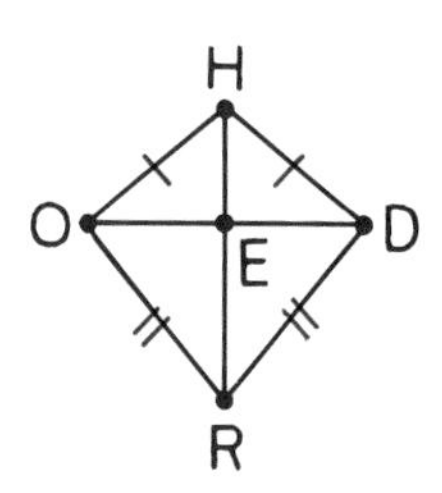

Given: HO = HD and OR = DR; O-E-D.
Prove: E is the midpoint of $\overline{OD}$.

Proof.

Statements	***Reasons***
1. HO = HD and OR = DR.	a) ______
2. b) ______	Reflexive.
3. c) △OHR ≅ ______.	d) ______
4. e) ∠OHR = ______.	f) ______
5. HE = HE.	g) ______
6. h) △OHE ≅ ______.	i) ______
7. j) ______	If two triangles are congruent, their corresponding sides are equal.
8. k) ______	Given.
9. l) ______	m) ______

Write complete proofs for each of the following exercises. Copy the figure, "given," and "prove" before writing your statements and reasons.

22.

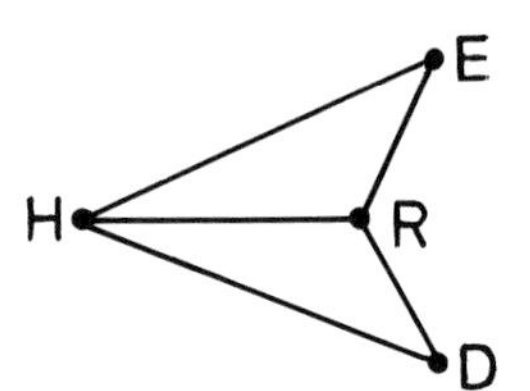

Given: HE = HD and RE = RD.
Prove: ∠E = ∠D.

23.

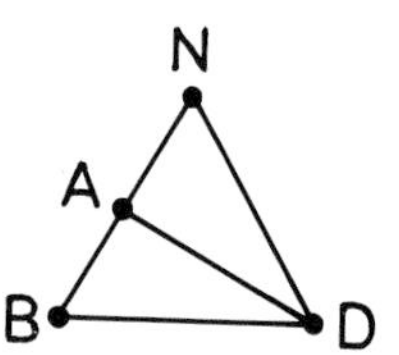

Given: △BND is equilateral; A is the midpoint of $\overline{BN}$.
Prove: △BAD ≅ △NAD.

24.

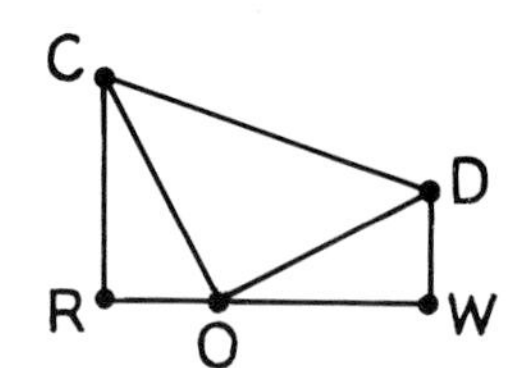

Given: CR = OW, RO = WD, and ∠OCD = ∠ODC.
Prove: ∠R = ∠W.

25.

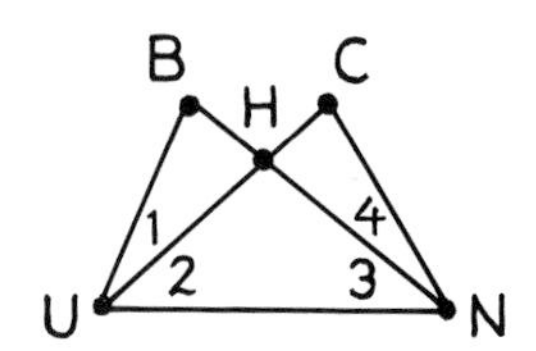

Given: △BUN ≅ △CNU.
Prove: △HUN is isosceles.

26.

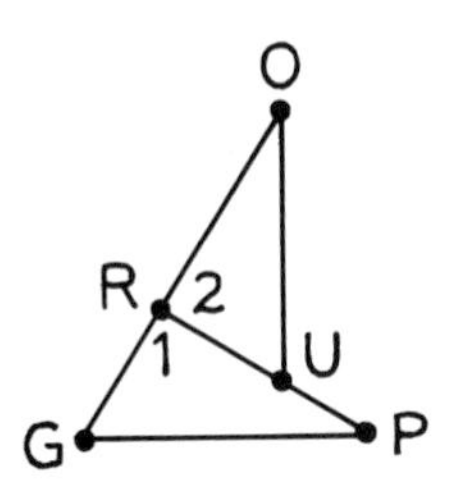

Given: GR = RU, RP = RO, and GP = OU; $\measuredangle 1$ and $\measuredangle 2$ are a linear pair.
Prove: $\triangle$GRP is a right triangle.

27.

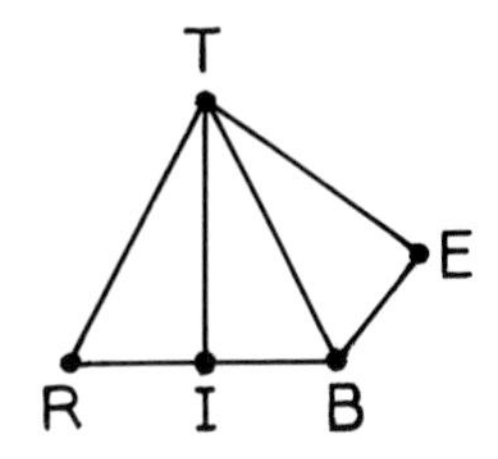

Given: TR = TB and RI = IB; TI = TE and $\angle$ITB = $\angle$BTE.
Prove: $\angle$R = $\angle$TBE.

Set III

Structures made of triangles owe their strength to the fact that the triangle is a rigid form. If three drinking straws are pinned together at their ends to form a triangle, the triangle will hold its shape, even if picked up and moved around.

Four straws pinned together to form a quadrilateral, however, result in a structure that is flexible. It can easily be changed into a variety of shapes.

Several frameworks are shown below. Tell whether you think each one is rigid or flexible. You might build some straw-and-pin models to check your answers.

1.

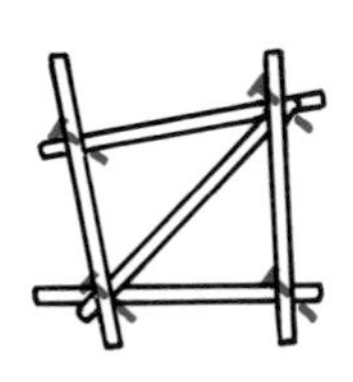

2.

3.

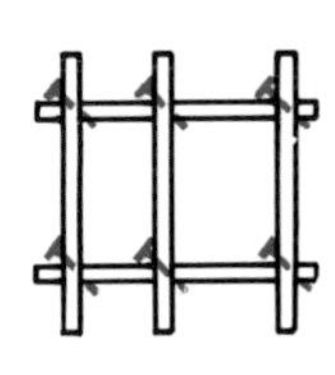

4.

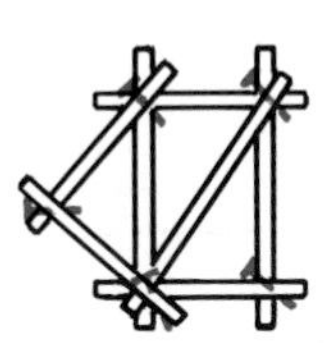

5.

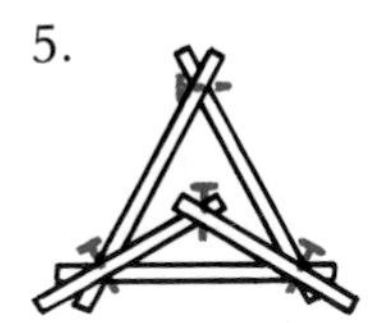

Lesson 7
Constructions

Ever since the time of Euclid, two tools have been used in making geometric drawings: the straightedge and the compass. The compass is used to draw *circles* or parts of circles called *arcs*.

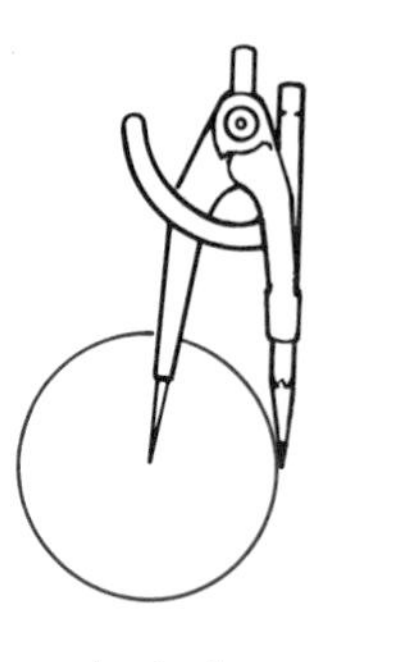

A circle

An arc

A circle is the set of all points in a plane that are at a given distance from a given point in the plane. The given distance is called the *radius* of the circle and the given point is called its *center*.

The arcs drawn with the compass are used to locate points. The points are used to determine lines, which are drawn with the straightedge. Drawings made with just a straightedge and compass are called *constructions* to distinguish them from those made with other tools such as a ruler and a protractor.

In this lesson, we will learn how to use a straightedge and compass to copy some simple figures.

► Construction 1
To construct a line segment equal in length to a given line segment.

Method.

1. Let $\overline{AB}$ be the given line segment. Adjust the radius of the compass to the length of $\overline{AB}$ by putting the metal point on A and the pencil point on B.

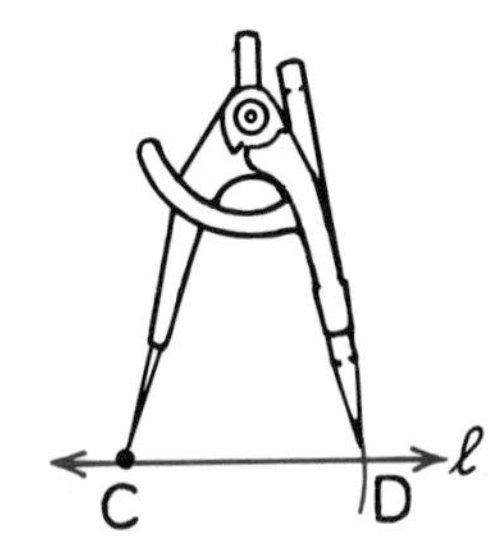

2. Draw a line (ℓ) and mark a point on it (point C in the figure). Draw an arc with center C and radius AB that intersects the line (in the figure, the arc intersects ℓ in point D.). CD = AB.

► Construction 2
To construct an angle equal in measure to a given angle.

Method.

1. Let ∡A be the given angle.

2. Draw a ray $\overrightarrow{BC}$ as one side of the angle to be constructed.

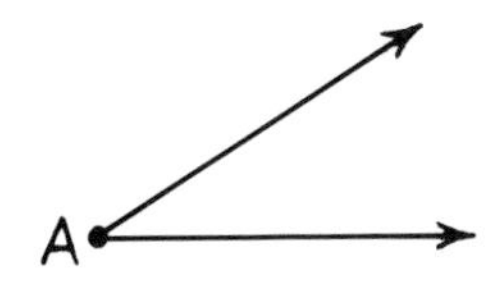

3. With A as center, draw an arc that intersects the sides of ∡A (points D and E in the figure at the left below).

4. Draw an arc with the same radius and with B as center that intersects $\overrightarrow{BC}$ (point F in the figure at the right below).

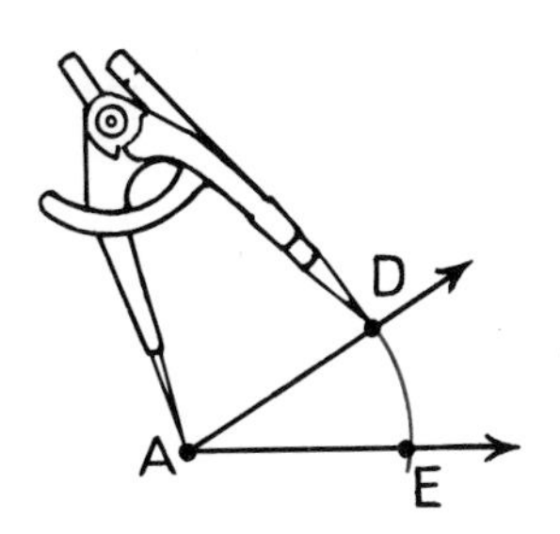

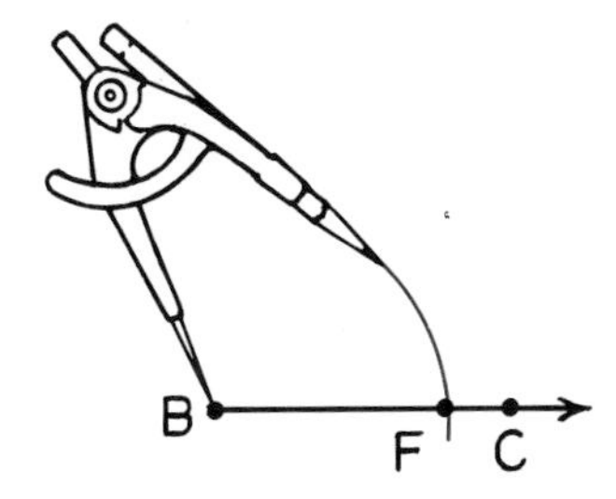

5. Adjust the radius of the compass to the distance between points D and E.

6. With F as center, draw an arc having this radius so that it intersects the previous one (point G in the figure).

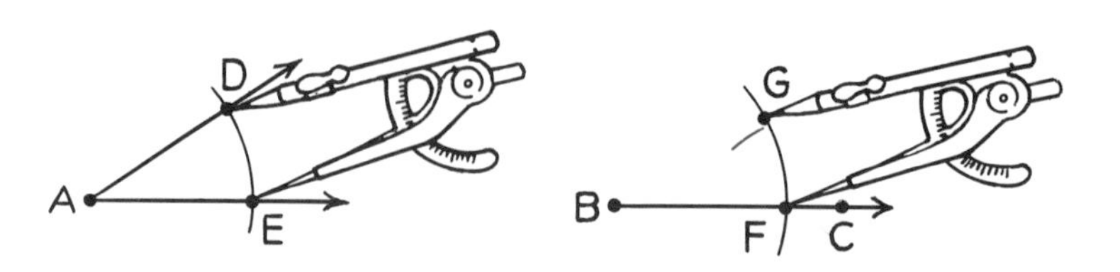

7. Draw $\overrightarrow{BG}$. $\angle B = \angle A$.

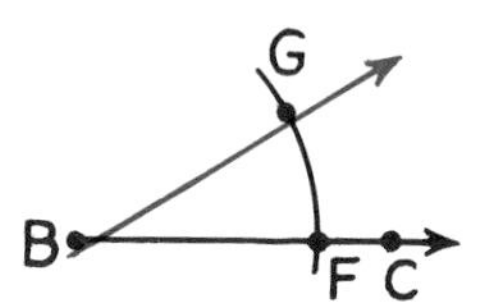

The figures below show why this method works. By the procedure described above, BG = BF = AD = AE. Also GF = DE. As a result, $\triangle BFG \cong \triangle AED$ (S.S.S.) and, because $\measuredangle B$ and $\measuredangle A$ are corresponding angles of these triangles, $\angle B = \angle A$.

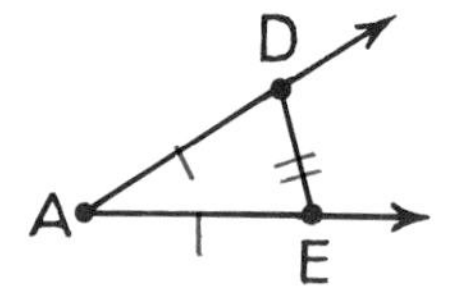

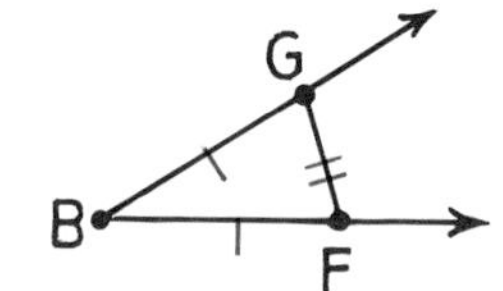

► **Construction 3**

To construct a triangle congruent to a given triangle.

Method.

1. Let $\triangle ABC$ be the given triangle.

2. Draw a line (ℓ) and copy $\overline{AB}$ on it ($\overline{DE}$ in the figure).

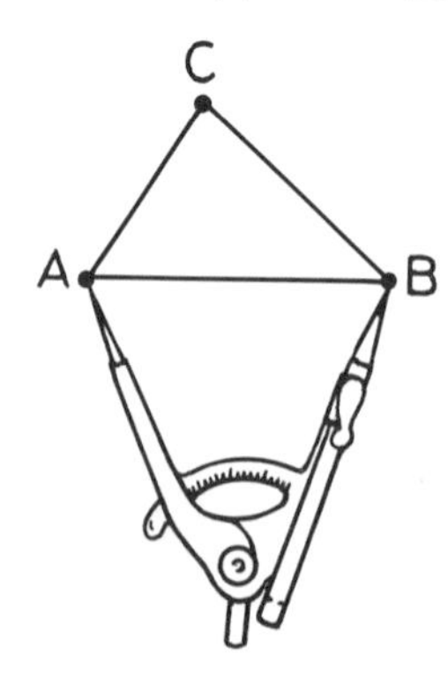

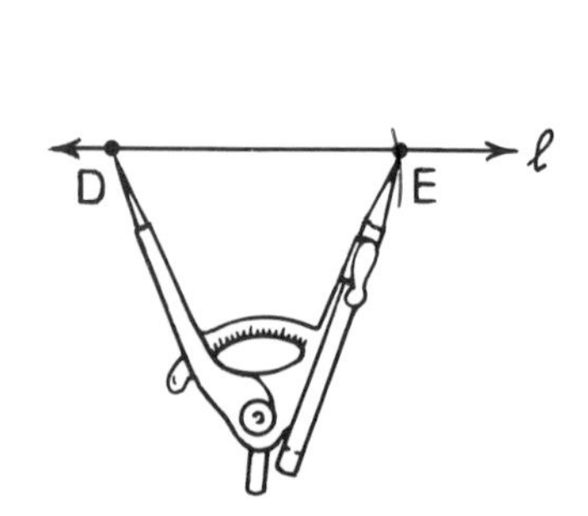

3. Adjust the radius of the compass to the length of $\overline{AC}$.

4. With D as center, draw an arc having this radius.

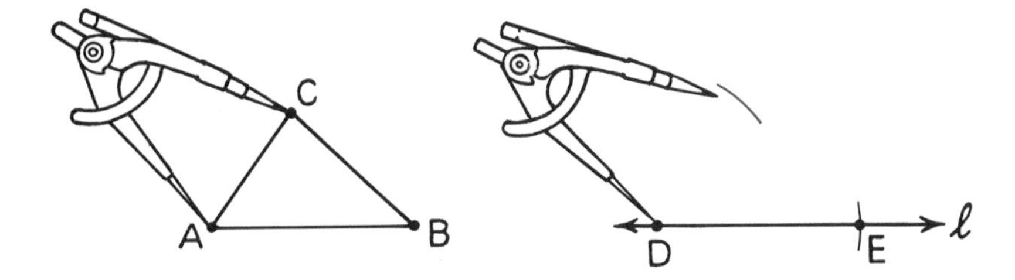

5. Adjust the radius of the compass to the length of $\overline{BC}$.

6. With E as center, draw a second arc having this radius so that it intersects the first arc (the arcs intersect in point F in the figure).

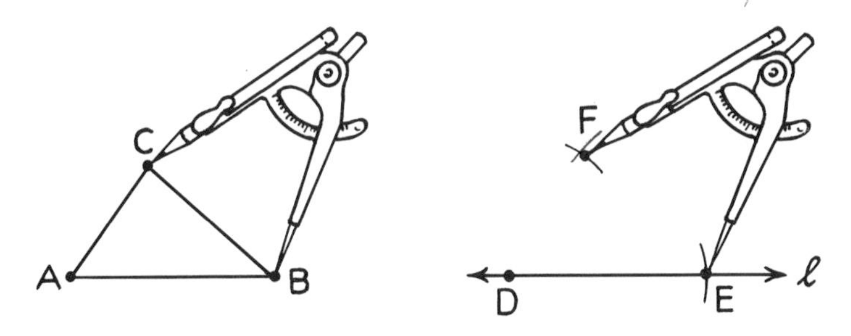

7. Use the straightedge to draw $\overline{DF}$ and $\overline{EF}$. $\triangle DEF \cong \triangle ABC$.

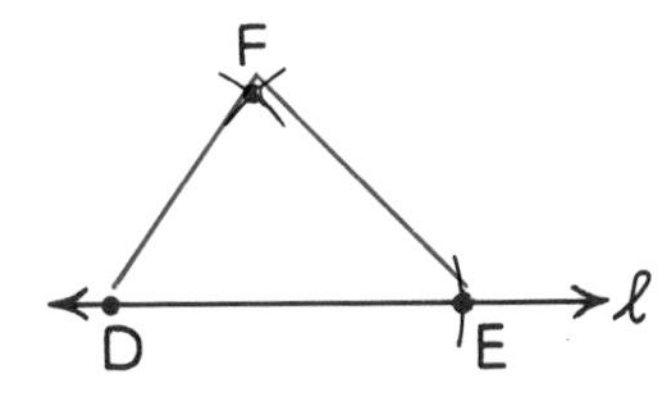

The figures below show why this method works. By the procedure described above, DE = AB, DF = AC, and EF = BC. As a result, $\triangle DEF \cong \triangle ABC$ (S.S.S.).

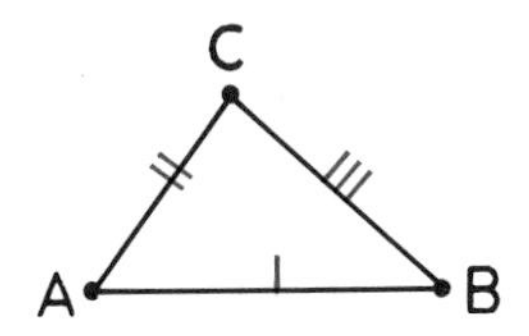

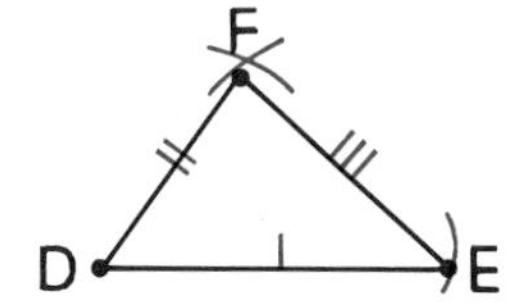

Exercises

Set I

In the figure below, $\overline{GE}$ and $\overline{ER}$ have been constructed on line ℓ so that they are each equal in length to $\overline{TI}$.

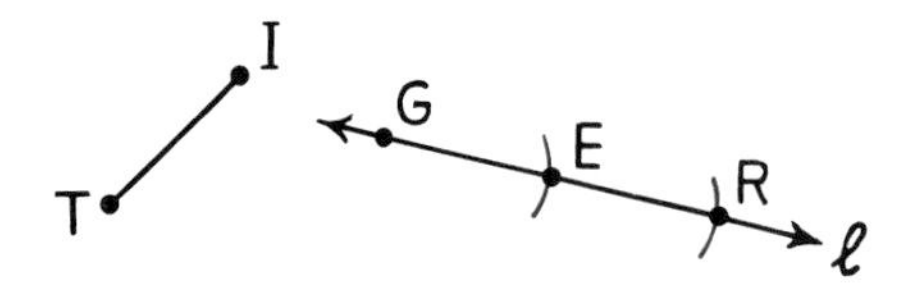

1. What is the name of the tool used to draw line ℓ?
2. What tool was used to draw the arcs that intersect ℓ in points E and R?
3. What is point E called with respect to $\overline{GR}$?
4. How does $\overline{GR}$ compare in length to $\overline{TI}$?
5. Draw a line on your paper and construct on it a line segment that is three times as long as $\overline{TI}$. Label the line segment $\overline{XY}$.

In the figure below, ∡E has been constructed so that it is equal in measure to ∡O.

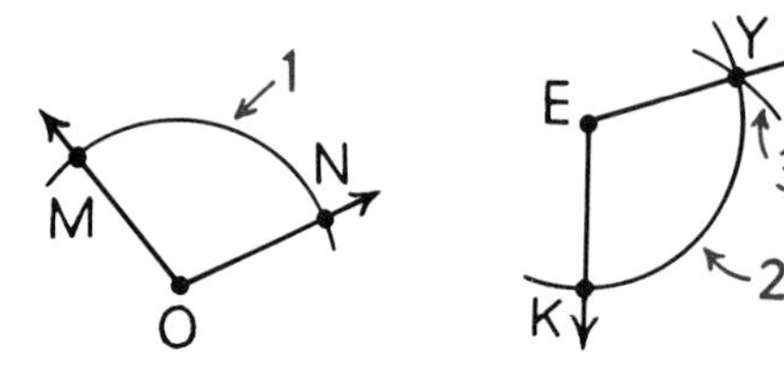

6. Which point is the center of the arc numbered 1?
7. What two lengths on ∡O are equal?
8. Which arc in the figure has the same radius as the arc numbered 1?
9. Which arc has its radius equal to the distance between M and N?

In the figure below, the equal lengths in the construction above have been marked.

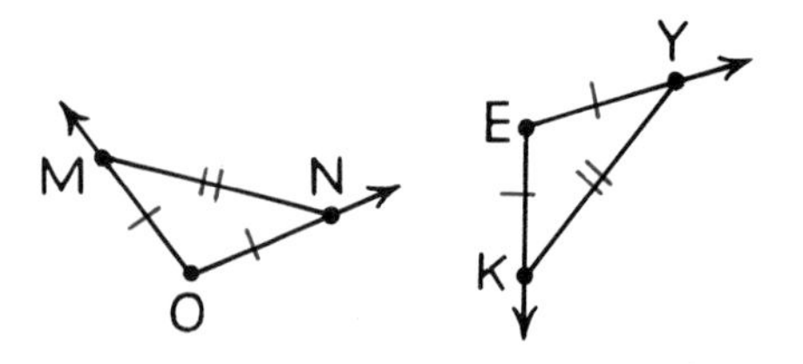

10. Why is △KEY ≅ △MON?
11. Why is ∠E = ∠O?

In the figure below, △BAT has been constructed so that it is congruent to △WOM.

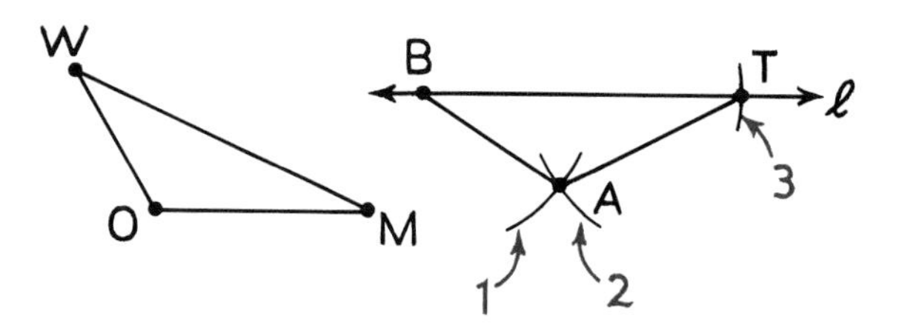

12. Which side of △WOM was copied on line ℓ?
13. Which arc was drawn with its radius equal to WO?
14. Which arc was drawn with its radius equal to MO?
15. Why are the triangles congruent?

Trace the line segments shown below and use your tracings to construct line segments having each of the following lengths. Shade the answers on your figures.

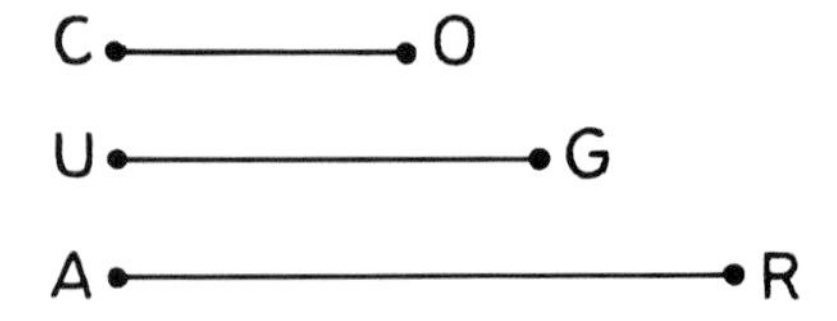

Example: Construct AR − CO.
Solution:

On line ℓ, $\overline{XY}$ has been constructed equal in length to $\overline{AR}$. $\overline{YZ}$ has been constructed equal in length to $\overline{CO}$. XZ = XY − YZ = AR − CO.

16. CO + AR.
17. AR − UG.
18. 2AR.
19. 3CO − AR.
20. UG − CO + AR.

Example: Construct $2\angle X$.
Solution:

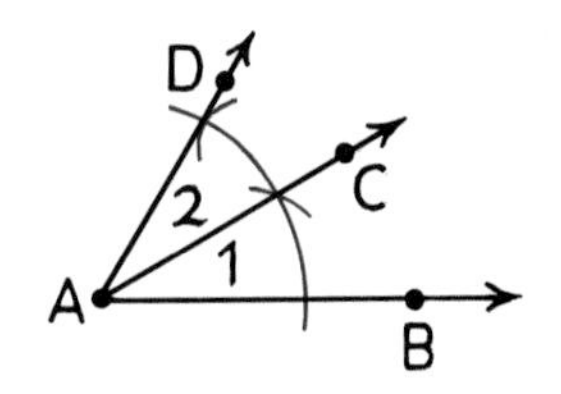

Starting on $\overrightarrow{AB}$, $\measuredangle 1$ and $\measuredangle 2$ have each been constructed equal in measure to $\measuredangle X$. $\angle BAD = \angle 1 + \angle 2 = 2\angle X$.

Trace the angles shown here and use your tracings to construct angles having each of the following measures.

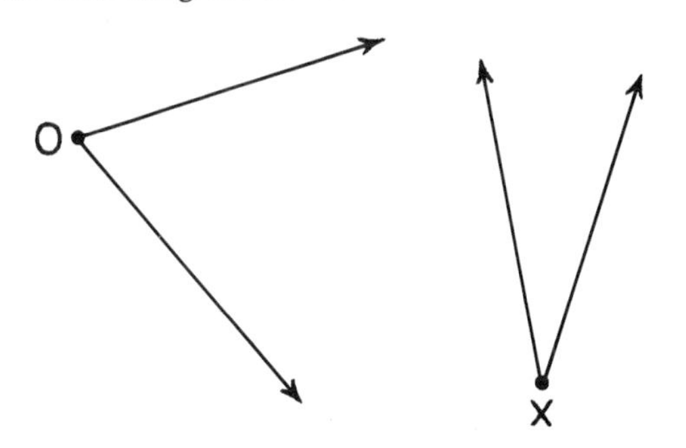

21. $\angle O$.
22. $3\angle X$.
23. $\angle O + \angle X$.
24. $2\angle O - \angle X$.

Set II

Trace the line segments shown here and use them to construct each of the following triangles.

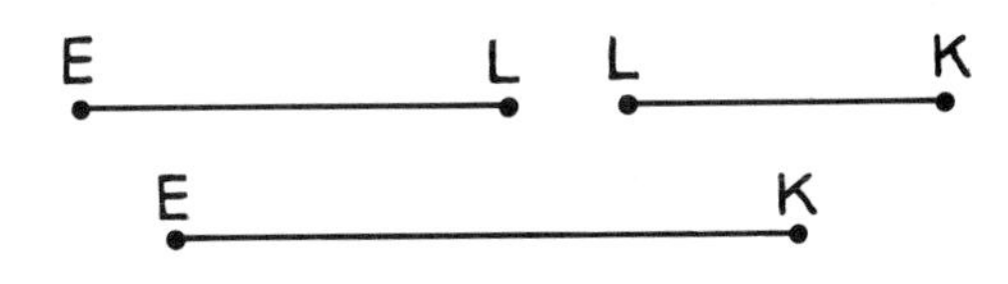

25. A triangle with sides $\overline{EL}$, $\overline{LK}$, and $\overline{EK}$.
26. An equilateral triangle with all three sides equal to $\overline{EL}$.
27. An isosceles triangle with one side equal to $\overline{EK}$ and the other two sides equal to $\overline{LK}$.
28. An isosceles triangle with one side equal to $\overline{LK}$ and the other two sides equal to $\overline{EK}$.

Trace the parts shown here.

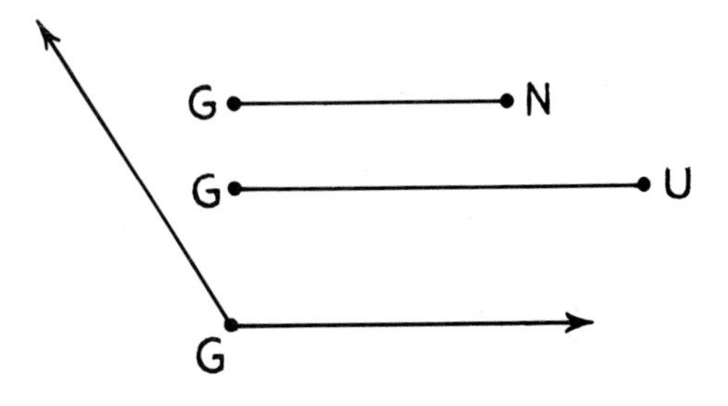

29. Use them to construct △GNU.

Trace the parts shown here.

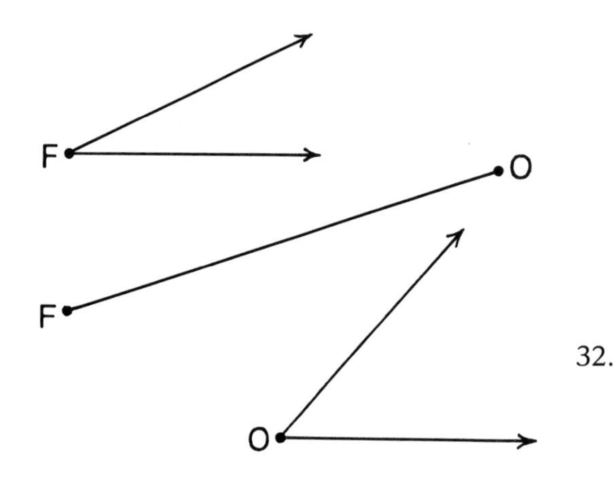

30. Use them to construct $\triangle$FOX.

Write complete proofs for each of the following exercises. Copy the figure, "given," and "prove" before writing your statements and reasons.

31.

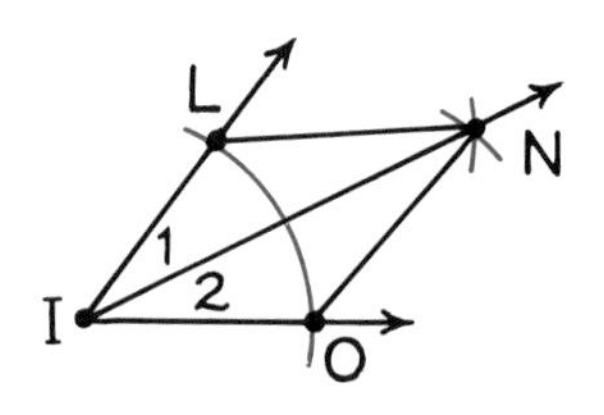

Given: IL = IO and LN = ON; $\overrightarrow{IL}$-$\overrightarrow{IN}$-$\overrightarrow{IO}$.
Prove: $\overrightarrow{IN}$ bisects ∡LIO.

32.

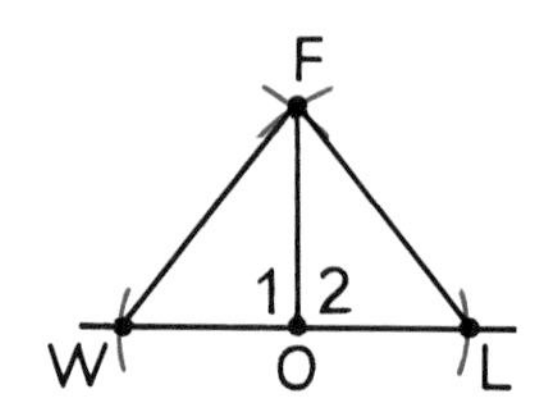

Given: OW = OL and WF = LF; ∡1 and ∡2 are a linear pair.
Prove: $\overline{OF} \perp \overline{WL}$.

Set III

Trace the line segments shown here.

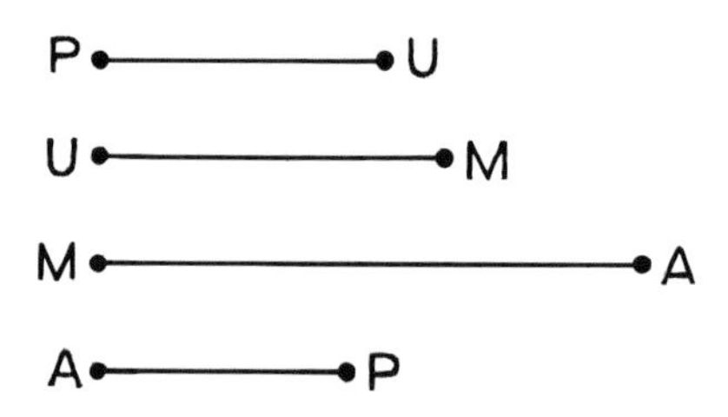

1. Use them to construct a quadrilateral PUMA.
2. Would you expect the quadrilateral you drew to be congruent to quadrilateral PUMA drawn by someone else? Explain.

Lesson 8

More Constructions

In this lesson, we will learn methods for constructing lines that bisect line segments and angles.

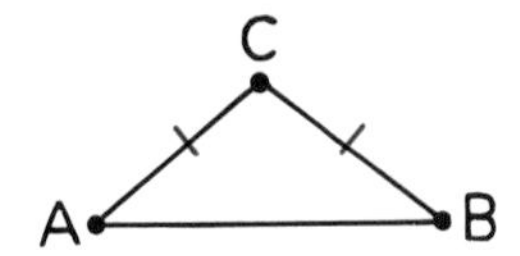

The figure above shows a line segment, $\overline{AB}$, and a point, C, whose distances from its endpoints, CA and CB, are equal. Because CA = CB, point C is said to be *equidistant* from points A and B.

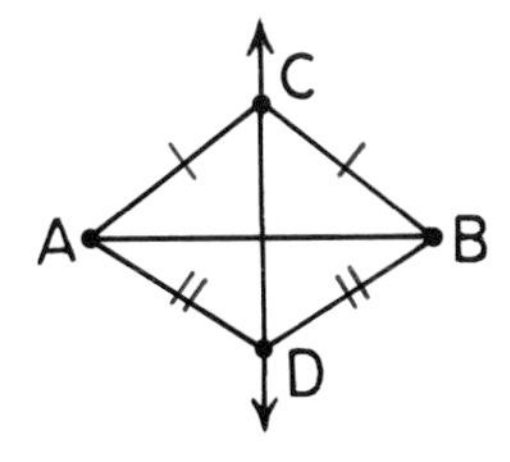

The figure above shows a line segment, $\overline{AB}$, and *two* points that are equidistant from its endpoints; that is, CA = CB and DA = DB. If a line is drawn through these two points, we can prove that it is perpendicular to $\overline{AB}$ and that it also bisects it. As a result, the line is called the *perpendicular bisector* of the line segment.

► **Theorem 16**
In a plane, two points each equidistant from the endpoints of a line segment determine the perpendicular bisector of the line segment.

The proof below is given for the case in which the two points are on opposite sides of the line segment (illustrated by the figure at the left below) and for the case in which they are on the same side of the line segment (illustrated by the figure at the right below).

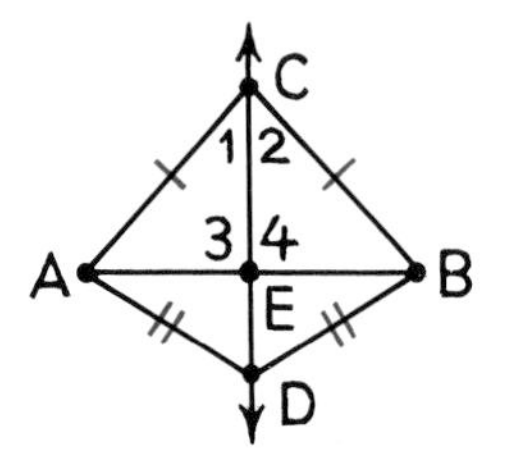

Given: CA = CB and DA = DB.

Prove: $\overleftrightarrow{CD}$ is the perpendicular bisector of $\overline{AB}$.

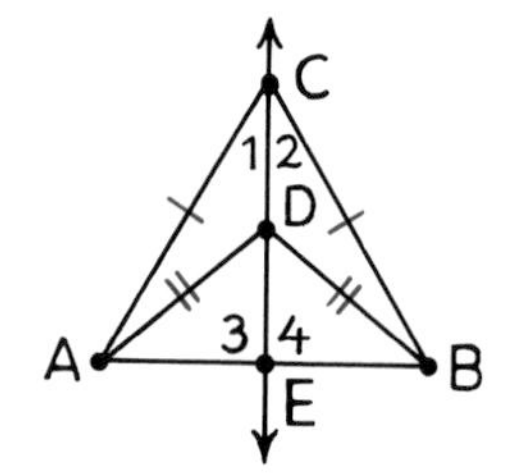

Proof.

Statements	Reasons
1. CA = CB and DA = DB.	Given.
2. CD = CD.	Reflexive.
3. $\triangle ACD \cong \triangle BCD$.	S.S.S.
4. $\angle 1 = \angle 2$.	If two triangles are congruent, their corresponding angles are equal.
5. CE = CE.	Reflexive.
6. $\triangle ACE \cong \triangle BCE$.	S.A.S.
7. $\angle 3 = \angle 4$.	Same as step 4.
8. $\measuredangle 3$ and $\measuredangle 4$ are right angles.	If two angles in a linear pair are equal, each is a right angle.
9. $\overleftrightarrow{CD} \perp \overline{AB}$	If two lines form right angles, they are perpendicular.
10. AE = BE.	If two triangles are congruent, their corresponding sides are equal.
11. $\overleftrightarrow{CD}$ bisects $\overline{AB}$.	If a line segment is divided into two equal segments, it is bisected.

This theorem is the basis for the following construction.

► **Construction 4**
To bisect a line segment.

Method.

1. Let $\overline{AB}$ be the given line segment.
2. With A and B as centers, draw two arcs that have the same radius and that intersect each other in two points (C and D in the figure).

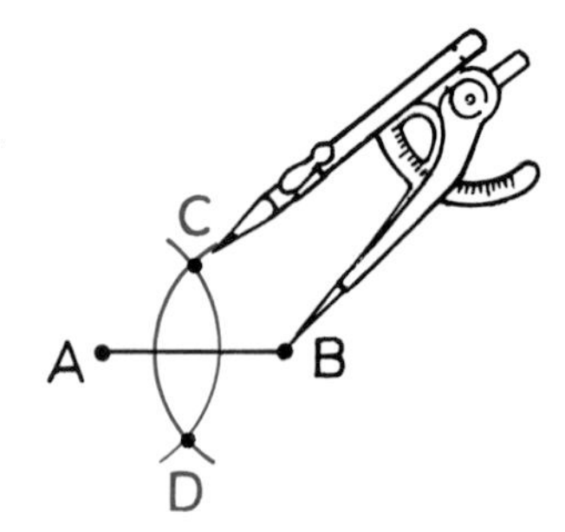

3. Draw $\overleftrightarrow{CD}$. $\overleftrightarrow{CD}$ bisects $\overline{AB}$.

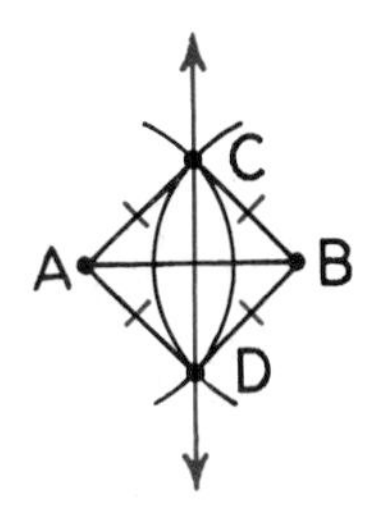

The figure at the left shows why this method works. By the procedure described, AC = BC and AD = BD. This means that points C and D are equidistant from the endpoints of $\overline{AB}$. Therefore, they determine the perpendicular bisector of $\overline{AB}$.

Angles can also be bisected with a straightedge and compass.

Construction 5
To bisect an angle.

Method.

1. Let ∡A be the given angle.

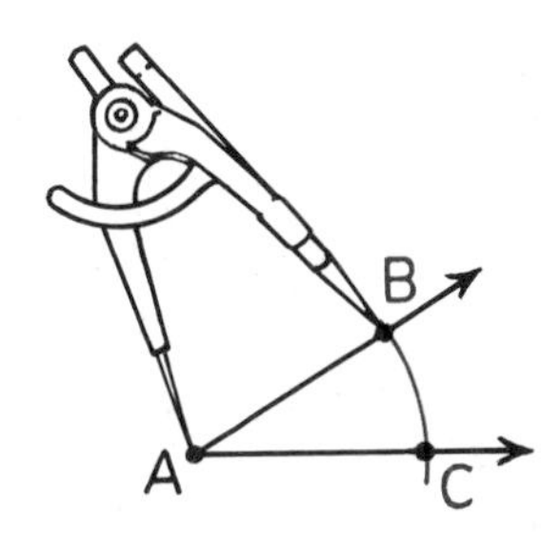

2. With A as center, draw an arc that intersects the sides of the angle (points B and C in the figure).

3. With B and C as centers, draw two arcs that have the same radius and that intersect each other in a point inside the angle (point D in the figure).

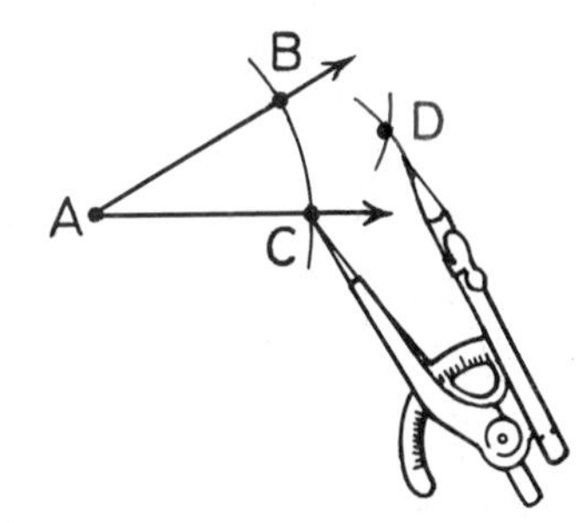

4. Draw $\overrightarrow{AD}$. $\overrightarrow{AD}$ bisects the angle.

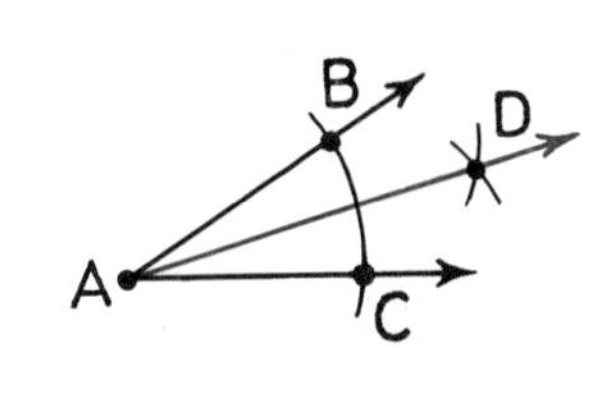

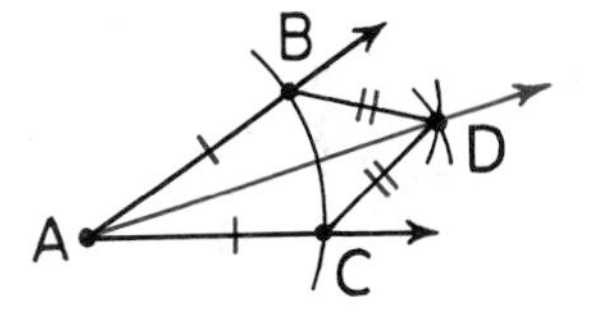

The figure at the right shows why this method works. By the procedure described above, AB = AC and BD = CD. Also, AD = AD, so △ABD ≅ △ACD (S.S.S.) and, because ∡BAD and ∡CAD are corresponding angles of these triangles, ∠BAD = ∠CAD. Because $\overrightarrow{AD}$ divides ∡BAC into two equal angles, it bisects it.

Exercises

Set I

Use your ruler to draw line segments with the following lengths. Then, use your straightedge and compass to bisect each line segment. Finally, use your ruler to check the accuracy of your construction.

1. 6 cm.
2. 4.8 cm.

Use your protractor to draw angles with the following measures. Then, use your straightedge and compass to bisect each angle. Finally, use your protractor to check the accuracy of your construction.

3. 50°.
4. 120°.

In the figure below, point C is equidistant from points A and O.

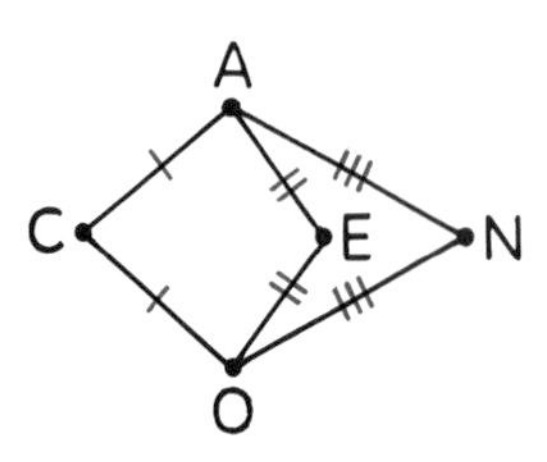

5. What other points are equidistant from A and O?
6. What relation do the points that are equidistant from A and O appear to have to each other?
7. What do the points that are equidistant from A and O determine?

In the figure below, point A is equidistant from points R and K.

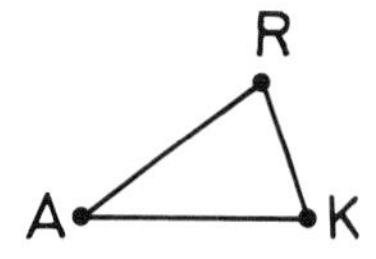

8. Write an equation expressing this fact.
9. Why does it follow that $\angle R = \angle K$?

In the figure below, KE = TE and KC = TC.

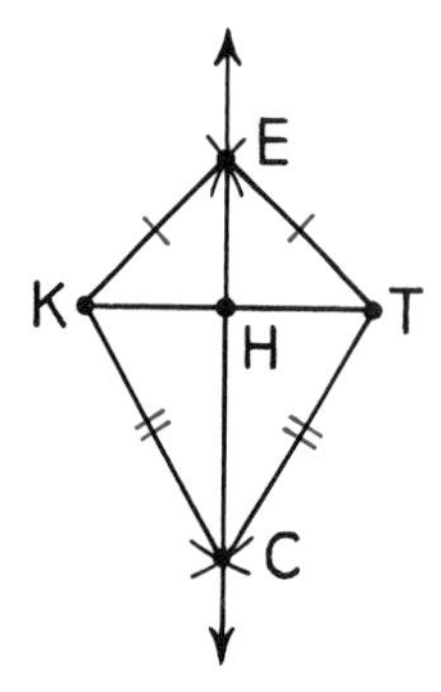

10. What relation do points E and C have to the endpoints of $\overline{KT}$?
11. Why does $\overleftrightarrow{EC}$ bisect $\overline{KT}$?
12. Do points T and K appear to be equidistant from the endpoints of $\overline{EC}$?
13. Does $\overline{KT}$ appear to bisect $\overline{EC}$?

In the figure below, OD = OR, and DY = RY.

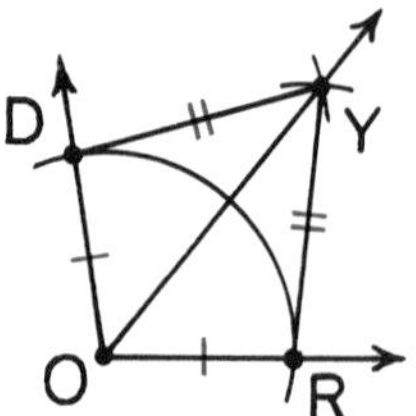

14. Which two points in the figure are equidistant from the other two points?
15. Why is $\triangle DOY \cong \triangle ROY$?
16. Why is $\angle DOY = \angle ROY$?
17. Why does $\overrightarrow{OY}$ bisect $\measuredangle DOR$?

Trace $\triangle TUG$ on your paper.

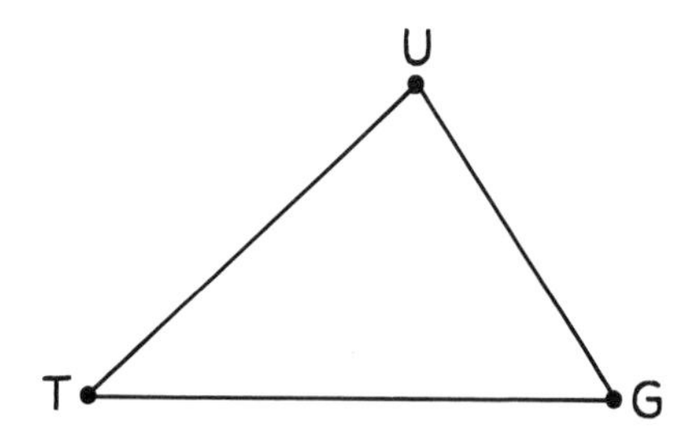

18. Use your straightedge and compass to bisect $\measuredangle G$.
19. Does the bisector of $\measuredangle G$ also bisect $\overline{TU}$?

Trace the figure below on your paper.

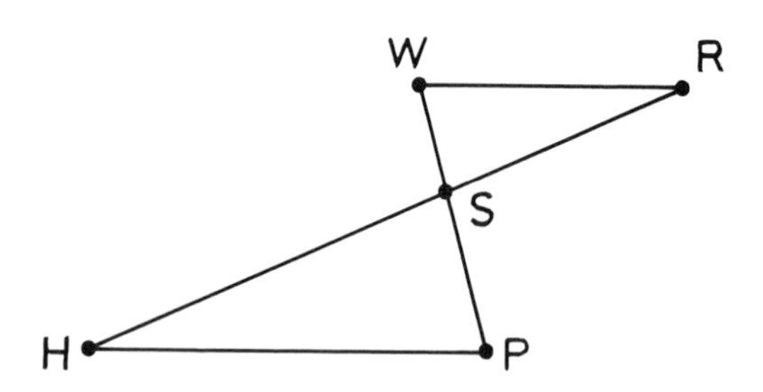

20. Use your straightedge and compass to bisect $\overline{WR}$ and $\overline{HP}$. Label the midpoints A and I respectively.
21. What seems to be true about points I, S, and A?
22. Draw $\overline{IA}$. Does $\overleftrightarrow{IA}$ bisect $\measuredangle WSR$ and $\measuredangle HSP$?

Set II

Trace the figure below on your paper.

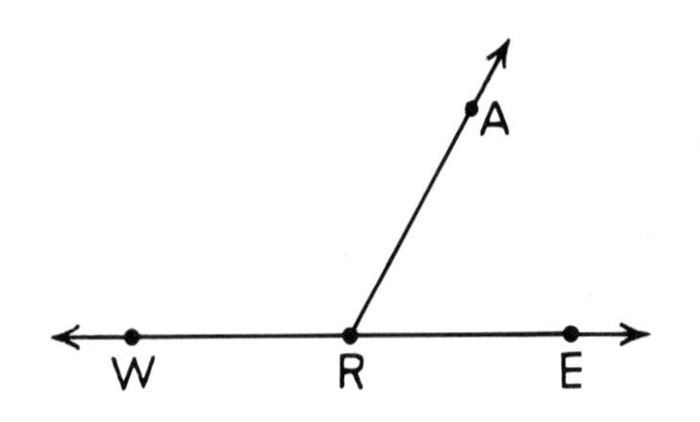

23. Use your straightedge and compass to construct the ray that bisects ∡WRA; label it $\overrightarrow{RH}$. Also construct the ray that bisects ∡ARE and label it $\overrightarrow{RL}$.
24. What seems to be true about ∡HRL?
25. What seems to be true about $\overrightarrow{RH}$ and $\overrightarrow{RL}$?

Trace quadrilateral LFEO on your paper.

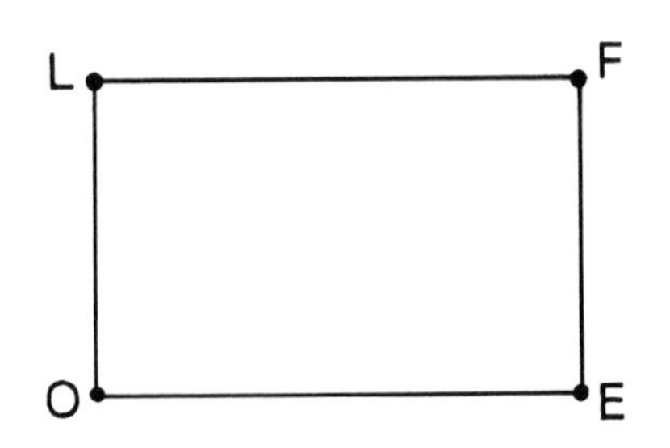

26. Use your straightedge and compass to bisect $\overline{LF}$ and $\overline{OE}$. Label the midpoints I and B respectively. Draw $\overline{LB}$ and $\overline{IE}$.
27. What seems to be true about $\overline{LB}$ and $\overline{IE}$?
28. Draw $\overline{OF}$. Label the points in which $\overline{LB}$ and $\overline{IE}$ intersect it A and T respectively. What seems to be true?

Write complete proofs for each of the following exercises.

29.

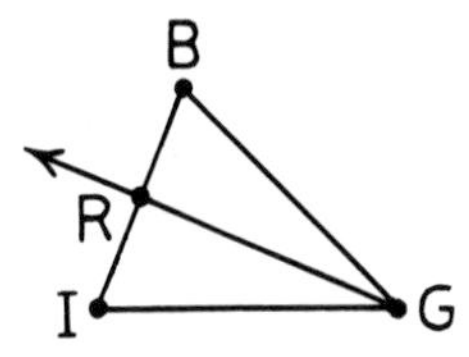

Given: $\overrightarrow{GR}$ bisects ∡IGB;
$\overrightarrow{GR} \perp \overline{IB}$.
Prove: BG = IG.

30.

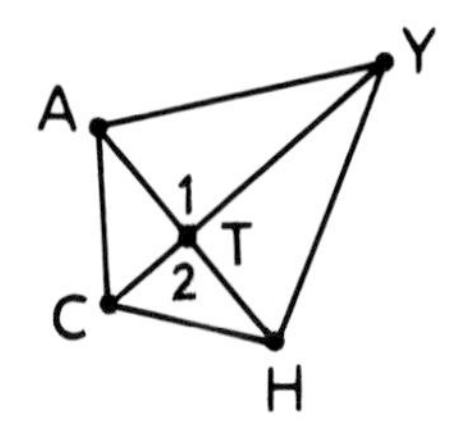

Given: C and Y are equidistant from A and H.
Prove: ∡1 and ∡2 are supplementary.

Set III

Obtuse Ollie left his compass out in the rain and it rusted so that he can no longer change its radius. He wanted to bisect $\overline{AB}$ but when he used his compass to draw arcs centered at A and B, they didn't intersect.

Can you help Ollie out by showing how he could still use his compass to bisect $\overline{AB}$, even though its radius cannot be changed? (Trace the figure and set your compass to the radius shown.)

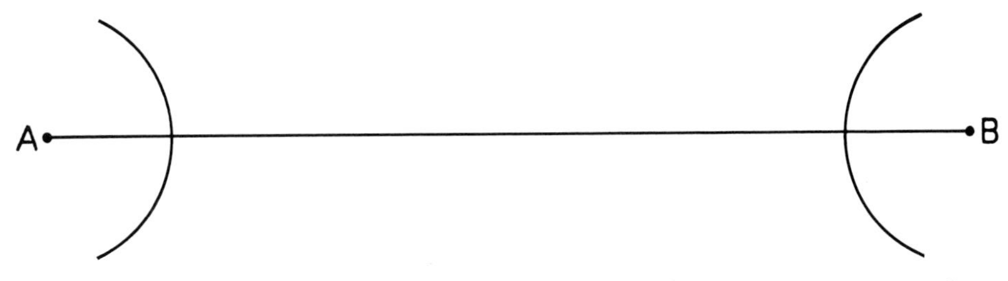

Chapter 4 / Summary and Review

Basic Ideas

Congruent polygons 126
Corollary 141
Equidistant points 157
Included and opposite sides and angles 120
Isosceles triangle, legs of 121
Perpendicular bisector of line segment 157
Right triangle, legs and hypotenuse of 121
Triangle 120
  classification according to angles 121
  classification according to sides 121

Postulates

8. *The S.A.S. Congruence Postulate.* If two sides and the included angle of one triangle are equal to two sides and the included angle of another triangle, the triangles are congruent. 126

9. *The A.S.A. Congruence Postulate.* If two angles and the included side of one triangle are equal to two angles and the included side of another triangle, the triangles are congruent. 127

Theorems

12. Two triangles congruent to a third triangle are congruent to each other. 137

13. If two sides of a triangle are equal, the angles opposite them are equal. 140
    Corollary. If a triangle is equilateral, it is also equiangular. 141

14. If two angles of a triangle are equal, the sides opposite them are equal. 141
    Corollary. If a triangle is equiangular, it is also equilateral. 141

15. *The S.S.S. Congruence Theorem.* If the three sides of one triangle are equal to the three sides of another triangle, the triangles are congruent. 146

16. In a plane, two points each equidistant from the endpoints of a line segment determine the perpendicular bisector of the line segment. 157

Constructions

1. To construct a line segment equal in length to a given line segment. 151
2. To construct an angle equal in measure to a given angle. 151
3. To construct a triangle congruent to a given triangle. 152
4. To bisect a line segment. 158
5. To bisect an angle. 159

Exercises

Set I

Complete the statements of the following theorems.

1. Two triangles congruent ||||||||||.
2. If a triangle is equiangular, then ||||||||||.
3. If two sides of a triangle are equal, ||||||||||.
4. In a plane, two points each equidistant from the endpoints of a line segment ||||||||||.

In the figure below, N-I-L and ∠N = ∠NEL.

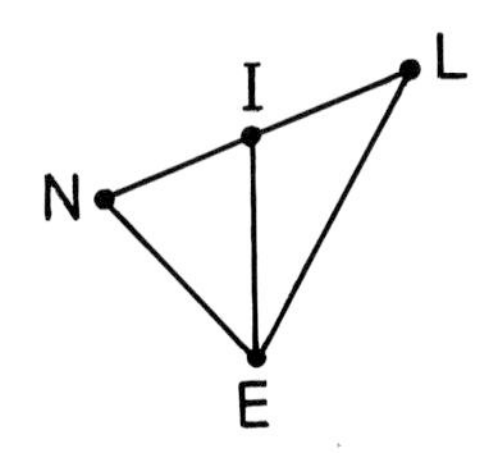

5. Which triangle in the figure is obtuse?
6. Which triangle in the figure must be isosceles?
7. State the theorem upon which your answer is based.

The two polygons below are congruent.

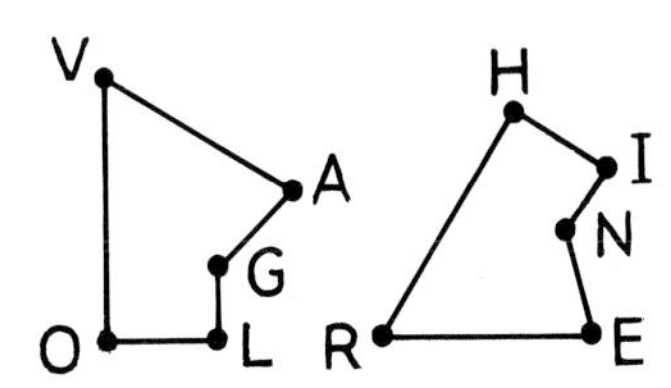

8. Define congruent polygons.
9. If a tracing of the first polygon were placed on the second, which of its vertices would fall on R?
10. Copy and complete the following congruence correspondence between the vertices of the polygons: VOLGA ↔ ||||||||||.

In the figure below, ∠1 = ∠2.

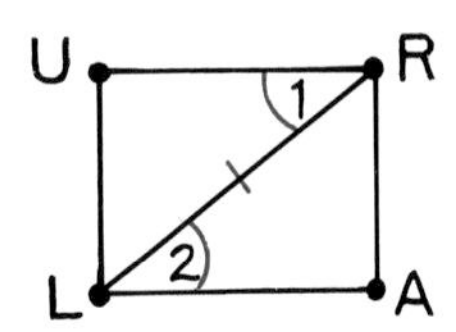

11. Name the other pair of parts needed to show that the triangles are congruent by the S.A.S. postulate.
12. State the S.A.S. postulate as a complete sentence.
13. If you knew that ∠U = ∠A, could you prove the triangles congruent by the A.S.A. postulate?
14. Explain why or why not.

In △DON, ∠D = ∠N.

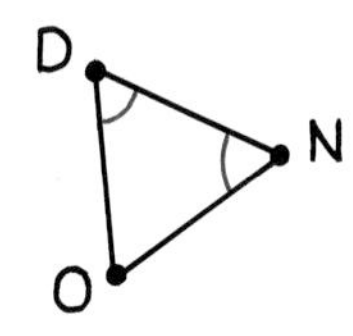

15. Does it follow that DO = DN?
16. Does it follow that △DON is isosceles?
17. Does it follow that △DON is equiangular?
18. Can △DON be scalene?
19. Can △DON be equilateral?

The figure below illustrates a construction.

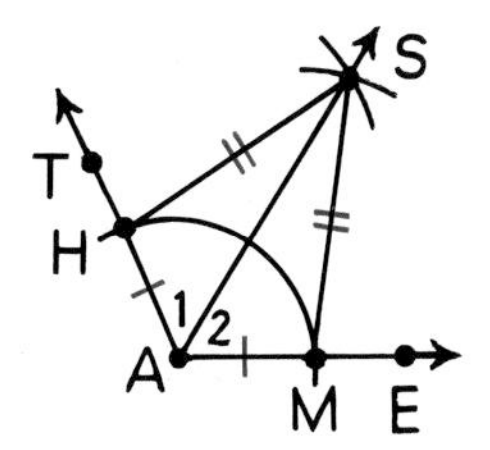

20. What has been constructed?

21. Why is $\triangle HAS \cong \triangle MAS$?

22. Why is $\angle 1 = \angle 2$?

Trace $\triangle OKA$ on your paper.

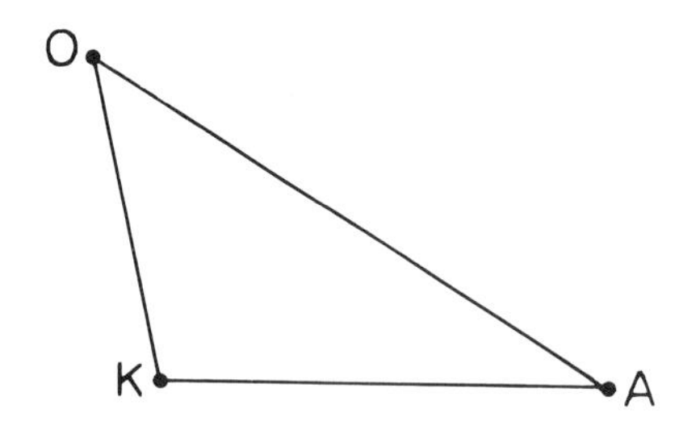

23. Use your straightedge and compass to construct the ray that bisects $\measuredangle A$.

24. Use your straightedge and compass to construct the line that is the perpendicular bisector of $\overline{OA}$.

Trace the parts shown below and use them to construct each of the following triangles.

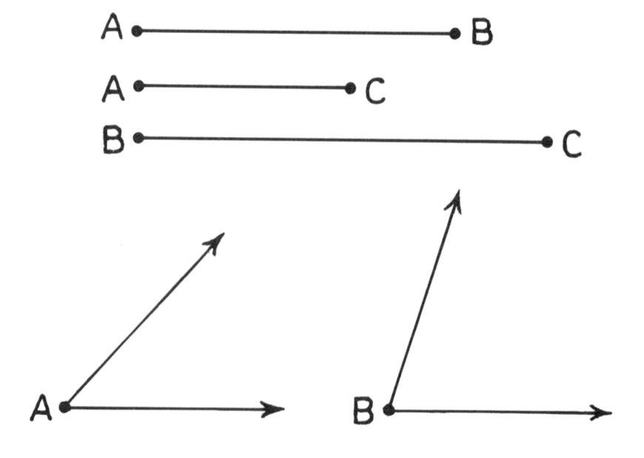

25. A triangle with sides $\overline{AB}$, $\overline{AC}$, and $\overline{BC}$.

26. A triangle with side $\overline{AB}$, angle A, and side $\overline{AC}$.

27. A triangle with side $\overline{AB}$, angle A, and angle B.

Set II

Write complete proofs for each of the following exercises.

28.

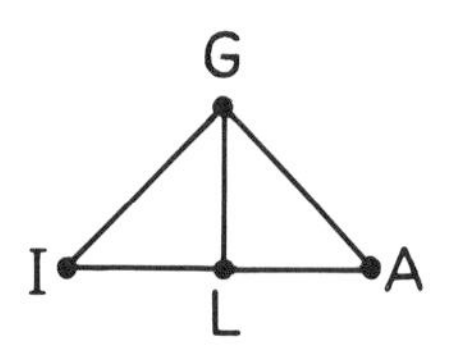

Given: L is the midpoint of $\overline{IA}$; IG = AG.
Prove: $\triangle GIL \cong \triangle GAL$.

29.

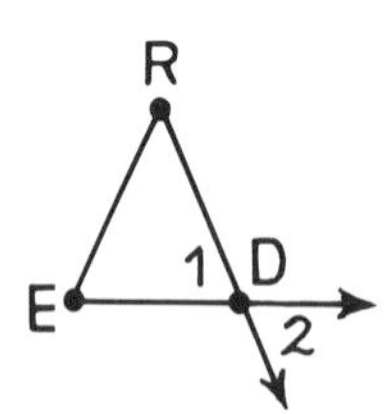

Given: RE = RD; $\measuredangle 1$ and $\measuredangle 2$ are vertical angles.
Prove: $\angle E = \angle 2$.

30.

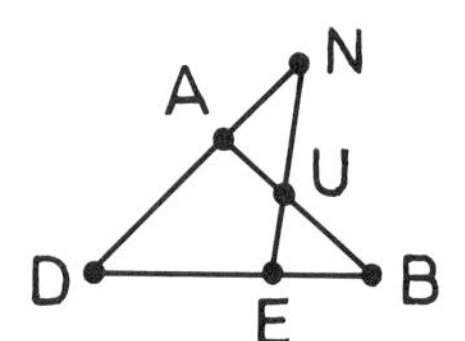

Given: DN = DB and $\angle N = \angle B$.
Prove: NE = BA.

31.

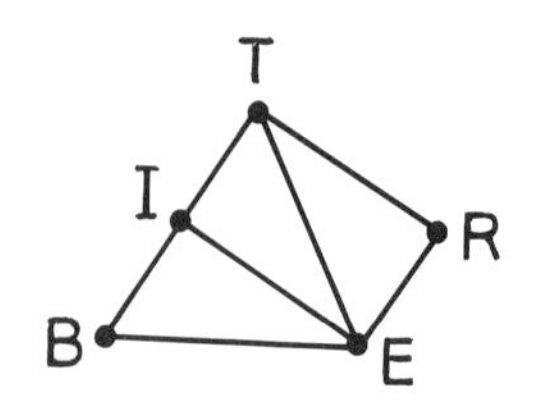

Given: $\triangle BIE \cong \triangle ERT$ and $\triangle TIE \cong \triangle ERT$; $\measuredangle BIE$ and $\measuredangle TIE$ are a linear pair.
Prove: $\overline{BT} \perp \overline{IE}$.

32.

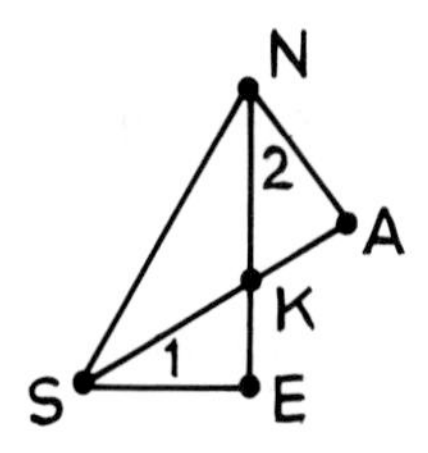

Given: $\measuredangle 1$ and $\measuredangle 2$ are complements of $\measuredangle NSE$;
SE = NA;
$\overline{NE} \perp \overline{SE}$ and $\overline{NA} \perp \overline{SA}$.
Prove: $\triangle SNK$ is isosceles.

33.

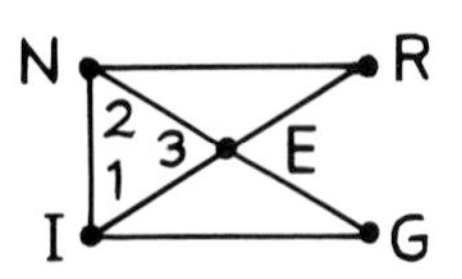

Given: $\angle NIG = \angle INR$;
NR = IG;
$\angle 1 = \angle 3$.
Prove: $\triangle NIE$ is equilateral.

Algebra Review

More on Solving Simultaneous Equations

Another way to solve a pair of simultaneous equations is to combine the equations to form an equation that contains only one of the variables. The solution to this equation can then be substituted into either of the original equations to find the other variable.

Example 1.
Solve this pair of simultaneous equations.

$$\begin{aligned} 4x + 3y &= 34 \\ x - 3y &= 1 \end{aligned}$$

Solution.
We can eliminate y by adding the two equations:

$$\begin{array}{rl} 4x + 3y &= 34 \\ \underline{x - 3y} &\underline{= 1} \\ 5x \quad &= 35 \end{array}$$

Dividing both sides of the resulting equation by 5, we get

$$x = 7.$$

Substituting 7 for x in the first equation, we get

$$\begin{aligned} 4(7) + 3y &= 34 \\ 28 + 3y &= 34 \\ 3y &= 6 \\ y &= 2. \end{aligned}$$

The solution is the ordered pair (7, 2).

Example 2.
Solve this pair of simultaneous equations.

$$\begin{aligned} 3x + 7y &= 25 \\ 5x + 4y &= 11 \end{aligned}$$

Solution.
We can eliminate x by multiplying the first equation by 5, the second equation by -3, and then adding the resulting equations:

$$\begin{array}{rcrl} 5(3x + 7y) = 5(25) & \rightarrow & 15x + 35y &= 125 \\ -3(5x + 4y) = -3(11) & \rightarrow & \underline{-15x - 12y} &\underline{= -33} \\ & & 23y &= 92 \\ & & y &= 4 \end{array}$$

$$\begin{aligned} 3x + 7(4) &= 25 \\ 3x + 28 &= 25 \\ 3x &= -3 \\ x &= -1. \end{aligned}$$

The solution is (-1, 4).

Exercises

Solve the following pairs of simultaneous equations.

1. $x + y = 35$, $x - y = 67$
2. $x - y = 8$, $x + y = 43$
3. $6x + 11y = 21$, $6x + y = -9$
4. $5x - 7y = 92$, $5x + y = 4$
5. $3x - 5y = 51$, $x + 5y = 23$
6. $9x + 2y = 5$, $x - 2y = -15$
7. $8x - 7y = 62$, $4x - 7y = 66$
8. $3x - 11y = 43$, $12x - 11y = 7$
9. $x + y = 7$, $3x + 2y = 25$
10. $4x + 3y = 31$, $2x - 9y = 5$
11. $2x + 5y = 29$, $4x - y = 25$
12. $5x + 4y = 53$, $x - 2y = 5$
13. $8x - 3y = 32$, $7x + 9y = 28$
14. $6x + 6y = 24$, $10x - y = -15$
15. $5x - 7y = 54$, $2x - 3y = 22$
16. $7x - 5y = 40$, $3x - 2y = 16$

Chapter 5

INEQUALITIES

Drawing by C. Barsotti; © 1972 The New Yorker Magazine, Inc.

"I wonder, sir, if you would indulge me in a rather unusual request?"

Lesson 1

Properties of Inequality

By the permission of The New Yorker Magazine, Inc.

The bartender in this cartoon likes to keep everything in order and one of his customers is in the wrong place. In the diagram at the left, the heights of the little fellow and the two men standing next to him have been represented by a, b, and c. To show that a is greater than b, we write $a > b$ and to show that b is less than c, we write $b < c$. Given that $b < c$, we also know that $c > b$, which is the order that the bartender would prefer because he wants the taller fellow to be at the left.

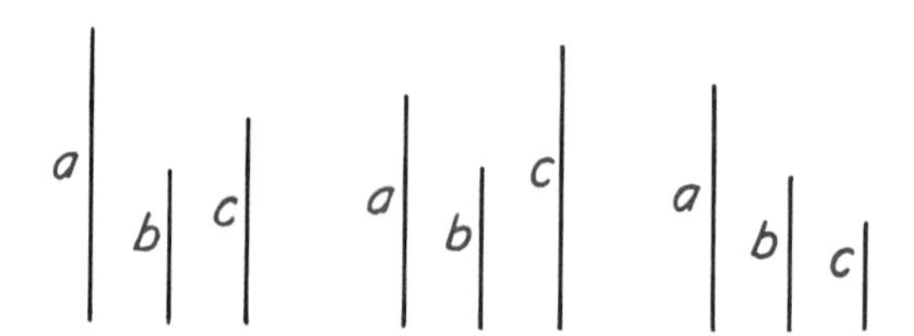

If $a > b$ and $c > b$, can we conclude anything about the relative sizes of a and c? The first two figures above reveal that the answer to this is no. What if $a > b$ and $b > c$? The third figure suggests that it follows that $a > c$. We will assume this conclusion as an algebraic postulate about inequality. It is stated on the next page along with some other properties of inequality. The letters represent real numbers.

► **The Three Possibilities Property**
Either $a > b$, $a = b$, or $a < b$.

► **The Transitive Property***
If $a > b$ and $b > c$, then $a > c$.

► **The Substitution Property**
If $a = b$, then a can be substituted for b in any inequality.

► **The Addition Properties***
If $a > b$, then $a + c > b + c$.
If $a > b$ and $c > d$, then $a + c > b + d$.

► **The Subtraction Property***
If $a > b$, then $a - c > b - c$.

► **The Multiplication Property**
If $a > b$ and $c > 0$, then $ac > bc$.

► **The Division Property**
If $a > b$ and $c > 0$, then $\frac{a}{c} > \frac{b}{c}$.

► **The "Whole Greater than Its Part" Property**
If $c = a + b$ and $b > 0$, then $c > a$.

The "whole greater than its part" property gets its name from the way that it is applied to geometric figures. For example, look at the figure at the right. The "whole" angle is $\measuredangle$APC and its "parts" are $\measuredangle$APB and $\measuredangle$BPC. Clearly, $\angle$APC > $\angle$APB and $\angle$APC > $\angle$BPC. The "whole greater than its part" property can be used to prove this.

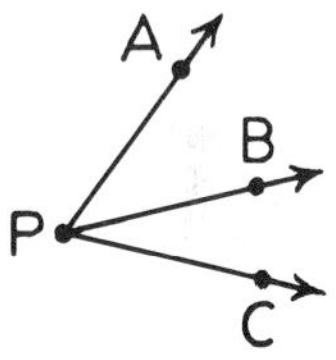

Because $\overrightarrow{PA}$-$\overrightarrow{PB}$-$\overrightarrow{PC}$, it follows from the Betweenness of Rays Theorem that

$$\angle APC = \angle APB + \angle BPC.$$

Now $\angle$BPC > 0 because the measure of every angle is greater than zero. By the "whole greater than its part" property, we can conclude that

$$\angle APC > \angle APB.$$

In a similar fashion, we can also conclude that

$$\angle APC > \angle BPC.$$

*Although these properties are expressed in terms of the symbol >, they are equally valid in terms of the symbol <.

Exercises

Set I

Refer to the figure below to copy and complete the following inequalities. Base your answers on appearances.

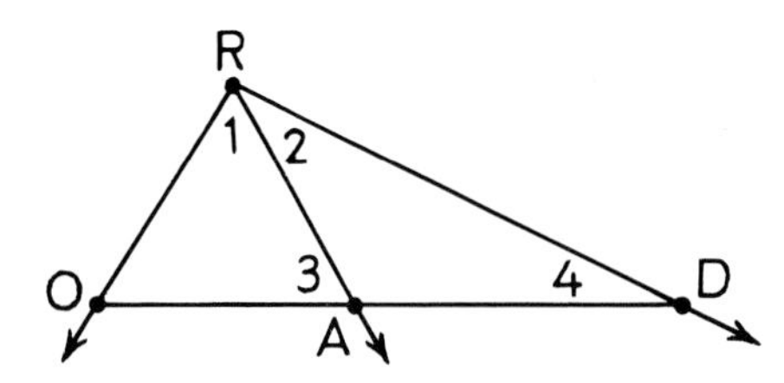

Example: $\angle 2$ |||||||| $\angle 3$.
Answer: $\angle 2 < \angle 3$ (because $\measuredangle 2$ looks smaller than $\measuredangle 3$).

1. $\angle 1$ |||||||| $\angle 2$.
2. OA |||||||| AD.
3. OD |||||||| RD.
4. $\angle 3$ |||||||| $\angle 4$.

In the figure below, UL = LE = EV.

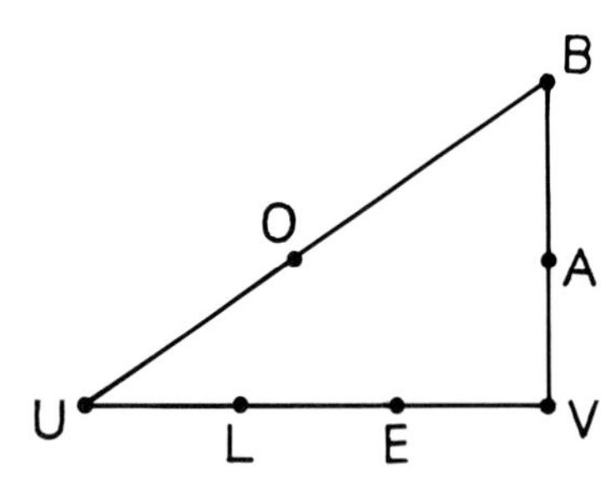

5. Trace the figure and draw $\overline{BL}$, $\overline{BE}$, and $\overline{OA}$. Label the points in which $\overline{BL}$ and $\overline{BE}$ intersect $\overline{OA}$ D and R respectively.

Refer to your figure to copy and complete the following equations and inequalities. Base your answers on appearances.

6. $\angle$UBL |||||||| $\angle$LBE |||||||| $\angle$EBV.
7. BO |||||||| BD |||||||| BR |||||||| BA.
8. OD |||||||| DR |||||||| RA.
9. $\angle$ULB |||||||| $\angle$UEB |||||||| $\angle$UVB.

Name the property of inequality illustrated by each of the following statements.

10. If $x < y$ and $y < 0$, then $x < 0$.
11. If $x = y + z$ and $z > 0$, then $x > y$.
12. If $x > y$ and $y = z$, then $x > z$.
13. Either $x > y$, $x = y$, or $x < y$.
14. If $x < 3$ and $y < 7$, then $x + y < 10$.
15. If $x > 4$, then $2x > 8$.

The following exercises refer to the figure below. Which fact (or facts) and what property of inequality can be used to prove the conclusion stated in each exercise?

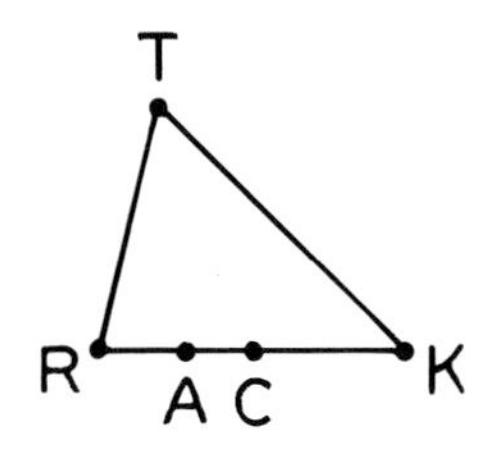

Fact 1. TK > RK.
Fact 2. RA < CK.
Fact 3. CK < TR.
Fact 4. RK = RA + AK.

16. RA < TR.
17. RK > RA.
18. RA + AC < CK + AC.
19. TK > RA + AK.

The following exercises refer to the figure below. Which fact (or facts) and what property of inequality can be used to prove the conclusion stated in each exercise?

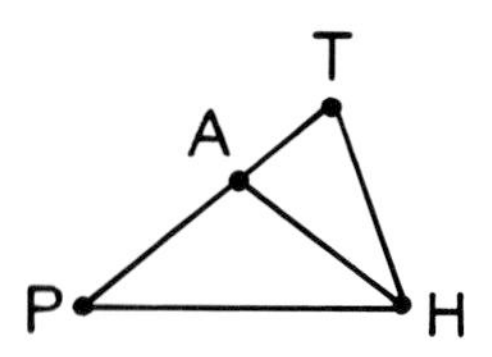

Fact 1. $\angle$TAH > $\angle$P.
Fact 2. $\angle$PHT = $\angle$PHA + $\angle$AHT.
Fact 3. $\angle$PAH > $\angle$TAH.
Fact 4. $\angle$P = $\angle$PHA.

20. $\frac{1}{2}\angle TAH > \frac{1}{2}\angle P$.

21. $\angle TAH > \angle PHA$.

22. $\angle PAH > \angle P$.

23. $\angle PHT > \angle AHT$.

Set II

Use your ruler and protractor to draw the following triangles.

24. Draw $\triangle RUN$ in which RU = 6 cm, $\angle R = 45°$, and $\angle U = 60°$.

25. Which angle of $\triangle RUN$ is the largest? (Use your protractor to find out.)

26. Which side of $\triangle RUN$ is the longest? (Use your ruler to find out.)

27. What is the relation of the largest angle of $\triangle RUN$ to the longest side?

28. Draw $\triangle WAY$ in which WA = 5 cm, $\angle W = 25°$, and WY = 8 cm.

29. Which side of $\triangle WAY$ is the shortest?

30. Which angle of $\triangle WAY$ is the smallest?

31. What is the relation of the shortest side of $\triangle WAY$ to the smallest angle?

The "whole greater than its part" property can be proved by means of other properties. Copy and complete the following proof showing this.

32.

Given: $c = a + b$ and $b > 0$.
Prove: $c > a$.

Proof.

Statements — **Reasons**

1. $c = a + b$. — ||||||||
2. $c - a = b$. — Subtraction property of equality.
3. $b > 0$. — ||||||||
4. |||||||| — Substitution property of inequality.
5. |||||||| — ||||||||

Use the "whole greater than its part" property to write the inequalities that follow from each of the following equations.

Example: $12 = 3 + 9$.
Answer: $12 > 3$ (because $9 > 0$) *and* $12 > 9$ (because $3 > 0$).

33. $7 = -4 + 11$.

34. $AC = AB + BC$, $AB > 0$, and $BC > 0$.

35. $x = y + z$, $y > 0$.

Give the missing statements and reasons in the following proof.

36.

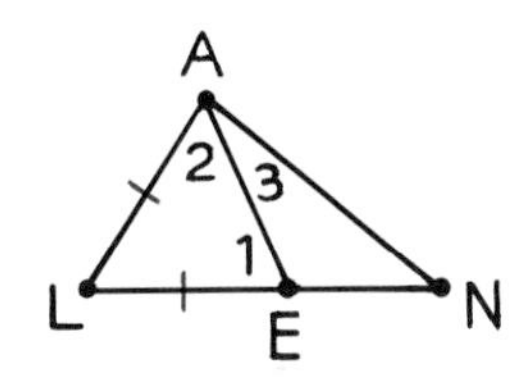

Given: LA = LE.
Prove: $\angle LAN > \angle 1$.

Proof.

Statements — **Reasons**

1. a) |||||||| — b) ||||||||
2. $\angle 1 = \angle 2$. — c) ||||||||
3. d) $\angle LAN =$ ||||||||. — Betweenness of Rays Theorem.
4. $\angle LAN > \angle 2$. — e) ||||||||
5. f) |||||||| — g) ||||||||

Write complete proofs for each of the following exercises.

37.

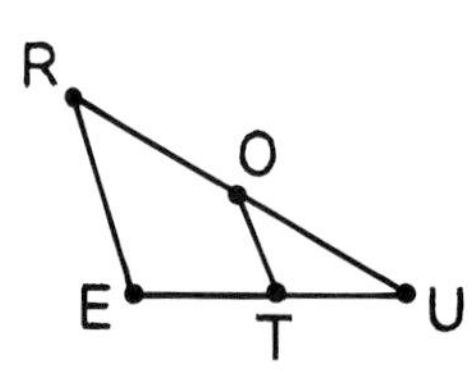

Given: Points O and T are the midpoints of $\overline{RU}$ and $\overline{EU}$ respectively; RU > EU.
Prove: RO > ET.

38.

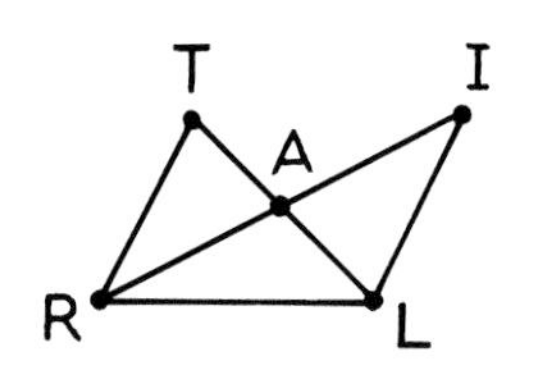

Given: △TRA ≅ △LIA.
Prove: ∠TRL > ∠I.

Set III

A bricklayer has nine bricks. Eight of the bricks weigh the same amount and one is a little heavier than the others. If the man has a balance scale, how can he find the heaviest brick in only two weighings?

Broom-Hilda © 1973 The Chicago Tribune

Lesson 2

The Exterior Angle Theorem

Suppose that an unidentified flying object is passing overhead and that it is being watched by two observers. Several of its different positions with respect to the two observers, at A and B, are shown in the diagrams.

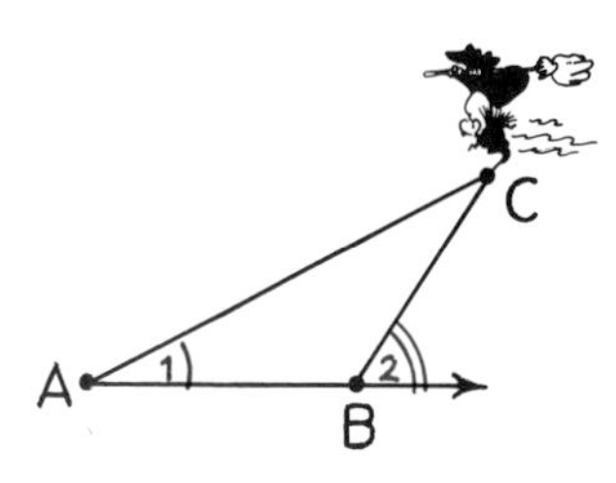

As the object moves past the observers, their lines of sight, $\overleftrightarrow{AC}$ and $\overleftrightarrow{BC}$, continually change direction. As a result, the angles that these lines of sight make with the ground, $\measuredangle 1$ and $\measuredangle 2$, are steadily changing.

The angle for observer B seems in every diagram to be larger than that for observer A. In symbols, $\angle 2 > \angle 1$. Do you think that this would continue to be true until the object had flown out of sight?

Angle 1 is an angle of the triangle ABC. It is sometimes called an *interior* angle of the triangle to distinguish it from the *exterior* angles of the triangle, of which $\measuredangle 2$ is an example.

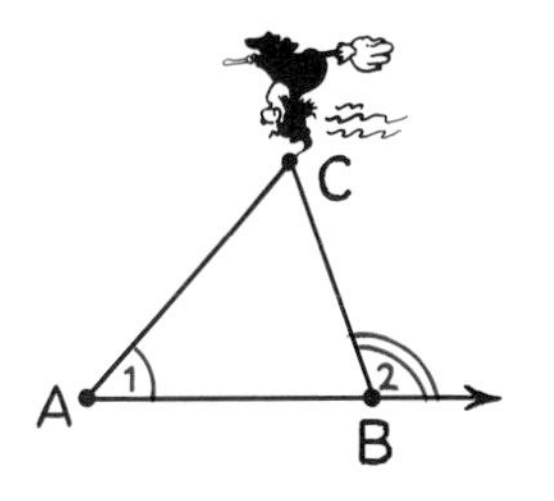

► **Definition**
An ***exterior angle of a polygon*** is an angle that forms a linear pair with one of the angles of the polygon.

The other two angles of the triangle are called *remote interior angles* with respect to the exterior angle under consideration. In the flying-object diagrams, $\measuredangle A$ and $\measuredangle C$ are the remote interior angles with respect to the exterior angle at B, $\measuredangle 2$.

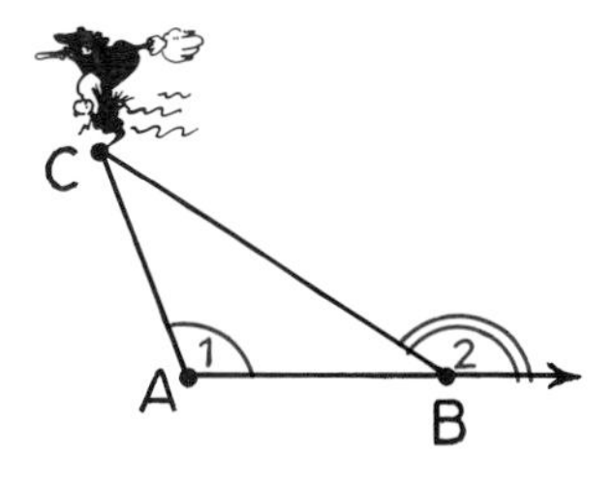

Look again at the diagrams and you will notice that it always seems that $\angle 2 > \angle A$ and that $\angle 2 > \angle C$. We can prove that these inequalities are always true; in other words, that an exterior angle of a triangle is always larger than either remote interior angle of the triangle.

► **Theorem 17** (The Exterior Angle Theorem)
An exterior angle of a triangle is greater than either remote interior angle.

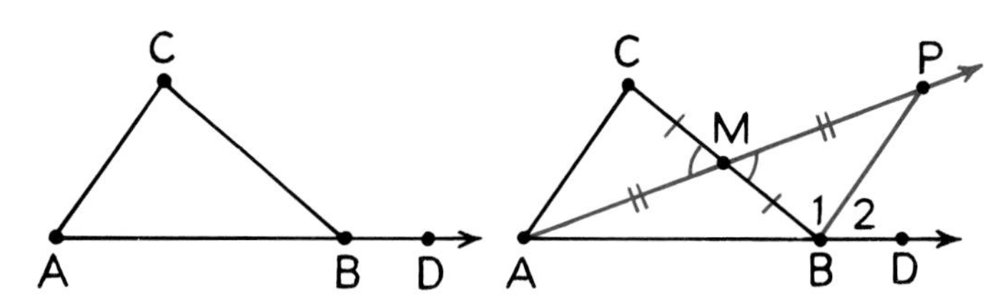

Given: $\measuredangle$CBD is an exterior angle of $\triangle$ABC.
Prove: $\angle$CBD > $\angle$A and $\angle$CBD > $\angle$C.

We will prove the theorem for just one of the two conclusions, because the proof of the other is similar. We begin by marking the midpoint of $\overline{CB}$, M, drawing a line through A and M such that MP = AM, and then drawing $\overline{BP}$. It is easy to prove that $\triangle$ACM $\cong$ $\triangle$PBM so that $\angle$C = $\angle$1. Because $\angle$CBD > $\angle$1, we can conclude that $\angle$CBD > $\angle$C.

Proof.

Statements	***Reasons***
1. $\measuredangle$CBD is an exterior angle of $\triangle$ABC.	Given.
2. Let M be the midpoint of $\overline{CB}$.	A line segment has exactly one midpoint.
3. CM = MB.	The midpoint of a line segment divides it into two equal segments.
4. Draw $\overleftrightarrow{AM}$.	Two points determine a line.
5. Choose P so that MP = AM.	The Ruler Postulate.
6. Draw $\overline{BP}$.	Two points determine a line.
7. $\angle$AMC = $\angle$BMP.	Vertical angles are equal.
8. $\triangle$ACM $\cong$ $\triangle$PBM.	S.A.S.
9. $\angle$C = $\angle$1.	Corresponding parts of congruent triangles are equal.
10. $\angle$CBD = $\angle$1 + $\angle$2.	Betweenness of Rays Theorem.
11. $\angle$CBD > $\angle$1.	The "whole greater than its part" property.
12. $\angle$CBD > $\angle$C.	Substitution.

Exercises

Set I

In the figure below, $\overrightarrow{MA}$ and $\overrightarrow{MO}$ are opposite rays.

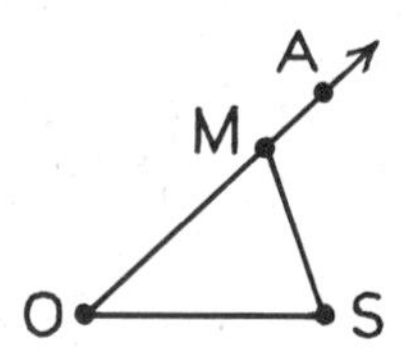

1. Name the exterior angle of the triangle shown.
2. Name the two remote interior angles with respect to that angle.
3. Write two inequalities for the figure that follow from the Exterior Angle Theorem.

In the figure below, $\measuredangle$JON and $\measuredangle$HOA are exterior angles of $\triangle$ONA.

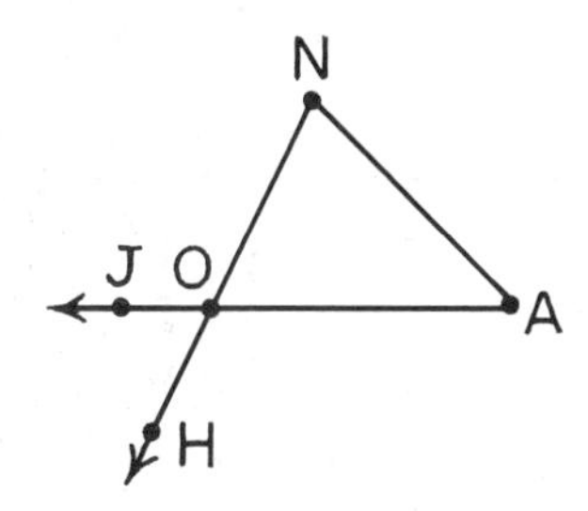

4. Name two relations that exist between $\measuredangle$JON and $\measuredangle$NOA.
5. Name two relations that exist between $\measuredangle$JON and $\measuredangle$HOA.
6. Is $\measuredangle$JOH an exterior angle of the triangle?
7. How many exterior angles does a triangle have at one vertex?
8. How many exterior angles does a triangle have in all?

In the figure below, $\measuredangle$MIH is an exterior angle of quadrilateral ICAH.

9. How many exterior angles does a quadrilateral have in all?
10. Does an exterior angle of a quadrilateral appear to be greater than any remote interior angle?

Use the three figures below to help decide whether the following statements are true or false.

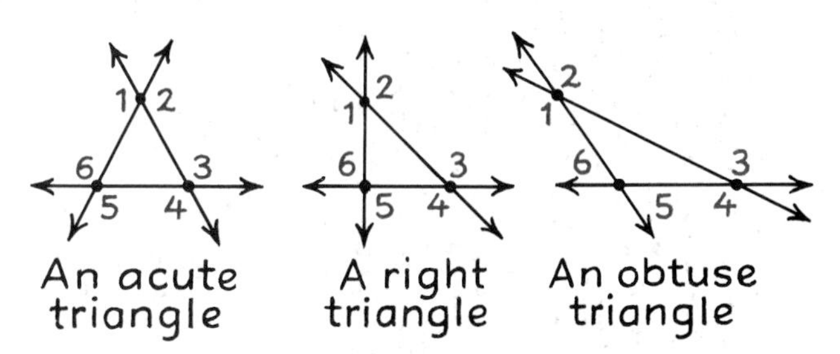

11. All six exterior angles of a triangle may be obtuse.
12. Some of the exterior angles of an obtuse triangle are acute.
13. All six exterior angles of a triangle may have different measures.
14. An exterior angle of a triangle may be smaller than one of the remote interior angles of the triangle.
15. An exterior angle of a triangle may be equal to one of the interior angles of the triangle.

In the figure below, J-A-M and S-E-M. Name all of the angles in the figure that are exterior angles of each of the following triangles. For each exterior angle, write the inequalities that follow from the Exterior Angle Theorem.

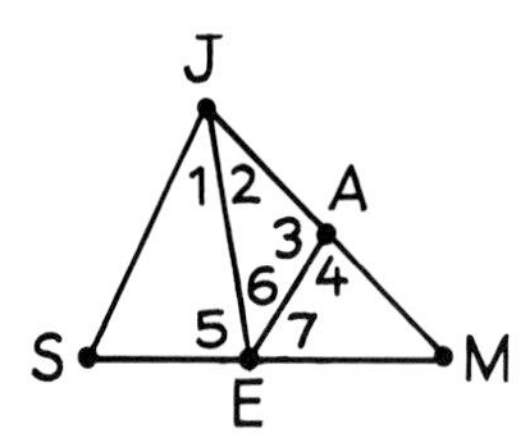

Example: △JSE.
Answer: ∡JEM; ∠JEM > ∠1 and ∠JEM > ∠S.

16. △JEA.

17. △JEM.

18. △AEM.

In the figure below, ∡JDE is an exterior angle of △JUD; ∠JDE = $(3x - 4)°$ and ∠JDU = $(x + 40)°$.

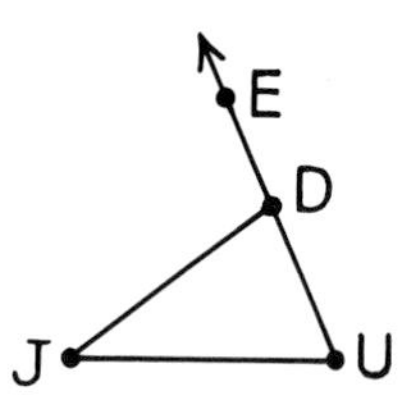

19. Find x.

20. Find ∠JDE.

21. What can you conclude about ∠U?

22. Find ∠JDU.

Give the missing statements and reasons in the following proof.

23.

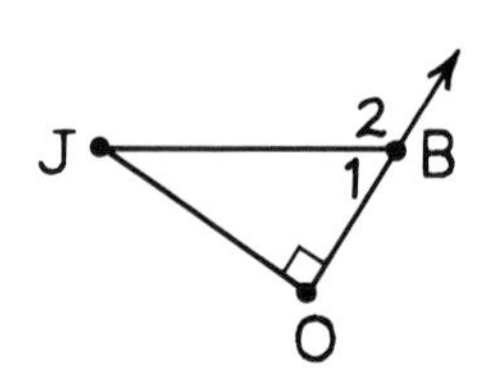

Given: ∡2 is an exterior angle of △JOB; $\overline{JO} \perp \overrightarrow{OB}$.
Prove: ∡2 is obtuse.

Proof.

Statements	***Reasons***
1. a) ________	b) ________
2. ∠2 > ∠O.	c) ________
3. d) ________	Given.
4. ∡O is a right angle.	e) ________
5. ∠O = 90°.	f) ________
6. ∠2 > 90°.	g) ________
7. h) ________	i) ________

Set II

Write complete proofs for each of the following exercises.

24.

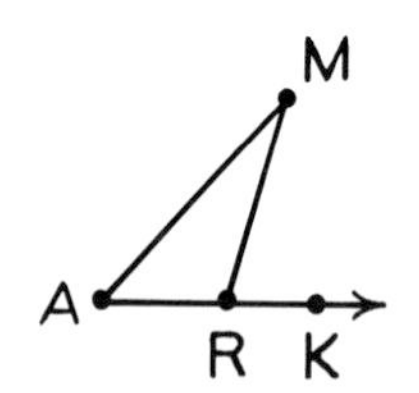

Given: ∡MRK is an exterior angle of △MAR.
Prove: ∠MRK − ∠M > 0.

25.

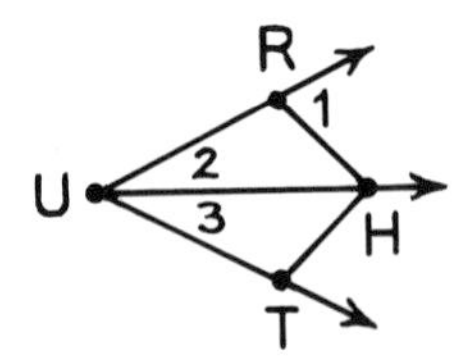

Given: ∡1 is an exterior angle of △RUH; UR = UT and RH = TH.
Prove: ∠1 > ∠3.

26.

Given: ∡1 is an exterior angle of △SEA; ∡2 is an exterior angle of △HOE.
Prove: ∠1 > ∠H.

27.

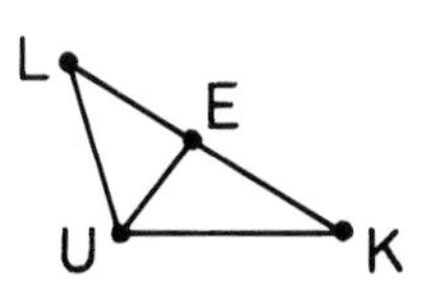

Given: $\overrightarrow{UE}$ bisects ∡LUK; ∡LEU is an exterior angle of △UKE.
Prove: ∠LEU > ∠LUE.

28.

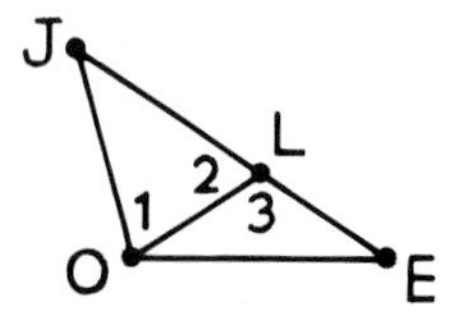

Given: JO = JL; ∡2 and ∡3 are a linear pair.
Prove: ∠1 > ∠E.

29.

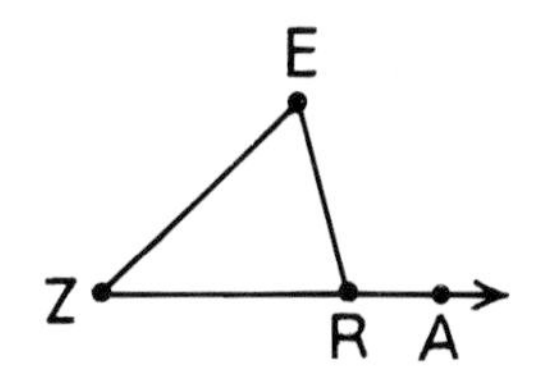

Given: ∡ERA is an exterior angle of △EZR.
Prove: 180° > ∠Z + ∠ERZ.

Set III

Proclus, a Greek philosopher and mathematician who lived in the fifth century A.D., used the Exterior Angle Theorem to prove that from one point there cannot be drawn to the same line three line segments equal in length.

He proved this by drawing the figure shown here and assuming that $\overline{AB}$, $\overline{AC}$, and $\overline{AD}$ are equal in length.

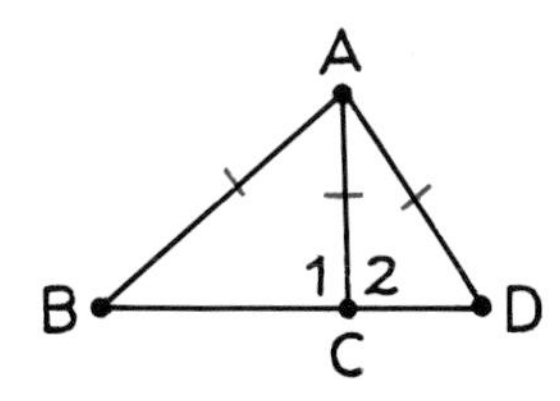

1. If AB = AC, which angles in the figure must be equal?
2. If AB = AD, which angles in the figure must be equal?
3. What conclusion follows from your answers to the previous two questions?
4. Why does this conclusion contradict the Exterior Angle Theorem?
5. What kind of proof did Proclus write?

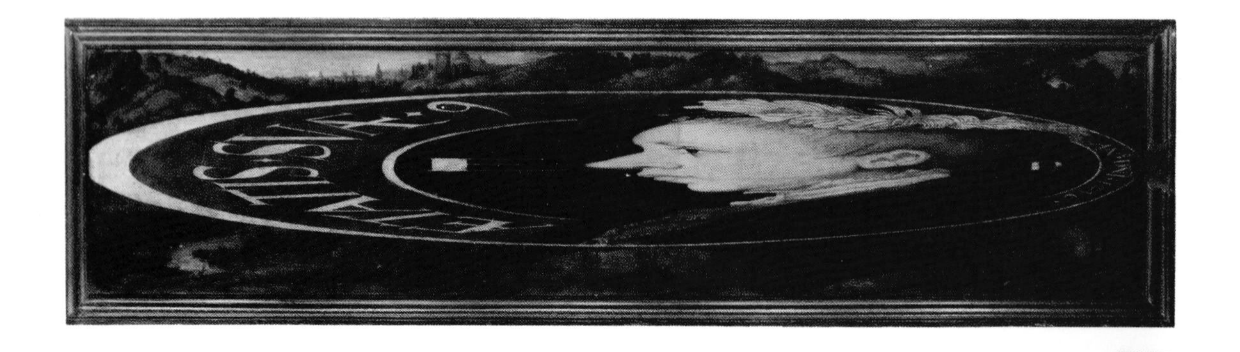

National Portrait Gallery, London

Lesson 3

Triangle Side and Angle Inequalities

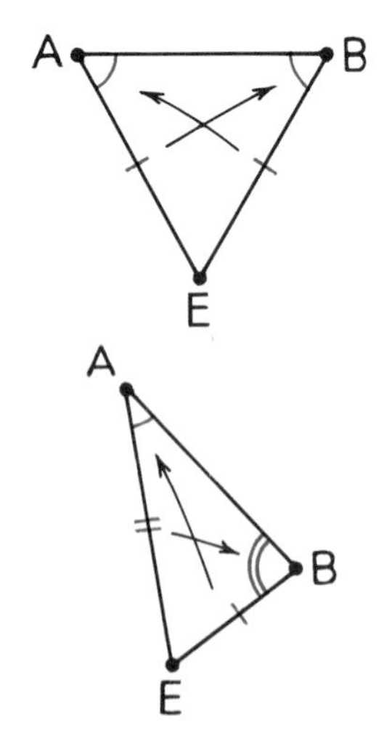

This strange-looking painting is a sixteenth-century portrait of one of the kings of England, Edward VI. It is drawn with a peculiar perspective, so that it is almost unrecognizable when viewed from the front. When viewed from the edge, however, the perspective is such that the distortion disappears and the portrait looks normal.

The diagrams at the left show the two ways of looking at the picture as seen from overhead. The top edge of the picture is represented by $\overline{AB}$ and the eyes by point E. In the first diagram, if our eyes are centered with respect to the edges of the picture, $EA = EB$. We can conclude by means of the Isosceles Triangle Theorem that $\angle A = \angle B$. So the angles formed by the picture and our lines of sight to its left and right edges are equal.

In the second diagram, on the other hand, we are looking at the picture from a point much closer to one edge than the other: $EA > EB$. How do the angles at A and B compare in measure now? Evidently, $\angle B > \angle A$. Notice that of the two angles the larger angle, $\angle B$, is opposite the longer side, $\overline{EA}$.

This relation of unequal sides and unequal angles holds true for all triangles.

► Theorem 18
If two sides of a triangle are unequal, the angles opposite them are unequal and the larger angle is opposite the longer side.

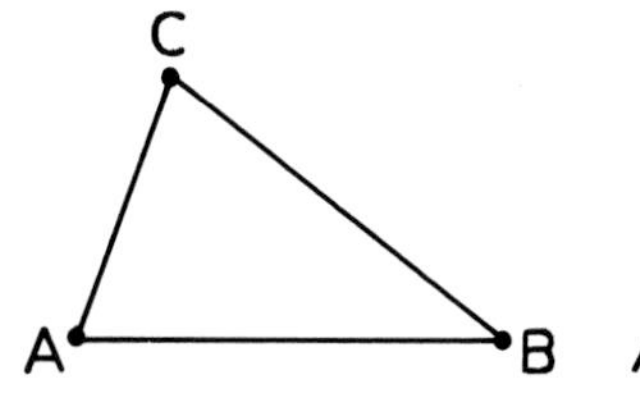

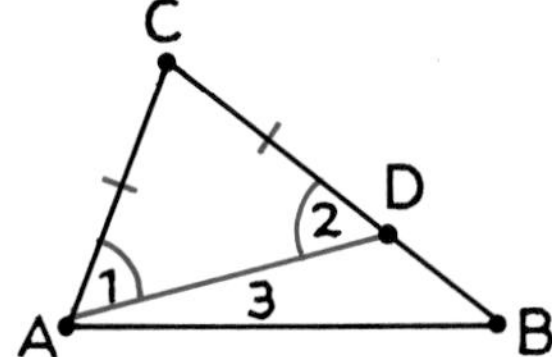

Given: $\triangle ABC$ with $BC > AC$.
Prove: $\angle A > \angle B$.

To prove this theorem, we will construct an isosceles triangle within the given triangle as shown in the second figure above. Because $\angle CAB > \angle 1$ and $\angle 1 = \angle 2$, we can conclude that $\angle CAB > \angle 2$. Also $\angle 2 > \angle B$, because $\measuredangle 2$ is an exterior angle of $\triangle ADB$. Therefore, $\angle CAB > \angle B$.

Proof.

Statements	***Reasons***
1. $\triangle ABC$ with $BC > AC$.	Given.
2. Choose point D on $\overrightarrow{CB}$ so that $CD = CA$.	The Ruler Postulate.
3. Draw $\overline{AD}$.	Two points determine a line.
4. $\angle 1 = \angle 2$.	If two sides of a triangle are equal, the angles opposite them are equal.
5. $\angle CAB = \angle 1 + \angle 3$.	Betweenness of Rays Theorem.
6. $\angle CAB > \angle 1$.	The "whole greater than its part" property.
7. $\angle CAB > \angle 2$.	Substitution.
8. $\angle 2 > \angle B$.	An exterior angle of a triangle is greater than either remote interior angle.
9. $\angle CAB > \angle B$.	Transitive property of inequality.

The converse of this theorem is also true:

► Theorem 19
If two angles of a triangle are unequal, the sides opposite them are unequal and the longer side is opposite the larger angle.

Exercises

Set I

Copy and complete the statements of the following theorems.

1. An exterior angle of a triangle ||||||||.
2. If two sides of a triangle are unequal, ||||||||.
3. If two angles of a triangle are unequal, ||||||||.

Copy and complete the statements about the following properties of inequality.

4. The transitive property: If $x < y$ and $y < z$, ||||||||.
5. The three possibilities property: Either $x < y$, ||||||||.
6. The substitution property: If $x = y$, ||||||||.
7. The "whole greater than its part" property: If $x = y + z$ and $z > 0$, ||||||||.

In the figure below, TA > EA and $\angle$A > $\angle$E.

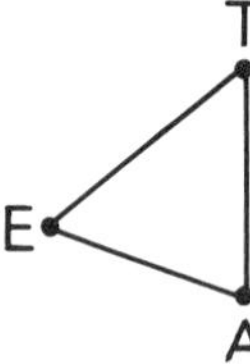

8. What can you conclude about $\angle$E and $\angle$T?
9. What can you conclude about $\angle$A and $\angle$T?
10. What can you conclude about TE and EA?

In $\angle$ALE (not shown here), AL = 4 cm, AE = 5 cm, and LE = 3 cm.

11. Which angle of the triangle must be the largest?
12. Which angle of the triangle must be the smallest?
13. Use your straightedge and compass to construct $\triangle$ALE.
14. What kind of triangle does it appear to be?

In $\triangle$RUM (not shown here), $\angle$R = 50°, $\angle$U = 30°, and $\angle$M = 100°.

15. Which side of the triangle must be the longest?
16. Which side of the triangle must be the shortest?
17. Use your ruler and protractor to draw $\triangle$RUM. (Choose any lengths you wish for the sides.)
18. What kind of triangle is it?

Theorem 19 can be proved indirectly.

If two angles of a triangle are unequal, the sides opposite them are unequal and the longer side is opposite the larger angle.

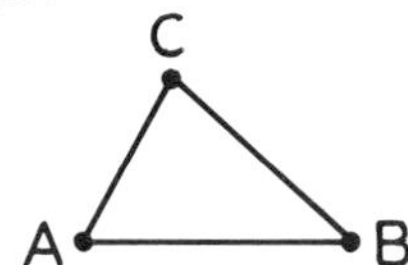

Given: $\triangle$ABC with $\angle$A > $\angle$B.
Prove: BC > AC.

19. Answer the following questions about the indirect proof of this theorem.
    a) With what assumption does the proof begin?
    b) If this assumption is true, what are the other two ways in which BC and AC might be related? (Hint: Use the Three Possibilities Property.)
    c) If BC = AC, it follows that $\angle$A = $\angle$B. Why?
    d) What does this conclusion contradict?
    e) If BC < AC, it follows that $\angle$A < $\angle$B. Why?
    f) What does this conclusion contradict?
    g) What do these contradictions tell us about the assumption with which we started the proof?
    h) What conclusion follows?

Set II

On the basis of the measures of the angles indicated, answer the following questions about the figure below.

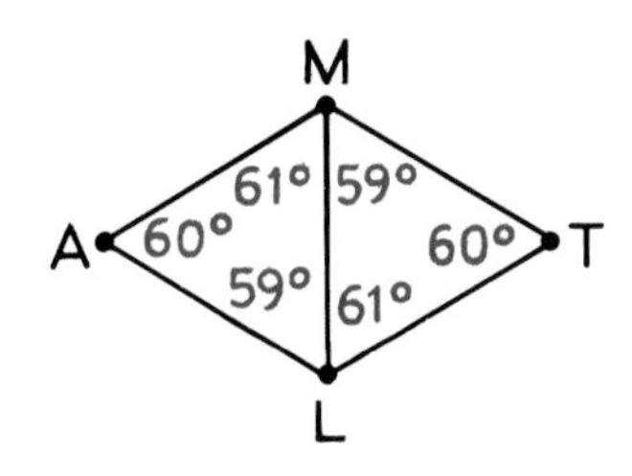

20. Are the triangles congruent?
21. Which side of $\triangle$MAL is the longest?
22. Which side of $\triangle$MLT is the longest?
23. Do these two segments necessarily have equal lengths?

On the basis of the measures of the angles indicated, answer the following questions about the figure below.

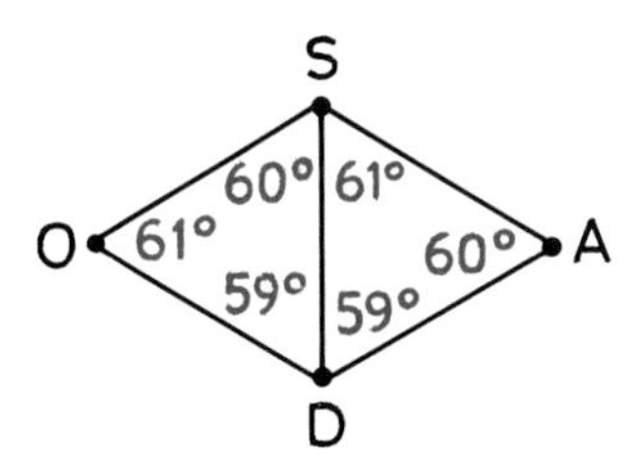

24. Which side of $\angle$SOD is the longest?
25. Which side of $\triangle$SDA is the longest?
26. Can you draw any conclusion about the relative lengths of these two segments? If so, what is it?
27. Are the triangles congruent?

Write complete proofs for each of the following exercises.

28.

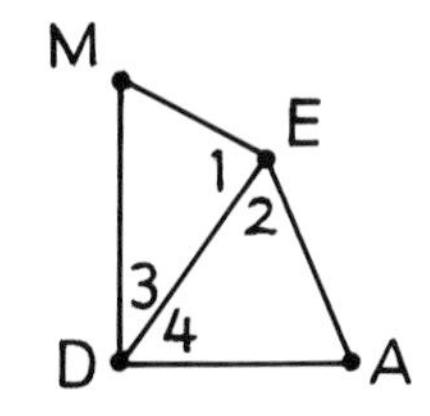

Given: $\angle 1 > \angle M$ and $\angle A > \angle 4$.
Prove: MD > EA.

29.

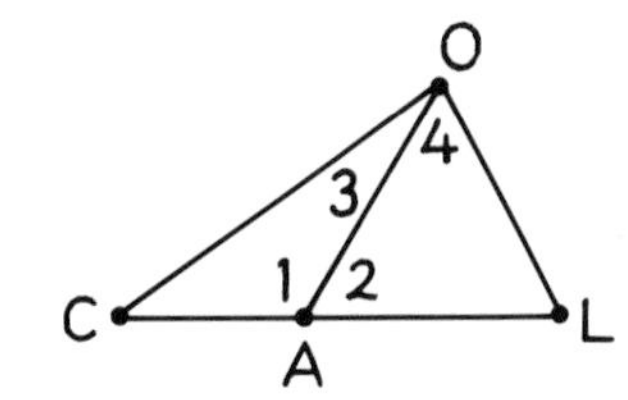

Given: $\measuredangle 1$ is an exterior angle of $\triangle$OAL; AL = OL.
Prove: $\angle 1 > \angle 2$.

30.

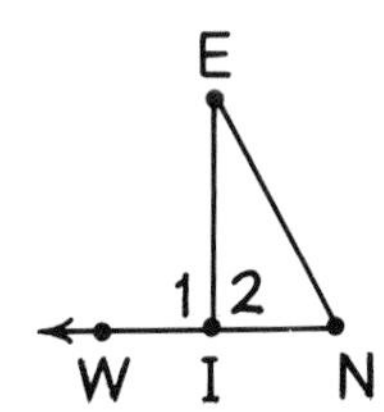

Given: $\overline{EI} \perp \overleftrightarrow{WN}$; $\measuredangle 1$ is an exterior angle of $\triangle$INE.
Prove: EN > EI.

31.

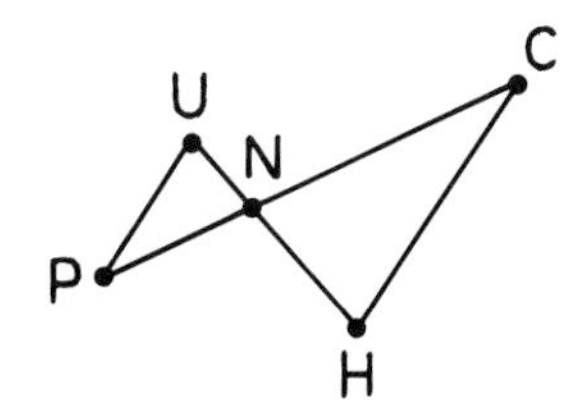

Given: $\angle U > \angle P$ and $\angle H > \angle C$; P-N-C and U-N-H.
Prove: PC > UH.

32.

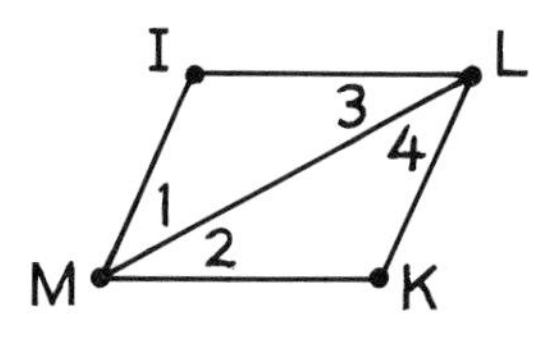

Given: MI = LK and IL = MK; $\overrightarrow{MI}$-$\overrightarrow{ML}$-$\overrightarrow{MK}$.

Prove: ∠IMK > ∠4.

33.

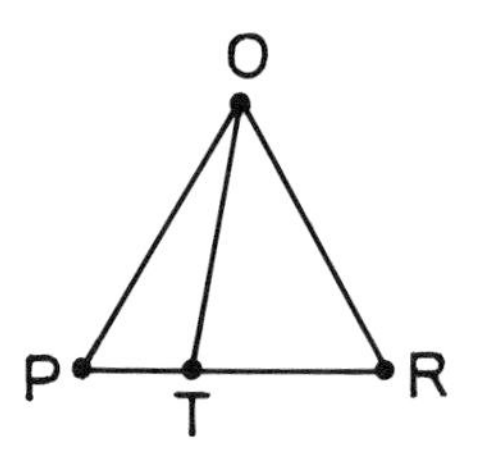

Given: △POR is equilateral; $\overrightarrow{OP}$-$\overrightarrow{OT}$-$\overrightarrow{OR}$.

Prove: OT > TR.

Set III

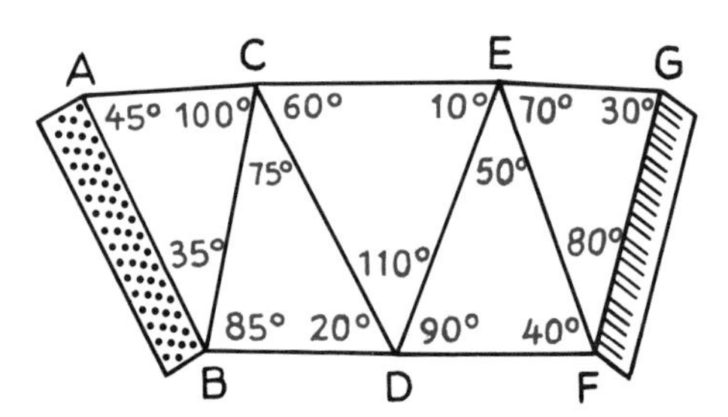

If this accordion were drawn so that the angles had the measures indicated, which segment would be the shortest? Explain the basis for your answer. (Hint: The answer is not $\overline{CD}$.)

Courtesy of the Calaveras County Jumping Frog Jubilee

Lesson 4

The Triangle Inequality Theorem

The Calaveras Frog Jump is held each spring in Angels Camp, California. Hundreds of frogs compete in trying to jump the farthest.

Suppose that a frog jumps 4 feet and then jumps 6 feet. Is it possible that it could end up 8 feet from its starting point? The figure at the right shows that the answer is yes.

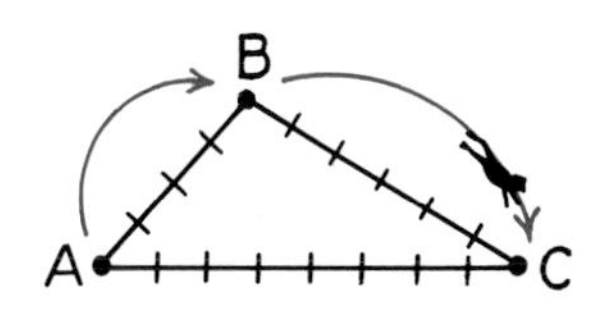

Could the frog jump 4 feet, then jump 6 feet, and end up 12 feet from its starting point? The answer to this question is no. If the frog does not change direction in making its second jump, it will land 10 feet from its starting point. If it does change direction, it will end up *less* than 10 feet from its starting point. This can be proved by means of a useful fact known as the Triangle Inequality Theorem.

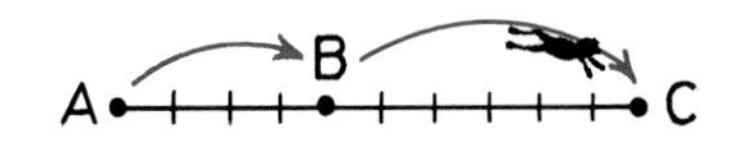

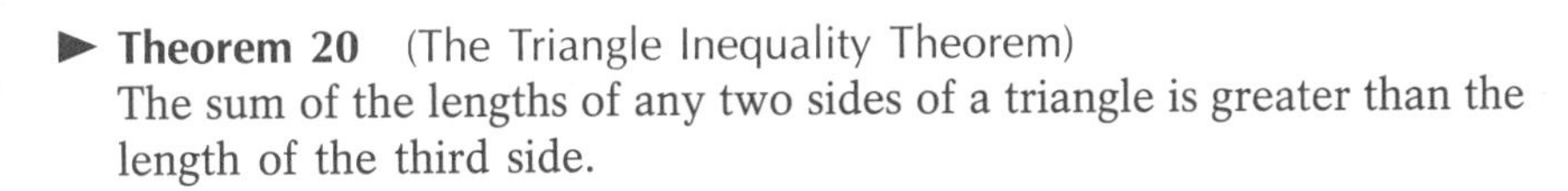

► **Theorem 20** (The Triangle Inequality Theorem)
The sum of the lengths of any two sides of a triangle is greater than the length of the third side.

The proof of this theorem, like those of the other inequality theorems that we have proved, requires the addition of some extra parts to the figure. This time we will construct an isosceles triangle next to the given triangle and then relate its equal angles to other angles in the figure.

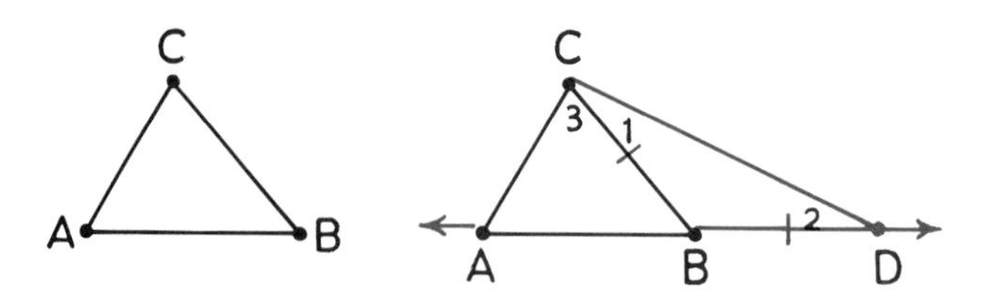

Given: △ABC.
Prove: AB + BC > AC.*

Proof.

Statements	Reasons
1. ABC is a triangle.	Given.
2. Draw $\overleftrightarrow{AB}$.	Two points determine a line.
3. Choose point D on $\overleftrightarrow{AB}$ so that BD = BC.	The Ruler Postulate.
4. Draw $\overline{CD}$.	Two points determine a line.
5. ∠1 = ∠2.	If two sides of a triangle are equal, the angles opposite them are equal.
6. ∠ACD = ∠3 + ∠1.	Betweenness of Rays Theorem.
7. ∠ACD > ∠1.	The "whole greater than its part" property.
8. ∠ACD > ∠2.	Substitution (steps 5 and 7).
9. In △ACD, AD > AC.	If two angles of a triangle are unequal, the sides opposite them are unequal and the longer side is opposite the larger angle.
10. AD = AB + BD.	Betweenness of Points Theorem.
11. AB + BD > AC.	Substitution (steps 9 and 10).
12. AB + BC > AC.	Substitution (steps 3 and 11).

*Also, AC + CB > AB and BA + AC > BC. These inequalities can be proved in the same way.

Exercises

Set I

The Triangle Inequality Theorem is true for all triangles.

1. State it as a complete sentence.

Use it to copy and complete the following inequalities for △JOE.

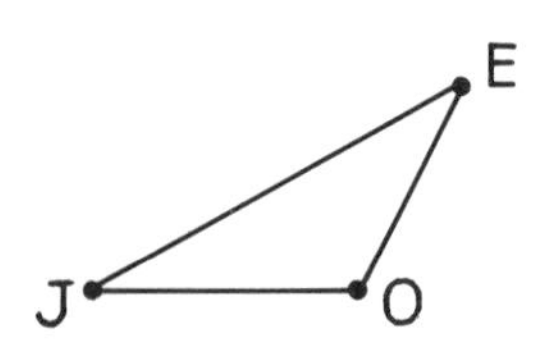

2. JO + OE > ||||||||.
3. JE + |||||||| > JO.
4. |||||||| + |||||||| > OE.

Use your ruler and compass to try to construct triangles having each of the following sets of sides. If you cannot construct a triangle, use the Triangle Inequality Theorem to explain why not.

5. △TED with TE = 7 cm, TD = 4 cm, and ED = 4 cm.

6. △JIM with JI = 5 cm, JM = 6 cm, and IM = 3 cm.

7. △KEN with KE = 8 cm, KN = 2 cm, and EN = 5 cm.

8. △ROY with RO = 3 cm, RY = 7 cm, and OY = 4 cm.

In the figure below, DA = 14, AV = 16, and DV = 20. Use this information to copy and complete the following inequalities by replacing each |||||||| with a number.

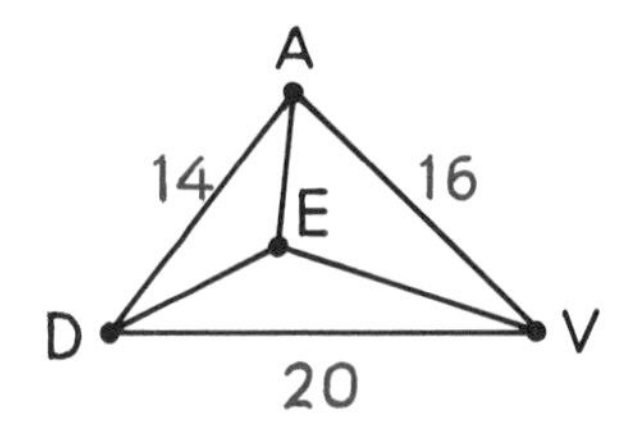

9. ED + EA > ||||||||.

10. EA + EV > ||||||||.

11. EV + ED > ||||||||.

12. (ED + EA) + (EA + EV) + (EV + ED) > ||||||||.

13. 2ED + 2EA + 2EV > ||||||||.

14. ED + EA + EV > ||||||||.

Refer to the figure below to answer each of the following questions.

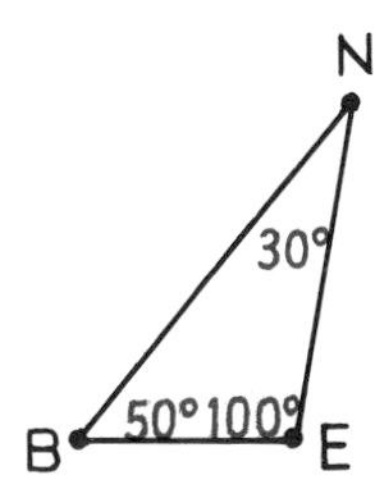

15. Can you conclude that BN < BE + EN?

16. Is it true that the length of any side of a triangle is less than the sum of the lengths of the other two sides?

17. Is it true that ∠B < ∠E + ∠N?

18. Is it true that ∠B + ∠N > ∠E?

19. Is it true that the sum of the measures of any two angles of a triangle is greater than the measure of the third angle?

Give the missing statements and reasons in the following proof.

20.

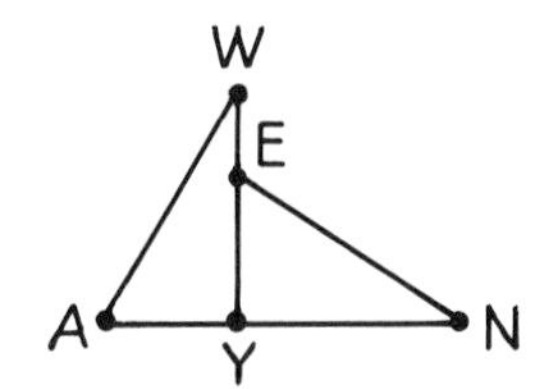

Given: △WAY ≅ △NEY.
Prove: AN > AW.

Proof.

<table>
<tr><th>Statements</th><th>Reasons</th></tr>
<tr><td>1. a) ||||||||</td><td>b) ||||||||</td></tr>
<tr><td>2. YW = YN.</td><td>c) ||||||||</td></tr>
<tr><td>3. AY + YW > AW.</td><td>d) ||||||||</td></tr>
<tr><td>4. e) ||||||||</td><td>Substitution (steps 2 and 3).</td></tr>
<tr><td>5. f) AN = |||||||| + ||||||||.</td><td>Betweenness of Points Theorem.</td></tr>
<tr><td>6. g) ||||||||</td><td>h) ||||||||</td></tr>
</table>

The following statement can be proved indirectly.

If AB + BC = AC, then
A, B, and C are collinear.

21. Answer the following questions about its proof.
    a) With what assumption does the proof begin?

b) If this assumption is true, what figure do the segments $\overline{AB}$, $\overline{BC}$, and $\overline{AC}$ form?
c) If this figure is formed, it follows that AB + BC > AC. Why?
d) What does this conclusion contradict?
e) What does this contradiction tell us about the assumption with which we started the proof?
f) What conclusion follows?

Set II

Write complete proofs for each of the following exercises.

22.

Given: $\angle O = \angle 1$.
Prove: JE + ES > JO.

23.

Given: $\angle 1 = \angle 2$ and $\angle 3 = \angle 4$.
Prove: RI + IE > EC.

24.

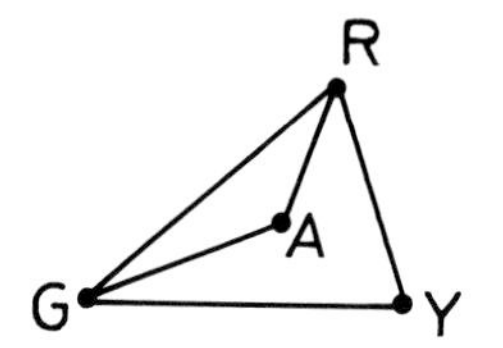

Given: $\angle Y > \angle RGY$.
Prove: GA + AR > RY.

25.

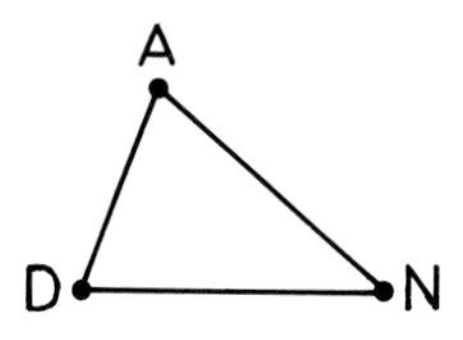

Given: In $\triangle DAN$, DN = AN.
Prove: $DN > \frac{1}{2}DA$.

26.

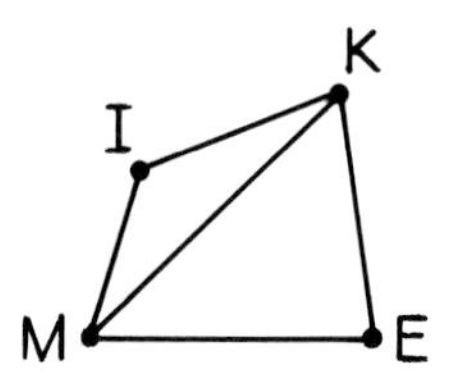

Given: $\triangle MIK$ and $\triangle MKE$.
Prove: MI + IK + KE > ME.

27.

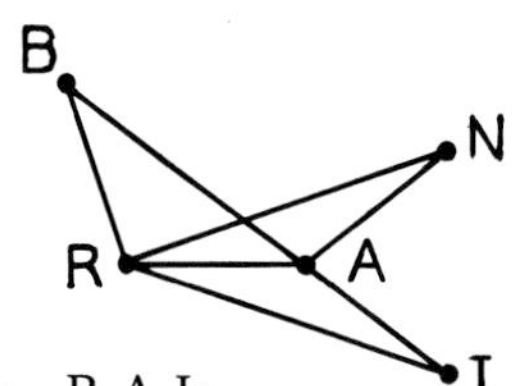

Given: B-A-I; RN = RI and AN = AI.
Prove: BR + RN > BA + AN.

Set III

The distances from Hobbiton to Bucklebury, Whitfurrows, and Frogmorton are 3, 4, and 5 kilometers, respectively. Frodo begins at Whitfurrows and drives along the outside roads to Bucklebury, Frogmorton, and back to Whitfurrows. He figures at the end of the trip that he has traveled 25 kilometers altogether.

Is this possible? Explain why or why not.

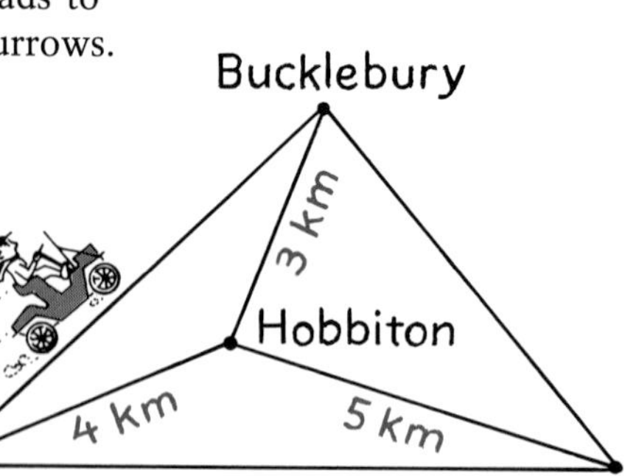

Chapter 5 / Summary and Review

Basic Ideas

Addition properties of inequality 169
Division property of inequality 169
Exterior angle of a polygon 173
Multiplication property of inequality 169
Substitution property of inequality 169
Subtraction property of inequality 169
Three possibilities property 169
Transitive property of inequality 169
"Whole greater than its part" property 169

Theorems

17. *The Exterior Angle Theorem.* An exterior angle of a triangle is greater than either remote interior angle. 174
18. If two sides of a triangle are unequal, the angles opposite them are unequal and the larger angle is opposite the longer side. 179
19. If two angles of a triangle are unequal, the sides opposite them are unequal and the longer side is opposite the larger angle. 179
20. *The Triangle Inequality Theorem.* The sum of the lengths of any two sides of a triangle is greater than the length of the third side. 183

Exercises

Set I

The figure below is an old optical illusion of inequality.

1. How does the height of the top hat (CD) *seem* to compare with the width of its brim (AB)?
2. How do they actually compare? (Measure them.)

Tell whether each of the following statements is true or false.

3. If a triangle is not equilateral, it is not equiangular.

Figure from *Illusions*, by Ebi Lanners. Copyright © 1973 by Verlag C. J. Bucher, Lauzern and Frankfurt/M. Copyright © 1977 by Thames and Hudson Ltd., London. Reprinted by the permission of Henry Holt and Company, Inc.

4. If a triangle is not isosceles, it is scalene.
5. An exterior angle of a triangle is greater than any interior angle of the triangle.
6. The length of each side of a triangle is less than the sum of the lengths of the other two sides.
7. If two sides of a triangle are unequal, the angles opposite them are unequal.
8. The smallest angle of a triangle must be opposite its shortest side.

The following exercises refer to this figure.

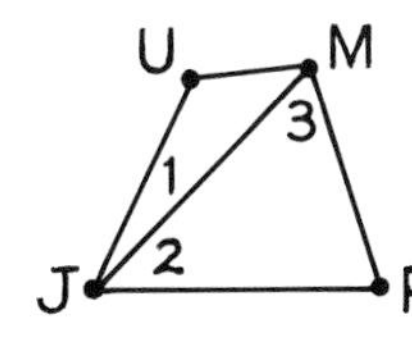

Fact 1. $\angle U > \angle UMP$.
Fact 2. $\angle P > \angle UJP$.
Fact 3. $\angle UJP = \angle 1 + \angle 2$.
Fact 4. $\angle UJP > \angle 3$.

Tell which fact (or facts) and what property of inequality can be used to prove the conclusion stated in each exercise.

9. $\angle U - \angle UMP > 0$.
10. $\angle P > \angle 3$.
11. $\angle 1 + \angle 2 > \angle 3$.
12. $\angle UJP > \angle 2$.

In the figure below, $\measuredangle$RIP is an exterior angle of $\triangle$TRI.

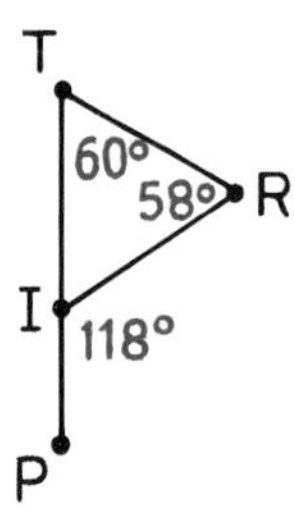

13. Find $\angle$TIR.
14. What kind of triangle is $\triangle$TRI?
15. Which side of $\triangle$TRI is longest?
16. Which side of $\triangle$TRI is shortest?

In the figure below, L-E-A, $\angle 1 > \angle 2$, and LA < LP. Use this information to tell whether each of the following statements *must be true, may be true,* or *is false.*

17. LE > EA.
18. LP = AP.
19. $\angle$LPA < $\angle$A.
20. $\triangle$LAP is equilateral.
21. LE + LP > EP.
22. $\overline{LA} \perp \overline{EP}$.
23. $\measuredangle 1$ and $\measuredangle 2$ are a linear pair.
24. $\angle 3 > \angle 1$.

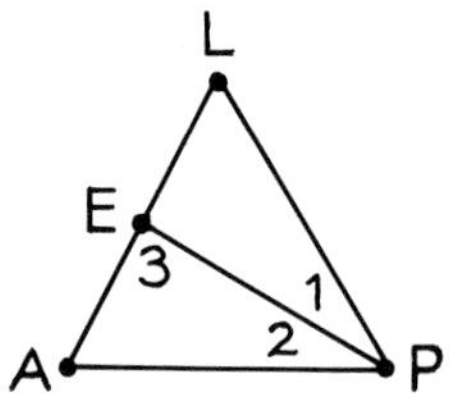

Set II

Write complete proofs for each of the following exercises.

25.

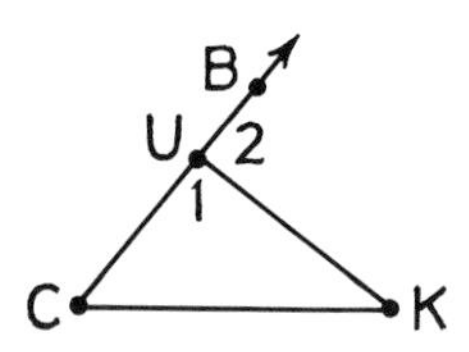

Given: $\angle 1 > \angle 2$;
$\measuredangle 2$ is an exterior angle of $\triangle$UKC.
Prove: CK > UK.

26.

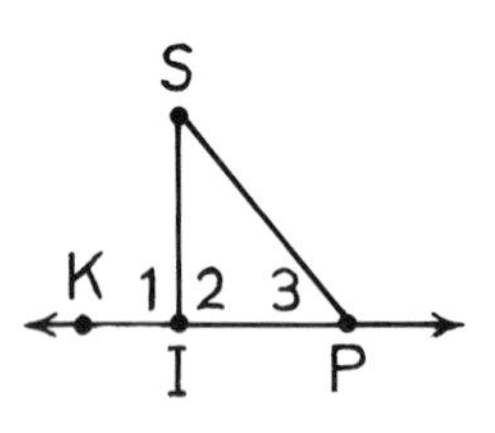

Given: $\overline{SI} \perp \overleftrightarrow{KP}$;
$\measuredangle 1$ is an exterior angle of $\triangle$SIP.
Prove: $\measuredangle 3$ is acute.

27.

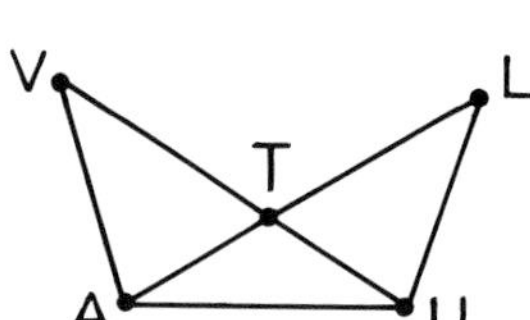

Given: $\angle$VAU = $\angle$LUA and $\angle$VUA = $\angle$LAU.
Prove: AV + AU > AL.

28.

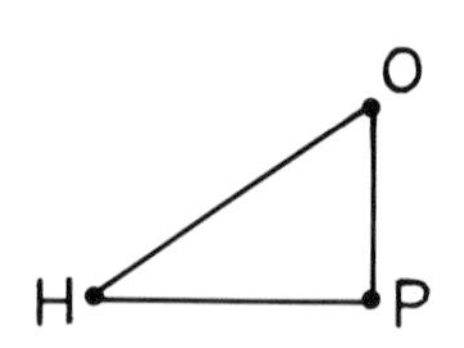

Given: HO > PO;
$\measuredangle$H and $\measuredangle$O are complementary.
Prove: $\angle P + \angle O > 90°$.

29.

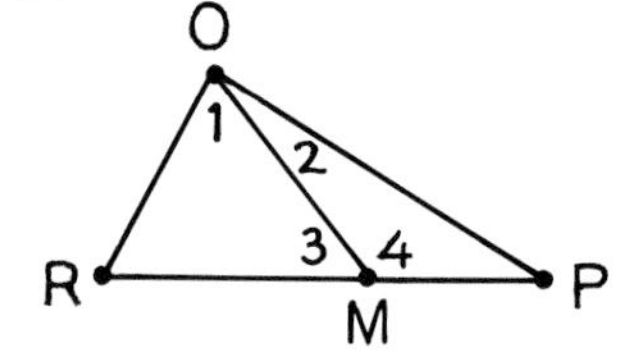

Given: RM > RO;
$\measuredangle 3$ and $\measuredangle 4$ are a linear pair.
Prove: $\angle 1 > \angle$P.

Algebra Review

Polynomials and Factoring

Adding and Subtracting Polynomials
To add or subtract two polynomials, add or subtract like terms (those having the same degree and variable).

Multiplying Polynomials
To multiply two polynomials, multiply each term of one polynomial by each term of the other and then add the resulting terms.

Factoring Polynomials
To factor a polynomial, check to see if its terms have a greatest common factor other than 1. If so, factor it out first. Then factor the remaining factor if possible.

Exercises

Add.

Example: $5x^2 + 3x - 10$ and $x^2 + 8$.

Solution:

$$\begin{array}{r} 5x^2 + 3x - 10 \\ +\ x^2 + 8 \\ \hline 6x^2 + 3x - 2 \end{array}$$

1. $5x + 9$ and $x - 8$.
2. $x - 3y$ and $x - y$.
3. $3x^2 - 2$ and $x^2 + 10$.
4. $x^2 + 6x - 1$ and $2x^2 - 4$.

Subtract the second polynomial from the first.

Example: $9x - 2y$ and $4x + y$.

Solution:

$$\begin{array}{r} 9x - 2y \\ -\ 4x + y \\ \hline 5x - 3y \end{array}$$

5. $11x + 3$ and $4x - 1$.
6. $7x - 7y$ and $x - 7y$.
7. $x^2 - 8x$ and $2x^2 + 3x$.
8. $5x^2 + 9x + 6$ and $3x^2 - x + 7$.

Multiply.

Example: $7x - 4$ and $3x + 5$.

Solution:

$$\begin{array}{r} 7x - 4 \\ \times\ 3x + 5 \\ \hline 35x - 20 \\ 21x^2 - 12x \\ \hline 21x^2 + 23x - 20 \end{array}$$

Another way to write the solution is

$$\begin{aligned} (7x - 4)(3x + 5) &= 21x^2 + 35x - 12x - 20 \\ &= 21x^2 + 23x - 20. \end{aligned}$$

9. $x + 12$ and $x + 2$.
10. $3x + 4$ and $2x - 5$.
11. $10x - 1$ and $10x - 1$.
12. $4x + 3$ and $4x - 3$.

Factor.

Example 1: $20x^2 + 36x$.
Solution: The greatest common factor of $20x^2$ and $36x$ is $4x$, and so $20x^2 + 36x = 4x(5x + 9)$.

Example 2: $x^2 + 2x - 35$.

Solution: Although x^2, $2x$ and -35 do not have a greatest common factor other than 1, we can factor x^2 into x and x and -35 into 7 and -5:

$$x^2 + 2x - 35 = (x + 7)(x - 5).$$

Check: $(x + 7)(x - 5) = x^2 - 5x + 7x - 35$
$= x^2 + 2x - 35$.

13. $6x - 21$.
14. $x^2 + 10x$.
15. $3x^3 - 2x^2$.
16. $x^2 - 49$.
17. $x^2 + 8x + 16$.
18. $x^2 + 16x + 39$.
19. $x^2 + x - 42$.
20. $2x^2 + 15x + 7$.
21. $3x^2 + x - 10$.
22. $6x^3 - 6x$.
23. $x^2 + 2xy$.
24. $9x^2 - y^2$.
25. $x^2 + 6xy + 5y^2$.

Chapter 6
PARALLEL LINES

By the permission of Johnny Hart and News America Syndicate

Lesson 1

Proving Lines Parallel

Parallel lines play an important role in Euclidean geometry. Peter's remark that they never meet is, as you may recall, the basis for our definition of such lines.

Definition

Two lines are ***parallel*** iff they lie in the same plane and do not intersect.

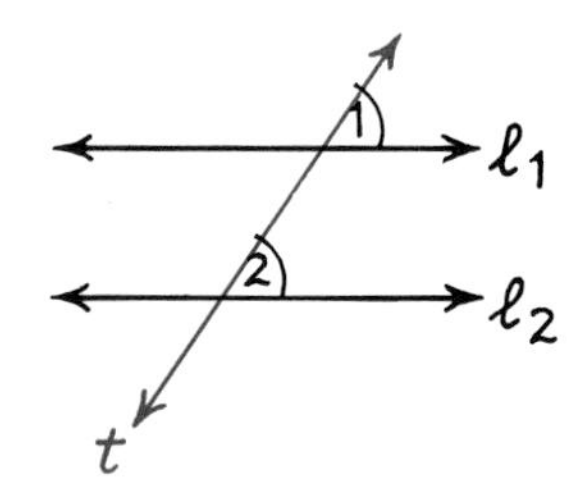

Peter is upset because he knows that parallel lines never meet and yet, from wherever he looks, it seems that they do. This predicament illustrates the fact that our definition of parallel lines is difficult to use. Because lines are infinite in extent, how can we prove that there is no point somewhere in which they intersect?

The figure at the left suggests a way to deal with this. Lines ℓ_1 and ℓ_2 appear to be parallel. A third line, t, has been drawn across them, forming a number of angles. If some of these angles are equal, such as the pair that has been numbered, then it is easy to prove that the lines must be parallel.

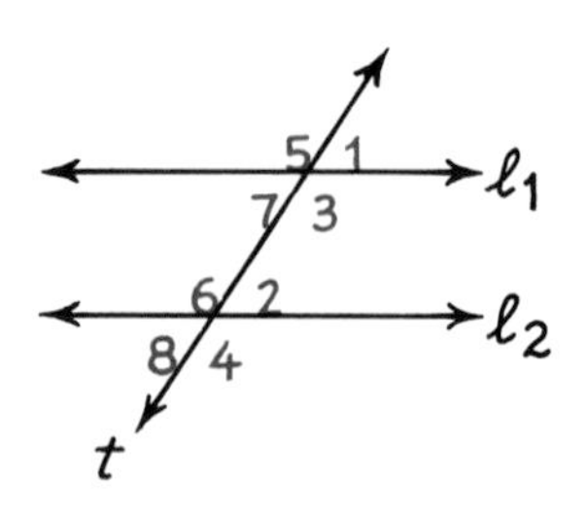

Before doing this, we will illustrate a few of the terms that we will be using. A *transversal* is a line that intersects two or more lines that lie in the same plane in different points. In the two figures shown here, line t is a transversal with respect to lines ℓ_1 and ℓ_2. A transversal that intersects two lines forms eight angles; certain pairs of these angles are given special names. In the second figure, $\measuredangle 1$ and $\measuredangle 2$ are called *corresponding angles*. Other pairs of corresponding angles are $\measuredangle 3$ and $\measuredangle 4$, $\measuredangle 5$ and $\measuredangle 6$, and $\measuredangle 7$ and $\measuredangle 8$.

Angles 7 and 2 are called *alternate interior angles*. The other pair of alternate interior angles is ∡3 and ∡6. Angles 7 and 6 are called *interior angles on the same side of the transversal*. The other pair of such angles is ∡3 and ∡2.

We have claimed that, if ℓ_1 and ℓ_2 form equal corresponding angles with transversal t, then $\ell_1 \parallel \ell_2$. We can prove this by means of the indirect method.

► **Theorem 21**
If two lines form equal corresponding angles with a transversal, then the lines are parallel.

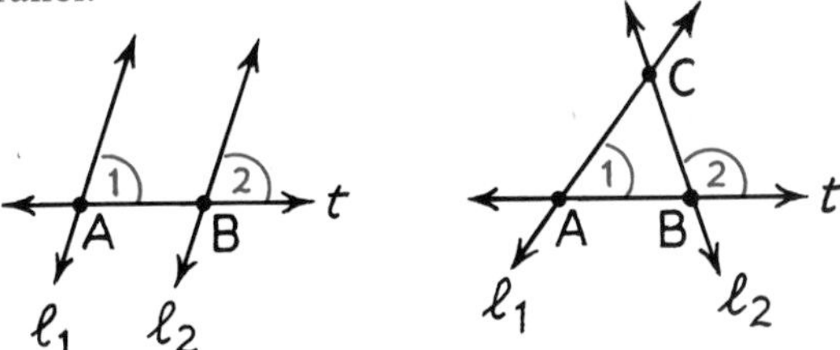

Given: Lines ℓ_1 and ℓ_2 with transversal t, $\angle 1 = \angle 2$.
Prove: $\ell_1 \parallel \ell_2$.

Remember that in an indirect proof, we begin by assuming the opposite of what we are trying to prove. To prove that the lines are parallel, we assume that they are not parallel and show that this assumption leads to a contradiction.

Proof.
Suppose that ℓ_1 and ℓ_2 are not parallel. Then they must intersect (because they lie in the same plane), and the second figure above illustrates this. If ℓ_1 and ℓ_2 intersect in some point C, then they form a triangle, $\triangle$ABC. Because ∡2 is an exterior angle of this triangle and ∡1 is a remote interior angle, $\angle 2 > \angle 1$. But this contradicts the hypothesis that $\angle 1 = \angle 2$. Therefore, our assumption that the lines are not parallel is false, and so $\ell_1 \parallel \ell_2$.

Using the angles formed by two lines and a transversal is such a convenient way to prove the lines parallel that we will prove three more theorems of this type.

► **Corollary 1**
If two lines form equal alternate interior angles with a transversal, then the lines are parallel.

► **Corollary 2**
If two lines form supplementary interior angles on the same side of a transversal, then the lines are parallel.

► **Corollary 3**
In a plane, two lines perpendicular to a third line are parallel to each other.

Exercises

Set I

To prove that lines are parallel, we usually refer to the angles that they form with a transversal.

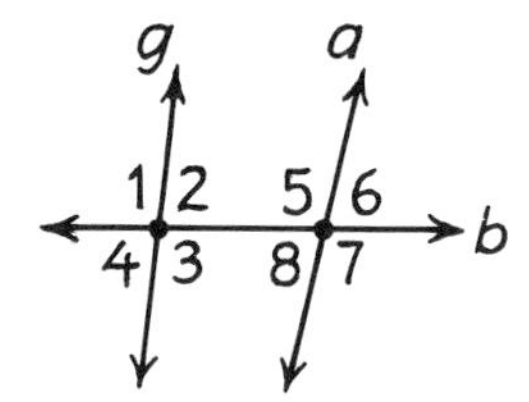

1. What is a transversal?
2. Which line in the figure above is the transversal?

One pair of corresponding angles in the figure above is $\measuredangle 1$ and $\measuredangle 5$.

3. Which angle corresponds to $\measuredangle 2$?
4. Which angle corresponds to $\measuredangle 8$?

One pair of alternate interior angles in the figure above is $\measuredangle 2$ and $\measuredangle 8$.

5. Name the other pair of alternate interior angles.

One pair of interior angles on the same side of the transversal in the figure above is $\measuredangle 2$ and $\measuredangle 5$.

6. Name the other pair of interior angles on the same side of the transversal.

In the figure below, lines y and a are intersected by lines c and k.

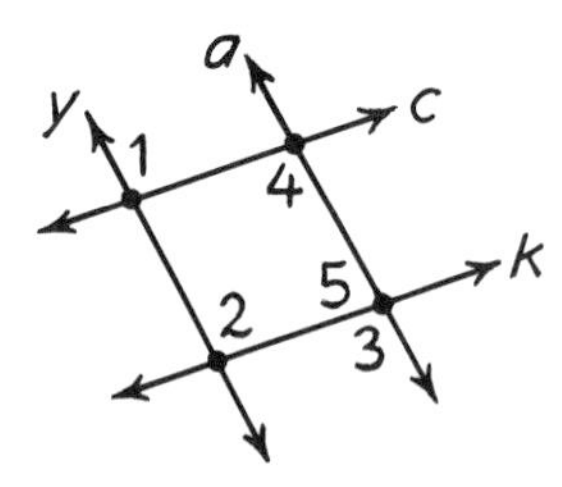

7. Name the three lines that form $\measuredangle 1$ and $\measuredangle 2$.
8. If $\angle 1 = \angle 2$, which of these lines must be parallel?
9. Why?
10. Name the three lines that form $\measuredangle 2$ and $\measuredangle 3$.
11. If $\angle 2 = \angle 3$, which of these lines must be parallel?
12. Why?
13. Name the three lines that form $\measuredangle 4$ and $\measuredangle 5$.
14. If $\measuredangle 4$ and $\measuredangle 5$ are supplementary, which of these lines must be parallel?
15. Why?

In the figure below, $\overline{AB}$ and $\overline{BE}$ intersect the sides of $\triangle WRL$. Remember that line segments are parallel iff the lines that contain them are parallel.

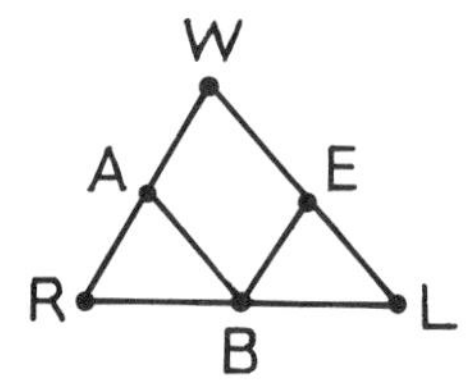

16. Which line segments in the figure are parallel if $\angle ABE = \angle BEL$?
17. Why?
18. Which line segments are parallel if $\measuredangle W$ and $\measuredangle WEB$ are supplementary?
19. Why?
20. Which line segments are parallel if $\angle R = \angle EBL$?
21. Why?

The figure below represents six lines and some of the angles that they form. If possible, name a pair of lines in the figure that must be parallel if each of the following is true.

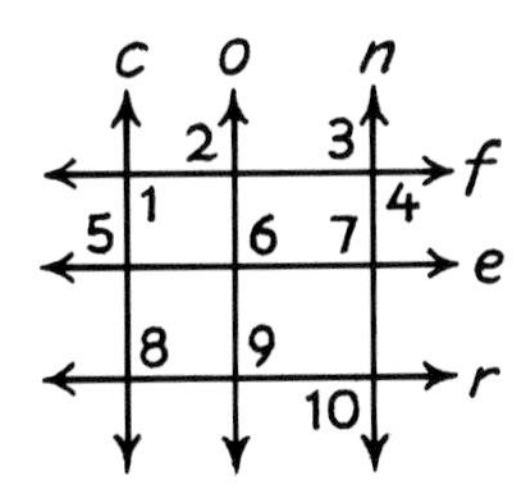

Example 1: $\angle 1 = \angle 3$.
Answer: $c \parallel n$.

Example 2: $\angle 2 = \angle 5$.
Answer: No lines.

22. $\angle 3 = \angle 7$.

23. $o \perp r$ and $r \perp n$.

24. $\angle 5 = \angle 6$.

25. $\measuredangle 1$ and $\measuredangle 8$ are supplementary.

26. $\angle 3 = \angle 4$.

27. $\measuredangle 5$ and $\measuredangle 7$ are supplementary.

28. $\angle 8 = \angle 10$.

29. $c \perp f$ and $o \perp e$.

Give the missing statements and reasons in the following proofs of the corollaries to Theorem 21.

30. *Corollary 1.* If two lines form equal alternate interior angles with a transversal, then the lines are parallel.

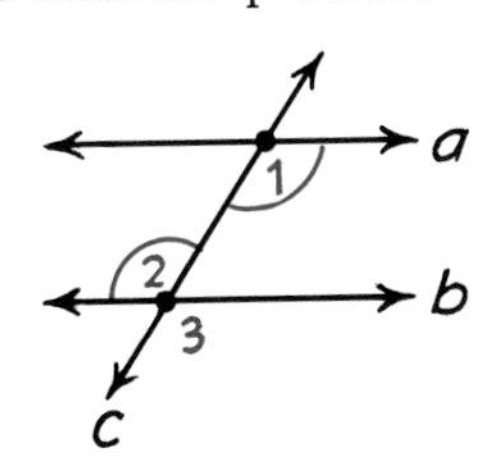

Given: Lines a and b with transversal c, $\angle 1 = \angle 2$.
Prove: $a \parallel b$.

Proof.

Statements — **Reasons**

1. Lines a and b with transversal c, $\angle 1 = \angle 2$. — a) ||||||||
2. b) |||||||| — Vertical angles are equal.
3. c) |||||||| — Transitive.
4. d) |||||||| — If two lines form equal corresponding angles with a transversal, the lines are parallel.

31. *Corollary 2.* If two lines form supplementary interior angles on the same side of a transversal, then the lines are parallel.

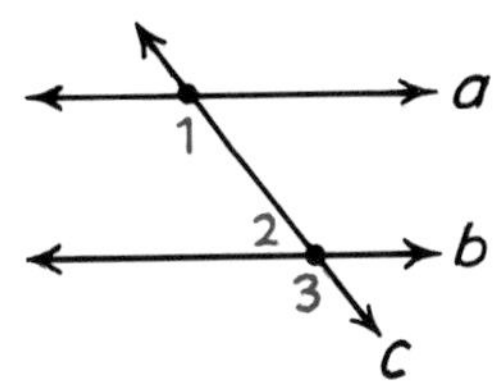

Given: Lines a and b with transversal c, $\measuredangle 1$ and $\measuredangle 2$ are supplementary.
Prove: $a \parallel b$.

Proof.

Statements — **Reasons**

1. Lines a and b with transversal c, $\measuredangle 1$ and $\measuredangle 2$ are supplementary. — a) ||||||||
2. b) |||||||| — If two angles are a linear pair, they are supplementary.
3. c) |||||||| — Supplements of the same angle are equal.
4. d) |||||||| — e) ||||||||

32. *Corollary 3.* In a plane, two lines perpendicular to a third line are parallel to each other.

Given: Lines a, b, and c lie in a plane, $a \perp c$ and $b \perp c$.
Prove: $a \parallel b$.

Proof.

Statements	Reasons
1. Lines a, b, and c lie in a plane, $a \perp c$ and $b \perp c$.	a) ▮▮▮
2. ∡1 and ∡2 are right angles.	b) ▮▮▮
3. c) ▮▮▮	d) ▮▮▮
4. e) ▮▮▮	f) ▮▮▮

Set II

33.

Given: CH = HA; $\angle 2 = \angle 3$; $\overleftrightarrow{CT}$ is a transversal of $\overline{CH}$ and $\overline{AN}$.
Prove: $\overline{CH} \parallel \overline{AN}$.

34.

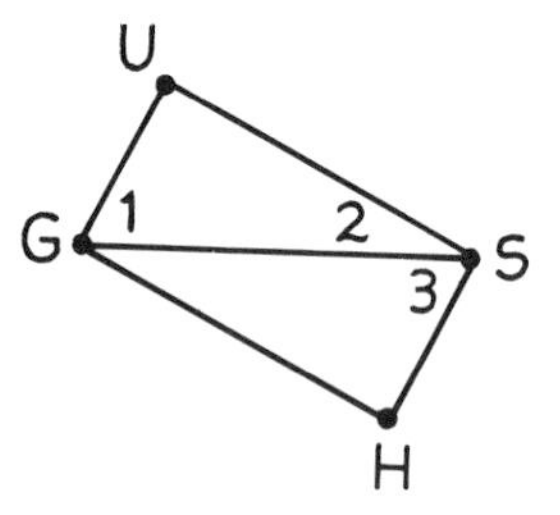

Given: ∡1 and ∡2 are complementary and ∡2 and ∡3 are complementary.
Prove: $\overline{GU} \parallel \overline{HS}$.

35.

Given: ∡A and ∡T are right angles.
Prove: $\overline{AL} \parallel \overline{TK}$.

36.

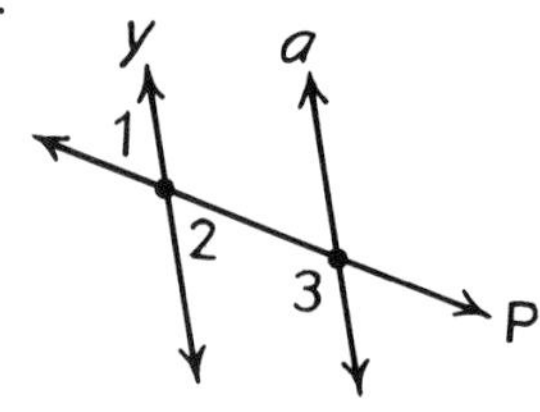

Given: ∡1 and ∡2 are vertical angles; $\angle 1 + \angle 3 = 180°$.
Prove: $y \parallel a$.

37.

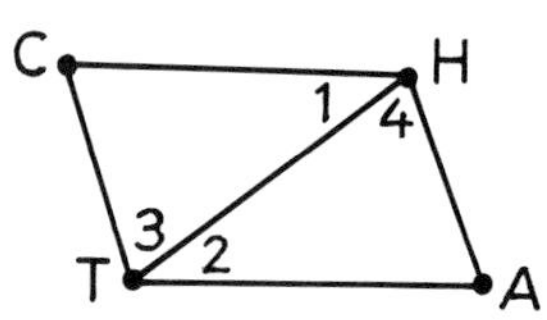

Given: CH = TA and CT = HA.
Prove: $\overline{CH} \parallel \overline{TA}$.

38.

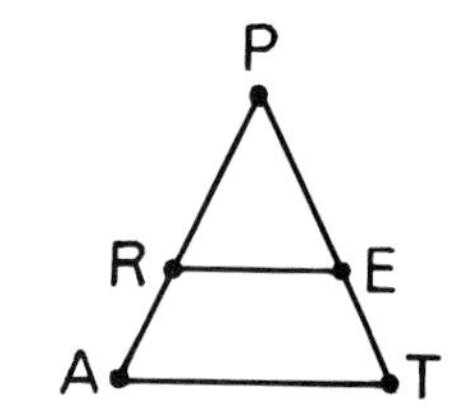

Given: PA = PT; $\angle PRE = \angle T$.
Prove: $\overline{RE} \parallel \overline{AT}$.

Set III

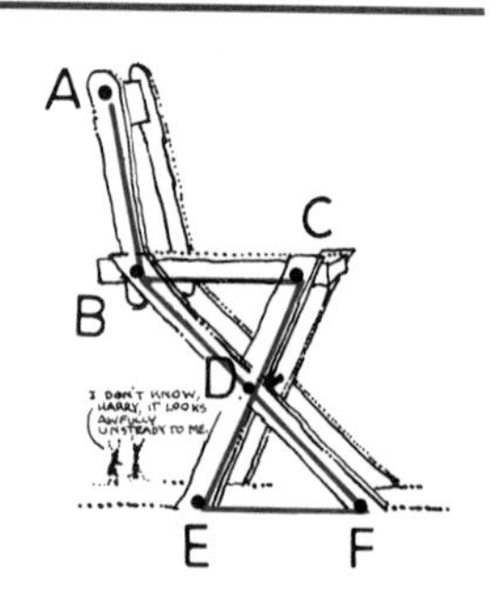

Dilcue built a folding chair in which the legs, $\overline{BF}$ and $\overline{CE}$, are attached at their midpoints as shown in this picture. Even though the legs are unequal in length, the seat of the chair, $\overline{BC}$, is parallel to the floor, $\overline{EF}$.

Why is this so?

By the permission of Johnny Hart and News America Syndicate

Lesson 2

Perpendicular Lines

The Leaning Tower of Pisa has been standing for eight hundred years, and every year it leans a little more. Built with a shallow foundation on sandy ground, it started to tip even before it was finished and, if nothing is done to prevent it, the tower will eventually collapse. Many people have come up with ideas on how to save it. One man, perhaps in the same condition as the fellow in this cartoon, decided to straighten it by putting a rope around the tower and pulling it with his car. When he drove off, he left not only the tower behind but also his bumper.

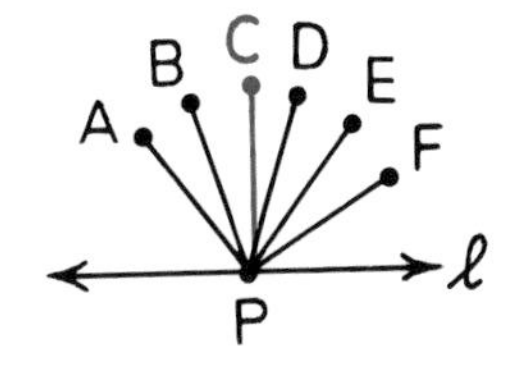

The problem, stated geometrically, is that the tower is not perpendicular to the ground. In the figure at the right, line ℓ represents the ground and the line segments represent some of the many possible positions of the tower. Notice that only one of them, $\overline{CP}$, is perpendicular to line ℓ. In other words, through a point on a line, there is exactly one line perpendicular to the line. We can prove this as a theorem.

► **Theorem 22**
In a plane, through a point on a line there is exactly one line perpendicular to the line.

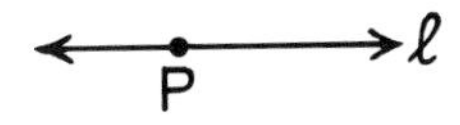

Given: Line ℓ and point P on it.
Prove: There is exactly one line through P perpendicular to ℓ.

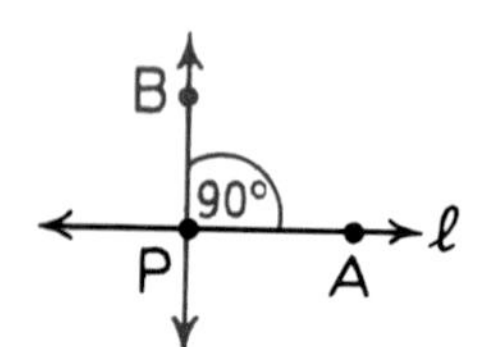

Proof.
We can choose a second point, A, on line ℓ because a line contains at least two points. Because of the Protractor Postulate, we know that there is exactly one ray, $\overrightarrow{PB}$, above line ℓ such that $\angle APB = 90°$. Because $\measuredangle APB$ has a measure of 90°, it is a right angle. So, $\overrightarrow{PB}$ (and hence, $\overleftrightarrow{PB}$) is perpendicular to line ℓ.

We can also prove that there is exactly one line perpendicular to a line through a given point *not* on it.

► **Theorem 23**
Through a point not on a line, there is exactly one line perpendicular to the line.

Given: Line ℓ and point P not on it.
Prove: There is exactly one line through P perpendicular to ℓ.

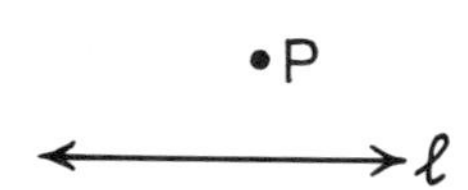

To prove this theorem, we will do two things. First, we will show that there is *at least* one perpendicular through P to the line. Second, we will show that there is *no more* than one perpendicular to the line through P.

Proof (that there is at least one perpendicular).

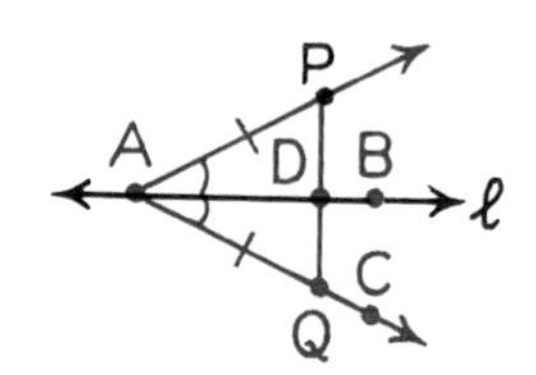

Choose two points A and B on line ℓ. Draw $\overrightarrow{AP}$ (two points determine a line). Draw $\overrightarrow{AC}$ such that $\angle CAB = \angle PAB$ (Protractor Postulate). Choose point Q on $\overrightarrow{AC}$ such that AQ = AP (Ruler Postulate). Draw $\overline{PQ}$, labeling the point in which it intersects line ℓ D. Because AQ = AP, $\angle CAB = \angle PAB$, and AD = AD (reflexive), $\triangle APD \cong \triangle AQD$ (S.A.S.). So $\angle ADP = \angle ADQ$. Because $\measuredangle ADP$ and $\measuredangle ADQ$ are also a linear pair, they are right angles (if the angles in a linear pair are equal, each is a right angle). Because $\overline{PQ}$ and line ℓ form right angles, they are perpendicular. This establishes the fact that through P there is at least one line perpendicular to ℓ.

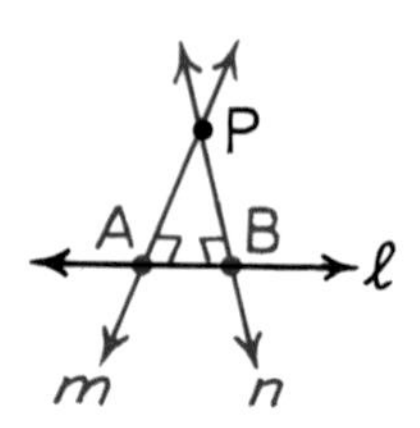

Proof (that there is no more than one perpendicular).
We will use the indirect method. Suppose that there is more than one line through P perpendicular to ℓ: for example, $\overleftrightarrow{PA} \perp \ell$ and $\overleftrightarrow{PB} \perp \ell$. Then $\overleftrightarrow{PA} \parallel \overleftrightarrow{PB}$ because, in a plane, two lines perpendicular to a third line are parallel to each other. But this contradicts the fact that $\overleftrightarrow{PA}$ and $\overleftrightarrow{PB}$ intersect at point P. Therefore, the assumption that there is more than one perpendicular is false and the conclusion is established.

We will now consider the method by which perpendicular lines can be constructed.

► **Construction 6**
To construct the line perpendicular to a given line through a point on it.

Method.

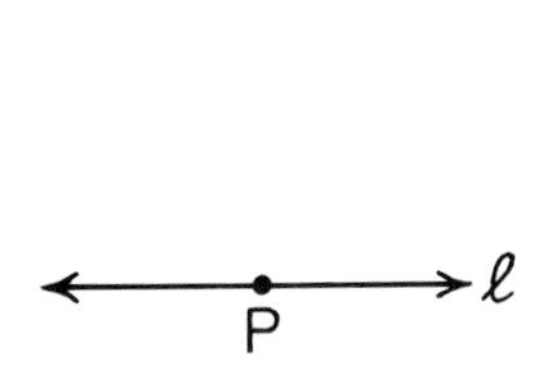

1. Let ℓ be the given line and P be a point on it.

2. With P as center, draw an arc that intersects ℓ in two points (A and B in the figure). Note that P is equidistant from A and B.

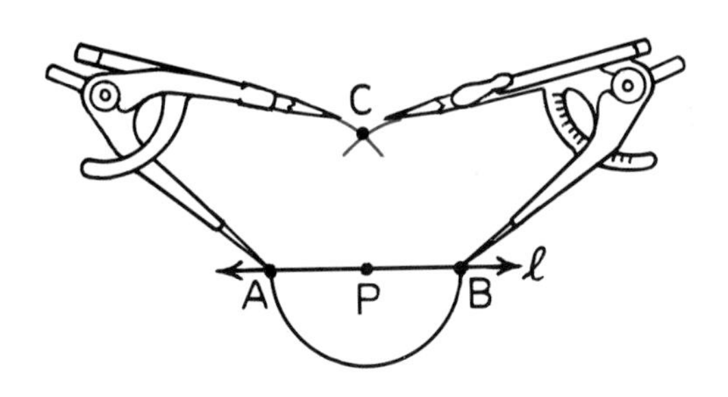

3. Use your compass to locate another point that is equidistant from A and B (C in the figure).

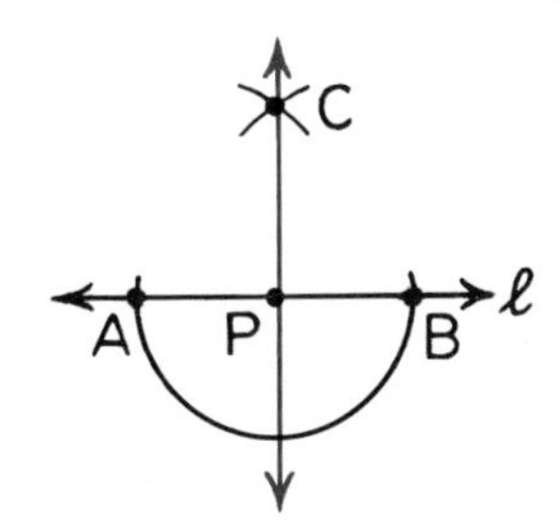

4. Draw $\overleftrightarrow{CP}$.

Because in a plane, two points each equidistant from the endpoints of a line segment determine the perpendicular bisector of the line segment, $\overleftrightarrow{CP}$ is the perpendicular bisector of $\overline{AB}$; therefore, $\overleftrightarrow{CP} \perp \ell$.

► **Construction 7**
To construct the line perpendicular to a given line through a point not on it.

Method.
Let ℓ be the given line and P be a point not on it. The steps in the procedure, illustrated below, are similar to those illustrated on page 199 for Construction 6.*

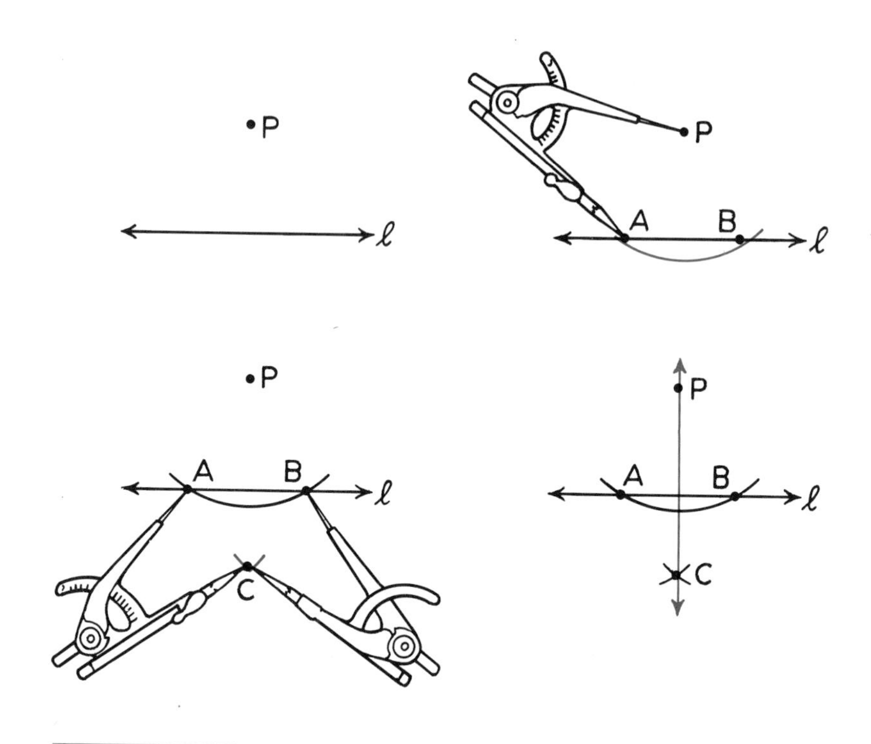

*Point C has been chosen so that it is on the other side of line ℓ from point P. It could also have been chosen on the same side.

Exercises

Set I

One way to prove that lines are parallel is to

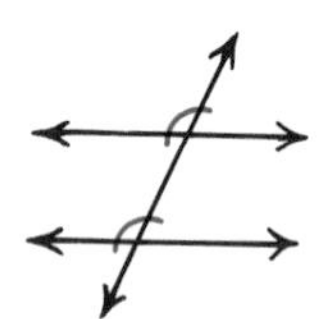

show that they form equal corresponding angles with a transversal. What other ways to prove lines parallel are illustrated by the following figures?

1. 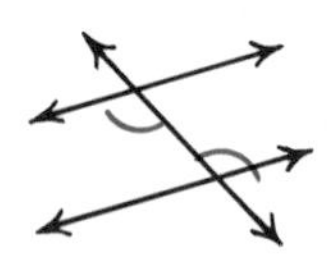

2. 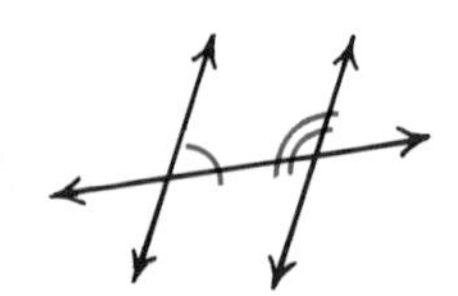

3. 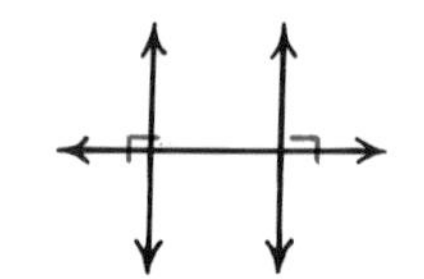

The figure below shows a wheel connected to an axle. The line of the axle is perpendicular to all of the spokes of the wheel: $\ell \perp a$, $\ell \perp b$, $\ell \perp c$, etc.

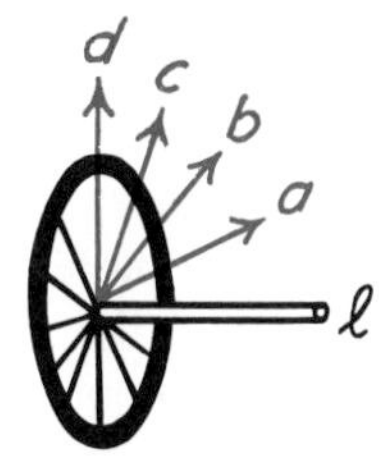

4. What does the figure suggest about the number of lines in *space* that are perpendicular to a line through a point on the line?
5. How many lines in a *plane* are perpendicular to a line through a point on the line?

Tell whether you think each of the following statements about the figure below is true or false.

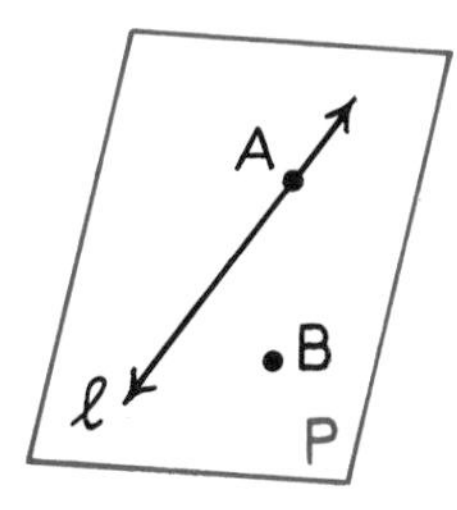

6. There is at least one line through point A that is perpendicular to line ℓ.
7. There is no more than one line through point A that is perpendicular to line ℓ.
8. There is at least one line through point B that is perpendicular to line ℓ.
9. There is no more than one line through point B that is perpendicular to line ℓ.
10. There is at least one line through A that is parallel to line ℓ.
11. There is at least one line through B that is parallel to line ℓ.

In the figure below, $\angle 1 = \angle 2$, $\measuredangle 2$ and $\measuredangle 3$ are complementary, and O-C-E. Use this information to tell whether each of the following statements *must be true, may be true,* or *appears to be false.*

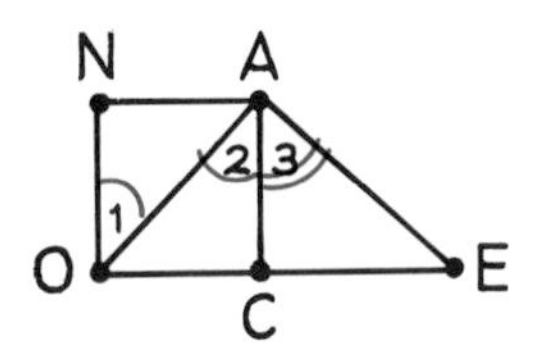

12. NA = OC.
13. AO = AE.
14. $\measuredangle$OAE is a right angle.
15. $\overline{OA} \perp \overline{AE}$.
16. $\overline{AC} \perp \overline{OE}$.
17. $\overline{ON} \parallel \overline{CA}$.
18. $\overline{NA} \parallel \overline{OC}$.

Set II

Trace this figure.

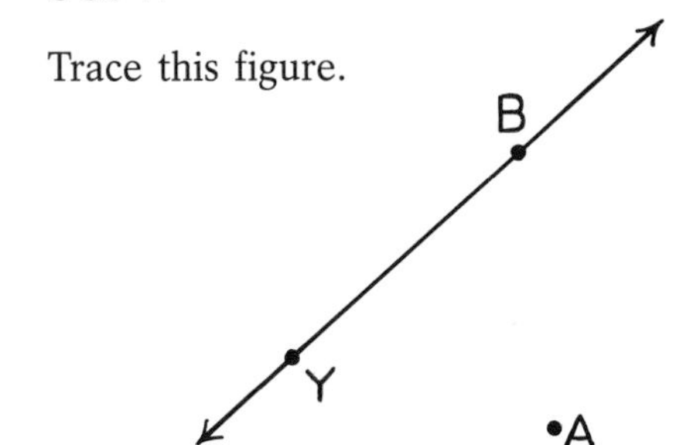

19. Use your straightedge and compass to construct two lines perpendicular to $\overleftrightarrow{BY}$: one through point B and one through point A.
20. What relation do the two lines that you have drawn have to each other?
21. Why?

Trace this figure.

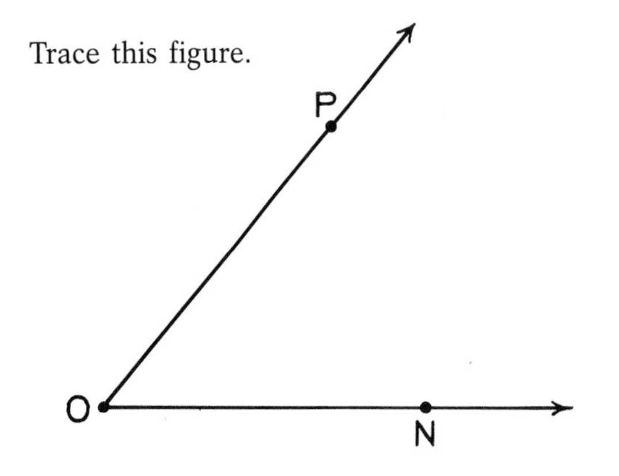

22. Use your straightedge and compass to construct a line through point P that is perpendicular to $\overrightarrow{OP}$ and a line through point N that is perpendicular to $\overrightarrow{ON}$. Label the point in which these lines intersect each other D.

23. Use your protractor to measure $\measuredangle O$ and $\measuredangle PDN$.

24. What relation do $\measuredangle O$ and $\measuredangle PDN$ seem to have?

Use your straightedge and compass to construct $\triangle SEA$ in which SE = 10 cm, SA = 9 cm, and EA = 8 cm.

25. Construct lines through the three vertices of the triangle so that each line is perpendicular to the opposite side.

26. Use your ruler to measure the distances along the lines you have constructed from the vertices to the opposite sides. Give each measurement to the nearest 0.1 cm.

27. Use your protractor to measure the angles of the triangle, each to the nearest degree.

Use your results to answer the following questions.

28. Which vertex of a triangle is farthest from the opposite side?

29. Which vertex of a triangle is closest to the opposite side?

Set III

Obtuse Ollie and Acute Alice rented a rowboat and took off from the dock as shown in the figure below. They decided to row the boat on a course such that their lines of sight back to the dock and to their favorite fishing spot on the shore were always perpendicular to each other.

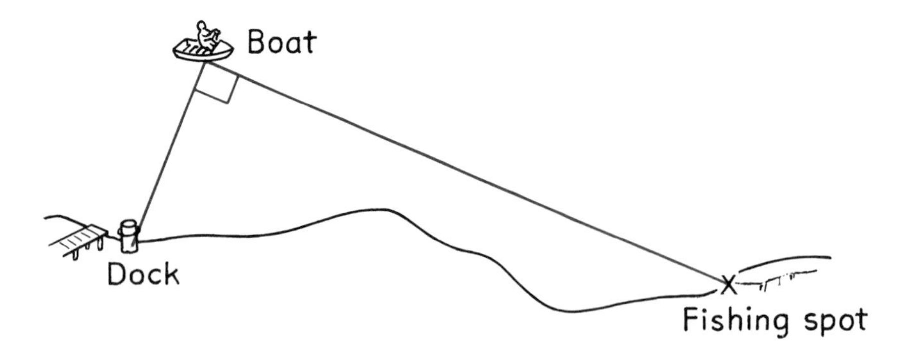

1. Trace the figure on your paper and use a right angle corner of a sheet of paper to mark the various positions of the boat as it sailed from the dock to the fishing spot. (Slide the paper around, keeping the sides of the right angle on the dock and fishing spot.)

2. What kind of a path on the lake do you think Ollie and Alice traveled?

Photograph by Friedrich Rauch in *Laughing Camera 2*, Hanns Reich Verlag

Lesson 3
The Parallel Postulate

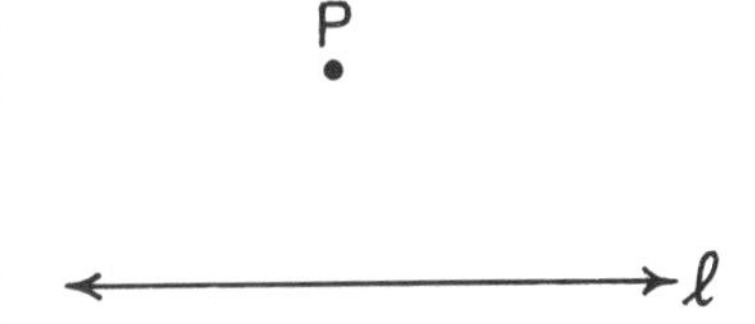

One of the most famous and, for mathematicians before the nineteenth century, most aggravating problems in geometry is about the illustration at the right. The problem concerns this question: through point P, how many lines can be drawn that are parallel to line ℓ?

We have already considered a similar question. In the previous lesson, we proved that through point P exactly one line can be drawn that is *perpendicular* to line ℓ. We did this by first proving that there is at least one perpendicular and then proving that there is no more than one. It seems reasonable to think that we should also be able to prove that through point P, exactly one line can be drawn that is *parallel* to line ℓ. To do this, we would have to first prove that there is at least one parallel and then prove that there is no more than one.

It is easy to prove that there is at least one parallel. Look at the figures below.

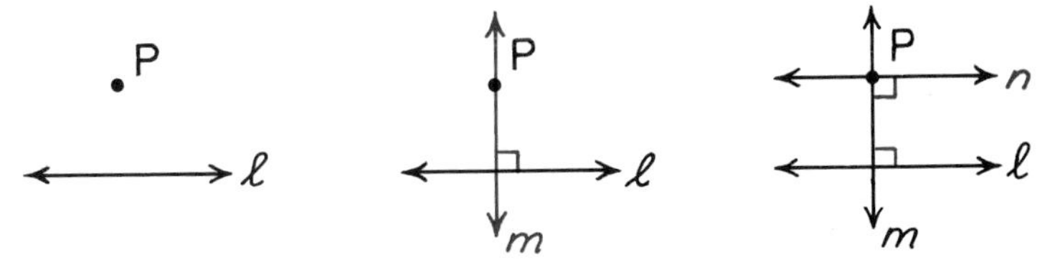

The first figure shows a line ℓ and a point P not on it. We can draw, through P, a line $m \perp \ell$ as shown in the second figure because, through a point not on a line, there is exactly one line perpendicular to the line. Then, through P, we can draw a line $n \perp m$ as shown in the third figure because, in a plane, through a point on a line there is exactly one line perpendicular to the line. It follows that $n \parallel \ell$, because in a plane two lines perpendicular to a third line are parallel to each other. Therefore, there is at least one line through P parallel to ℓ.

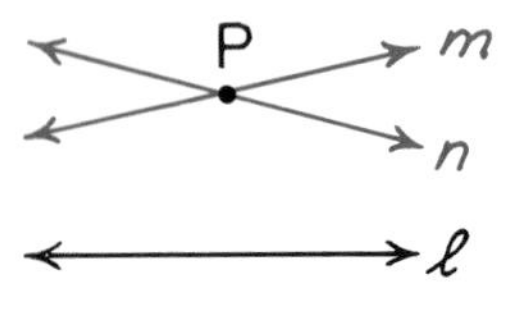

Now that we have proved that there is at least one line through P parallel to line ℓ, it seems reasonable to prove that there is no more than one. To do this, we might assume that there are two such lines parallel to the line, as the figure at the left illustrates, and show that this assumption leads to a contradiction with something we already know.

The trouble with this plan is that *we won't be able to come up with a contradiction.* That there is only one line *cannot be proved* by means of the postulates and theorems we already have.* If we wish to use the idea that there is exactly one line parallel to a line through a given point not on the line, we must assume it without proof. We will call it the Parallel Postulate.

► **Postulate 10** (The Parallel Postulate)
Through a point not on a line, there is exactly one line parallel to the line.

The parallel to a given line through a point not on it can be constructed in several different ways. One of them is described in the following construction.

► **Construction 8**
To construct the line parallel to a given line through a point not on it.

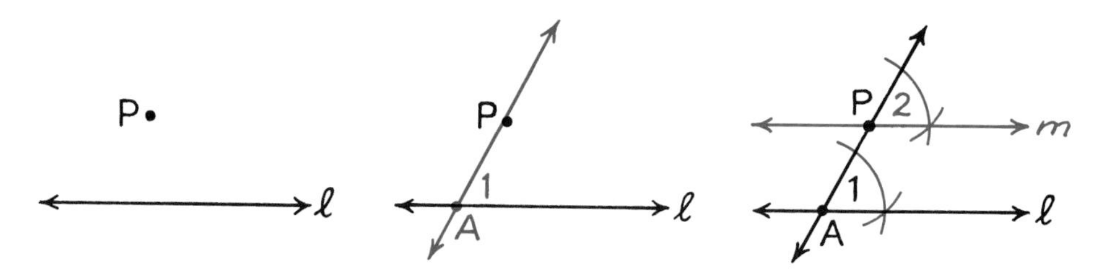

Method.
Let ℓ be the given line and P a point not on it. Choose any point A on line ℓ and draw $\overleftrightarrow{PA}$. Name one of the angles formed at A $\measuredangle 1$ as shown in the second figure. At P, construct an angle equal in measure to $\measuredangle 1$ as shown in the third figure. Name it $\measuredangle 2$. Because $\measuredangle 1$ and $\measuredangle 2$ are equal corresponding angles formed by lines m and ℓ and transversal $\overleftrightarrow{PA}$, $m \parallel \ell$.

The Parallel Postulate can be used to prove the following theorem.

► **Theorem 24**
In a plane, two lines parallel to a third line are parallel to each other.

---

*If this surprises you, you will be interested to know that it took more than two thousand years for mathematicians to realize this. The assumption that there can be more than one line through a point parallel to a given line led to one of the so-called non-Euclidean geometries that we will study later in the course.

Exercises

Set I

We now have a definition, a postulate, and several theorems about parallel lines.

1. State the definition of parallel lines.
2. State the Parallel Postulate.
3. State the theorem about corresponding angles that can be used to prove lines parallel.

The following exercises are about this statement:

If two angles are complementary, then the sum of their measures is 90°.

4. Is this statement a *definition,* a *postulate,* or a *theorem?*
5. State its converse.
6. Is its converse true?
7. If a statement is a definition, does it follow that its converse must be true?

The following exercises are about this statement:

If two angles are vertical angles, then they are equal.

8. Is this statement a *definition,* a *postulate,* or a *theorem?*
9. State its converse.
10. Is its converse true?
11. If we have proved that a statement is true, does it follow that its converse must also be true?

State or name the postulate or theorem that justifies each of the following assertions about the figure below.

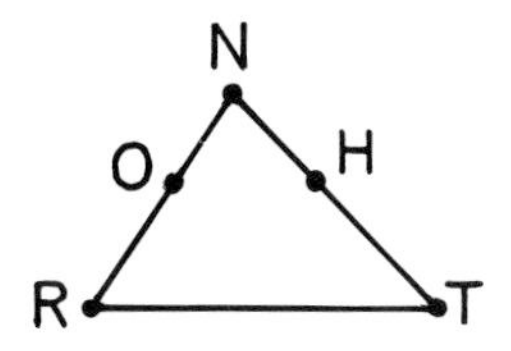

Example: It is possible to draw, through O, a line perpendicular to $\overline{RT}$.

Answer: Through a point not on a line, there is exactly one line perpendicular to the line.

12. It is possible to draw $\overleftrightarrow{OH}$.
13. It is possible to draw, through O, a line parallel to $\overline{NT}$.
14. It is possible to draw, through H, a line perpendicular to $\overline{NT}$.
15. It is possible to draw a ray $\overrightarrow{RX}$ below $\overline{RT}$ so that $\angle TRX = \angle TRN$.
16. It is possible to choose a point Y on $\overline{RT}$ so that RY = RO.
17. It is possible to choose a point Z so that it is the midpoint of $\overline{RT}$.

Theorem 24 can be proved indirectly.

18. Answer the questions included in the following proof.
    In a plane, two lines parallel to a third line are parallel to each other.

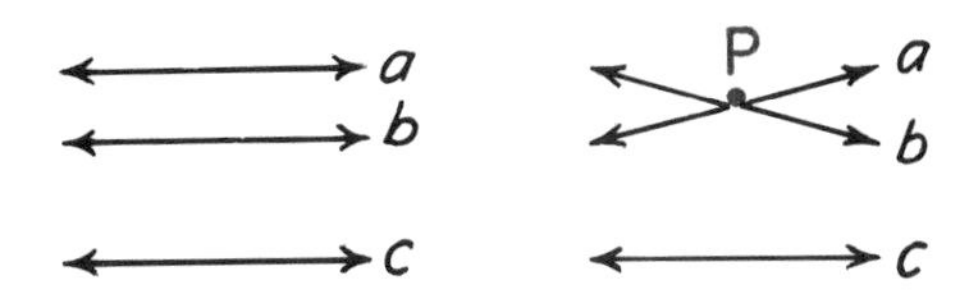

Given: Lines a, b, and c lie in a plane; $a \parallel c$ and $b \parallel c$.
Prove: $a \parallel b$.

Proof.
Suppose that a and b are not parallel.
a) The second figure illustrates what we can then conclude. What is it?
This means that two lines through the same point (P in the figure) have the same relation to line c.
b) The relation is stated in the given. What is it?
c) What fact does this contradict?
d) What does this contradiction tell us?

Set II

Trace this figure.

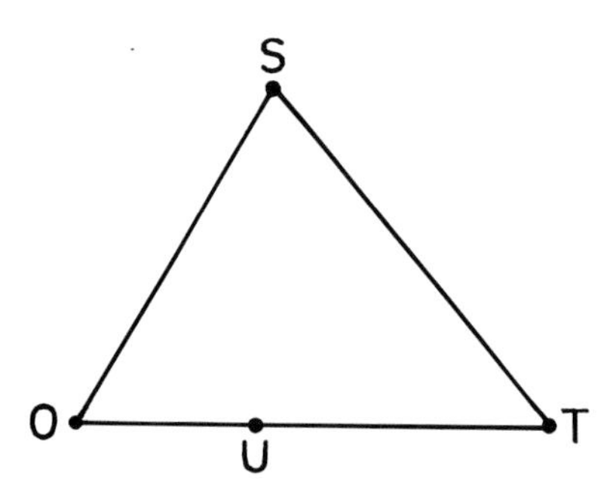

19. At point U, construct an angle that is both equal in measure to ∡O and corresponding to it with respect to transversal $\overline{OT}$. Label the point in which its other side intersects $\overline{ST}$ H.
20. What relation does $\overline{UH}$ have to $\overline{OS}$?
21. Why?

Trace this figure.

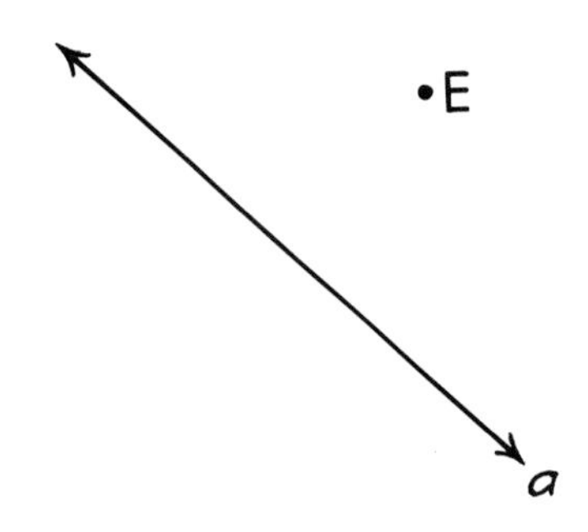

22. Construct a line through point E that is perpendicular to line *a*. Label the line *s*. Now, construct a line through point E that is perpendicular to line *s*. Label the line *t*.
23. What relation does line *t* have to line *a*?
24. Why?

Trace this figure.

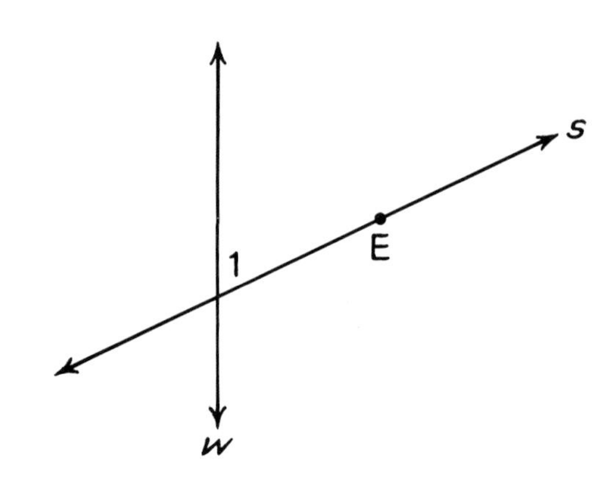

25. At point E, construct an angle that is both equal in measure to ∡1 and alternate interior to it. Label it ∡2 and label the line that contains the other side of ∡2 line *t*.
26. What relation does line *t* have to line *w*?
27. Why?

In the figure below, $a \parallel b$ and line *c* intersects line *a* at point P.

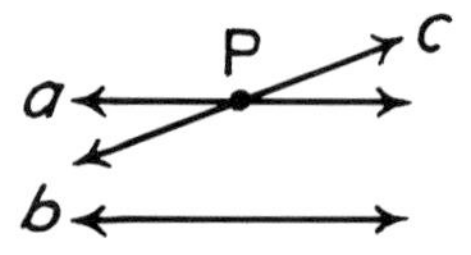

28. Does it follow that line *c* must also intersect line *b*? Explain why or why not.

Set III

In the test on this chapter, one of the problems is to construct a line through point A that is parallel to line ℓ.

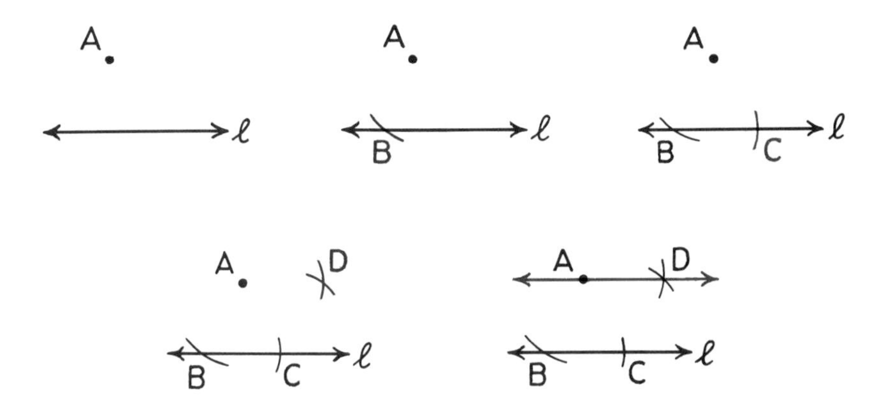

Obtuse Ollie can't remember what to do, and so he puts the metal point of his compass on A and draws an arc intersecting ℓ at B as shown in the second figure. Then, with B as center, he draws an equal arc intersecting ℓ at C as shown in the third figure. Next, with A and C as centers, he draws two more equal arcs that intersect each other at D as shown in the fourth figure. He draws $\overleftrightarrow{AD}$, hoping that it is parallel to ℓ.

Ollie's teacher says that his method does result in $\overleftrightarrow{AD} \parallel \ell$, but that he will not receive any credit unless he can explain why.

Why is Ollie's method correct? (Hint: The explanation is based on congruent triangles.)

Lesson 4

Some Consequences of the Parallel Postulate

The fact that the world is round was known to the ancient Greeks. One of them, Eratosthenes, who lived in the third century B.C., determined the earth's circumference without sailing around it or, in fact, traveling more than 500 miles. How did he do it?

The diagram below illustrates the method that Eratosthenes used. He knew that the city of Alexandria in Egypt is about 500 miles due north of the city of Syene (now called Aswan). By comparing the apparent directions of the sun in the sky as observed at the same time from each of these cities, he was able to figure out the distance around the earth.

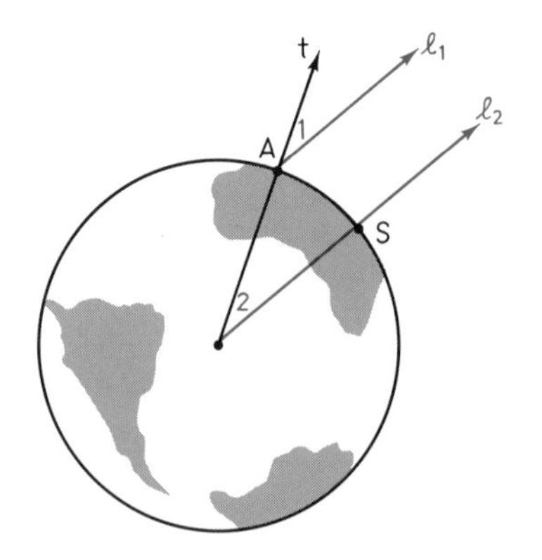

The points A and S in the diagram represent Alexandria and Syene, respectively, and the lines ℓ_1 and ℓ_2 represent the direction of the sun as seen from each city. At noon on a certain day of the year, the sun was directly overhead in Syene, as shown by ℓ_2. In Alexandria at the same time, the direction of the sun was along ℓ_1, in contrast with the overhead direction, t. Eratosthenes measured the angle between these two directions and found that it was about 7.5°. This, together with the distance between Alexandria and Syene (500 miles), is sufficient to determine the earth's circumference.

Because the sun is so far away from the earth, the lines ℓ_1 and ℓ_2 are very close to being parallel. For simplicity, Eratosthenes assumed that they were. Angles 1 and 2 are corresponding angles formed by the parallel lines ℓ_1 and ℓ_2 and the transversal t. If $\angle 1 = \angle 2$, then $\angle 2 = 7.5°$. Because 7.5° is $\frac{1}{48}$ of 360°, 500 miles (the distance from A to S along the circle) is apparently $\frac{1}{48}$ of the earth's circumference (the entire distance around the circle):

$$48 \times 500 = 24{,}000 \text{ miles.}$$

The modern value for the circumference of the earth is about 25,000 miles, so Eratosthenes' estimate was remarkably accurate. His method depends on the following fact: If two parallel lines are cut by a transversal, the corresponding angles are equal. Although it seems as if we already know this, it is the *converse* of this statement that we have proved. If we want to use it, we must prove it as a new theorem.

► **Theorem 25**
If two parallel lines are cut by a transversal, then the corresponding angles are equal.

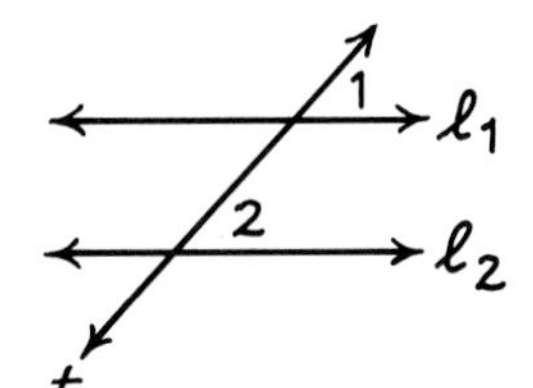

Given: Lines ℓ_1 and ℓ_2 with transversal t; $\ell_1 \parallel \ell_2$.
Prove: $\angle 1 = \angle 2$.

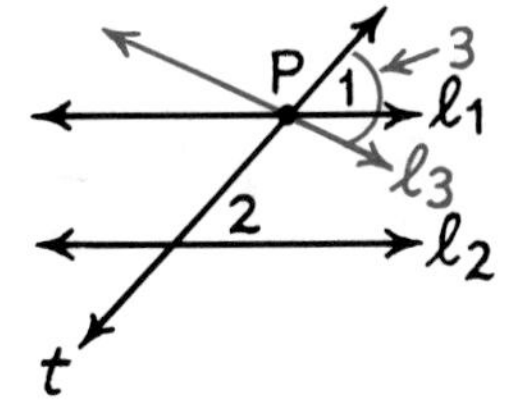

Proof.
Suppose that $\angle 1 \neq \angle 2$. Then we can draw a line ℓ_3 through P such that $\angle 3 = \angle 2$ (the Protractor Postulate). Now, $\ell_3 \parallel \ell_2$ (because two lines that form equal corresponding angles with a transversal are parallel) and $\ell_1 \parallel \ell_2$ (by hypothesis). Therefore, through point P we have two lines parallel to ℓ_2. Because this contradicts the Parallel Postulate, our assumption that $\angle 1 \neq \angle 2$ is false. Therefore, $\angle 1 = \angle 2$.

Now that we have established that parallel lines form equal corresponding angles with a transversal, it is easy to prove some more theorems about the angles formed by parallel lines. You will recognize the first two as the converses of theorems we have already proved.

► **Corollary 1**
If two parallel lines are cut by a transversal, then the alternate interior angles are equal.

► **Corollary 2**
If two parallel lines are cut by a transversal, then the interior angles on the same side of the transversal are supplementary.

► **Corollary 3**
In a plane, if a line is perpendicular to one of two parallel lines, it is also perpendicular to the other.

Exercises

Set I

In the figure below, lines a and r are cut by transversal m, and $a \parallel r$.

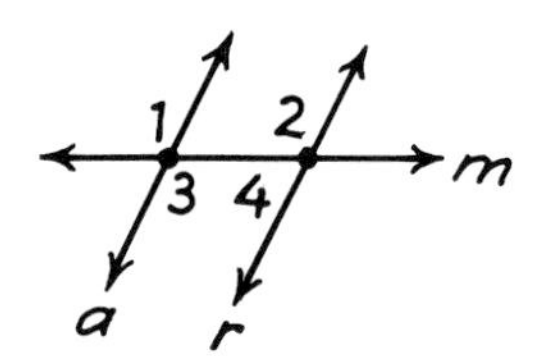

1. What are ∡1 and ∡2 called with respect to a and r and the transversal?
2. Why is $\angle 1 = \angle 2$?
3. What are ∡2 and ∡3 called with respect to a and r and the transversal?
4. Why is $\angle 2 = \angle 3$?
5. What are ∡3 and ∡4 called with respect to a and r and the transversal?
6. Why are ∡3 and ∡4 supplementary?

The following exercises refer to the figure below.

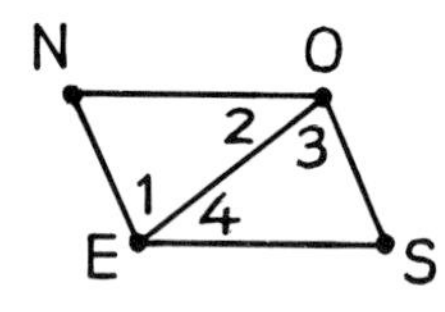

7. Which angles are equal if $\overline{NO} \parallel \overline{ES}$?
8. Which angles are equal if $\overline{NE} \parallel \overline{OS}$?
9. Which angle is supplementary to ∡NES if $\overline{NO} \parallel \overline{ES}$?
10. Which angle is supplementary to ∡N if $\overline{NE} \parallel \overline{OS}$?

In quadrilateral CHIN, $\overline{CN} \parallel \overline{HI}$. Also, $\angle C = (6x + 1)°$, $\angle H = (9x - 1)°$, and $\angle I = 5(x + 3)°$.

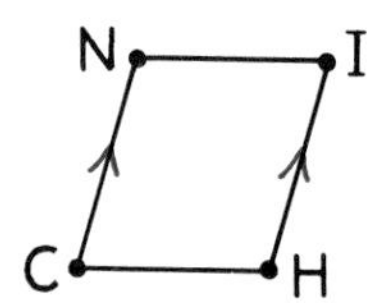

11. Find x.
12. Find $\angle C$.
13. Find $\angle H$.
14. Find $\angle I$.
15. Is $\overline{CH} \parallel \overline{NI}$?

In $\triangle MUH$, $\overline{OT} \parallel \overline{MH}$. Also, $\angle M = (x + 80)°$, $\angle H = (25 - 2x)°$, and $\angle TOU = (\text{-}4x)°$.

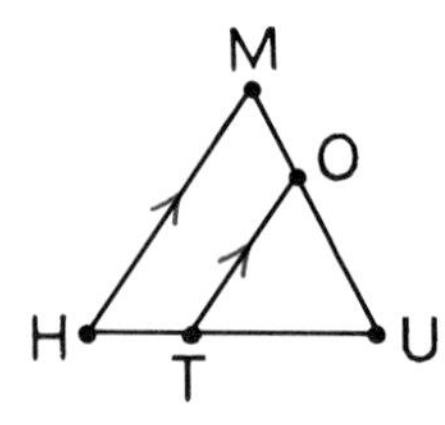

16. Find x.
17. Find $\angle M$.
18. Find $\angle H$.
19. Find $\angle TOU$.
20. Find $\angle OTU$.

Trace this figure.

21. Construct, through point E, a line parallel to $\overline{HC}$. Label the point in which it intersects $\overline{CS}$ T.
22. Name the angle in your figure that is equal in measure to $\measuredangle$C.
23. State the theorem in this lesson that is the basis for your answer.
24. Name the angle in your figure that is supplementary to $\measuredangle$H.
25. State the theorem in this lesson that is the basis for your answer.

Set II

Give the missing statements and reasons in the following proofs of the corollaries to Theorem 25.

26. *Corollary 1.* If two parallel lines are cut by a transversal, then the alternate interior angles are equal.

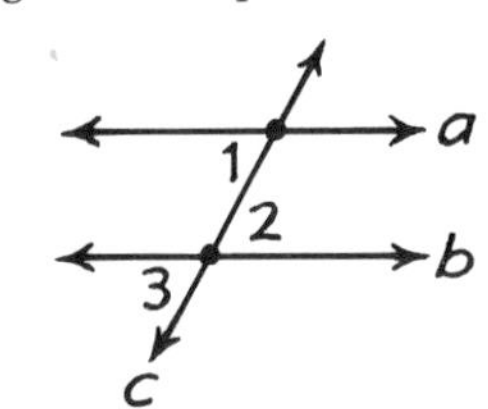

Given: Lines a and b with transversal c, $a \parallel b$.
Prove: $\angle 1 = \angle 2$.

Proof.

Statements	Reasons
1. Lines a and b with transversal c, $a \parallel b$.	a) ________
2. b) ________	If two parallel lines are cut by a transversal, the corresponding angles are equal.
3. c) ________	Vertical angles are equal.
4. d) ________	e) ________

27. *Corollary 2.* If two parallel lines are cut by a transversal, then the interior angles on the same side of the transversal are supplementary.

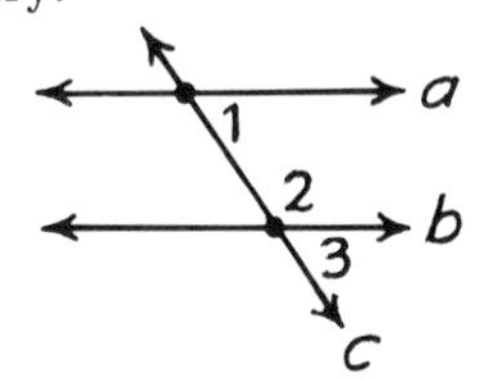

Given: Lines a and b with transversal c, $a \parallel b$.
Prove: $\measuredangle 1$ and $\measuredangle 2$ are supplementary.

Proof.

Statements	Reasons
1. Lines a and b with transversal c, $a \parallel b$.	a) ________
2. b) ________	If two parallel lines are cut by a transversal, the corresponding angles are equal.
3. c) ________	If two angles are a linear pair, they are supplementary.

4. $\angle 2 + \angle 3 = 180°$.	d) ________
5. e) ________	Substitution.
6. f) ________	g) ________

28. *Corollary 3.* In a plane, if a line is perpendicular to one of two parallel lines, it is also perpendicular to the other.

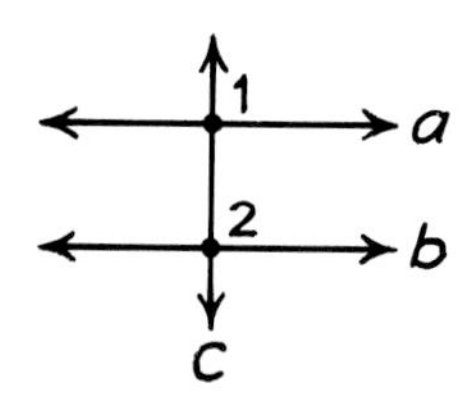

Given: $a \parallel b$ and $c \perp a$.
Prove: $c \perp b$.

Proof.

Statements	**Reasons**
1. $a \parallel b$.	a) ________
2. b) ________	If two parallel lines are cut by a transversal, the corresponding angles are equal.
3. $c \perp a$.	c) ________
4. d) ________	Perpendicular lines form right angles.
5. $\angle 1 = 90°$.	e) ________
6. f) ________	Substitution (steps 2 and 5).
7. g) ________	An angle whose measure is 90° is a right angle.
8. h) ________	i) ________

29.

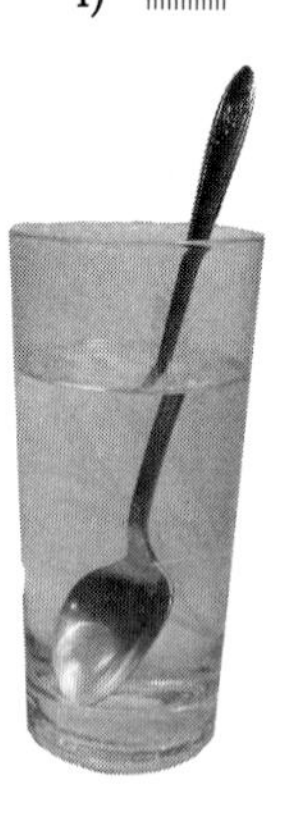

When a spoon is put in a glass of water, it appears to be bent. This is because light rays are bent when they go from water into air. Parallel light rays are bent by the same amount; therefore, in the figure below, if $\overline{AB} \parallel \overline{DE}$, then $\angle ABC = \angle DEF$.

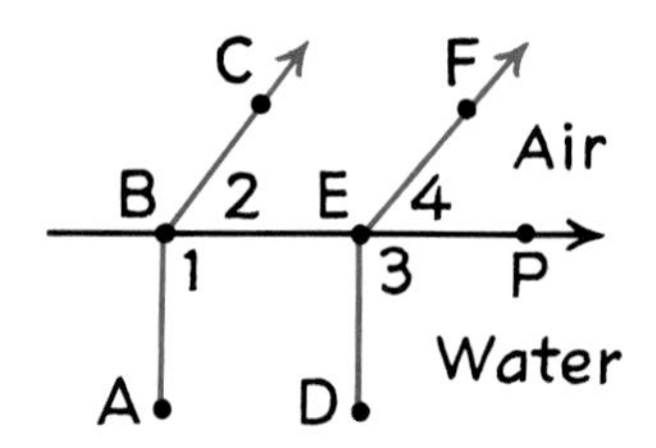

Refer to this information and the figure to show why light rays that are parallel in the water are still parallel when they emerge into the air.

a) Why is $\angle ABC = \angle 1 + \angle 2$ and $\angle DEF = \angle 3 + \angle 4$?
b) Given that $\angle ABC = \angle DEF$, why does it follow that $\angle 1 + \angle 2 = \angle 3 + \angle 4$?
c) Given that $\overline{AB} \parallel \overline{DE}$, why is $\angle 1 = \angle 3$?
d) Why does it follow from *b* and *c* that $\angle 2 = \angle 4$?
e) Why does it follow that $\overline{BC} \parallel \overline{EF}$?

30.

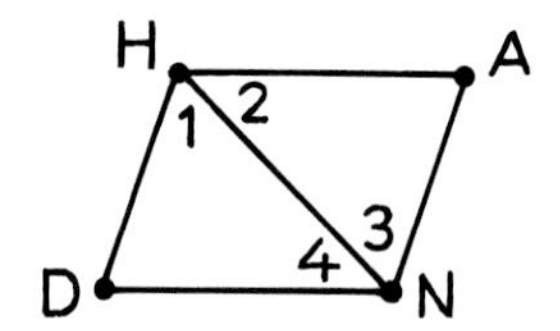

Given: $\overline{HA} \parallel \overline{DN}$; $HA = DN$.
Prove: $\triangle HAN \cong \triangle NDH$.

31.

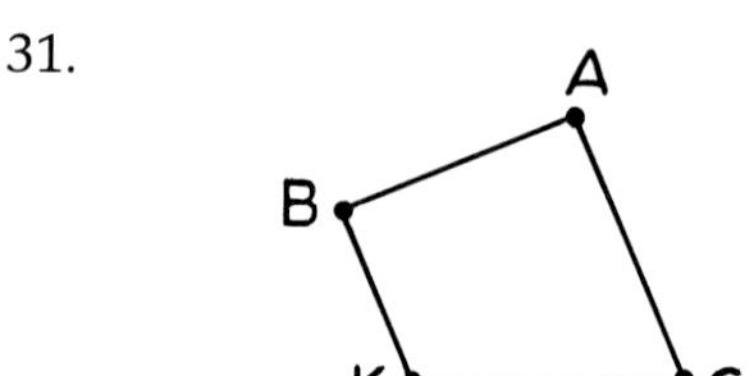

Given: $\overline{BA} \perp \overline{AC}$ and $\overline{BK} \parallel \overline{AC}$.
Prove: $\measuredangle B$ is a right angle.

32.

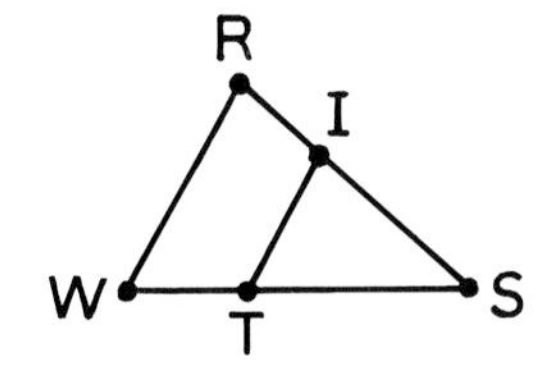

Given: $\measuredangle$WTI is an exterior angle of $\triangle$IST; $\overline{TI} \parallel \overline{WR}$.
Prove: $\angle WTI > \angle R$.

33.

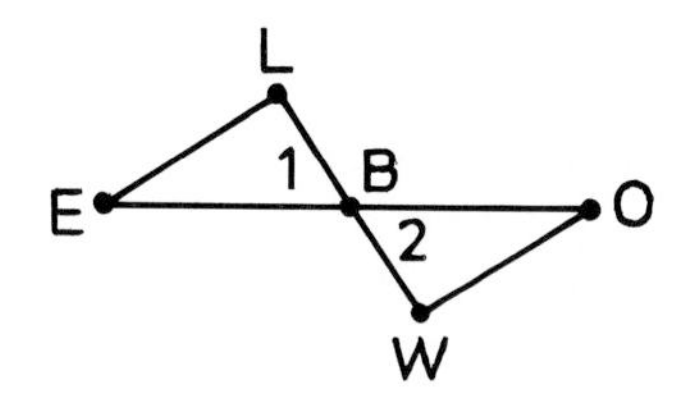

Given: $\measuredangle 1$ and $\measuredangle 2$ are vertical angles; B is the midpoint of $\overline{LW}$; $\overline{EL} \parallel \overline{WO}$.
Prove: EL = OW.

34.

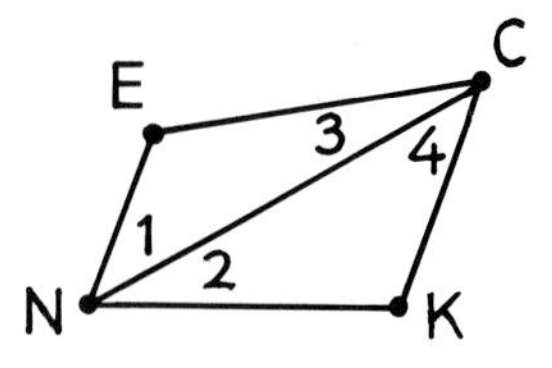

Given: NK > CK; $\overline{NE} \parallel \overline{KC}$.
Prove: $\angle 1 > \angle 2$.

35.

Given: KA = KL; $\overline{NE} \parallel \overline{KL}$.
Prove: $\triangle$ANE is isosceles.

Set III

Which fishpole caught the fish?

Adapted from *Challenge!* by Charles Rice, Hallmark Editions, 1968.

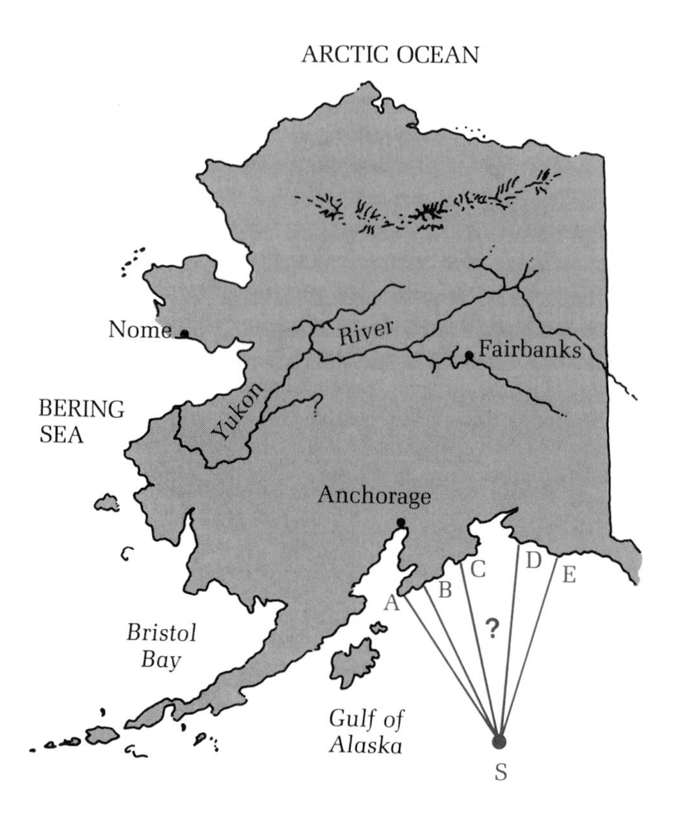

Lesson 5

More on Distance

Most countries, including the United States, prohibit foreign ships from fishing within 200 miles of their shore lines. In the map above, point S represents the location of a ship off the coast of Alaska. How would its distance from the shore line be determined?

Because the distance between two points is the length of the shortest path between them, it seems reasonable to associate other distances with shortest paths as well. If $\overline{SB}$, for example, is the shortest line segment that can be drawn from S to the shore line, then its length measures the distance of the ship from shore.

If the shore line is a straight line, then it is easy to find the shortest segment that can be drawn from a point to it: we can prove that it is the perpendicular segment.

► **Theorem 26**
The perpendicular segment from a point to a line is the shortest segment joining them.

In the figure at the right, we will show that if $\overline{SA} \perp \ell$, then $\overline{SA}$ is shorter than any other segment joining S to ℓ.

Given: $\overline{SA} \perp \ell$ and $\overline{SB}$ is any other segment joining S to ℓ.
Prove: SA < SB.

Proof.

Statements	*Reasons*
1. $\overline{SA} \perp \ell$.	Given.
2. $\measuredangle 1$ and $\measuredangle 2$ are right angles.	Perpendicular lines form right angles.
3. $\angle 1 = \angle 2$.	Any two right angles are equal.
4. $\angle 1 > \angle 3$.	An exterior angle of a triangle is greater than either remote interior angle.
5. $\angle 2 > \angle 3$.	Substitution.
6. SB > SA, and so SA < SB.	If two angles of a triangle are unequal, the sides opposite them are unequal and the longer side is opposite the larger angle.

This theorem is the basis for defining the distance from a point to a line, something that we took for granted in the problem of the surfer in the introduction to this book.

► Definition
The ***distance from a point to a line*** is the length of the perpendicular segment from the point to the line.

Note that it is assumed in this definition that the point is not on the line. If the point is on the line, then there is no perpendicular segment, and the distance from the point to the line is zero.

Parallel lines are often referred to as lines that are everywhere equidistant. The following theorem is the basis for this statement.

► Theorem 27
If two lines are parallel, every perpendicular segment joining one line to the other line has the same length.

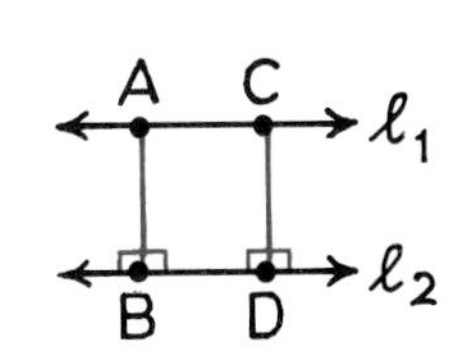

In terms of the figure at the right, this means that, if $\ell_1 \parallel \ell_2$, then AB = CD. Because of this theorem, we can make the following definition.

► Definition
The ***distance between two parallel lines*** is the length of any perpendicular segment joining one line to the other.

Exercises

Set I

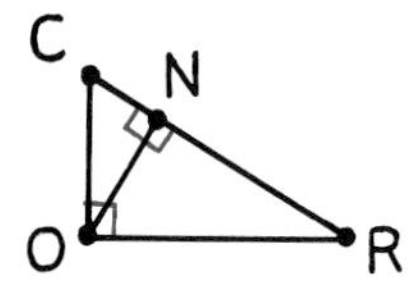

In this figure, $\overline{CO} \perp \overline{OR}$ and $\overline{ON} \perp \overline{CR}$. Which line segment would you measure to find each of the following distances?

1. The distance from C to $\overline{OR}$.
2. The distance from O to $\overline{CR}$.
3. The distance from R to $\overline{CO}$.
4. What is the definition of the distance from a point to a line?

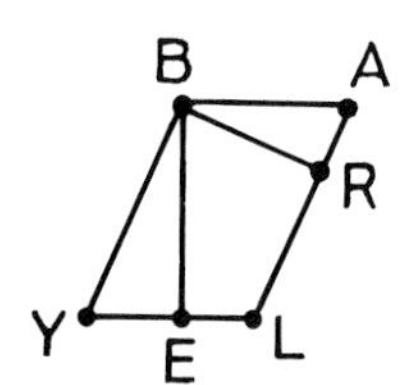

In the figure above, $\overline{BA} \parallel \overline{YL}$, $\overline{YB} \parallel \overline{LA}$, $\overline{BE} \perp \overline{YL}$, and $\overline{BR} \perp \overline{LA}$. Which line segment would you measure to find each of the following distances?

5. The distance between $\overline{BA}$ and $\overline{YL}$.
6. The distance between $\overline{YB}$ and $\overline{LA}$.
7. What is the definition of the distance between two parallel lines?

In the figure below, $\overline{WH} \parallel \overline{TE}$, $\overline{WT} \perp \overline{TE}$, and $\overline{HA} \perp \overline{TE}$. State the theorem that is the basis for each of the following statements.

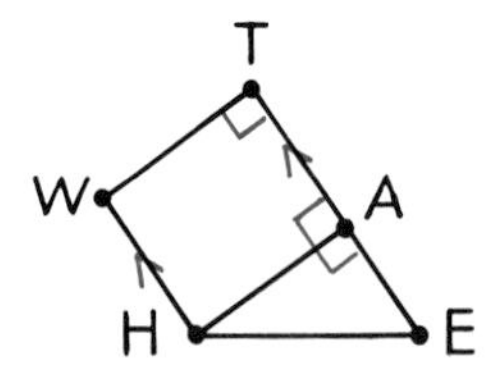

Example: $\overleftrightarrow{WT}$ is the only line at T perpendicular to $\overleftrightarrow{TE}$.

Answer: In a plane, through a point on a line, there is exactly one line perpendicular to the line.

8. $\overleftrightarrow{TE}$ is the only line from T perpendicular to $\overleftrightarrow{HA}$.
9. $\overline{WT} \parallel \overline{HA}$.
10. $\overline{WT} \perp \overline{WH}$.
11. $WT = HA$.
12. $HA < HE$.

Trace this figure.

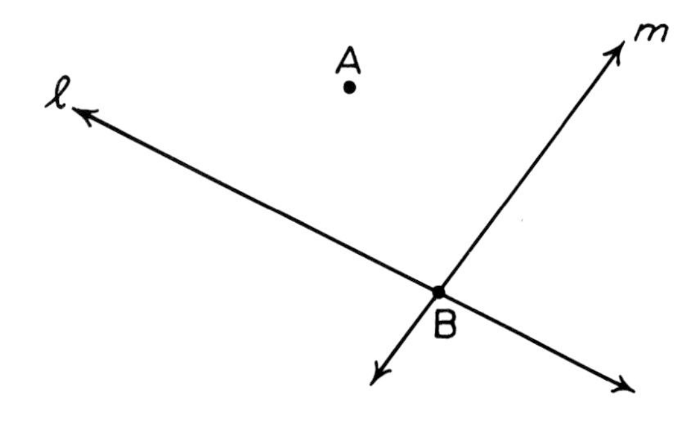

13. Use your straightedge and compass to construct a line through point A perpendicular to line ℓ and a line through point A perpendicular to line m.

Use your ruler to find the following distances, each to the nearest 0.1 cm.

14. The distance from point A to line ℓ.
15. The distance from point A to line m.
16. The distance from point A to point B.

Draw a horizontal line on your paper and label it ℓ.

17. Use your straightedge and compass to construct a line m parallel to line ℓ and 2 cm above it. Also, construct a line n parallel to line ℓ and 3 cm below it.
18. What relation do lines m and n have to each other?
19. State the theorem that is the basis for your answer.
20. What is the distance between lines m and n?

Set II

Give the missing statements and reasons in the following proof of Theorem 27.

21. If two lines are parallel, every perpendicular segment joining one line to the other has the same length.

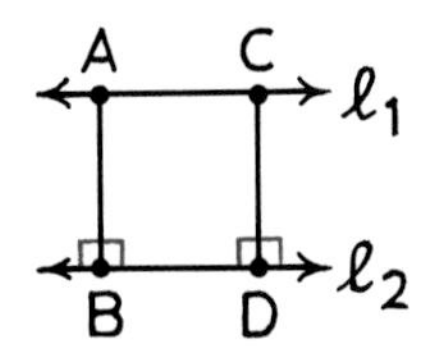

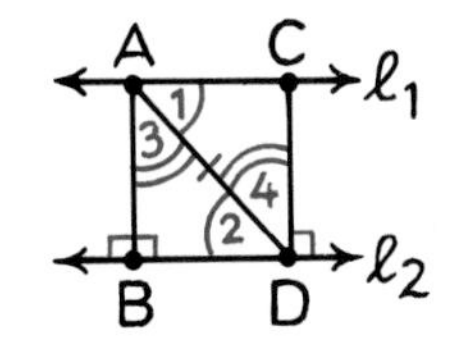

Given: $\ell_1 \parallel \ell_2$; $\overline{AB} \perp \ell_2$ and $\overline{CD} \perp \ell_2$.
Prove: AB = CD.

Proof.

Statements	Reasons
1. $\ell_1 \parallel \ell_2$.	Given.
2. Draw $\overline{AD}$.	a) ▮▮▮▮
3. b) ▮▮▮▮	If two parallel lines are cut by a transversal, the alternate interior angles are equal.
4. $\overline{AB} \perp \ell_2$ and $\overline{CD} \perp \ell_2$.	c) ▮▮▮▮
5. d) ▮▮▮▮	In a plane, two lines perpendicular to a third line are parallel to each other.
6. e) ▮▮▮▮	Same as step 3.
7. f) ▮▮▮▮	Reflexive.
8. △ABD ≅ △DCA.	g) ▮▮▮▮
9. h) ▮▮▮▮	i) ▮▮▮▮

22.

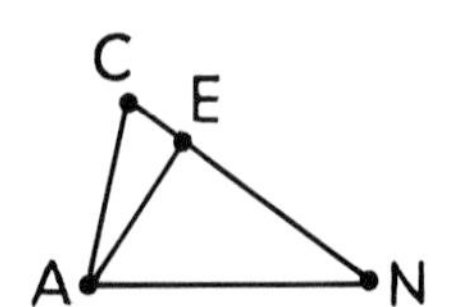

Given: △AEN is a right triangle with right ∡AEN and C-E-N.
Prove: AE < AC.

23.

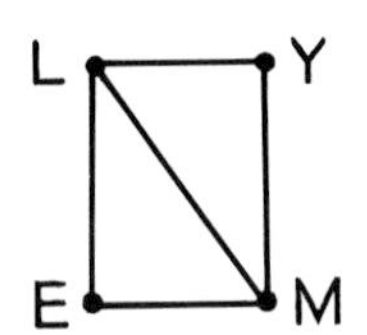

Given: ∠ELM = ∠LMY; $\overline{LY} \perp \overline{YM}$ and $\overline{EM} \perp \overline{YM}$.
Prove: LY = EM.

24.

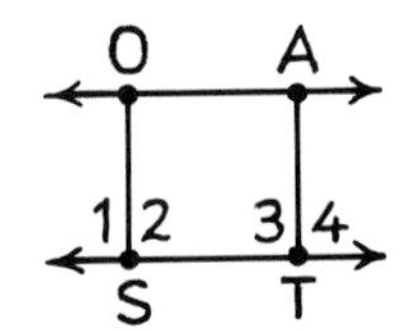

Given: $\overleftrightarrow{OA} \parallel \overleftrightarrow{ST}$; ∠1 = ∠2 and ∠3 = ∠4.
Prove: OS = AT.

25.

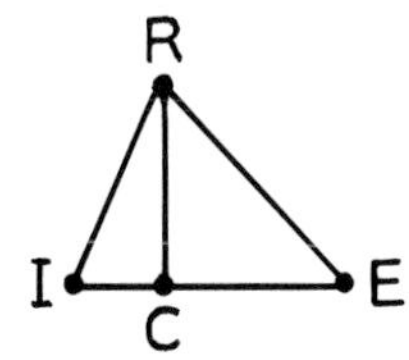

Given: In △RIE, $\overline{RC} \perp \overline{IE}$; ∠E > ∠CRE.
Prove: RI > CE.

Set III

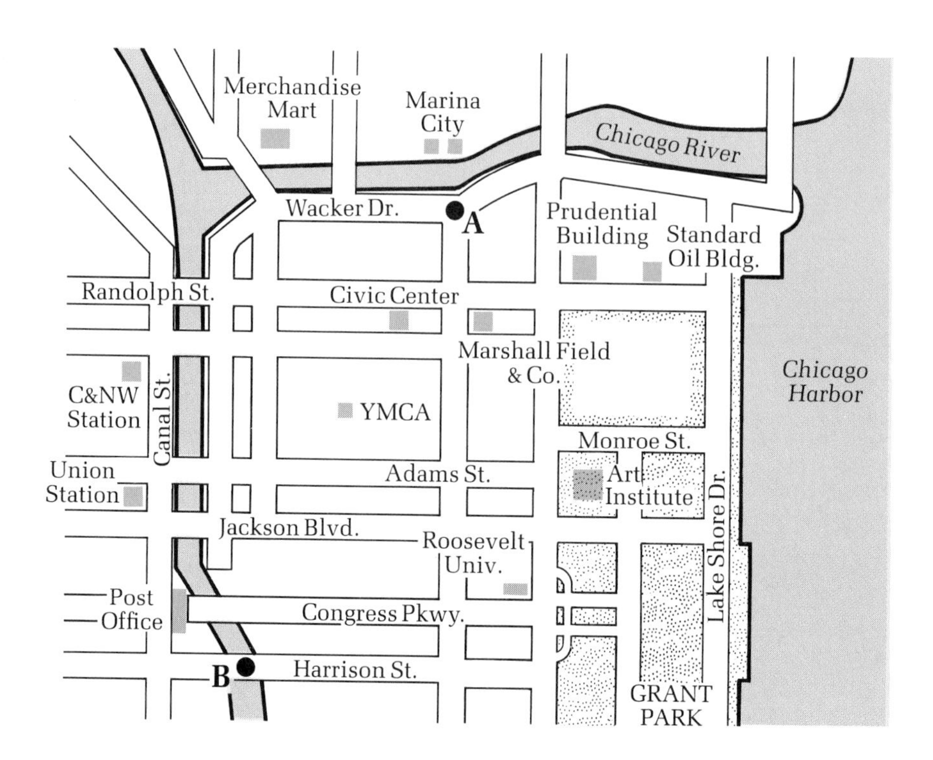

This map shows part of downtown Chicago. On it, 1 cm represents 0.25 km. Estimate the distance, in kilometers, from A, the intersection of State Street and Wacker Drive, to B, the Harrison Street Bridge of the Chicago River, from the viewpoint of

1. a bird.

2. a taxi driver.

3. a boat captain (assume that the boat sails down the center of the river).

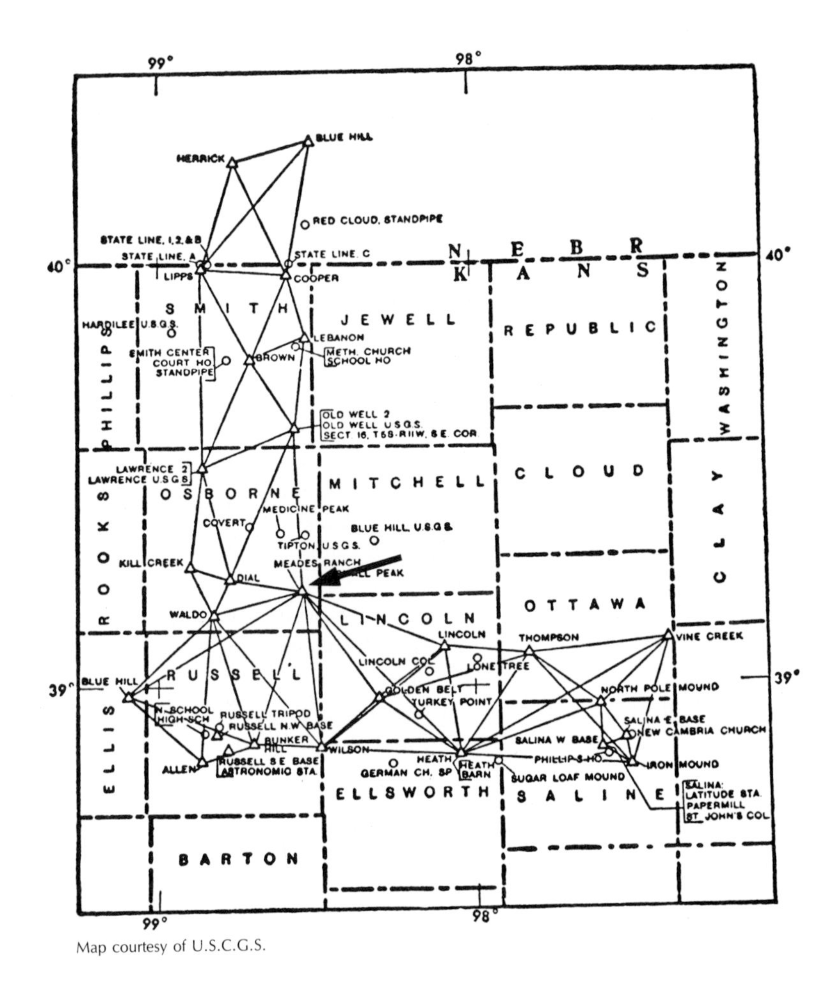

Map courtesy of U.S.C.G.S.

Lesson 6

The Angles of a Triangle

In a pasture in Meades Ranch, Kansas, there is a concrete block with a bronze disk marked with a small cross. This cross mark is the initial point from which the entire North American continent has been mapped.

To map the United States, surveyors used a method called "triangulation"—so named because it locates points by means of a network of triangles. Some of these triangles are visible in the map above, which shows part of Kansas. Meades Ranch is marked with an arrow.

A basic procedure in triangulation is to measure the angles of these triangles. One check of how accurately a given triangle has been determined is to find the sum of the measures of its angles: the sum should be exactly 180°. This is easy to prove by means of the Parallel Postulate.

► **Theorem 28**
The sum of the measures of the angles of a triangle is 180°.

Given: $\triangle ABC$.
Prove: $\angle A + \angle B + \angle C = 180°$.

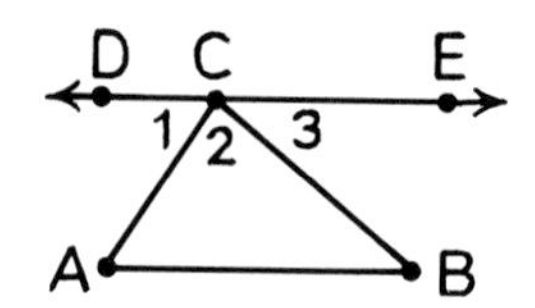

Proof.

Statements	Reasons
1. $\triangle ABC$.	Given.
2. Through point C, draw $\overleftrightarrow{DE} \parallel \overline{AB}$.	Through a point not on a line, there is exactly one line parallel to the line.
3. $\angle 1 = \angle A$ and $\angle 3 = \angle B$.	If two parallel lines are cut by a transversal, the alternate interior angles are equal.
4. $\measuredangle DCB$ and $\measuredangle 3$ are supplementary.	If two angles are a linear pair, they are supplementary.
5. $\angle DCB + \angle 3 = 180°$.	If two angles are supplementary, the sum of their measures is 180°.
6. $\angle DCB = \angle 1 + \angle 2$.	Betweenness of Rays Theorem.
7. $\angle 1 + \angle 2 + \angle 3 = 180°$.	Substitution (steps 5 and 6).
8. $\angle A + \angle B + \angle C = 180°$.	Substitution (steps 3 and 7).

This theorem has some useful corollaries:

► **Corollary 1**
If two angles of one triangle are equal to two angles of another triangle, then the third pair of angles are equal.

► **Corollary 2**
The acute angles of a right triangle are complementary.

► **Corollary 3**
Each angle of an equilateral triangle has a measure of 60°.

Another theorem that can be proved by means of Theorem 28 concerns the measure of an exterior angle of a triangle. You will recall that we have already proved that an exterior angle of a triangle is greater than either remote interior angle. We can now prove that it is *equal to the sum of their measures.*

► **Theorem 29**
An exterior angle of a triangle is equal in measure to the sum of the measures of the remote interior angles.

Given: $\triangle ABC$ with exterior $\measuredangle 2$.
Prove: $\angle 2 = \angle A + \angle B$.

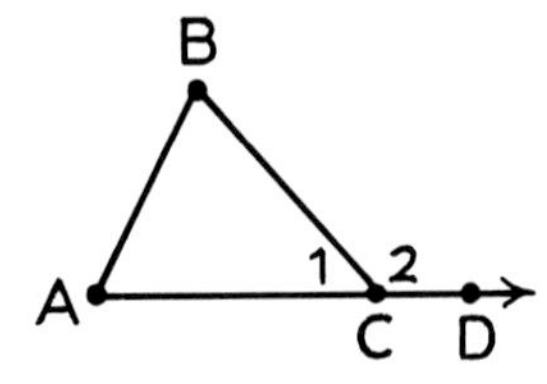

Proof.
In $\triangle ABC$,

$$\angle A + \angle B + \angle 1 = 180°.$$

Because $\measuredangle 2$ is an exterior angle of $\triangle ABC$, it forms a linear pair with $\measuredangle 1$. Hence $\measuredangle 1$ and $\measuredangle 2$ are supplementary, and so

$$\angle 1 + \angle 2 = 180°.$$

From the two equations above, it follows by substitution that

$$\angle 1 + \angle 2 = \angle A + \angle B + \angle 1.$$

Subtracting $\angle 1$ from each side of this equation gives

$$\angle 2 = \angle A + \angle B.$$

Exercises

Set I

In $\triangle OAR$, $\angle O = 5°$ and $\angle A = 25°$.

1. Find $\angle R$.
2. State the theorem that is the basis for your answer.
3. What kind of triangle is $\triangle OAR$ with respect to its angles?

In $\triangle RIG$, $\angle R = 30°$ and $\angle I = 75°$.

4. Find $\angle G$.
5. What kind of triangle is $\triangle RIG$ with respect to its sides?
6. State the theorem that is the basis for your answer to Exercise 5.

In $\triangle BOW$, $BO = OW = BW$.

7. What kind of triangle is $\triangle BOW$ with respect to its sides?
8. What kind of triangle is $\triangle BOW$ with respect to its angles?
9. What is the measure of each angle of $\triangle BOW$?
10. State the theorem that is the basis for your answer to Exercise 9.

In $\triangle TAC$ and $\triangle KLE$, $\angle T = \angle K$ and $\angle A = \angle L$.

11. What can you conclude about $\angle C$ and $\angle E$?
12. State the theorem that is the basis for your answer.

In $\triangle JIB$, $\measuredangle J$ is a right angle.

13. What relation do $\measuredangle I$ and $\measuredangle B$ have?
14. State the theorem that is the basis for your answer.

Use the theorems of this lesson and deductive reasoning to find the measure of the indicated angle in each of the following figures.

15.

16.

17.

18.

19.

20.

In the figure below, $\overline{UN} \perp \overline{RN}$. Also, $\angle R = (3x + 5)°$ and $\angle U = (4x - 6)°$.

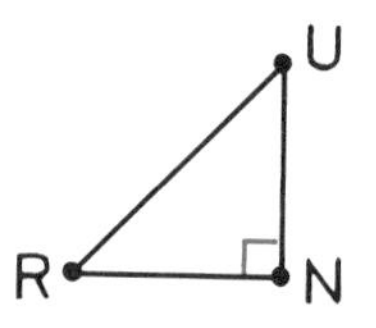

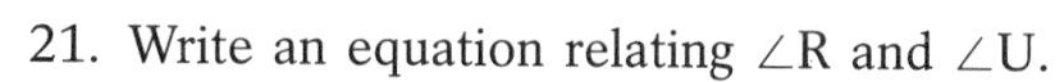

21. Write an equation relating $\angle R$ and $\angle U$.
22. Find x.
23. Find $\angle R$.
24. Find $\angle U$.
25. Is $\triangle RUN$ isosceles?

In the figure below, S-A-I. Also, $\angle SAL = (-25x)°$, $\angle ALI = (35 - x)°$, and $\angle I = (x + 65)°$.

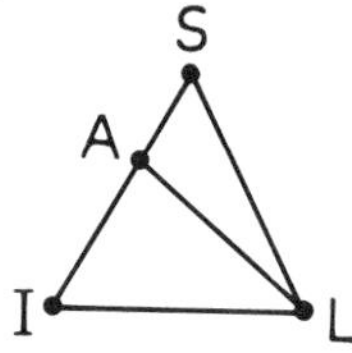

26. Write an equation relating $\angle SAL$, $\angle ALI$, and $\angle I$.
27. Find x.
28. Find $\angle SAL$.
29. Find $\angle ALI$.
30. Find $\angle I$.
31. Is $\triangle SIL$ equilateral?

Set II

Give the missing statements and reasons in the following proofs of the corollaries to Theorem 28.

32. *Corollary 1.* If two angles of one triangle are equal to two angles of another triangle, then the third pair of angles are equal.

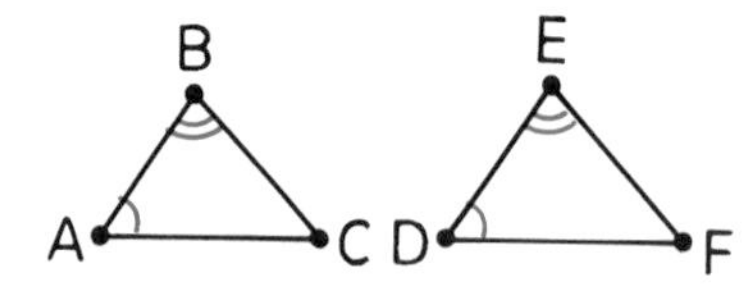

Given: $\triangle ABC$ and $\triangle DEF$; $\angle A = \angle D$ and $\angle B = \angle E$.
Prove: $\angle C = \angle F$.

Proof.

Statements	***Reasons***
1. $\triangle ABC$ and $\triangle DEF$.	Given.
2. $\angle A + \angle B + \angle C = 180°$ and $\angle D + \angle E + \angle F = 180°$.	a) ______
3. b) ______	Substitution.
4. $\angle A = \angle D$ and $\angle B = \angle E$.	c) ______
5. $\angle A + \angle B + \angle C = \angle A + \angle B + \angle F$.	d) ______
6. e) ______	f) ______

33. *Corollary 2.* The acute angles of a right triangle are complementary.

Given: Right $\triangle ABC$ with right $\measuredangle A$.
Prove: $\measuredangle B$ and $\measuredangle C$ are complementary.

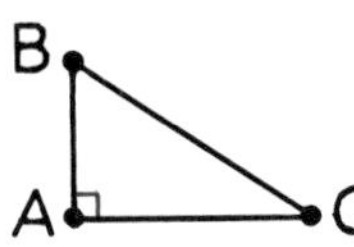

Proof.

Statements	***Reasons***
1. Right $\triangle ABC$ with right $\measuredangle A$.	Given.
2. a) __________	A right angle has a measure of 90°.
3. b) __________	The sum of the measures of the angles of a triangle is 180°.
4. c) __________	Subtraction.
5. d) __________	e) __________

34. *Corollary 3.* Each angle of an equilateral triangle has a measure of 60°.

Given: $\triangle ABC$ is equilateral.
Prove: $\angle A = 60°$, $\angle B = 60°$, and $\angle C = 60°$.

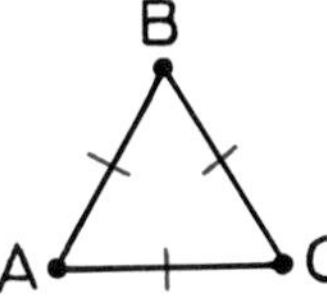

Proof.

Statements	***Reasons***
1. $\triangle ABC$ is equilateral.	Given.
2. $\triangle ABC$ is equiangular.	a) __________
3. b) __________	If a triangle is equiangular, all of its angles are equal.
4. c) __________	The sum of the measures of the angles of a triangle is 180°.
5. $\angle A + \angle A + \angle A = 180°$, and so $3\angle A = 180°$.	d) __________
6. $\angle A = 60°$.	e) __________
7. $\angle B = 60°$ and $\angle C = 60°$.	f) __________

35.

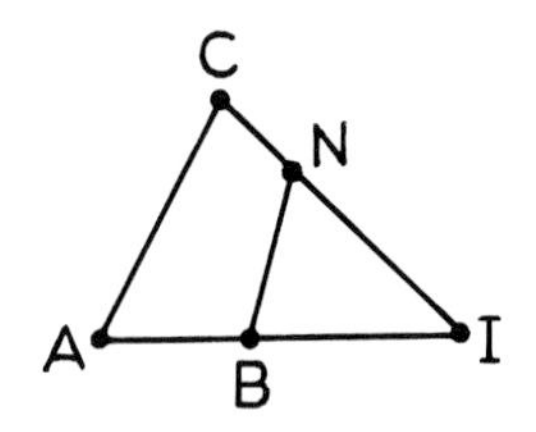

Given: In $\triangle CAI$ and $\triangle BIN$, $\angle C = \angle NBI$.
Prove: $\angle A = \angle BNI$.

36.

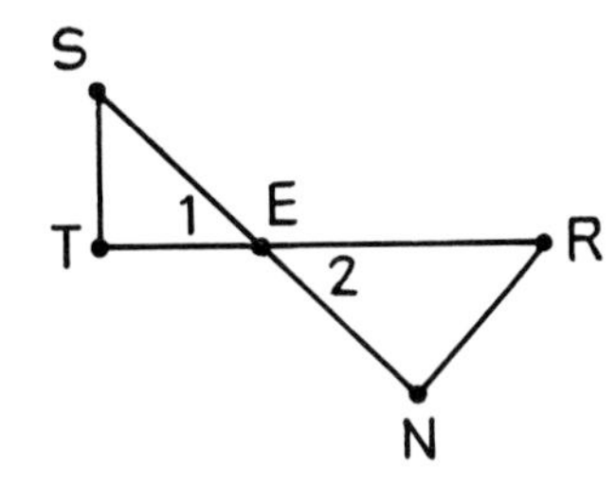

Given: $\overline{ST} \perp \overline{TE}$ and $\overline{EN} \perp \overline{NR}$; $\measuredangle 1$ and $\measuredangle 2$ are vertical angles.
Prove: $\angle S = \angle R$.

37.

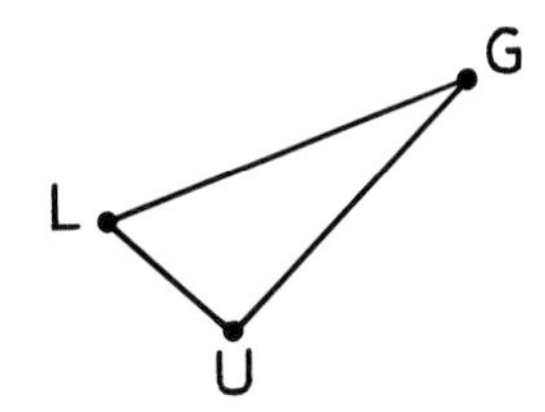

Given: $\overline{LU} \perp \overline{UG}$.
Prove: $\angle L + \angle G = 90°$.

38.

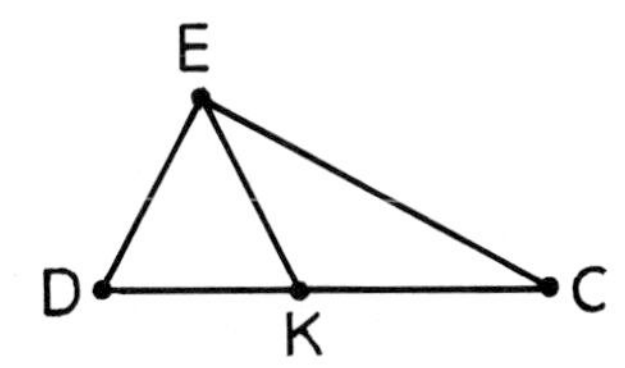

Given: $\triangle DEC$ is a right triangle; $\measuredangle DKE$ and $\measuredangle C$ are complementary.
Prove: $DE = EK$.

39.

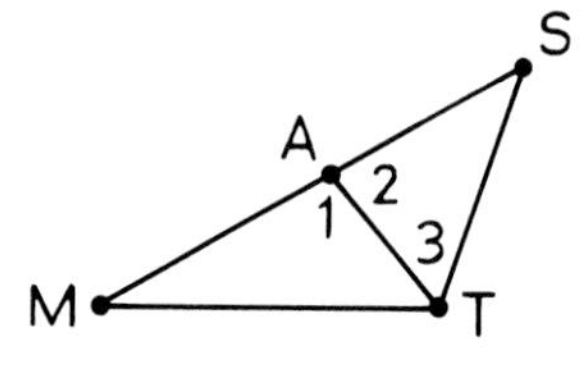

Given: $\measuredangle 1$ and $\measuredangle 2$ are a linear pair.
Prove: $\angle S = \angle 1 - \angle 3$.

40.

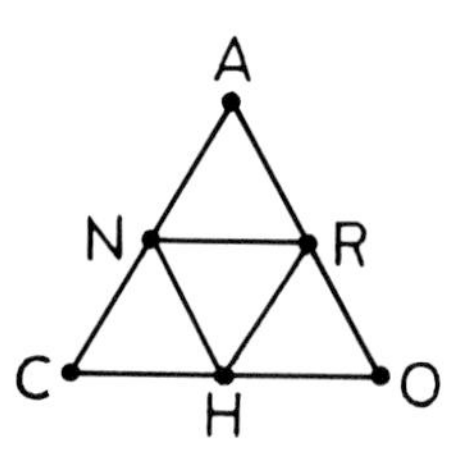

Given: $\triangle ACO$ and $\triangle NHR$ are equilateral.
Prove: $\angle A = \angle NHR$.

Set III

In this lesson, we proved that the sum of the measures of the three angles of every triangle is the same number: 180°.

Every triangle has six exterior angles, as illustrated in this figure.

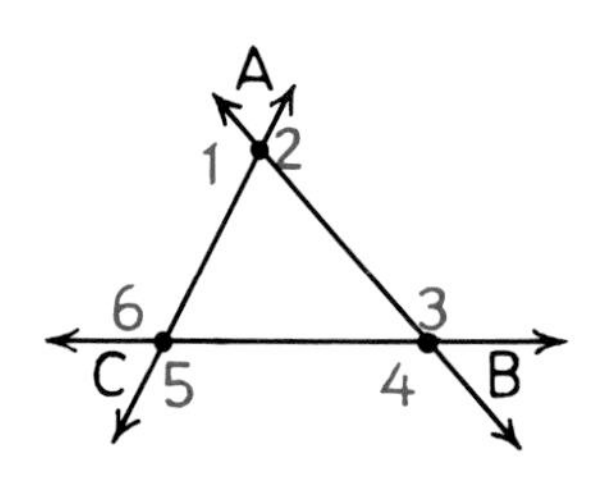

1. Do you think that the sum of the measures of the six exterior angles of every triangle is the same number?
2. If you do, what number is it and how did you arrive at it? If you don't, give an example of two triangles for which the sums are different.

H. Armstrong Roberts

Lesson 7
Two More Ways to Prove Triangles Congruent

A popular type of hang glider is the Rogallo, named after a NASA engineer who invented the design in the 1940s as a potential space reentry vehicle. The sail of a Rogallo glider consists of two congruent triangles that share a common side as shown in the figure below.

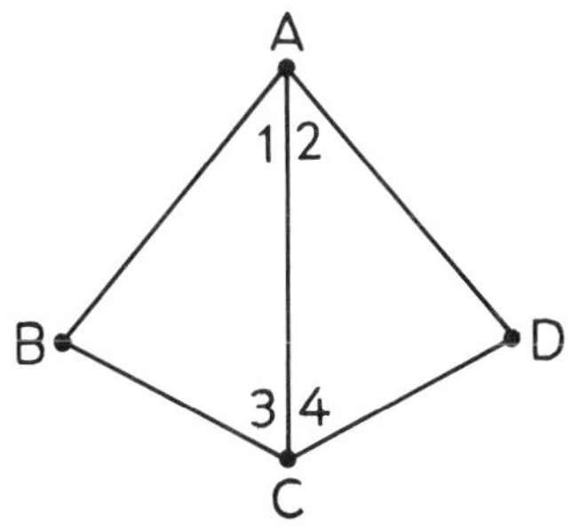

You already know several sets of conditions that are sufficient to insure that the triangles are congruent. For example, if AB = AD and $\angle 1 = \angle 2$, then $\triangle ABC \cong \triangle ADC$ by S.A.S. If $\angle 1 = \angle 2$ and $\angle 3 = \angle 4$, the triangles are congruent by A.S.A., and if AB = AD and BC = DC, the triangles are congruent by S.S.S. Are there any other sets of parts that can be used to prove triangles congruent? The answer is yes, and we will consider two of them in this lesson.

▶ **Theorem 30** (The A.A.S. Congruence Theorem)
If two angles and the side opposite one of them in one triangle are equal to the corresponding parts of another triangle, then the triangles are congruent.

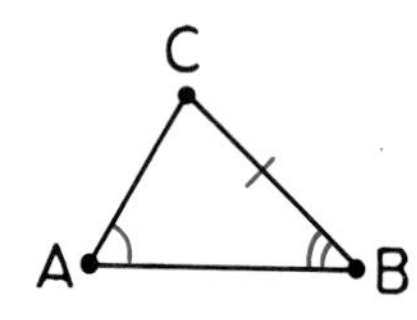

Given: $\triangle ABC$ and $\triangle DEF$ with $\angle A = \angle D$, $\angle B = \angle E$, and $BC = EF$.
Prove: $\triangle ABC \cong \triangle DEF$.

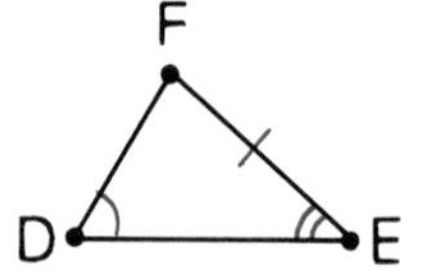

Proof.
Because $\angle A = \angle D$ and $\angle B = \angle E$, it must also be true that $\angle C = \angle F$. (If two angles of one triangle are equal to two angles of another triangle, then the third pair of angles are equal.) Because $BC = EF$, it follows that $\triangle ABC \cong \triangle DEF$ by the A.S.A. Congruence Postulate.

The A.A.S. Congruence Theorem, like S.A.S., A.S.A., and S.S.S., can be applied to all triangles, regardless of their type. The next theorem, called H.L., applies only to right triangles.

▶ **Theorem 31** (The H.L. Congruence Theorem)
If the hypotenuse and a leg of one right triangle are equal to the corresponding parts of another right triangle, then the triangles are congruent.

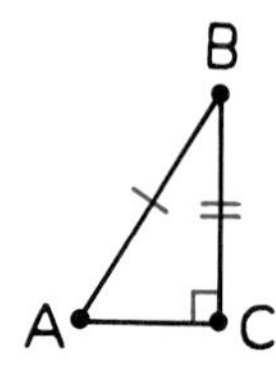

Given: $\triangle ABC$ and $\triangle DEF$ are right triangles (with right angles C and F); $AB = DE$, $BC = EF$.
Prove: $\triangle ABC \cong \triangle DEF$.

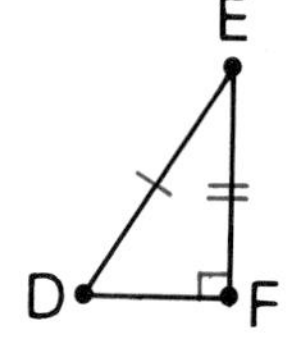

Proof.
Draw $\overleftrightarrow{DF}$ (two points determine a line) and choose point G on $\overleftrightarrow{DF}$ as shown in the figure below so that $FG = AC$ (the Ruler Postulate); also, draw $\overline{EG}$.

In $\triangle ABC$ and $\triangle GEF$,

$AC = FG$,
$\angle C = \angle 4$ (they are both right angles), and
$BC = EF$ (given),

and so $\triangle ABC \cong \triangle GEF$ by the S.A.S. Congruence Postulate.

Also

$GE = AB$ (they are corresponding sides of the congruent triangles), and
$AB = DE$ (given), and so
$GE = DE$ (transitive).

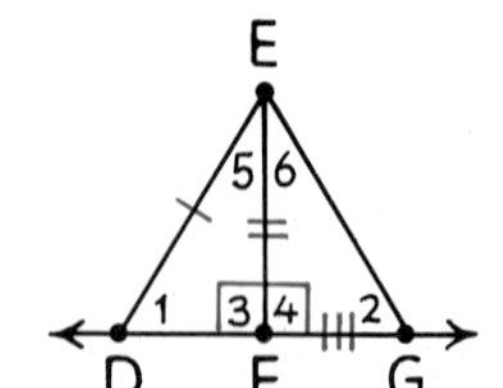

But $\overline{GE}$ and $\overline{DE}$ are sides of $\triangle DEG$, and so $\angle 1 = \angle 2$ (if two sides of a triangle are equal, the angles opposite them are equal). Also, because $\triangle ABC \cong \triangle GEF$, $\angle 2 = \angle A$. It follows that $\angle 1 = \angle A$ (transitive), and so $\triangle ABC \cong \triangle DEF$ by the A.A.S. Congruence Theorem.

Exercises

Set I

The following questions refer to the triangles below.

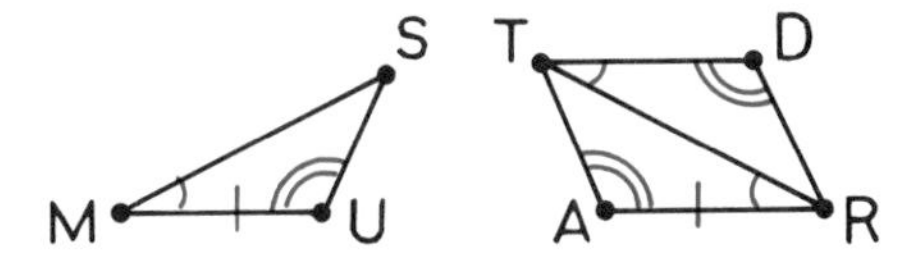

1. Why is $\triangle MUS \cong \triangle RAT$?
2. Why is $\triangle RAT \cong \triangle TDR$?
3. Why is $\triangle MUS \cong \triangle TDR$?

In $\triangle SAV$ and $\triangle ORY$, $\measuredangle A$ and $\measuredangle R$ are right angles and AV = RY. Name the third pair of parts in the triangles needed to prove them congruent by each of the following reasons.

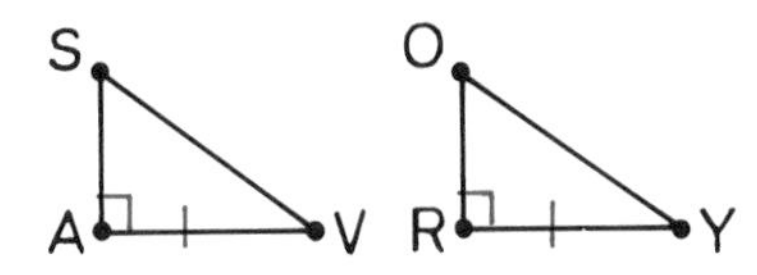

Example: The A.S.A. postulate.
Answer: $\angle V = \angle Y$.

4. The S.A.S. postulate.
5. The H.L. theorem.
6. The A.A.S. theorem.

The following questions refer to the triangles below.

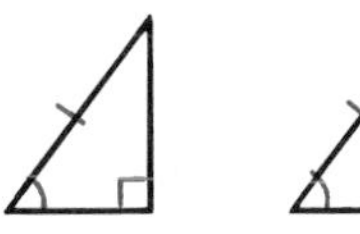

7. Do the triangles appear to be congruent?
8. If the hypotenuse and an acute angle of one right triangle are equal to the corresponding parts of another right triangle, are the triangles necessarily congruent?
9. Explain why or why not.

The following questions refer to the triangles below.

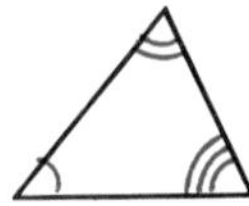

10. Do the triangles appear to be congruent?
11. If the three angles of one triangle are equal to the three angles of another triangle, are the triangles necessarily congruent?

The following questions refer to the triangles below.

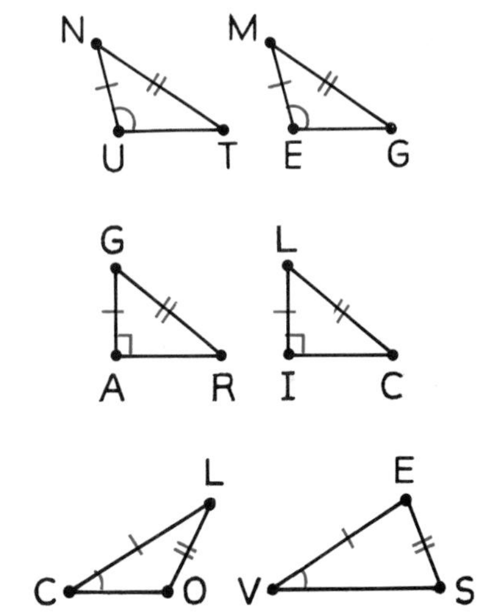

12. Do $\triangle NUT$ and $\triangle MEG$ appear to be congruent?
13. Do you know that they are congruent from the equal parts marked?
14. Do $\triangle GAR$ and $\triangle LIC$ appear to be congruent?
15. Do you know that they are congruent from the equal parts marked?
16. Do $\triangle CLO$ and $\triangle VES$ appear to be congruent?
17. Do you know that they are congruent from the equal parts marked?

18. If two sides and the angle opposite one of them in one triangle are equal to the corresponding parts of another triangle, does it follow that the triangles are congruent?

In the figure below, $\overline{IM} \perp \overline{MT}$, $\overline{IN} \perp \overline{NT}$, and IM = IN. Also, $\angle NIT = (x^2 + 10x)°$ and $\angle ITN = (50 - x^2)°$.

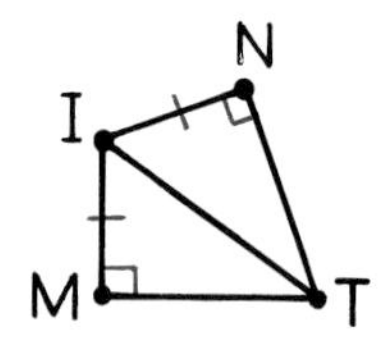

19. Write an equation relating ∠NIT and ∠ITN.

20. Find x.

21. Find ∠NIT.

22. Find ∠ITN.

23. Find ∠MIN.

Give the missing statements and reasons in the following proof.

24.

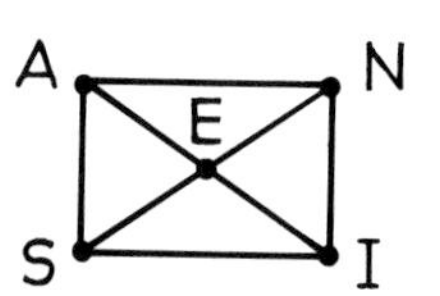

Given: ∡ASI and ∡NIS are right angles; AI = NS.
Prove: AS = NI.

Proof.

Statements	**Reasons**
1. ∡ASI and ∡NIS are right angles.	Given.
2. a) ||||||||	If a triangle has a right angle, it is a right triangle.
3. b) ||||||||	Given.
4. c) ||||||||	Reflexive.
5. △ASI ≅ △NIS.	d) ||||||||
6. e) ||||||||	f) ||||||||

Set II

25.

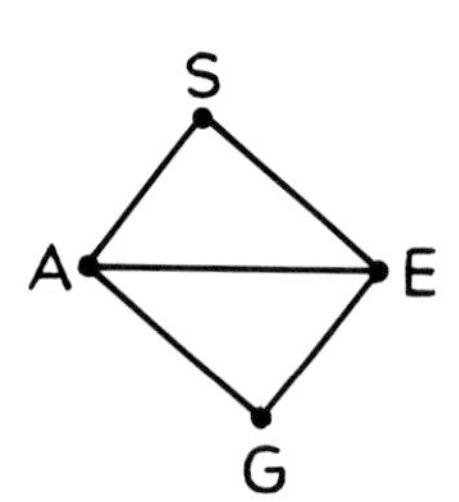

Given: △ASE and △AGE are right triangles and AS = GE.
Prove: $\overline{SE} \parallel \overline{AG}$.

26.

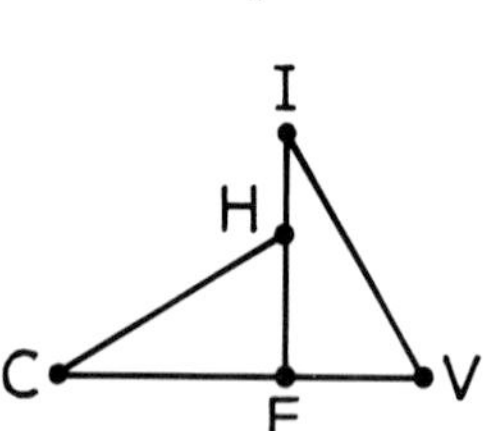

Given: $\overline{IE} \perp \overline{CV}$; CH = IV and CE = IE.
Prove: ∠C = ∠I.

27.

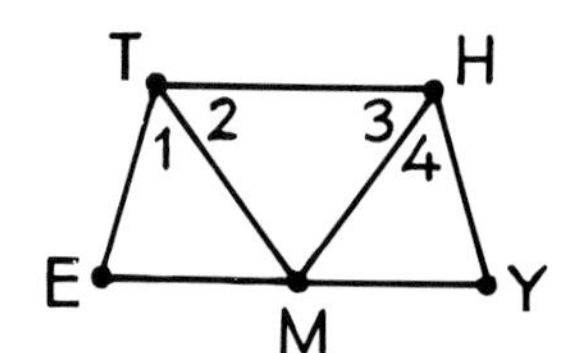

Given: ∠E = ∠Y, ∠1 = ∠4, ∠2 = ∠3, and E-M-Y.
Prove: M is the midpoint of $\overline{EY}$.

28.

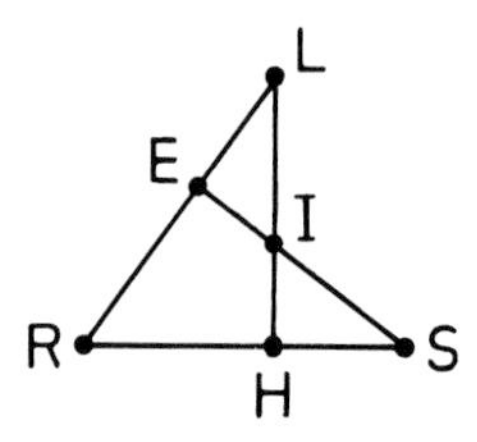

Given: ∠L = ∠S and RE = RH.
Prove: LH = SE.

29.

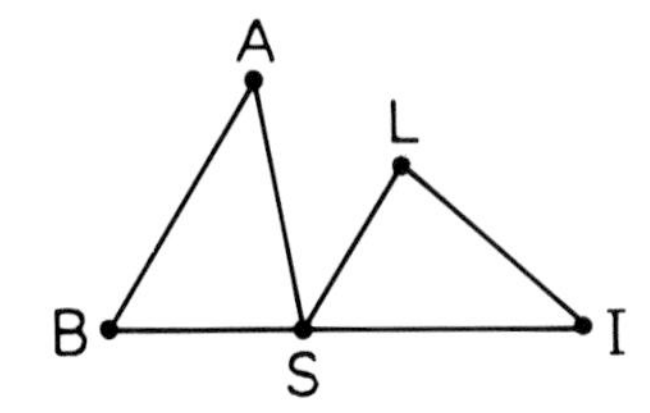

Given: $\overline{BA} \parallel \overline{SL}$ with transversal $\overline{BI}$; $\angle A = \angle I$ and BS = SL.
Prove: $\triangle BAS \cong \triangle SIL$.

30.

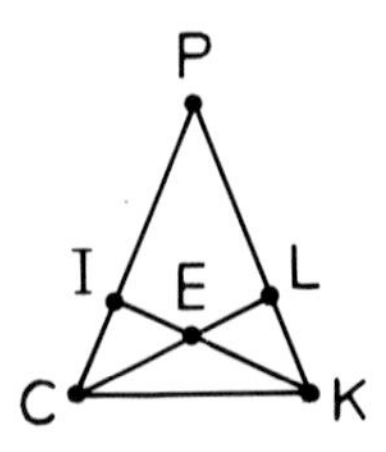

Given: $\triangle CPK$ with $\overline{KI} \perp \overline{CP}$ and $\overline{CL} \perp \overline{KP}$; KI = CL.
Prove: $\triangle CPK$ is isosceles.

Set III

Three pairs of equal parts, if they have the right relation, are sufficient to prove two triangles congruent. Nevertheless, it is a rather remarkable fact that two triangles can have more than three pairs of equal parts and yet not be congruent.

1. Use your ruler and compass to construct $\triangle ABC$ with AB = 5.4 cm, AC = 3.6 cm, and BC = 2.4 cm.
2. Construct $\triangle DEF$ with DE = 8.1 cm, DF = 5.4 cm, and EF = 3.6 cm.
3. Are $\triangle ABC$ and $\triangle DEF$ congruent?

It can be proved that all of the parts of $\triangle ABC$ and $\triangle DEF$ that *appear* to be equal *are* equal.

4. How many pairs of parts of the triangles are equal?

Chapter 6 / Summary and Review

Photograph by René Maltete in *Laughing Camera 2*, Hanns Reich Verlag

Basic Ideas

Angles formed by a transversal 192–193
Distance
  between two parallel lines 215
  from a point to a line 215
Parallel lines 192
Transversal 192

Postulate

10. *The Parallel Postulate.* Through a point not on a line, there is exactly one line parallel to the line. 204

Theorems

21. If two lines form equal corresponding angles with a transversal, then the lines are parallel. 193
    Corollary 1. If two lines form equal alternate interior angles with a transversal, then the lines are parallel. 193
    Corollary 2. If two lines form supplementary interior angles on the same side of a transversal, then the lines are parallel. 193
    Corollary 3. In a plane, two lines perpendicular to a third line are parallel to each other. 193
22. In a plane, through a point on a line there is exactly one line perpendicular to the line. 197
23. Through a point not on a line, there is exactly one line perpendicular to the line. 198
24. In a plane, two lines parallel to a third line are parallel to each other. 204
25. If two parallel lines are cut by a transversal, then the corresponding angles are equal. 209
    Corollary 1. If two parallel lines are cut by a transversal, then the alternate interior angles are equal. 209
    Corollary 2. If two parallel lines are cut by a transversal, then the interior angles on the same side of the transversal are supplementary. 210
    Corollary 3. In a plane, if a line is perpendicular to one of two parallel lines, it is also perpendicular to the other. 210
26. The perpendicular segment from a point to a line is the shortest segment joining them. 215
27. If two lines are parallel, every perpendicular segment joining one line to the other has the same length. 215

28. The sum of the measures of the angles of a triangle is 180°. 220
    Corollary 1. If two angles of one triangle are equal to two angles of another triangle, then the third pair of angles are equal. 220
    Corollary 2. The acute angles of a right triangle are complementary. 220
    Corollary 3. Each angle of an equilateral triangle has a measure of 60°. 220
29. An exterior angle of a triangle is equal in measure to the sum of the measures of the remote interior angles. 221
30. *The A.A.S. Congruence Theorem.* If two angles and the side opposite one of them in one triangle are equal to the corresponding parts of another triangle, then the triangles are congruent. 226
31. *The H.L. Congruence Theorem.* If the hypotenuse and a leg of one right triangle are equal to the corresponding parts of another right triangle, then the triangles are congruent. 226

Constructions

6. To construct the line perpendicular to a given line through a point on it. 199
7. To construct the line perpendicular to a given line through a point not on it. 199–200
8. To construct the line parallel to a given line through a point not on it. 204

Exercises

Set I

Tell whether each of the following statements is true or false.

1. No triangle can have two right angles.
2. If two lines are cut by a transversal, the interior angles on the same side of the transversal are supplementary.
3. The H.L. Congruence Theorem applies only to right triangles.
4. If two lines form unequal corresponding angles with a transversal, then the lines are not parallel.
5. A corollary is a theorem.
6. If two angles and a side of one triangle are equal to two angles and a side of another triangle, the triangles are congruent.
7. Through any point, there is exactly one line parallel to a given line.
8. If the sum of the measures of two angles of a triangle is equal to the measure of the third angle, the triangle is a right triangle.

Trace this figure.

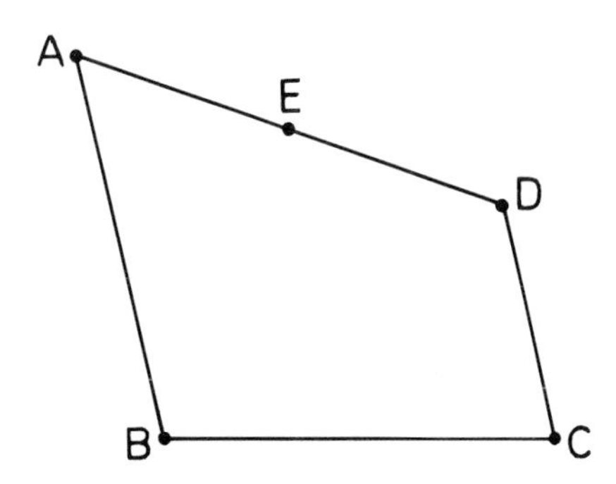

9. Construct, through E, a line parallel to $\overline{AB}$. Label the point in which it intersects $\overline{BC}$ F.
10. What relation does $\overline{EF}$ appear to have to $\overline{DC}$?
11. What relation does F appear to have to $\overline{BC}$?

Trace this figure, in which $\measuredangle B$ is a right angle.

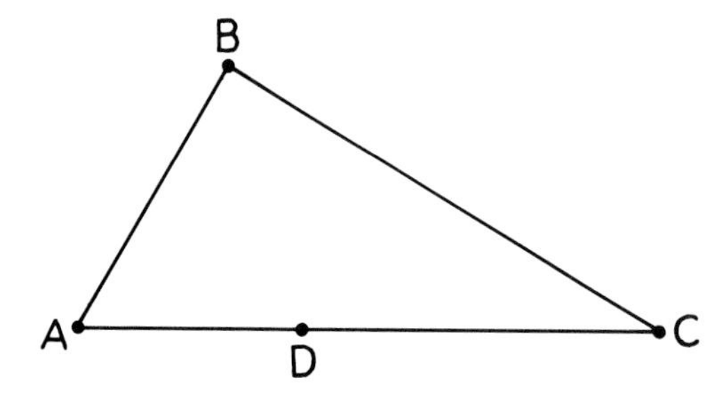

12. Construct, through D, a line perpendicular to $\overline{BC}$. Label the point in which it intersects $\overline{BC}$ E.
13. What relation does $\overline{AB}$ have to $\overline{BC}$?
14. State the definition that is the basis for your answer.
15. What relation does $\overline{DE}$ have to $\overline{AB}$?
16. State the theorem that is the basis for your answer.

In the figure below, $\overleftrightarrow{DS} \parallel \overleftrightarrow{PE}$. Find the measures of the following angles.

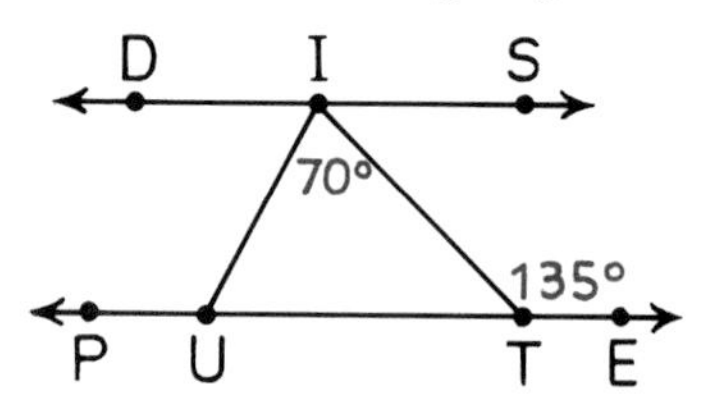

17. $\measuredangle SIT$.
18. $\measuredangle IUT$.
19. $\measuredangle PUI$.

In the figure below, G-I-F and G-H-T. Find the measures of the following angles.

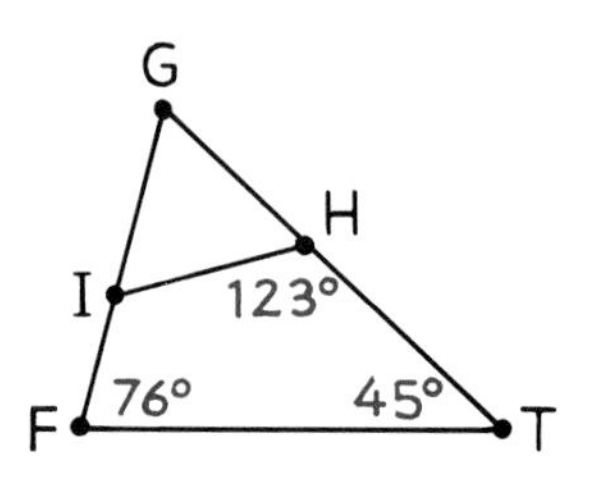

20. $\measuredangle G$.
21. $\measuredangle GHI$.
22. $\measuredangle GIH$.

Set II

If the lens of a magnifying glass is held so that it is perpendicular to the light rays of the sun, the ray passing through its center is not bent. The rest of the rays passing through the lens are bent so that they meet at a single point.

Suppose that, in the figure below, $\overline{EP}$ is

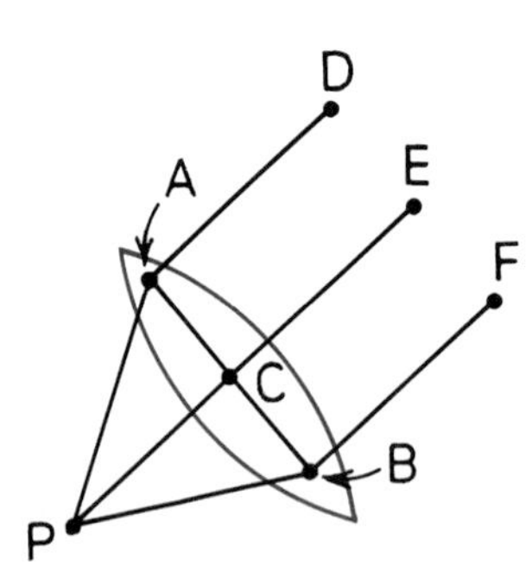

the perpendicular bisector of $\overline{AB}$ and that $\overline{DA} \parallel \overline{EP}$. Use this information to tell whether each of the following statements *must be true, may be true,* or *is false.*

23. C is the midpoint of $\overline{AB}$.
24. $\overline{DA} \perp \overline{AB}$.
25. $\overline{FB} \parallel \overline{EP}$.
26. $AP = BP$.
27. $PC < PB$.
28. $AC + CP < AP$.
29. $\angle PAB = 60°$.
30. $\angle ACE = \angle PAC + \angle APC$.
31.

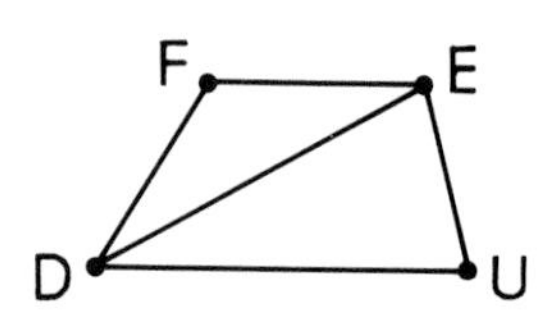

Given: $\overrightarrow{DE}$ bisects $\measuredangle FDU$;
$DF = FE$.

Prove: $\overline{FE} \parallel \overline{UD}$.

By the permission of Johnny Hart and News America Syndicate

32.

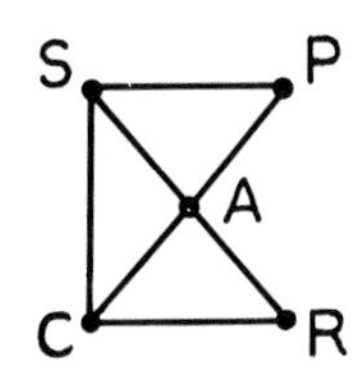

Given: $\overline{SC} \perp \overline{SP}$ and $\overline{SC} \perp \overline{CR}$; SR = CP.
Prove: $\angle P = \angle R$.

33.

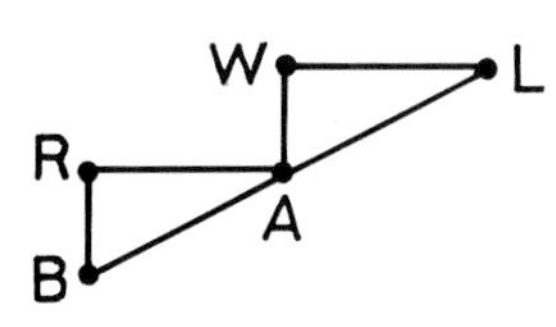

Given: $\measuredangle R$ and $\measuredangle W$ are right angles; $\overline{RA} \parallel \overline{WL}$ with transversal $\overline{BL}$; BA = AL.
Prove: $\triangle BRA \cong \triangle AWL$.

34.

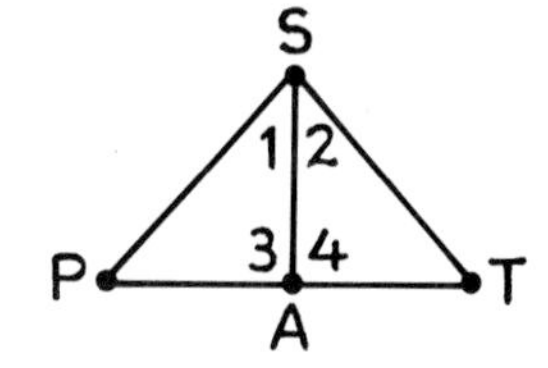

Given: $\overrightarrow{SA}$ bisects $\measuredangle PST$; $\angle P = \angle T$; $\measuredangle 3$ and $\measuredangle 4$ are a linear pair.
Prove: $\overline{SA} \perp \overline{PT}$ without proving triangles congruent.

Algebra Review

Square Roots

Simple Radical Form
An expression containing a square root is in simple radical form if the number or expression under the square root sign has no perfect square factors other than 1.

The Square Root of a Product
If x and y are positive, $\sqrt{xy} = \sqrt{x}\sqrt{y}$.

The Square Root of a Quotient
If x and y are positive, $\sqrt{\frac{x}{y}} = \frac{\sqrt{x}}{\sqrt{y}}$.

Exercises

Write in simple radical form.

Example: $\sqrt{150}$
Solution: $\sqrt{150} = \sqrt{25 \cdot 6} = \sqrt{25}\sqrt{6} = 5\sqrt{6}$

1. $\sqrt{28}$
2. $\sqrt{98}$
3. $\sqrt{360}$
4. $\sqrt{47^2}$

Write in simple radical form, assuming that x represents a positive number.

Example: $\sqrt{x^7}$
Solution: $\sqrt{x^7} = \sqrt{x^6 \cdot x} = \sqrt{x^6}\sqrt{x} = x^3\sqrt{x}$

5. $\sqrt{25x}$
6. $\sqrt{x^{16}}$

Simplify each expression.

Example: $\sqrt{32} + \sqrt{18}$
Solution: $\sqrt{32} + \sqrt{18} = \sqrt{2 \cdot 16} + \sqrt{2 \cdot 9} = 4\sqrt{2} + 3\sqrt{2} = 7\sqrt{2}$

7. $\sqrt{48} + \sqrt{3}$
8. $\sqrt{48 - 3}$
9. $\sqrt{700} - \sqrt{175}$
10. $\sqrt{700 - 175}$
11. $\sqrt{20 + 20 + 20}$
12. $\sqrt{20} + \sqrt{20} + \sqrt{20}$
13. $\sqrt{6^2} + \sqrt{8^2}$
14. $\sqrt{6^2 + 8^2}$
15. $\sqrt{15^2} - \sqrt{10^2}$
16. $\sqrt{15^2 - 10^2}$

Multiply or divide and simplify.

Example: $\sqrt{13}\sqrt{26}$
Solution: $\sqrt{13}\sqrt{26} = \sqrt{13 \cdot 26} = \sqrt{13 \cdot 13 \cdot 2} = 13\sqrt{2}$

17. $\sqrt{14}\sqrt{21}$
18. $\sqrt{500}\sqrt{45}$
19. $\frac{\sqrt{242}}{\sqrt{2}}$
20. $\frac{\sqrt{156}}{\sqrt{3}}$
21. $(4\sqrt{5})(6\sqrt{2})$
22. $(3\sqrt{7})^2$
23. $\frac{12\sqrt{15}}{4\sqrt{3}}$
24. $\frac{\sqrt{200}}{2}$

Do as indicated.

Example: $(1 + 2\sqrt{5}) + (7 + \sqrt{5})$
Solution: $(1 + 2\sqrt{5}) + (7 + \sqrt{5}) = 8 + 3\sqrt{5}$

25. $(5 - \sqrt{2}) + (5 + \sqrt{2})$
26. $(5 - \sqrt{2})(5 + \sqrt{2})$
27. $\sqrt{3}(\sqrt{27} + 1)$
28. $\sqrt{3} + (\sqrt{27} + 1)$
29. $(\sqrt{x} + \sqrt{y}) + (\sqrt{x} + \sqrt{y})$
30. $(\sqrt{x} + \sqrt{y})^2$

Chapter 7

QUADRILATERALS

Menil Foundation Collection, Houston

Lesson 1

Quadrilaterals

This picture of a partly open window is a painting by the Belgian artist René Magritte. At first glance, it seems that we are looking through the window at the sky. The partly open pane on the right, however, reveals that behind the window is darkness. Perhaps the "sky" is a picture fastened to the glass or even a reflection of actual sky on our side of the window. But the top of the right-hand pane seems to rule out both of these possibilities. The painting represents an unreal window in an unreal world.

The frames of the window panes are physical models of quadrilaterals. In this lesson we will consider the names of various types of quadrilaterals, and in the following lessons we will study their properties.

► **Definition**

A ***quadrilateral*** is a polygon that has four sides.

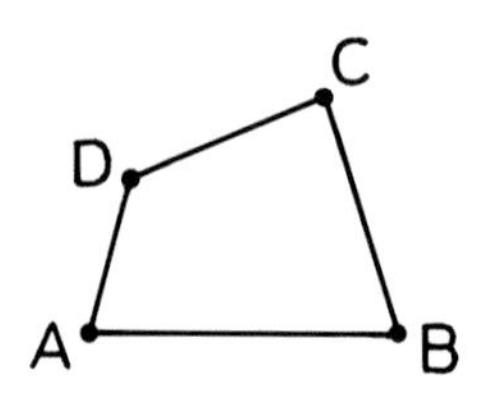

Quadrilateral ABCD, shown at the left, has four vertices, points A, B, C, and D; four sides, $\overline{AB}$, $\overline{BC}$, $\overline{CD}$, and $\overline{DA}$; and four angles, ∡A, ∡B, ∡C, and ∡D. Two sides that intersect each other, such as $\overline{AB}$ and $\overline{BC}$, are called *consecutive*. A pair of sides that do *not* intersect, such as $\overline{AB}$ and $\overline{DC}$, are called *opposite*. The same terms are applied to the vertices and angles of a quadrilateral; ∡A and ∡D, for example, are *consecutive angles;* points A and C are *opposite vertices.*

Several special types of quadrilaterals are illustrated below.

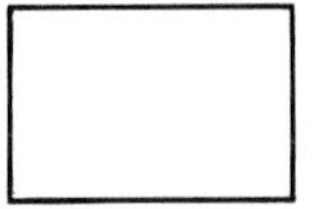

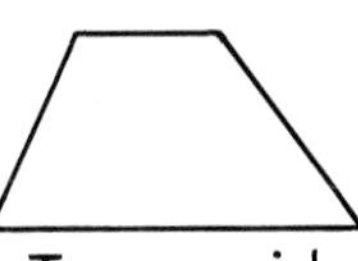

► **Definitions**
A quadrilateral is
a ***parallelogram*** iff both pairs of opposite sides are parallel.
a ***rhombus*** iff all of its sides are equal. (It is equilateral.)
a ***rectangle*** iff all of its angles are equal. (It is equiangular.)
a ***square*** iff it is both equilateral and equiangular.
a ***trapezoid*** iff it has exactly one pair of parallel sides.

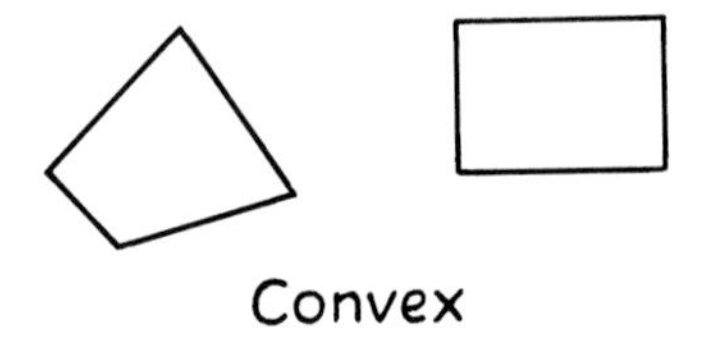

Although all triangles are convex, quadrilaterals may be convex or concave, as the figures at the right illustrate. Because all of the quadrilaterals that have special names are convex, you may assume that the quadrilaterals referred to in this book are convex unless mentioned otherwise. It is easy to prove that the sum of the measures of the angles of every such quadrilateral is the same number.

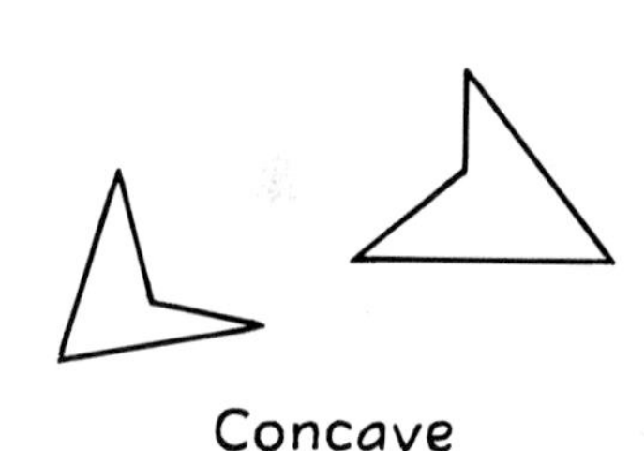

► **Theorem 32**
The sum of the measures of the angles of a quadrilateral is 360°.

Given: Quadrilateral ABCD.
Prove: $\angle A + \angle B + \angle C + \angle D = 360°$.

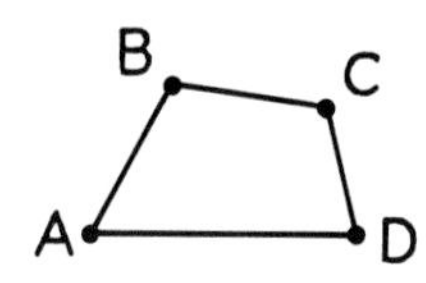

To prove this theorem, we will draw a diagonal of the quadrilateral to form two triangles.

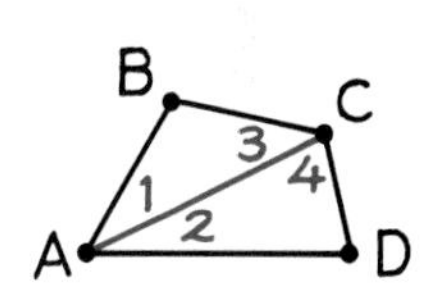

Proof.

Statements	***Reasons***
1. Quadrilateral ABCD.	Given.
2. Draw $\overline{AC}$.	Two points determine a line.
3. $\angle 1 + \angle B + \angle 3 = 180°$ and $\angle 2 + \angle D + \angle 4 = 180°$.	The sum of the measures of the angles of a triangle is 180°.
4. $\angle 1 + \angle B + \angle 3 + \angle 2 + \angle D + \angle 4 = 360°$.	Addition.
5. $\angle 1 + \angle 2 = \angle BAD$ and $\angle 3 + \angle 4 = \angle BCD$.	Betweenness of Rays Theorem.
6. $\angle BAD + \angle B + \angle BCD + \angle D = 360°$.	Substitution.

► **Corollary**
Each angle of a rectangle is a right angle.

Given: ABCD is a rectangle.
Prove: Each of its angles is a right angle.

Proof.
Because $\angle A = \angle B = \angle C = \angle D$ (all of the angles of a rectangle are equal) and $\angle A + \angle B + \angle C + \angle D = 360°$, it follows that each angle has a measure of 90°. Therefore, each angle is a right angle.

Exercises

Set I

In the figure below, the arrowheads indicate parallel sides. From them, we can conclude that the figure is a parallelogram.

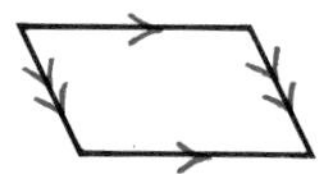

Use the marked parts and the definitions in this lesson to identify each of the following figures.

1. 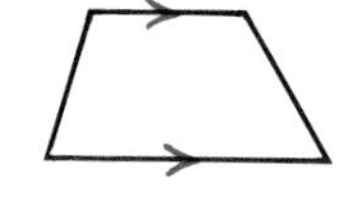

2. 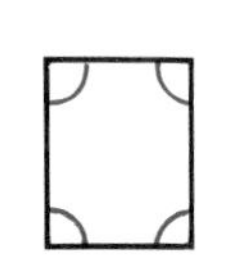

3. 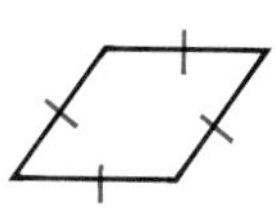

4. 

5. 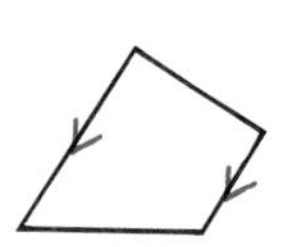

6. 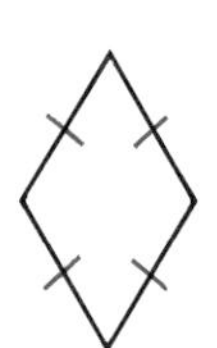

7. 

Use your ruler to make accurate drawings of the following quadrilaterals. Draw both diagonals in each figure.

Example: A rhombus that is not a square.
Answer:

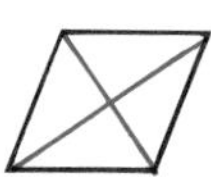

8. A square.
9. A rectangle that is not a square.
10. A parallelogram that is not a rectangle or rhombus.
11. A trapezoid that has no equal sides.

Refer to your drawings for Exercises 8 through 11 to name the types of quadrilaterals that seem to have each of the following properties.

Example: Both pairs of opposite sides are equal.
Answer: Parallelograms, rhombuses, rectangles, and squares.

12. Both pairs of opposite sides are parallel.
13. Both pairs of opposite angles are equal.
14. The diagonals are equal.
15. The diagonals are perpendicular to each other.
16. The diagonals bisect each other.

Find the measure of the indicated angle in each of the following figures.

17. 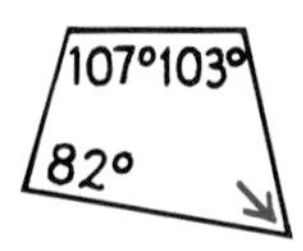

18. 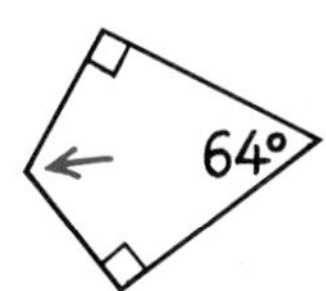

19. 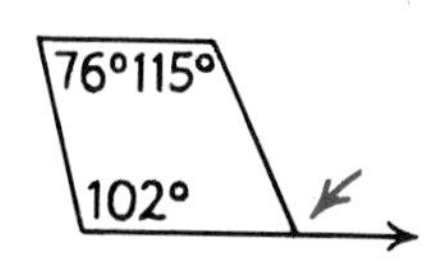

20. 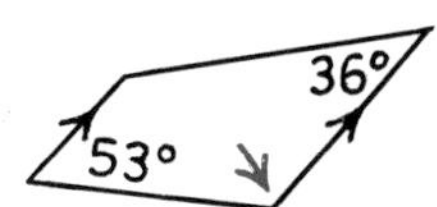

21. 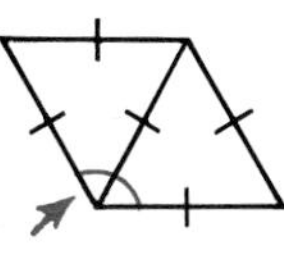

22. 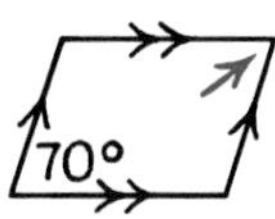

Set II

In quadrilateral BOLT, ∠B = ∠O = ∠L = ∠T.

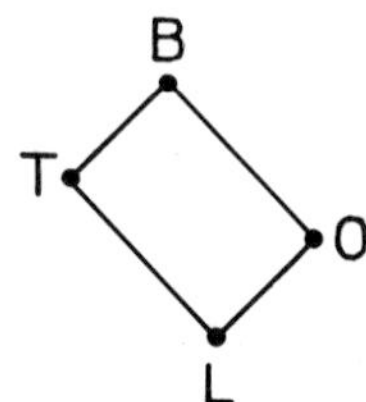

23. What kind of quadrilateral is BOLT?
24. What can you conclude about its angles?

In quadrilateral LOCK, ∠O = ∠1 = ∠2 = ∠3 = ∠4 = ∠K.

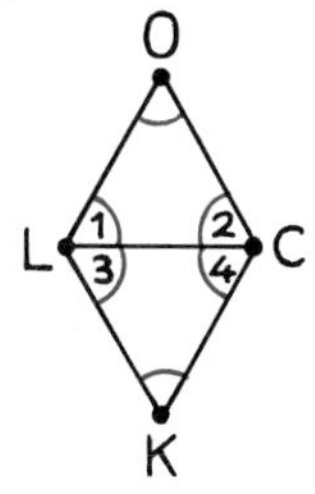

25. What kind of triangles are △LOC and △LKC?
26. Why are the triangles congruent?
27. What kind of quadrilateral is LOCK?

In quadrilateral BAND, $\overline{AN} \perp \overline{ND}$ and $\overline{BD} \perp \overline{ND}$.

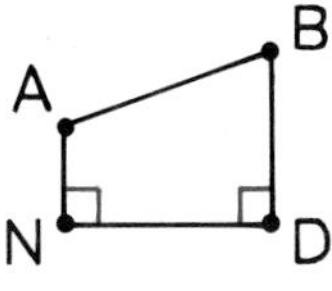

28. Why is $\overline{AN} \parallel \overline{BD}$?
29. What kind of quadrilateral is BAND?
30. What are ∡A and ∡B called with respect to transversal $\overline{AB}$?
31. Why are ∡A and ∡B supplementary?
32. 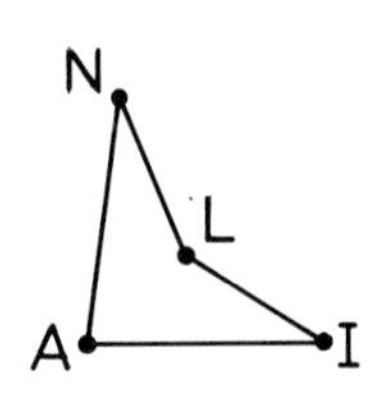

Given: In concave quadrilateral NAIL, NA = AI and NL = LI.

Prove: ∠N = ∠I.
(Hint: Draw diagonal $\overline{AL}$.)

33. 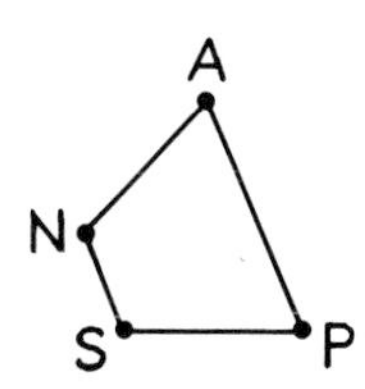

Given: In quadrilateral SNAP, $\overline{SN} \parallel \overline{AP}$; ∡N and ∡P are supplementary.

Prove: ∠S = ∠N.

34.

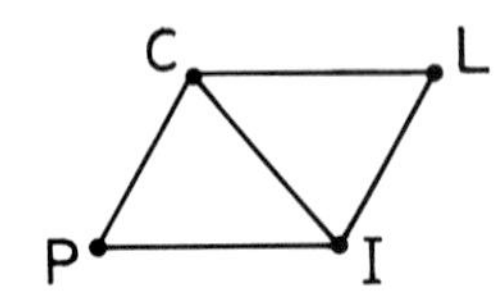

Given: In quadrilateral CLIP,
$\overline{CL} \parallel \overline{PI}$;
CL = PI.

Prove: $\triangle CLI \cong \triangle IPC$.

35.

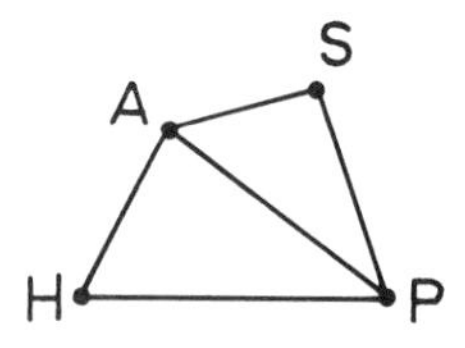

Given: In quadrilateral HASP,
$\overrightarrow{PA}$ bisects $\measuredangle HPS$;
HP > HA.

Prove: $\angle HAP > \angle APS$.

Set III

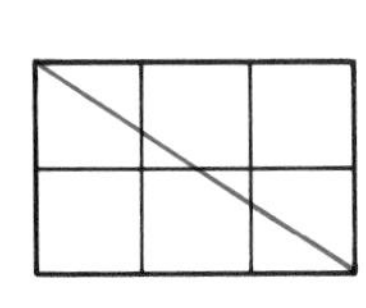

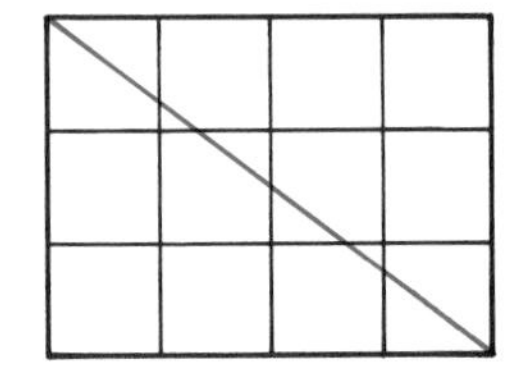

When asked to count the number of squares through which the diagonal of the first figure above passes, Obtuse Ollie got four and Acute Alice got six.

What answers do you think they got for the second figure?

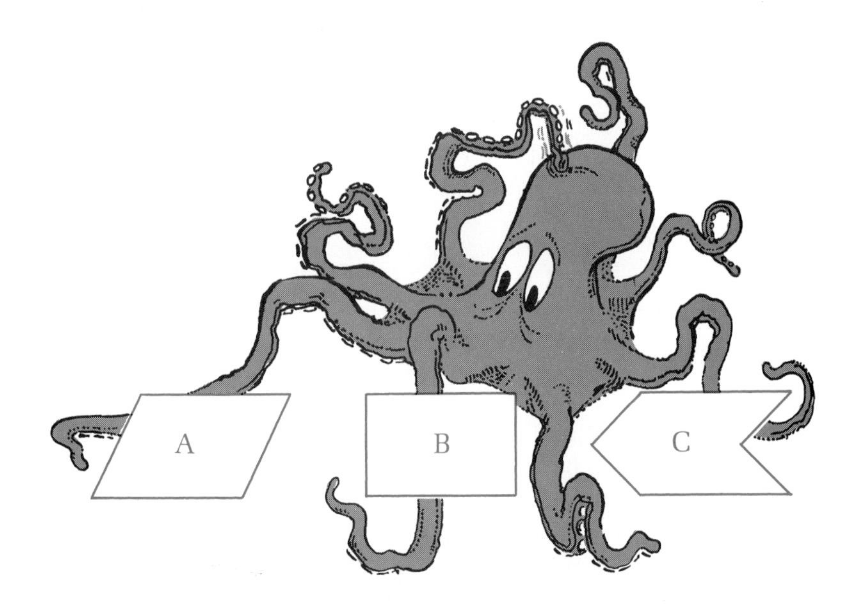

Lesson 2
Parallelograms

Experiments with octopuses show that they can distinguish between certain types of geometric shapes but not between others.* For example, of the three figures shown above, two look alike to an octopus and one looks different. Which do you suppose looks different?

If the octopus recognized a geometric figure on the basis of the number of sides it has, the answer would be C. However, from the way in which the eyes of an octopus work, it is shapes A and C that look alike and shape B that looks different!

This is rather surprising because, from a geometric point of view, figures A and B are the most nearly alike. They are parallelograms, whereas figure C is a concave hexagon. Because figures A and B are parallelograms, they have several properties in common. Their opposite sides are not only parallel but equal. We can also show that their opposite angles are equal. These properties, together with others, are proved in this lesson. In writing the proofs, we will use the symbol ▱ to mean parallelogram.

► **Theorem 33**
The opposite sides of a parallelogram are equal.

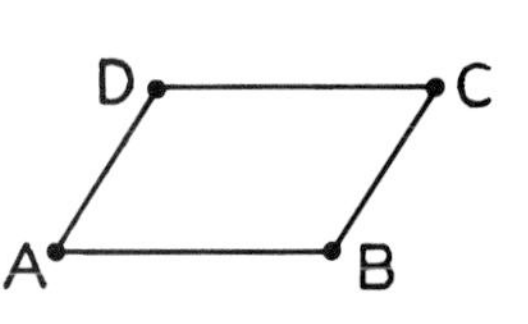

Given: ▱ABCD.
Prove: DC = AB and AD = BC.

Proof.
Draw a diagonal of ▱ABCD. Because $\overline{DC} \parallel \overline{AB}$ and $\overline{AD} \parallel \overline{BC}$ (the opposite sides of a parallelogram are parallel), $\angle 2 = \angle 3$ and $\angle 1 = \angle 4$ (parallel lines form equal alternate interior angles with a transversal). Also DB = DB (reflexive), and so $\triangle DCB \cong \triangle BAD$ (A.S.A.). Therefore, DC = AB and AD = BC (corresponding sides of congruent triangles are equal).

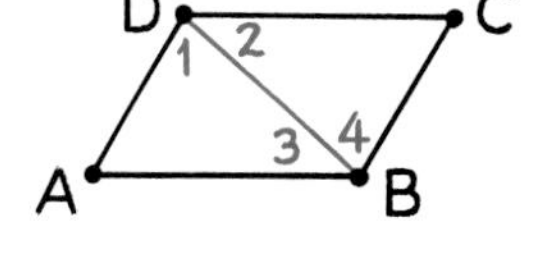

*Niko Tinbergen, *Animal Behavior,* a book in the Life Nature Library (Time, Inc., 1965), pp. 40–41.

► **Theorem 34**
The opposite angles of a parallelogram are equal and its consecutive angles are supplementary.

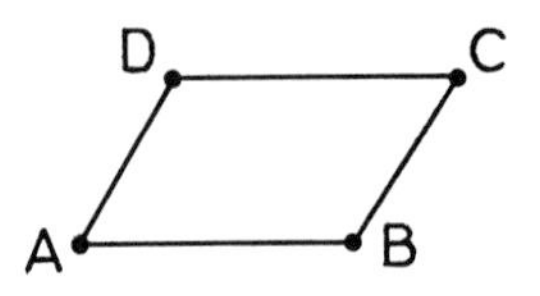

Given: ▱ABCD.

Prove: $\angle A = \angle C$, $\angle B = \angle D$,
$\measuredangle A$ and $\measuredangle B$ are supplementary,
and so forth.

Proof.
Because $\overline{DC} \parallel \overline{AB}$, $\measuredangle A$ and $\measuredangle D$ are supplementary and $\measuredangle B$ and $\measuredangle C$ are supplementary (if two parallel lines are cut by a transversal, the interior angles on the same side of the transversal are supplementary). Likewise, because $\overline{AD} \parallel \overline{BC}$, $\measuredangle A$ and $\measuredangle B$ are supplementary as are $\measuredangle C$ and $\measuredangle D$.

Because we have just proved that $\measuredangle A$ and $\measuredangle C$ are both supplementary to $\measuredangle D$, it follows that $\angle A = \angle C$ (supplements of the same angle are equal). Also, $\measuredangle B$ and $\measuredangle D$ are both supplementary to $\measuredangle C$; so $\angle B = \angle D$.

► **Theorem 35**
The diagonals of a parallelogram bisect each other.

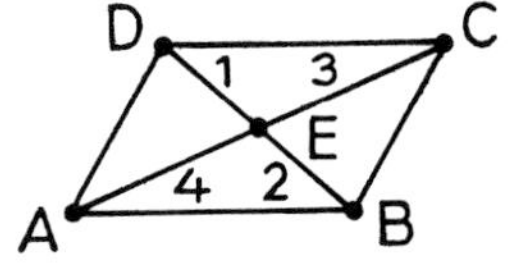

Given: ▱ABCD with diagonals $\overline{AC}$ and $\overline{DB}$.

Prove: $\overline{AC}$ and $\overline{DB}$ bisect each other.

Proof.
Because $\overline{DC} \parallel \overline{AB}$, $\angle 1 = \angle 2$ and $\angle 3 = \angle 4$ (parallel lines form equal alternate interior angles with a transversal). Also, DC = AB (the opposite sides of a parallelogram are equal), and so $\triangle DEC \cong \triangle BEA$ (A.S.A.). This means that DE = BE and EC = EA (corresponding sides of congruent triangles are equal), and so $\overline{AC}$ and $\overline{DB}$ bisect each other.

Exercises

Set I

Complete the statements of the following theorems.

1. The opposite sides of a parallelogram ▯▯▯▯.
2. The consecutive angles of a parallelogram ▯▯▯▯.
3. The opposite angles of a parallelogram ▯▯▯▯.
4. The diagonals of a parallelogram ▯▯▯▯.

In ▱ABCD and ▱EFGH, AD = EH, $\angle A = \angle E$, and AB = EF.

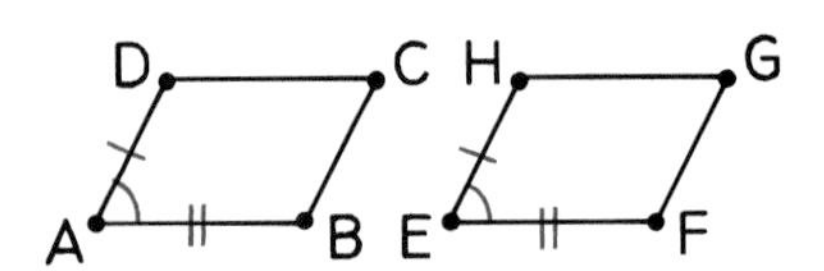

5. Name the other pairs of parts that would have to be equal in order for the parallelograms to be congruent.

6. Can you conclude that these parts are equal?

7. If two sides and the included angle of one parallelogram are equal to two sides and the included angle of another parallelogram, does it follow that the parallelograms are congruent?

In ▱ABCD and ▱EFGH, $\angle A = \angle E$, AB = EF, and $\angle B = \angle F$.

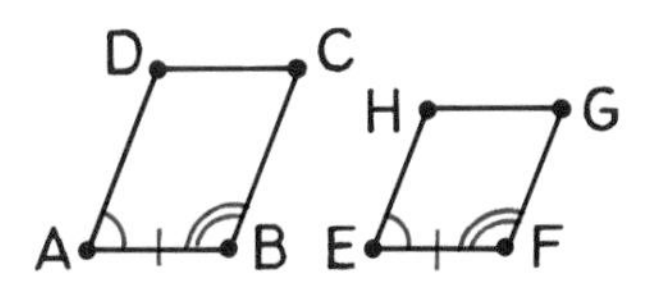

8. What other pairs of parts of these parallelograms can you conclude must also be equal?

9. If two angles and the included side of one parallelogram are equal to two angles and the included side of another parallelogram, does it follow that the parallelograms are congruent?

In ▱ABCD and ▱EFGH, AD = EH, AB = EF, and BC = FG.

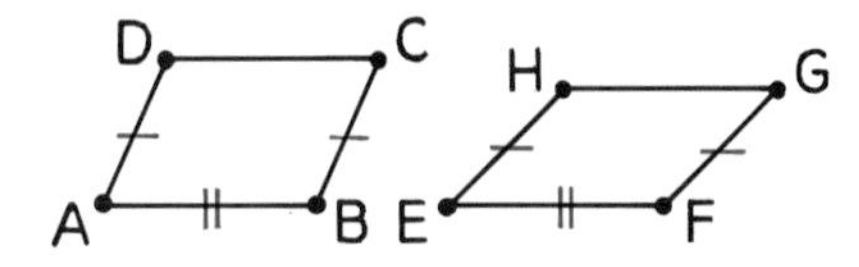

10. What other pairs of parts of these parallelograms can you conclude must also be equal?

11. If three sides of one parallelogram are equal to three sides of another parallelogram, does it follow that the parallelograms are congruent?

Set II

In ▱SLED, SL = $2(x + 5)$, LE = $x^2 + 6$, and ED = $3(10 - x)$.

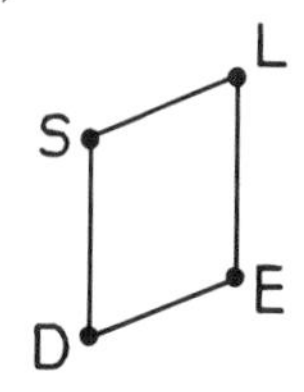

12. Find x.

13. Find SL.

14. Find LE.

15. Find ED.

In ▱CART, $\angle C = 3x°$ and $\angle A = (x + 40)°$.

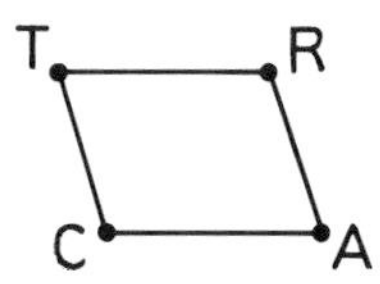

16. Find x.

17. Find $\angle C$.

18. Find $\angle A$.

19. Find $\angle R$.

20. Find $\angle T$.

$\overline{SA}$ and $\overline{GT}$ are diagonals of ▱STAG; GE = $x + 20$, SA = $2(6 - x)$, and ET = $2 - x$.

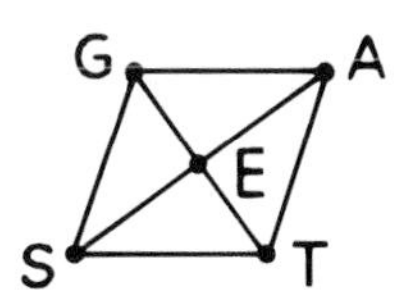

21. Find x.

22. Find GE.

23. Find SA.

24. Find SE.

25. Find GT.

26.

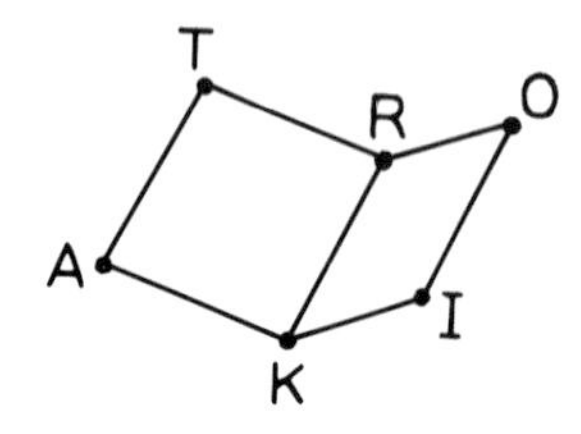

Given: TRKA and ROIK are parallelograms.

Prove: TA = OI.

27.

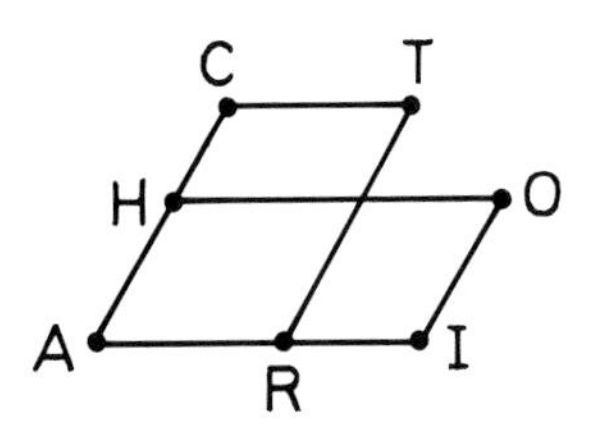

Given: CART and HAIO are parallelograms.

Prove: $\angle T = \angle O$.

28.

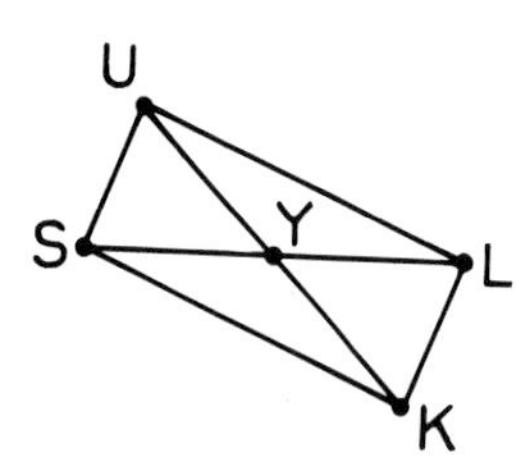

Given: $\overline{SU} \parallel \overline{KL}$ and $\overline{UL} \parallel \overline{SK}$.

Prove: SY = YL.

29.

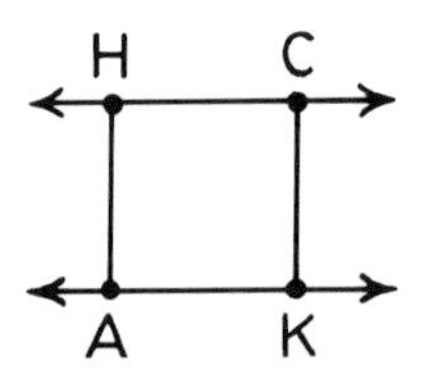

Given: $\overline{HA} \perp \overline{AK}$, $\overline{CK} \perp \overline{AK}$, and $\overline{HC} \parallel \overline{AK}$.

Prove: HC = AK.

30.

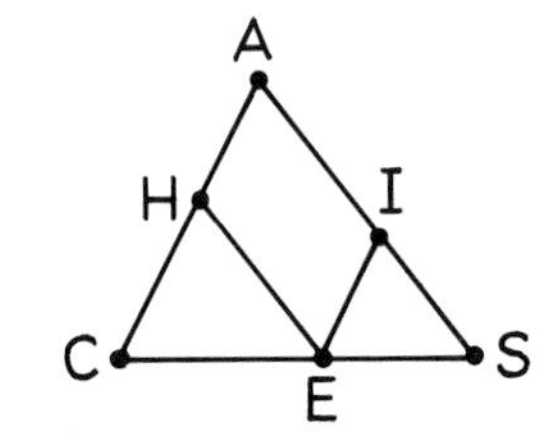

Given: AS = CS; $\overline{HE} \parallel \overline{AI}$ and $\overline{IE} \parallel \overline{AH}$.

Prove: $\angle C = \angle HEI$.

31.

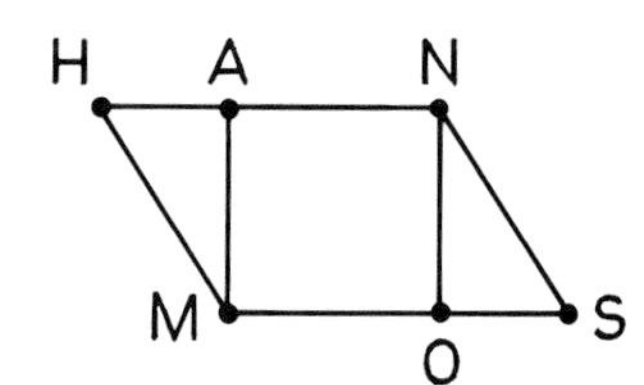

Given: HNSM is a parallelogram; $\overline{MA} \perp \overline{HN}$ and $\overline{NO} \perp \overline{MS}$.

Prove: $\triangle HAM \cong \triangle SON$.

Set III

Escher's animal mosaics are among his most remarkable pictures. You have already seen one of them on page 125.

In several of his mosaics, Escher began by dividing the plane into parallelograms. One of them is based on the two figures shown at the top of the next page. Trace the grid on the facing page and see if you can show how Escher filled it with just these two shapes. The position of one of them is shown to help you in getting started. Be sure to fill the entire grid.

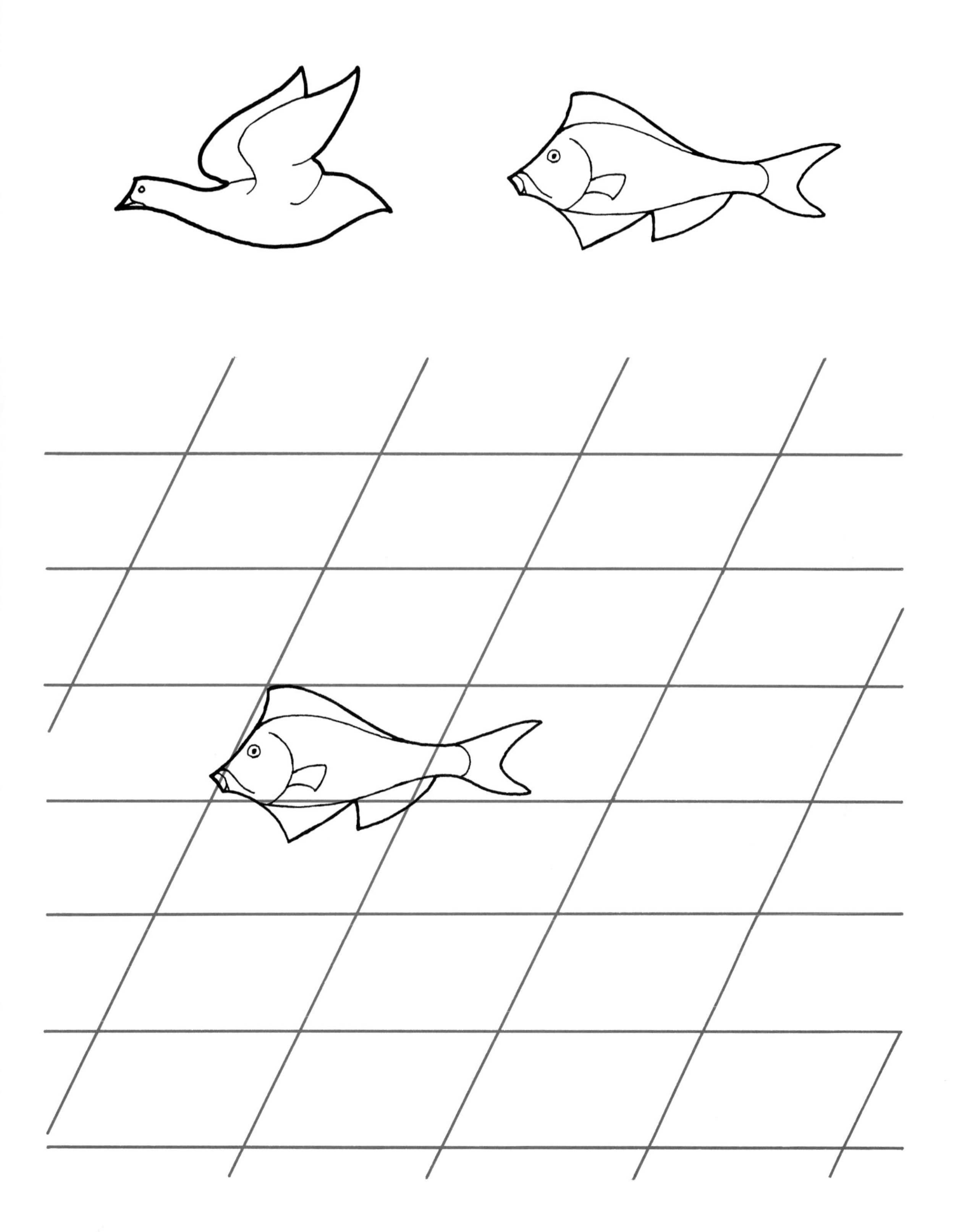

William Vandivert and Scientific American, Inc.

Lesson 3

Quadrilaterals That Are Parallelograms

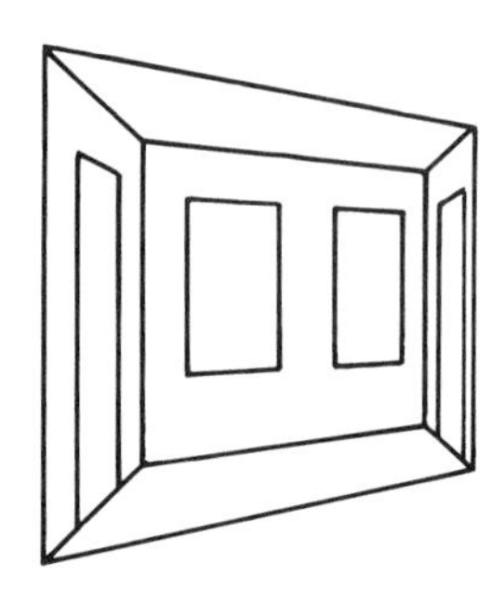

Room as "seen"

Actual room

Camera

Both photographs show the same person in the same room. How can this be?

We are fooled by the design of the room. Although the walls and windows seem to be rectangular, they are not. The first diagram at the left reveals the room's actual shape. The back wall is not rectangular because, although its left and right edges are parallel, the edges along the ceiling and floor are not. They slope toward each other at the right side of the room, so that the right wall is smaller than the left. Furthermore, the right edge of the back wall is much closer to us than the left edge. The actual floor plan of the room, in contrast with the one that we think we are seeing, is shown in the second diagram. Because we assume that the room is a normal one, we interpret the larger size of someone at the right to mean that that person *is* larger rather than merely closer to us.

An important part of the illusion is our presumption that all of the quadrilaterals in the room are *parallelograms*, which they are not. What do we need to know about a quadrilateral in order to be sure that it is a parallelogram?

There are five basic conditions, any one of which is sufficient to prove that a quadrilateral is a parallelogram:

1. Both pairs of opposite sides are parallel.
2. Both pairs of opposite sides are equal.
3. Two opposite sides are parallel and equal.
4. Both pairs of opposite angles are equal.
5. Its diagonals bisect each other.

The first of these conditions is the definition of a parallelogram. The proofs of the rest are given below and on the next page.

► Theorem 36
A quadrilateral is a parallelogram if both pairs of opposite sides are equal.

Given: In ABCD, DC = AB and AD = BC.
Prove: ABCD is a parallelogram.

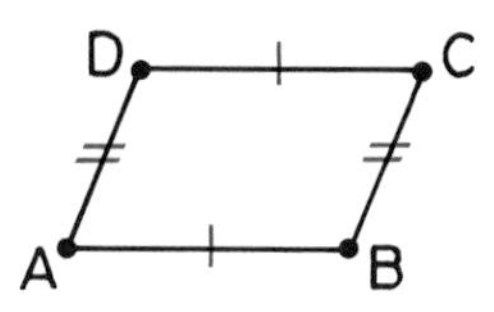

Proof.
Draw $\overline{DB}$. Because DC = AB and AD = BC (given), and DB = DB (reflexive), $\triangle DCB \cong \triangle BAD$ (S.S.S.). Therefore, $\angle 2 = \angle 3$ and $\angle 1 = \angle 4$ (corresponding angles of congruent triangles are equal).

So $\overline{DC} \parallel \overline{AB}$ and $\overline{AD} \parallel \overline{BC}$ (lines that form equal alternate interior angles with a transversal are parallel). ABCD is a parallelogram because both pairs of opposite sides are parallel.

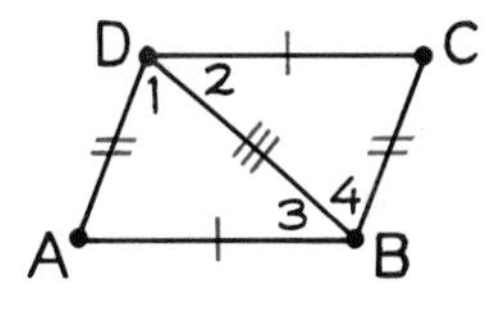

► Theorem 37
A quadrilateral is a parallelogram if two opposite sides are both equal and parallel.

Given: In ABCD, DC = AB and $\overline{DC} \parallel \overline{AB}$.
Prove: ABCD is a parallelogram.

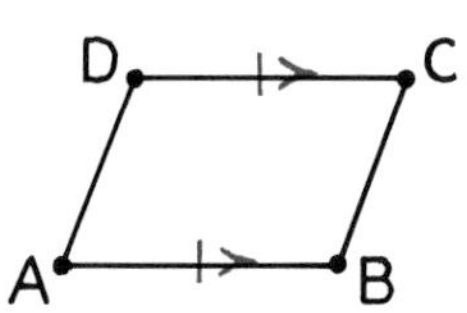

Proof.
Draw $\overline{DB}$. Because $\overline{DC} \parallel \overline{AB}$, $\angle 1 = \angle 2$. We also know that DC = AB and DB = DB, and so $\triangle DCB \cong \triangle BAD$ (S.A.S.). Therefore, AD = BC; so ABCD is a parallelogram (a quadrilateral is a parallelogram if both pairs of opposite sides are equal).

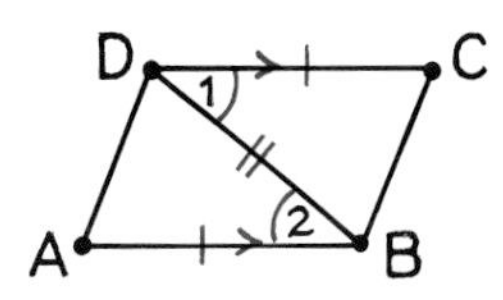

► **Theorem 38**
A quadrilateral is a parallelogram if both pairs of opposite angles are equal.

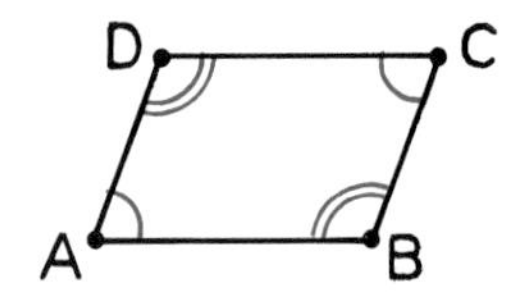

Given: In ABCD, $\angle A = \angle C$ and $\angle B = \angle D$.

Prove: ABCD is a parallelogram.

Proof.
$\angle A + \angle B + \angle C + \angle D = 360°$ (the sum of the measures of the angles of a quadrilateral is 360°). Because $\angle A = \angle C$ and $\angle B = \angle D$, we can substitute into this equation, getting $\angle A + \angle B + \angle A + \angle B = 360°$. Simplifying the left side of the equation, we get $2\angle A + 2\angle B = 360°$. Dividing both sides by 2 gives $\angle A + \angle B = 180°$.

This means that $\measuredangle A$ and $\measuredangle B$ are supplementary, and so $\overline{AD} \parallel \overline{BC}$ (lines that form supplementary interior angles on the same side of a transversal are parallel). Because $\angle B = \angle D$, it follows that $\measuredangle A$ and $\measuredangle D$ are also supplementary. Hence, $\overline{DC} \parallel \overline{AB}$; so ABCD is a parallelogram.

► **Theorem 39**
A quadrilateral is a parallelogram if its diagonals bisect each other.

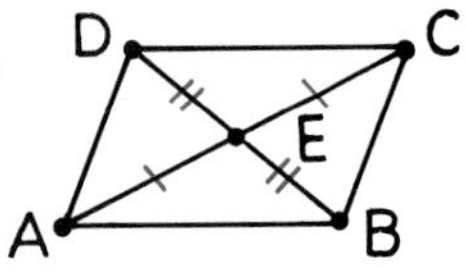

Given: In ABCD, $\overline{AC}$ and $\overline{DB}$ bisect each other.

Prove: ABCD is a parallelogram.

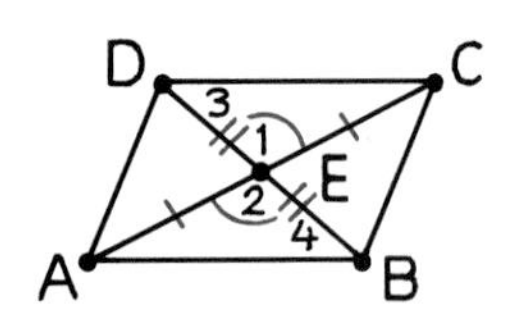

Proof.
Because $\overline{AC}$ and $\overline{DB}$ bisect each other, DE = EB and AE = EC. Also, $\angle 1 = \angle 2$, and so $\triangle DEC \cong \triangle BEA$ (S.A.S.). Therefore, DC = AB and $\angle 3 = \angle 4$. Because $\angle 3 = \angle 4$, $\overline{DC} \parallel \overline{AB}$. It follows that ABCD is a parallelogram because we have proved two opposite sides both equal and parallel.

Exercises

Set I

Without looking them up, write the definitions of the following geometric terms.

1. Quadrilateral.
2. Parallelogram.
3. Rhombus.
4. Rectangle.

Use the marked parts of the figures below to tell whether each one must be a parallelogram. If it is a parallelogram, state a definition or theorem that explains why.

5. 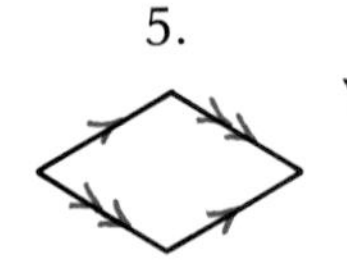

6. 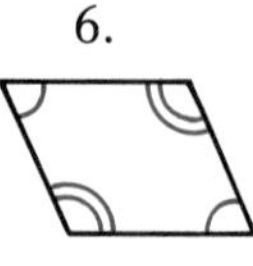

7. 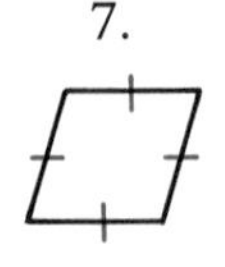

8. 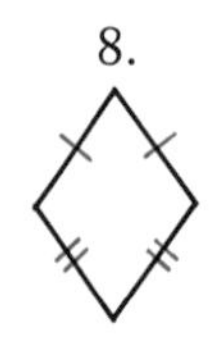

9. 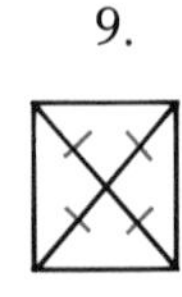

10. 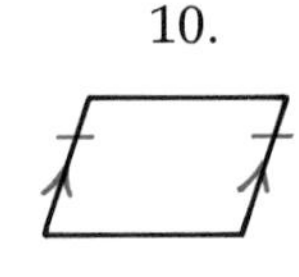

11. 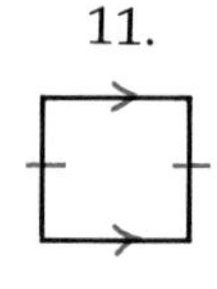

12. 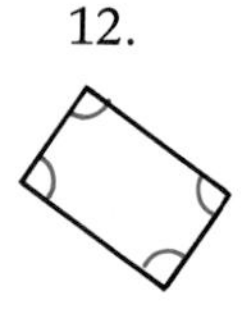

In the figure below, $\overline{BI}$ and $\overline{CU}$ are diagonals of quadrilateral BUIC. What else would you have to know in addition to each of the following statements in order to conclude that BUIC is a parallelogram?

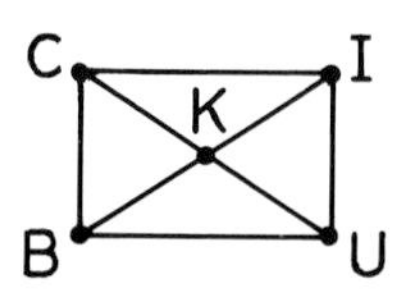

Example: BC = UI.

Answer: $\overline{BC} \parallel \overline{UI}$ or CI = BU.

13. $\angle$CBU = $\angle$CIU.
14. $\overline{BC} \parallel \overline{UI}$.
15. BK = KI.

In the figure below, PONT is a rectangle, NTIA is a rhombus, and ONAC is a parallelogram.

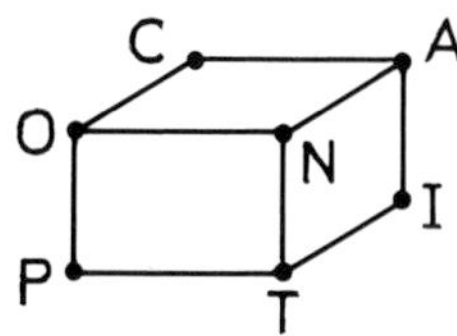

16. What can you conclude about the angles of PONT from the fact that it is a rectangle?
17. Why is PONT a parallelogram?
18. What can you conclude about the sides of NTIA from the fact that it is a rhombus?
19. Why is NTIA a parallelogram?

Tell whether each of the following statements is true or false.

20. If a quadrilateral is a parallelogram, then at least two opposite sides are parallel.
21. A quadrilateral is a parallelogram if at least two opposite sides are parallel.
22. If a quadrilateral is a trapezoid, then exactly two opposite sides are parallel.
23. A quadrilateral is a trapezoid if exactly two opposite sides are parallel.
24. If a quadrilateral is a parallelogram, then its diagonals bisect each other.
25. If a quadrilateral is not a parallelogram, then its diagonals do not bisect each other.
26. If a quadrilateral is a parallelogram, then it is equilateral.
27. A quadrilateral is a parallelogram if it is equilateral.

Set II

28.

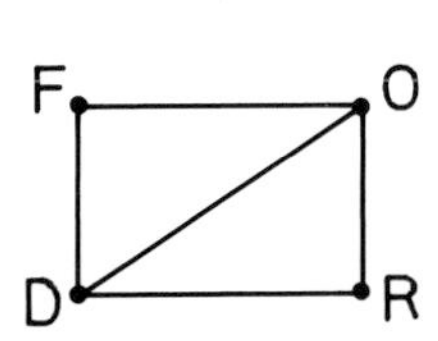

Given: $\triangle$DFO $\cong$ $\triangle$ORD.

Prove: FORD is a parallelogram.

29.

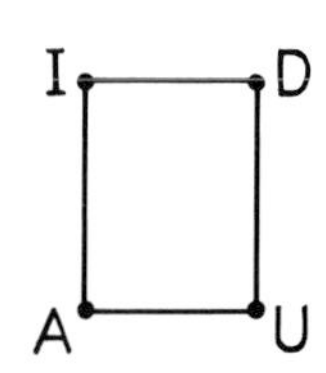

Given: $\overline{ID} \perp \overline{DU}$ and $\overline{AU} \perp \overline{DU}$; ID = AU.

Prove: AUDI is a parallelogram.

30.

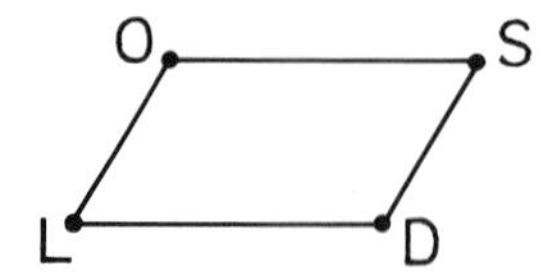

Given: $\angle O = \angle D$ and $\angle L = \angle S$.

Prove: OL = DS.

31.

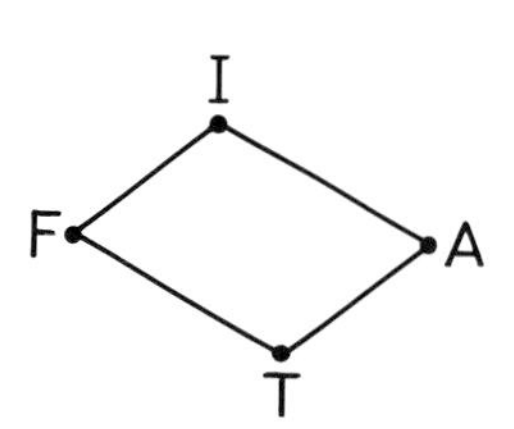

Given: FI = TA and IA = FT.

Prove: ∡F and ∡I are supplementary.

32.

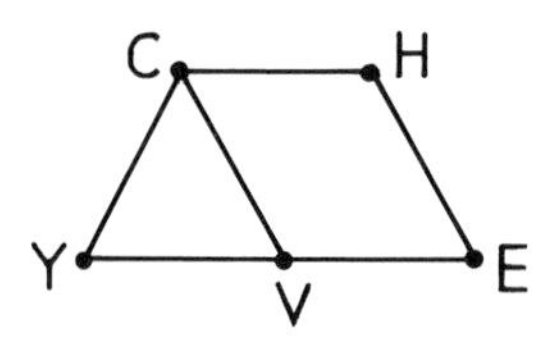

Given: Quadrilateral CHEY with $\overline{CV} \parallel \overline{HE}$ and CV = HE.

Prove: CHEY is a trapezoid.

33.

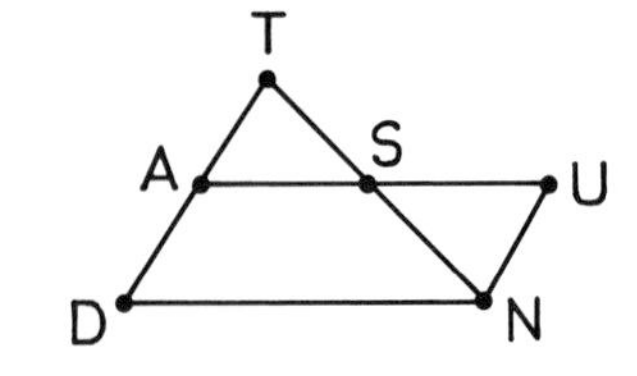

Given: $\triangle ATS \cong \triangle UNS$; A is the midpoint of $\overline{DT}$.

Prove: DAUN is a parallelogram.

Set III

Obtuse Ollie invented the following theorem and proof.

A diagonal of a parallelogram bisects two of its angles.

Given: ▱ABCD with diagonal $\overline{AC}$.

Prove: $\overline{AC}$ bisects ∡DAB and ∡DCB.

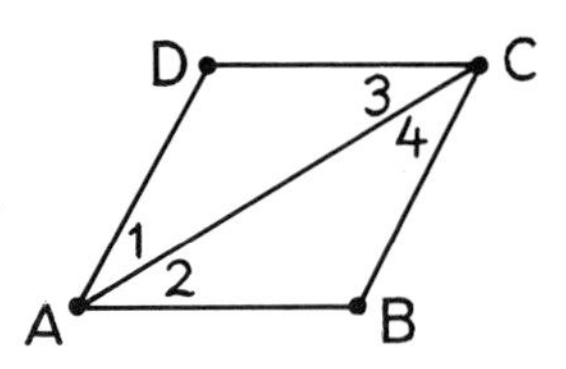

Proof.
Because ABCD is a parallelogram, AD = BC and DC = AB. Also, AC = AC; so $\triangle ADC \cong \triangle ABC$. Therefore, $\angle 1 = \angle 2$ and $\angle 3 = \angle 4$.

1. Do you think that the theorem is true?

2. Is Ollie's proof correct? Explain.

Courtesy of Bell Laboratories

Lesson 4
Rectangles, Rhombuses, and Squares

This pattern of dark and light squares is actually a picture of a famous man. If you look at it from several feet away, the portrait looks quite different. It was created at Bell Laboratories in an experiment to determine how much detail a picture must contain in order to be recognizable.

The basic shape used within the picture is the square and the picture itself is in the shape of a rectangle. The rectangle is undoubtedly the most commonly used of all quadrilateral shapes. It is easy to prove that a rectangle is a parallelogram.

► **Theorem 40**
All rectangles are parallelograms.

Given: ABCD is a rectangle.
Prove: ABCD is a parallelogram.

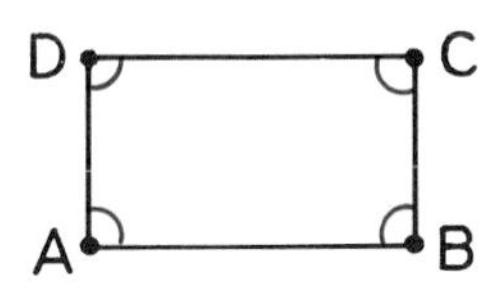

Proof.
Because ABCD is a rectangle, $\angle A = \angle C$ and $\angle B = \angle D$ (a quadrilateral is a rectangle iff all of its angles are equal). So ABCD must also be a parallelogram (a quadrilateral is a parallelogram if both pairs of opposite angles are equal).

Because every rectangle is a parallelogram, it follows that the diagonals of a rectangle bisect each other. We can also prove that they are equal.

► **Theorem 41**
The diagonals of a rectangle are equal.

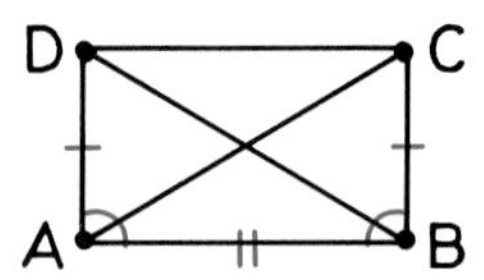

Given: ABCD is a rectangle.
Prove: DB = CA.

Proof.
Because ABCD is a rectangle, ∠DAB = ∠CBA (a quadrilateral is a rectangle iff all of its angles are equal). ABCD is also a parallelogram (all rectangles are parallelograms), and so DA = CB (the opposite sides of a parallelogram are equal). Also, AB = AB. Therefore, △DAB ≅ △CBA (S.A.S.), and so DB = CA.

As we did for rectangles, we can prove that all rhombuses also are parallelograms.

► **Theorem 42**
All rhombuses are parallelograms.

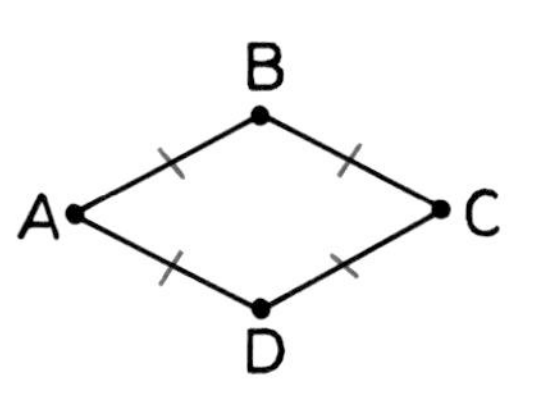

Given: ABCD is a rhombus.
Prove: ABCD is a parallelogram.

Proof.
Because ABCD is a rhombus, AB = DC and BC = AD (a quadrilateral is a rhombus iff all of its sides are equal). So ABCD must also be a parallelogram (a quadrilateral is a parallelogram if both pairs of opposite sides are equal).

Because rhombuses, like rectangles, are parallelograms, it follows that the diagonals of a rhombus bisect each other. It is also easy to prove that they are perpendicular.

► **Theorem 43**
The diagonals of a rhombus are perpendicular.

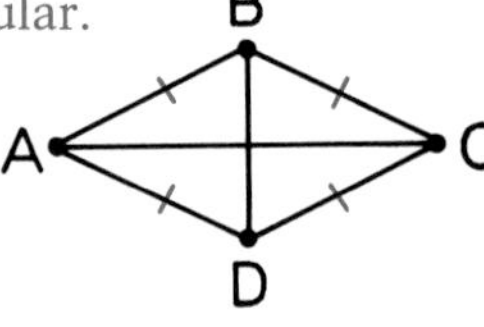

Given: ABCD is a rhombus.
Prove: $\overline{BD} \perp \overline{AC}$.

Proof.
Because ABCD is a rhombus, BA = BC and DA = DC. This means that points B and D are equidistant from the endpoints of $\overline{AC}$. Therefore, $\overline{BD} \perp \overline{AC}$ (in a plane, two points each equidistant from the endpoints of a line segment determine the perpendicular bisector of the line segment).

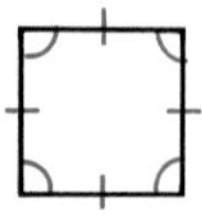

We do not need to prove any theorems about squares due to the fact that, because a square is a quadrilateral that is both equilateral and equiangular, every square is also a rhombus, rectangle, and parallelogram. Therefore, squares possess every property that we have proved for these other figures.

Exercises

Set I

Complete the following definitions and theorems.

1. The opposite sides of a parallelogram are ||||||||. (Definition.)
2. The consecutive angles of a parallelogram are ||||||||.
3. All of the sides of a rhombus are ||||||||.
4. Each angle of a rectangle is a ||||||||.
5. A quadrilateral is a parallelogram if two opposite sides are ||||||||.
6. The diagonals of a parallelogram ||||||||.

Quadrilateral BYRO is a rhombus, and $\overline{BR}$ and $\overline{OY}$ are its diagonals.

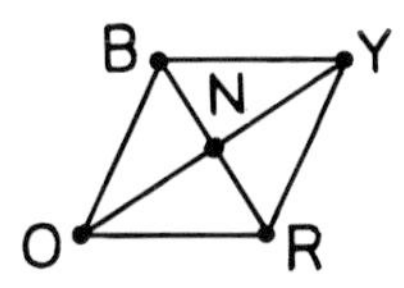

7. What kind of triangle is $\triangle$BYR?
8. How do you know?
9. What kind of triangle is $\triangle$BYN?
10. How do you know?

Quadrilateral KEAT is a rectangle, and $\overline{KA}$ and $\overline{TE}$ are its diagonals.

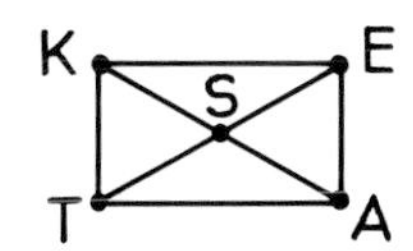

11. What kind of triangle is $\triangle$TEA?
12. How do you know?
13. What kind of triangle is $\triangle$TSA?
14. How do you know?

In quadrilateral TWAI, $\overline{TA}$ and $\overline{IW}$ bisect each other.

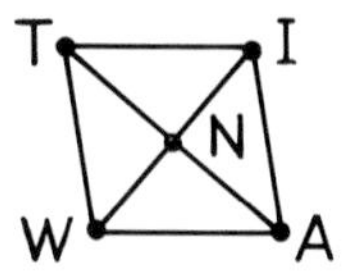

15. What kind of quadrilateral is TWAI?
16. How do you know?
17. Does it follow that $\overline{TA} \perp \overline{IW}$?
18. Does it follow that TA = IW?

Quadrilateral HUGO is a square.

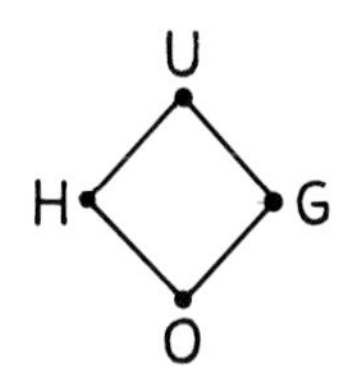

19. Why is HUGO a rhombus?
20. Why is HUGO a parallelogram?
21. Why is HUGO a rectangle?

Quadrilateral JAME is a rectangle with diagonals $\overline{JM}$ and $\overline{AE}$; JA = 10, JS = 13, and JE = 24. Use this information to find each of the following perimeters.

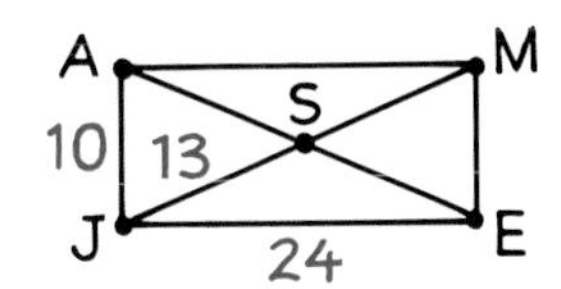

22. The perimeter of rectangle JAME.
23. The perimeter of $\triangle$JME.
24. The perimeter of $\triangle$JSE.
25. The perimeter of $\triangle$MSE.

Set II

Use your ruler to draw each of the following figures. (Draw the diagonals first.)

26. A quadrilateral whose diagonals bisect each other.

27. A quadrilateral with perpendicular diagonals that is not a rhombus.

28. A quadrilateral with equal diagonals that is not a rectangle.

29. A quadrilateral whose diagonals are perpendicular, equal, and bisect each other.

30.

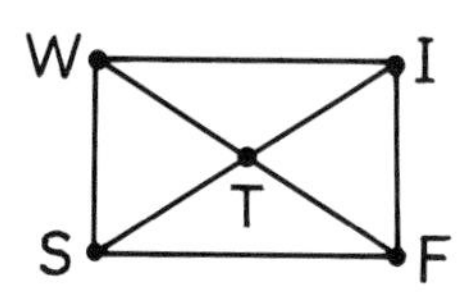

Given: SWIF is a parallelogram; WF = IS.

Prove: ∠SWI = ∠WIF.

31.

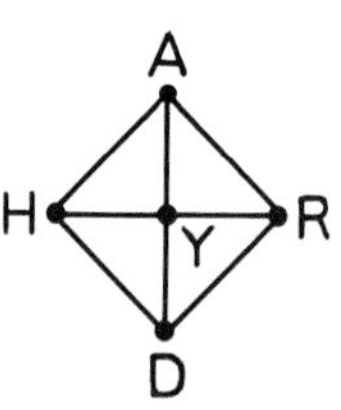

Given: HARD is a square.

Prove: HR = AD.

32.

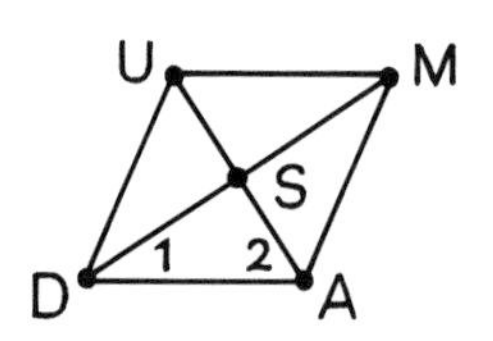

Given: DUMA is a rhombus with diagonals $\overline{DM}$ and $\overline{UA}$.

Prove: ∡1 and ∡2 are complementary.

33.

Given: ASTN is a rectangle; AU = ET.

Prove: NU = ES.

34.

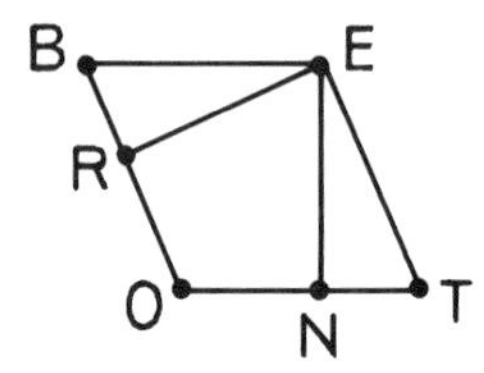

Given: BOTE is a rhombus; $\overline{ER} \perp \overline{BO}$ and $\overline{EN} \perp \overline{OT}$.

Prove: ER = EN.

35.

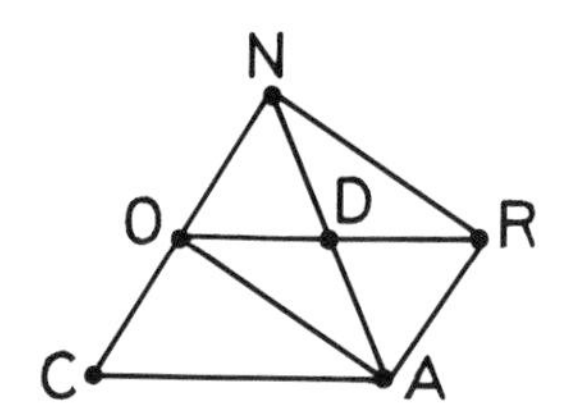

Given: NRAO is a rectangle; CORA is a parallelogram.

Prove: △CNA is isosceles.

Set III

Dilcue, a backward geometry student, had a nightmare because of studying too hard for his geometry final. He dreamed that four congruent right triangles came marching toward him and surrounded him.

The triangles lined up their sides so that ABCD was a square. Dilcue noticed that EFGH also seemed to be a square but was too scared to be able to think clearly enough to figure out why. Can you?

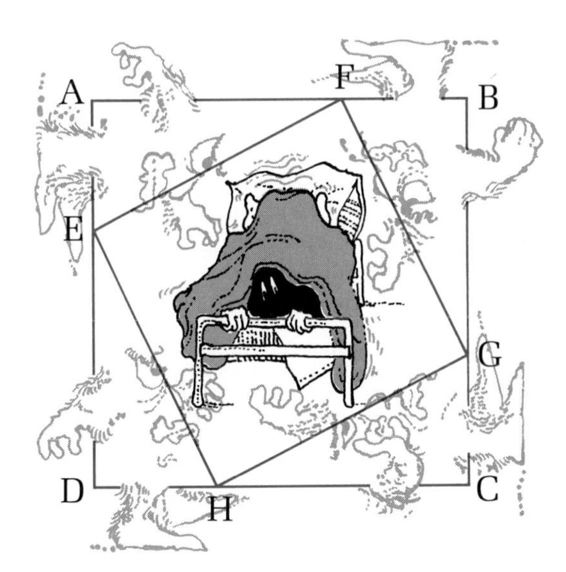

Lesson 5

Trapezoids

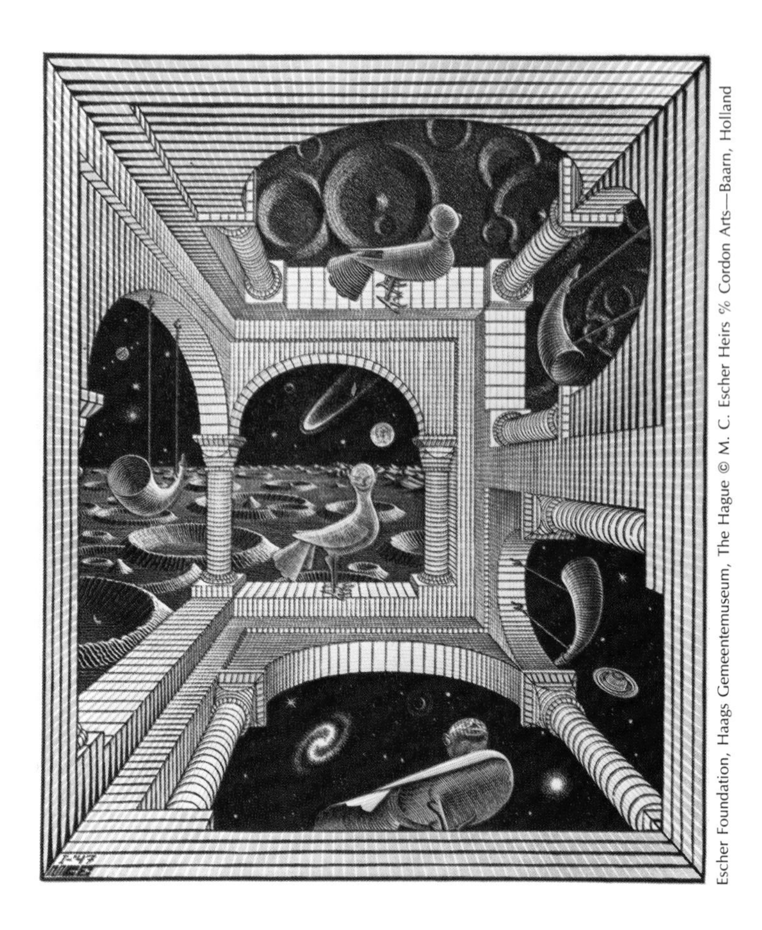

Escher Foundation, Haags Gemeentemuseum, The Hague © M. C. Escher Heirs % Cordon Arts—Baarn, Holland

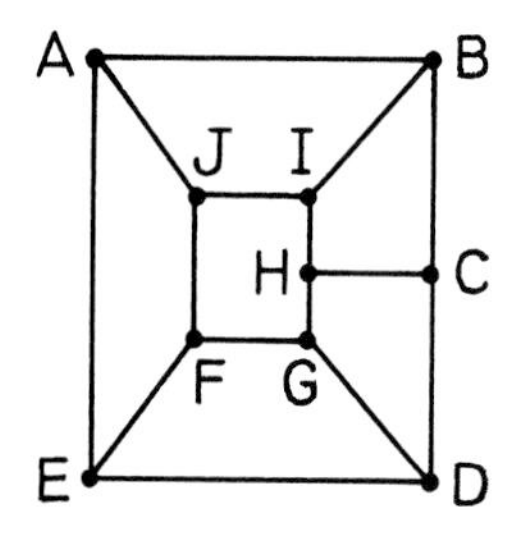

This woodcut by Escher titled *Another World* shows the interior of a strange building in which the walls, floor, and ceiling are interchangeable. The view through the windows at the top and upper right indicates that we are looking down from the ceiling. The view through the windows at the bottom and lower right, on the other hand, is from the floor up toward the sky. And through the center and left-hand windows we are looking at the horizon. Is the rectangular region in the center a *wall, floor,* or *ceiling?* It depends on your point of view.

The design of the drawing reveals that rectangles seen in perspective may look like another type of quadrilateral instead: the trapezoid, in which just one pair of sides are parallel.

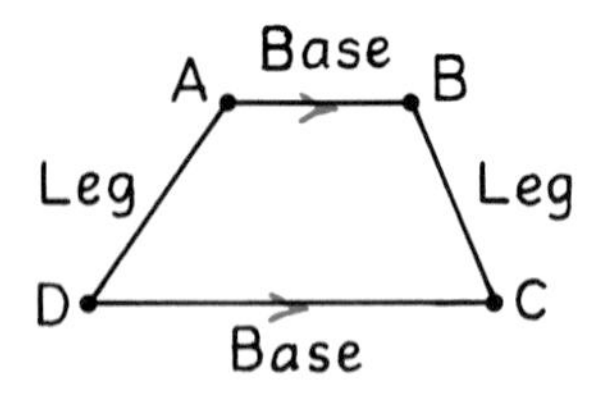

The parallel sides of a trapezoid are called its *bases* and the nonparallel sides are called its *legs.* The pairs of angles that include each base are called *base angles:* one pair of base angles in trapezoid ABCD is ∡A and ∡B and the other pair is ∡C and ∡D.

The base angles of a trapezoid do not ordinarily have any special relation. If the legs of a trapezoid are equal, however, then we can prove that the base angles are equal.

▶ **Definition**

An ***isosceles trapezoid*** is a trapezoid whose legs are equal.

▶ Theorem 44
The base angles of an isosceles trapezoid are equal.

Given: Isosceles trapezoid ABCD with bases $\overline{AB}$ and $\overline{DC}$.

Prove: $\angle A = \angle B$ and $\angle C = \angle D$.

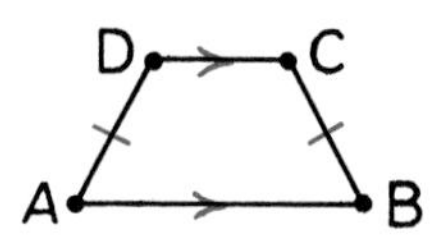

Proof.
Because $\overline{AB}$ and $\overline{DC}$ are the bases of the trapezoid, $\overline{AB} \parallel \overline{DC}$. Through C, draw $\overline{CE} \parallel \overline{DA}$ (through a point not on a line, there is exactly one parallel to the line). AECD is a parallelogram, and so DA = CE (the opposite sides of a parallelogram are equal). Also, DA = CB (the legs of an isosceles trapezoid are equal), and so CE = CB (substitution).

It follows that $\angle CEB = \angle B$ (if two sides of a triangle are equal, the angles opposite them are equal). Also, because $\overline{CE} \parallel \overline{DA}$, $\angle CEB = \angle A$ (parallel lines form equal corresponding angles with a transversal), and so $\angle A = \angle B$ (substitution).

In the original figure, $\measuredangle D$ and $\measuredangle A$ are supplementary, as are $\measuredangle C$ and $\measuredangle B$ (parallel lines form supplementary interior angles on the same side of a transversal). Therefore, $\angle D = \angle C$ (supplements of equal angles are equal).

By using this theorem, it is easy to prove that the diagonals of an isosceles trapezoid, like those of a rectangle, are equal.

▶ Theorem 45
The diagonals of an isosceles trapezoid are equal.

Given: Isosceles trapezoid ABCD with bases $\overline{AB}$ and $\overline{DC}$.

Prove: DB = CA.

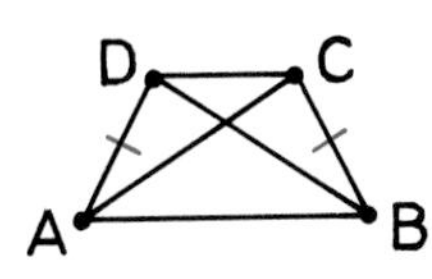

Proof.
Because ABCD is an isosceles trapezoid, $\angle DAB = \angle CBA$ (the base angles of an isosceles trapezoid are equal). Also, DA = CB (the legs of an isosceles trapezoid are equal), and AB = AB. Therefore, $\triangle DAB \cong \triangle CBA$ (S.A.S.), and so DB = CA.

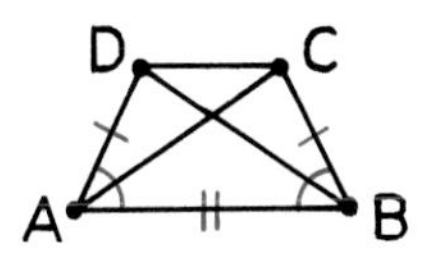

Exercises

Set I

Quadrilateral OKAY is a trapezoid in which $\overline{OK} \parallel \overline{YA}$.

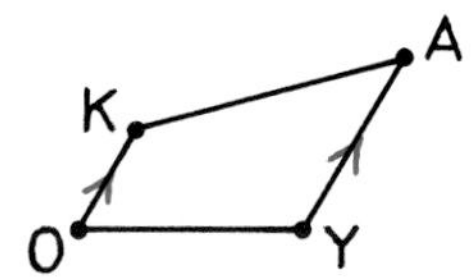

1. Which sides are its bases?
2. Which sides are its legs?
3. Does OKAY appear to be isosceles?

Quadrilateral FINE is an isosceles trapezoid with bases $\overline{FI}$ and $\overline{EN}$.

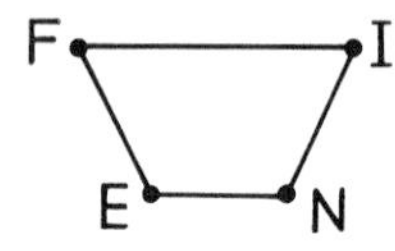

4. Why is $\overline{FI} \parallel \overline{EN}$?
5. Why is FE = IN?
6. Why is $\angle F = \angle I$?
7. Why are $\measuredangle I$ and $\measuredangle N$ supplementary?

In quadrilateral NICE, $\angle N = 93°$, $\angle I = 84°$, and $\angle C = 96°$.

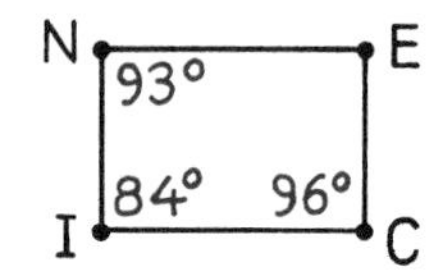

8. Find $\angle E$.
9. Which sides of NICE are parallel?
10. Why?
11. What kind of quadrilateral is NICE?

Quadrilateral PRIZ is an isosceles trapezoid with diagonals $\overline{PI}$ and $\overline{ZR}$. Use this information to tell whether each of the following statements *must be true, may be true,* or *appears to be false.*

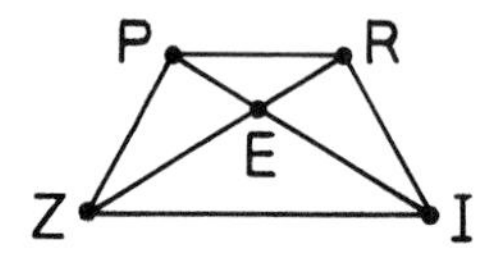

12. PI = ZR.
13. $\overline{PI}$ and $\overline{ZR}$ bisect each other.
14. $\overline{PI} \perp \overline{ZR}$.
15. $\triangle PZI \cong \triangle RIZ$.
16. $\triangle PER \cong \triangle ZEI$.
17. $\triangle ZPR$ is an isosceles triangle.
18. $\triangle ZEI$ is an isosceles triangle.

Tell whether each of the following statements is true or false.

19. If a quadrilateral is a trapezoid, its base angles are equal.
20. A trapezoid can have three equal sides.
21. A trapezoid can have four equal angles.
22. A trapezoid can have three right angles.
23. If the diagonals of a quadrilateral bisect each other, it is not a trapezoid.

Quadrilateral BEST is a trapezoid with bases $\overline{TS}$ and $\overline{BE}$; $\angle B = 5(x - 4)°$, $\angle E = 4(x + 13)°$, and $\angle S = 7(x - 10)°$.

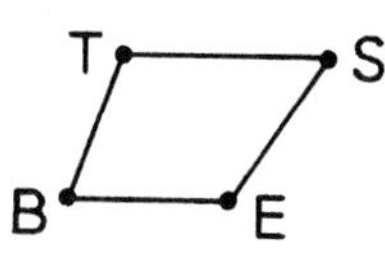

24. Find x.
25. Find $\angle B$.
26. Find $\angle E$.
27. Find $\angle S$.
28. Find $\angle T$.

Set II

29.

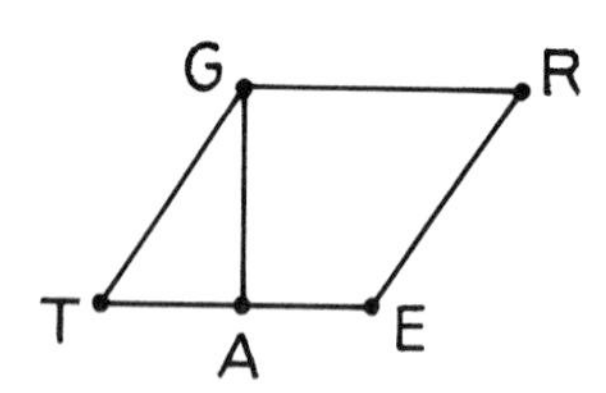

Given: GREA is a trapezoid with bases $\overline{GR}$ and $\overline{AE}$; GR = TE.

Prove: GRET is a parallelogram.

30.

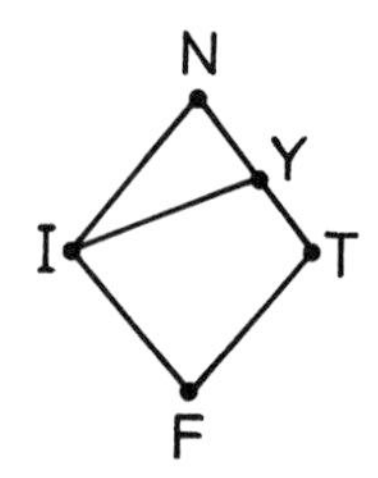

Given: NIFT is a rhombus.

Prove: IFTY is a trapezoid.

31.

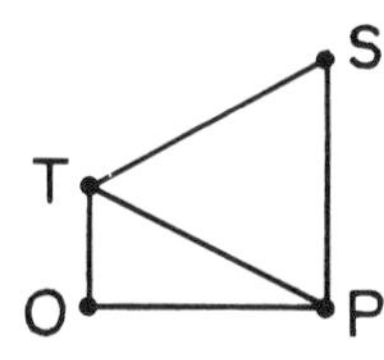

Given: TS = TP;
$\angle$S = $\angle$OTP.

Prove: TOPS is a trapezoid.

32.

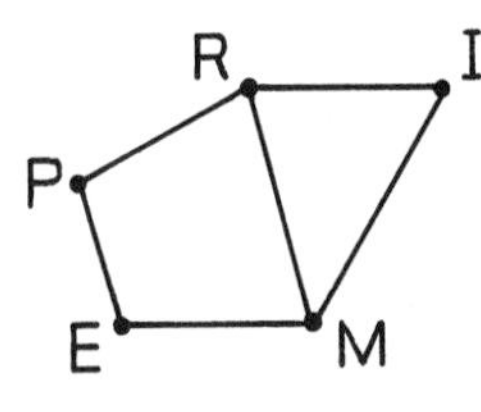

Given: PRME is an isosceles trapezoid with bases $\overline{PE}$ and $\overline{RM}$;
$\overrightarrow{RM}$ bisects $\measuredangle$PRI.

Prove: $\overline{RI} \parallel \overline{EM}$.

33.

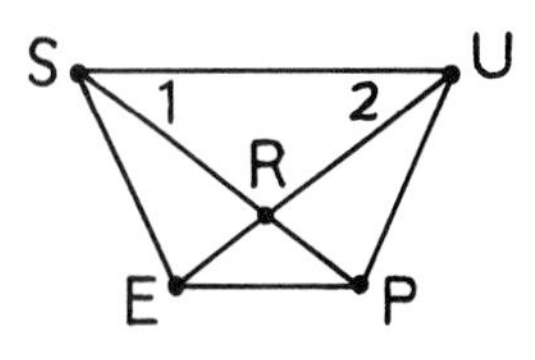

Given: SUPE is an isosceles trapezoid with bases $\overline{SU}$ and $\overline{EP}$ and diagonals $\overline{SP}$ and $\overline{EU}$.

Prove: $\angle 1 = \angle 2$.

34.

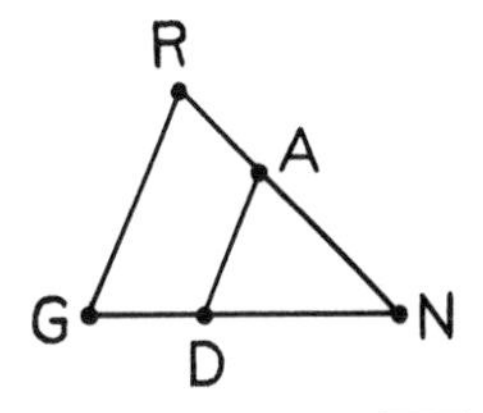

Given: $\triangle$GRN with $\overline{DA} \parallel \overline{GR}$ and RA = GD.

Prove: $\triangle$GRN is isosceles.

Set III

The following problem is from a popular puzzle book published in the Soviet Union.*

An equilateral triangle can be separated into four congruent triangles as shown in the figures below.

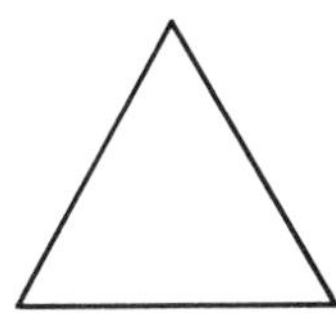

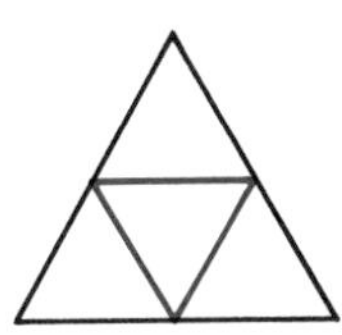

Without the top triangle, the three remaining triangles form a trapezoid. Trace the copy of it shown below. Can you figure out how to separate it into four congruent parts?

*An English language edition of *The Moscow Puzzles,* by Boris A. Kordemsky, was edited by Martin Gardner (Scribners, 1972).

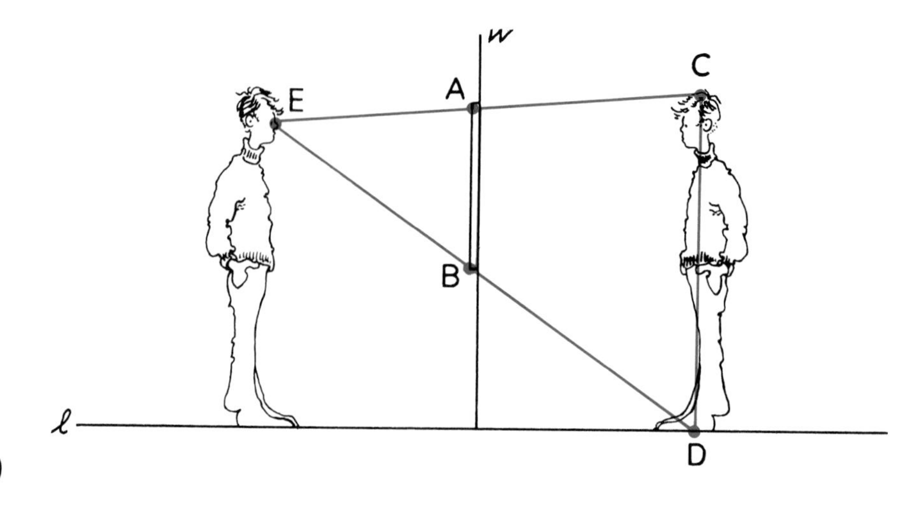

Lesson 6

The Midsegment Theorem

Suppose that you are looking at a reflection of yourself in a large mirror on a wall. How tall would the mirror have to be in order for you to see all of yourself from the top of your head to the bottom of your feet?

When you stand before a mirror, your reflection seems to be standing an equal distance behind it. In the diagram above, line w represents the wall with the mirror on it and line ℓ represents the floor. The person looking in the mirror is shown at the left with his eyes at point E. His reflection, which seems to stand behind the mirror, is represented by $\overline{CD}$ and the mirror itself is represented by $\overline{AB}$. Points A and B lie on the sides of $\triangle ECD$; so AB, the length of the mirror, is determined by this triangle.

It can be shown, using the reflection properties of a mirror, that A is the midpoint of $\overline{EC}$ and that B is the midpoint of $\overline{ED}$. Because of this, $\overline{AB}$ is called a *midsegment* of $\triangle ECD$.

► **Definition**
A ***midsegment*** of a triangle is a line segment that joins the midpoints of two of its sides.

We can use this definition and our knowledge of parallelograms to prove the following theorem.

► **Theorem 46** (The Midsegment Theorem)
A midsegment of a triangle is parallel to the third side and half as long.

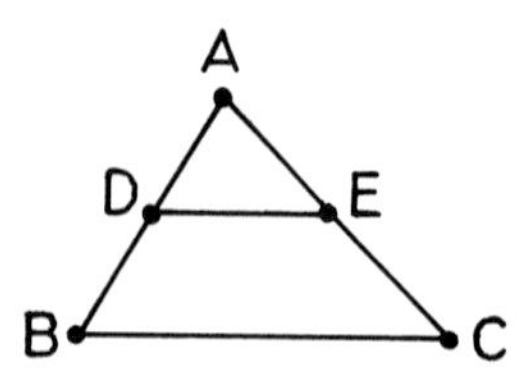

Given: $\triangle ABC$ with midsegment $\overline{DE}$.

Prove: $\overline{DE} \parallel \overline{BC}$ and $DE = \frac{1}{2}BC$.

It is impossible to prove either part of the conclusion without adding something to the figure. Our method will be to draw $\triangle CFE \cong \triangle ADE$ as shown in the figure at the right. We can then show that BCFD is a parallelogram. One side of this parallelogram contains the midsegment $\overline{DE}$; both conclusions about the midsegment of a triangle can be derived from this fact.

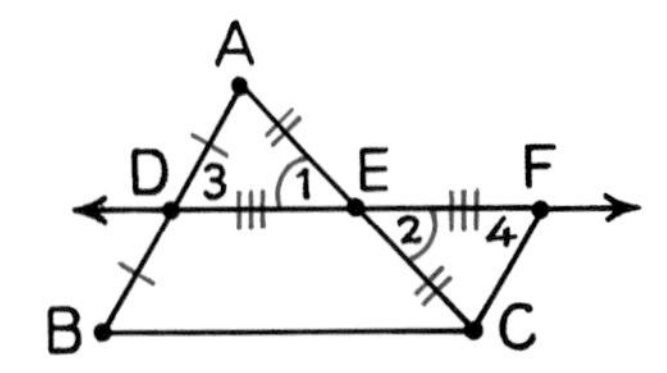

Proof.
Draw $\overleftrightarrow{DE}$ (two points determine a line), choose point F on $\overleftrightarrow{DE}$ so that EF = DE (the Ruler Postulate), and draw $\overline{CF}$.

In $\triangle ADE$ and $\triangle CFE$, DE = EF, $\angle 1 = \angle 2$ (vertical angles are equal), and AE = EC (because E is the midpoint of $\overline{AC}$, it divides it into two equal segments). So $\triangle ADE \cong \triangle CFE$ (S.A.S.).

Now BD = DA (because D is the midpoint of $\overline{BA}$), and DA = CF (corresponding sides of congruent triangles are equal), and so BD = CF (transitive). Also, $\angle 3 = \angle 4$ (corresponding angles of congruent triangles are equal); so $\overline{BA} \parallel \overline{CF}$ (two lines that form equal alternate interior angles with a transversal are parallel). Hence $\overline{BD} \parallel \overline{CF}$.

Because BD = CF and $\overline{BD} \parallel \overline{CF}$, BCFD is a parallelogram (a quadrilateral is a parallelogram if two sides are parallel and equal). Therefore, $\overline{DF} \parallel \overline{BC}$ and DF = BC (the opposite sides of a parallelogram are both parallel and equal). Hence $\overline{DE} \parallel \overline{BC}$.

We also know that DE = $\frac{1}{2}$DF because E is the midpoint of $\overline{DF}$, and the midpoint of a line segment divides it into segments half as long. Because DF = BC, we have DE = $\frac{1}{2}$BC (substitution).

The midsegment theorem can be used to discover how long the mirror on the wall must be in order for you to see your complete reflection. In the figure at the beginning of this lesson, $\overline{AB}$ is a midsegment of $\triangle ECD$, and so AB = $\frac{1}{2}$CD. The segment $\overline{CD}$, your image in the mirror, is just as tall as you are. So it is sufficient for the mirror to be half as long as you are tall.

Exercises

Set I

$\overline{AR}$, $\overline{AL}$, and $\overline{RL}$ connect points on the sides of $\triangle TYO$.

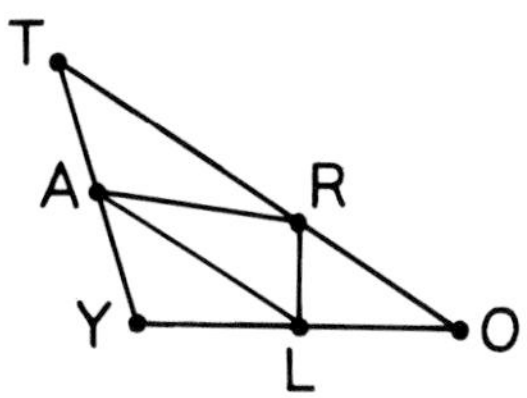

1. Which segment looks like it might be a midsegment of $\triangle TYO$?
2. Define midsegment.
3. How many midsegments can a triangle have?
4. State the Midsegment Theorem.

In the figure below, $\overline{RN}$ and $\overline{NM}$ are midsegments of $\triangle TUA$. $\angle T = 40°$, $\angle U = 70°$, TR = 25, and RN = 17.

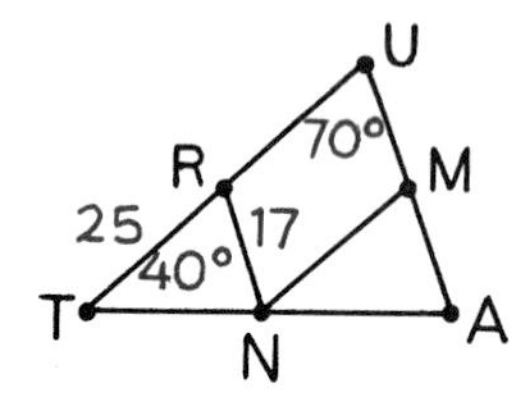

5. What kind of quadrilateral is RUMN?
6. Find $\angle RNM$.
7. Find $\angle A$.
8. What kind of triangle is $\triangle TUA$?
9. Find UM.
10. Find TU.
11. Find TA.
12. Find the perimeter of RUMN.
13. Find the perimeter of $\triangle TUA$.

In the figure below, $\overline{TA}$ is a midsegment of $\triangle GRN$. $\angle G = 60°$, $\angle R = 60°$, and GR = 16.

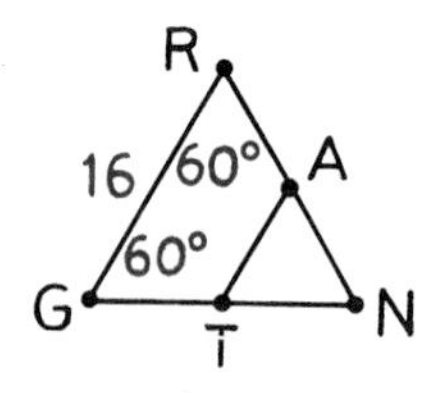

14. Find $\angle N$.
15. What kind of triangle is $\triangle GRN$?
16. What kind of triangle is $\triangle TAN$?
17. What kind of quadrilateral is GRAT?
18. Find AT.
19. Find RA.
20. Find the perimeter of $\triangle GRN$.
21. Find the perimeter of $\triangle TAN$.
22. Find the perimeter of GRAT.

In the figure below, $\overline{RL}$ is a midsegment of $\triangle TYE$. $TY = x + 40$, $YL = x^2 - 15$, and $LR = 3x + 35$.

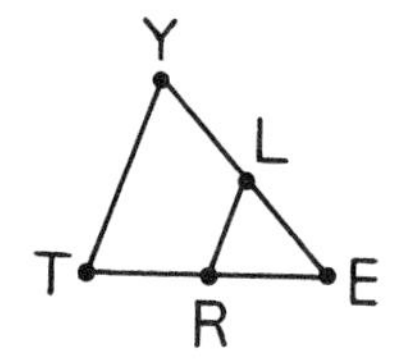

23. Find x.
24. Find TY.
25. Find YL.
26. Find LR.
27. Find YE.

Set II

Trace quadrilateral ABCD.

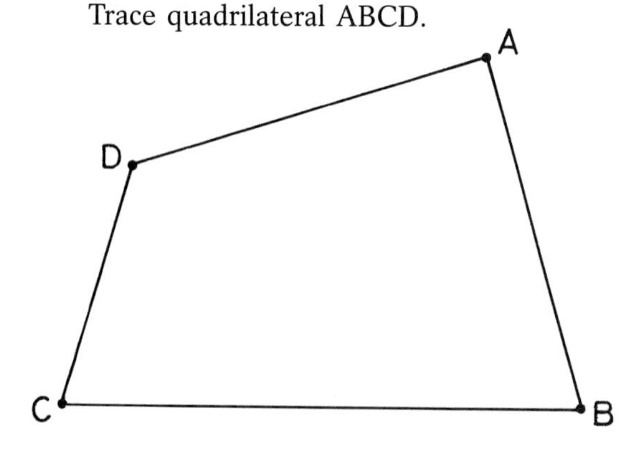

28. Use your straightedge and compass to bisect each of its sides. Starting with the midpoint of side $\overline{AB}$, label the midpoints (moving clockwise around the figure) E, F, G, and H, respectively. Draw quadrilateral EFGH.
29. What kind of quadrilateral does EFGH seem to be?
30. Why is $\overline{HE} \parallel \overline{DB}$ and $\overline{GF} \parallel \overline{DB}$?
31. Why is $\overline{HE} \parallel \overline{GF}$?

32. Why is $HE = \frac{1}{2}DB$ and $GF = \frac{1}{2}DB$?

33. Why is HE = GF?

34. Why is EFGH a parallelogram?

35.

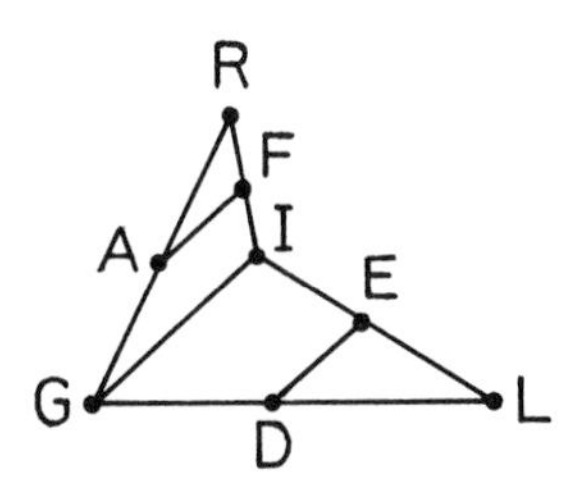

Given: $\overline{AF}$ and $\overline{DE}$ are midsegments of $\triangle GRI$ and $\triangle GIL$.

Prove: AF = DE.

36.

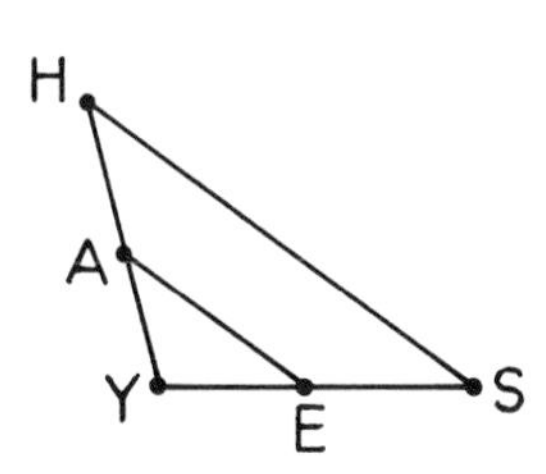

Given: A and E are the midpoints of sides $\overline{HY}$ and $\overline{YS}$ of $\triangle HYS$.

Prove: HAES is a trapezoid.

37.

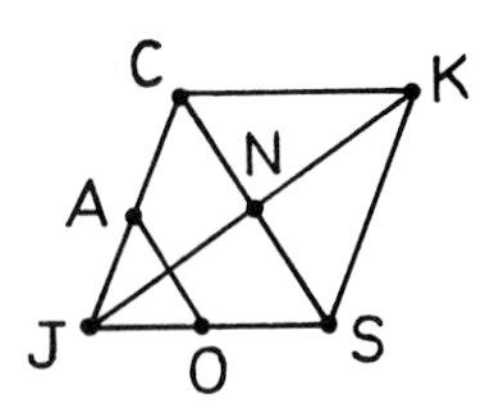

Given: JCKS is a rhombus; $\overline{AO}$ is a midsegment of $\triangle JCS$.

Prove: $\overline{AO} \perp \overline{JK}$.

38.

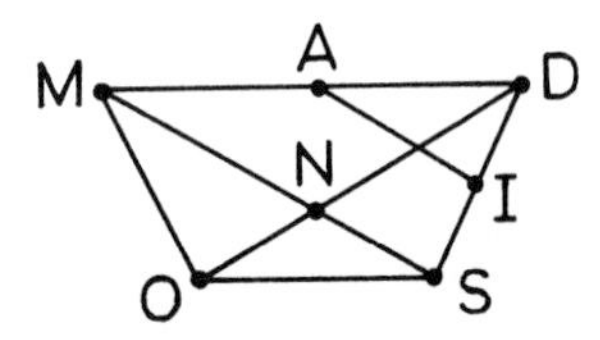

Given: MDSO is an isosceles trapezoid; $\overline{AI}$ is a midsegment of $\triangle MDS$.

Prove: $AI = \frac{1}{2}OD$.

39.

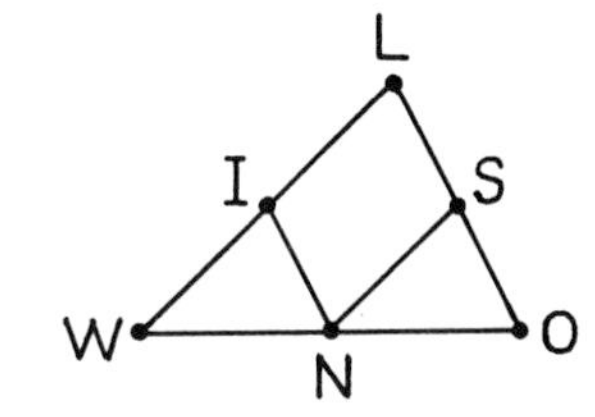

Given: $\overline{IN}$ and $\overline{NS}$ are midsegments of $\triangle WLO$; $\angle O > \angle W$.

Prove: NS > IN.

Set III

Obtuse Ollie says that, if you draw a quadrilateral and connect the midpoints of its opposite sides with two line segments, they will always bisect each other. He drew the figure shown here to support his assertion.

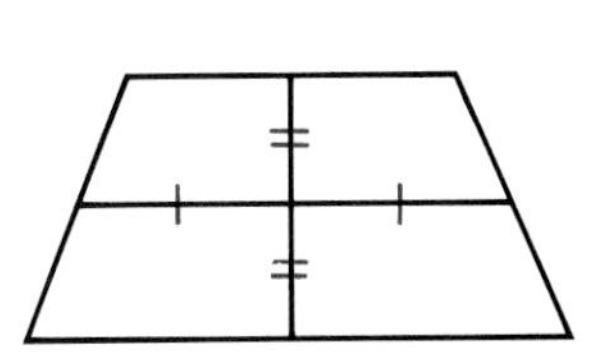

1. Does it? Explain why or why not.

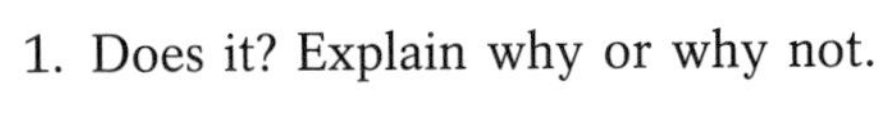

2. Can you draw a figure that contradicts Ollie's conclusion?

3. Do you think that he is correct?

Chapter 7 / Summary and Review

Courtesy of Sunkist Growers, Inc.

Basic Ideas

Convex and concave quadrilaterals 237
Isosceles trapezoid 256
Midsegment of a triangle 260
Parallelogram 237
Quadrilateral 236
Rectangle 237
Rhombus 237
Square 237
Trapezoid 237

Theorems

32. The sum of the measures of the angles of a quadrilateral is 360°. 237
    Corollary. Each angle of a rectangle is a right angle. 238
33. The opposite sides of a parallelogram are equal. 241
34. The opposite angles of a parallelogram are equal and its consecutive angles are supplementary 242
35. The diagonals of a parallelogram bisect each other. 242
36. A quadrilateral is a parallelogram if both pairs of opposite sides are equal. 247
37. A quadrilateral is a parallelogram if two opposite sides are both equal and parallel. 247
38. A quadrilateral is a parallelogram if both pairs of opposite angles are equal. 248
39. A quadrilateral is a parallelogram if its diagonals bisect each other. 248
40. All rectangles are parallelograms. 251
41. The diagonals of a rectangle are equal. 252
42. All rhombuses are parallelograms. 252
43. The diagonals of a rhombus are perpendicular. 252
44. The base angles of an isosceles trapezoid are equal. 257
45. The diagonals of an isosceles trapezoid are equal. 257
46. *The Midsegment Theorem.* A midsegment of a triangle is parallel to the third side and half as long. 260

Exercises

Set I

If you have ever driven by an orchard in which the trees are evenly spaced in parallel rows, you may have noticed that the trees seem to line up in other parallel rows as your perspective of them changes.

The figure below represents an overhead view of some of these rows. In it, AB = BC = CD = DE = EF = GH = HI = IJ = JK = KL = MN = NO = OP = PQ = QR and AG = GM = BH = HN = CI = IO = DJ = JP = EK = KQ = FL = LR.

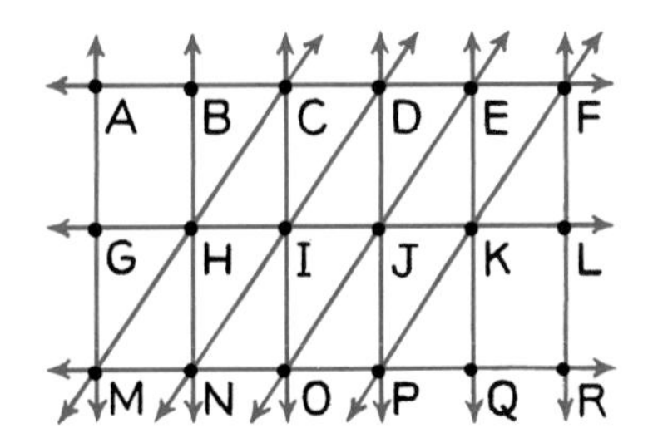

1. What kind of quadrilaterals are ABHG, BCIH, CDJI, and so on?
2. How do you know?
3. Why are $\overleftrightarrow{AF} \parallel \overleftrightarrow{GL} \parallel \overleftrightarrow{MR}$ and $\overleftrightarrow{AM} \parallel \overleftrightarrow{BN} \parallel \overleftrightarrow{CO} \parallel \overleftrightarrow{DP} \parallel \overleftrightarrow{EQ} \parallel \overleftrightarrow{FR}$?
4. What kind of quadrilaterals are CDIH, DEJI, EFKJ, HINM, and so on?
5. How do you know?
6. Why are $\overleftrightarrow{CM} \parallel \overleftrightarrow{DN} \parallel \overleftrightarrow{EO} \parallel \overleftrightarrow{FP}$?

Tell whether each of the following statements is true or false.

7. If both pairs of opposite angles of a quadrilateral are equal, the quadrilateral must be a parallelogram.
8. One of the diagonals of a parallelogram divides it into two congruent triangles.
9. If one of the diagonals of a quadrilateral divides it into two congruent triangles, the quadrilateral must be a parallelogram.
10. If two consecutive sides of a parallelogram are equal, the parallelogram must be a rhombus.
11. All squares are rhombuses.
12. All parallelograms are rectangles.
13. All four angles of a trapezoid can have different measures.
14. The opposite angles of an isosceles trapezoid are supplementary.
15. The diagonals of an isosceles trapezoid bisect each other.

In quadrilateral TALC, ∠T = 92°, ∠A = 89°, and ∠L = 91°.

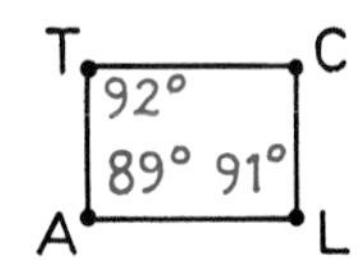

16. Find ∠C.
17. What kind of quadrilateral is TALC?
18. What are sides $\overline{TA}$ and $\overline{CL}$ of TALC called?
19. Is TALC isosceles?

Quadrilateral CLAY is a rhombus. $CL = 3(x + 12)$, $CA = x^2 - 25$, and $CY = 1 - 2x$.

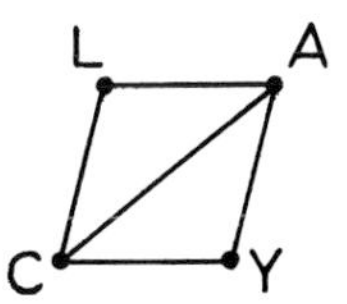

20. Find x.
21. Find CL.
22. Find CA.
23. Find the perimeter of △CLA.
24. Find the perimeter of CLAY.

Trace parallelogram FLIN.

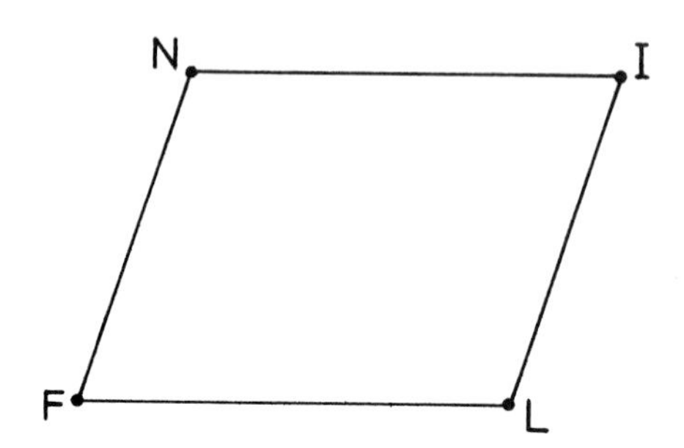

25. Use your straightedge and compass to bisect ∡F and ∡L. Label the point in which the two bisectors intersect T.
26. To what number is ∠NFL + ∠FLI equal?
27. How do you know?
28. Write equations relating ∠TFL to ∠NFL and ∠FLT to ∠FLI.
29. To what number is ∠TFL + ∠FLT equal?
30. What can you conclude about ∡FTL?
31. What can you conclude about $\overline{FT}$ and $\overline{LT}$?

Set II

32.

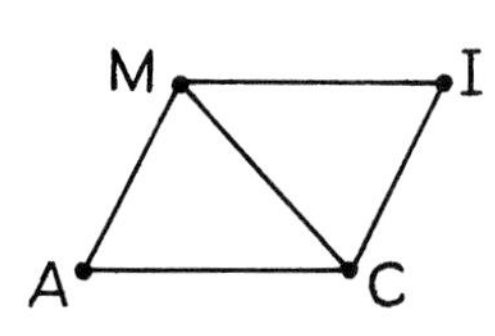

Given: MA = IC;
∠AMC = ∠MCI.

Prove: MICA is a parallelogram.

33.

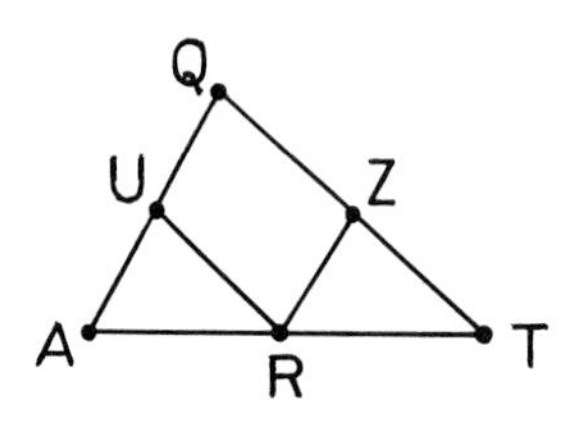

Given: U, Z, and R are the midpoints of the sides of △QAT.

Prove: △ARU ≅ △RTZ.

34.

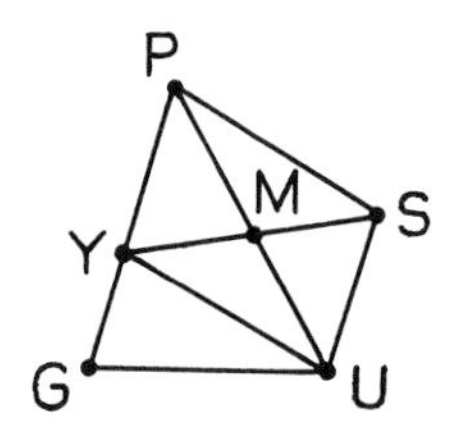

Given: In quadrilateral GPSU, $\overline{PU}$ and $\overline{YS}$ bisect each other; PS = GU.

Prove: ∠G = ∠GPS.

35.

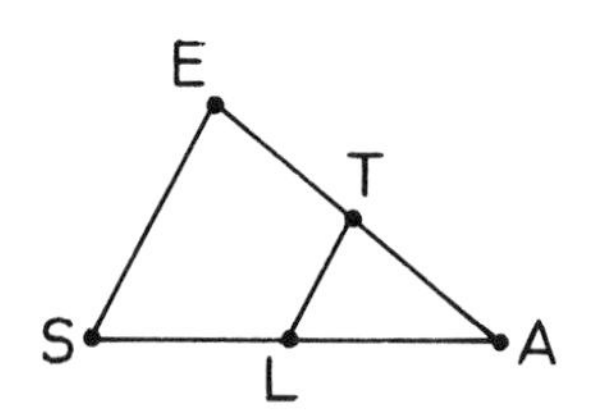

Given: $\overline{LT}$ is a midsegment of △SAE;
∠TLA > ∠A.

Prove: EA > ES.

Algebra Review

Fractions

Fractions
A fraction is the indicated quotient of two numbers or algebraic expressions called its numerator and denominator. A fraction consisting of two numbers can be changed into decimal form by dividing the numerator by the denominator.

Equivalent Fractions
Fractions that represent the same number are equivalent. Whenever the numerator and denominator of a fraction are multiplied or divided by the same number other than zero, the result is an equivalent fraction: in general, if $n \neq 0$, $\frac{a}{b} = \frac{na}{nb}$ and $\frac{na}{nb} = \frac{a}{b}$. To reduce a fraction to lowest terms, divide its numerator and denominator by their common factors.

Operations with Fractions
Addition and Subtraction
Same denominator

$$\frac{a}{b} \pm \frac{c}{b} = \frac{a \pm c}{b}$$

Different denominators

$$\frac{a}{b} \pm \frac{c}{d} = \frac{ad}{bd} \pm \frac{bc}{bd} = \frac{ad \pm bc}{bd}$$

Multiplication

$$\frac{a}{b} \cdot \frac{c}{d} = \frac{ac}{bd}$$

Division

$$\frac{a}{b} \div \frac{c}{d} = \frac{a}{b} \cdot \frac{d}{c} = \frac{ad}{bc}$$

Exercises

Change to decimal form.

Example: $\frac{5}{8}$.

Solution:

$$\begin{array}{r} 0.625 \\ 8\overline{)5.000} \\ \underline{4\,8} \\ 20 \\ \underline{16} \\ 40 \\ \underline{40} \\ 0 \end{array}$$

1. $\frac{3}{4}$
2. $\frac{1}{20}$
3. $\frac{9}{5}$
4. $\frac{7}{12}$

Reduce to lowest terms.

Example: $\frac{21}{35}$.

Solution: $\frac{21}{35} = \frac{7 \cdot 3}{7 \cdot 5} = \frac{3}{5}$.

5. $\frac{12}{54}$
6. $\frac{x^3}{3x}$
7. $\frac{4x + 4}{7x + 7}$
8. $\frac{x - 5}{x^2 - 25}$

Write as a single fraction in lowest terms.

Example: $\frac{3}{x} + \frac{1}{4}$.

Solution: $\frac{3}{x} + \frac{1}{4} = \frac{3 \cdot 4}{x \cdot 4} + \frac{1 \cdot x}{4 \cdot x}$
$= \frac{12}{4x} + \frac{x}{4x} = \frac{12 + x}{4x}$.

9. $\frac{1}{2} + \frac{2}{3}$

10. $\frac{x}{4} - \frac{x}{5}$

11. $\frac{x + 5}{2x} + \frac{5}{2x}$

12. $\frac{4}{x^2} - \frac{3}{x^3}$

Example: $5x \cdot \frac{x}{5}$.

Solution: $5x \cdot \frac{x}{5} = \frac{5x}{1} \cdot \frac{x}{5} = \frac{5x^2}{5} = x^2$.

13. $\frac{3}{4} \cdot \frac{8}{9}$

14. $\frac{x^2}{2} \cdot \frac{x^3}{3}$

15. $\frac{1}{2} \div \frac{7}{10}$

16. $\frac{x + 2}{x} \div \frac{2}{x}$

17. $\frac{5x}{16} + \frac{7x}{16}$

18. $\frac{x + 4}{10} - \frac{x + 2}{10}$

19. $\frac{x + 3}{9} \cdot \frac{3}{x}$

20. $\frac{10}{x} \div \frac{x}{5}$

21. $\frac{1}{x} + \frac{1}{y}$

22. $\frac{6}{2x - 3} \cdot \frac{x - 3}{3}$

23. $\frac{x}{y} - 1$

24. $\frac{y}{xy - y^2} \div \frac{x}{x - y}$

Chapter 8
TRANSFORMATIONS

Escher Foundation, Haags Gemeentemuseum, The Hague © M. C. Escher Heirs % Cordon Arts—Baarn, Holland

Lesson 1

Reflections

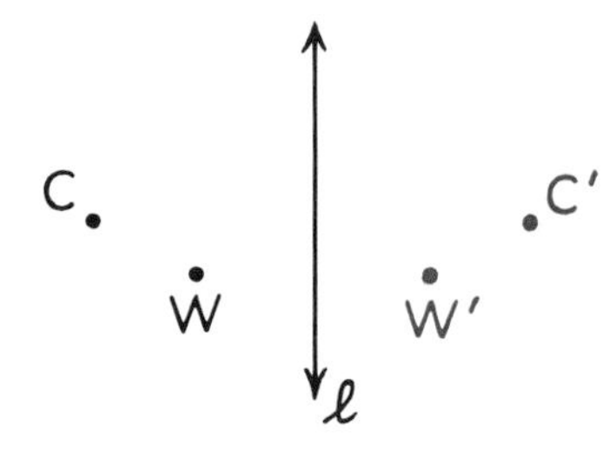

This woodcut by the artist Maurits Escher is titled *Day and Night.* The interlocking black and white birds at the top of the picture are flying over two different views of the same landscape. The two views, except for the reversal of dark and light, are mirror images of each other.

Each point of the river and village on the left corresponds to an "image point" of the river and village on the right. The diagram at the left shows some of these points. The church steeple on the left, C, corresponds to the church steeple on the right, C′. The windmill on the left, W, corresponds to the windmill on the right, W′.

If a mirror were placed on line ℓ so that it faced toward the left, the reflections of points C and W would appear to be where C′ and W′ are. Because of this, we will refer to points C′ and W′ as the *images* of points C and W *reflected through line* ℓ. We will refer to line ℓ as the *mirror* of the reflection.

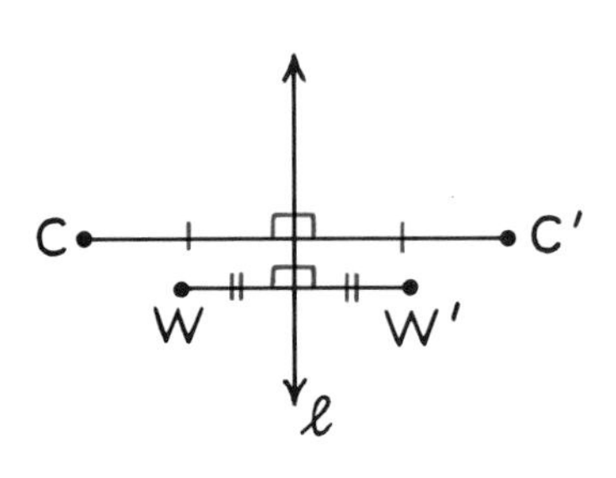

In the figure at the left, a segment has been drawn from each point to its image. The mirror of the reflection, ℓ, is related to each of these segments in the same way: it is the *perpendicular bisector* of each segment.

It is easy to see that, as a point moves toward the reflection line, so does its image. In fact, if the point moves onto the reflection line, it and its image merge.

► **Definition**
The ***reflection*** of point P through line ℓ is:

point P itself if P is on ℓ or
the point P′ such that ℓ is the perpendicular bisector of $\overline{PP'}$ if P is not on ℓ.

One way to find the reflection of a point through a line is to use a mirror—a Plexiglas mirror called a MIRA is especially convenient for this purpose. Another way to find the reflection of a point is by construction.

► **Construction 9**
To construct the reflection of a point through a line.

Method.

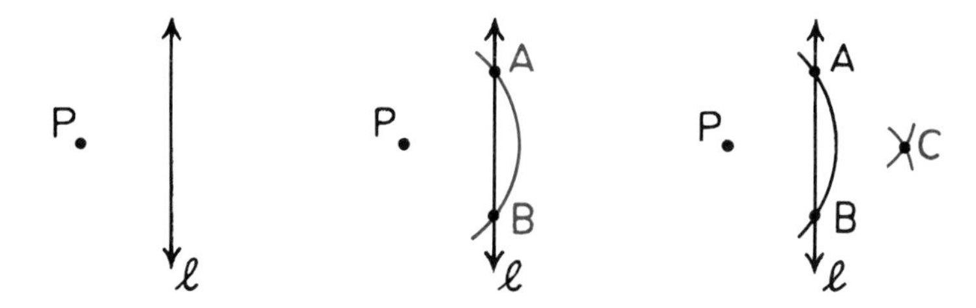

Let the point be P and the line be ℓ. First, with P as center, draw an arc that intersects ℓ in two points, A and B. Then, with A and B as centers, draw two more arcs with the same radius as the first arc. The point in which they intersect, C, is the reflection of P through ℓ.

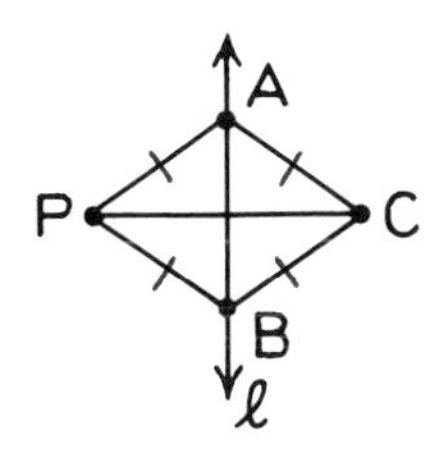

The figure at the right reveals why this method works. Because PA = PB = AC = BC, points A and B are equidistant from P and C. In a plane, two points each equidistant from the endpoints of a line segment determine the perpendicular bisector of the line segment, and so ℓ is the perpendicular bisector of $\overline{PC}$. It follows from the definition of the reflection of a point through a line that C is the reflection of P through ℓ.

It is easy to prove that, whenever a pair of points is reflected through a line, the distance between the points is always equal to the distance between their reflections.

► **Transformation Theorem 1**
Reflection of a pair of points through a line preserves distance.

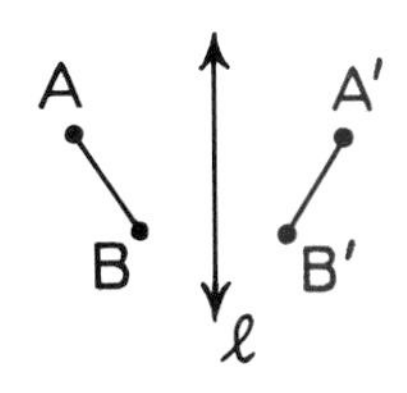

With regard to the figure at the right, this theorem says that, if A′ and B′ are the reflections of A and B through line ℓ, then AB = A′B′. The proof of this is left as an exercise.

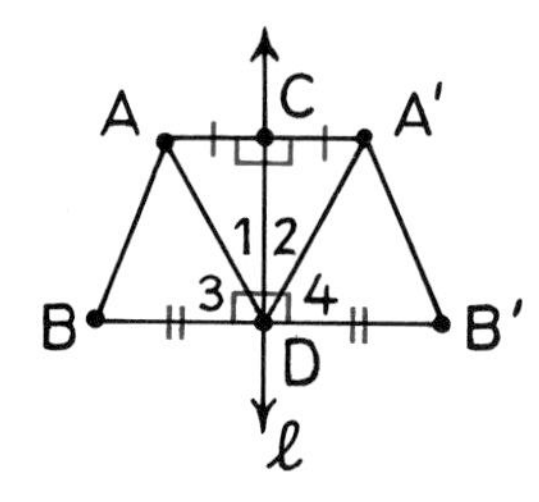

Proof.
Draw $\overline{AA'}$, $\overline{BB'}$, $\overline{AD}$, and $\overline{A'D}$.

26. What relation does line ℓ have to $\overline{AA'}$ and $\overline{BB'}$?
27. Why is $\triangle ACD \cong \triangle A'CD$?
28. Why is $AD = A'D$ and $\angle 1 = \angle 2$?
29. What relation does $\measuredangle 1$ have to $\measuredangle 3$ and $\measuredangle 2$ have to $\measuredangle 4$?
30. Why is $\angle 3 = \angle 4$?
31. Why is $\triangle ABD \cong \triangle A'B'D$?
32. Why is $AB = A'B'$?

Use Transformation Theorem 1 in each of the following proofs.

33.

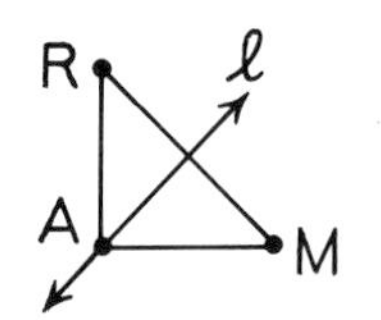

Given: Point M is the reflection of point R through line ℓ.

Prove: $\triangle RAM$ is isosceles.

34.

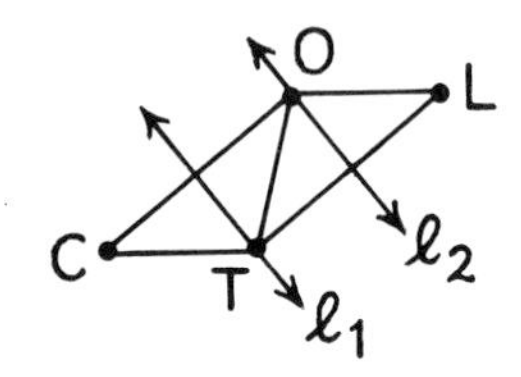

Given: Point O is the reflection of point C through ℓ_1; point L is the reflection of point T through ℓ_2; $CO = TL$.

Prove: COLT is a parallelogram.

Set III

In the murder mystery *The House of the Arrow* by A. E. W. Mason, one of the characters opens a door and sees a clock without realizing that it is actually the clock's reflection in a mirror that is being seen.

If the time on the clock appeared to be 12:20, what was the actual time? Make accurate drawings to support your answer.

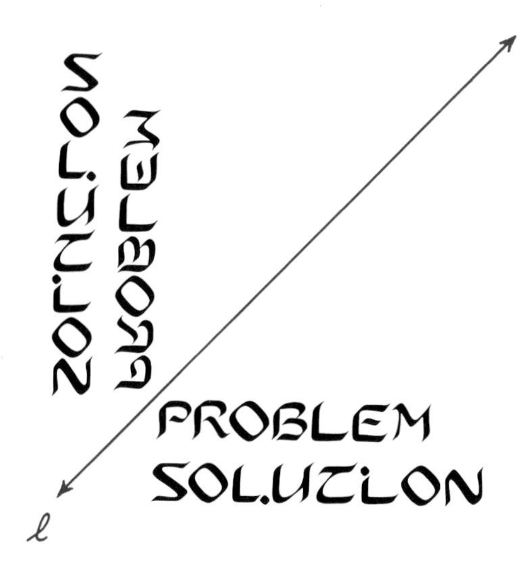

By the permission of Scott Kim

Lesson 2

Properties of Isometries

In this design by Scott Kim, the words PROBLEM and SOLUTION at the lower right are reflected through line ℓ to form the images at the upper left. Although there is nothing special about the reflection of PROBLEM, the design of the letters in the word SOLUTION makes its reflection quite remarkable.

The reflections that we have been studying are a special type of *transformation.*

▶ **Definition**
A ***transformation*** is a one-to-one correspondence between two sets of points.

If the transformation is a reflection, such as the one pictured above, this correspondence can be easily seen by imagining that the figure is folded along line ℓ, the mirror of the reflection. If the figure were folded in this way, the points in the two sets that correspond to each other would touch each other.

The reflection of a set of points through a line is also an example of an *isometry.*

▶ **Definition**
An ***isometry*** is a transformation that preserves distance.

In the preceding lesson, we proved that reflection through a line preserves distance. Therefore reflections are isometries.

The definition of an isometry can be used to prove that an isometry also preserves collinearity and betweenness.

► Transformation Theorem 2
An isometry preserves collinearity and betweenness.

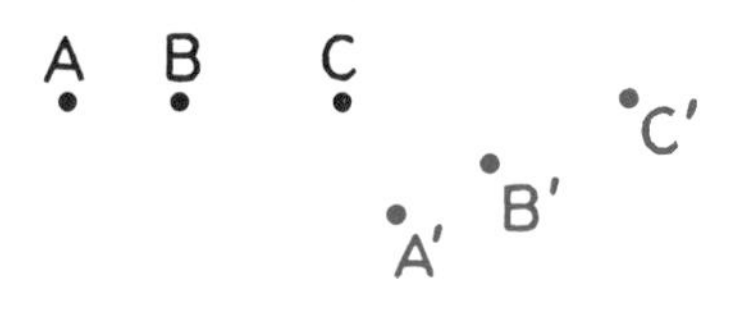

Given: Points A, B, and C are collinear with A-B-C; their images under a given isometry are A′, B′, and C′.

Prove: Points A′, B′, and C′ are collinear and A′-B′-C′.

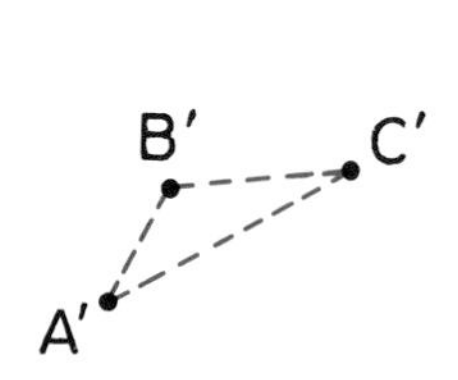

Proof.
First, we will prove that collinearity is preserved. Suppose that A′, B′, and C′ are not collinear. Then they determine a triangle and, by the Triangle Inequality Theorem,

$$A'B' + B'C' > A'C'. \tag{1}$$

Because A-B-C, we can conclude from the Betweenness of Points Theorem that

$$AB + BC = AC. \tag{2}$$

Also, because an isometry preserves distance,

$$AB = A'B', \quad BC = B'C', \quad \text{and} \quad AC = A'C'. \tag{3}$$

By substitution of A′B′, B′C′, and A′C′ into equation 2,

$$A'B' + B'C' = A'C'. \tag{4}$$

This contradicts inequality 1; so A′, B′, and C′ must be collinear.

To prove that betweenness is preserved, we must show that A′-B′-C′. Suppose that B′ is not between A′ and C′. Then, because collinearity is preserved, either A′ is between the other two points or C′ is between them.

If B′-A′-C′, then

$$A'B' + A'C' = B'C'. \tag{5}$$

Substituting from equation 4 into equation 5, we get

$$A'B' + (A'B' + B'C') = B'C',$$

or

$$2A'B' = 0,$$

$$A'B' = 0.$$

This contradicts the fact that the distance between two distinct points is a positive number.

The possibility that B′-C′-A′ can be eliminated in the same way, and so our proof is complete.

Two more properties of isometries that are easy to prove are stated as the following theorems.

► **Transformation Theorem 3**
An isometry preserves angle measure.

► **Transformation Theorem 4**
A triangle and its image under an isometry are congruent.

Exercises

Set I

In the figure below, RMDA is the reflection image of COEA through line ℓ.

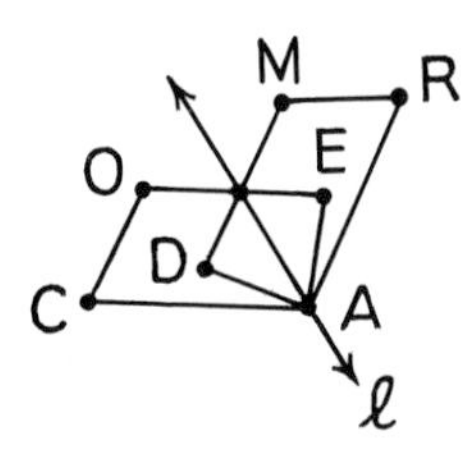

1. Which point is the reflection of point O?
2. Which segment is the reflection of $\overline{DM}$?
3. Which angle is the reflection of ∡CAE?
4. Which segment is equal in length to $\overline{EA}$?
5. Which angle is equal in measure to ∡C?

In the figure below, △P′A′L′ is the image of △PAL under a certain transformation.

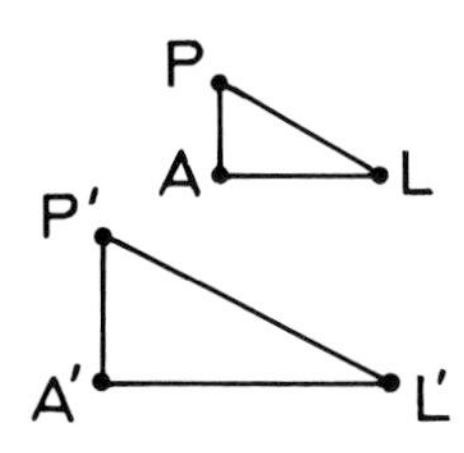

6. Is this transformation an isometry?
7. Does it appear to preserve distance?
8. Does it appear to preserve angle measure?
9. Does it appear to preserve collinearity?
10. Are the two triangles congruent?

In the figure below, trapezoid C′H′U′M′ is the image of square CHUM under a certain transformation.

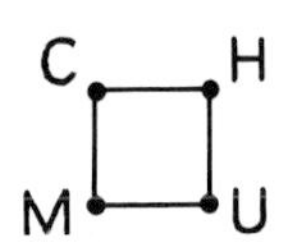

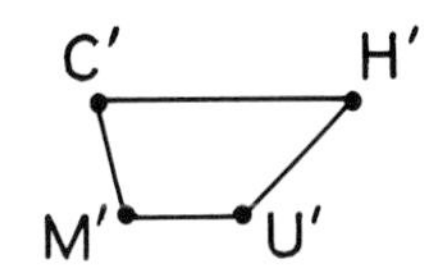

11. Is this transformation an isometry?
12. Does it appear to preserve distance?
13. Does it appear to preserve angle measure?
14. Does it appear to preserve collinearity?

The following questions refer to the figure below.

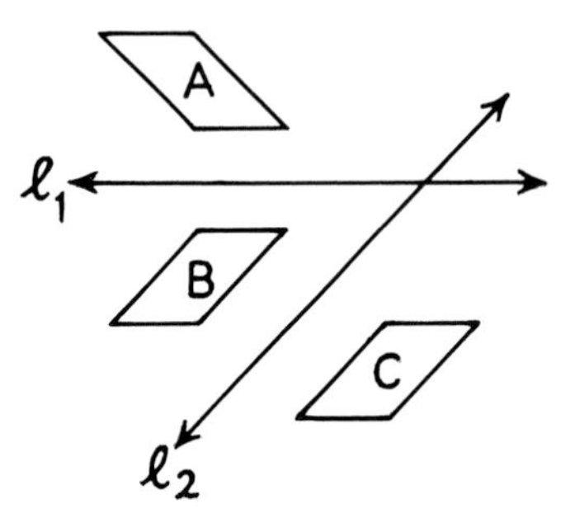

15. Does figure B appear to be a reflection image of figure A through line ℓ_1?
16. Does figure C appear to be a reflection image of figure B through line ℓ_2?
17. Does figure C look as if it could be a reflection image of figure A?
18. Does figure C look as if it could be an image of figure A under some isometry?

Set II

Trace the following figures. Then sketch, as accurately as you can, the reflection of each figure through the given line. Draw the reflection images in a different color.

19.

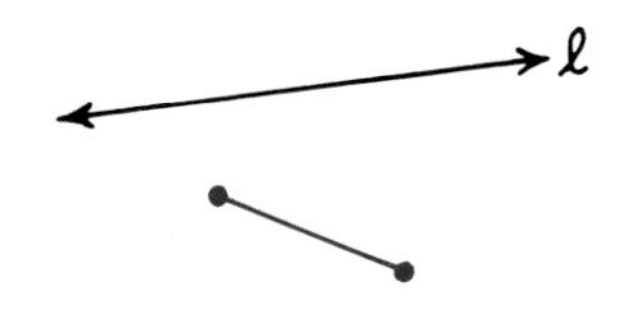

20.

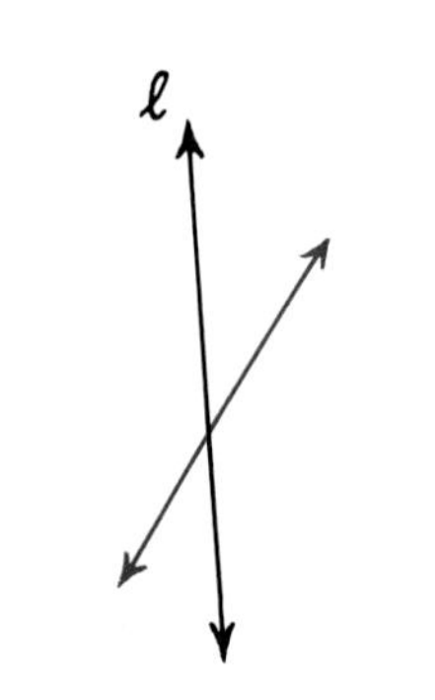

21.

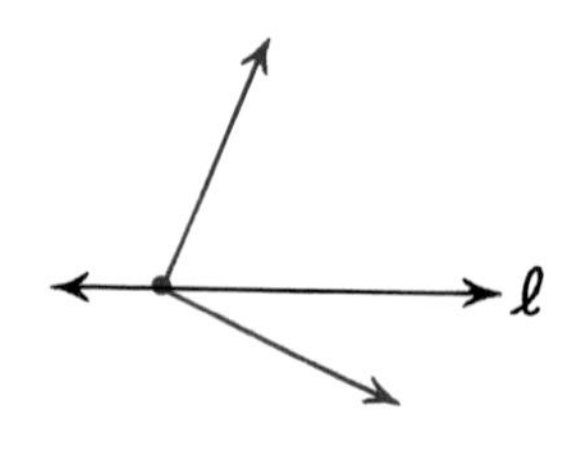

22.

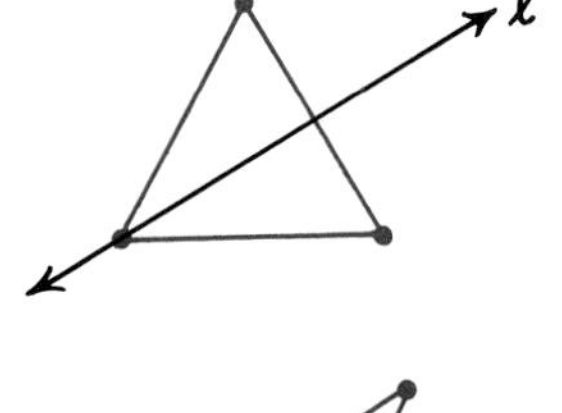

23.

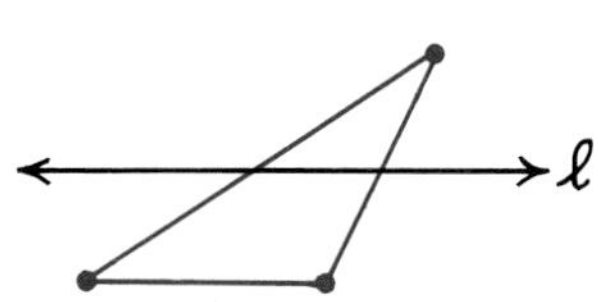

24.

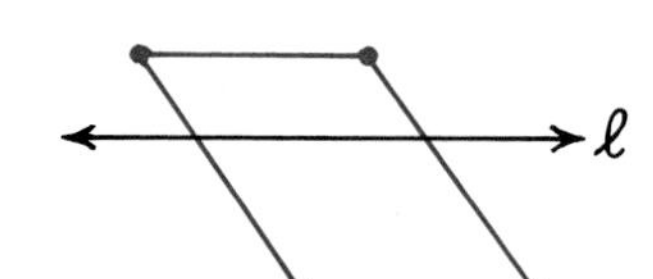

25.

26.

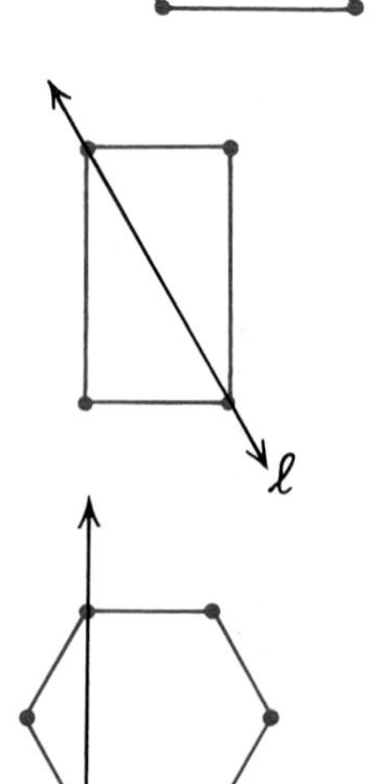

27. Give the missing statements and reasons in this proof of Transformation Theorem 3.

Theorem.
An isometry preserves angle measure.

Given: ∡A′ is the image of ∡A under a certain isometry.*

Prove: $\angle A = \angle A'$.

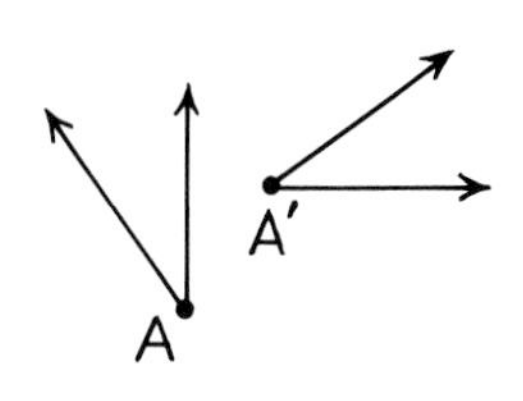

Proof.

Statements	**Reasons**
1. ∡A′ is the image of ∡A under a certain isometry.	Given.
2. Choose points B and C on the sides of ∡A so that AB = AC.	a) ||||||||||
3. Let B′ and C′ be their images on the sides of ∡A′.	An isometry is a transformation and a transformation is a one-to-one correspondence between two sets of points.
4. Draw $\overline{BC}$ and $\overline{B'C'}$.	b) ||||||||||
5. AB = A′B′, AC = A′C′, and BC = B′C′.	c) ||||||||||
6. d) ||||||||||	e) ||||||||||
7. f) ||||||||||	g) ||||||||||

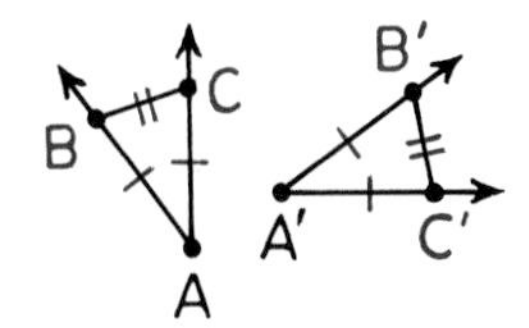

*This proof is for the case in which the angles are not straight angles.

28. Give the missing statements and reasons in this proof of Transformation Theorem 4.

Theorem.
A triangle and its image under an isometry are congruent.

Given: △A′B′C′ is the image of △ABC under a certain isometry.

Prove: $\triangle ABC \cong \triangle A'B'C'$.

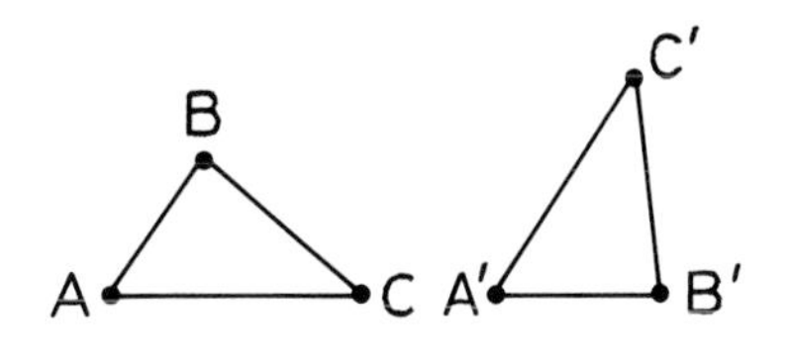

Proof.

Statements	**Reasons**
1. △A′B′C′ is the image of △ABC under a certain isometry.	Given.
2. a) ||||||||||	An isometry is a transformation that preserves distance.
3. b) ||||||||||	c) ||||||||||

29.

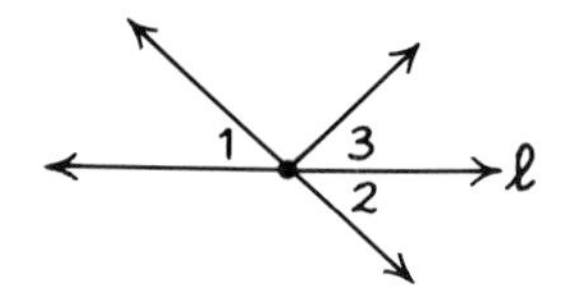

Given: ∡1 and ∡2 are vertical angles; ∡3 is the reflection image of ∡2 through ℓ.

Prove: $\angle 1 = \angle 3$.

30.

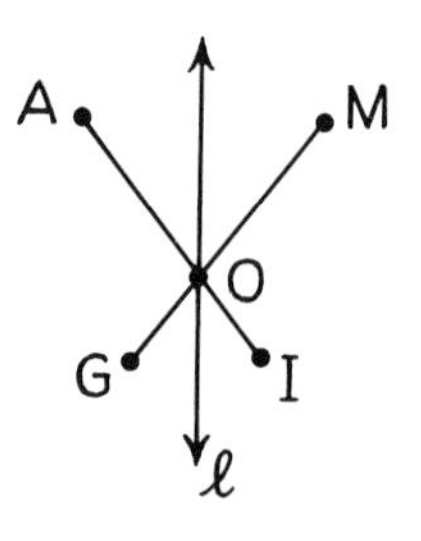

Given: Points M and G are reflection images of points A and I through line ℓ; A-O-I.

Prove: M-O-G.

Set III

A Construction That Does Not Require a Compass

Trace this figure, in which point B′ is the reflection of point B through line ℓ. Can you figure out a way to find the reflection of point A through ℓ that uses only a *straightedge*?

> Hint: Because a reflection is an isometry, it preserves collinearity.

Drawing by Chas. Addams; © 1957, 1985 The New Yorker Magazine, Inc.

Lesson 3
Translations

Some barber shops have mirrors on the side walls that face each other. The result is a series of successive reflections.

The cartoon reveals that the resulting images are not all alike. For example, some of the images of the man in the barber chair face in the same direction and some face in the opposite direction. The second image, which faces in the same direction as the man in the chair, is an example of a *translation.* A translation can be thought of as the result of *sliding, without turning,* a figure from one position to another.

The figure at the right shows that a translation is the result of two successive reflections in parallel mirrors. We will call a transformation that is the result of two or more successive transformations their *composite.*

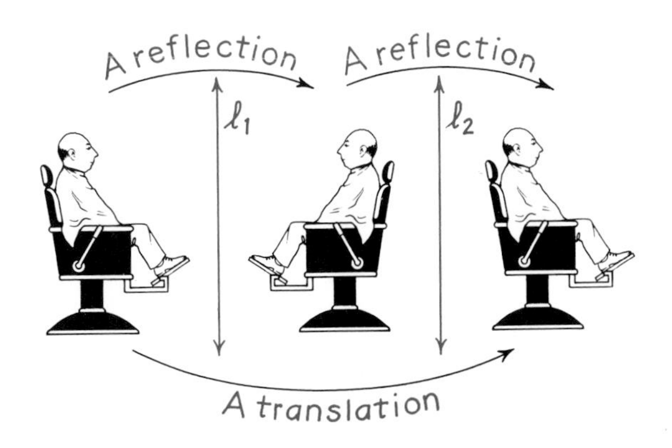

► **Definition**

A transformation is a ***translation*** iff it is the composite of two successive reflections through parallel lines.

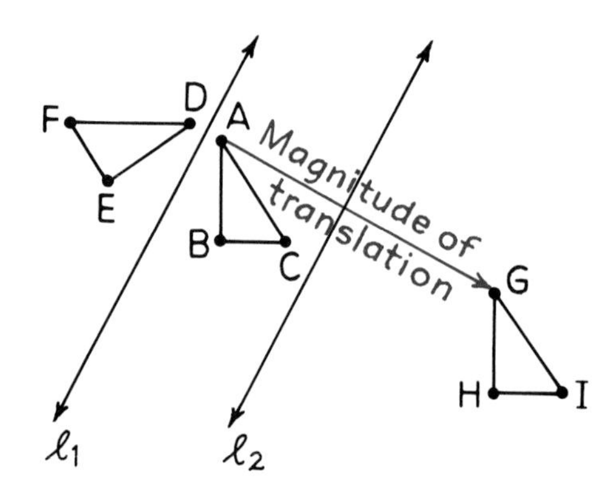

Another example of a translation is shown here. The reflection of △ABC through ℓ_1 is △DEF, and the reflection of △DEF through ℓ_2 is △GHI. If $\ell_1 \parallel \ell_2$, then △GHI is a translation image of △ABC. If a tracing of △ABC were made, it could slide onto △GHI without being turned. This means that the distance between each point of △ABC and the corresponding point of △GHI (for example, the distance from A to G) is the same. This distance is called the *magnitude* of the translation.

Because translations are composites of reflections, it follows that translations, like reflections, are isometries. As we proved in the preceding lesson, isometries preserve not only distance but also collinearity, betweenness of points, and angle measure.

Exercises

Set I

In the figure below, △DEF is the reflection of △ABC through ℓ_1, and △GHI is the reflection of △DEF through ℓ_2; $\ell_1 \parallel \ell_2$.

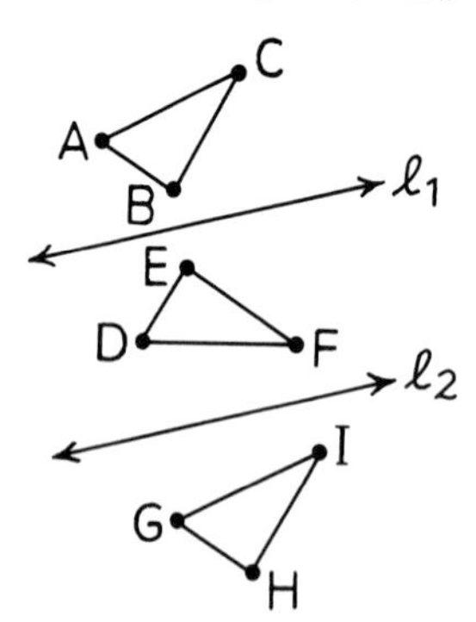

1. Through what transformation is △GHI the image of △ABC?

2. Why is △ABC ≅ △DEF and △DEF ≅ △GHI?

3. Why is △ABC ≅ △GHI?

In the figure below, EFGH and IJKL are images of ABCD under certain transformations.

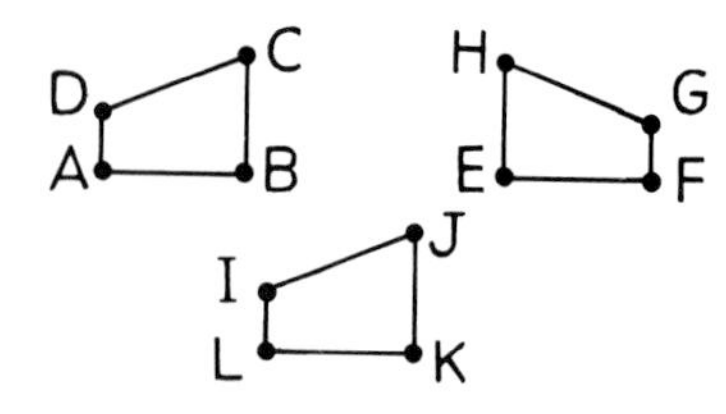

4. What transformation seems to relate ABCD and EFGH?

5. Which point of EFGH seems to be the image of A?

6. What transformation seems to relate ABCD and IJKL?

7. Which point of IJKL seems to be the image of B?

In the figure below, $\ell_1 \parallel \ell_2$. $\triangle$DEF, $\triangle$GHI, $\triangle$JKL, and $\triangle$MNO are reflection images of $\triangle$ABC through either or both of the lines. Which triangle is the reflection image of

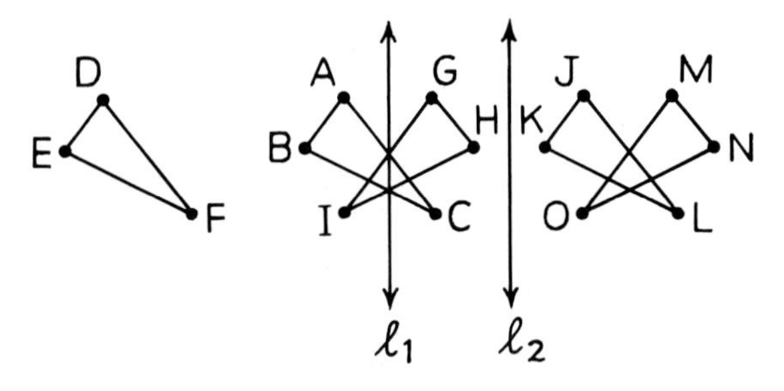

8. $\triangle$ABC through ℓ_1?

9. $\triangle$GHI through ℓ_2?

10. $\triangle$ABC through ℓ_2?

11. $\triangle$MNO through ℓ_1?

Which triangle is the translation image of $\triangle$ABC as a result of successive reflections through

12. ℓ_1 and ℓ_2?

13. ℓ_2 and ℓ_1?

You know that the composite of two successive reflections through parallel lines is a translation. Look again at the cartoon at the beginning of this lesson. Using it as a clue, what do you think is the composite of

14. *three* successive reflections through parallel lines?

15. *four* successive reflections through parallel lines?

Set II

Trace this figure, in which $\ell_1 \parallel \ell_2$.

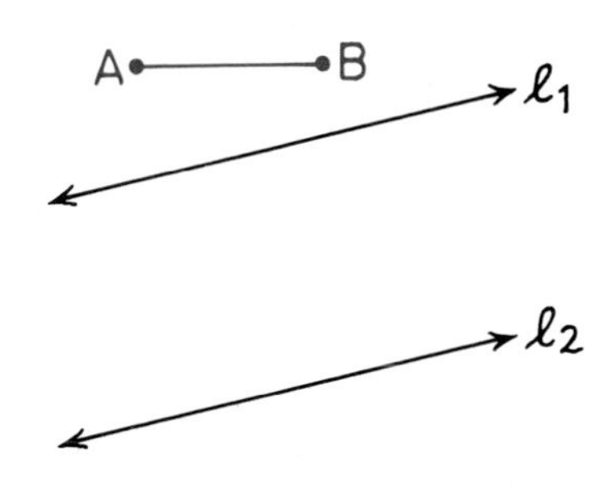

16. Reflect $\overline{AB}$ through ℓ_1 and name its image $\overline{CD}$. Then, reflect $\overline{CD}$ through ℓ_2 and name its image $\overline{EF}$.

17. Explain how you know that AB = EF.

18. What other relation do $\overline{AB}$ and $\overline{EF}$ seem to have?

Trace this figure, in which $\ell_1 \parallel \ell_2$.

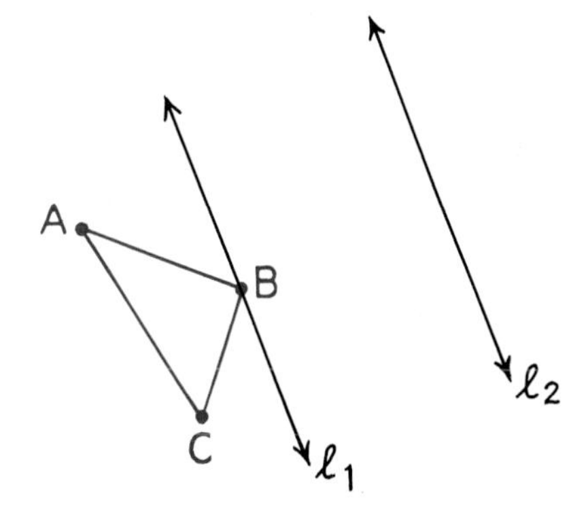

19. Reflect $\triangle$ABC through ℓ_1 and name its reflection $\triangle$DBE so that $\triangle DBE \cong \triangle ABC$. Then, reflect $\triangle$DBE through ℓ_2 and name its reflection $\triangle$FGH so that $\triangle FGH \cong \triangle DBE$. Draw $\overline{AF}$, $\overline{BG}$, and $\overline{CH}$.

20. What relation do points A, D, and F seem to have?

21. Do points C, E, and H seem to have the same relation?

If a tracing of $\triangle ABC$ were made and slid onto $\triangle FGH$, its vertices would move along the "tracks" $\overline{AF}$, $\overline{BG}$, and $\overline{CH}$.

22. Name two relations that these tracks seem to have to each other.

23. What relation do these tracks seem to have to ℓ_1 and ℓ_2?

Trace this figure, in which $\ell_1 \parallel \ell_2$, so that ℓ_1 and ℓ_2 are centered on your paper.

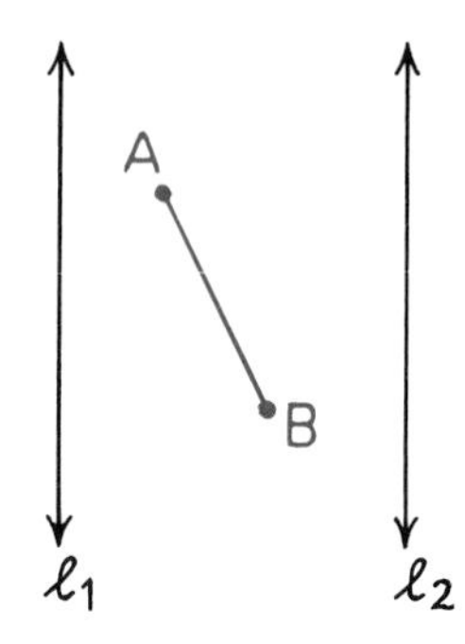

24. Measure the distance between ℓ_1 and ℓ_2 in centimeters.

25. Translate $\overline{AB}$ by reflecting it through ℓ_1 and its image through ℓ_2. Label the translation image $\overline{CD}$. Then, translate $\overline{AB}$ by reflecting it through ℓ_2 and its image through ℓ_1. Label the translation image $\overline{EF}$.

26. Measure the magnitude of the translation relating $\overline{AB}$ and $\overline{CD}$ in centimeters.

27. Measure the magnitude of the translation relating $\overline{AB}$ and $\overline{EF}$ in centimeters.

28. If the distance between two parallel lines is d, what do you think is the magnitude of a translation through the two lines?

In the figure below, point C is the translation image of point A resulting from successive reflections through ℓ_1 and ℓ_2.

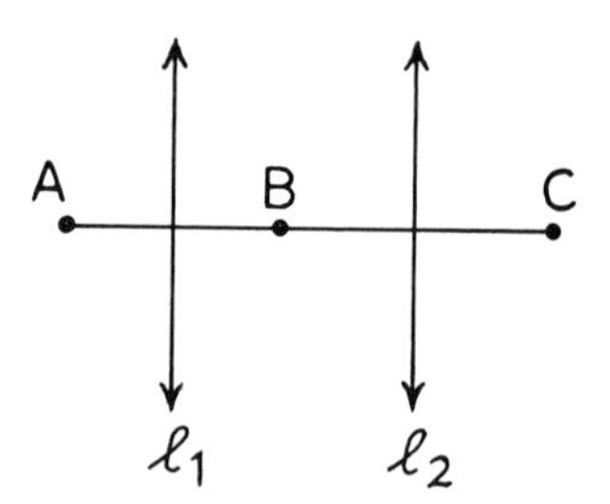

29. Trace the figure and label the distance from A to ℓ_1 x and the distance from B to ℓ_2 y. Also, label the distances from B to ℓ_1 and C to ℓ_2 in terms of x and y.

30. What is the distance between ℓ_1 and ℓ_2 in terms of x and y?

31. What is the distance between A and C in terms of x and y?

32. How does the magnitude of this translation compare to the distance between ℓ_1 and ℓ_2?

Set III

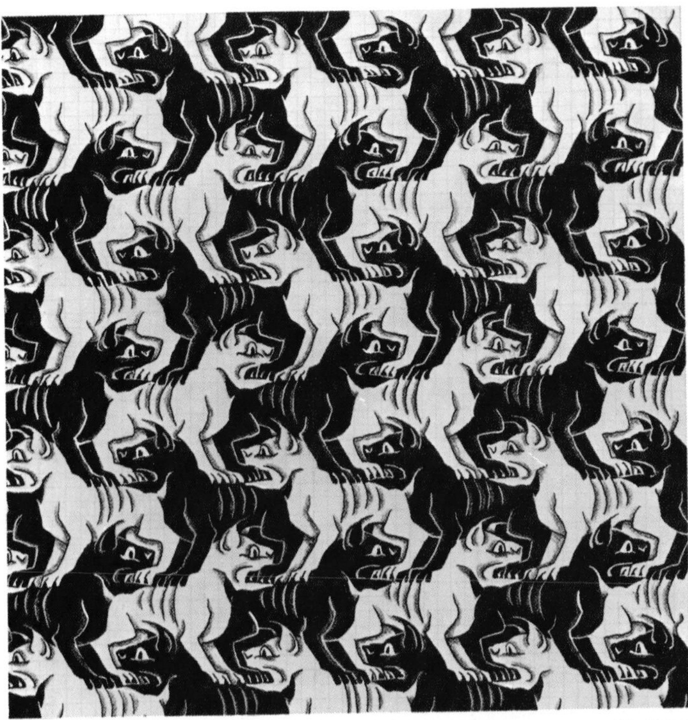

The bulldogs in this mosaic by Escher are related through several different translations.

An obvious translation is shown in the first pair of bulldogs below; a less obvious one is shown in the second pair of bulldogs.

Trace the second pair of bulldogs. Can you find a line through which to reflect the bulldog at the upper left and a second line through which to reflect its image so that the result is the bulldog at the lower right?

By the permission of Johnny Hart and News America Syndicate

Lesson 4

Rotations

You have become acquainted with two geometric transformations: reflections and translations. A third transformation is illustrated by the two figures below. There is a one-to-one correspondence between the points of the rightside-up tortoise and the upside-down tortoise in the first figure, yet it is neither a reflection nor a translation.

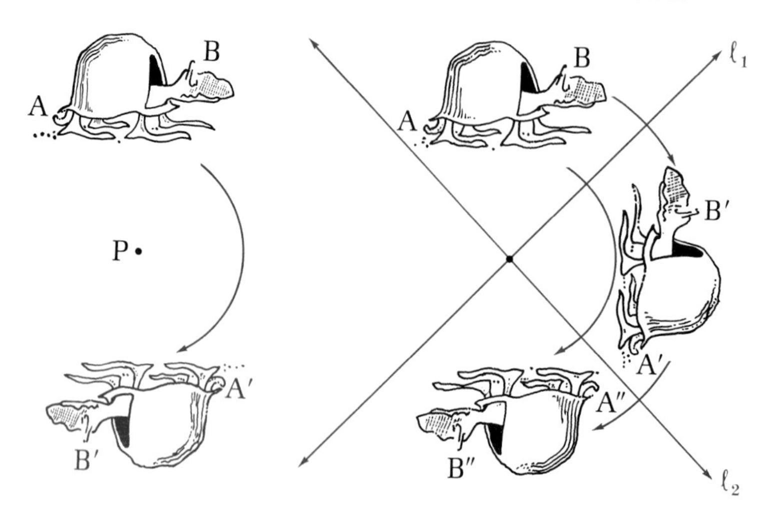

By the permission of Johnny Hart and News America Syndicate

One way to illustrate this correspondence would be to put a sheet of tracing paper over the figure and pin it at point P. If a tracing were made of the top tortoise and the paper were then turned about the pin, the tracing would eventually coincide with the drawing of the bottom tortoise. For this reason, the bottom tortoise is called a *rotation* image of the top one.

Another way to illustrate the correspondence is by reflections. You know that two successive reflections through parallel lines result in a translation. If the lines intersect, then two successive reflections through them result in a rotation instead. The second figure on the facing page shows that the tortoise rotation is the composite of two such reflections. For this reason, we can define "rotation" in the same way that we defined "translation," with the exception of just one word.

► **Definition**
A transformation is a ***rotation*** iff it is the composite of two successive reflections through intersecting lines.

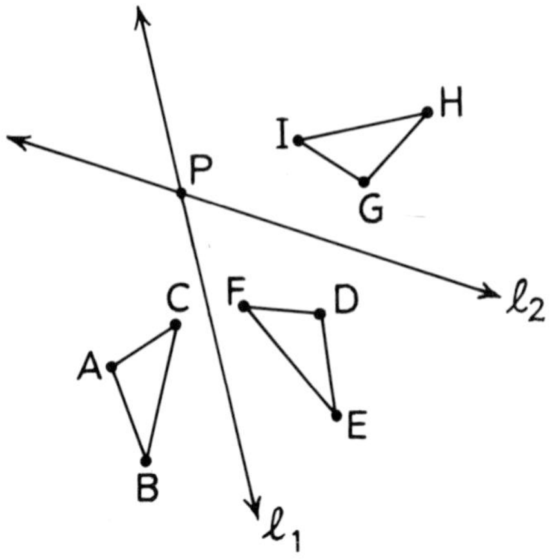

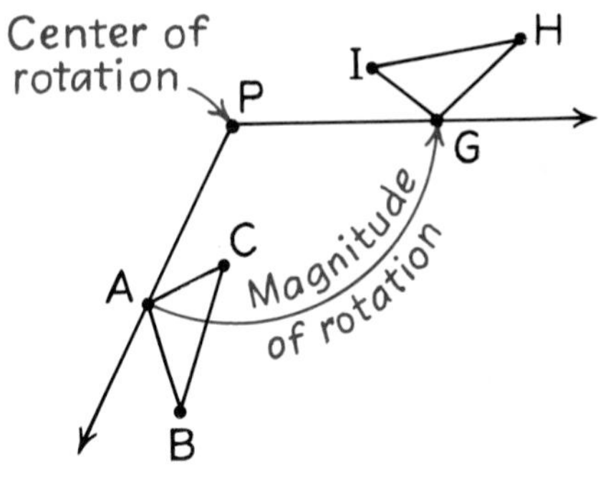

Another example of a rotation is shown in the figures above. The reflection of △ABC through ℓ_1 is △DEF, and the reflection of △DEF through ℓ_2 is △GHI. Because ℓ_1 and ℓ_2 intersect in point P, △GHI is a rotation image of △ABC. If a tracing of △ABC were made, it could be turned about point P so that it coincides with △GHI. This means that the angle through which each point of △ABC rotates about P (for example, ∡APG) is the same. Point P is called the *center* of the rotation and the measure of the angle of the rotation is called its *magnitude*.

Because rotations, like translations, are composites of reflections, it follows that they also are isometries.

Exercises

Set I

In the figure below, $\ell_1 \parallel \ell_2$.

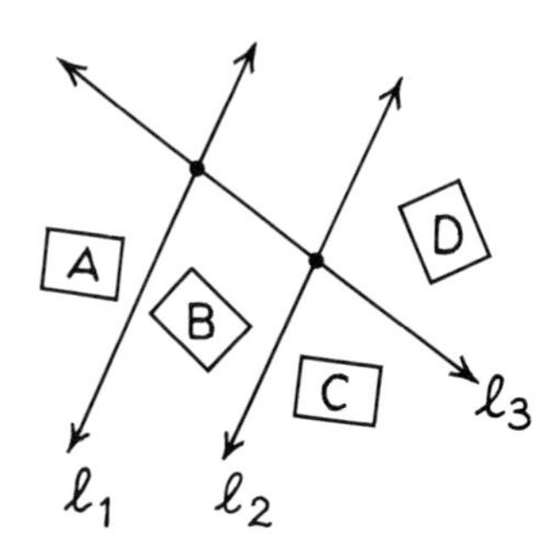

1. Through what transformation does figure B appear to be the image of figure A?
2. Through what transformation does figure C appear to be the image of figure A?
3. Through what transformation does figure D appear to be the image of figure B?
4. What does the transformation that relates figure D to figure A have in common with the transformations that you have named in Exercises 1 through 3?

In the figure below, ℓ_1 intersects ℓ_2 at O. $\overline{CD}$, $\overline{EF}$, $\overline{GH}$, and $\overline{IJ}$ are reflection images of $\overline{AB}$ through either or both of the lines.

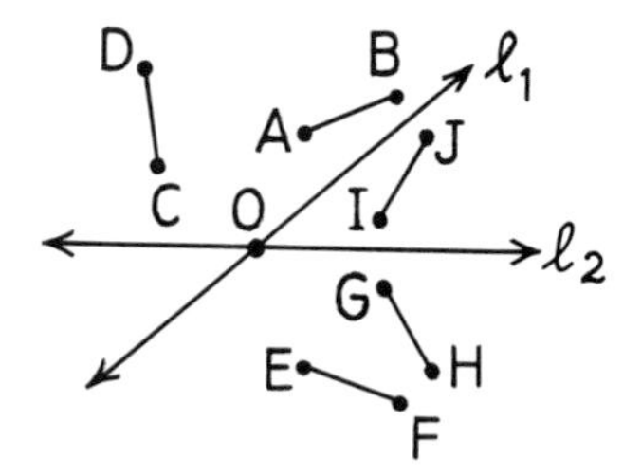

Which line segment is the reflection image of

5. $\overline{AB}$ through ℓ_1?
6. $\overline{IJ}$ through ℓ_2?
7. $\overline{AB}$ through ℓ_2?
8. $\overline{EF}$ through ℓ_1?

Which line segment is the rotation image of $\overline{AB}$ as a result of successive reflections through

9. ℓ_1 and ℓ_2?
10. ℓ_2 and ℓ_1?
11. What is point O called with respect to these rotations?

In the figure below, each of the three lines

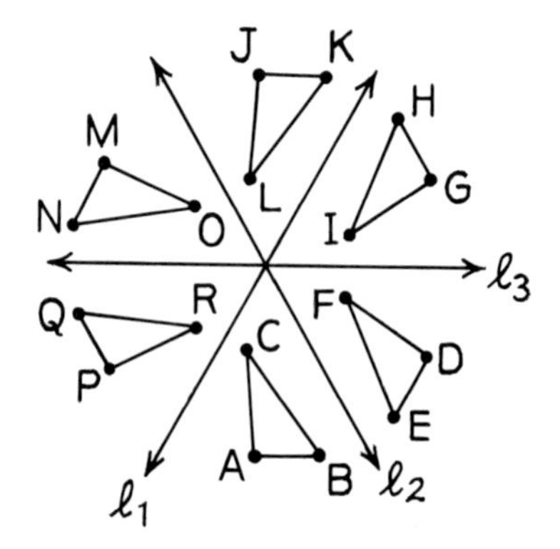

is a mirror of reflection with respect to the triangles on each side of it. For example, $\triangle PQR$ is a reflection image of $\triangle ABC$ through ℓ_1.

12. Which triangle is a reflection image of $\triangle MNO$ through ℓ_2?
13. Which triangle is a reflection image of $\triangle JKL$ through ℓ_3?
14. Which triangle is a rotation image of $\triangle ABC$ resulting from successive reflections through ℓ_1 and ℓ_2?
15. Which triangle is a rotation image of $\triangle DEF$ resulting from successive reflections through ℓ_1 and ℓ_3?

Set II

Trace this figure, in which A-B-C.

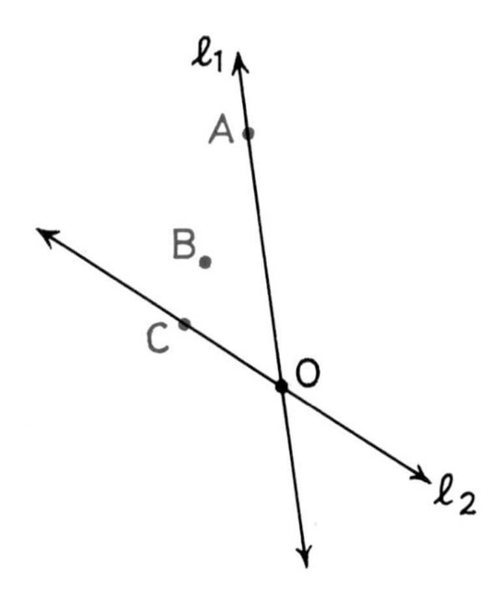

16. Rotate points A, B, and C by reflecting them through ℓ_1 and their images through ℓ_2. Label the rotation images A′, B′, and C′.

17. What properties are preserved in this rotation?

Trace this figure.

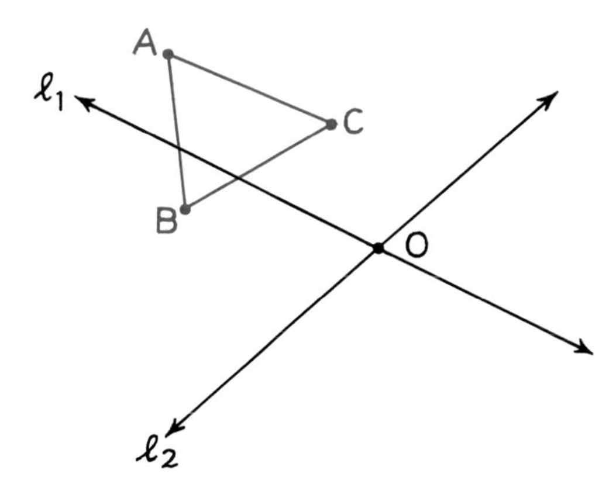

18. Reflect △ABC through ℓ_1 and name its reflection △DEF so that △DEF ≅ △ABC. Reflect △DEF through ℓ_2 and name its reflection △GHI so that △GHI ≅ △DEF. Draw three circles centered at O so that one circle goes through A, one goes through B, and one goes through C.

19. What do you notice about the three circles?

20. If a tracing of △ABC were made, about what point could it be rotated so that it would coincide with △GHI?

21. What is this point called with respect to the rotation?

Trace this figure.

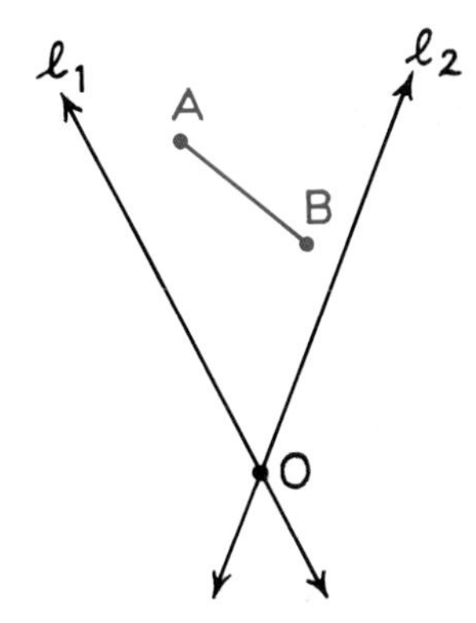

22. Measure one of the acute angles formed by ℓ_1 and ℓ_2.

23. Rotate $\overline{AB}$ by reflecting it through ℓ_1 and its image through ℓ_2. Label the rotation image $\overline{CD}$ so that C is the image of A. Draw ∡AOC.

24. Measure ∡AOC.

25. What is the measure of ∡AOC called with respect to the rotation?

26. How does it seem to compare to the measure of one of the acute angles formed by ℓ_1 and ℓ_2?

In the figure below, point C is the rotation image of point A resulting from successive reflections through ℓ_1 and ℓ_2.

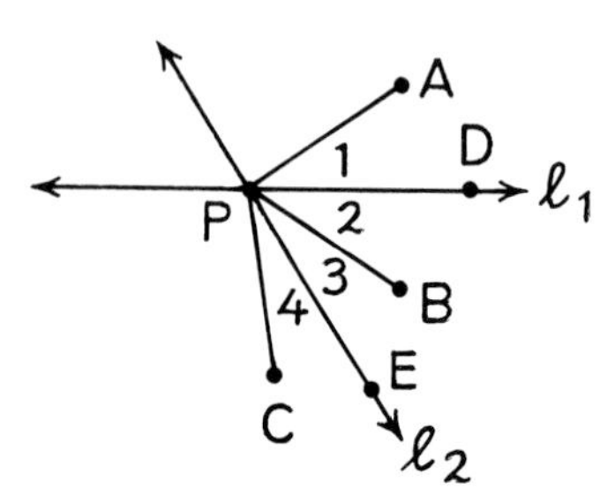

27. What is $\angle APC$ called?
28. Why is $\angle 1 = \angle 2$ and $\angle 3 = \angle 4$?
29. What relation does $\angle APC$ have to the measure of $\measuredangle DPE$, one of the acute angles formed by ℓ_1 and ℓ_2?

Trace this figure, in which $\ell_1 \perp \ell_2$.

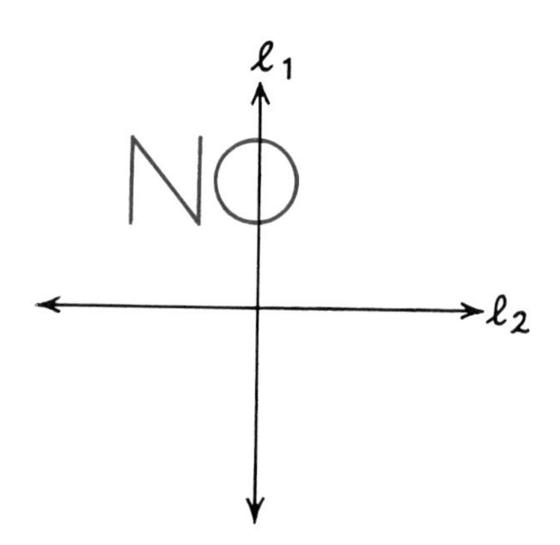

30. Reflect NO through ℓ_1 and reflect the resulting image through ℓ_2.
31. What is the measure of the angles formed by ℓ_1 and ℓ_2?
32. What is the magnitude of the rotation of NO?

Set III

Little children like being rotated in a swing.

1. Trace the two drawings below. Can you find a line through which

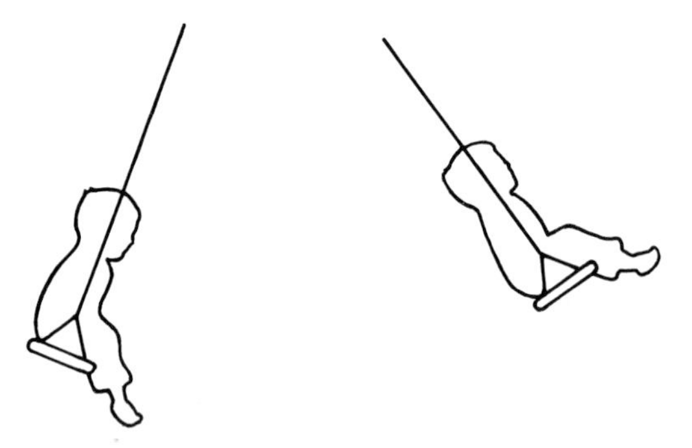

to reflect one of them and a second line through which to reflect its image so that the result is the same as the rotation shown?

2. Describe your method in words.

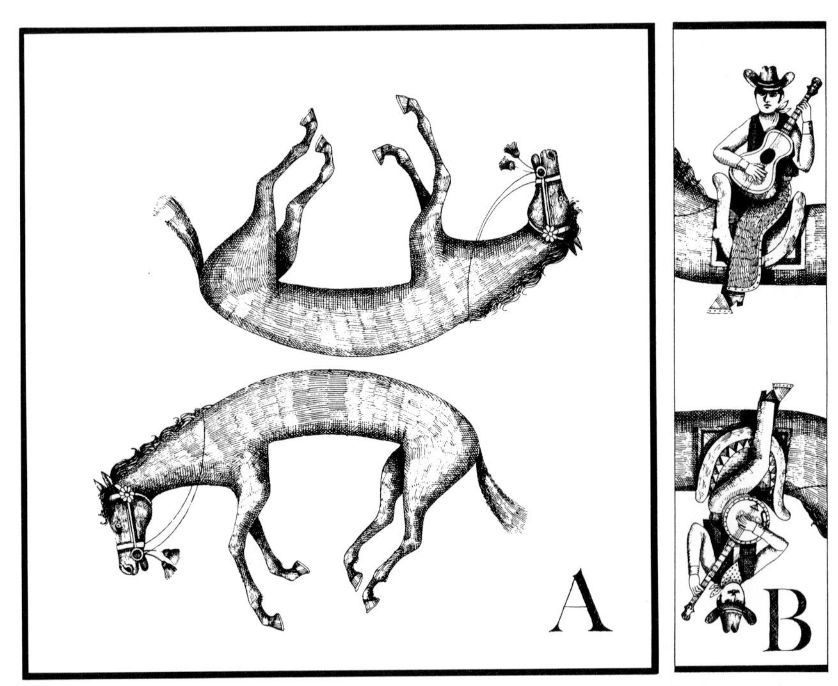

Lesson 5

Congruence and Isometries

An interesting puzzle consists of the two drawings shown here. The problem is to place drawing B on drawing A in such a way that each rider is astride a horse. Although it appears from the positions and spacing of the horses that this is impossible, the puzzle does have a solution.*

The drawings of the two horses in figure A look as if one might be a rotation image of the other. If this is the case, it follows that the figures of the horses must be congruent. We have already defined congruence but only in terms of polygons. These horses, and many other geometric figures, are not polygons. The transformation that seems to relate them, however, suggests a general definition of congruence.

Definition
Two figures are ***congruent*** iff there is an isometry such that one figure is the image of the other.

---

*From "Problem-Solving," by Martin Scheerer, *Scientific American,* April 1963.

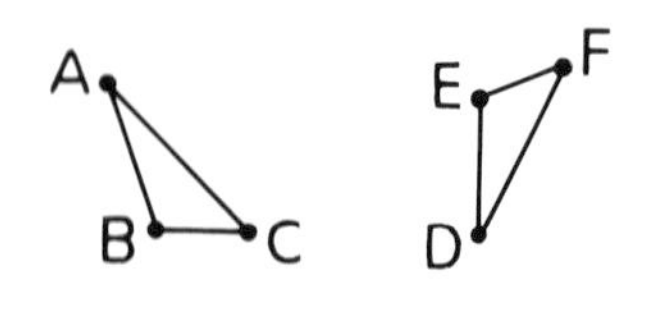

Look at the two triangles at the left. The definition of congruence in terms of isometries says that $\triangle ABC \cong \triangle DEF$ iff there is an isometry such that $\triangle DEF$ is the image of $\triangle ABC$. We have learned that reflections, and hence translations and rotations because they are composites of reflections, are isometries. The reasoning that led us to these conclusions can be extended to show that *any* transformation that is a composite of reflections is an isometry.

How we might find such a composite of reflections relating $\triangle ABC$ and $\triangle DEF$ is shown here. First, we observe that D appears to be the

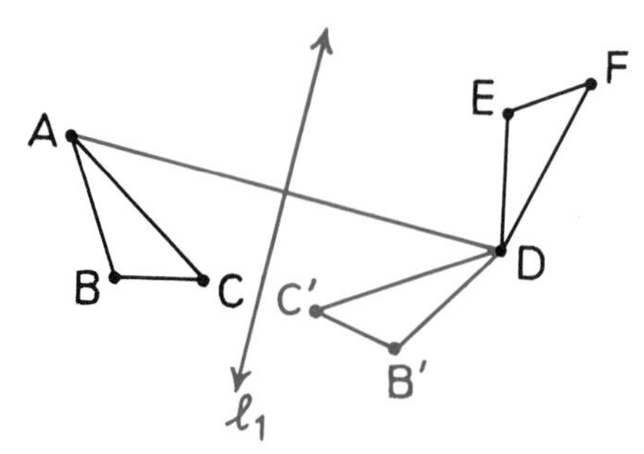

image of A. Let ℓ_1 be the perpendicular bisector of $\overline{AD}$. Reflect $\triangle ABC$ through ℓ_1 to produce $\triangle DB'C'$.

Next, because E appears to be the image of B, and hence B′, let ℓ_2 be the perpendicular bisector of $\overline{EB'}$. Reflect $\triangle DB'C'$ through ℓ_2 to produce $\triangle DEC''$.

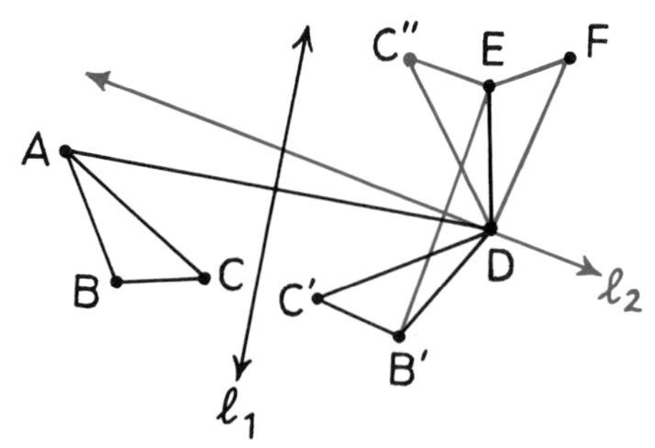

Finally, because F appears to be the image of C, and hence C′ and C″, let ℓ_3 be the perpendicular bisector of $\overline{FC''}$. Reflect $\triangle DEC''$ through ℓ_3 to produce $\triangle DEF$.

The composite of the successive reflections through ℓ_1, ℓ_2, and ℓ_3 is an isometry in which $\triangle DEF$ is the image of $\triangle ABC$.

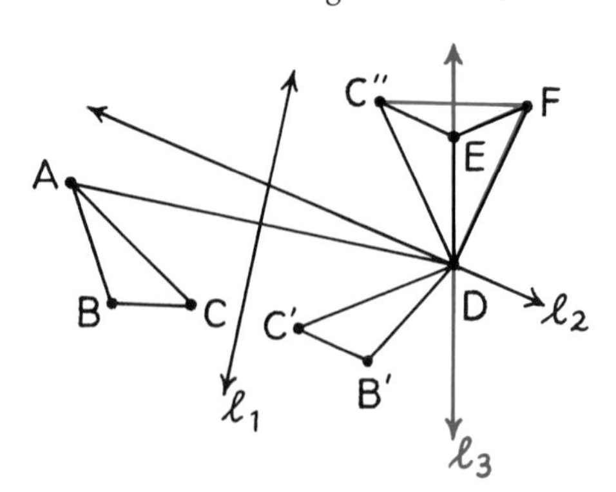

Exercises

Set I

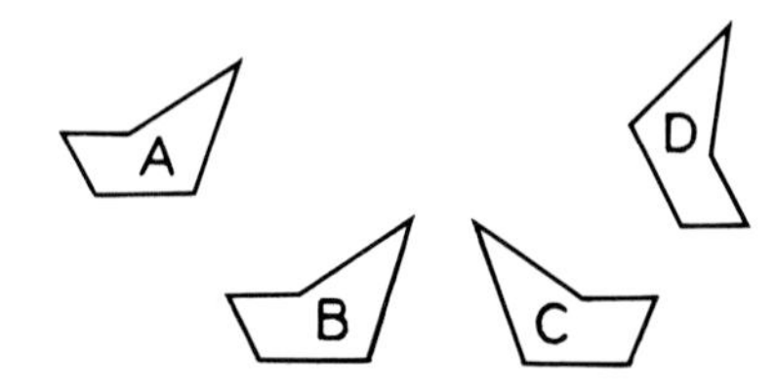

All of the polygons in the figure above are congruent. Name the isometry through which

1. polygon B appears to be the image of polygon A.
2. polygon C appears to be the image of polygon B.
3. polygon D appears to be the image of polygon C.
4. Is there an isometry through which polygon D is the image of polygon A?
5. Use the definition of congruent figures in this lesson to explain why or why not.

In the figure below, ℓ_1 and ℓ_2 are lines of reflection for the polygons shown.

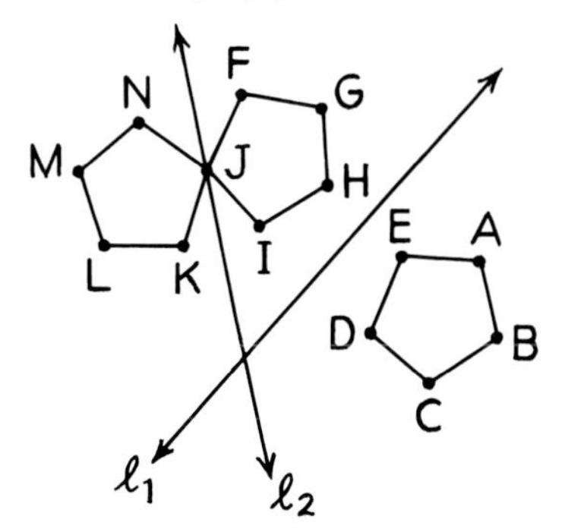

6. Which point is the reflection image of A through ℓ_1?
7. Copy and complete the following congruence correspondence between the vertices of ABCDE and FGHIJ:

ABCDE $\leftrightarrow$ ||||||||.

8. Through what transformation is JKLMN the image of ABCDE?
9. Copy and complete the following congruence correspondence between the vertices of ABCDE and JKLMN:

ABCDE $\leftrightarrow$ ||||||||.

In the figure below, ABCD $\cong$ EFGH. The

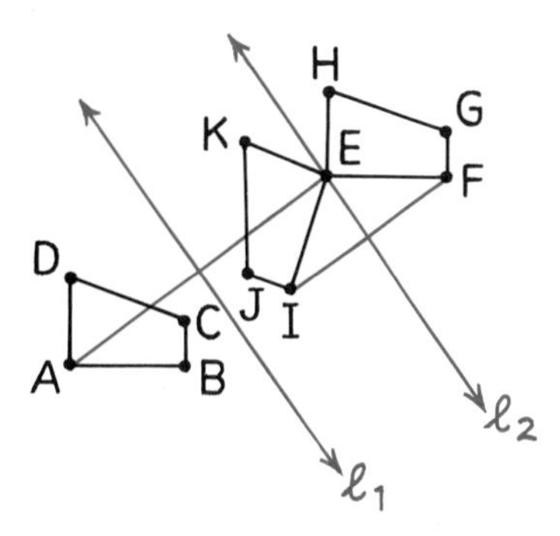

lines and line segments illustrate a procedure for showing that EFGH is an image of ABCD resulting from a given isometry: $\overline{AE}$ was drawn first, followed by ℓ_1.

10. What assumption permits us to draw $\overline{AE}$?
11. What relation does ℓ_1 have to $\overline{AE}$?
12. What line or line segment was drawn after $\overline{AE}$ and ℓ_1?
13. What relation does ℓ_2 appear to have to ℓ_1?
14. If this relation is true, through what transformation is EFGH the image of ABCD?

In the figure below, $\measuredangle$B is the reflection image of $\measuredangle$A through ℓ_1 and $\measuredangle$C is the reflection image of $\measuredangle$B through ℓ_2.

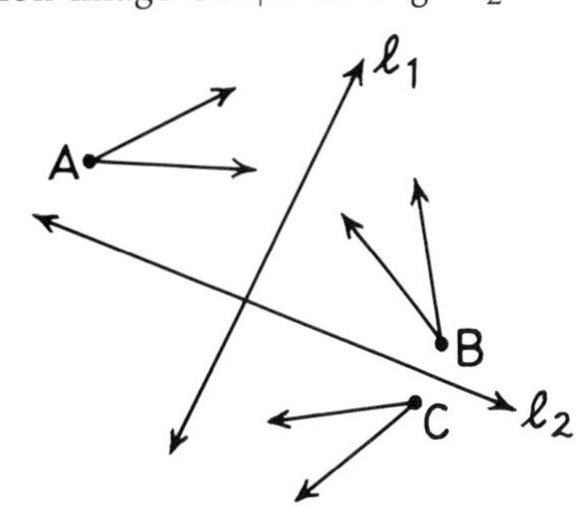

15. Does it follow that $\measuredangle A \cong \measuredangle C$?

16. Explain why or why not.

17. If two angles are congruent, what must be true about their measures?

Use the definition of congruent figures given in this lesson to decide whether each of the following statements is true or false.

18. Any two points are congruent.

19. Any two line segments are congruent.

20. Any two lines are congruent.

Set II

Trace this figure, in which $\triangle ABC \cong \triangle DEF$.

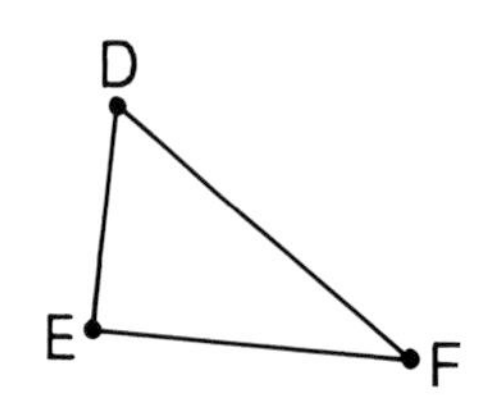

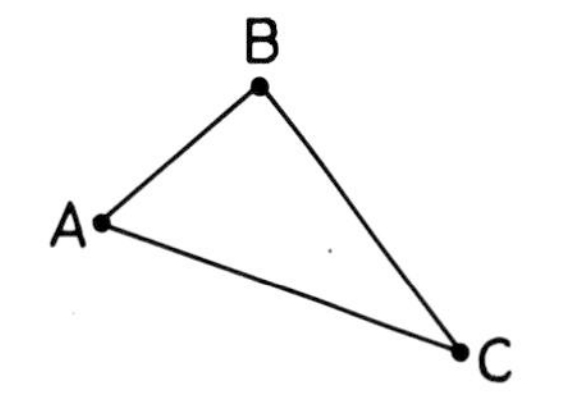

21. Draw $\overline{AD}$. Draw its perpendicular bisector and label it ℓ. Draw $\overline{BE}$ and $\overline{CF}$.

22. What relation does ℓ appear to have to $\overline{BE}$ and $\overline{CF}$?

23. Given that this relation is true, what can you conclude about $\triangle DEF$ with respect to $\triangle ABC$?

Trace this figure, in which ABCD $\cong$ EFGH.

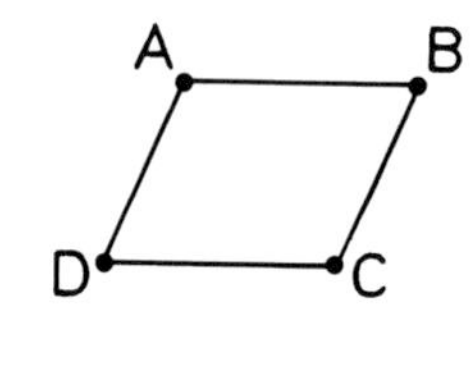

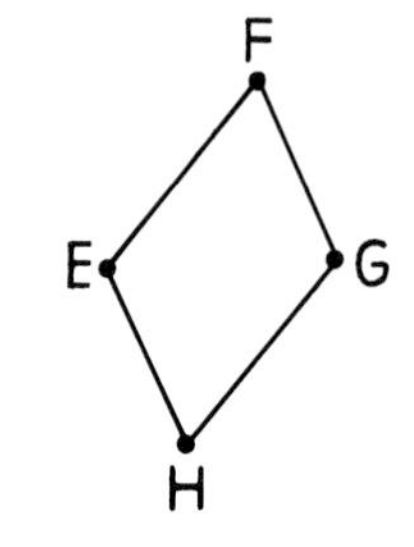

24. Draw $\overline{AE}$. Draw its perpendicular bisector and label it ℓ_1. Reflect ABCD through ℓ_1 and name its reflection EB′C′D′. Draw $\overline{B'F}$. Draw its perpendicular bisector and label it ℓ_2.

25. What relation does EFGH appear to have to EB′C′D′?

26. Given that this relation is true, what can you conclude about EFGH with respect to ABCD?

27. Where is the center of the transformation relating ABCD and EFGH?

Set III

Trace the two drawings at the beginning of this lesson, and then use your tracings to try to solve the puzzle.

Lesson 6

Symmetry

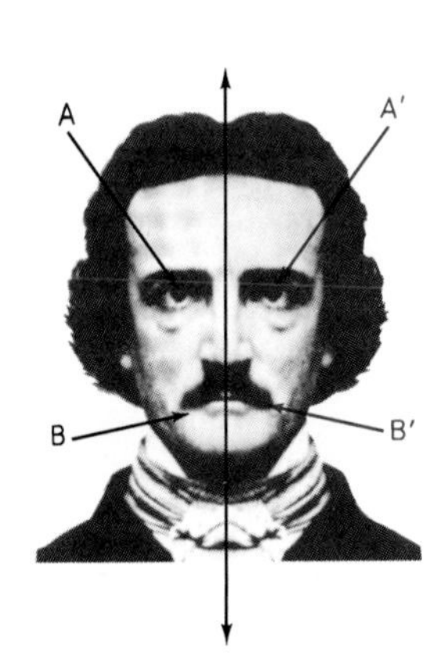

The left side of a human face is close to being the mirror image of the right side. But not exactly. The first photograph above is of Edgar Allan Poe as he actually looked. The second and third photographs are composite pictures made by merging each side of the first picture with its corresponding reflection.

If a vertical line is drawn through the center of each of the composite pictures, each point of the picture on one side of the line has an image point on the other side of the line. Such a figure is said to be *symmetric* with respect to the line.

► **Definition**
A figure has ***reflection symmetry*** with respect to a line iff it coincides with its reflection image through the line.

The biologist's term for reflection symmetry is "bilateral symmetry." A simple test to determine whether a figure has it is to fold the figure along the supposed symmetry line and see if the two halves of the figure coincide.

An example of a geometric figure that has reflection symmetry is an isosceles triangle. In the adjoining figure, $\triangle ABC$ is isosceles with $AC = BC$. The triangle is symmetric with respect to a line through its vertex, C, and the midpoint of its base, M.

Although the parallelogram shown here does not have reflection symmetry, it possesses another type of symmetry called *rotation symmetry*.

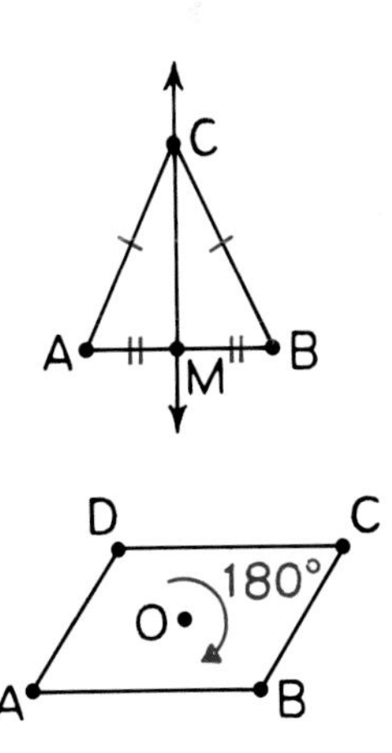

Photograph of Edgar Allan Poe reprinted with the permission of the American Antiquarian Society. Composite photographs from *Symmetry*, by I. Bernal, W. C. Hamilton, and J. S. Ricci. W. H. Freeman and Company. Copyright © 1972.

► **Definition**
A figure has ***rotation symmetry*** with respect to a point iff it coincides with its rotation image about the point.

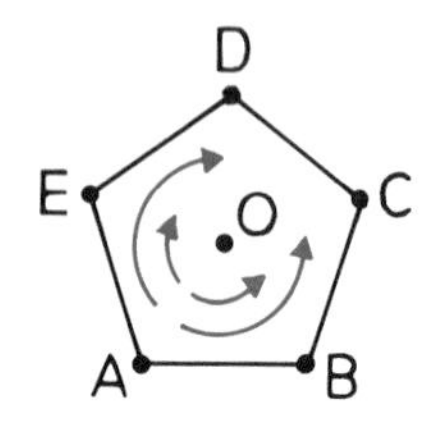

The point, which is the center of the rotation, is called the center of the symmetry. If ▱ABCD is rotated 180° about the point labeled O, each vertex will fall on the opposite vertex (for example, D will fall on B), causing the parallelogram and its rotation image to coincide.

Another example of a figure that has rotation symmetry is this pentagon. It can be rotated 72° or 144° in either direction about the point labeled O, and the pentagon and its resulting rotation image will coincide.

Exercises

Set I

In the figure below, lines ℓ_1, ℓ_2, ℓ_3, and ℓ_4 intersect at the midpoint of $\overline{AB}$.

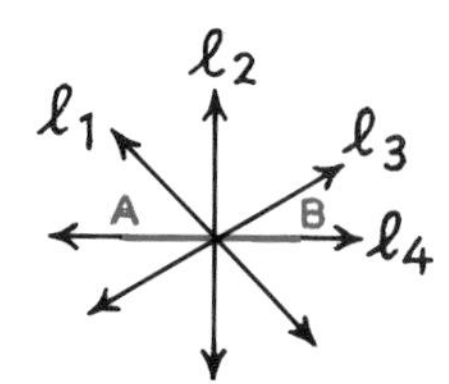

1. Which line or lines look like lines of symmetry with respect to $\overline{AB}$?
2. In general, how many lines of symmetry does a line segment have that lie in a given plane that contains the line segment?

An equilateral triangle has three lines of symmetry, as shown in the figure below.

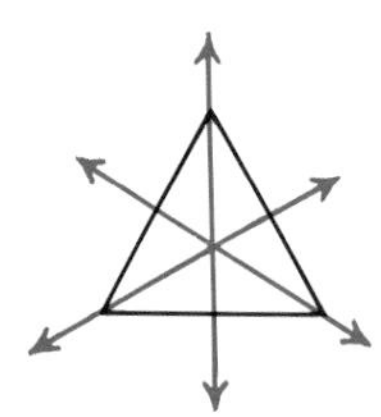

Trace the following figures and draw every line that you think is a symmetry line for each figure.

3. 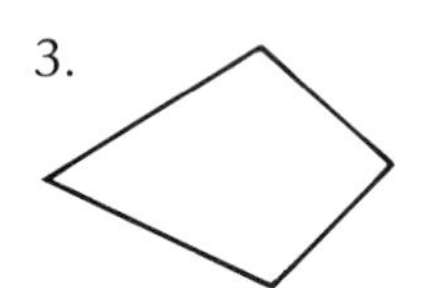

4. 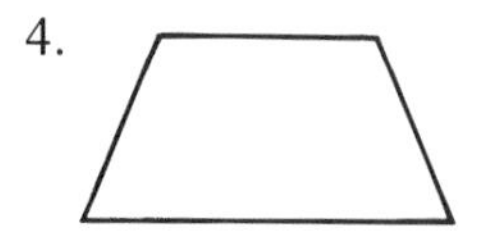

5. 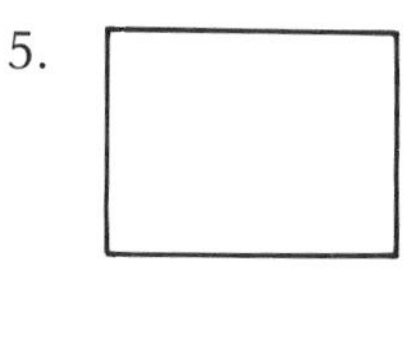

6. 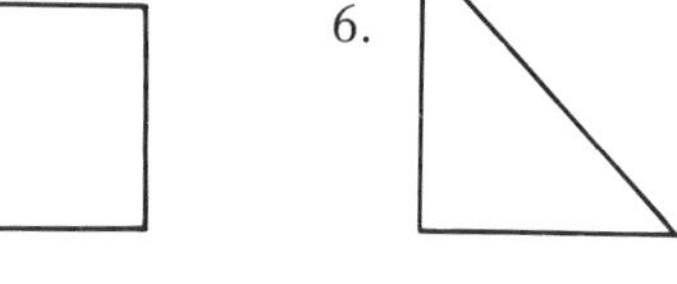

7. 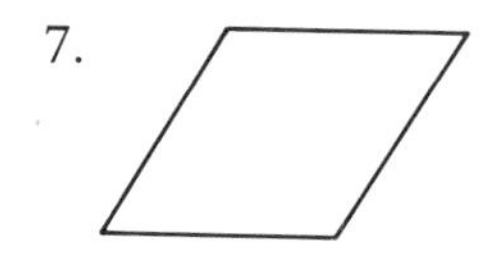

8. 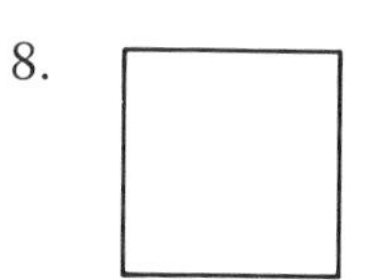

9. 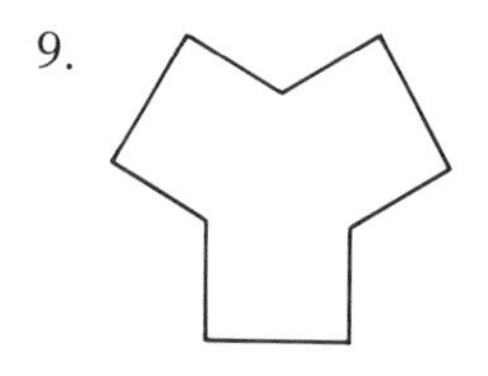

10. 

In the alphabet shown below, the letter A has a vertical line of symmetry, the letter B has a horizontal line of symmetry, and the letter Z has rotation symmetry.

ABCDEFGHI
JKLMNOPQR
STUVWXYZ

11. Which letters, in addition to A, have a vertical line of symmetry?
12. Which letters, in addition to B, have a horizontal line of symmetry?

13. Which letters, in addition to Z, have rotation symmetry?

Describe any symmetry that you see in the objects in the following photographs.

14.

Butterfly

15.

Snowflake

16.

Lick Observatory

Galaxy

Set II

A kaleidoscope is a simple optical instrument that uses two mirrors to produce symmetrical patterns. This photograph shows the face

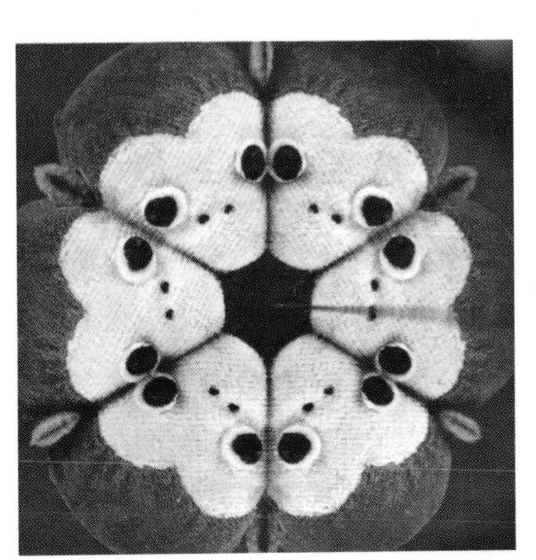

Al Freni—Time-Life Picture Agency

of a toy monkey and five of its images in a kaleidoscope with mirrors that meet at a 60° angle. The pattern of the monkey faces appears to be very symmetrical.

17. How many lines of symmetry does it seem to have?

18. Does it seem to have rotation symmetry?

This figure has rotation symmetry.

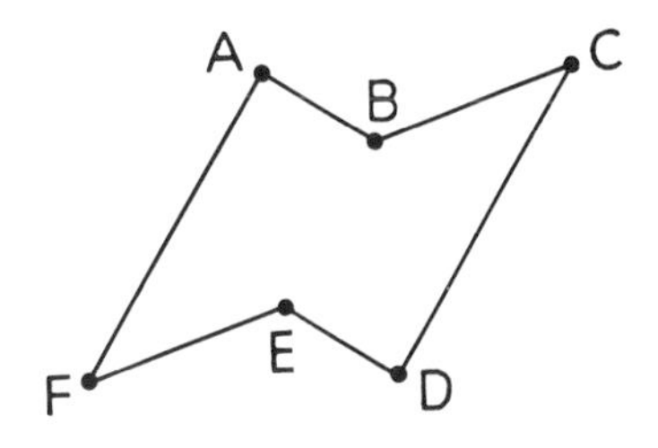

19. Trace the figure and use your straightedge and compass to find the point that is the center of the symmetry; name the point P. Draw three circles centered at P so that one circle goes through A, one circle goes through B, and one circle goes through C.

20. What do you notice about the three circles?

21. Through how many degrees would the figure be rotated so that it and its rotation image coincide?

22. Which point of the figure is the rotation image of A?

Trace the following figures, and then add enough parts to each so that it is symmetric with respect to the lines in the figure.

23.

24.

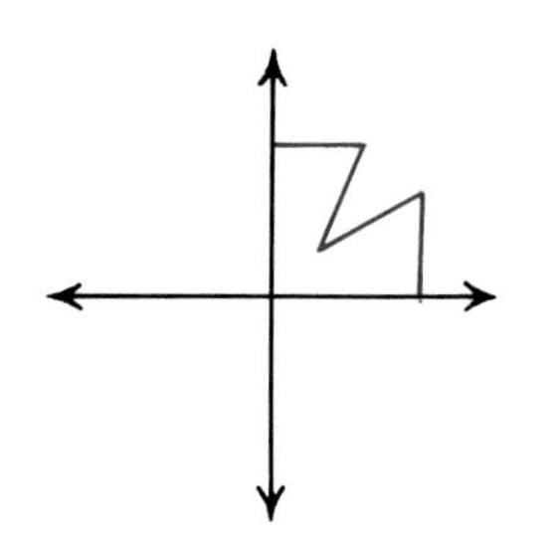

25.

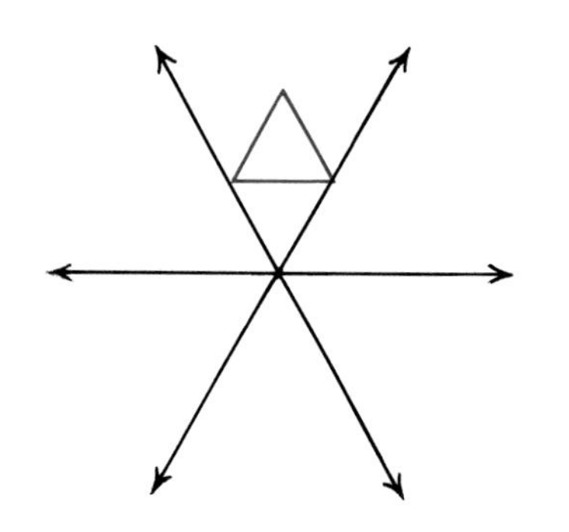

Trace the following figures, and then add as few parts as possible to each figure so that it has rotation symmetry with respect to the point named P.

26. 27.

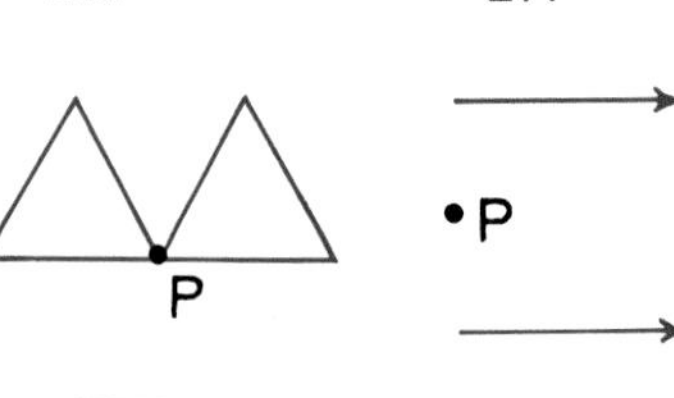

28.

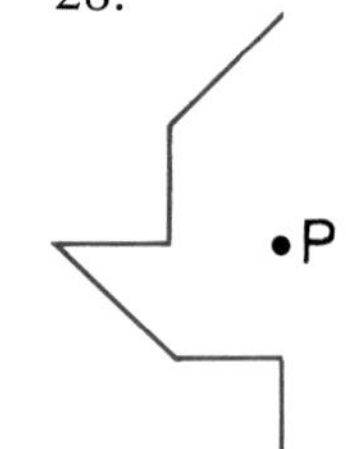

If it is possible, draw a figure fitting each of the following descriptions. Otherwise, write *not possible.*

29. A triangle that has rotation symmetry.
30. A quadrilateral that has exactly one line of symmetry.
31. A parallelogram that has exactly two lines of symmetry.
32. A rhombus that has four lines of symmetry.
33. A quadrilateral that has rotation symmetry but does not have reflection symmetry.

Set III

Courtesy of John McClellan

The swimming pool in this cartoon is closed on Mondays. There is something rather remarkable in the picture other than the fact that there are mice swimming in the pool. What is it?

Chapter 8 / Summary and Review

Basic Ideas

Congruent figures 291
Isometry 275
Magnitude of rotation 287
Magnitude of translation 282
Reflection 271
Reflection symmetry 295
Rotation 287
Rotation symmetry 296
Transformation 275
Translation 282

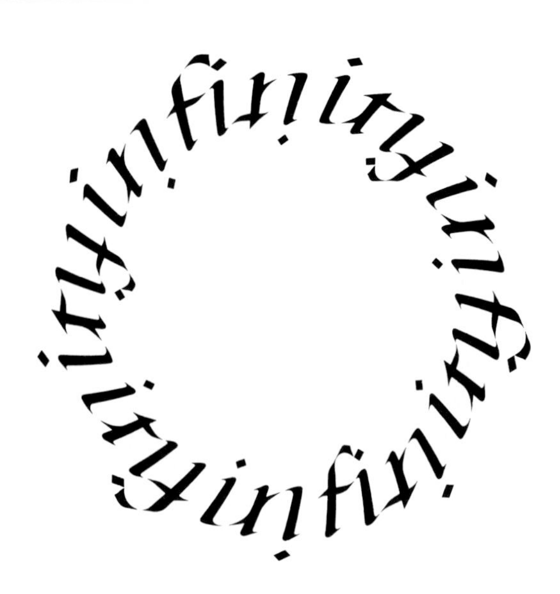

By the permission of Scott Kim

Constructions

9. To construct the reflection of a point through a line. 271

Transformation Theorems

1. Reflection of a pair of points through a line preserves distance. 271
2. An isometry preserves collinearity and betweenness. 276
3. An isometry preserves angle measure. 277
4. A triangle and its image under an isometry are congruent. 277

Exercises

Set I

Tell whether each of the following statements is true or false.

1. A transformation is a one-to-one correspondence between two sets of points.
2. If one point is the reflection image of another point through a given line, then the line is the perpendicular bisector of the line segment that joins the two points.
3. The composite of two successive reflections through intersecting lines is a translation.
4. If a triangle has reflection symmetry, it must be isosceles.
5. Every parallelogram has rotation symmetry.
6. If a figure has rotation symmetry, then it looks the same when turned upside-down.
7. If a transformation is an isometry, then it must be either a translation or a rotation.
8. If there is a transformation such that one figure is the image of another, then the two figures must be congruent.

In the figure below, A′B′C′D′ is the image of ABCD under a certain transformation.

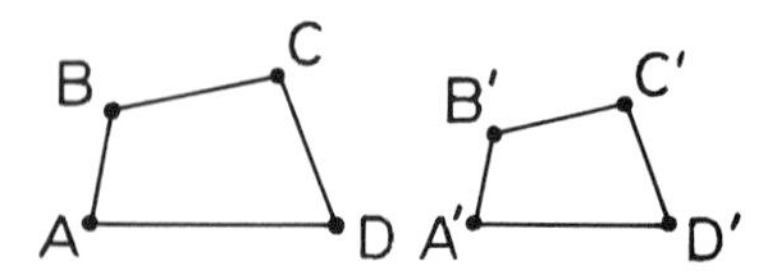

9. What is a transformation?

10. Is this transformation an isometry?

11. What property is not preserved by this transformation?

In the figure below, rectangle BFDE is the reflection image of rectangle ABCD through line ℓ.

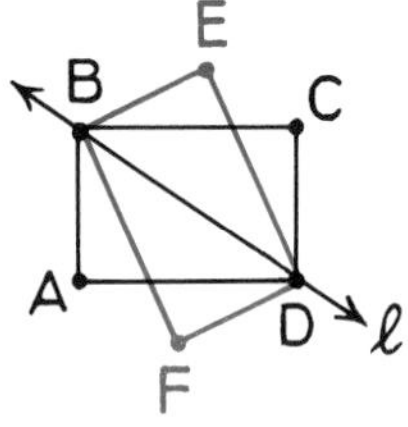

12. Which points of ABCD are their own images in this transformation?

13. What properties are preserved by this transformation?

14. Copy and complete the following congruence correspondence between the vertices of ABCD and BFDE: ABCD $\leftrightarrow$ ______.

Trace these figures and draw every line that you think is a symmetry line for each figure.

15.

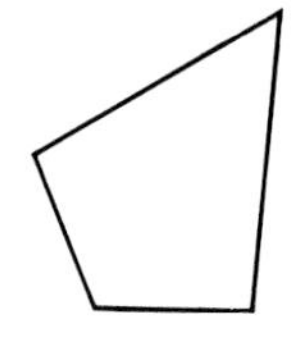

16.

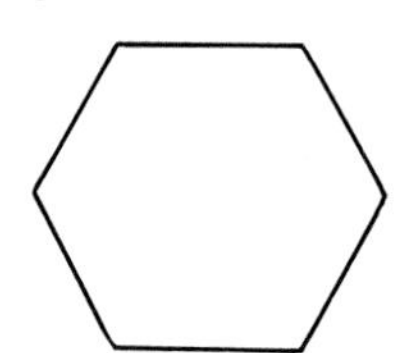

Set II

Two more mosaics by Escher based on animal shapes are shown here.* Imagine that each pattern continues indefinitely without any borders.

Beetles

Birds

*An entire book has been written about Escher's mosaics. It is *Fantasy and Symmetry: The Periodic Drawings of M. C. Escher*, by Caroline H. MacGillavry (Abrams, 1976).

17. What transformations are illustrated by the beetle mosaic?

18. What transformations are illustrated by the bird mosaic?

This polygon is symmetrical.

19. What kind of polygon is it? (Name it with two words.)

20. Does it have reflection symmetry?

21. Does it have rotation symmetry?

22. Trace the figure, and draw any lines or points about which the polygon is symmetrical.

Trace this figure, in which ℓ_1 and ℓ_2 intersect in point P.

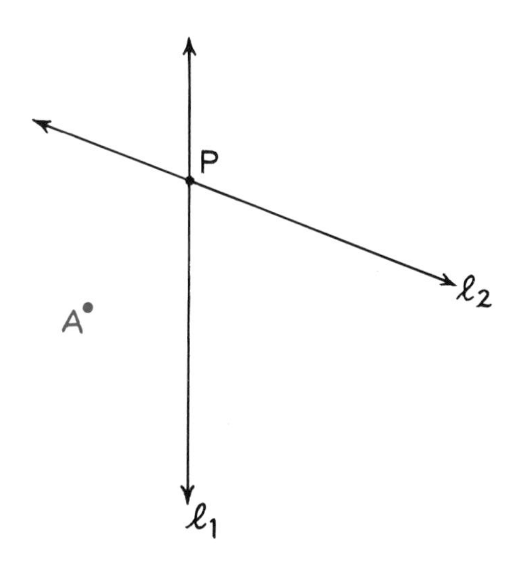

23. Use your straightedge and compass to construct the reflection of A through ℓ_1; name it B. Construct the reflection of B through ℓ_2; name it C.
24. Under what transformation is C the image of A?
25. How do you know?
26. What relation do points A, B, and C have to point P?
27. Why?

Trace this figure, in which $\ell_1 \parallel \ell_2$.

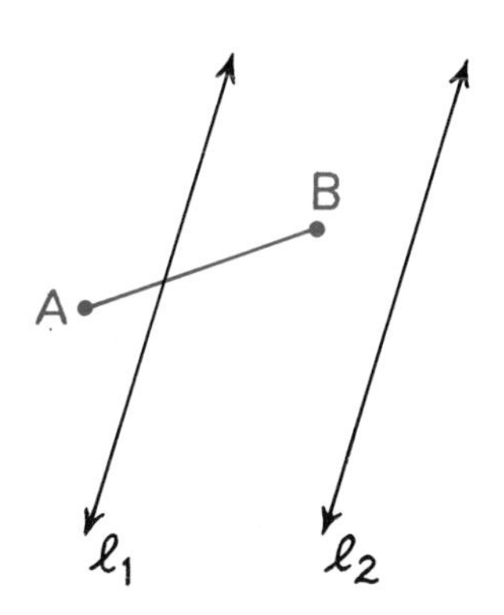

28. Reflect $\overline{AB}$ through ℓ_1, and name its image $\overline{CD}$. Then reflect $\overline{CD}$ through ℓ_2, and name its image $\overline{EF}$.
29. Under what transformation is $\overline{EF}$ the image of $\overline{AB}$?
30. How do you know?
31. Can you conclude that $\overline{AB} \cong \overline{CD}$? Explain why or why not.

Make two tracings of this figure.

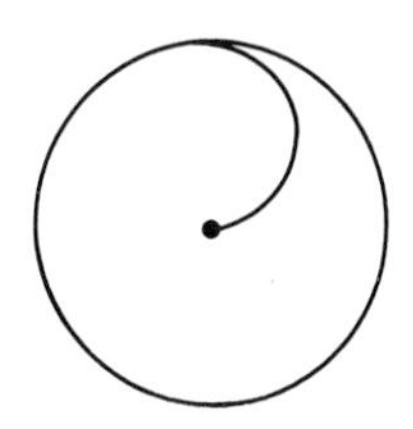

32. Add as few parts as possible to your first tracing so that the resulting figure has reflection symmetry.
33. Add as few parts as possible to your second tracing so that the resulting figure has rotation symmetry.

Trace this figure, in which $\triangle DEF$ is a rotation image of $\triangle ABC$.

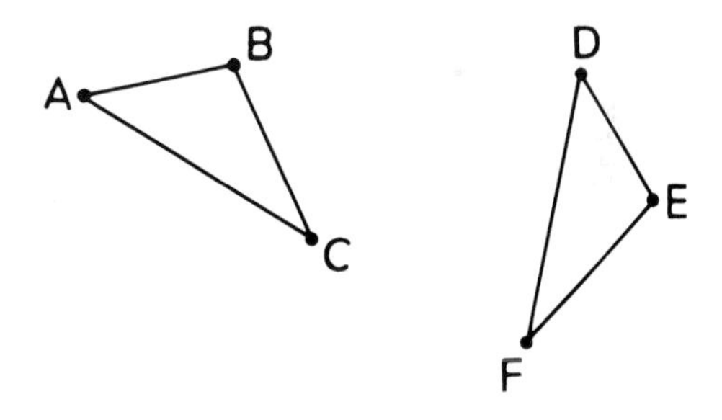

34. Draw ℓ_1 so that the reflection image of A through ℓ_1 is D. Reflect $\triangle ABC$ through ℓ_1, and name its reflection $\triangle DB'C'$. Draw ℓ_2 so that the reflection image of $\triangle DB'C'$ through ℓ_2 is $\triangle DEF$.
35. Where is the center of the rotation?
36. Find the magnitude of the rotation.

MIDTERM REVIEW

Set I

Do the following exercises as indicated.

1. Find the sum of 5 and -12.
2. Simplify: $(x^2)(x^3)$.
3. Use the distributive property to eliminate the parentheses: $9(x + 7)$.
4. Solve for x: $3(x - 11) = 2x + 1$.
5. Solve this pair of simultaneous equations:

$$x - 4y = 13$$
$$x + 4y = 5.$$

6. From $8x - 1$ subtract $x + 1$.
7. Factor $x^2 - 9y^2$.
8. Write $\sqrt{75}$ in simple radical form.

The following questions refer to this statement:

All elephants have ivory tusks.

9. Write it in "if-then" form.
10. Does it follow that, if an animal does not have ivory tusks, it is not an elephant?
11. What relation does this idea have to the original statement?
12. Does it follow that an animal that has ivory tusks is an elephant?
13. What relation does this idea have to the original statement?

Write the letter of the correct answer.

14. If the sides of a triangle have lengths x, x, and y, its perimeter is
    a) $x^2 + y$.
    b) $2xy$.
    c) x^2y.
    d) $2x + y$.
15. Two lines are parallel if the interior angles that they form on the same side of a transversal
    a) are right angles.
    b) are equal.
    c) are complementary.
    d) form a linear pair.
16. Which one of the following parts of the Protractor Postulate is stated incorrectly?
    The rays in a half-rotation can be numbered so that
    a) to every ray there corresponds exactly one real number.
    b) to every real number there corresponds exactly one ray.
    c) to every pair of rays there corresponds exactly one real number.
    d) the measure of an angle is the absolute value of the difference between the coordinates of its rays.
17. If all of the exterior angles of a triangle are obtuse, the triangle must be
    a) obtuse.
    b) equiangular.
    c) acute.
    d) scalene.
18. Which one of the following statements about the diagonals of a parallelogram is always true?
    a) They are equal.
    b) They are longer than the sides.
    c) They are perpendicular.
    d) They bisect each other.
19. If $a > b$ and $c > d$, which of the following must be true?
    a) $a > d$.
    b) $a + c > b + d$.
    c) $a - c > b - d$.
    d) $ac > bd$.

20. If a quadrilateral is equilateral,
    a) it is also equiangular.
    b) it is a square.
    c) its consecutive angles are equal.
    d) it is a parallelogram.

21. What did the acorn say when it grew up? (This is a joke.)
    a) Arithmetic.
    b) Algebra.
    c) Geometry.
    d) Calculus.

Read the following statements carefully. If a statement is always true, write *true*. If it is not, *do not write* false. Instead, write a word or words that could replace the underlined word to make the statement true. Some of the questions in this section may have more than one correct answer; however, do not make a change in any statement that is already true.

22. For any two points, there is exactly one plane that contains them.

23. According to the Betweenness of Points Theorem, if A-B-C, then AB = BC.

24. To prove a theorem indirectly, we begin by assuming that the opposite of its hypothesis is true.

25. If a triangle has two equal sides, it must be equilateral.

26. Two lines that do not intersect must be parallel.

27. The complement of an acute angle is obtuse.

28. If the two angles in a linear pair are supplementary, then each is a right angle.

29. The legs of an isosceles trapezoid are parallel.

30. In a plane, through a point on a line there is exactly one line parallel to the line.

31. All rectangles are convex.

32. The measure of an angle is found by adding the coordinates of its sides and taking the absolute value of the result.

33. The "whole greater than its part" property says that if $c = a + b$ and $b > 0$, then $c > a$.

34. A line segment has exactly one line that bisects it.

35. If a quadrilateral has four equal angles, it must be a square.

36. In a right triangle, the hypotenuse must be the longest side.

Make four tracings of the figure below.

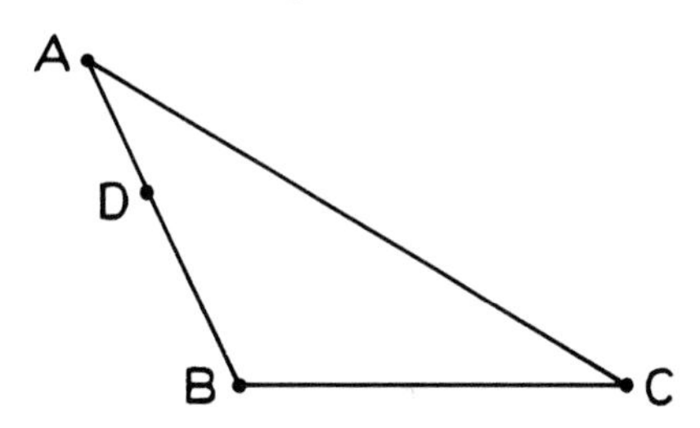

Use your straightedge and compass to make the following constructions.

37. Through D in your first figure, construct a line perpendicular to $\overline{AB}$.

38. Through D in your second figure, construct a line parallel to $\overline{AC}$.

39. Find the midpoint of $\overline{AC}$ in your third figure.

40. Bisect $\measuredangle C$ in your fourth figure.

Set II

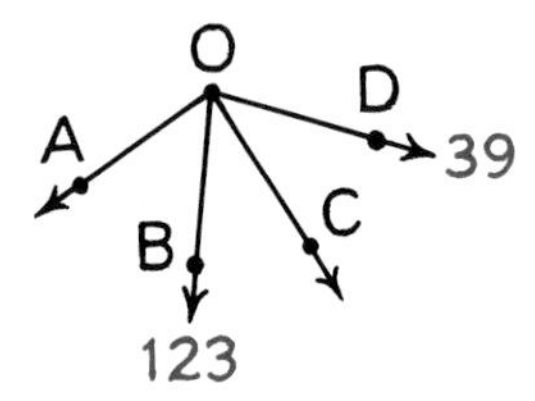

Copy this figure and mark it as necessary to find each of the following numbers.

41. Find ∠BOD.

42. Find ∠BOC, given that $\overrightarrow{OC}$ bisects ∡BOD.

43. Find the coordinate of $\overrightarrow{OC}$.

44. Find ∠AOB, given that ∡AOB and ∡BOC are complementary.

45. Find the coordinate of $\overrightarrow{OA}$.

In the figure below, $\overline{AC} \perp \overline{BD}$ and AD > DC.

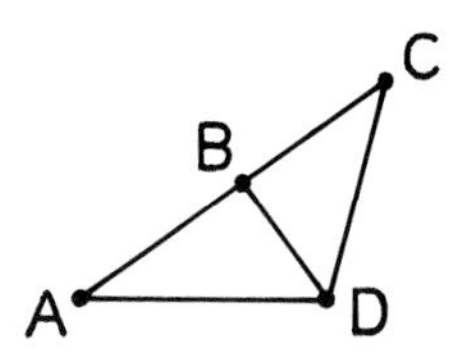

Write the appropriate symbol (>, =, or <) that should replace each |||||||| to make the following statements true.

46. AD |||||||| AB.

47. ∠ADB |||||||| ∠ADC.

48. AB + BD |||||||| AD.

49. AB + BC |||||||| AC.

50. ∠C |||||||| ∠A.

51. ∠A + ∠ADB |||||||| ∠DBC.

Copy the figure below, in which $\overline{BC} \parallel \overline{AD}$, $\overline{DE} \perp \overline{AB}$, ∠B = 112° and ∠EDC = 50°.

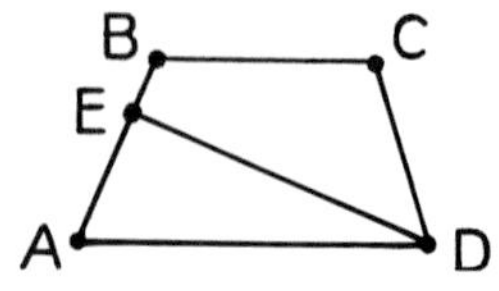

Find the measures of the following angles.

52. ∡AED.

53. ∡A.

54. ∡EDA.

55. ∡CDA.

In the figure below, ∠A = ∠D, A-C-E, and D-C-B.

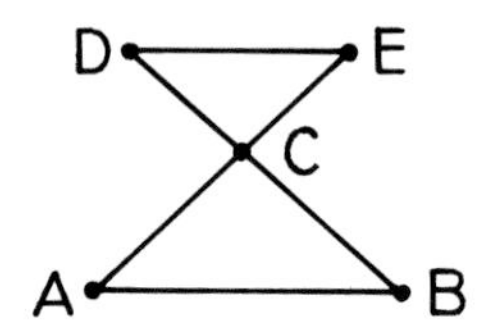

Use this information to tell whether each of the following statements *must be true, may be true,* or *appears to be false.*

56. ∠DCE = ∠ACB.

57. $\overline{AE} \perp \overline{DB}$.

58. ∠E = ∠B.

59. △ACB ≅ △DCE.

60. $\overline{DE} \parallel \overline{AB}$.

61. △ACB is isosceles.

62. ∡D and ∡DCE are supplementary.

63. A, C, and E are collinear.

64. Give the missing statements and reasons in this proof.

Given: Quadrilateral ABCD with diagonals $\overline{AC}$ and $\overline{BD}$; AB = BC = CD = DA = BD.

Prove: $\angle 1 = 30°$.

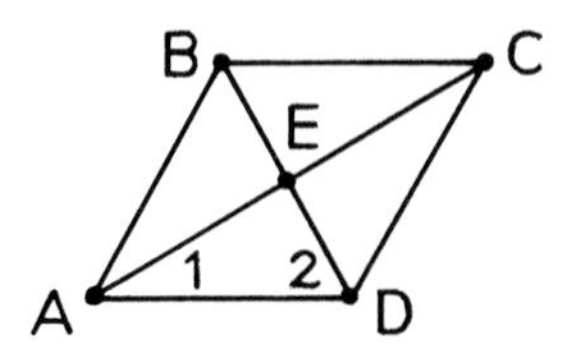

Proof.

Statements	***Reasons***
1. Quadrilateral ABCD with diagonals $\overline{AC}$ and $\overline{BD}$; AB = BC = CD = DA = BD.	Given. 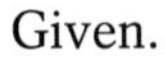
2. ABCD is a rhombus.	a) ______
3. $\overline{BD} \perp \overline{AC}$.	b) ______
4. c) ______	d) ______
5. $\triangle$AED is a right triangle.	A triangle that has a right angle is a right triangle.
6. ∡1 and ∡2 are complementary.	e) ______
7. f) ______	g) ______
8. h) ______	A triangle that has three equal sides is equilateral.
9. $\angle 2 = 60°$.	i) ______
10. j) ______	Substitution.
11. k) ______	l) ______

65. Write a complete proof for this exercise.

Given: $\overline{AB} \parallel \overline{DC}$; AD = DC.

Prove: $\overrightarrow{AC}$ bisects ∡BAD.

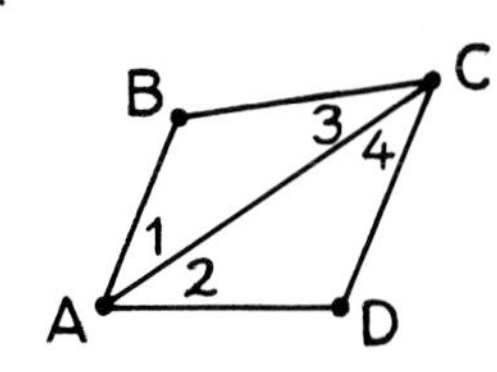

Chapter 9
AREA

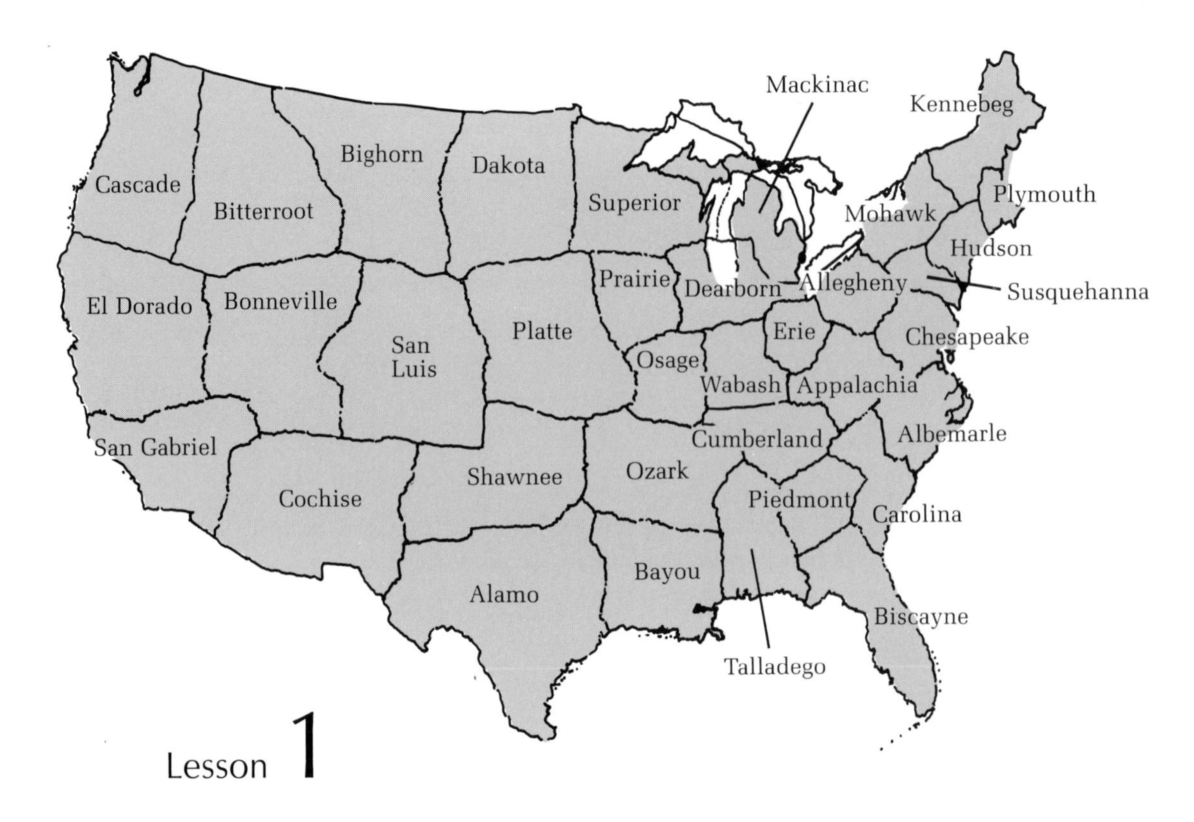

Lesson 1

Polygonal Regions and Area

Several years ago it was suggested that the United States be reorganized into thirty-eight states. The only state that would remain intact after the rearrangement is Hawaii. Alaska would become two states, and the rest of the country would be divided as shown in the map above. Among the proposed benefits of this reorganization are a better balance in the population and size of states. The size of each state is measured with a number called its *area.*

In this chapter we will learn how to determine the areas of geometric figures called polygonal regions. Polygonal regions are defined in terms of triangular regions.

► Definition
A ***triangular region*** is the union of a triangle and its interior.

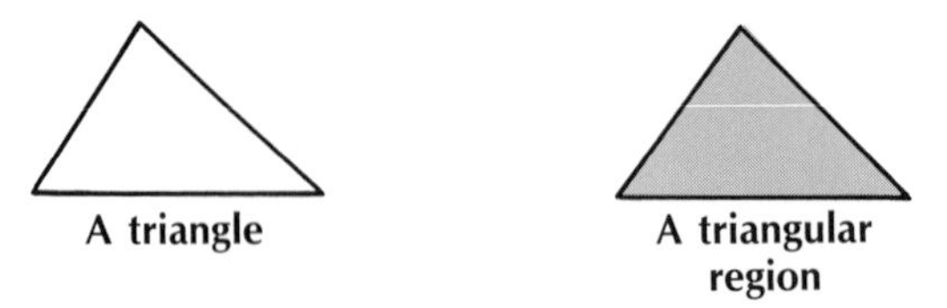

Polygonal regions can be divided into triangular regions. Some examples of polygonal regions are shown on the next page.

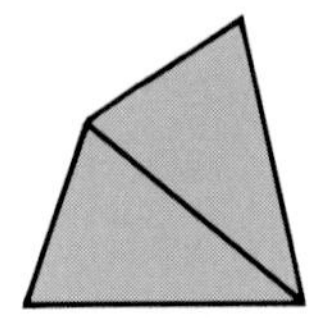 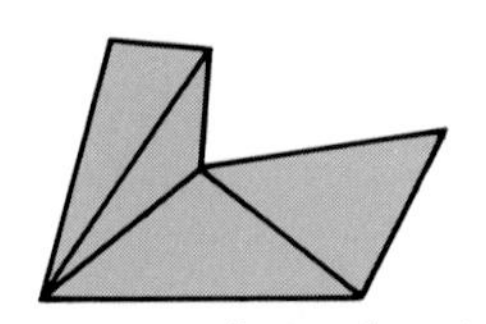 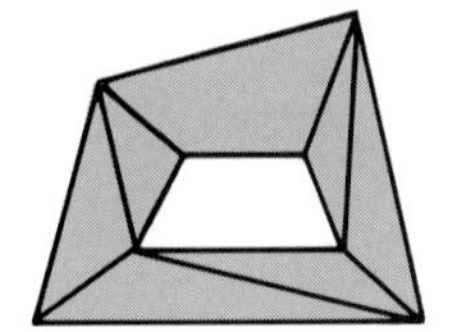

Some polygonal regions

► **Definition**
A ***polygonal region*** is the union of a finite number of nonoverlapping triangular regions in a plane.

It seems reasonable to assume that every polygonal region has an area. The first figure above suggests that triangular regions bounded by congruent triangles have equal areas. All three figures suggest that, if a polygonal region is divided into smaller regions, its area is equal to the sum of their areas. These ideas are summarized in the following postulate.

► **Postulate 11** (The Area Postulate)
To every polygonal region there corresponds a positive number called its area such that
a) triangular regions bounded by congruent triangles have equal areas,
b) and polygonal regions consisting of two or more nonoverlapping polygonal regions are equal in area to the sum of their areas.

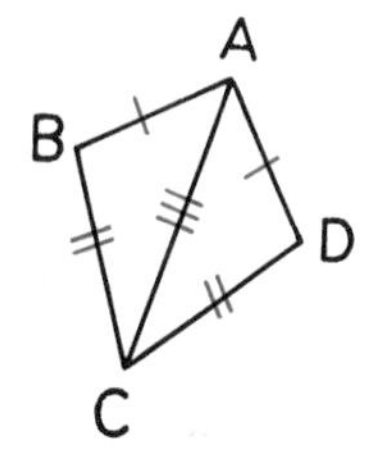

We will represent the word "area" by α (alpha), the first letter in the Greek alphabet. The figure at the right is a quadrilateral that is divided by a diagonal into two congruent triangles. On the basis of the Area Postulate, we can conclude that

$$\alpha\triangle ABC = \alpha\triangle ADC \quad \text{and} \quad \alpha ABCD = \alpha\triangle ABC + \alpha\triangle ADC.$$

Exercises

Set I

Trace the figures below and then show that each is a polygonal region by dividing it into triangular regions. In each case, try to form *as few triangular regions as possible.*

1. 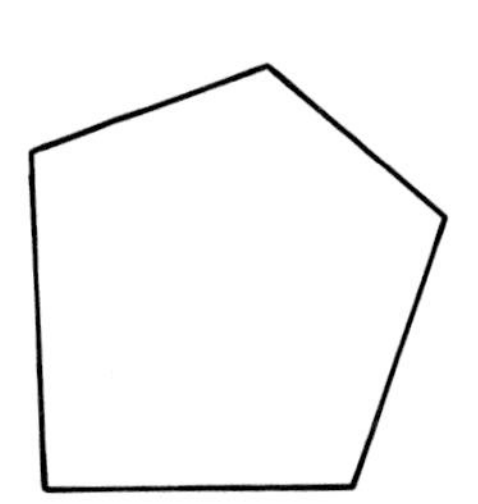

2. 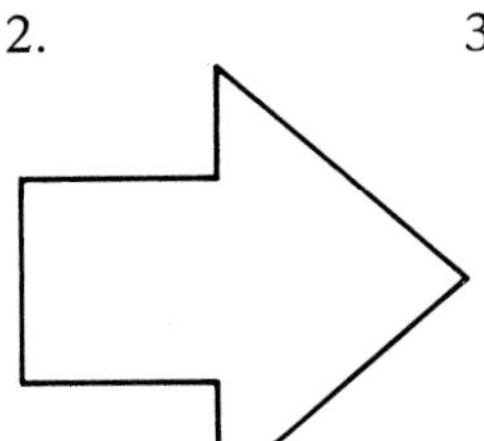

3. 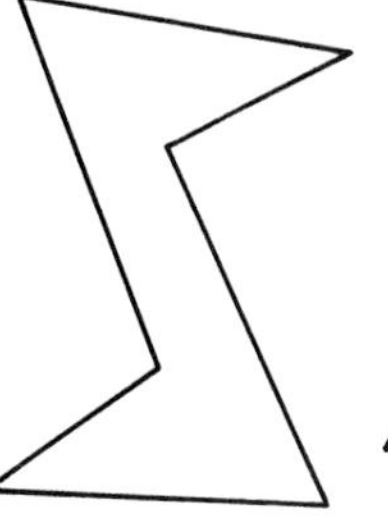

4. 

5. 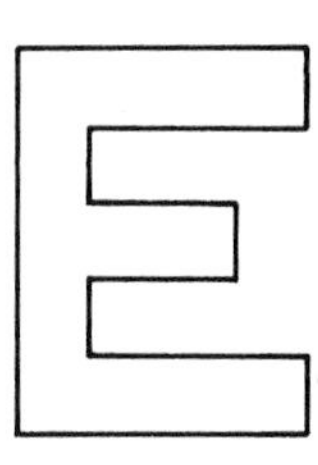

In the figure below, $\triangle PRI \cong \triangle ENC$. Copy and complete the following equations.

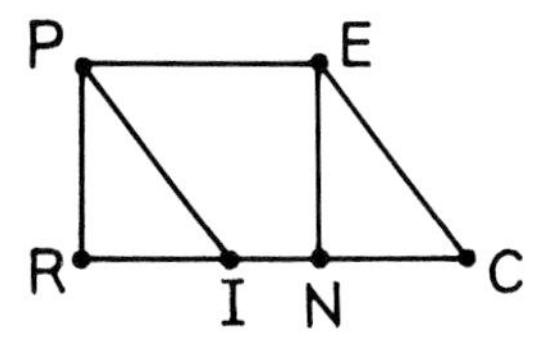

6. $\alpha\triangle PRI = \alpha$ ____.
7. $\alpha PRCE = \alpha$ ____ $+ \alpha$ ____ $+ \alpha$ ____.

State the part of the Area Postulate that is the basis for your answer to

8. Exercise 6.
9. Exercise 7.

In the figure below, $\alpha\triangle SUL = \alpha\triangle TAN$.

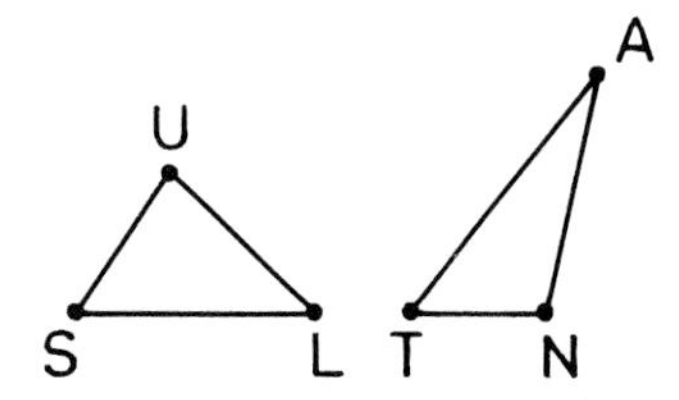

10. Is $\triangle SUL \cong \triangle TAN$?
11. If two triangular regions have equal areas, does it follow that they are bounded by congruent triangles?

In the figure below, $\triangle MIK$ and $\triangle ADO$ are isosceles and $MI = IK = AD = DO$.

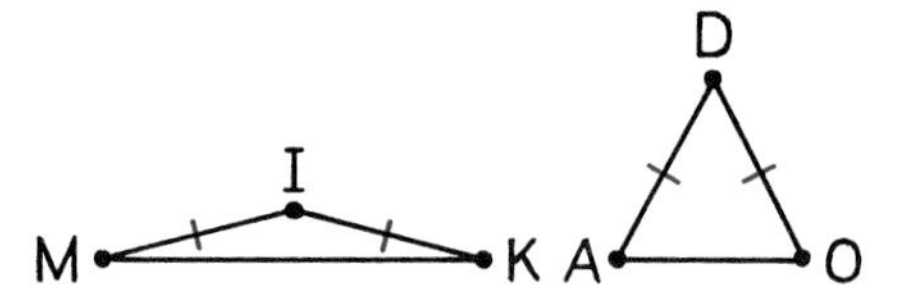

12. Which triangle appears to have the greater area?
13. Which triangle has the greater perimeter?

In the figure below, MOGU is an isosceles trapezoid with diagonals $\overline{MG}$ and $\overline{OU}$. Use this information to tell whether each of the following statements *must be true, may be true,* or *is false.*

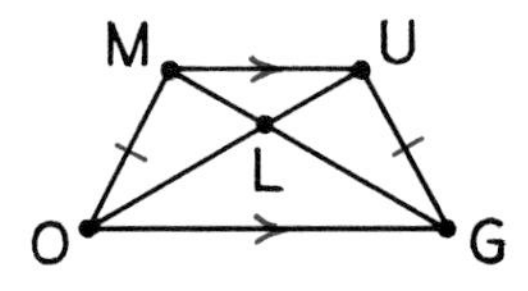

14. $\triangle MOG \cong \triangle UGO$.
15. $\triangle MUL \cong \triangle OLG$.
16. $\alpha\triangle MOG = \alpha\triangle UGO$.
17. $\alpha\triangle MOU = \alpha\triangle MOL + \alpha\triangle MLU$.
18. $\alpha MOGU = \alpha\triangle MOG + \alpha\triangle UGO + \alpha\triangle MLU$.
19. $\alpha MOGUL = \alpha\triangle MOL + \alpha\triangle UGO$.

Set II

Quadrilateral SHEI is a rhombus with diagonals $\overline{SE}$ and $\overline{HI}$.

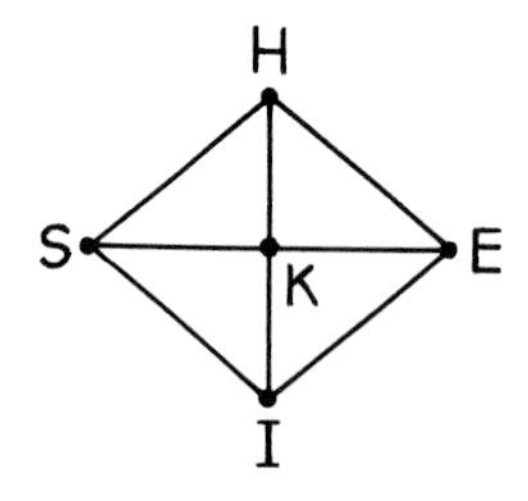

20. State two relations that $\overline{SE}$ and $\overline{HI}$ have to each other.
21. Copy the figure and mark these relations on it.
22. Why is $\triangle SHK \cong \triangle EHK \cong \triangle EIK \cong \triangle SIK$?
23. Why do these triangles have equal areas?
24. What fraction of $\alpha SHEI$ is $\alpha\triangle SHK$?

CLP is a triangle with midsegments $\overline{AI}$, $\overline{IH}$, and $\overline{HA}$.

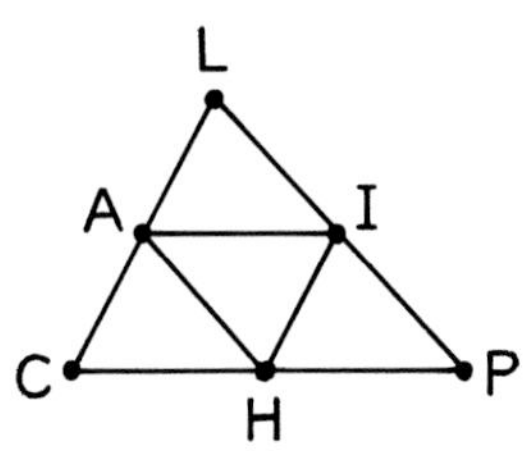

25. What kind of quadrilaterals are CAIH, LIHA, and AIPH?
26. How do you know?
27. Copy the figure and mark the parts that can be used to prove that $\triangle ACH \cong \triangle LAI \cong \triangle IHP \cong \triangle HIA$.
28. What fraction of αCAIP is $\alpha\triangle$LAI?

The figure below consists of six equilateral triangles; $\alpha\text{MCRH} = 8x + 6$ and $\alpha\text{CNAR} = 15x - 9$.

29. Find x.
30. Find αMCRH.
31. Find αCNAR.

32.

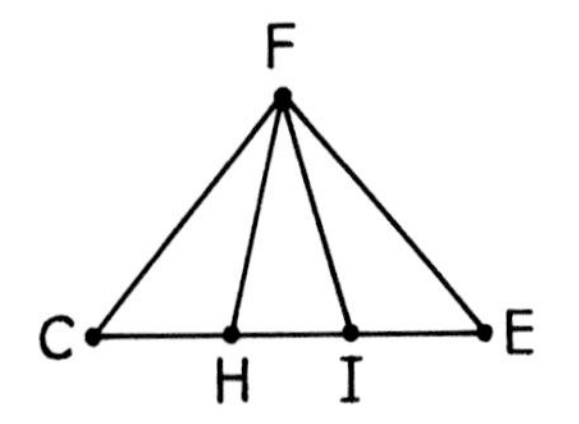

Given: In $\triangle$CFE, CF = EF; $\angle$CFI = $\angle$EFH.

Prove: $\alpha\triangle\text{CFI} = \alpha\triangle\text{EFH}$.

33.

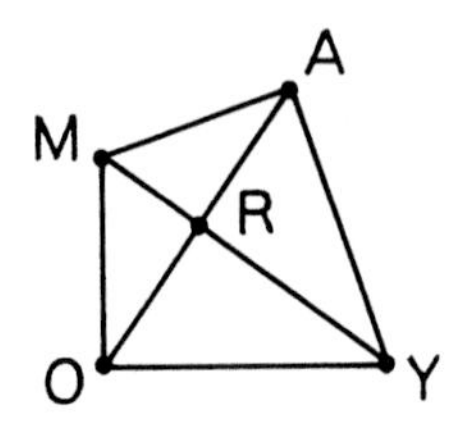

Given: $\triangle$MAY and $\triangle$MOY are right triangles with right angles ∡MAY and ∡MOY; $\angle$MOA = $\angle$MAO.

Prove: $\alpha\triangle\text{MAY} = \alpha\triangle\text{MOY}$.

Set III

Carefully trace this figure and cut it apart. Can you rearrange the pieces to form a symmetrical six-pointed star that has the same area?

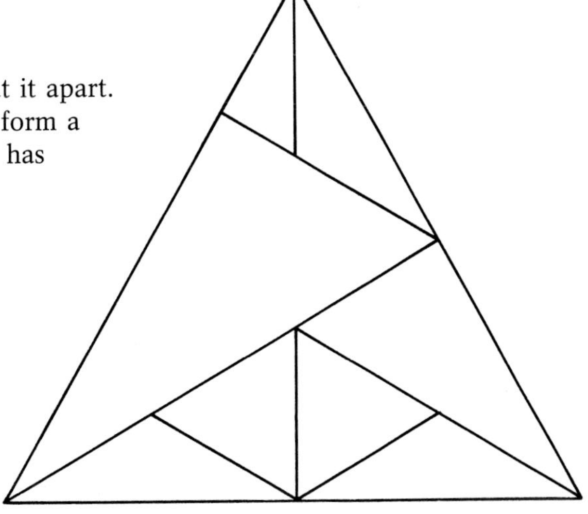

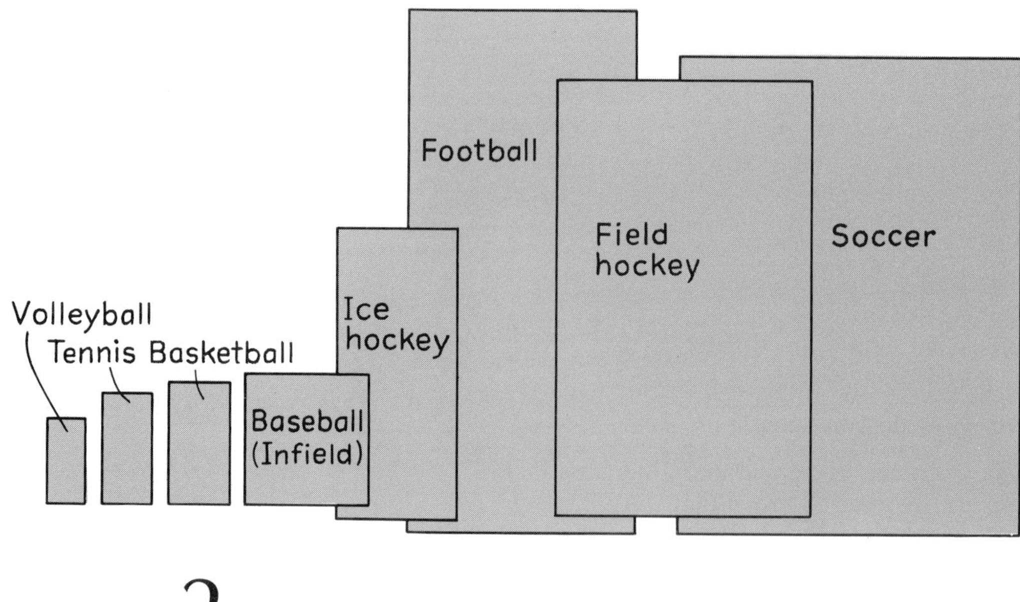

Lesson 2

Squares and Rectangles

The playing area of a sport is defined by its rules. According to the rules of tennis, for example, the court on which the game is played has an area of 312 square yards. The rules of soccer specify a field having an area of 8,800 square yards. The figures above, which are drawn to scale, show the playing areas of these sports and several others.

Notice that all of them are rectangular in shape. If a rectangle is divided into a set of squares that are all equal in area, its area can be found by using one of them as the unit and counting their number. For example, a tennis court is 12 yards wide and 26 yards long. It can be divided into squares measuring 1 yard along each side as shown in the figure below. One of the small squares is said to have an area of *1 square yard;*

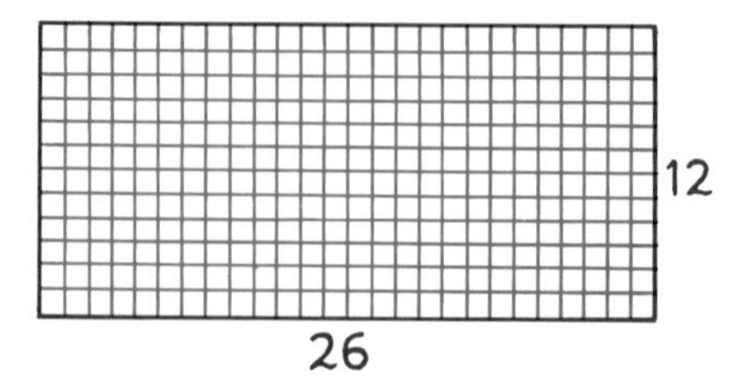

the tennis court contains

$$12 \times 26 = 312$$

of these squares, and so it has an area of 312 square yards.

Two consecutive sides of a rectangle may be called its base and altitude. The letters b and h are usually used to represent their respective lengths (h for "height," another word for "altitude"). The example of the tennis court suggests that the area of every rectangle can be found by multiplying these two lengths. We will assume that this is always true, even for rectangles whose dimensions are not integers.

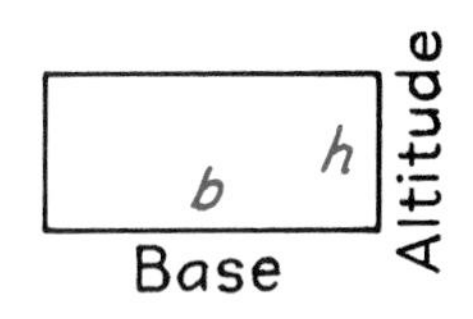

► **Postulate 12**
The area of a rectangle is the product of the lengths of its base and its altitude.

Because a square is a rectangle whose base and altitude are equal, the following corollary is obvious.

► **Corollary**
The area of a square is the square of the length of its side.

The figure at the right shows, for example, that the area of a square whose sides are 5 units long is

$$5^2 = 25 \text{ square units.}$$

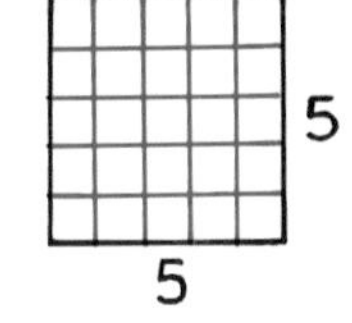

The following table summarizes the relations of the perimeters and areas of squares and rectangles to their dimensions.

Quadrilateral	*Perimeter*	*Area*
Square, side s	$4s$	s^2
Rectangle, base b, altitude h	$2b + 2h$	bh

Exercises

Set I

Find the perimeter and area of each of the following rectangles.

Example:

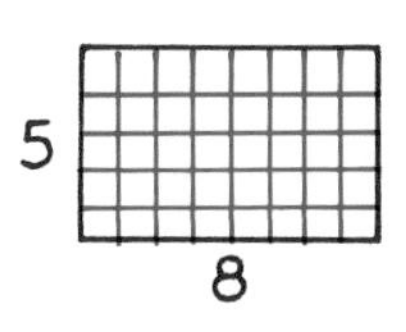

Solution: $2 \cdot 8 + 2 \cdot 5 = 16 + 10 = 26$.
The perimeter is 26.
$8 \cdot 5 = 40$.
The area is 40.

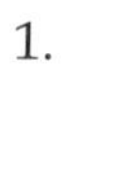

1.

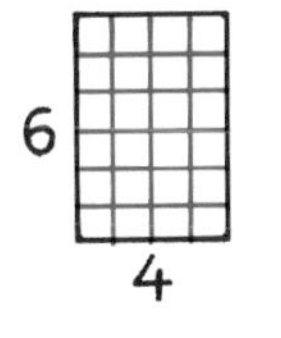

2.

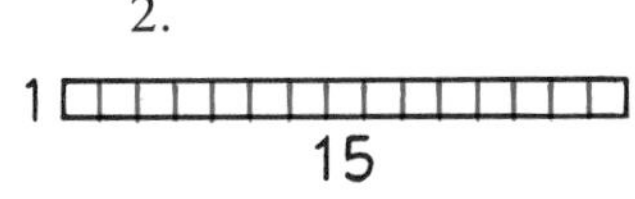

3.

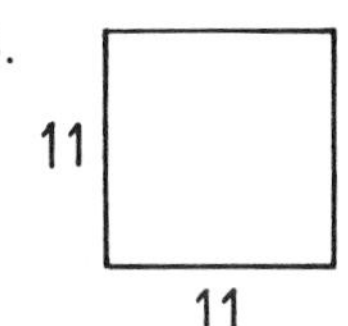

4.

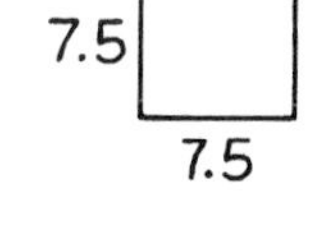

5.

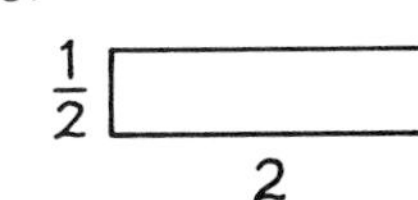

6.

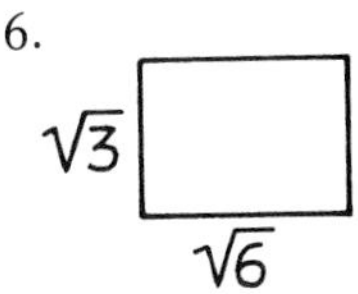

A basketball court has a length of 28 yards and a width of 15 yards.

7. What is its perimeter in yards?
8. What is its perimeter in feet?
9. What is its area in square yards?
10. What is its area in square feet?

The figure below shows the relation between square inches and square feet.

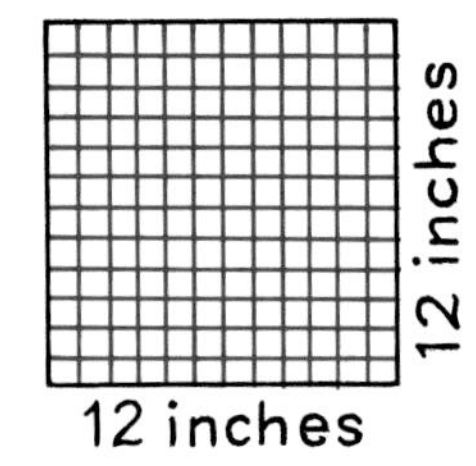

11. How many square inches are equal to one square foot?
12. How many square inches are equal to one square yard?

A pool table has a length of 10 feet and a width of 5 feet.

13. What is its perimeter in inches?
14. What is its area in square inches?

If the area of a square is 100 square inches, then its sides must be 10 inches long because $\sqrt{100} = 10$.

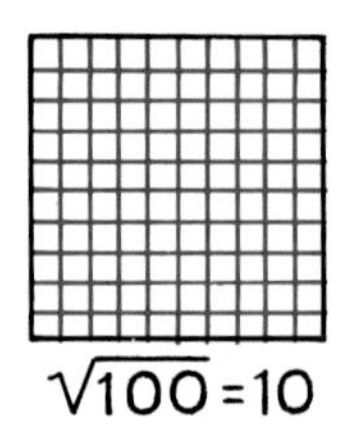

15. How long are the sides of a square whose area is 49 square inches?
16. How long are the sides of a square whose area is 50 square inches? (Express your answer in simple radical form.)
17. Express your answer to Exercise 16 in decimal form ($\sqrt{2} \approx 1.41$).
18. Find the area of a square whose sides are exactly 7.05 inches long.

A boxing ring is square and has an area of 400 square feet.

19. How long is one of its sides?
20. What is its perimeter?

Set II

Draw figures to illustrate the following exercises and find their areas.

Example: A square whose perimeter is 12 feet.

Solution: Each side of the square must be 3 feet, and so its area is 9 square feet.

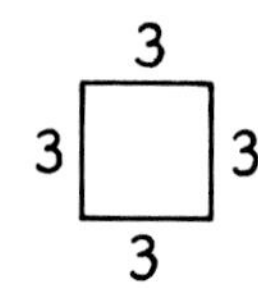

21. A square whose perimeter is 20 feet.
22. A rectangle whose perimeter is 20 feet and whose altitude is 4 feet.
23. A rectangle whose perimeter is 20 feet and whose altitude is 8 feet.
24. A rectangle whose perimeter is 48 feet and whose base is twice its altitude.
25. A rectangle whose perimeter is 48 feet and whose base is one-third its altitude.
26. A square whose perimeter is equal to its area.

Find the area of the shaded region in each of the following figures.

27.

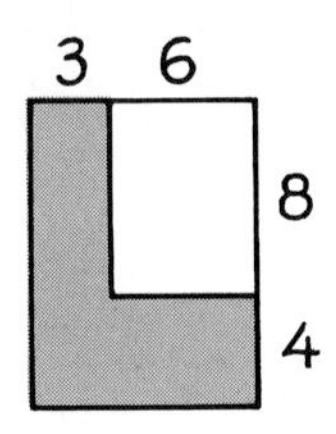

Both quadrilaterals are rectangles.

28.

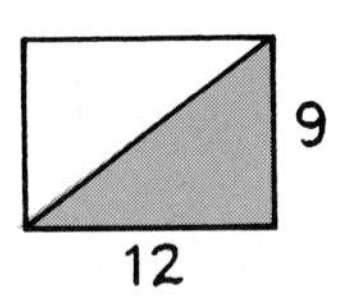

The quadrilateral is a rectangle.

29.

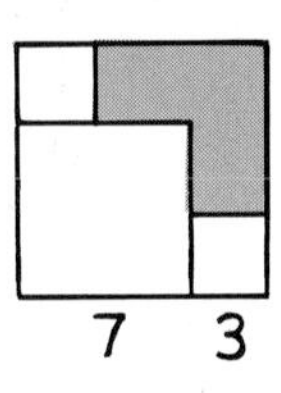

All of the quadrilaterals are squares.

30.

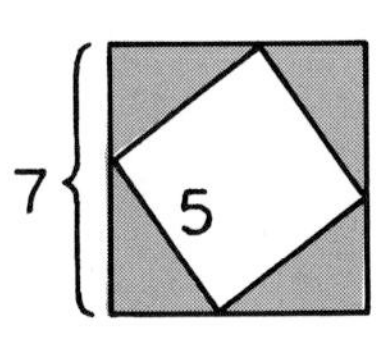

Both quadrilaterals are squares.

31.

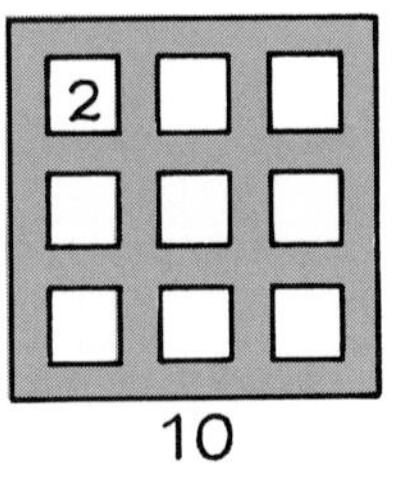

The "frame" and "holes" are squares; all the holes are alike.

32.

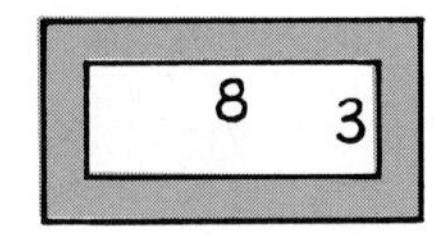

The "hole" is a rectangle surrounded by a "frame" 1 unit wide.

The figure below represents a floorplan of a room. Each of its angles is a right angle and each length is given in feet.

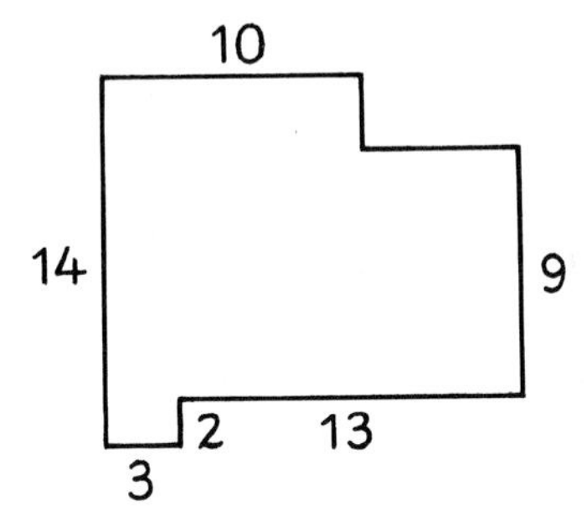

33. What kind of polygon is the floorplan?

34. Copy the figure and write in the missing lengths.

35. Find the perimeter of the room in feet.

36. Find the area of the room in square feet.

Set III

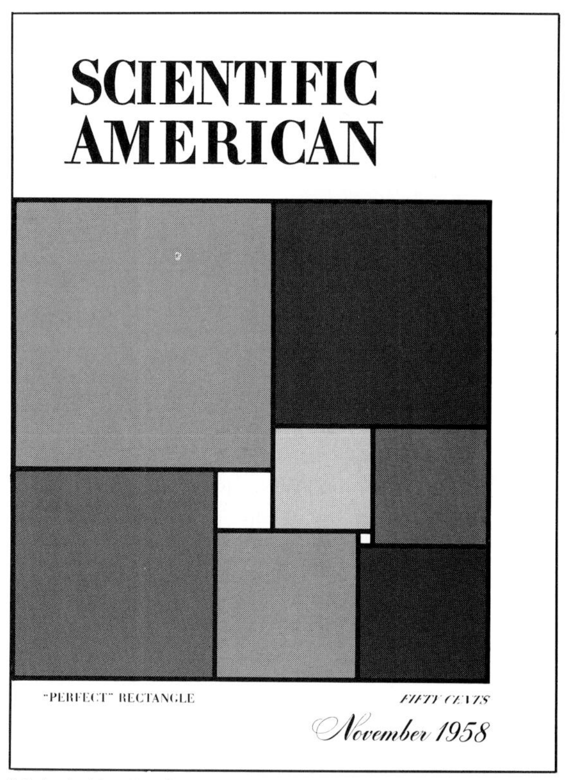

Painting by Mary Russel

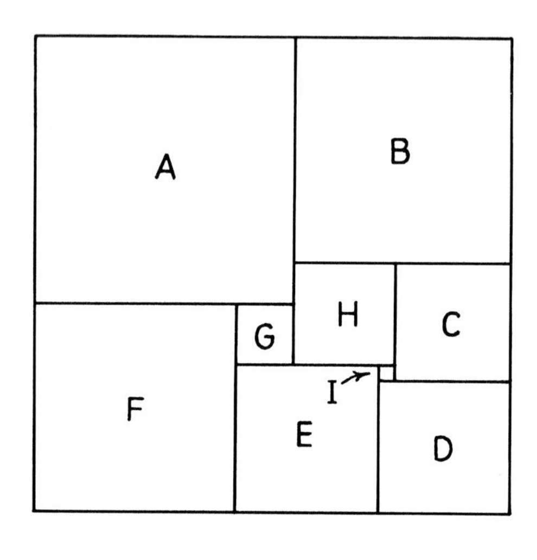

The picture on this magazine cover seems to show a square that has been divided into nine smaller squares, all having different areas. Suppose that the areas of squares C and D as shown in the figure at the right are 64 and 81 respectively.

1. Can you figure out the areas of the other seven squares? (Trace the figure on your paper so that you can mark it as you find each area.)

2. Is the figure containing the nine squares a square? Explain why or why not.

Alastair Black

Lesson 3

Triangles

One of the fastest racing yachts in the world is the Crossbow II, which is reported to have attained a speed of more than 50 miles per hour. Yachts used in racing are divided into classes based on factors affecting their design and speed. One of these factors is the area of their sails. Since 1920, the sails of most yachts have been triangular in shape.

The figure at the right represents the mainsail of a yacht: it has the shape of a right triangle. The area of a right triangle can be found from the lengths of its legs, as we will establish in the proof of the following theorem.

Theorem 47

The area of a right triangle is half the product of the lengths of its legs.

Given: Right $\triangle ABC$ with right $\angle B$.

Prove: $\alpha\triangle ABC = \frac{1}{2}AB \cdot BC.$

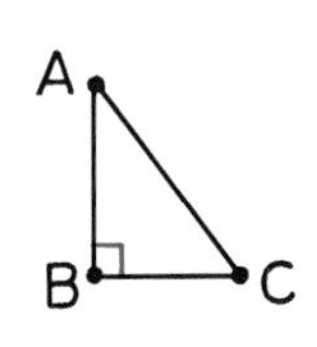

Proof.

Through A draw a line parallel to $\overline{BC}$ and through C draw a line parallel to $\overline{AB}$. ABCD is a parallelogram, and so its consecutive angles are supplementary and its opposite angles are equal. It follows from this that, because $\angle B$ is a right angle, $\angle BAD$, $\angle BCD$, and $\angle D$ must also be right

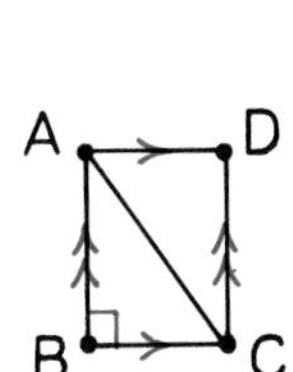

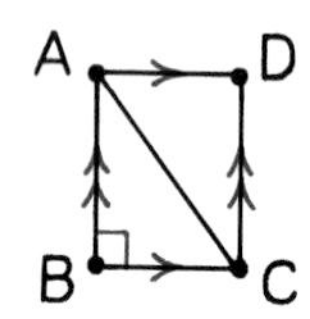

angles. This means that ABCD is equiangular, and so it is also a rectangle. Therefore, $\alpha ABCD = AB \cdot BC$. But $\triangle ABC \cong \triangle CDA$, and so $\alpha\triangle ABC = \alpha\triangle CDA$. It follows that $\alpha\triangle ABC = \frac{1}{2}\alpha ABCD$, and so $\alpha\triangle ABC = \frac{1}{2}AB \cdot BC$ (substitution).

Although Theorem 47 is true only for right triangles, it can be used to derive a formula for the area of any triangle, regardless of its shape. To do so, we need to define the *base* and *altitude* of a triangle.

► **Definition**
An ***altitude*** of a triangle is a perpendicular line segment from a vertex of the triangle to the line of the opposite side. The side is called the corresponding ***base*** of the triangle.

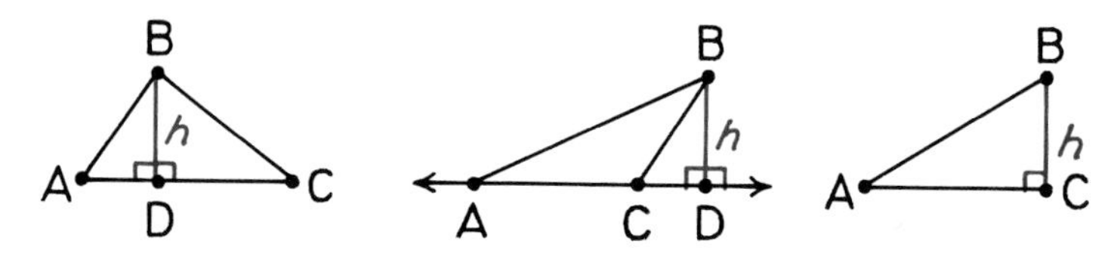

The figures above show that according to this definition, an altitude of a triangle may lie inside the triangle, lie outside it, or be one of its sides. Regardless of where the altitude may be, the following theorem is true.

► **Theorem 48**
The area of a triangle is half the product of the lengths of any base and corresponding altitude.

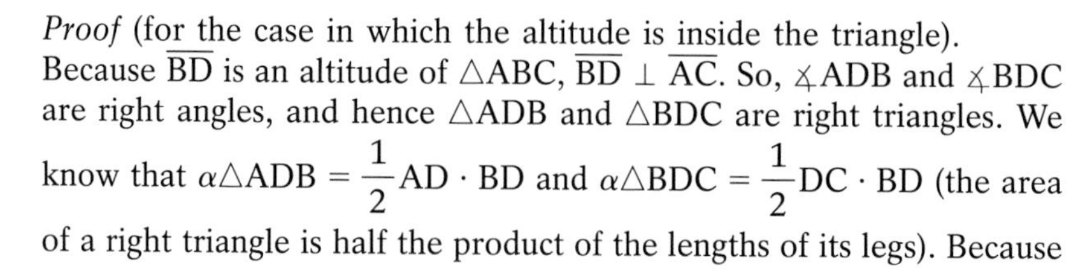

Given: $\triangle ABC$ with altitude $\overline{BD}$.

Prove: $\alpha\triangle ABC = \frac{1}{2}AC \cdot BD$.

Proof (for the case in which the altitude is inside the triangle).
Because $\overline{BD}$ is an altitude of $\triangle ABC$, $\overline{BD} \perp \overline{AC}$. So, $\measuredangle ADB$ and $\measuredangle BDC$ are right angles, and hence $\triangle ADB$ and $\triangle BDC$ are right triangles. We know that $\alpha\triangle ADB = \frac{1}{2}AD \cdot BD$ and $\alpha\triangle BDC = \frac{1}{2}DC \cdot BD$ (the area of a right triangle is half the product of the lengths of its legs). Because

$$\alpha\triangle ABC = \alpha\triangle ADB + \alpha\triangle BDC \text{ (Area Postulate)},$$

$$\alpha\triangle ABC = \frac{1}{2}AD \cdot BD + \frac{1}{2}DC \cdot BD \text{ (substitution)}.$$

By factoring the right side of this equation, we get

$$\alpha\triangle ABC = \frac{1}{2}BD(AD + DC).$$

Because AD + DC = AC (Betweenness of Points Theorem),

$$\alpha\triangle ABC = \frac{1}{2}BD \cdot AC.$$

A similar proof can be written for the case in which the altitude is outside the triangle.

An immediate consequence of this theorem is the following corollary.

► **Corollary**
Triangles with equal bases and equal altitudes have equal areas.

The fact that the diagonals of a rhombus are perpendicular makes it possible to derive a formula for its area in terms of their lengths.

► **Theorem 49**
The area of a rhombus is half the product of the lengths of its diagonals.

Given: Rhombus ABCD with diagonals $\overline{AC}$ and $\overline{BD}$.

Prove: $\alpha ABCD = \frac{1}{2}BD \cdot AC.$

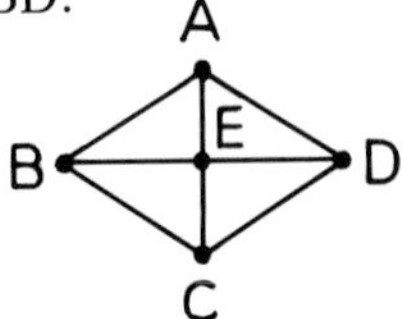

Proof.
Because ABCD is a rhombus, $\overline{AC} \perp \overline{BD}$. So $\overline{AE}$ is an altitude of $\triangle ABD$ and $\overline{EC}$ is an altitude of $\triangle BCD$. It follows that

$$\alpha\triangle ABD = \frac{1}{2}BD \cdot AE \quad \text{and} \quad \alpha\triangle BCD = \frac{1}{2}BD \cdot EC,$$

and so

$$\alpha\triangle ABD + \alpha\triangle BCD = \frac{1}{2}BD \cdot AE + \frac{1}{2}BD \cdot EC$$

$$= \frac{1}{2}BD(AE + EC).$$

But $\alpha ABCD = \alpha\triangle ABD + \alpha\triangle BCD$ and AE + EC = AC, and so

$$\alpha ABCD = \frac{1}{2}BD \cdot AC.$$

Exercises

Set I

Find the perimeter and area of each of the following triangles.

1.

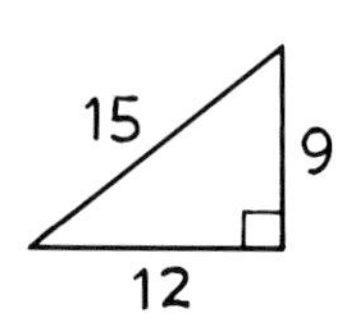

2.

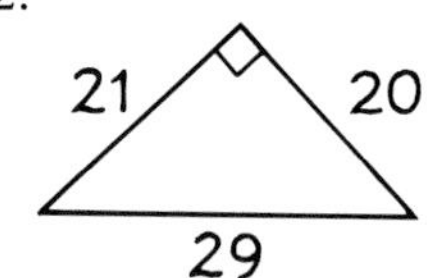

3.

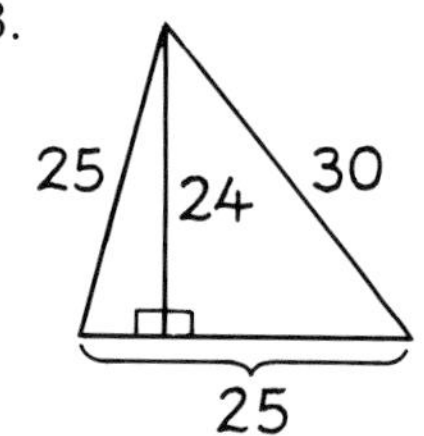

4.

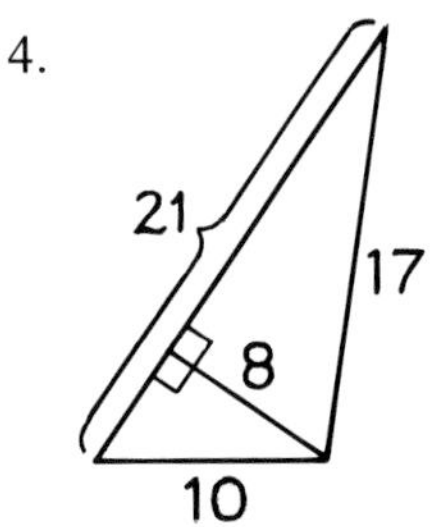

5.

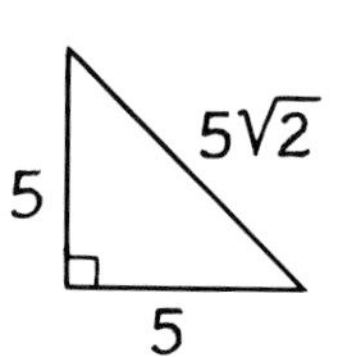

6.

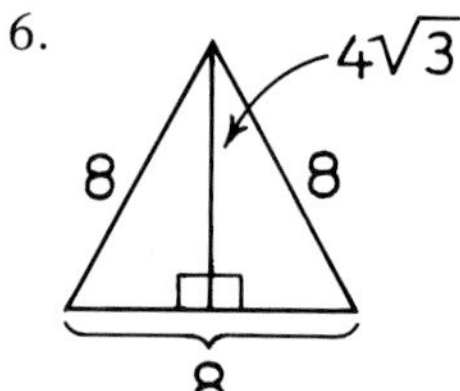

Write expressions for the perimeter and area of each of the following triangles.

Example:

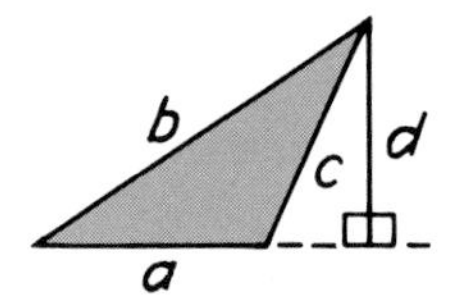

Answer: The perimeter is $a + b + c$.

The area is $\frac{1}{2}ad$.

7.

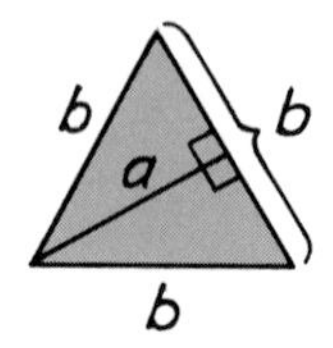

8.

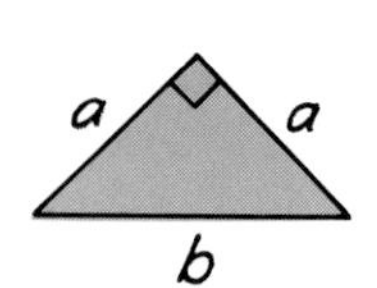

9.

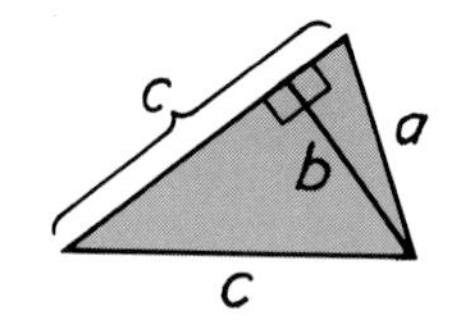

In the figure below, $\triangle$ORC and $\triangle$HID are right triangles.

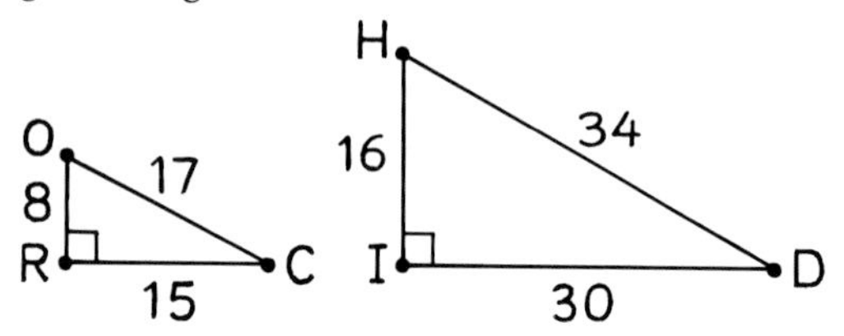

10. How do the sides of $\triangle$HID compare in length with the sides of $\triangle$ORC?
11. Find the perimeter of each triangle.
12. How does the perimeter of $\triangle$HID compare with the perimeter of $\triangle$ORC?
13. Find the area of each triangle.
14. How does the area of $\triangle$HID compare with the area of $\triangle$ORC?
15. If the sides of a triangle are doubled, is its perimeter doubled?
16. If the sides of a triangle are doubled, is its area doubled?

Quadrilaterals GERA and NIUM are rhombuses.

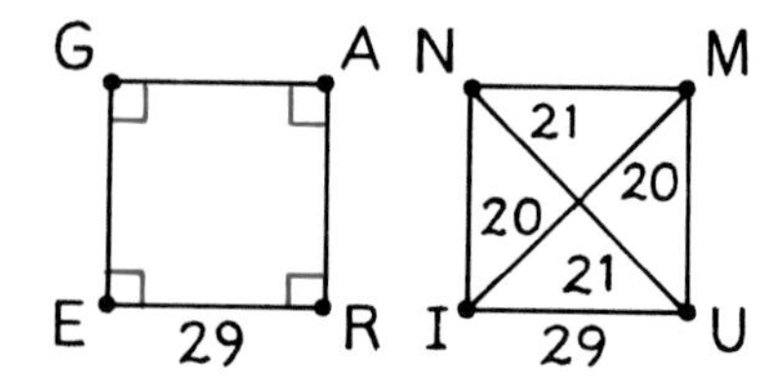

17. Find the perimeter of each rhombus.
18. Find the area of GERA.
19. Find the area of NIUM.
20. If two rhombuses have equal perimeters, does it follow that they also have equal areas?

Set II

Find the area of the shaded region in each of the following figures.

21. 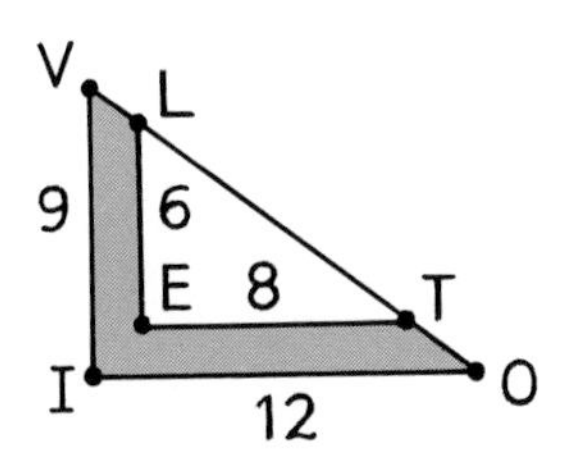

$\triangle VIO$ and $\triangle LET$ are right triangles.

22. 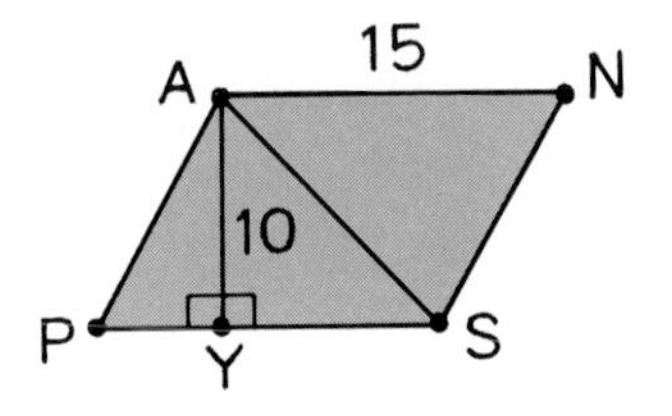

PANS is a parallelogram.

23. 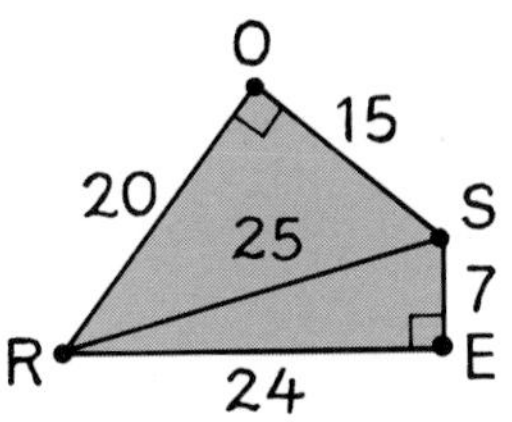

$\overline{RO} \perp \overline{OS}$ and $\overline{RE} \perp \overline{ES}$.

24. 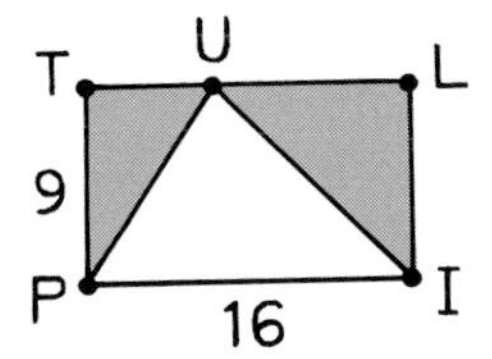

TLIP is a rectangle.

25. 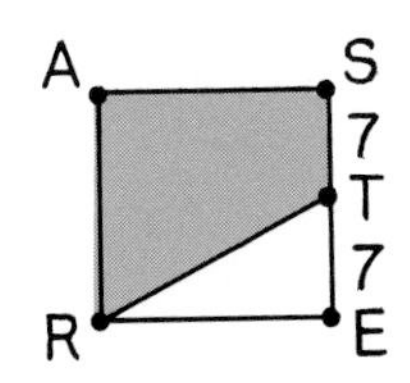

ASER is a square.

26. 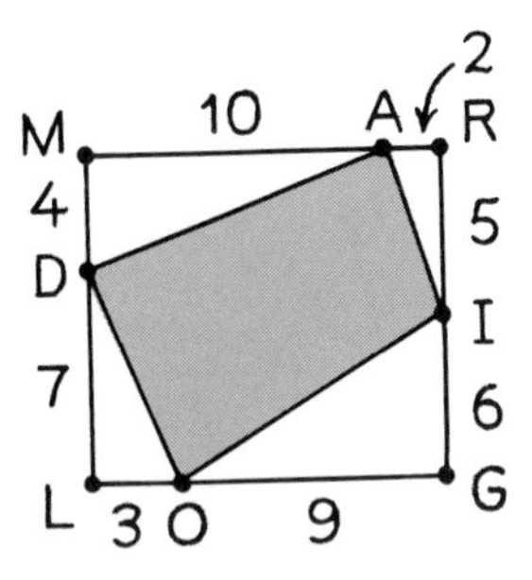

MRGL is a rectangle.

27. 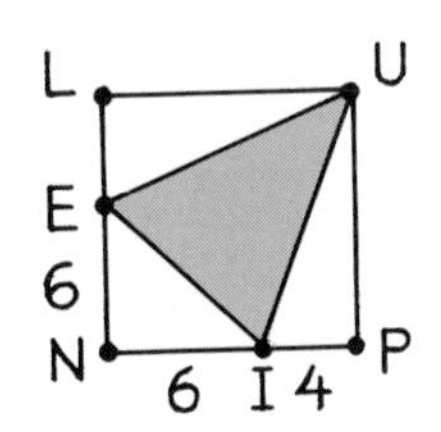

LUPN is a square.

Quadrilateral MYTL is a trapezoid with bases $\overline{YT}$ and $\overline{ML}$; $\overline{YE} \perp \overline{ML}$ and $\overline{RL} \perp \overline{YT}$.

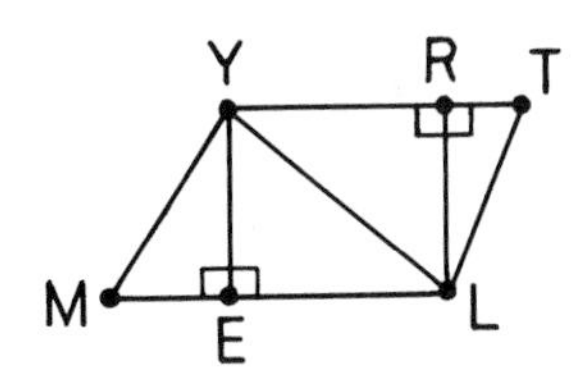

28. Why is $\overline{YT} \parallel \overline{ML}$?
29. Why is YE = RL?
30. Write an expression for $\alpha\triangle MYL$. Let ML = a and YE = h.
31. Write an expression for $\alpha\triangle YLT$. Let YT = b and RL = h.
32. Write an expression for αMYTL.

In the figure below, $\overline{LT}$ and $\overline{OS}$ are altitudes of $\triangle LOU$; $LT = 24$.

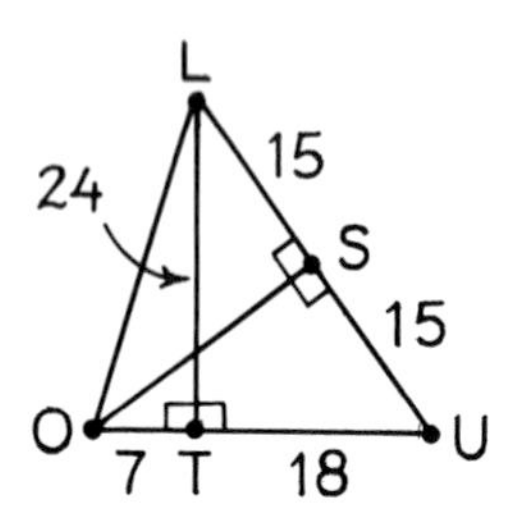

33. Find the perimeter of $\triangle LOU$.

34. Find the area of $\triangle LOU$.

35. Find OS.

Set III

Three "triangles" have sides of lengths 5 cm, 5 cm, and 6 cm; 5 cm, 5 cm, and 8 cm; and 5 cm, 5 cm, and 10 cm.

1. Guess which "triangle" has the largest area.

2. Use your straightedge and compass to construct the three "triangles."

3. Did you guess correctly? If not, what do you think your drawings indicate?

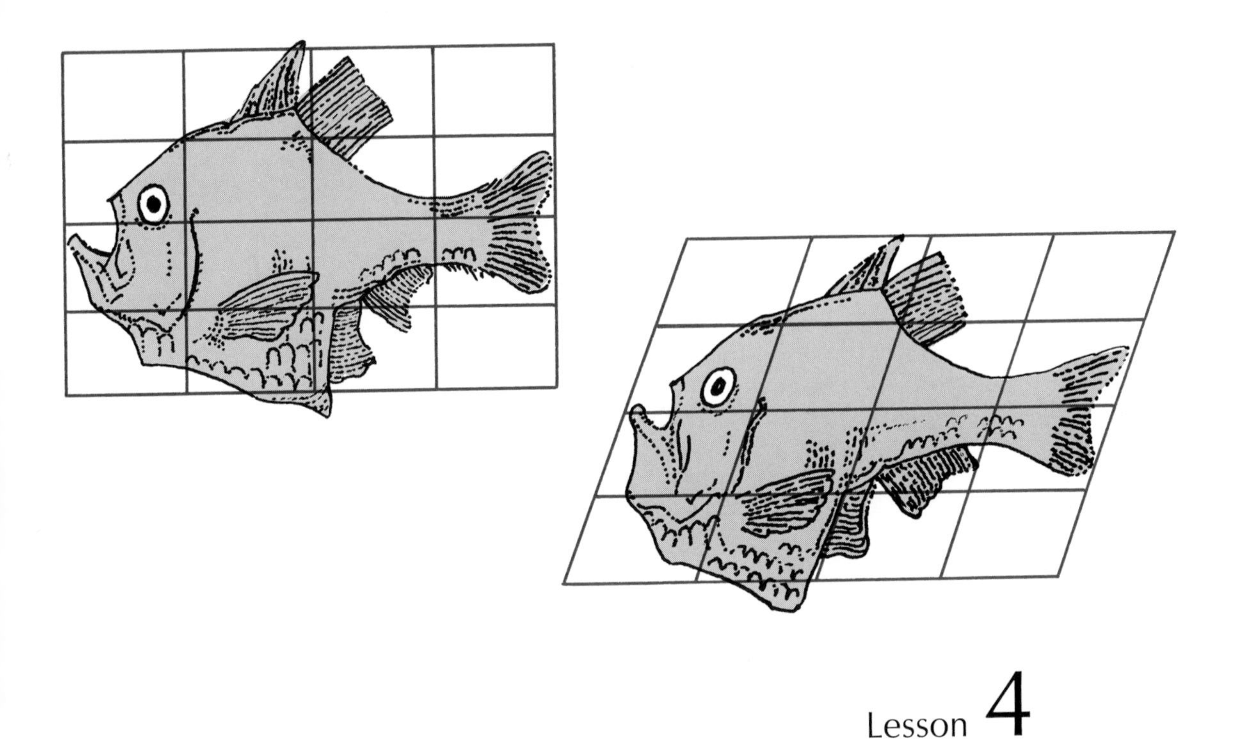

Lesson 4

Parallelograms and Trapezoids

One of the ancient Greeks, perhaps Plato or Pythagoras, once said that the book of nature is written in characters of geometry. A remarkable application of geometry to the study of biological shapes was made by the great British scientist D'Arcy Thompson. In his book *On Growth and Form*, he showed many examples of how one species of animal can be considered to be a geometric transformation of another.* For example, the fish shown on the rectangular grid at the left above is of the species *Argyropelecus olfersi*. If the rectangles are transformed into the parallelograms shown at the right, the shape of the fish becomes that of a species in an entirely different genus, *Sternoptyx diaphana*.

How do you think the pictures of the two fish compare in size? The answer to this question depends on how the rectangles in the first grid compare in area with the parallelograms in the second. In this lesson we will prove that the formulas for the areas of these two figures are the same.

The figure at the right shows a parallelogram that has been divided by a diagonal into two triangles. The base and altitude of $\triangle ABD$ marked b and h can also be considered to be a base and altitude of the parallelogram.

*D'Arcy Thompson, "On the theory of transformations, or the comparison of related forms," *On Growth and Form*, edited by J. T. Bonner, Cambridge University Press © 1961.

► **Definition**

An ***altitude*** of a quadrilateral that has parallel sides is a perpendicular line segment that joins a point on one of the parallel sides to the line that contains the other side.

Any side of a parallelogram may be chosen as its base. An altitude that joins this side to the line containing the opposite side is a corresponding altitude.

► **Theorem 50**

The area of a parallelogram is the product of the lengths of any base and corresponding altitude.

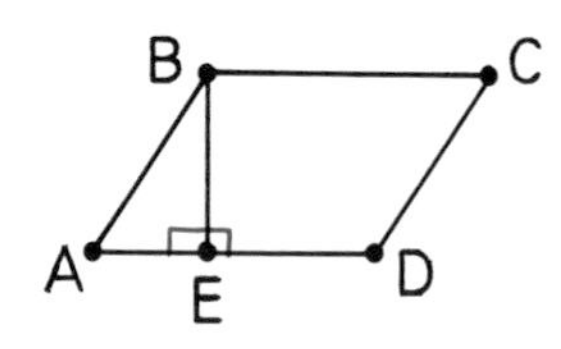

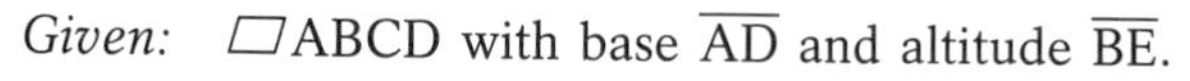

Given: $\square$ABCD with base $\overline{AD}$ and altitude $\overline{BE}$.

Prove: $\alpha ABCD = AD \cdot BE$.

Proof.

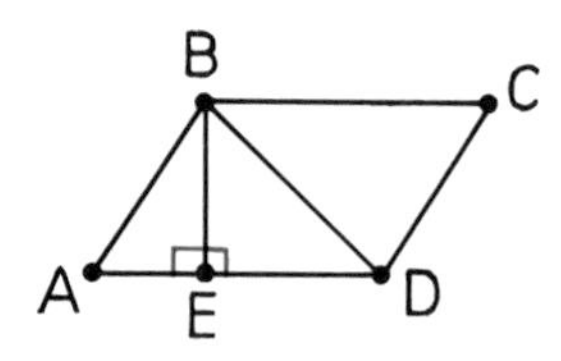

Draw $\overline{BD}$. Because AB = DC and BC = AD (the opposite sides of a parallelogram are equal) and BD = BD, $\triangle ABD \cong \triangle CDB$ (S.S.S.). So $\alpha\triangle ABD = \alpha\triangle CDB$.

Also, $\alpha\square ABCD = \alpha\triangle ABD + \alpha\triangle CDB$ (the Area Postulate), and so

$$\alpha\square ABCD = \alpha\triangle ABD + \alpha\triangle ABD = 2\alpha\triangle ABD.$$

Because $\alpha\triangle ABD = \frac{1}{2}AD \cdot BE$,

$$\alpha\square ABCD = 2\left(\frac{1}{2}AD \cdot BE\right) = AD \cdot BE.$$

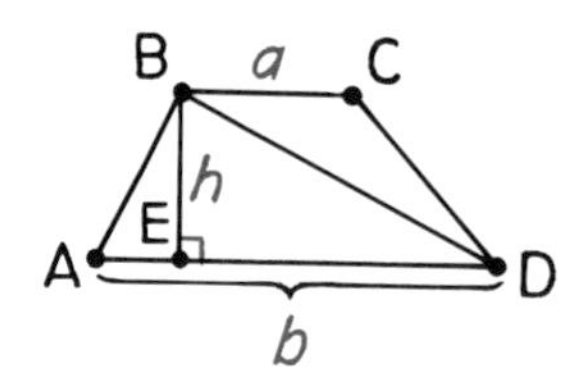

A formula for the area of a trapezoid can be derived in essentially the same way. The figure at the left shows a trapezoid that has been divided by a diagonal into two triangles. The bases of $\triangle CDB$ and $\triangle ABD$ marked a and b are also bases of the trapezoid. The altitude of $\triangle ABD$ marked h is also an altitude of the trapezoid.

► **Theorem 51**

The area of a trapezoid is half the product of the length of its altitude and the sum of the lengths of its bases.

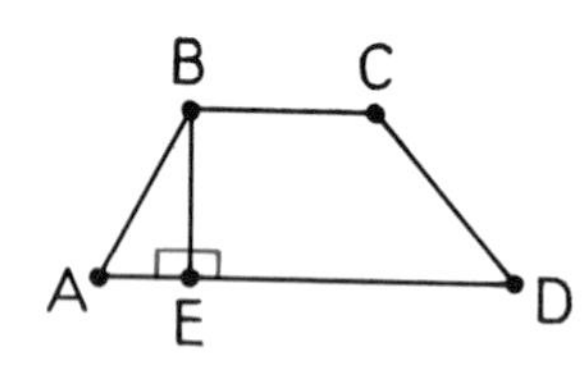

Given: Trapezoid ABCD with bases $\overline{AD}$ and $\overline{BC}$ and altitude $\overline{BE}$.

Prove: $\alpha ABCD = \frac{1}{2}BE(AD + BC)$.

Proof.

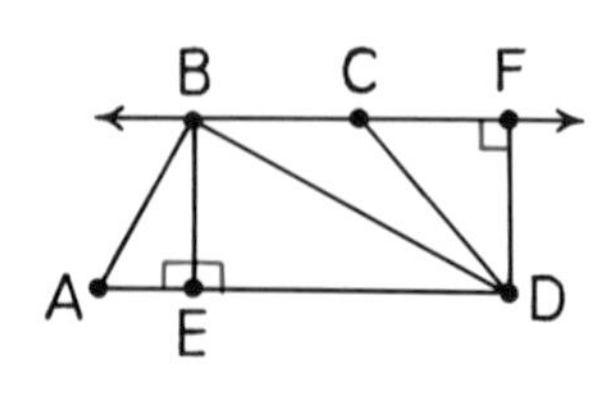

Draw $\overline{BD}$. Altitude $\overline{BE}$ of trapezoid ABCD is also an altitude of $\triangle ABD$, and so $\alpha\triangle ABD = \frac{1}{2}AD \cdot BE$. Draw $\overleftrightarrow{BC}$ and, through D, draw $\overline{DF} \perp \overleftrightarrow{BC}$; $\overline{DF}$ is the altitude corresponding to base $\overline{BC}$ of $\triangle CDB$, and so $\alpha\triangle CDB = \frac{1}{2}BC \cdot DF$.

Because $\overline{AD} \parallel \overline{BC}$ (the bases of a trapezoid are parallel), BE = DF (if two lines are parallel, every perpendicular segment joining one line to the other line has the same length).

Because $\alpha ABCD = \alpha\triangle ABD + \alpha\triangle CDB$ (the Area Postulate),

$$\alpha ABCD = \frac{1}{2}AD \cdot BE + \frac{1}{2}BC \cdot DF.$$

Substituting BE for DF in this equation gives

$$\alpha ABCD = \frac{1}{2}AD \cdot BE + \frac{1}{2}BC \cdot BE = \frac{1}{2}BE(AD + BC).$$

Exercises

Set I

Quadrilaterals RAVE, HOLS, and VERD are parallelograms.

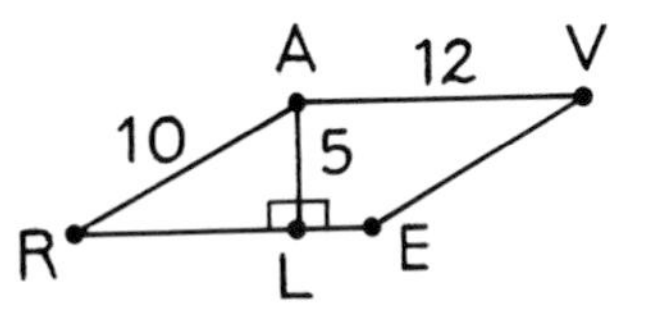

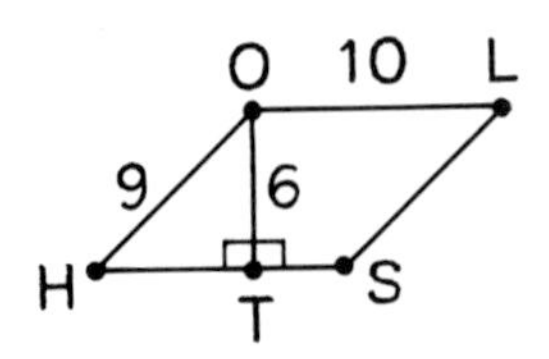

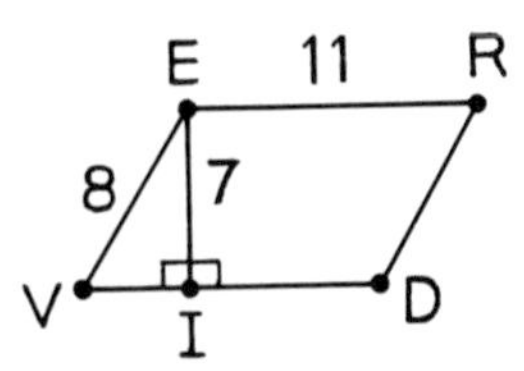

1. Find the perimeter of each parallelogram.
2. Find the area of each parallelogram.
3. If two parallelograms have equal perimeters, does it follow that they have equal areas?
4. If two parallelograms have equal areas, does it follow that they have equal perimeters?

Find the perimeter and area of each of the following quadrilaterals.

5.

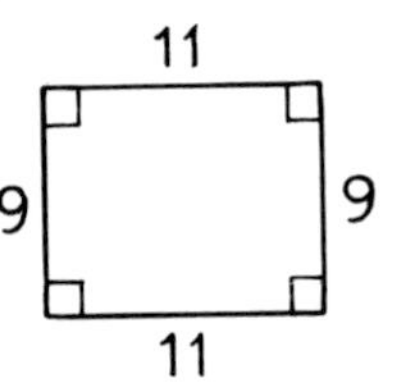

6.

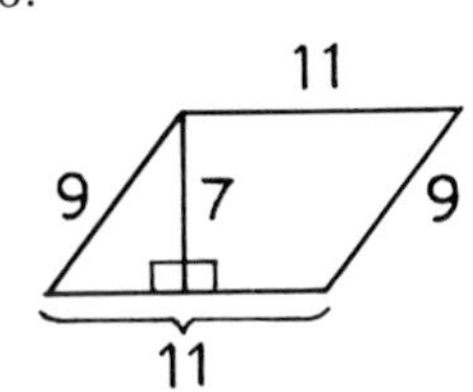

7.

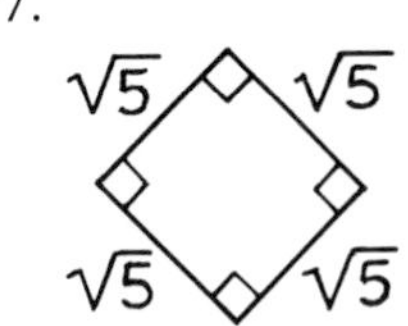

8.

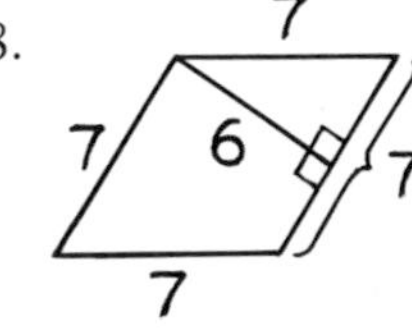

9.

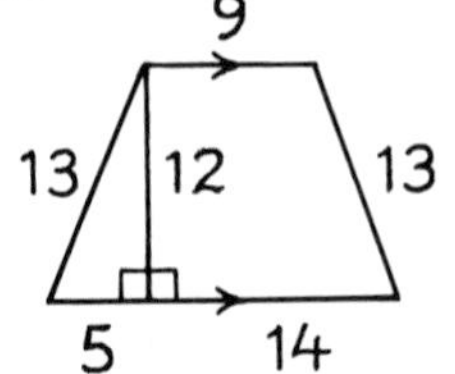

10.

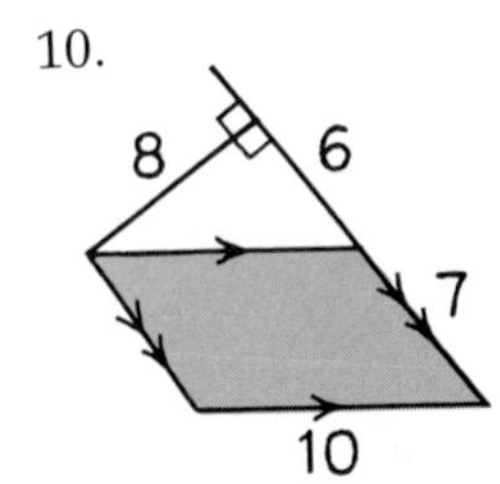

11.

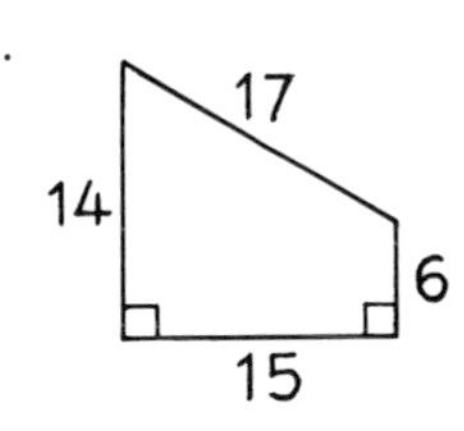

Write expressions for the perimeter and area of each of the following figures.

Example:

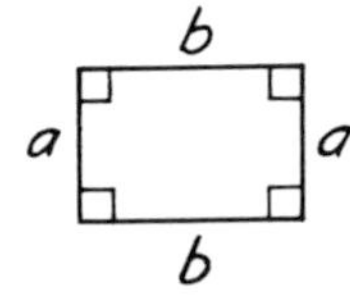

Answer: Its perimeter is $2a + 2b$, and its area is ab.

12.

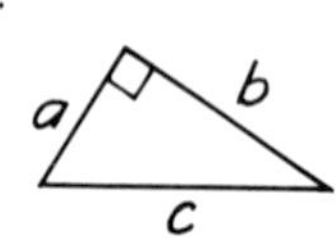

13.

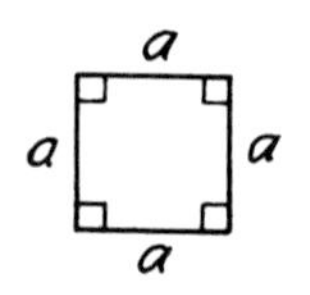

14.

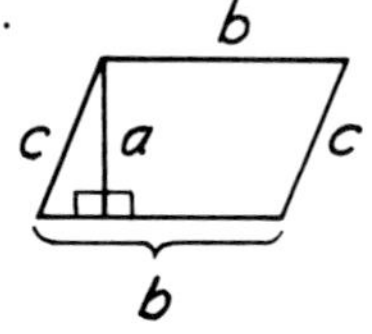

15.

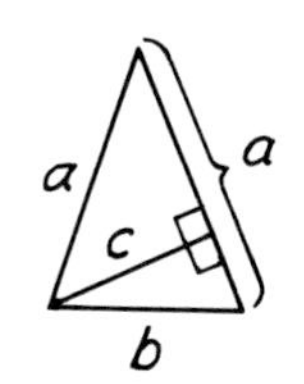

16.

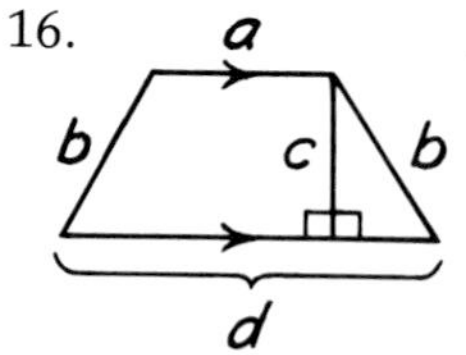

17.

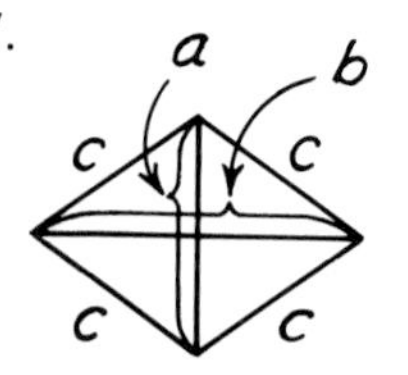

The figure below shows a square that has been divided into two smaller squares and two rectangles. It illustrates the fact that

$$(a + b)^2 = a^2 + 2ab + b^2.$$

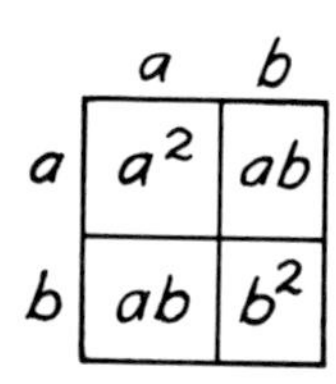

Refer to the figures below to complete the following equations.

18.

	c	5
c	c^2	$5c$
5	$5c$	25

$(c + 5)^2 =$ ____.

19.

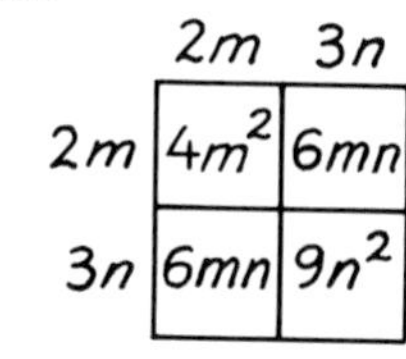

$(2m + 3n)^2 =$ ____.

20.

	x	$-y$
x	x^2	$-xy$
$-y$	$-xy$	y^2

$(x - y)^2 =$ ____.

Complete the following equations without referring to figures.

21. $(a + 4b)^2 =$ ____.

22. $m^2 + 14m + 49 = (\_\_\_\_)^2$.

23. $9x^2 - 6xy + y^2 = (\_\_\_\_)^2$.

Set II

In the figure below, WGER is a parallelogram with altitudes $\overline{RA}$ and $\overline{RN}$.

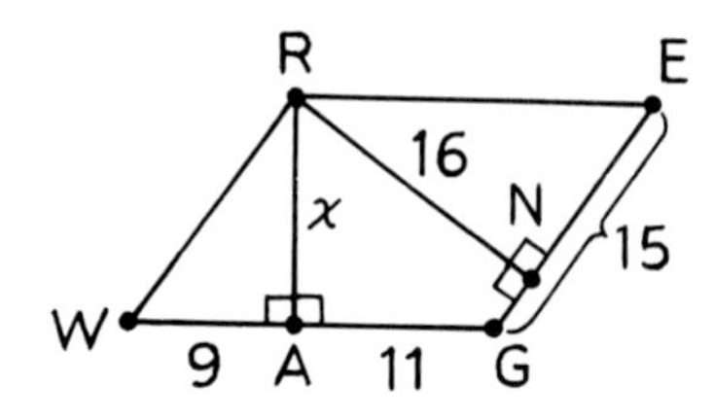

24. Find αWGER.

25. Write an expression for αWGER in terms of x.

26. Find x.

27. Find αAGER.

Find the colored area in each of the following figures.

28.

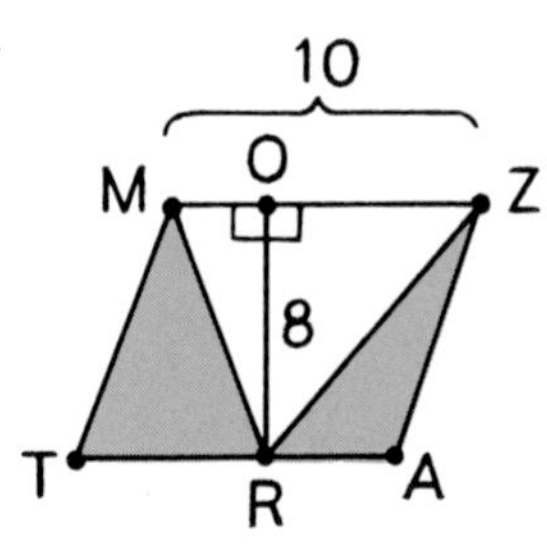

MZAT is a parallelogram.

29.

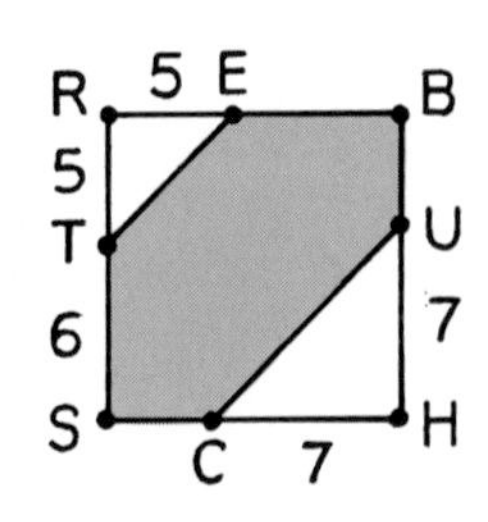

SHBR is a square.

30.

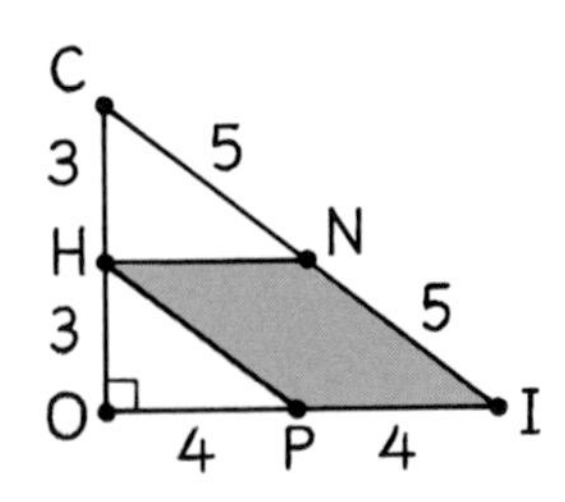

△COI is a right triangle.

31.

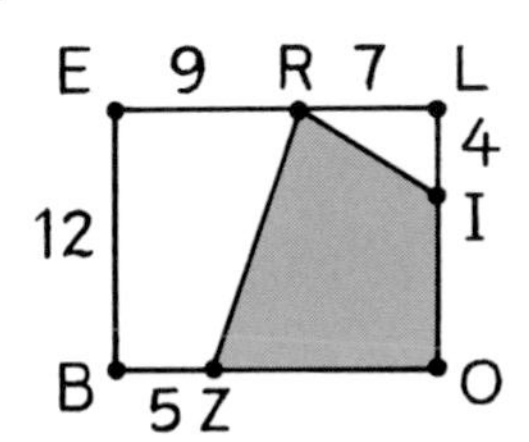

BELO is a rectangle.

32.

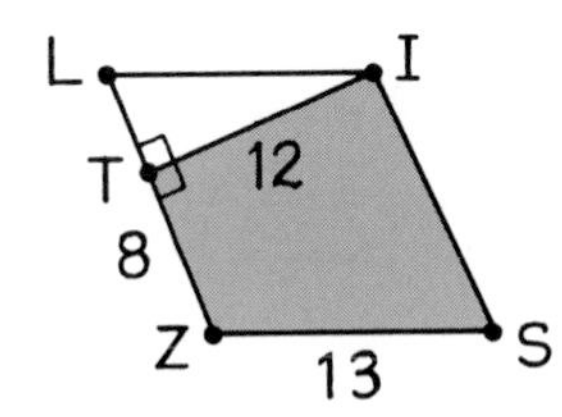

LISZ is a rhombus.

33.

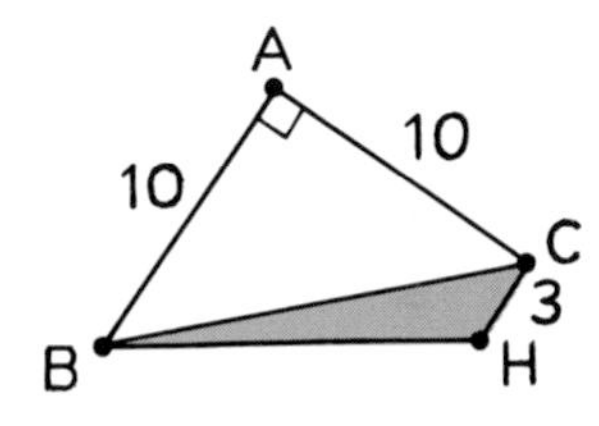

BACH is a trapezoid.

Set III

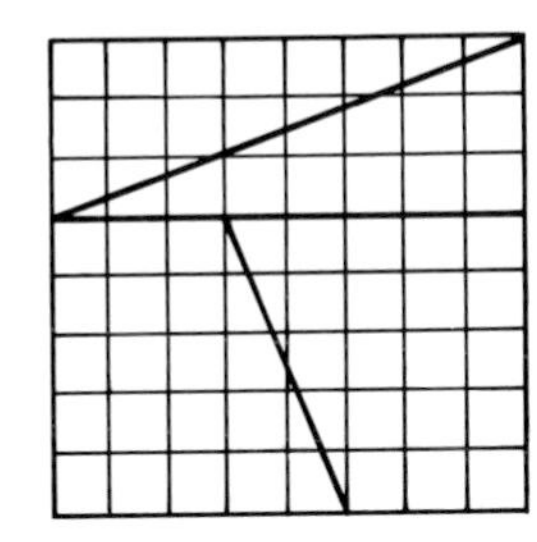

This square "checkerboard" has been divided into two trapezoids and two right triangles. Make a large copy of it, cut out the four pieces, and try to rearrange them to form a rectangle having a different shape. If you succeed, tape the pieces to your paper.

1. What seems to be the area of the rectangle you have formed?
2. What is the area of the original square?
3. Can you explain why the area of the rectangle and the area of the square seem to be different?

Lesson 5

The Pythagorean Theorem

At the beginning of this century, many scientists believed that intelligent creatures might live on Mars. Among them was the American astronomer Percival Lowell, whose work led to the discovery of Pluto. Mr. Lowell thought that he could see canals on the surface of Mars through his telescope and speculated that they might have been dug by a Martian civilization to irrigate their dry land with water melted from polar ice caps.

To let the Martians know that there was also intelligent life on the earth (in a time long before any kind of space travel was possible), it was proposed that gigantic geometric figures be used to convey a message. For example, broad lanes of trees might be planted in Siberia to form a huge right triangle. Or canals might be dug in the Sahara desert to do the same thing; kerosene could be poured on the water in them and set on fire at night for the Martians to see through their telescopes. A geometric figure felt to be especially appropriate for this is shown here. It illustrates what is perhaps the most famous theorem in all of geometry—the Pythagorean Theorem.

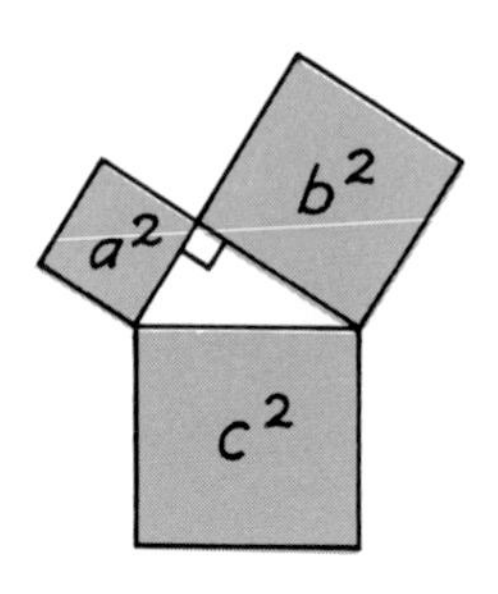

The theorem says that, if squares are constructed on the three sides of a right triangle, the area of the square on the hypotenuse is equal to the sum of the areas of the squares on the two legs. If a and b are the lengths of the legs and c is the length of the hypotenuse, then

$$c^2 = a^2 + b^2.$$

The following statement of this theorem emphasizes the relation of the lengths of the triangle's sides rather than the areas of the squares on them.

► **Theorem 52** (The Pythagorean Theorem)
In a right triangle, the square of the hypotenuse is equal to the sum of the squares of the legs.

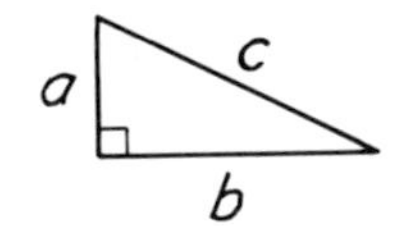

Given: A right triangle with hypotenuse c and legs a and b.

Prove: $c^2 = a^2 + b^2$.

Many different proofs have been developed for the Pythagorean Theorem—more, in fact, than for any other theorem of geometry. One proof is based on the two figures shown below. Each figure is a square

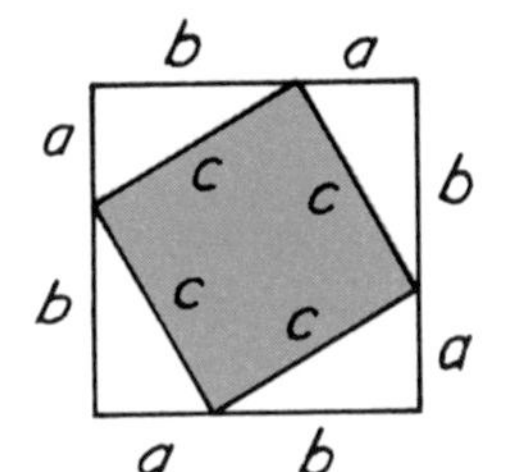

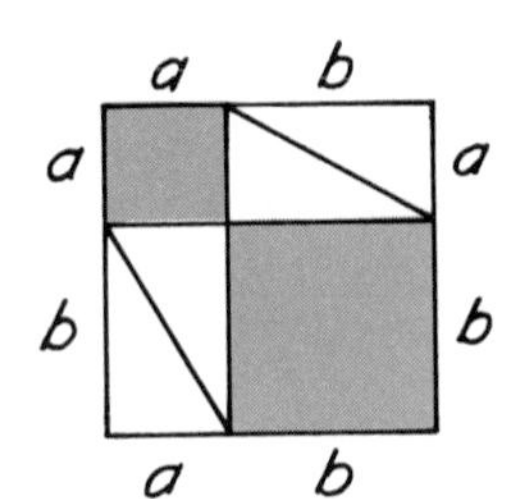

with sides of length $a + b$. The first has been subdivided into four right triangles congruent to the original triangle and a square whose sides are equal to its hypotenuse. The second figure also contains four right triangles congruent to the original triangle. The theorem follows from the fact that the rest of the figure consists of two squares whose sides are equal to the legs of the triangle.

Expressing this algebraically in terms of the total areas of the two figures, we have:

$$(a + b)^2 = 4\left(\frac{1}{2}ab\right) + c^2$$

and

$$(a + b)^2 = 4\left(\frac{1}{2}ab\right) + a^2 + b^2.$$

Hence,

$$4\left(\frac{1}{2}ab\right) + c^2 = 4\left(\frac{1}{2}ab\right) + a^2 + b^2$$

and

$$c^2 = a^2 + b^2.$$

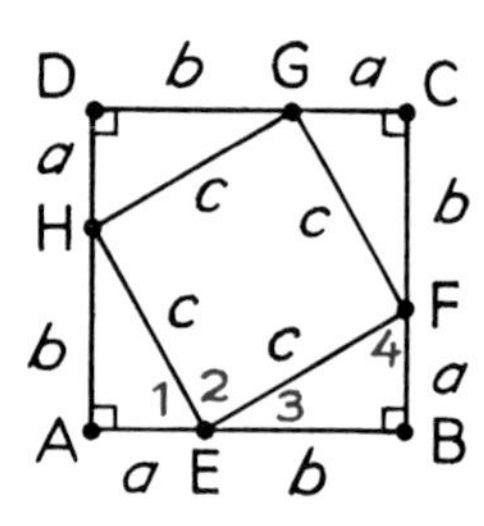

One assumption that we made about the first figure is not completely obvious. How do we know that the quadrilateral in the center is a square? If you figured out the problem about Dilcue's nightmare,* you already know the answer to this. It is reviewed below in terms of the figure shown here.

The four right triangles are congruent by S.A.S., and so EF = FG = GH = HE. Hence, EFGH is equilateral.

Angles 3 and 4 are complementary because they are the acute angles of right $\triangle$EBF, and so $\angle 3 + \angle 4 = 90°$. Because they are corresponding angles of congruent triangles, $\angle 1 = \angle 4$. Therefore, $\angle 3 + \angle 1 = 90°$ by substitution. Because $\angle 1 + \angle 2 + \angle 3 = 180°$, $\angle 2 = 90°$ by subtraction. In the same way, it can be shown that the measures of the other three angles of EFGH are 90°. Because EFGH is equiangular as well as equilateral, it is a square.

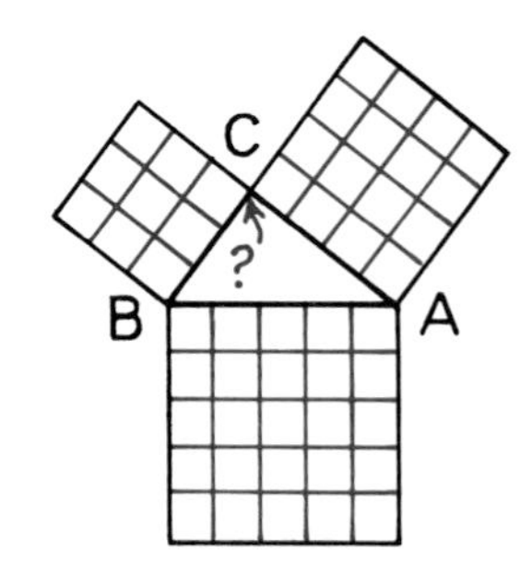

In the figure shown here, squares have been drawn on the sides of $\triangle$ABC and the squares into which each has been subdivided are all the same size. Is $\triangle$ABC a right triangle?

It is tempting to say yes, because

$$3^2 + 4^2 = 5^2.$$

However, to conclude that $\triangle$ABC must be a right triangle because the square of one of its sides is equal to the sum of the squares of the other two sides is to assume that the *converse* of the Pythagorean Theorem is true. Conveniently, it is.

► **Theorem 53**

If the square of one side of a triangle is equal to the sum of the squares of the other two sides, then the triangle is a right triangle.

*See page 255.

Exercises

Set I

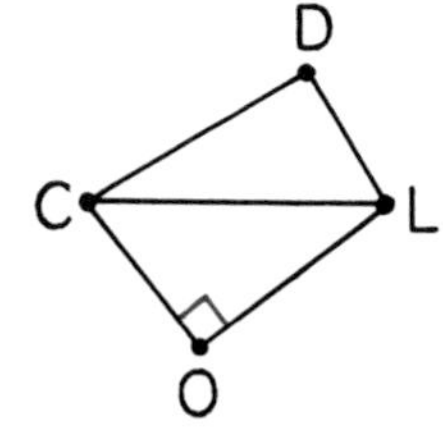

$\triangle$COL is a right triangle and $CD^2 + DL^2 = CL^2$. State, as a complete sentence, the theorem that is the basis for each of the following conclusions.

1. $CO^2 + OL^2 = CL^2$.
2. $\triangle$CLD is a right triangle.

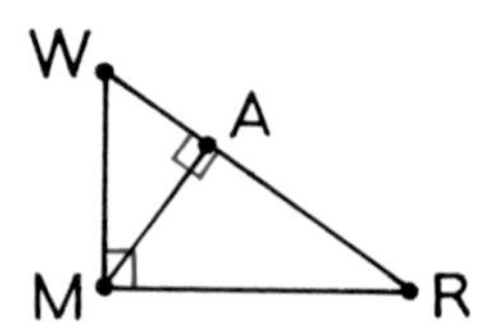

$\triangle$WRM is a right triangle and $\overline{MA} \perp \overline{WR}$. Copy and complete the following equations.

3. $WM^2 + ▒▒ = WR^2$.
4. $WM^2 = ▒▒ + ▒▒$.
5. $AR^2 + ▒▒ = ▒▒$.

Squares have been drawn on the sides of $\triangle FRO$ and $\triangle STY$.

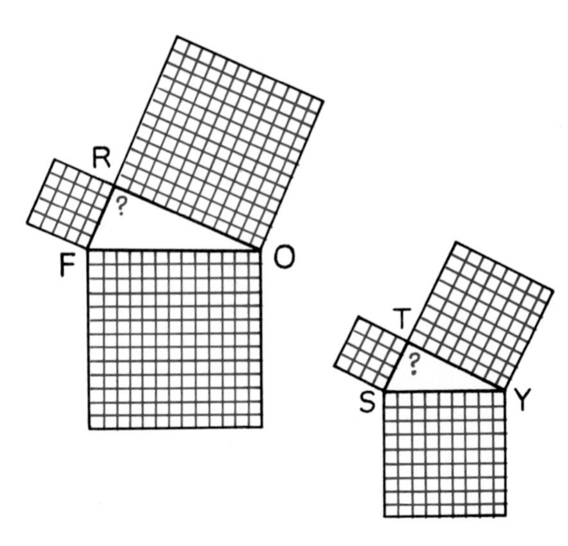

6. Is $\triangle FRO$ a right triangle?
7. Explain why or why not.
8. Is $\triangle STY$ a right triangle?
9. Explain why or why not.

Find the areas of the three squares and the length of the segment marked x in each of the following figures.

Example:

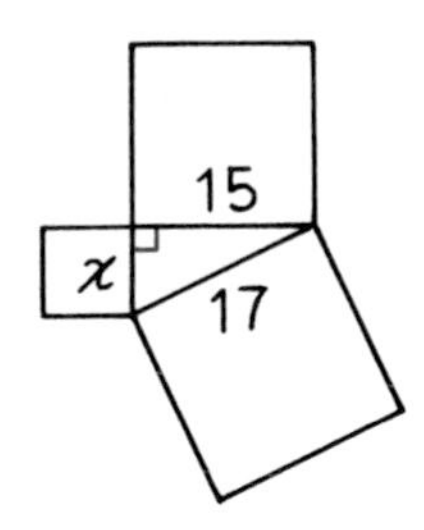

Solution: The areas of two squares are $15^2 = 225$ and $17^2 = 289$. $x^2 + 225 = 289$, and so $x^2 = 64$. The area of the third square is 64 and $x = \sqrt{64} = 8$.

10.

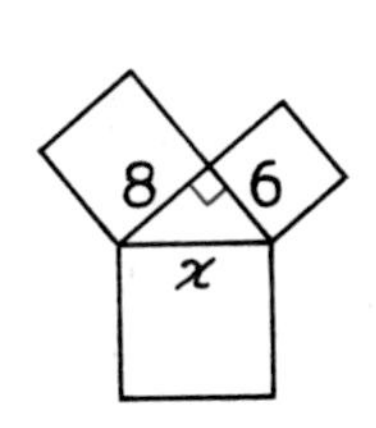

11.

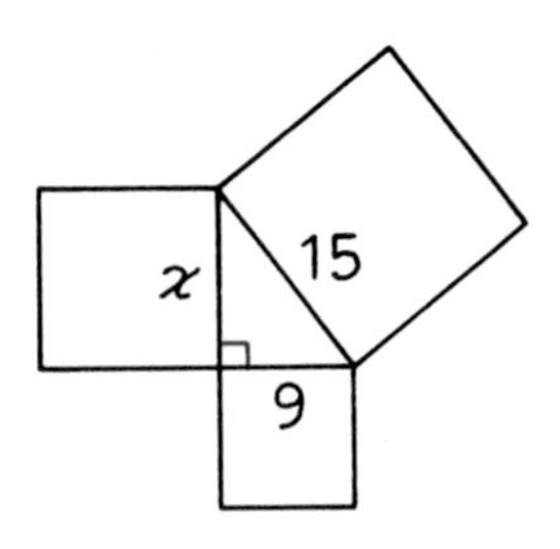

12.

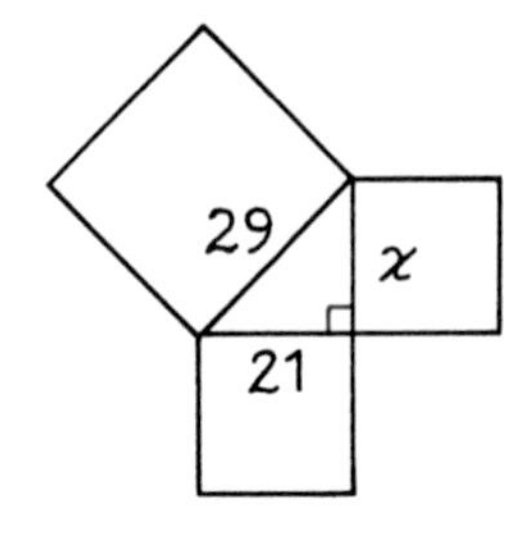

Find the length of the side marked x in each of the following right triangles. Express all irrational answers in simple radical form.

Example:

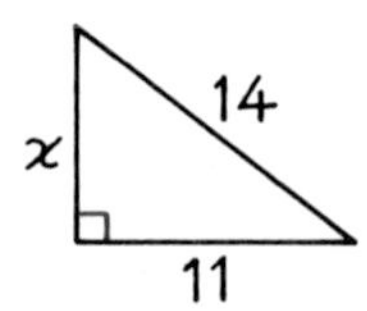

Solution:

$$x^2 + 11^2 = 14^2$$
$$x^2 + 121 = 196$$
$$x^2 = 75$$
$$x = \sqrt{75} = \sqrt{25 \cdot 3}$$
$$= \sqrt{25}\sqrt{3} = 5\sqrt{3}.$$

13.

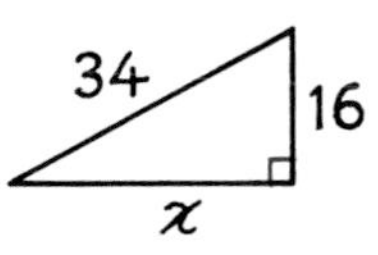

14.

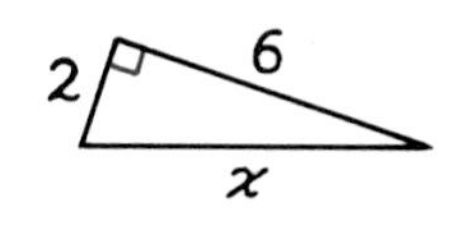

15.

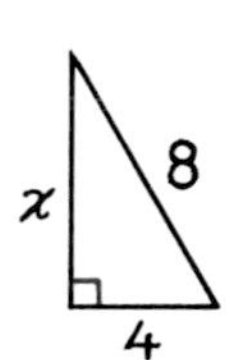

16.

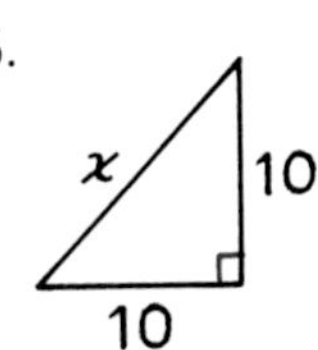

Give the missing statements and reasons in the following proof of the converse of the Pythagorean Theorem.

If the square of one side of a triangle is equal to the sum of the squares of the other two sides, then the triangle is a right triangle.

Given: $\triangle ABC$ with $c^2 = a^2 + b^2$.

Prove: $\triangle ABC$ is a right triangle.

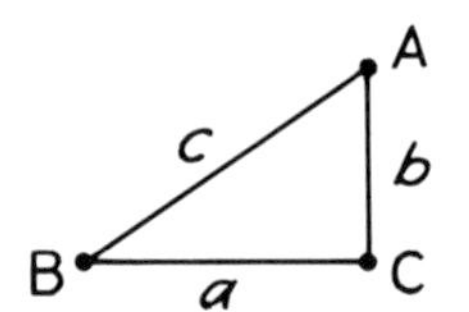

17. *Proof.*

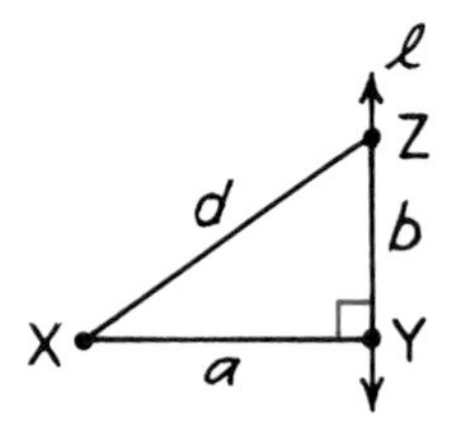

Statements	***Reasons***
1. $\triangle ABC$ with $c^2 = a^2 + b^2$.	Given.
2. Draw $\overline{XY}$ so that $XY = a$.	Ruler Postulate.
3. Through Y, draw $\ell \perp \overline{XY}$.	a) ______
4. b) ______	Perpendicular lines form right angles.
5. c) ______	A right angle has a measure of 90°.
6. Choose point Z on ℓ so that $YZ = b$.	d) ______
7. Draw $\overline{XZ}$.	e) ______
8. $\triangle XYZ$ is a right triangle.	f) ______
9. $d^2 = a^2 + b^2$.	g) ______
10. $c^2 = d^2$.	Substitution (steps 1 and 9).
11. h) ______	Square roots property.
12. i) ______	S.S.S.
13. j) ______	If two triangles are congruent, the corresponding angles are equal.
14. $\angle C = 90°$.	k) ______
15. l) ______	A 90° angle is a right angle.
16. m) ______	n) ______

Set II

Find the shaded area in each of the following figures.

18.

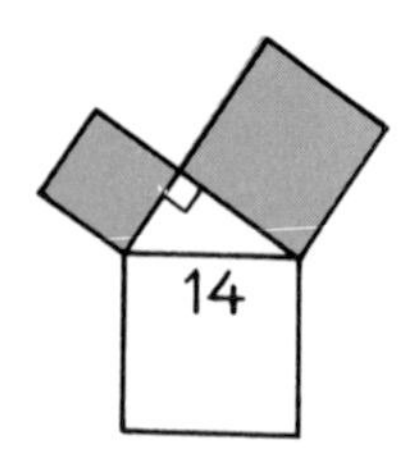

Squares are drawn on the sides of the right triangle.

19.

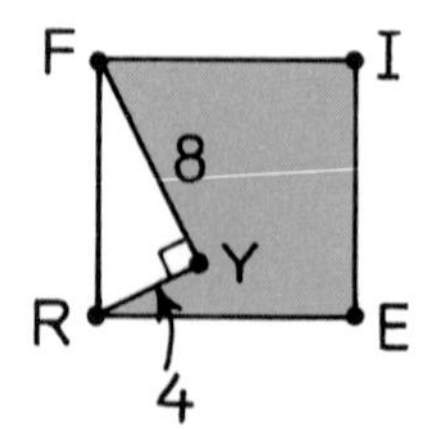

FIER is a square.

20.

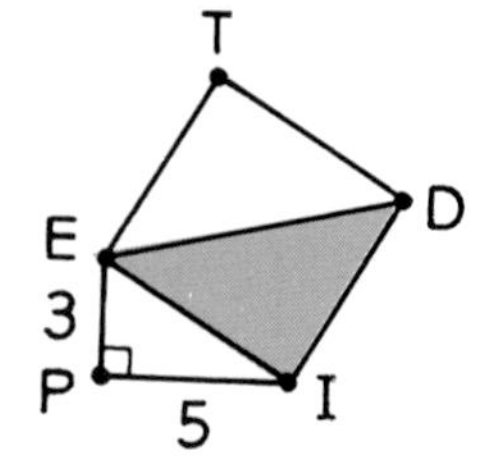

TEID is a square.

21.

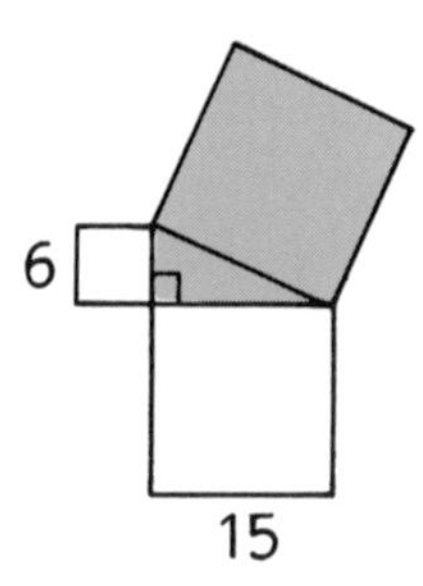

Squares are drawn on the sides of the right triangle.

Use the Pythagorean Theorem to solve each of the following problems.

22. The hypotenuse of a right triangle is two units longer than one of its legs. The other leg is 8 units long. Find the area of the triangle.

23. One leg of a right triangle is three times as long as the other leg. The hypotenuse is 10 units long. Find the area of the triangle.

24. The hypotenuse of a right triangle is twice as long as one of its legs. The other leg is 12 units long. Find the area of the triangle.

In the following exercises, two famous proofs of the Pythagorean Theorem are presented. The approach in each case is very informal, with many of the details omitted.

Courtesy of Library of Congress

President Garfield invented an original proof for the Pythagorean Theorem in 1876 when he was a member of the House of Representatives. It is based on the figure shown here.

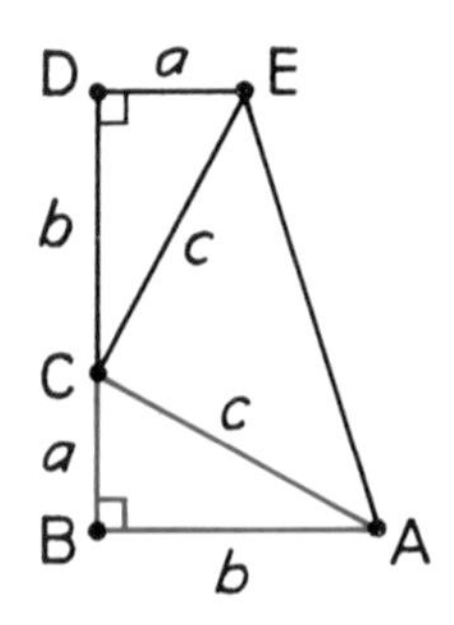

By an argument similar to that in the proof of the Pythagorean Theorem in this lesson, we can show that ∡ACE is a right angle so that △ACE is a right triangle.

25. Write expressions for the areas of the three triangles in terms of *a*, *b*, and *c*.

26. What kind of quadrilateral is ABDE?

27. Write an expression for αABDE in terms of *a* and *b*.

28. Write an equation relating the areas of the three triangles to αABDE.

29. Finish Garfield's proof by substituting your expressions for the areas of the triangles and the area of ABDE into this equation and simplifying it.

Bhaskara, an Indian mathematician of the twelfth century, created a proof of the Pythagorean Theorem based on the figure below. Three copies of the right triangle have been added to it to form two squares.

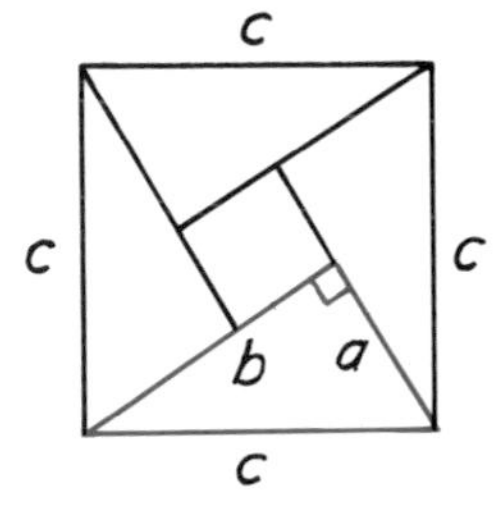

30. The length of a side of the large square is *c*. What is the length of a side of the small square?

31. What is the area of the small square?

32. What is the total area of the four triangles?

33. Use the Area Postulate to write an equation relating the areas of the small square and four triangles to the area of the large square.

34. Finish the proof by simplifying this equation.

Set III

The most famous proof of the Pythagorean Theorem is the one that originally appeared in Euclid's *Elements.* A reproduction of a manuscript of an Arabic translation of this proof made more than one thousand years ago is shown at the left.

The figure on which it is based contains a right triangle, $\triangle ABC$, with squares on its sides; to it have been added three line segments: $\overline{BK}$ (drawn parallel to $\overline{CH}$), $\overline{AG}$, and $\overline{BH}$.

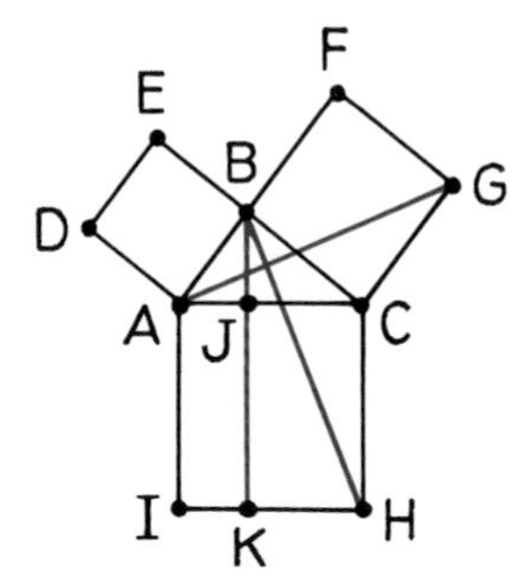

Euclid's proof is quite complex, and so we will consider only its main idea. As a test of your reasoning, you may want to try to justify some of the details.

It can be proved that $\alpha BFGC = 2\alpha\triangle AGC$, $\alpha JCHK = 2\alpha\triangle HBC$, and $\triangle AGC \cong \triangle HBC$.

1. What can you conclude about $\alpha BFGC$ and $\alpha JCHK$? Explain.

By forming some additional triangles, it can also be proved that $\alpha DEBA = \alpha AJKI$.

2. How does the Pythagorean Theorem follow from this fact and your conclusion to Question 1?

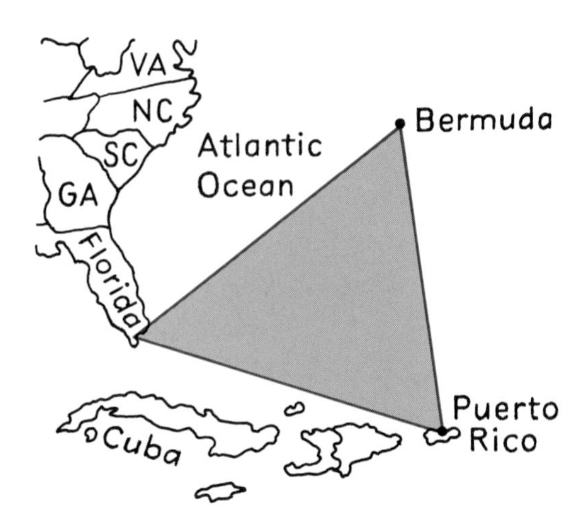

Lesson 6
Heron's Theorem

The Bermuda Triangle is a section of the Atlantic Ocean extending from Bermuda to southern Florida to Puerto Rico. More than 50 ships and 20 planes have disappeared in it, some of them under mysterious circumstances. Ships that had been abandoned for no apparent reason have been discovered within the triangle and airplanes have vanished within it without leaving any trace of wreckage.*

To determine the size of the Bermuda Triangle, it would be useful to have a formula for its area in terms of the lengths of its sides. Such a formula was derived by Heron of Alexandria, a mathematician of the first century A.D. Archimedes is thought to have actually discovered the formula, but it is Heron's proof that has survived.

The formula includes a number called the *semiperimeter* of the triangle.

► **Definition**
The ***semiperimeter*** of a triangle is the number that is half its perimeter.

Hence, if the sides of a triangle have lengths a, b, and c, and s is the semiperimeter of the triangle,

$$s = \frac{a + b + c}{2}.$$

The proof of Heron's theorem is quite complicated and requires some rather difficult algebra.

---

*Many of the stories about strange occurrences within the Bermuda Triangle, however, have very little basis in fact. See Chapter 17, "The Bermuda Triangle," in *Science and the Paranormal,* edited by George O. Abell and Barry Singer (Scribners, 1981).

► **Theorem 54** (Heron's Theorem)
The area of a triangle with sides of lengths a, b, and c and semiperimeter s is $\sqrt{s(s-a)(s-b)(s-c)}$.

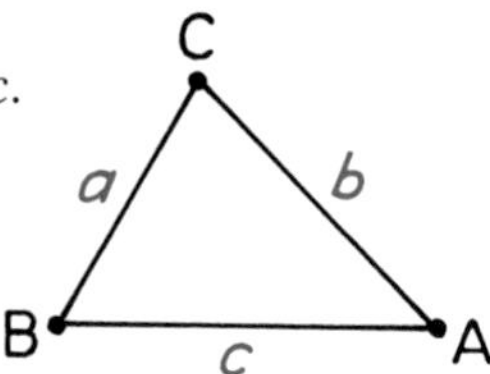

Given: $\triangle ABC$ with sides of lengths a, b, and c.

Prove: $\alpha\triangle ABC = \sqrt{s(s-a)(s-b)(s-c)}$.

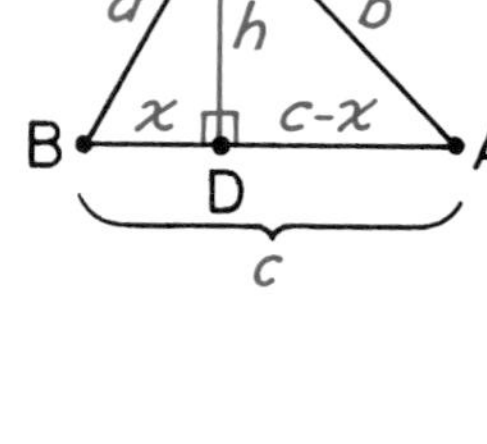

Proof.
Draw $\overline{CD} \perp \overline{BA}$, forming right triangles CBD and CAD. Let CD = h and BD = x, so that DA = $c - x$.

$$\alpha\triangle ABC = \frac{1}{2}ch.$$

To express h in terms of a, b, and c, we apply the Pythagorean Theorem:

$$h^2 + x^2 = a^2$$
$$h^2 = a^2 - x^2$$

and

$$\begin{aligned} h^2 + (c - x)^2 &= b^2 \\ h^2 &= b^2 - (c - x)^2 \\ &= b^2 - (c^2 - 2cx + x^2) \\ &= b^2 - c^2 + 2cx - x^2. \end{aligned}$$

Hence,

$$\begin{aligned} a^2 - x^2 &= b^2 - c^2 + 2cx - x^2 && \text{(Substitution)} \\ a^2 &= b^2 - c^2 + 2cx && \text{(Addition)} \\ a^2 - b^2 + c^2 &= 2cx && \text{(Subtraction)} \\ x &= \frac{a^2 - b^2 + c^2}{2c} && \text{(Division)} \end{aligned}$$

Because $h^2 = a^2 - x^2$,

$$\begin{aligned} h^2 &= (a - x)(a + x) \\ &= \left(a - \frac{a^2 - b^2 + c^2}{2c}\right)\left(a + \frac{a^2 - b^2 + c^2}{2c}\right) \\ &= \left(\frac{2ac - (a^2 - b^2 + c^2)}{2c}\right)\left(\frac{2ac + (a^2 - b^2 + c^2)}{2c}\right). \end{aligned}$$

$$4c^2h^2 = (2ac - a^2 + b^2 - c^2)(2ac + a^2 - b^2 + c^2)$$
$$= [b^2 - (a^2 - 2ac + c^2)][(a^2 + 2ac + c^2) - b^2]$$
$$= [b^2 - (a - c)^2][(a + c)^2 - b^2]$$
$$= [b - (a - c)][b + (a - c)][(a + c) - b][(a + c) + b]$$
$$= (b + c - a)(a + b - c)(a + c - b)(a + b + c).\dagger$$

Because s is the semiperimeter, $s = \dfrac{a + b + c}{2}$.

$$a + b + c = 2s.$$
$$(a + b + c) - 2a = 2s - 2a$$
$$b + c - a = 2(s - a)$$
$$(a + b + c) - 2b = 2s - 2b$$
$$a + c - b = 2(s - b)$$
$$(a + b + c) - 2c = 2s - 2c$$
$$a + b - c = 2(s - c).$$

By substitution of these results into the equation marked †,

$$4c^2h^2 = [2(s - a)][2(s - c)][2(s - b)][2s]$$
$$= 16s(s - a)(s - b)(s - c)$$
$$c^2h^2 = 4s(s - a)(s - b)(s - c).$$

Taking square roots,

$$ch = 2\sqrt{s(s - a)(s - b)(s - c)}.$$

But $\alpha\triangle ABC = \frac{1}{2}ch$, and so

$$\alpha\triangle ABC = \sqrt{s(s - a)(s - b)(s - c)}.$$

The sides of the Bermuda Triangle are 1,000, 1,000, and 1,100 miles in length. Its perimeter is 3,100 miles and its semiperimeter is 1,550 miles. Substituting these numbers into the formula of Heron's Theorem, we get

$$\sqrt{1550(1550 - 1000)(1550 - 1000)(1550 - 1100)} =$$
$$\sqrt{1550(550)(550)(450)} \approx$$
$$460{,}000.^*$$

* A calculator was used to find this figure.

The area of the Bermuda Triangle is approximately 460,000 square miles.

Heron's Theorem can be used to derive a formula for the area of an equilateral triangle.

► **Corollary**

The area of an equilateral triangle with sides of length a is $\frac{a^2}{4}\sqrt{3}$.

Proof.

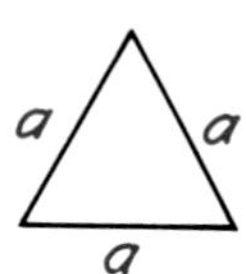

The perimeter of the triangle is $3a$, and so the semiperimeter is $\frac{3a}{2}$.

$$\begin{aligned}\text{Area} &= \sqrt{s(s-a)(s-b)(s-c)} \\ &= \sqrt{\frac{3a}{2}\left(\frac{3a}{2}-a\right)\left(\frac{3a}{2}-a\right)\left(\frac{3a}{2}-a\right)} \\ &= \sqrt{\frac{3a}{2}\cdot\frac{a}{2}\cdot\frac{a}{2}\cdot\frac{a}{2}} = \sqrt{\frac{3a^4}{16}} = \frac{a^2}{4}\sqrt{3}\end{aligned}$$

Exercises

Set I

In the figure below, △ABC is a right triangle.

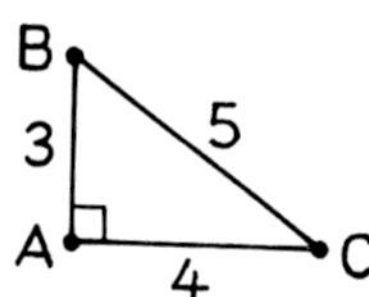

1. How many sides of △ABC are needed to find its area?
2. Use the formula for the area of a right triangle to find α△ABC.
3. Use Heron's Theorem to find α△ABC.

Use Heron's Theorem to find the areas of the following triangles.

Example:

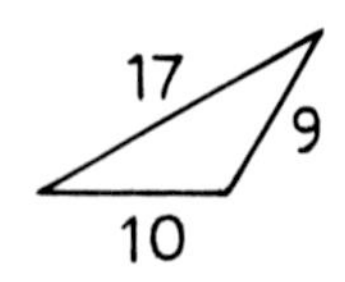

Solution:

Perimeter $= 17 + 10 + 9 = 36$

Semiperimeter $= s = \frac{36}{2} = 18$

$$\begin{aligned}\text{Area} &= \sqrt{18(18-17)(18-10)(18-9)} \\ &= \sqrt{18\cdot 1\cdot 8\cdot 9} \\ &= \sqrt{2\cdot 9\cdot 2\cdot 4\cdot 9} \\ &= \sqrt{2^2\cdot 9^2\cdot 2^2} \\ &= 2\cdot 9\cdot 2 = 36\end{aligned}$$

4.

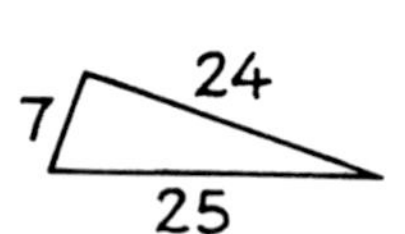

5.

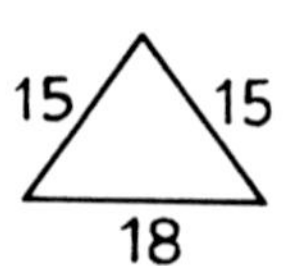

6.

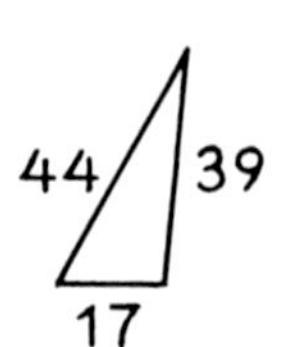

The areas of the following triangles are irrational. Use Heron's Theorem to find them. Express each answer in simple radical form and in decimal form to the nearest tenth.

Example:

Solution:

Perimeter $= 7 + 12 + 15 = 34$

Semiperimeter $= s = \frac{34}{2} = 17$

$$\begin{aligned} \text{Area} &= \sqrt{17(17-7)(17-12)(17-15)} \\ &= \sqrt{17 \cdot 10 \cdot 5 \cdot 2} \\ &= \sqrt{17 \cdot 10^2} \\ &= 10\sqrt{17} \\ &\approx 41.2 \end{aligned}$$

7.

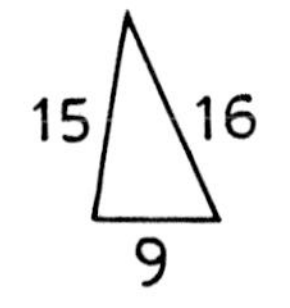

8.

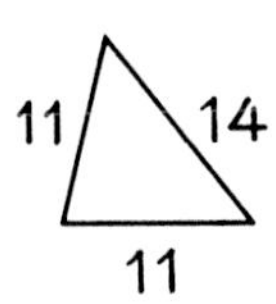

9.

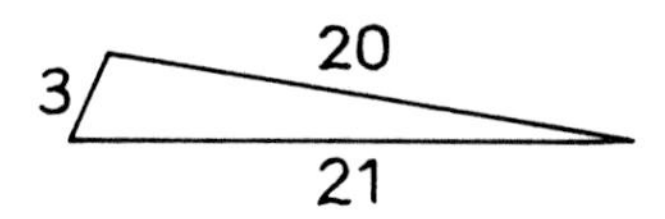

In the figure below, △DEF is an equilateral triangle.

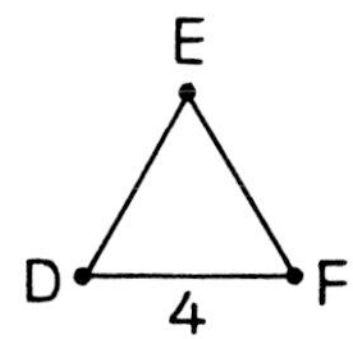

10. Use Heron's Theorem to find $\alpha\triangle$DEF. Express your answer in simple radical form.

11. Use the corollary to Heron's Theorem to find $\alpha\triangle$DEF.

12. Find the approximate area of the triangle. Express your answer to the nearest tenth.

Find the areas of the following equilateral triangles. Express your answers in simple radical form.

13.

14.

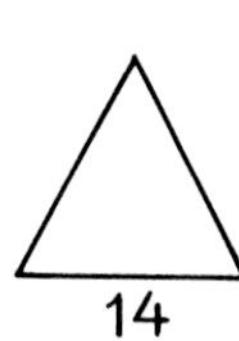

15.

Suppose that a triangle has sides of lengths 4, 6, and 10.

16. Try to use Heron's Theorem to find its area.

17. Why does the result turn out as it does?

18. State the theorem that is the basis for your answer.

Suppose that a triangle has sides of lengths 3, 5, and 12.

19. Try to use Heron's Theorem to find its area.

20. Why does the result turn out as it does?

Set II

Find the area of each of the following triangles. Express all irrational answers in simple radical form.

21. An equilateral triangle whose perimeter is 36.
22. A right triangle whose sides are 9, 12, and 15.
23. An isosceles triangle, two of whose sides are 4 and 10.

Find the perimeter and area of each of the following figures.

24.

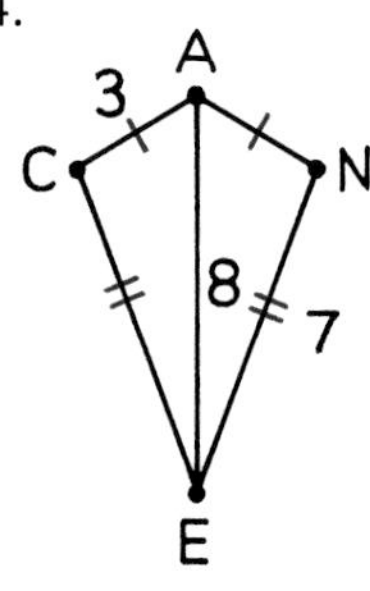

25.

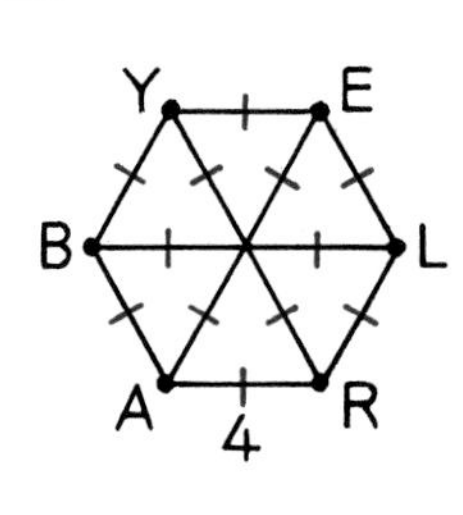

26.

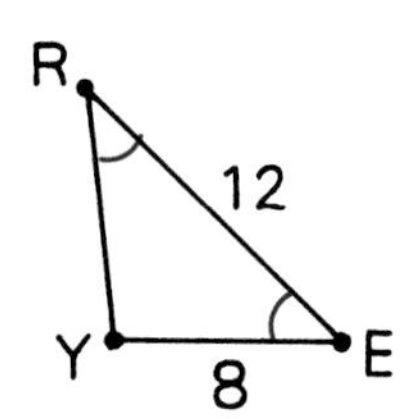

27.

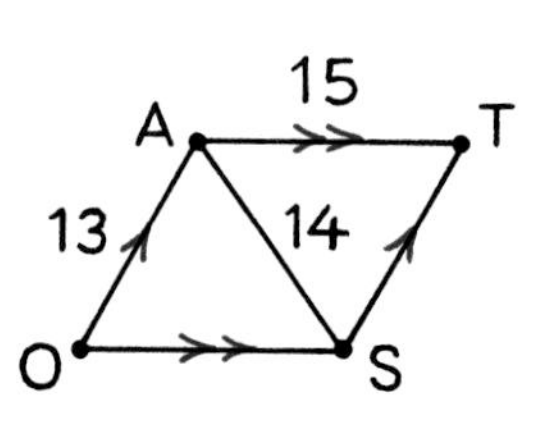

28.

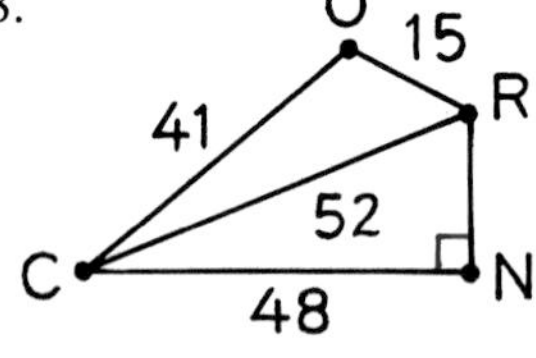

Find the colored area in each of the following figures.

29.

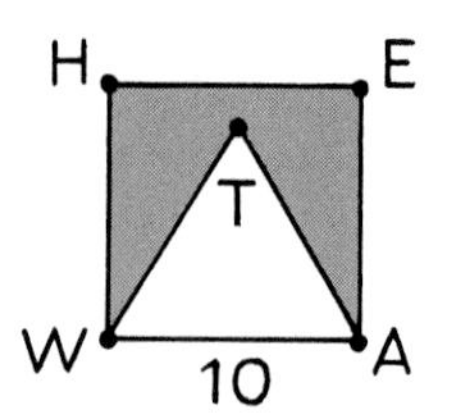

WHEA is a square and $\triangle$WAT is an equilateral triangle.

30.

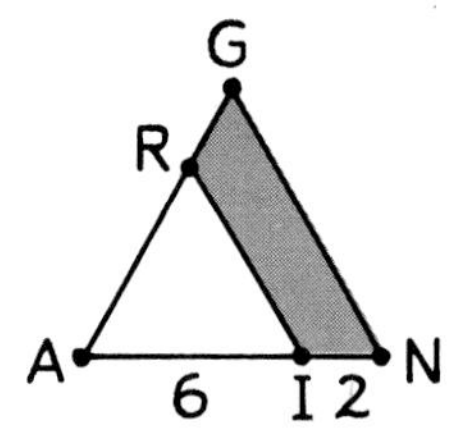

$\triangle$GAN and $\triangle$RAI are equilateral triangles.

31.

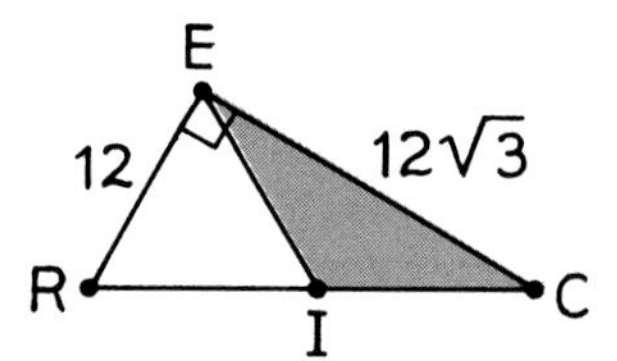

$\triangle$RCE is a right triangle and $\triangle$RIE is an equilateral triangle.

32.

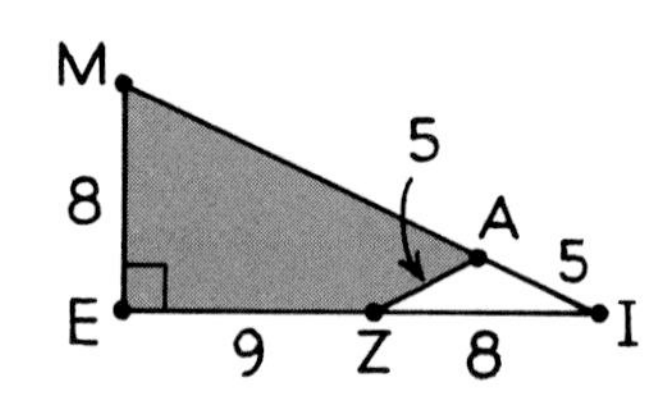

$\triangle$MIE is a right triangle.

Set III

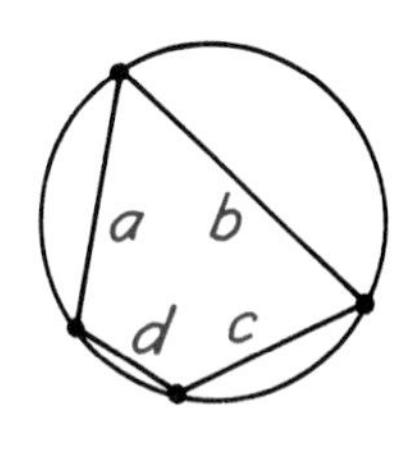

An Indian mathematician of the seventh century, named Brahmagupta, discovered a formula for the area of a quadrilateral whose vertices lie on a circle. It is

$$\text{Area} = \sqrt{(s-a)(s-b)(s-c)(s-d)},$$

where s is the semiperimeter of the quadrilateral, defined in the same way as it is for a triangle.

1. If the vertices of the quadrilateral move on the circle so that d shrinks to zero, what does the quadrilateral become?
2. If $d = 0$, what does Brahmagupta's formula become?

Use Brahmagupta's formula to find the areas of the following quadrilaterals.

3.

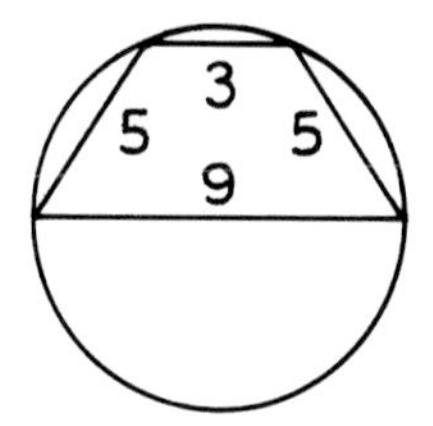

4.

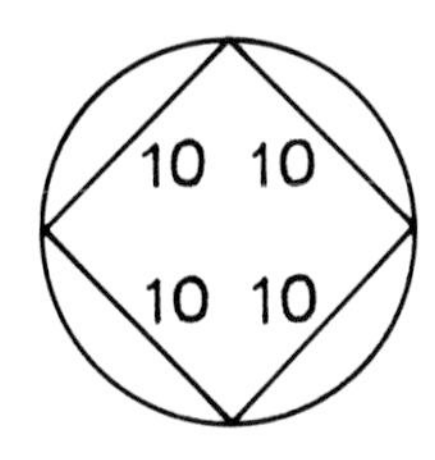

5. No one has ever discovered a formula for the area of a general quadrilateral in terms of the lengths of its sides and no one ever will. Why not?

Chapter 9 / Summary and Review

Basic Ideas

Altitude of a quadrilateral 324
Altitude of a triangle 318
Area 309
Polygonal region 309
Semiperimeter 335
Triangular region 308

Postulates

11. *The Area Postulate.* To every polygonal region there corresponds a positive number called its area such that
    a) triangular regions bounded by congruent triangles have equal areas,
    b) and polygonal regions consisting of two or more nonoverlapping polygonal regions are equal in area to the sum of their areas. 309

12. The area of a rectangle is the product of the lengths of its base and its altitude. 313

Theorems

Corollary to Postulate 12. The area of a square is the square of the length of its side. 313

47. The area of a right triangle is half the product of the lengths of its legs. 317

48. The area of a triangle is half the product of the lengths of any base and corresponding altitude. 318
    Corollary. Triangles with equal bases and equal altitudes have equal areas. 319

49. The area of a rhombus is half the product of the lengths of its diagonals. 319

50. The area of a parallelogram is the product of the lengths of any base and corresponding altitude. 324

51. The area of a trapezoid is half the product of the length of its altitude and the sum of the lengths of its bases. 324

52. *The Pythagorean Theorem.* In a right triangle, the square of the hypotenuse is equal to the sum of the squares of the legs. 329

53. If the square of one side of a triangle is equal to the sum of the squares of the other two sides, then the triangle is a right triangle. 330

54. *Heron's Theorem.* The area of a triangle with sides of lengths a, b, and c and semiperimeter s is $\sqrt{s(s-a)(s-b)(s-c)}$. 336
    Corollary. The area of an equilateral triangle with sides of length a is $\frac{a^2}{4}\sqrt{3}$. 338

Exercises

Set I

The dancing figures at the beginning of this review were made from tangrams, a Chinese puzzle made by dividing a square into seven pieces. The pieces are shown in the figure below. Given that the area of the square is 64 square units, find the area of each of the following pieces.

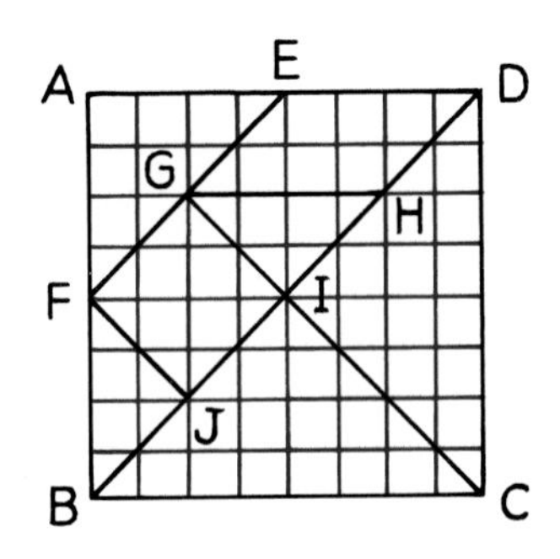

1. △BIC.
2. △AEF.
3. △BFJ.
4. FGIJ.
5. EDHG.

Find the areas of the following figures.

6. A square whose perimeter is 100.
7. A rectangle whose base is 5 and whose perimeter is 16.
8. A rhombus whose diagonals are 8 and 9.
9. A triangle whose sides are 10, 17, and 21.

△JAY has sides of lengths 21, 29, and 20.

10. Is △JAY a right triangle?
11. Find α△JAY.

In the figure below, $\overline{HO} \perp \overline{EN}$ and ER = ON.

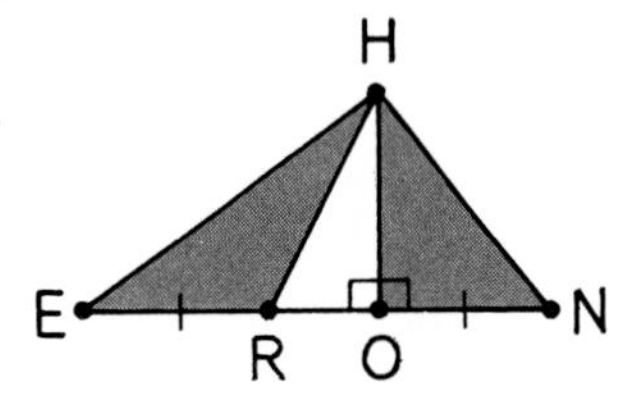

12. Can you conclude that △HRE ≅ △HON?
13. Can you conclude that α△HRE = α△HON?

△OWL is a right triangle whose hypotenuse is 41 and one of whose legs is 9.

14. Find the length of its other leg.
15. Find α△OWL.

△SOR is equilateral, with midsegment $\overline{TK}$.

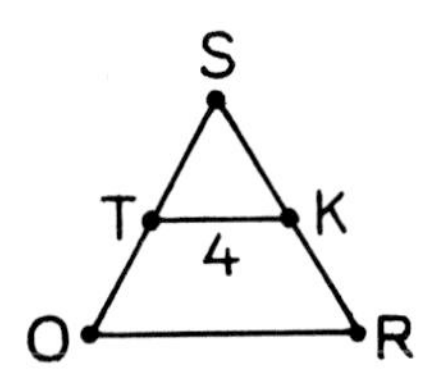

16. Find α△STK.
17. Find α△SOR.
18. Find αOTKR.
19. How does the area of △STK compare with the area of OTKR?

Set II

Acute Alice's mother wants new carpeting for her living room. The shape and dimensions of the room are shown here and the kind of carpeting she wants costs \$20 per square yard.

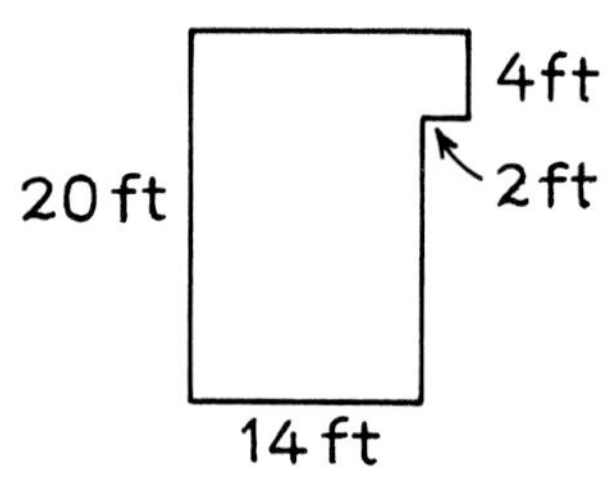

Obtuse Ollie figures that the area of the room is 288 square feet. Because 1 yard = 3 feet, he divides by 3 to get 96 square yards and because the carpet costs \$20 per square yard, he divides by 20 to get 4.8. He tells Alice's mother the carpeting will cost \$4.80.

20. What's wrong with Ollie's calculations?

21. What is the actual cost of the carpeting?

In the figure below, BLUE is a rectangle and BJYE is a parallelogram.

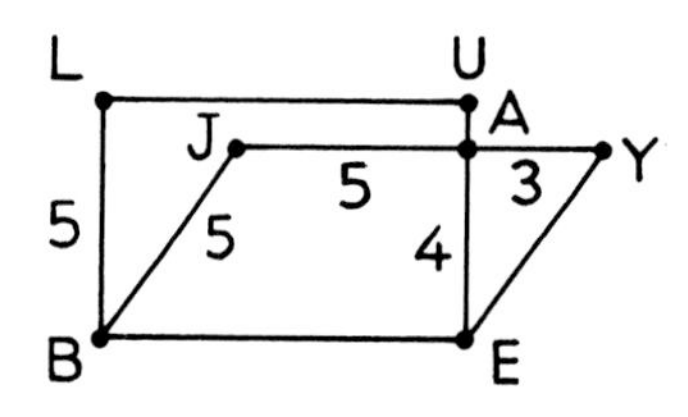

22. Find αBLUE.

23. Find αBJYE.

24. Find αBJAE.

Find the area of the shaded region in each of the following figures.

25.

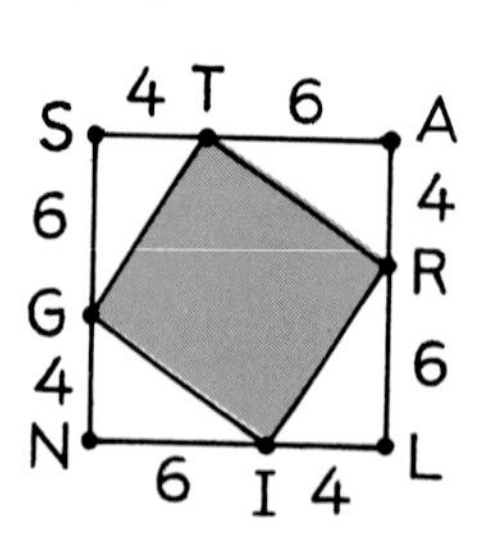

SALN is a square.

26.

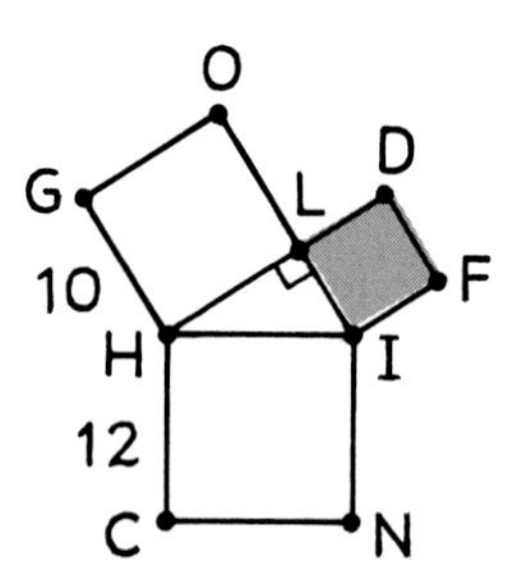

Squares are drawn on the sides of the right triangle.

27.

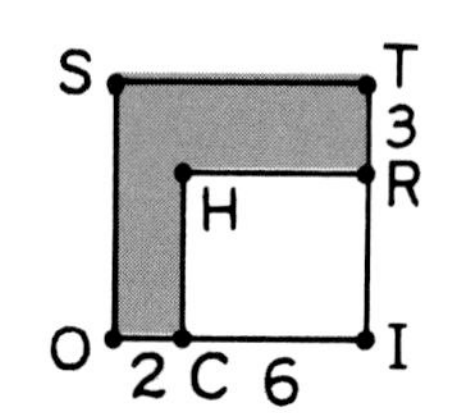

OSTI is a square and CHRI is a rectangle.

28.

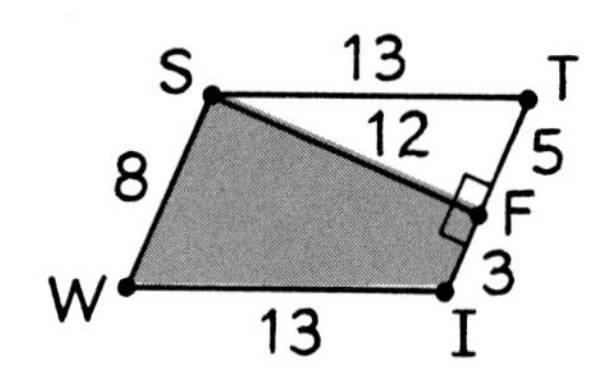

WSTI is a parallelogram.

29.

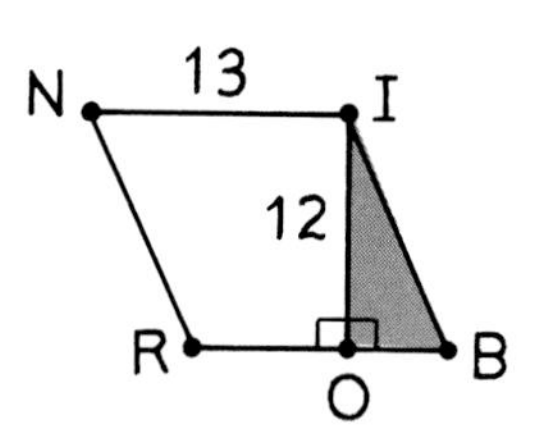

NIBR is a rhombus.

30.

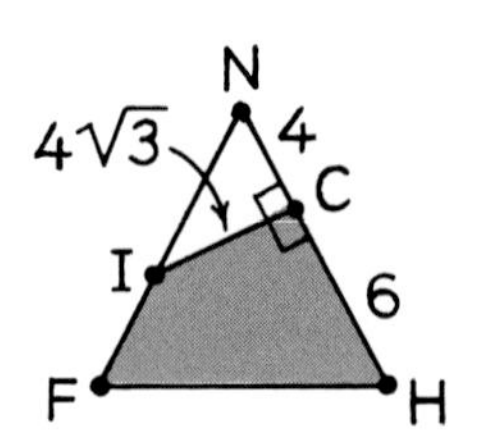

$\triangle$FNH is equilateral.

31.

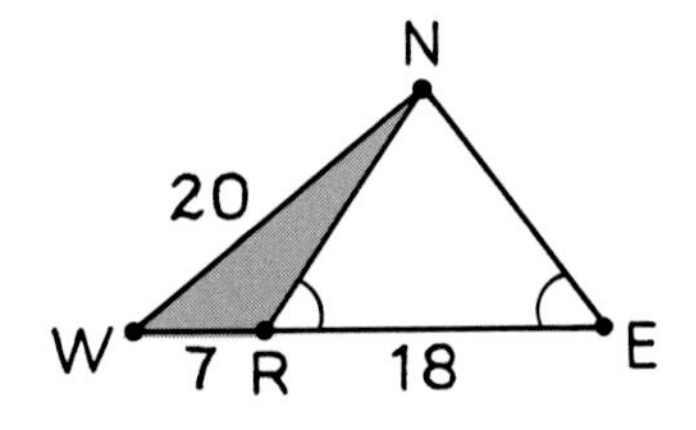

$\triangle$WNE is a right triangle.

32.

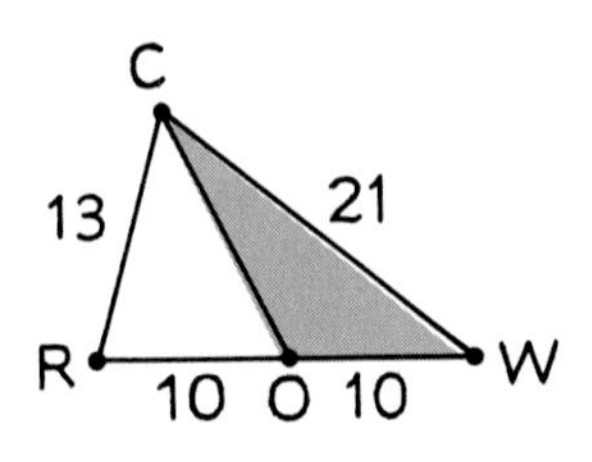

O is the midpoint of $\overline{RW}$.

Algebra Review

Fractional Equations

An equation that contains fractions can be solved by first multiplying both sides by a number that will clear the equation of all of the fractions. The simplest way to do this is to multiply by the least common denominator of the fractions.

Exercises

Solve the following equations. Check your answers.

Example 1: $\frac{x}{4} + \frac{x-1}{3} = 2$

Solution: The least common denominator of the fractions is 12. Multiplying both sides of the equation by 12, we get

$$12\left(\frac{x}{4}\right) + 12\left(\frac{x-1}{3}\right) = 12(2)$$

$$\frac{12x}{4} + \frac{12(x-1)}{3} = 24$$

$$3x + 4(x-1) = 24$$

$$3x + 4x - 4 = 24$$

$$7x = 28$$

$$x = 4$$

Checking this result in the original equation, we get

$$\frac{4}{4} + \frac{4-1}{3} = 1 + 1 = 2.$$

Example 2: $\frac{1}{6} = \frac{10}{x} - \frac{1}{2}$

Solution: The least common denominator of the fractions is $6x$. Multiplying both sides by $6x$, we get

$$6x\left(\frac{1}{6}\right) = 6x\left(\frac{10}{x}\right) - 6x\left(\frac{1}{2}\right)$$

$$\frac{6x}{6} = \frac{60x}{x} - \frac{6x}{2}$$

$$x = 60 - 3x$$

$$4x = 60$$

$$x = 15$$

Checking this result by substituting it in the right side of the original equation, we get

$$\frac{10}{15} - \frac{1}{2} = \frac{2}{3} - \frac{1}{2}$$

$$= \frac{4}{6} - \frac{3}{6} = \frac{1}{6}.$$

1. $\dfrac{x}{10} = 50$

2. $\dfrac{3}{x} = 15$

3. $\dfrac{6}{x - 1} = 4$

4. $\dfrac{4}{x} = \dfrac{x}{25}$

5. $\dfrac{11}{x} + 2 = \dfrac{5}{x}$

6. $\dfrac{x}{3} = \dfrac{x}{2} - 7$

7. $\dfrac{1}{3} - \dfrac{4}{x} = \dfrac{1}{x}$

8. $\dfrac{x}{4} = \dfrac{x}{5} + 1$

9. $\dfrac{x + 5}{2} = \dfrac{4x - 3}{7}$

10. $\dfrac{10}{x - 3} - \dfrac{6}{x - 3} = 8$

11. $\dfrac{5}{x + 3} - \dfrac{2}{3} = 1$

12. $\dfrac{7x - 2}{x} + \dfrac{x + 2}{x} = x$

13. $\dfrac{15}{x + 4} + \dfrac{1}{3} = 2$

14. $\dfrac{3x + 1}{4} - \dfrac{x}{5} = x - 2$

15. $\dfrac{x + 8}{2} - \dfrac{x - 4}{8} = \dfrac{x}{4}$

Solve for x in terms of the other variables. Simplify your answers as much as possible.

Example: $\dfrac{1}{a} = \dfrac{1}{x - b}$

Solution: Multiplying both sides of the equation by $a(x - b)$, we get

$$\frac{a(x - b)}{a} = \frac{a(x - b)}{x - b}$$

$$x - b = a$$

$$x = a + b$$

16. $\dfrac{x}{a} + b = 0$

17. $\dfrac{x}{a} = \dfrac{a}{b}$

18. $\dfrac{1}{x} - \dfrac{1}{a} = 0$

19. $\dfrac{a}{x} + \dfrac{b}{x} = c$

20. $x + \dfrac{1}{b} = \dfrac{a}{b}$

Chapter 10
SIMILARITY

Ben Rose—Time-Life Picture Agency

Lesson 1

Ratio and Proportion

The two cars parked on the street in this photograph look very much alike. The band of the crosswalk running underneath them, however, reveals that something is wrong. Because the band's width is so much narrower below the dark car than below the light one, the two cars must be parked quite a distance apart. Why doesn't the one closer to us, then, look much larger than the one parked across the street?

The closer car is in fact only a small model. The camera was carefully placed with respect to the two cars to give the illusion that they are the same size. Without the clues present in the photograph, it would be difficult to see that they are vastly different in size.

Suppose that the actual car is 200 cm wide and 560 cm long. To look like the actual car, how long must the model be if it is 10 cm wide? Because the actual car is 20 times as wide as the model, it must also be 20 times as long. Dividing 560 by 20, we find that the model is 28 cm long.

Another way to solve this problem is to write a proportion:

$$\frac{200}{10} = \frac{560}{x}.$$

This proportion has as its basis the fact that, because the car and the model have the *same shape,* their dimensions have the *same ratio.* In geometry, figures that have the same shape are said to be *similar* to each other.

Before studying the properties of similar figures, we will review the meaning of *ratio* and *proportion.*

► **Definitions**

The ***ratio*** of the numbers a to b is the number $\frac{a}{b}$. (Note that b cannot be 0 because division by 0 is undefined.)

A ***proportion*** is an equality between two ratios.

We can represent a proportion symbolically as

$$\frac{a}{b} = \frac{c}{d}.$$

The numbers a, b, c, and d are called the *first, second, third,* and *fourth terms* of the proportion, respectively. The second and third terms, b and c, are also called the *means,* and the first and fourth terms, a and d, are called the *extremes* of the proportion.

Proportions have many interesting properties. One of them concerns the means and extremes. If we multiply both sides of the proportion

$$\frac{a}{b} = \frac{c}{d}$$

by the common denominator bd, we get

$$bd\left(\frac{a}{b}\right) = bd\left(\frac{c}{d}\right)$$

$$\frac{abd}{b} = \frac{bcd}{d}$$

$$ad = bc.$$

This proves that the product of the means of a proportion is equal to the product of the extremes. We will refer to this as the "means-extremes" property. For example, to solve the proportion

$$\frac{200}{10} = \frac{560}{x},$$

we can multiply its means and extremes, getting

$$200x = 5600$$
$$x = 28.$$

We will encounter some proportions in geometry in which the means are the same number. An example of such a proportion is

$$\frac{4}{10} = \frac{10}{25}.$$

The number 10 is called the *geometric mean* between 4 and 25.

► **Definition**
The number b is the ***geometric mean*** between the numbers a and c iff a, b, and c are positive and

$$\frac{a}{b} = \frac{b}{c}.$$

Exercises

Set I

The following questions refer to this proportion: $\frac{7}{11} = \frac{21}{33}$.

1. What are the two ratios in this proportion?
2. Name the means of the proportion.
3. Show that the product of the means in this proportion is equal to the product of the extremes.

Solve for x in each of the following proportions.

4. $\frac{x}{3} = \frac{9}{10}$
5. $\frac{6}{17} = \frac{2x}{51}$
6. $\frac{x}{x+2} = \frac{3}{4}$
7. $\frac{1}{x} = \frac{3}{2x-1}$

Two ratios that are famous in the history of the number π are $\frac{22}{7}$ and $\frac{355}{113}$.

8. Express $\frac{22}{7}$ in decimal form to the nearest hundredth.
9. Express $\frac{355}{113}$ in decimal form to the nearest hundredth.
10. Is it true that $\frac{22}{7} = \frac{355}{113}$? Explain why or why not.

Find the geometric mean between each of the following pairs of numbers.

Example: 12 and 15.

Solution:

$$\frac{12}{x} = \frac{x}{15}$$
$$x^2 = 180$$
$$x = \sqrt{180} \text{ (because } x \text{ is positive)}$$
$$= \sqrt{36 \cdot 5}$$
$$= 6\sqrt{5}.$$

11. 3 and 27.
12. 5 and 80.
13. 8 and 10.
14. 1 and 21.

Set II

Suppose that Uncle Albert's arm span is 3 feet and that Alabaster's arm span is 15 inches.

15. Does it seem correct to say that the ratio of their respective arm spans is $\frac{3}{15}$ or $\frac{1}{5}$?

16. Change Uncle Albert's measurement to inches and find the ratio of their arm spans. Express the ratio in decimal form.

17. Change Alabaster's measurement to feet and find the ratio of their arm spans. Express the ratio in decimal form.

18. Does the ratio of two lengths depend on the unit of measure if both lengths are in terms of the same unit?

Write the proportion that results from dividing both sides of the equation

$$5x = 12y$$

by each of the following quantities.

Example: $5y$.

Solution: $\frac{5x}{5y} = \frac{12y}{5y}$, and so $\frac{x}{y} = \frac{12}{5}$.

19. 60

20. xy

21. $12x$

Tell what can be done to both sides of the equation

$$ad = bc$$

to get each of the following equations.

Example: $\frac{a}{b} = \frac{c}{d}$.

Answer: Divide by bd; $\frac{ad}{bd} = \frac{bc}{bd}$, and so $\frac{a}{b} = \frac{c}{d}$.

22. $\frac{d}{b} = \frac{c}{a}$

23. $\frac{a}{c} = \frac{b}{d}$

24. $\frac{d}{c} = \frac{b}{a}$

Solve for x and y.

25. $5 = \frac{60}{x} = \frac{y}{25}$

26. $\frac{x}{4} = 3 = \frac{18}{y}$

27. $\frac{x}{5} = \frac{20}{x} = y$

28. $\frac{x-1}{x} = \frac{2}{7} = \frac{y}{y+1}$

Give the missing statements and reasons in the following proofs.

29. *Given:* $\frac{a}{b} = \frac{c}{d}$.

Prove: $\frac{a+b}{b} = \frac{c+d}{d}$.

Proof.

Statements	**Reasons**
1. a) ▒▒▒	b) ▒▒▒
2. $\frac{a}{b} + 1 = \frac{c}{d} + 1$	c) ▒▒▒
3. $\frac{a}{b} + \frac{b}{b} = \frac{c}{d} + \frac{d}{d}$, and so $\frac{a+b}{b} = \frac{c+d}{d}$	d) ▒▒▒

30. *Given:* $\frac{a}{b} = k$, $\frac{c}{d} = k$, and $\frac{e}{f} = k$.

Prove: $\frac{a+c+e}{b+d+f} = k$.

Proof.

Statements	**Reasons**
1. a) ▒▒▒	b) ▒▒▒
2. $a = kb$, $c = kd$, and $e = kf$.	c) ▒▒▒
3. $a + c + e = kb + kd + kf$.	d) ▒▒▒
4. $a + c + e = k(b + d + f)$.	e) ▒▒▒
5. f) ▒▒▒	g) ▒▒▒

Write complete proofs for each of the following exercises.

31. *Given:* $\frac{a}{b} = \frac{b}{c}$

Prove: $b = \pm\sqrt{ac}$

32. *Given:* $\frac{a}{b} = \frac{c}{d}$.

Prove: $\frac{a-b}{b} = \frac{c-d}{d}$.

Set III

Photograph by S. H. Rosenthal, Jr., Rapho Guillumette Pictures

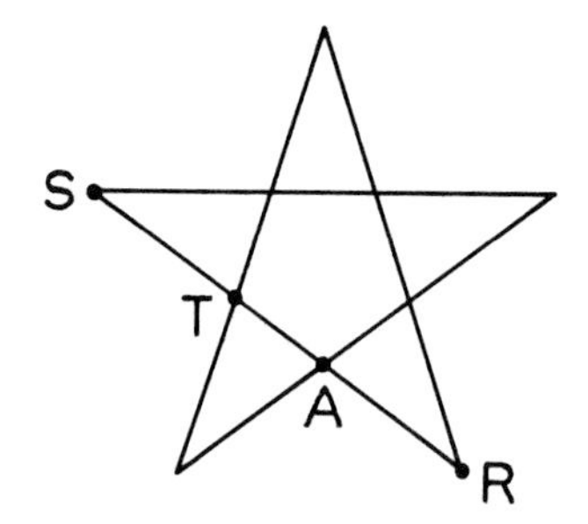

The geometric mean can be seen in nature in the starfish. Its shape, called a "pentagram," is shown in the figure at the left.

The length ST is the geometric mean between TA and SA; that is,

$$\frac{TA}{ST} = \frac{ST}{SA}.$$

Using the assumption that ST = AR, it is possible to show that SA is also the geometric mean between two lengths in the figure. Do this by doing each of the following exercises.

1. Add 1 to both sides of $\frac{TA}{ST} = \frac{ST}{SA}$.
2. Write each side of the resulting equation as a single fraction.
3. Substitute AR for ST in the right side of the resulting equation.
4. Use the Betweenness of Points Theorem to express the sums in the equation as single lengths.
5. Apply the symmetric property to the resulting equation.
6. Between which two lengths in the figure is SA the geometric mean?

Brown Brothers

Lesson 2

The Side-Splitter Theorem

In 1889, a dam above Johnstown, Pennsylvania, broke and sent down a torrent of water that destroyed most of the city and killed more than 2,300 people. The photograph above shows an overturned house through which an uprooted tree was hurled.

Before a modern dam is built, a lot of mathematics has to be worked out to insure that it will not collapse. Some of this mathematics is applied geometry.

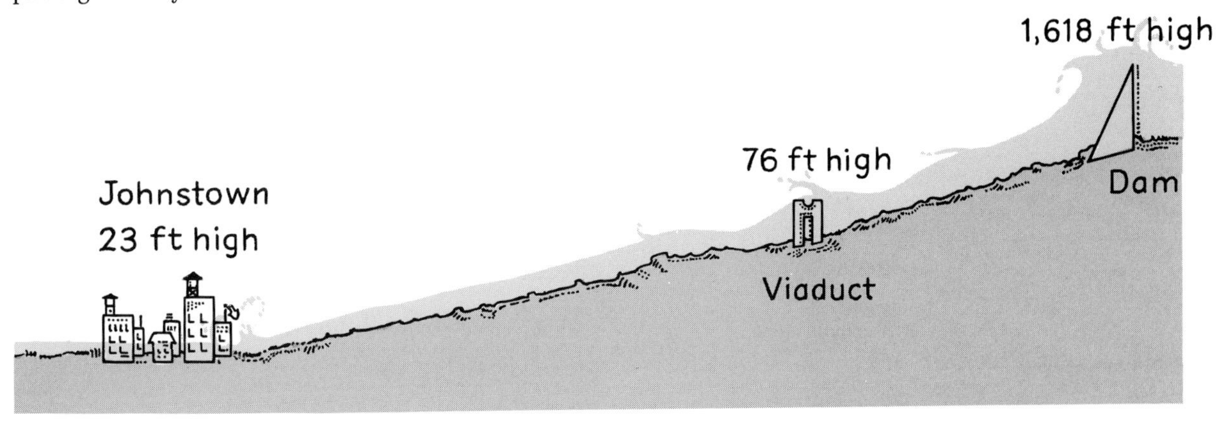

The drawing below represents a simplified side view of a dam. The cross section of the dam is shown as $\triangle DAM$; note that the line of the water level behind the dam, $\overleftrightarrow{XY}$, is parallel to the line of the ground, $\overleftrightarrow{AM}$.

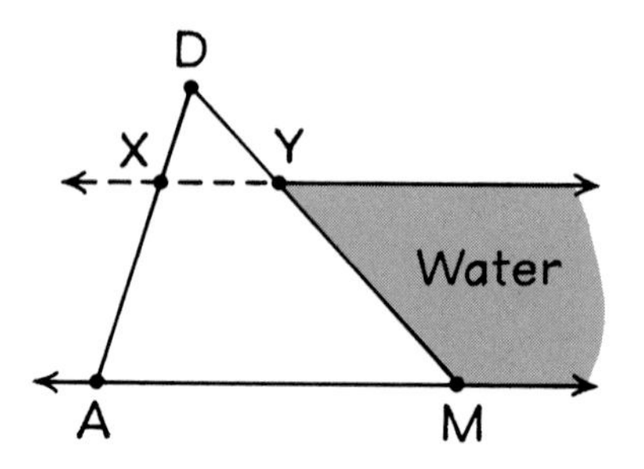

The amount of force of the water against the dam depends on several factors; one of them is the length of $\overline{YM}$, the segment that represents the surface of contact. Can the length of $\overline{YM}$ be determined indirectly from the lengths of other segments in the figure? For example, suppose that DX = 80 meters, XA = 160 meters, and DY = 100 meters; is it possible to find YM from these numbers?

We will show that, if $\overleftrightarrow{XY} \parallel \overline{AM}$, then

$$\frac{DX}{XA} = \frac{DY}{YM}.$$

By substituting into this equation,

$$\frac{80}{160} = \frac{100}{YM}.$$

Solving for YM,

$$80 \cdot YM = 100 \cdot 160,$$

$$YM = \frac{100 \cdot 160}{80} = 200,$$

we find that the length of $\overline{YM}$ is 200 meters.

This method depends on the following fact, which we will call the "Side-Splitter Theorem."

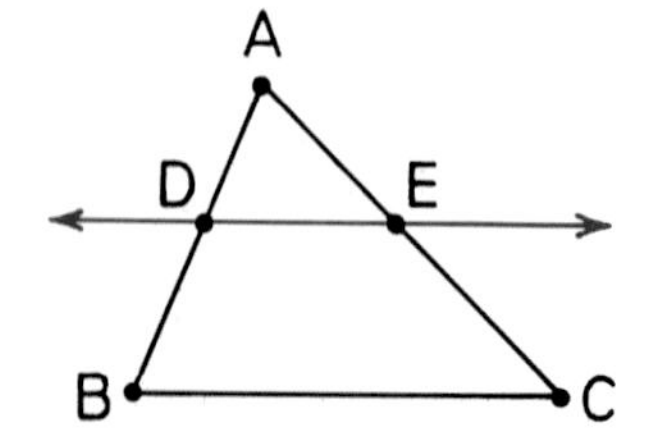

► **Theorem 55**

If a line parallel to one side of a triangle intersects the other two sides in different points, it divides the sides in the same ratio.

Given: $\triangle ABC$ with $\overleftrightarrow{DE} \parallel \overline{BC}$.

Prove: $\frac{AD}{DB} = \frac{AE}{EC}.$

To prove this theorem, we will add two line segments to the figure to form some triangles whose bases are the four segments in the theorem's conclusion.

Proof.
Draw $\overline{BE}$ and draw $\overline{EF} \perp \overline{AB}$. Because $\overline{EF}$ is an altitude of both $\triangle AED$ and $\triangle DEB$, we can write

$$\alpha\triangle AED = \frac{1}{2}AD \cdot EF \quad \text{and} \quad \alpha\triangle DEB = \frac{1}{2}DB \cdot EF.$$

Dividing, we get

$$\frac{\alpha\triangle AED}{\alpha\triangle DEB} = \frac{\frac{1}{2}AD \cdot EF}{\frac{1}{2}DB \cdot EF} = \frac{AD}{DB}. \tag{1}$$

Draw $\overline{DC}$ and draw $\overline{DG} \perp \overline{AC}$. Because $\overline{DG}$ is an altitude of both $\triangle AED$ and $\triangle DEC$, we can write

$$\alpha\triangle AED = \frac{1}{2}AE \cdot DG \quad \text{and} \quad \alpha\triangle DEC = \frac{1}{2}EC \cdot DG.$$

Dividing, we get

$$\frac{\alpha\triangle AED}{\alpha\triangle DEC} = \frac{\frac{1}{2}AE \cdot DG}{\frac{1}{2}EC \cdot DG} = \frac{AE}{EC}. \tag{2}$$

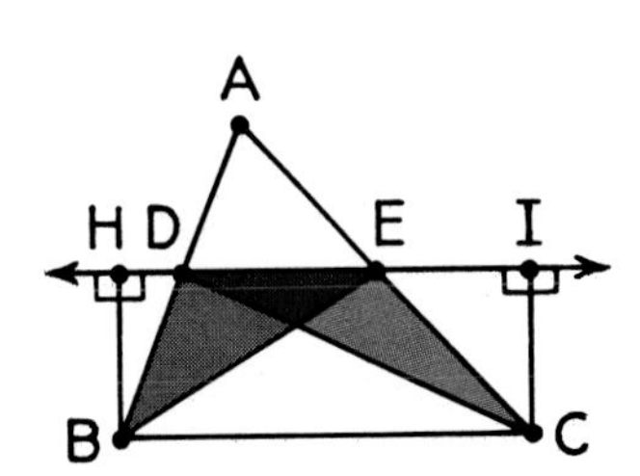

Draw $\overline{BH} \perp \overleftrightarrow{DE}$ and $\overline{CI} \perp \overleftrightarrow{DE}$. We know that $BH = CI$ because $\overleftrightarrow{DE} \parallel \overline{BC}$ and every perpendicular segment joining one of two parallel lines to the other has the same length. Because $\triangle DEB$ and $\triangle DEC$ have equal bases ($DE = DE$) and equal altitudes ($BH = CI$),

$$\alpha\triangle DEB = \alpha\triangle DEC.$$

Substituting this result in Equation 2, we have

$$\frac{\alpha\triangle AED}{\alpha\triangle DEB} = \frac{AE}{EC}. \tag{3}$$

We know from Equation 1 that

$$\frac{\alpha\triangle AED}{\alpha\triangle DEB} = \frac{AD}{DB},$$

and so we can substitute in Equation 3 to get

$$\frac{AD}{DB} = \frac{AE}{EC}.$$

A useful fact that follows directly from this theorem is:

► **Corollary**
If a line parallel to one side of a triangle intersects the other two sides in different points, it cuts off segments proportional to the sides.

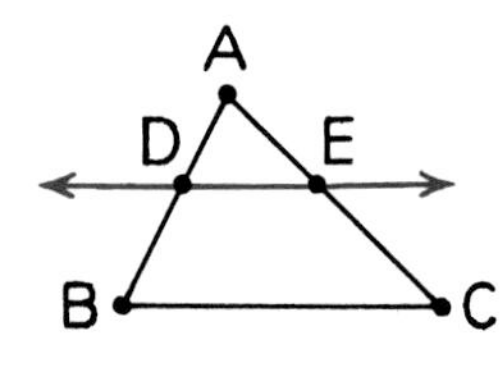

As illustrated in the figure at the left, this corollary permits us to conclude that, if $\overleftrightarrow{DE} \parallel \overline{BC}$, then

$$\frac{AD}{AB} = \frac{AE}{AC} \text{ and } \frac{DB}{AB} = \frac{EC}{AC}.$$

Exercises

Set I

In the figure below, $\overline{TI}$ is a midsegment of $\triangle SWF$.

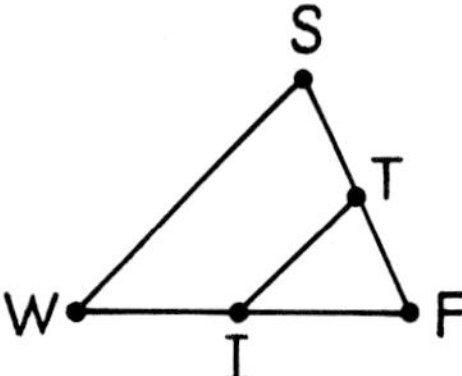

1. Why is $\overline{TI} \parallel \overline{SW}$?
2. Why is $\frac{ST}{TF} = \frac{WI}{IF}$?
3. Why is ST = TF and WI = IF?
4. To what number are $\frac{ST}{TF}$ and $\frac{WI}{IF}$ both equal?

Use the Side-Splitter Theorem and its corollary to complete the proportions for the following figures.
In $\triangle HST$, $\overline{AY} \parallel \overline{ST}$.

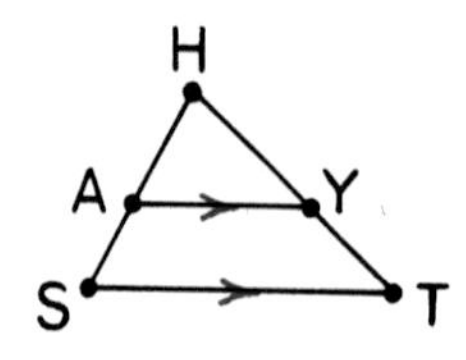

Example: $\frac{HA}{AS} = \text{||||||||||}.$

Answer: $\frac{HA}{AS} = \frac{HY}{YT}.$

5. $\frac{SA}{AH} = \text{||||||||||}.$
6. $\frac{HA}{HS} = \text{||||||||||}.$
7. $\frac{YT}{HT} = \text{||||||||||}.$

In $\triangle RPI$, $\overline{DA} \parallel \overline{IP}$.

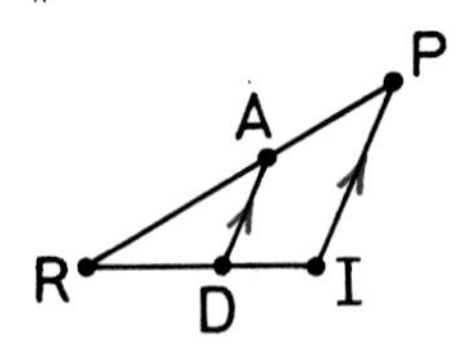

8. $\frac{RA}{AP} = \text{||||||||||}.$
9. $\frac{PA}{PR} = \text{||||||||||}.$
10. $\frac{RD}{RI} = \text{||||||||||}.$

The length marked x in the figure below can be found from several proportions.

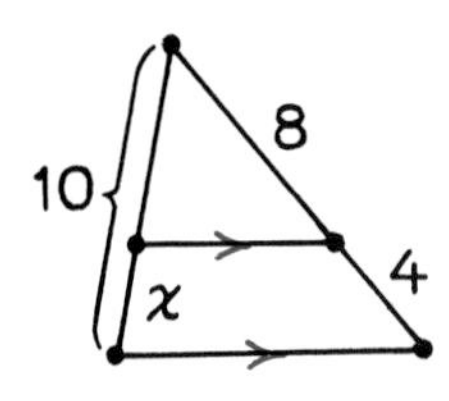

11. Which of the following proportions for finding x are correct?

a) $\dfrac{10 - x}{x} = \dfrac{8}{4}$

b) $\dfrac{x}{10} = \dfrac{4}{8}$

c) $\dfrac{x}{10 - x} = \dfrac{4}{8}$

d) $\dfrac{x}{10} = \dfrac{4}{12}$

12. Use one of the correct proportions to find x.

Find x in each of the following figures.

13.

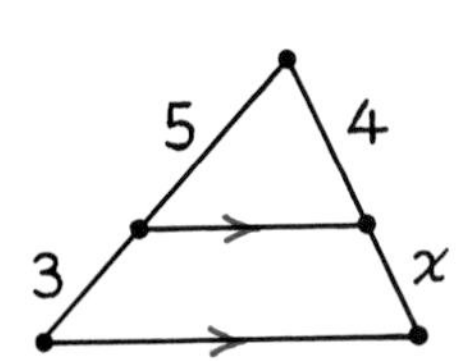

14.

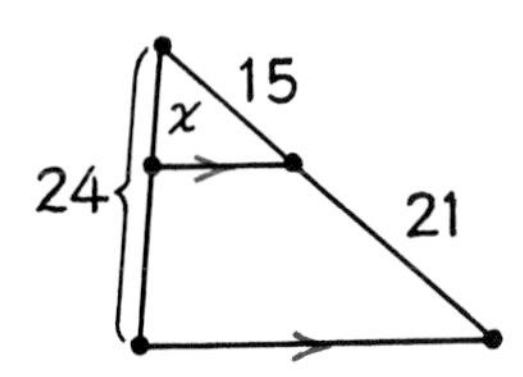

15.

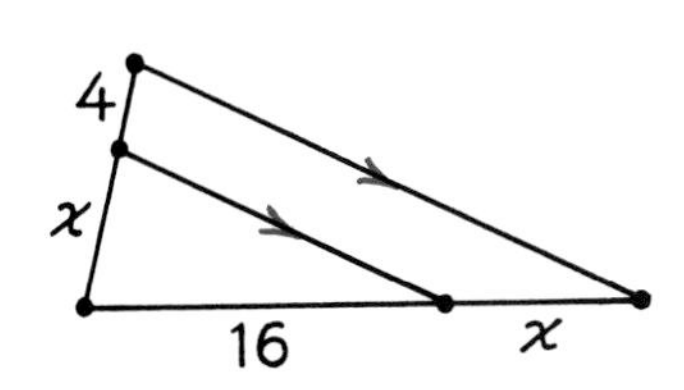

16.

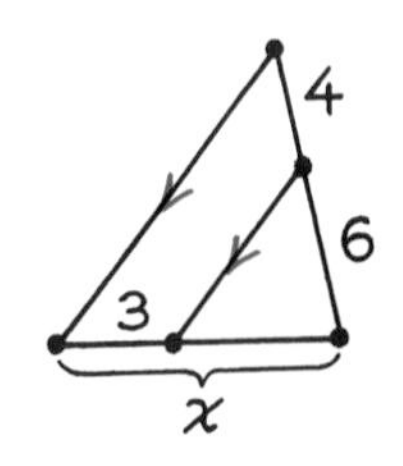

In $\triangle$WNE, $\overline{ID} \parallel \overline{NE}$ and $\overline{IG} \parallel \overline{WE}$.

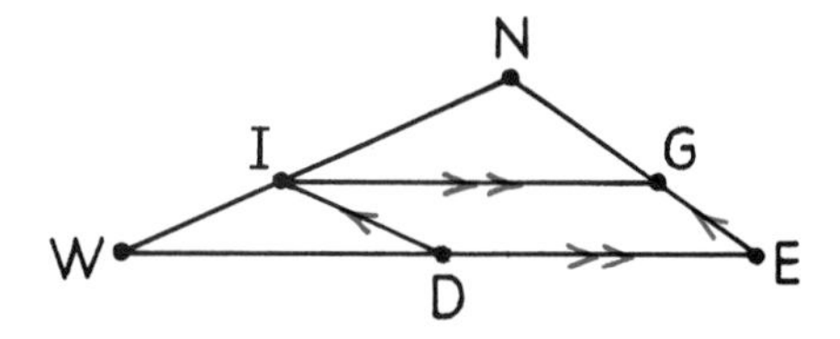

17. Copy the figure and mark the following lengths on it: WI = 12, NG = 12, WD = 18, and DE = 24.

18. Find IN.

19. Find GE.

20. Find IG.

In $\triangle$ARU, $\overline{BT} \parallel \overline{RP}$ and $\overline{BP} \parallel \overline{RU}$.

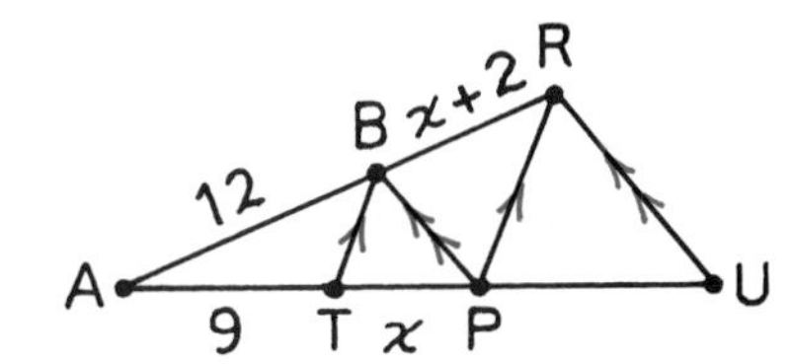

21. Find TP.

22. Find $\dfrac{AB}{BR}$.

23. Find PU.

Set II

Complete the following proof of the corollary to the Side-Splitter Theorem by giving the missing statements and reasons.

If a line parallel to one side of a triangle intersects the other two sides in different points, it cuts off segments proportional to the sides.

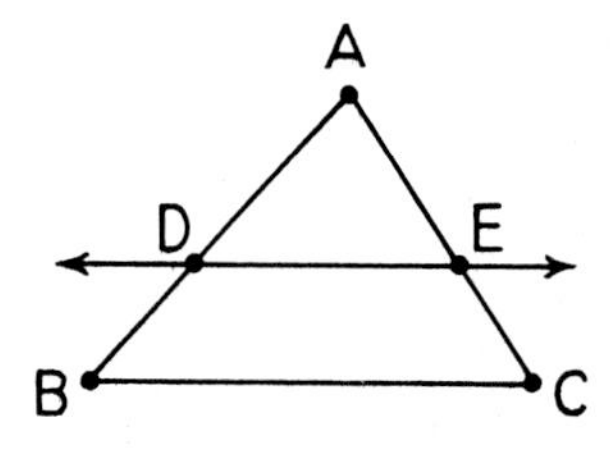

Given: $\triangle ABC$ with $\overleftrightarrow{DE} \parallel \overline{BC}$.

Prove: $\frac{DA}{BA} = \frac{EA}{CA}$.

24. *Proof.*

Statements	***Reasons***
1. $\triangle ABC$ with $\overleftrightarrow{DE} \parallel \overline{BC}$.	Given.
2. $\frac{BD}{DA} = \frac{CE}{EA}$.	a) ▯▯▯▯▯
3. $\frac{BD}{DA} + 1 = \frac{CE}{EA} + 1$.	b) ▯▯▯▯▯
4. $\frac{BD}{DA} + \frac{DA}{DA} = \frac{CE}{EA} + \frac{EA}{EA}$, and so $\frac{BD + DA}{DA} = \frac{CE + EA}{EA}$.	c) ▯▯▯▯▯
5. d) ▯▯▯▯▯ and ▯▯▯▯▯	Betweenness of Points Theorem.
6. e) ▯▯▯▯▯	Substitution.
7. $BA \cdot EA = CA \cdot DA$.	f) ▯▯▯▯▯
8. $\frac{EA}{CA} = \frac{DA}{BA}$.	g) ▯▯▯▯▯
9. h) ▯▯▯▯▯	i) ▯▯▯▯▯

25.

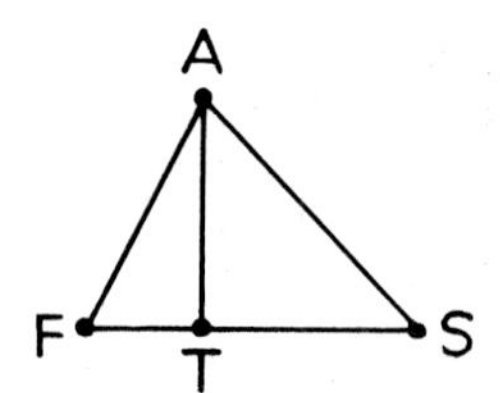

Given: $\triangle FAS$ with $\overline{AT} \perp \overline{FS}$.

Prove: $\frac{\alpha\triangle ATF}{\alpha\triangle ATS} = \frac{FT}{TS}$.

26.

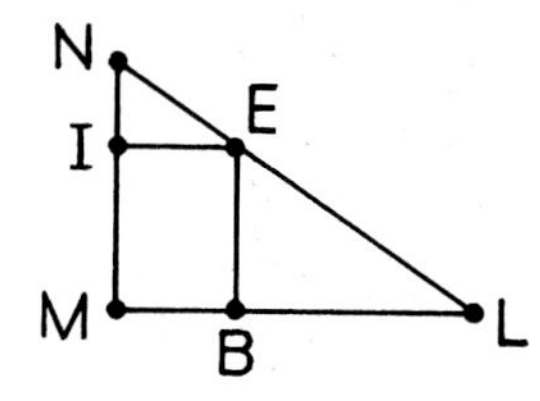

Given: $\triangle NML$ and rectangle IEBM.

Prove: $\frac{NI}{IM} = \frac{MB}{BL}$.

27.

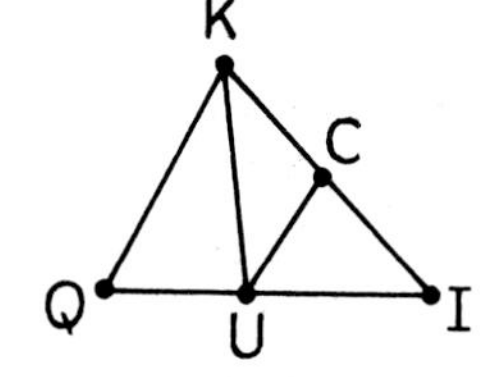

Given: In $\triangle QKI$, $\overrightarrow{KU}$ bisects $\measuredangle QKI$; $KC = CU$.

Prove: $\frac{KC}{KI} = \frac{QU}{QI}$.

28.

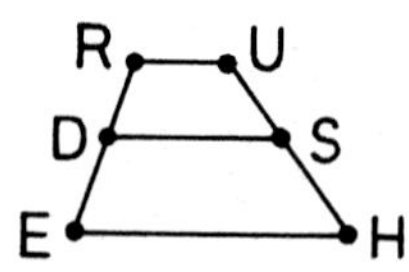

Given: Trapezoid RUHE with bases $\overline{RU}$ and $\overline{EH}$; $\overline{DS} \parallel \overline{EH}$.

Prove: $\frac{RD}{DE} = \frac{US}{SH}$.

Hint: Draw a diagonal of RUHE.

Set III

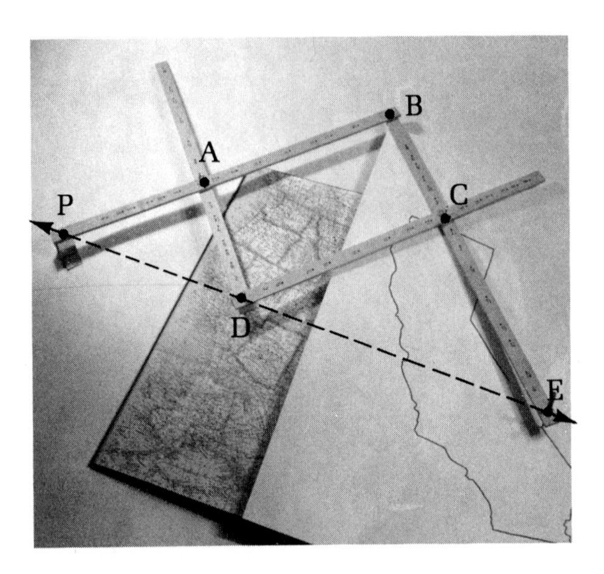

This photograph shows a device called a *pantograph* being used to enlarge a map. Its four bars are hinged together at A, B, C, and D so that AB = DC and AD = BC. As point D moves around the map, point E moves so that it is always in line with D and P. (Point P is attached to the drawing board and does not move.)

As the map is being enlarged, $\triangle PBE$ and quadrilateral ABCD continually change in shape but $\frac{PD}{PE}$ always stays the same.

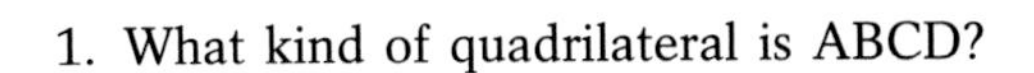

1. What kind of quadrilateral is ABCD?
2. Why?
3. What is always true about $\overline{DC}$ and $\overline{PB}$?
4. Why?

As the map is being enlarged, $\overline{PD}$ and $\overline{PE}$ continually change in length.

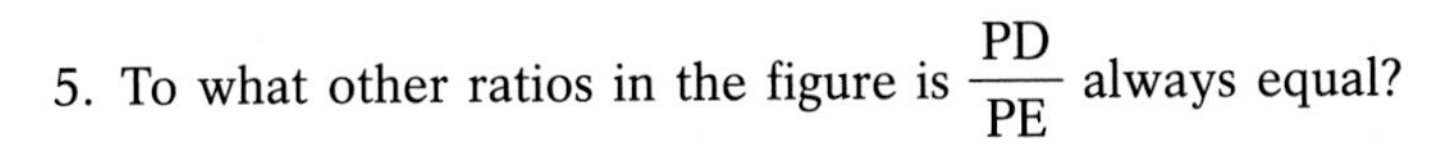

5. To what other ratios in the figure is $\frac{PD}{PE}$ always equal?
6. Why?
7. Why does $\frac{PD}{PE}$ always stay the same?

Lesson 3

Similar Polygons

4"
3"

Suppose that the fellow in this cartoon took in a photograph whose dimensions were 3 inches by 4 inches to be "blown up to poster size." If the store's equipment can make enlargements having dimensions nine times those of the original, the picture could be enlarged to 27 inches by 36 inches.

The picture and its enlargement have the *same shape* and are called *similar*.

► **Definition**
Two polygons are ***similar*** iff there is a correspondence between their vertices such that the corresponding sides of the polygons are proportional and the corresponding angles are equal.

The rectangles representing the picture and its enlargement are similar because their sides are proportional,

$$\frac{3}{27} = \frac{4}{36} = \frac{3}{27} = \frac{4}{36},$$

and their angles, because they are right angles, are equal.

If two polygons are congruent, we can usually write equations for their corresponding parts by looking at the figures and imagining which ones would fit together if they were made to coincide. For similar polygons, however, it is important to base the equations on the correspondence of their vertices. For example, to indicate that the two triangles at the right are similar, we write $\triangle ABC \sim \triangle EFD$. This notation indicates a special correspondence between the vertices of the two triangles, just as the notation used for congruence does. From this correspondence, we can write

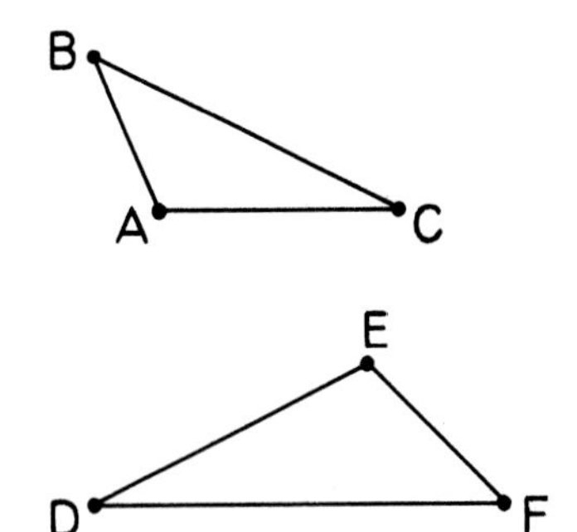

$$\angle A = \angle E, \quad \angle B = \angle F, \quad \angle C = \angle D, \quad \text{and} \quad \frac{AB}{EF} = \frac{BC}{FD} = \frac{AC}{ED}.$$

Exercises

Set I

The two pentagons below are similar.

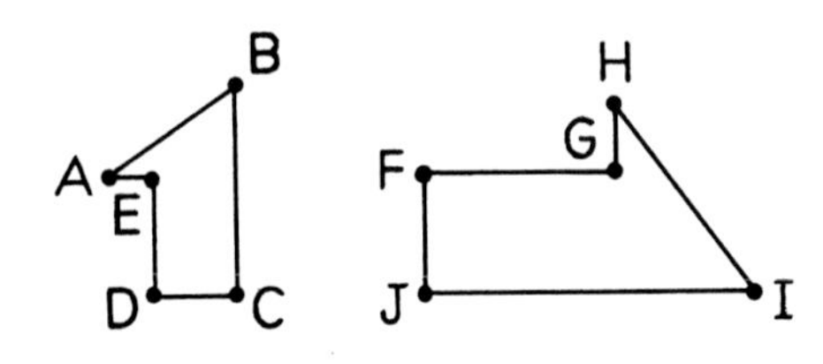

1. If an enlargement of ABCDE were placed on FGHIJ, which of its vertices would A fall on?
2. Which of its vertices would B fall on?
3. Copy and complete the following similarity correspondence between the vertices of the pentagons: ABCDE $\leftrightarrow$ ▮▮▮▮.

Use your correspondence to copy and complete the following statements.

4. $\angle C = \angle$▮▮▮▮.
5. $\frac{AB}{HI} = \frac{DE}{▮▮▮▮}$.
6. $\frac{BC}{▮▮▮▮} = \frac{EA}{▮▮▮▮}$.

Refer to the given correspondences to complete the following statements. If $\triangle BOG \sim \triangle ART$, then

7. $\frac{OG}{RT} = \frac{BG}{▮▮▮▮}$.
8. $\angle O = \angle$▮▮▮▮.

If $\triangle GRA \sim \triangle BLE$, then

9. $\frac{BE}{▮▮▮▮} = \frac{LB}{▮▮▮▮}$.
10. $\angle E = \angle$▮▮▮▮.

In the figure below, $\triangle HAR \sim \triangle LOW$.

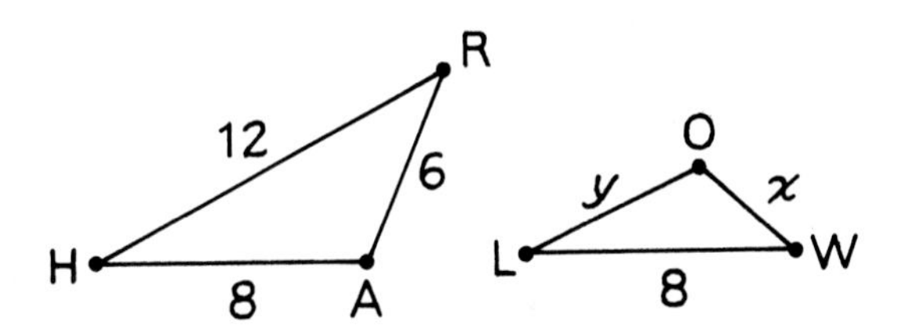

11. Refer to this correspondence to complete the following equation.
$\frac{HA}{▮▮▮▮} = \frac{AR}{▮▮▮▮} = \frac{HR}{▮▮▮▮}$.
12. Rewrite the equation in terms of the six lengths shown on the figure.
13. Solve for x and y.

The two triangles below are equilateral.

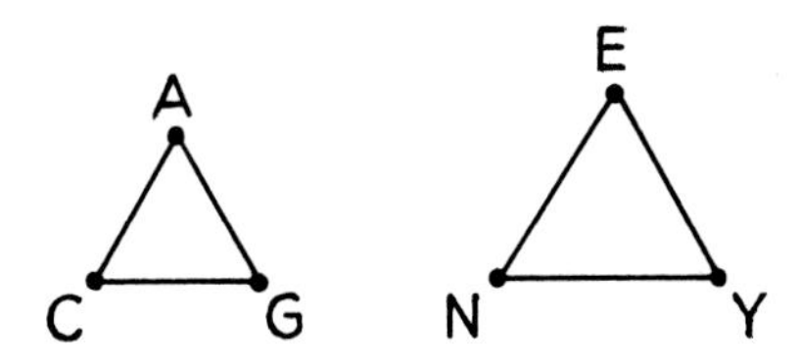

14. Why is CA = AG = GC and NE = EY = YN?

15. Why is $\frac{CA}{NE} = \frac{AG}{EY} = \frac{GC}{YN}$?

16. Why are the corresponding angles of the two triangles equal?

17. Why are the triangles similar?

PICK and FORD are rhombuses in which $\angle P = \angle F$. Use this information to tell why each of the following statements is true.

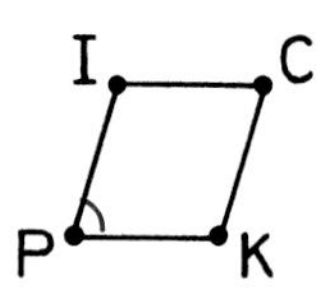

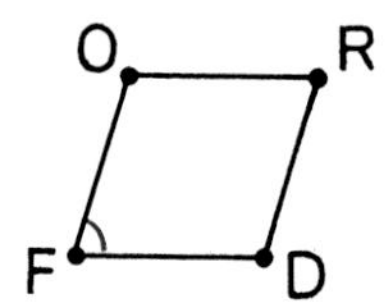

18. PICK and FORD are parallelograms.

19. $\measuredangle I$ and $\measuredangle P$ are supplementary and $\measuredangle O$ and $\measuredangle F$ are supplementary.

20. $\angle I = \angle O$.

21. $\angle C = \angle P$ and $\angle R = \angle F$.

22. $\angle C = \angle R$.

23. PI = IC = CK = KP and FO = OR = RD = DF.

24. $\frac{PI}{FO} = \frac{IC}{OR} = \frac{CK}{RD} = \frac{KP}{DF}$.

25. PICK ~ FORD.

STAN and WYCK are isosceles trapezoids.

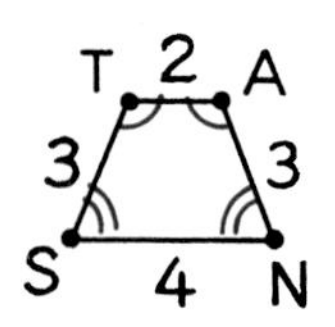

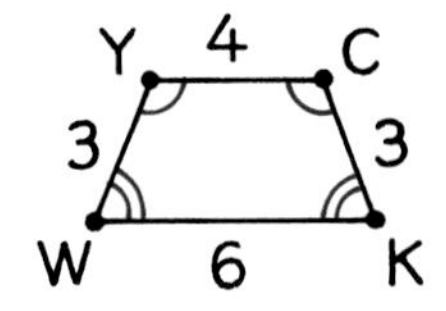

26. Is STAN ~ WYCK?

27. Use the definition of similar polygons to explain why or why not.

RATH and BONE are parallelograms.

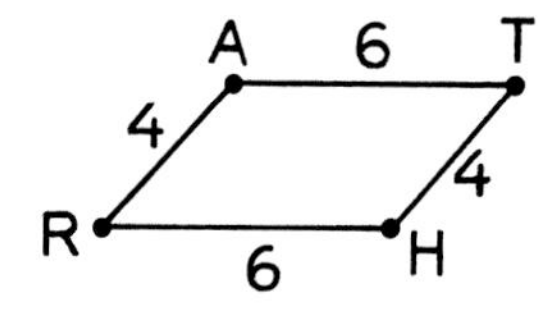

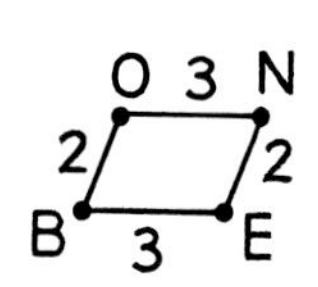

28. Is RATH ~ BONE?

29. Use the definition of similar polygons to explain why or why not.

Set II

30.

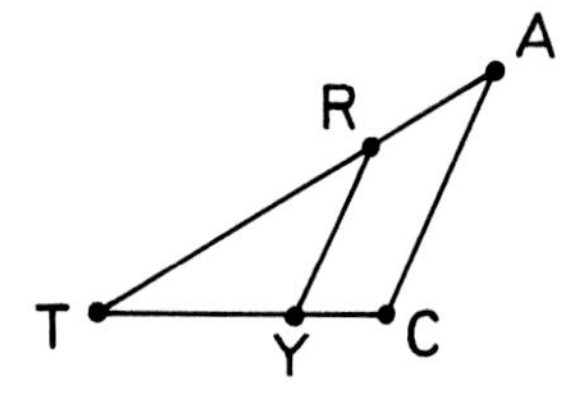

Given: $\triangle TAC \sim \triangle TRY$.
Prove: $\overline{RY} \parallel \overline{AC}$.

31.

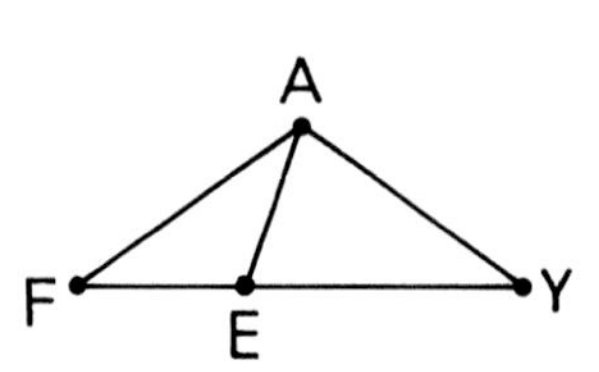

Given: $\triangle AEF \sim \triangle FAY$.
Prove: $\triangle AEF$ is isosceles.

32.

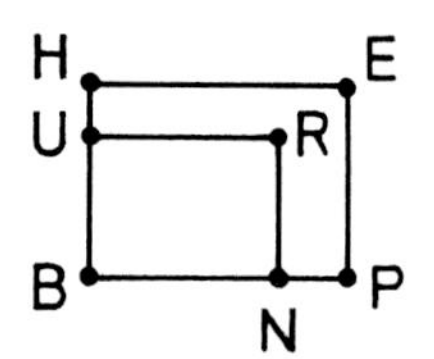

Given: Rectangle HEPB ~ rectangle URNB.
Prove: UB · BP = HB · BN.

33.

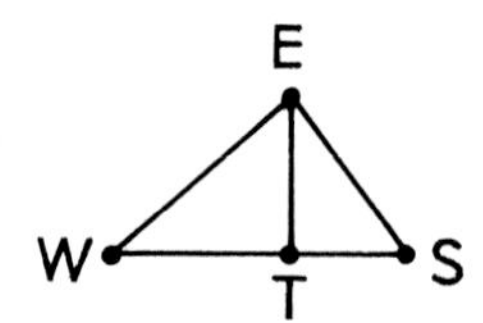

Given: $\triangle WTE \sim \triangle ETS$.
Prove: ET is the geometric mean between WT and TS.

Set III

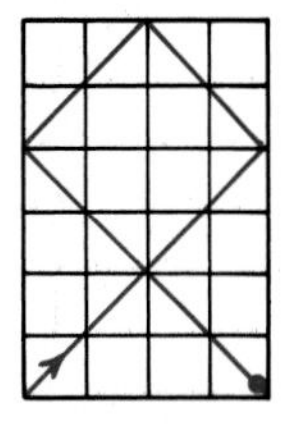

The figure at the right represents a billiard table on which a ball has been hit from the lower left-hand corner so that it travels at a 45° angle from the sides of the table. The path of the ball has been drawn to show that, each time it hits a cushion, the ball rebounds from it at the same angle. If it is hit hard enough, the ball will end up in the lower right-hand corner as shown.

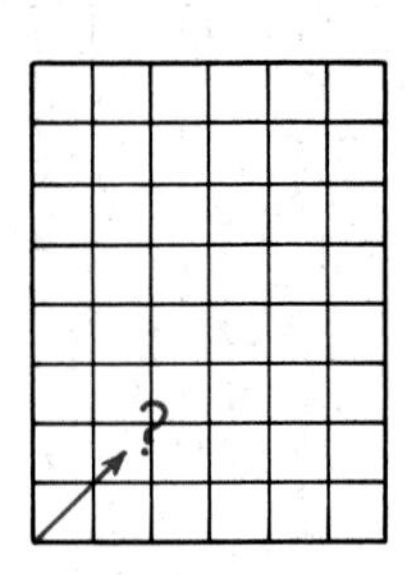

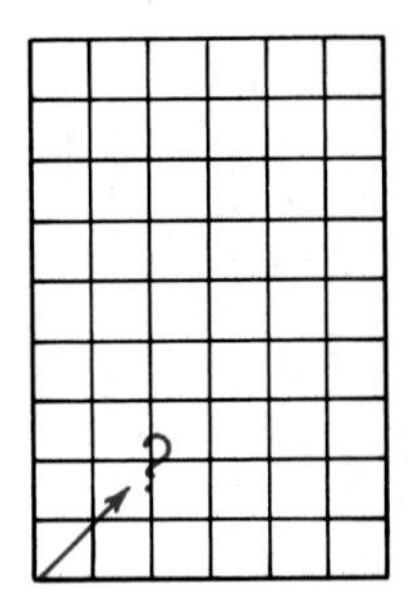

1. Copy the two tables above and draw the path of a ball hit in the same way from the lower left-hand corner of each table. Continue each path *until the ball hits a corner.*
2. What do you notice?
3. Explain.

Photograph by R. Kauffman in *Gentle Wilderness*, Sierra Club, © 1967

Lesson 4

The A.A. Similarity Theorem

The tallest trees in the world are the redwoods along the coast of northern California and southern Oregon. One way to measure one of these giants is to move some distance from the base of the tree and place a mirror faceup on the ground so that it is level. Then move still further away until the top of the tree can be seen reflected in the mirror.

If you know the height of your eye level above the ground, all you have to do is to measure your distance to the mirror and the mirror's distance to the tree in order to find the height of the tree.

To understand how the method works, we need the following theorem.

► **Theorem 56** (The A.A. Similarity Theorem)

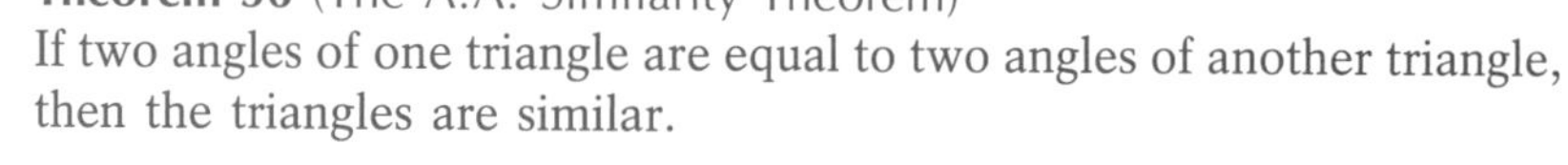

If two angles of one triangle are equal to two angles of another triangle, then the triangles are similar.

Given: $\triangle ABC$ and $\triangle DEF$ with $\angle A = \angle D$ and $\angle B = \angle E$.

Prove: $\triangle ABC \sim \triangle DEF$.

To prove this theorem, we must show that the third pair of angles in the triangles are equal and that all three pairs of corresponding sides are proportional. The angles are easy, and so most of the proof will deal with the sides. Our method will be to copy the smaller triangle in one corner of the larger one to form the figure shown at the top of the next page. We can then prove that we have a triangle with a line segment parallel to one side; this permits us to apply the corollary to the Side-Splitter Theorem to show that corresponding sides of the triangles are proportional.

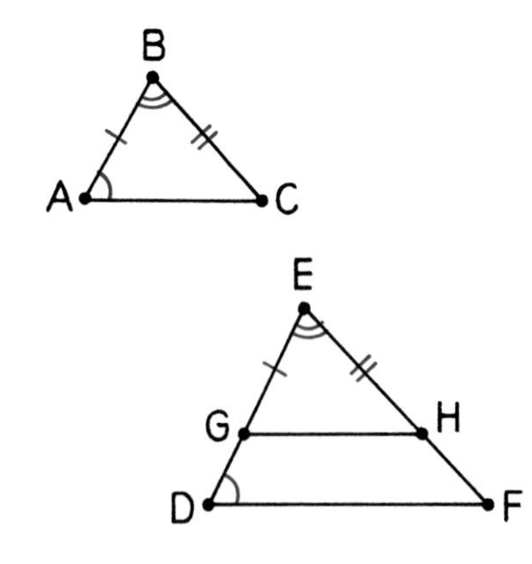

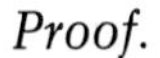
Proof.
Choose G and H on $\overline{ED}$ and $\overline{EF}$ so that EG = BA and EH = BC. Draw $\overline{GH}$. Because $\angle B = \angle E$, $\triangle GEH \cong \triangle ABC$ (S.A.S.). Therefore, $\angle EGH = \angle A$. Because $\angle A = \angle D$, $\angle EGH = \angle D$. So $\overline{GH} \parallel \overline{DF}$.

From the fact that a line parallel to one side of a triangle cuts off segments on the other two sides that are proportional to them, we have

$$\frac{EG}{ED} = \frac{EH}{EF}.$$

Because EG = BA and EH = BC,

$$\frac{BA}{ED} = \frac{BC}{EF}$$

by substitution.

By copying $\triangle ABC$ a second time as shown in the figure at the right, and by using the same reasoning, we can show that

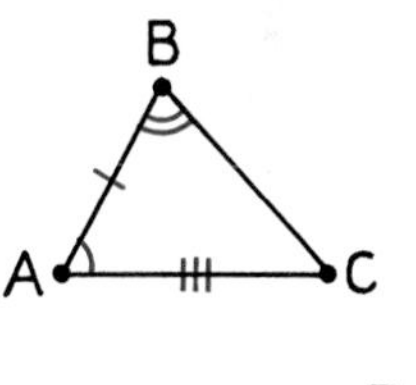

$$\frac{BA}{ED} = \frac{AC}{DF}.$$

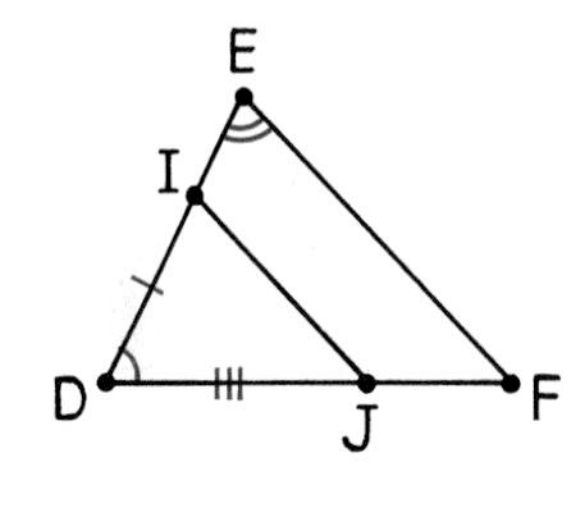

Hence, $\frac{BA}{ED} = \frac{BC}{EF} = \frac{AC}{DF}$, and so we have shown that all three pairs of corresponding sides of the triangles are proportional.

Because all three pairs of corresponding angles in the triangles are equal ($\angle C = \angle F$ because, if two angles of one triangle are equal to two angles of another, the third pair of angles are equal), the triangles are similar by definition.

Although we could have proved the following theorem directly from the definition of similar triangles, the A.A. Similarity Theorem makes its proof especially easy.

► **Corollary**
Two triangles similar to a third triangle are similar to each other.

Given: $\triangle ABC \sim \triangle GHI$ and $\triangle DEF \sim \triangle GHI$.

Prove: $\triangle ABC \sim \triangle DEF$.

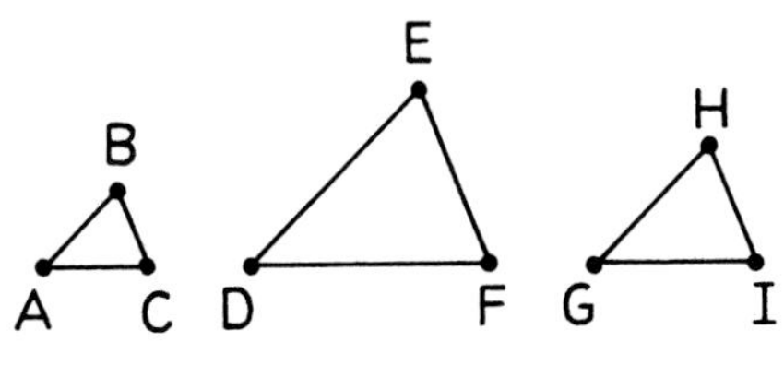

Proof.
Because $\triangle ABC \sim \triangle GHI$, $\angle A = \angle G$ and $\angle B = \angle H$, and because $\triangle DEF \sim \triangle GHI$, $\angle D = \angle G$ and $\angle E = \angle H$. Therefore, $\angle A = \angle D$ and $\angle B = \angle E$ (substitution). It follows from the A.A. Similarity Theorem that $\triangle ABC \sim \triangle DEF$.

Exercises

Set I

The figure below illustrates the method mentioned at the beginning of this lesson for finding the height of a tree. Point M represents the position of the mirror so that the

light from the top of the tree travels along $\overline{AM}$ and $\overline{MC}$ before reaching the measurer's eyes.

It is a law of physics that, when a beam of light strikes a mirror, the angles of incidence and reflection are equal, and so $\angle AMB = \angle CMD$.

1. Why is $\angle B = \angle D$?
2. Why is $\triangle AMB \sim \triangle CMD$?
3. Copy and complete the following proportion for the sides of these triangles:

$$\frac{AB}{\text{|||||||||}} = \frac{MB}{\text{|||||||||}}.$$

4. Suppose that CD = 6 ft, MB = 115 ft, and MD = 2 ft. Use the proportion that you wrote for Exercise 3 to find the height of the tree, AB.

Another way to measure the height of a tree uses its shadow. The measurer walks away from the tree along its shadow until her head is in line with the top of the tree and the tip of the shadow.

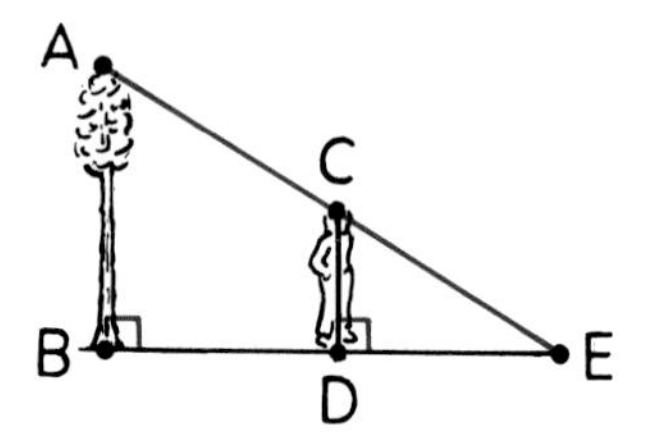

5. Why is $\triangle ABE \sim \triangle CDE$?
6. Copy and complete the following proportion:

$$\frac{AB}{\text{|||||||||}} = \frac{BE}{\text{|||||||||}}.$$

7. Suppose that CD = 6 ft and BD = 20 ft. What other distance would you need to know in order to determine the tree's height?
8. If this distance is 10 ft, how tall is the tree?

To find the height of a bridge between two buildings, the measurer can stand at one end and look down to the ground at the other end.

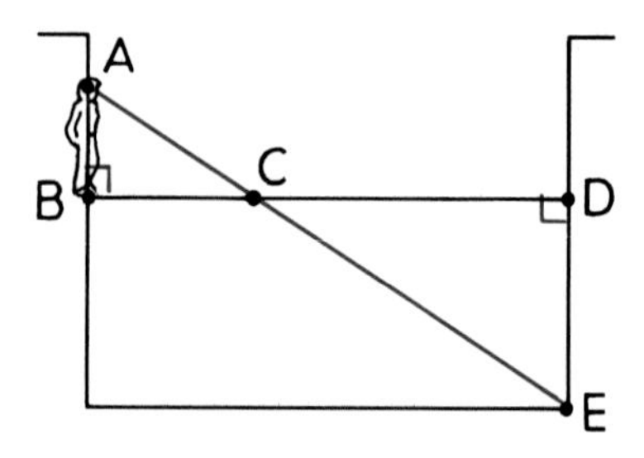

9. In the diagram, why is $\triangle ABC \sim \triangle EDC$?

10. Copy and complete the following proportion:

$$\frac{\text{|||||||||}}{BA} = \frac{\text{|||||||||}}{BC}.$$

11. If BC = 10 ft, CD = 25 ft, and BA = 6 ft, find DE.

In the figure below, $\triangle CAL \sim \triangle LIP$ and $\triangle LIP \sim \triangle ERS$.

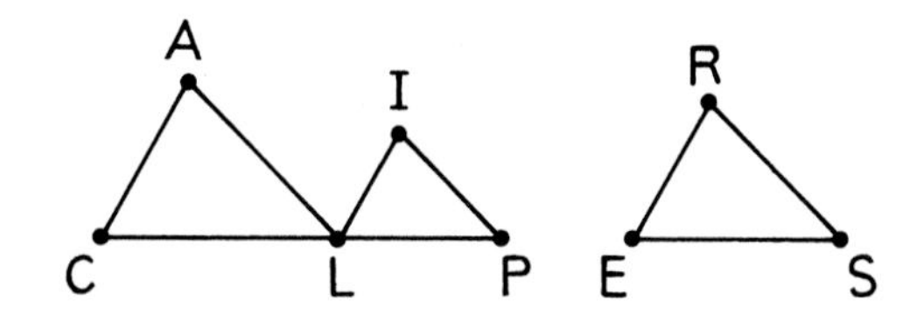

12. What can you conclude from this?

13. Why?

In the figure below, $\overline{EC}$ and $\overline{NR}$ are altitudes of $\triangle WEN$. Use this information to tell whether each of the following statements *must be true, may be true,* or *appears to be false.*

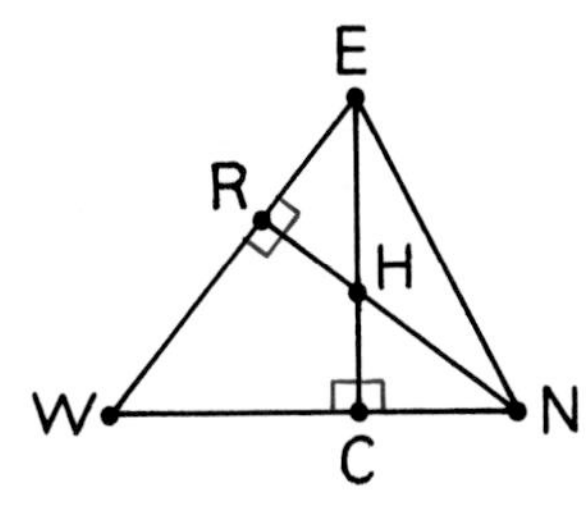

14. $\triangle WCE \sim \triangle WRN$.

15. $\triangle WCE \cong \triangle WRN$.

16. $\triangle RHE \sim \triangle CHN$.

17. $\triangle REN \sim \triangle CNE$.

18. $\triangle ECW \sim \triangle ECN$.

In right $\triangle PNC$, $\overline{UH} \perp \overline{PC}$. Find each of the following numbers.

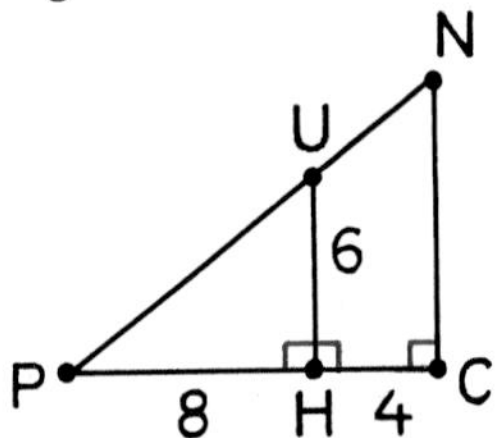

19. PU.

20. UN.

21. NC.

22. $\rho\triangle PHU$.

23. $\rho\triangle PCN$.

24. $\alpha\triangle PHU$.

25. $\alpha\triangle PCN$.

Find each of the following numbers as a common fraction in lowest terms.

26. $\dfrac{PH}{PC}$.

27. $\dfrac{\rho\triangle PHU}{\rho\triangle PCN}$.

28. $\dfrac{\alpha\triangle PHU}{\alpha\triangle PCN}$.

Set II

29.

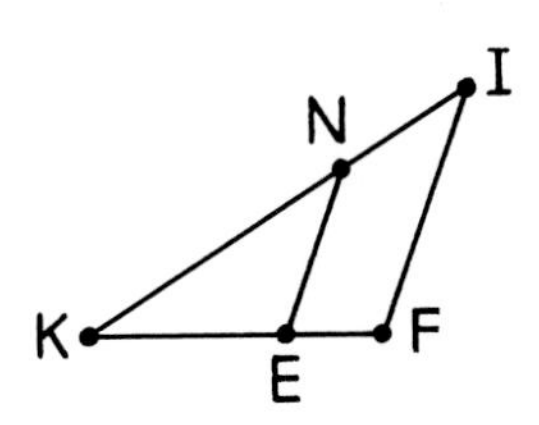

Given: $\triangle KIF$ with $\overline{NE} \parallel \overline{IF}$.
Prove: $\triangle KNE \sim \triangle KIF$.

30.

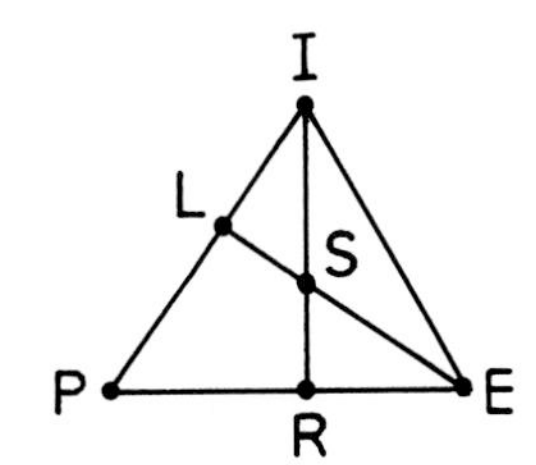

Given: $\triangle PIE$ with altitudes $\overline{IR}$ and $\overline{EL}$.
Prove: $\triangle ILS \sim \triangle ERS$.

31.

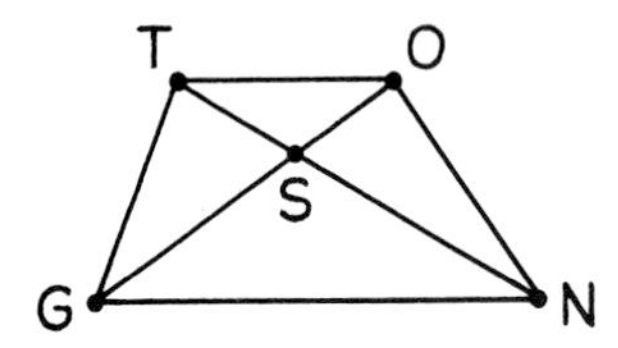

Given: Trapezoid TONG with bases $\overline{TO}$ and $\overline{GN}$ and diagonals $\overline{TN}$ and $\overline{OG}$.

Prove: $\frac{TS}{NS} = \frac{OS}{GS}$.

32.

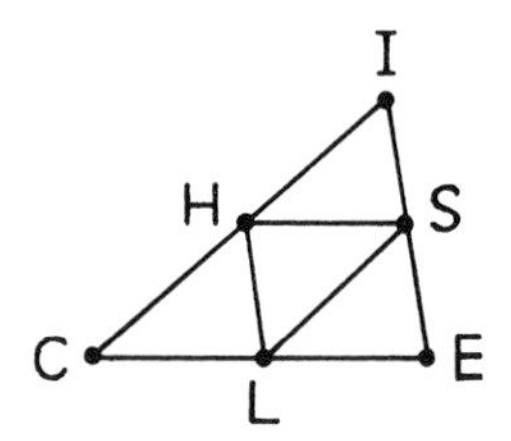

Given: $\triangle$ICE with midsegments $\overline{HS}$, $\overline{HL}$, and $\overline{LS}$.

Prove: $\triangle LSH \sim \triangle ICE$.

Set III

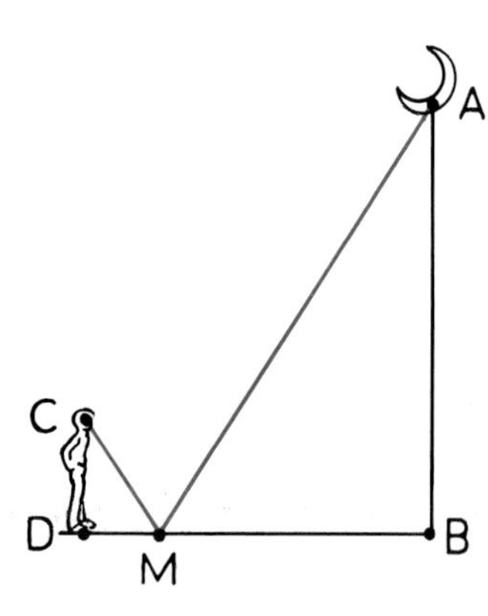

Dilcue has decided to try to measure the distance to the moon by the mirror method described in this lesson. He plans to mark the point on the ground at which the moon is directly overhead, place the mirror at a convenient distance from this point, and then move away from the mirror until he can see the moon's reflection in it.

Do you think the method will work? Explain why or why not. (This painting by the Belgian artist René Magritte contains a hint.)

Courtesy of L. Arnold Weissberger

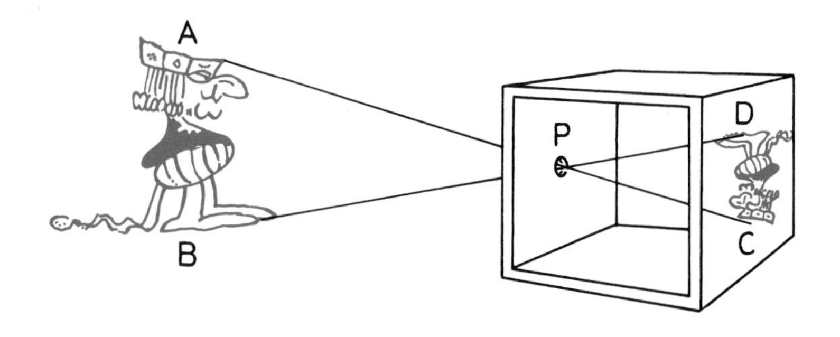

Lesson 5

The S.A.S. Similarity Theorem

Did you know that a camera capable of taking recognizable pictures can be made from an ordinary box simply by poking a pinhole in the center of one wall? The diagram above shows such a "pinhole camera" and its interior.

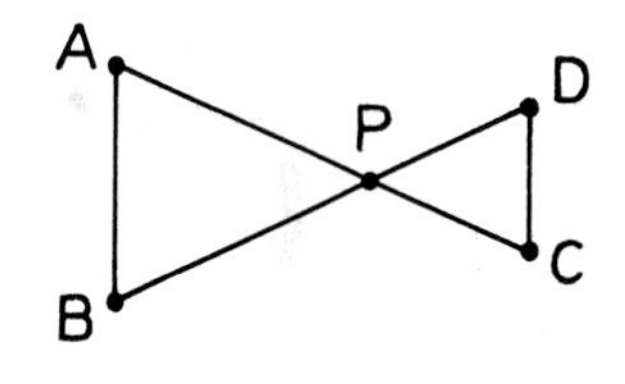

Notice that the camera produces an image that is upside-down with respect to the object being photographed. The size of the image depends on the distance of the object from the pinhole. In the diagram at the right, $\overline{AB}$ represents the object, $\overline{CD}$ represents the image, and P represents the pinhole.

The two triangles in the diagram appear to be similar. Because vertical angles are equal, we know that $\angle APB = \angle CPD$. If another pair of angles are equal, we can conclude that the triangles are similar by means of the A.A. Similarity Theorem. If instead we know that the sides including $\measuredangle APB$ and $\measuredangle CPD$ are proportional, the triangles can be proved similar by the S.A.S. Similarity Theorem.

► **Theorem 57** (The S.A.S. Similarity Theorem)
If an angle of one triangle is equal to an angle of another triangle and the sides including these angles are proportional, then the triangles are similar.

Given: $\triangle ABC$ and $\triangle DEF$ with $\angle B = \angle E$

and $\frac{BA}{ED} = \frac{BC}{EF}$.

Prove: $\triangle ABC \sim \triangle DEF$.

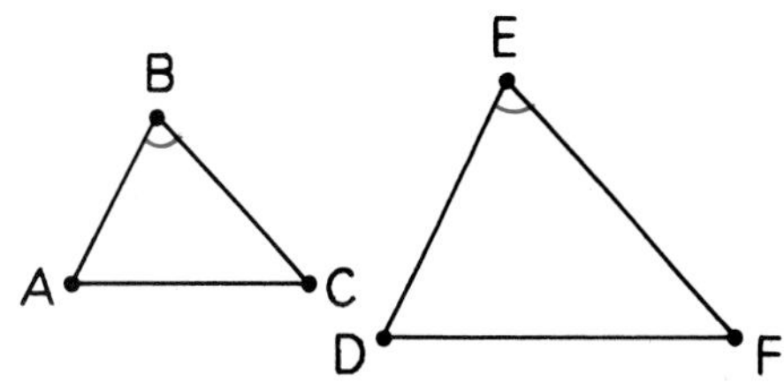

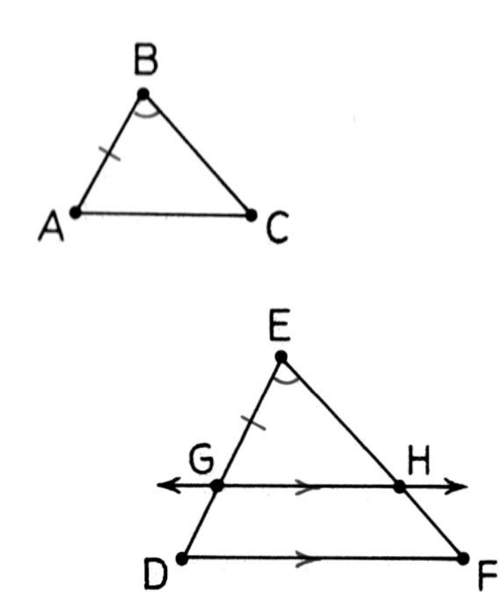

To prove this theorem, we will show that a second pair of angles of the triangles are equal and then apply the A.A. Similarity Theorem.

Proof.
Choose G on $\overline{ED}$ so that EG = BA. Through G, draw $\overleftrightarrow{GH} \parallel \overline{DF}$ (through a point not on a line, there is exactly one parallel to the line).

From the fact that a line parallel to one side of a triangle cuts off segments on the other two sides that are proportional to them, we have

$$\frac{EG}{ED} = \frac{EH}{EF}.$$

Because EG = BA,

$$\frac{BA}{ED} = \frac{EH}{EF}$$

by substitution.

Also, $\frac{BA}{ED} = \frac{BC}{EF}$ (given), and so $\frac{BC}{EF} = \frac{EH}{EF}$ (substitution). Multiplying both sides of this equation by EF, we get BC = EH. This establishes that $\triangle ABC \cong \triangle GEH$ (S.A.S.), and so $\angle A = \angle EGH$. Because it is also true that $\angle EGH = \angle D$ (if two parallel lines are cut by a transversal, the corresponding angles are equal), it follows that $\angle A = \angle D$ (transitive).

We have shown that a second pair of angles of the triangles are equal, and so $\triangle ABC \sim \triangle DEF$ by the A.A. Similarity Theorem.

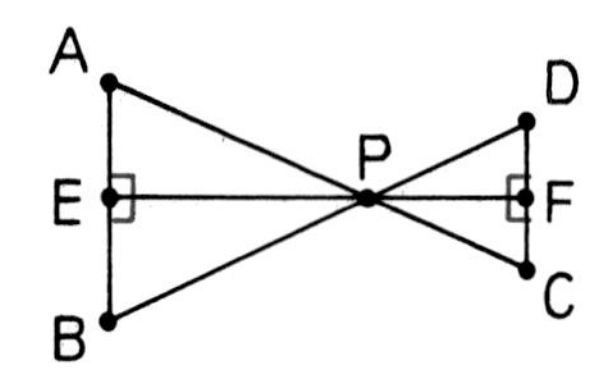

In the diagram at the left, the formation of an image in a pinhole camera is again represented. Altitudes have been added to the two triangles: their lengths represent the distance of the object from the pinhole (PE) and the distance of the image from the pinhole (PF). If $\triangle APB \sim \triangle CPD$, we can prove that these altitudes, called *corresponding altitudes* because they are drawn from corresponding vertices of the triangles, have the same ratio as a pair of corresponding sides:

$$\frac{PE}{PF} = \frac{AB}{CD}.$$

► **Theorem 58**
Corresponding altitudes of similar triangles have the same ratio as the corresponding sides.

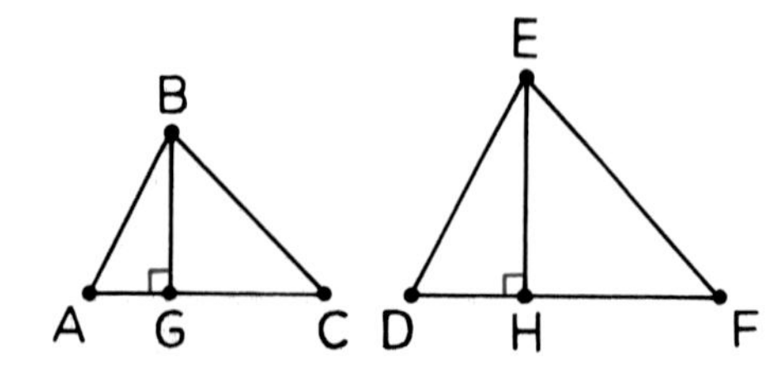

Given: $\overline{BG}$ and $\overline{EH}$ are altitudes of $\triangle ABC$ and $\triangle DEF$, respectively; $\triangle ABC \sim \triangle DEF$.

Prove: $\frac{BG}{EH} = \frac{AB}{DE} = \frac{BC}{EF} = \frac{AC}{DF}.$

*Proof.**
Because $\overline{BG}$ and $\overline{EH}$ are altitudes of $\triangle ABC$ and $\triangle DEF$, $\overline{BG} \perp \overline{AC}$ and $\overline{EH} \perp \overline{DF}$. It follows that ∡AGB and ∡DHE are right angles, and so $\angle AGB = \angle DHE$.

Also, because $\triangle ABC \sim \triangle DEF$, we know that $\angle A = \angle D$ (corresponding angles of similar triangles are equal). Hence $\triangle ABG \sim \triangle DEH$ (A.A.), and so

$$\frac{BG}{EH} = \frac{AB}{DE}.$$

Also, because $\triangle ABC \sim \triangle DEF$, $\frac{AB}{DE} = \frac{BC}{EF} = \frac{AC}{DF}$. It follows from the transitive property that

$$\frac{BG}{EH} = \frac{AB}{DE} = \frac{BC}{EF} = \frac{AC}{DF}.$$

*This proof is based on the assumption that the altitudes are inside the triangles. A similar proof can be written for the case in which they are not.

Exercises

Set I

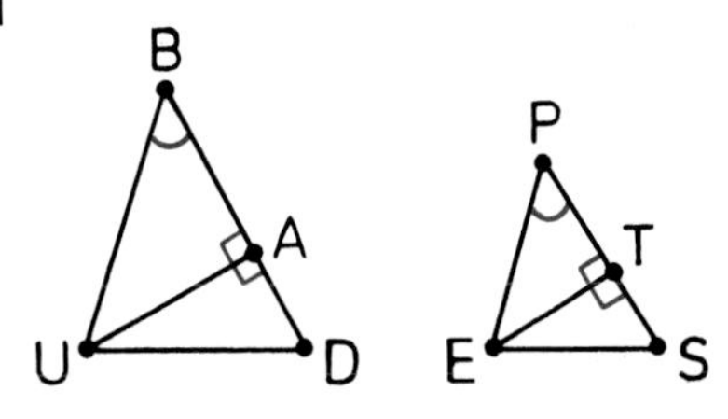

In $\triangle BUD$ and $\triangle PES$, $\angle B = \angle P$, $\frac{BU}{PE} = \frac{BD}{PS}$, $\overline{UA} \perp \overline{BD}$, and $\overline{ET} \perp \overline{PS}$.

1. Why is $\triangle BUD \sim \triangle PES$?
2. Why is $\angle D = \angle S$?
3. Why is $\frac{UA}{ET} = \frac{BD}{PS}$?

The following exercises are about $\triangle NAP$ and $\triangle LES$.

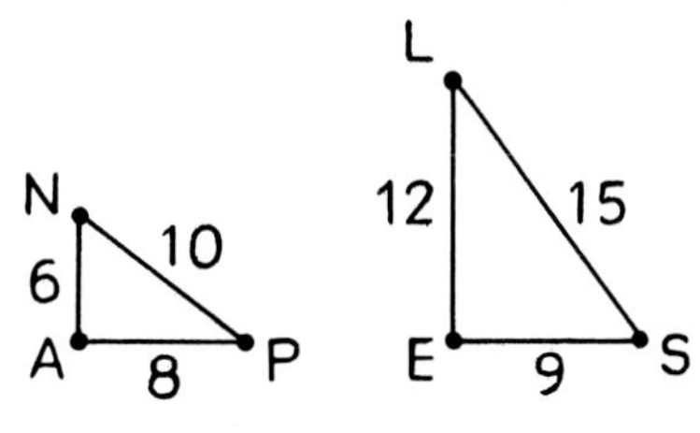

4. Show why ∡A and ∡E are right angles.
5. Are the triangles similar?
6. Explain why or why not.

The following exercises are about $\triangle DUB$ and $\triangle LIN$.

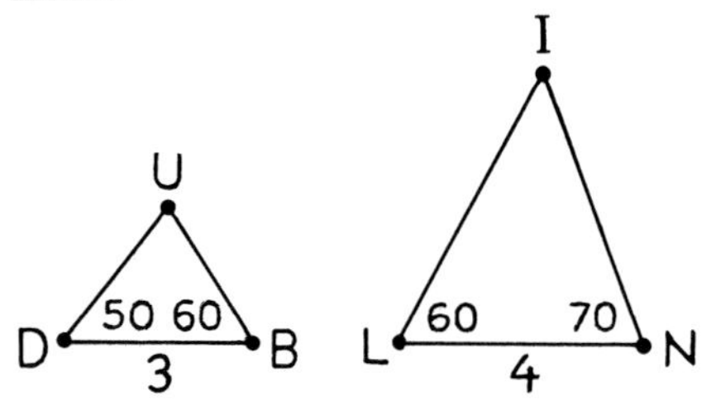

7. Find $\angle U$.

8. Are the triangles similar?

9. Explain why or why not.

In $\triangle LBN$, $\overline{NI} \perp \overline{LB}$, $\overline{OS} \perp \overline{LB}$, and $\overline{NL} \parallel \overline{OI}$.

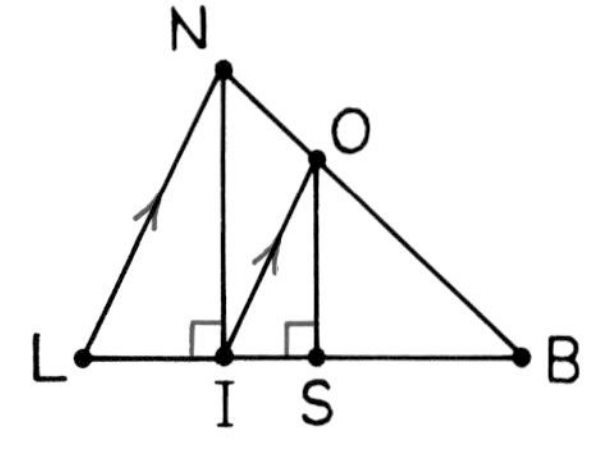

10. Why is $\angle L = \angle OIB$?

11. Why is $\triangle LBN \sim \triangle IBO$?

12. Why is $\dfrac{NI}{OS} = \dfrac{LB}{IB}$?

Solve for x in each of these figures.

13.

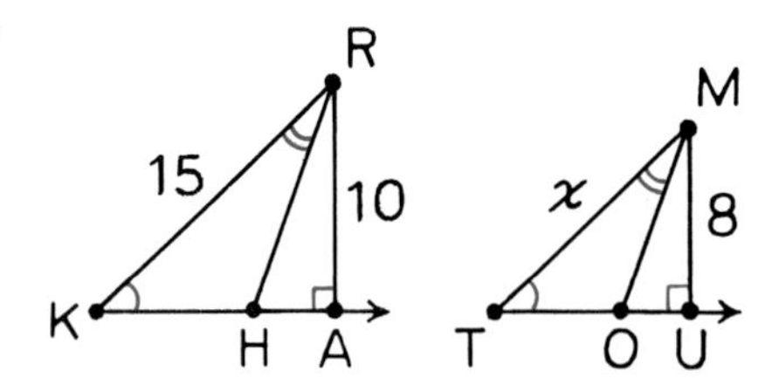

14.

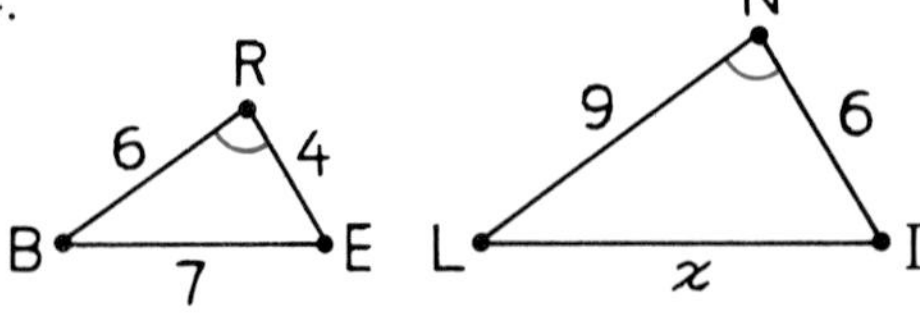

15.

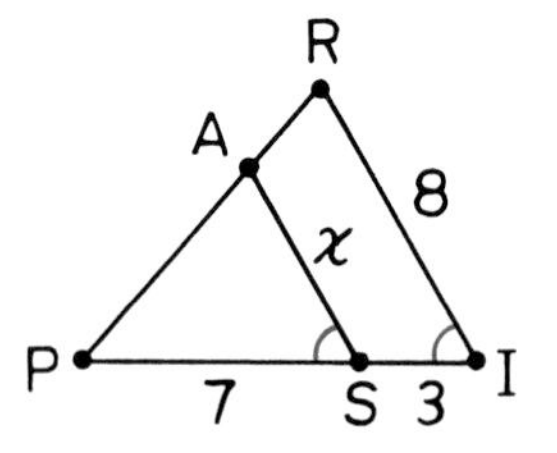

16.

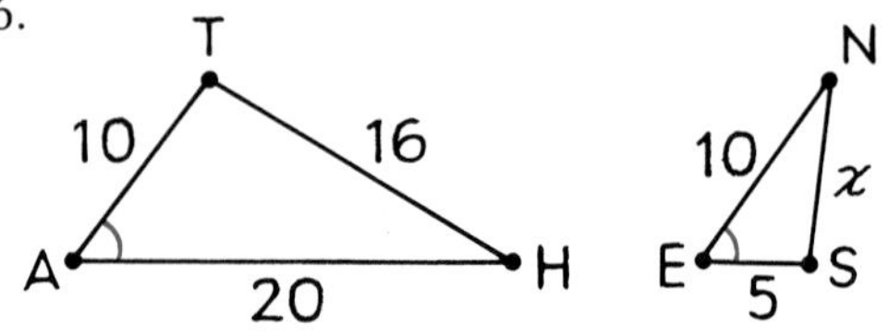

In the figure below, $\triangle IST \sim \triangle NBU$, $\overline{SA} \perp \overline{IT}$ and $\overline{BL} \perp \overline{NU}$. Find each of the following numbers.

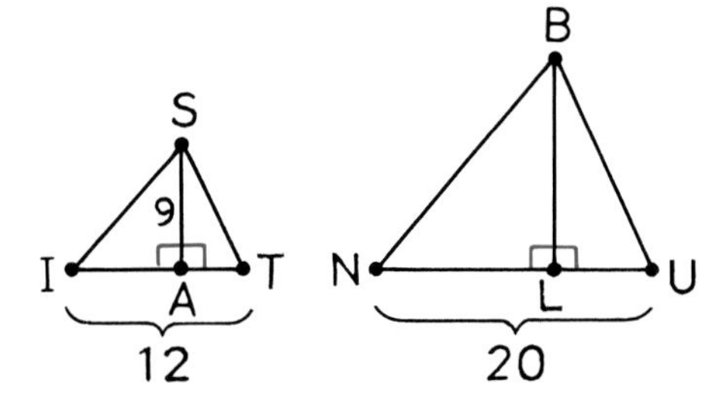

17. BL.

18. $\alpha\triangle IST$.

19. $\alpha\triangle NBU$.

Find each of the following ratios as a common fraction in lowest terms.

20. $\dfrac{IS}{NB}$.

21. $\dfrac{\alpha\triangle IST}{\alpha\triangle NBU}$.

Set II

22.

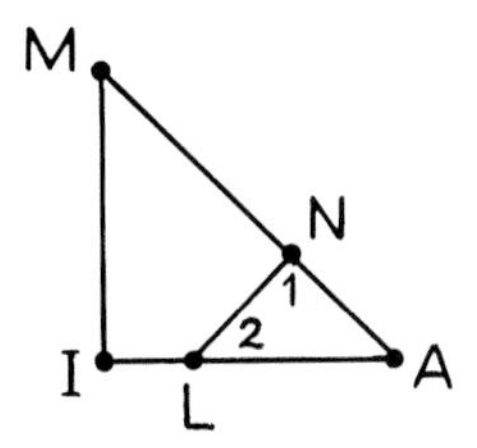

Given: $\triangle MIA$ with $MI = IA$ and $LN = NA$; $\angle I = \angle 1$.
Prove: $\triangle MIA \sim \triangle LNA$ by using the S.A.S. Similarity Theorem.

23. Using the information given in Exercise 22, prove $\triangle MIA \sim \triangle LNA$ by using the A.A. Similarity Theorem.

24.

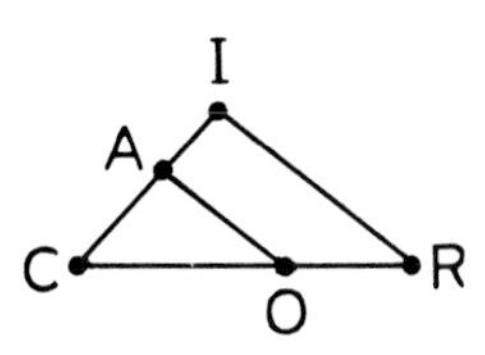

Given: $\frac{CA}{CI} = \frac{CO}{CR}$.
Prove: $\overline{AO} \parallel \overline{IR}$.

25.

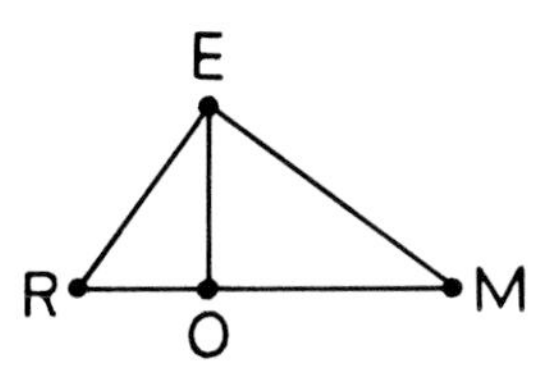

Given: RE is the geometric mean between RO and RM.
Prove: $\triangle ROE \sim \triangle REM$.

26.

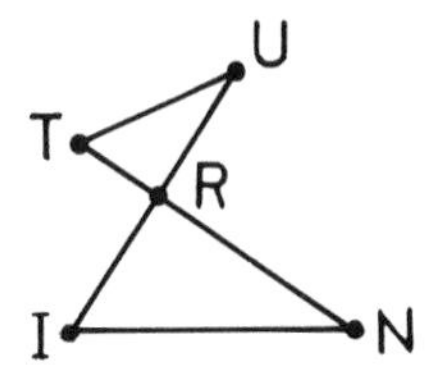

Given: $\measuredangle TRU$ and $\measuredangle IRN$ are vertical angles; $\frac{RT}{RI} = \frac{RU}{RN}$.
Prove: $\angle T = \angle I$.

27.

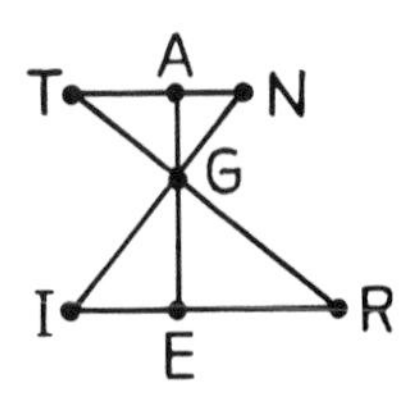

Given: $\triangle TGN$ and $\triangle IGR$ with $\overline{AE} \perp \overline{TN}$ and $\overline{AE} \perp \overline{IR}$.
Prove: $\frac{GA}{GE} = \frac{TN}{IR}$.

Set III

The following experiment will enable you to see an image made by a pinhole camera. All that you need is a stiff card (a file card is convenient), a candle, and a dark room.

Poke a hole through the center of the card with your compass point. The hole should have a diameter of about $\frac{1}{16}$ inch. Light the candle and hold it about 6 inches from a wall. Then hold the card between the candle and the wall so that the hole, flame, and wall are in line. You should observe an upside-down image of the flame on the wall.

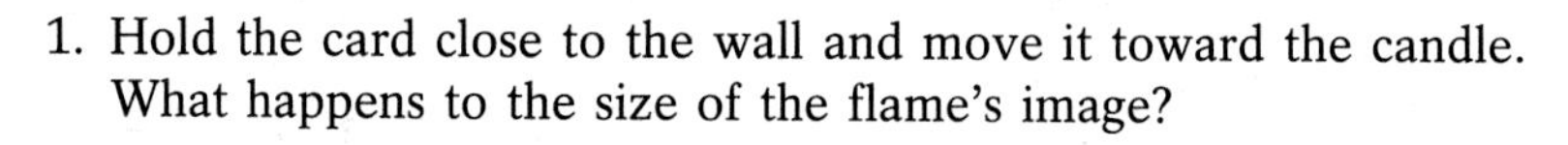

1. Hold the card close to the wall and move it toward the candle. What happens to the size of the flame's image?

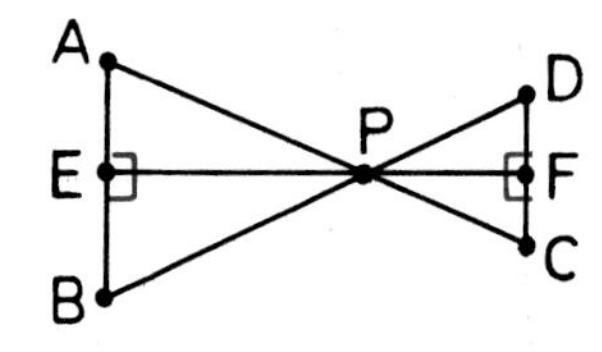

In the figure at the left, $\overline{AB}$ represents the candle flame, P represents the pinhole, and $\overline{CD}$ represents the flame's image on the wall.

2. Assuming that $\triangle ABP \sim \triangle CDP$, derive an equation for CD in terms of AB, PE, and PF.

3. When the card is moved from the wall toward the candle, what happens to PF?

4. What happens to PE?

5. What happens to $\frac{PF}{PE}$?

6. What happens to CD?

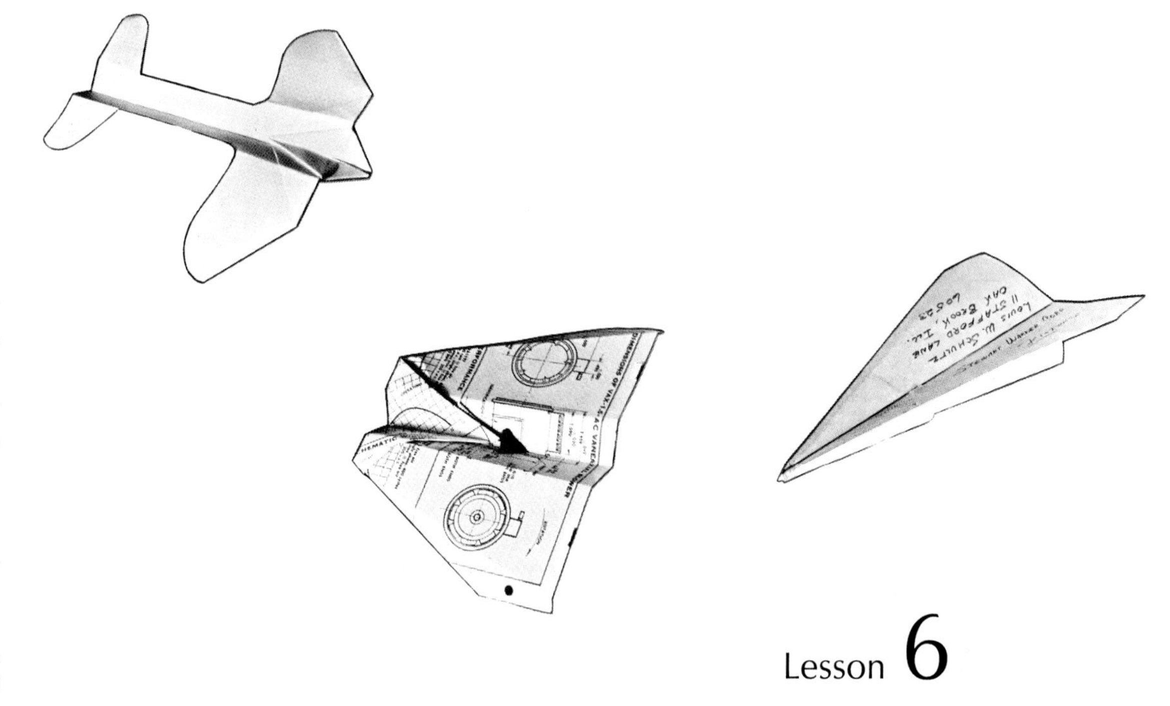

Lesson 6
The Angle Bisector Theorem

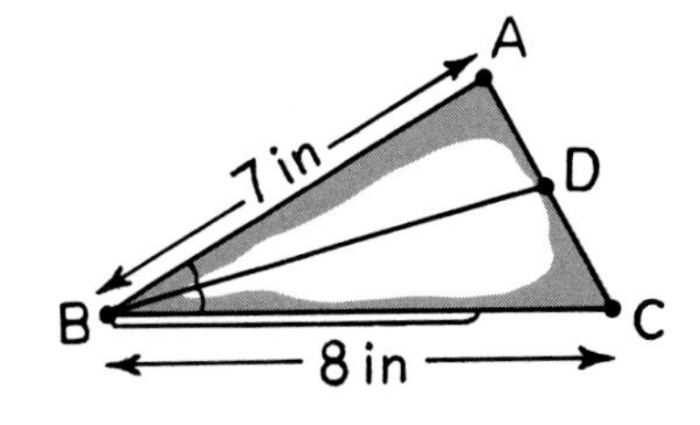

Three of the winners among more than eleven thousand entries in a paper airplane competition held by *Scientific American* are shown above. Which one do you suppose flew the farthest? Which one stayed in the air the longest?

So that they will fly straight, each of these planes is symmetrical; that is, their wings match in length and form equal angles with the center line of the plane. The figure at the right illustrates a lopsided airplane in which the wing angles are equal ($\angle ABD = \angle DBC$), but the lengths of their side edges are not (BA = 7 in and BC = 8 in). As a result of this, the back edges of the wings, $\overline{AD}$ and $\overline{DC}$, evidently do not have equal lengths either. If we assume that A, D, and C are collinear so that the two wings are coplanar (an especially appropriate word in this particular case), then it is possible to prove that the back edges must have the same length ratio as the side edges; that is,

$$\frac{AD}{DC} = \frac{AB}{BC} = \frac{7}{8}.$$

This result depends on the fact that, because $\angle ABD = \angle DBC$, $\overrightarrow{BD}$ bisects $\measuredangle ABC$. Stated more generally, if a line bisects one angle of a triangle, it divides the opposite side into segments that have the same ratio as the other two sides. We will prove this and call it the "Angle Bisector Theorem."

▶ **Theorem 59** (The Angle Bisector Theorem)
An angle bisector in a triangle divides the opposite side into segments that have the same ratio as the other two sides.

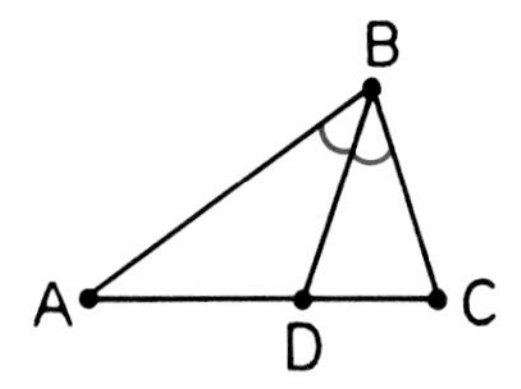

Given: In $\triangle ABC$, $\overrightarrow{BD}$ bisects $\measuredangle ABC$.

Prove: $\frac{AD}{DC} = \frac{AB}{BC}$.

One way to prove that line segments are proportional is to show that they are corresponding sides of similar triangles. The four segments in the conclusion of the theorem are sides of $\triangle ABD$ and $\triangle DBC$. These triangles, however, are not necessarily similar because we know only that one pair of angles in them are equal.

Another way to prove that segments are proportional is to use the Side-Splitter Theorem. It, however, applies only to a triangle in which there is a line parallel to one side that intersects the other two sides. The angle bisector in the figure illustrating the theorem is obviously not parallel to any side of the triangle. However, by carefully choosing certain lines to add to the figure, we can form a triangle to which the Side-Splitter Theorem will apply. By extending $\overline{AB}$ and drawing, through C, a line parallel to $\overline{DB}$, a triangle is produced in which a line is parallel to one side: $\triangle AEC$. At the same time, we can prove that an isosceles triangle has also been formed: $\triangle BEC$. The conclusion of the theorem can easily be derived from these results.

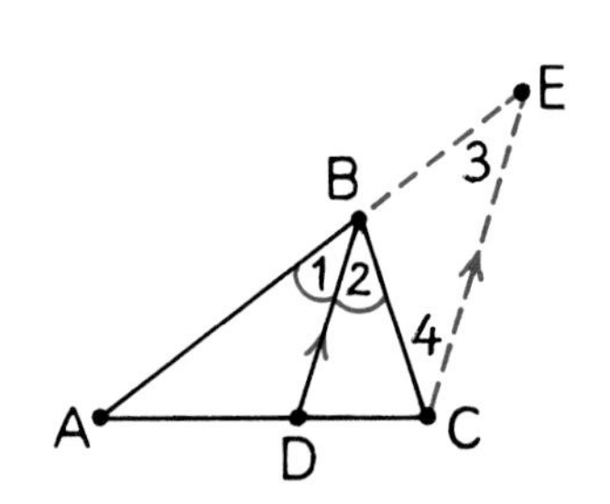

Proof.

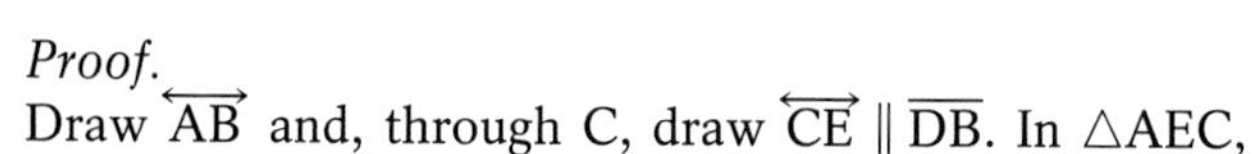

Draw $\overleftrightarrow{AB}$ and, through C, draw $\overleftrightarrow{CE} \parallel \overline{DB}$. In $\triangle AEC$,

$$\frac{AD}{DC} = \frac{AB}{BE} \qquad (1)$$

(if a line parallel to one side of a triangle intersects the other two sides in different points, it divides the sides in the same ratio).

Because $\overline{CE} \parallel \overline{DB}$, $\angle 3 = \angle 1$ (parallel lines form equal corresponding angles with a transversal) and $\angle 4 = \angle 2$ (parallel lines form equal alternate interior angles with a transversal). Because $\overrightarrow{BD}$ bisects $\measuredangle ABC$, $\angle 1 = \angle 2$. Substituting $\angle 3$ for $\angle 1$ and $\angle 4$ for $\angle 2$, we get $\angle 3 = \angle 4$. It follows that BC = BE (if two angles of a triangle are equal, the sides opposite them are equal). Substituting BC for BE in Equation 1, we get

$$\frac{AD}{DC} = \frac{AB}{BC}. \qquad (2)$$

Of the three paper airplanes at the beginning of this lesson, the second stayed in the air the longest and the third flew the farthest.

Exercises

Set I

State the reason that is the basis for concluding that

$$\frac{a}{b} = \frac{c}{d}$$

for each of the following figures.

1.

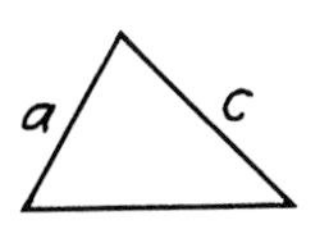

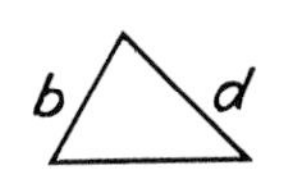

The triangles are similar.

2.

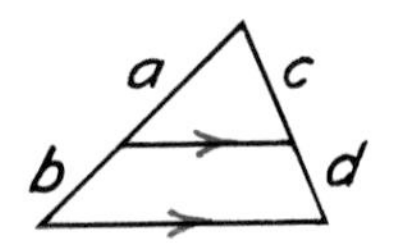

3.

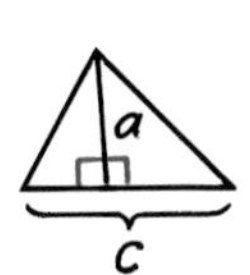

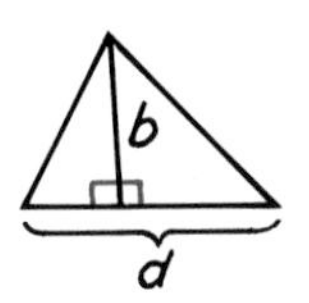

The triangles are similar.

4.

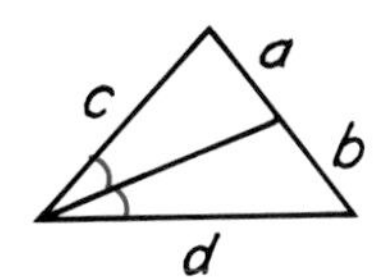

In $\triangle BLW$, $\overrightarrow{WE}$ bisects $\measuredangle BWL$ and $\overline{EO} \parallel \overline{BW}$. Copy and complete the following proportions.

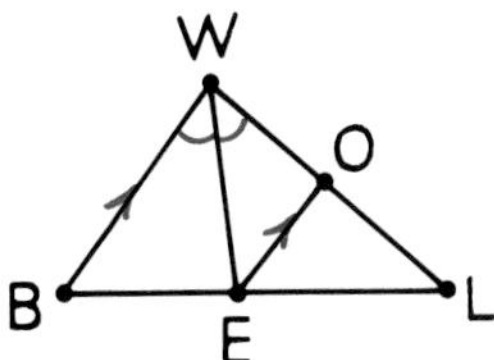

5. $\dfrac{BE}{EL} = \dfrac{WO}{\text{||||||||}}$

6. $\dfrac{BE}{EL} = \dfrac{BW}{\text{||||||||}}$

7. $\dfrac{OL}{WL} = \dfrac{\text{||||||||}}{\text{||||||||}}$

8. $\dfrac{LW}{WB} = \dfrac{\text{||||||||}}{\text{||||||||}}$

Use the Angle Bisector Theorem to solve for x in each of the following figures.

Example:

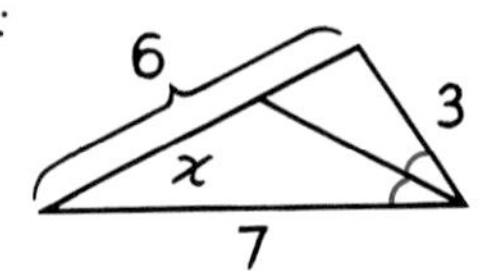

Solution:

$$\frac{x}{6 - x} = \frac{7}{3}$$

$$3x = 7(6 - x)$$

$$3x = 42 - 7x$$

$$10x = 42$$

$$x = \frac{42}{10} = 4.2$$

9.

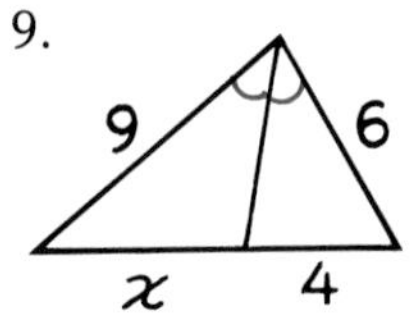

10.

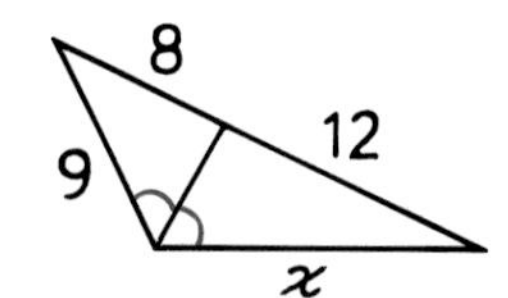

11.

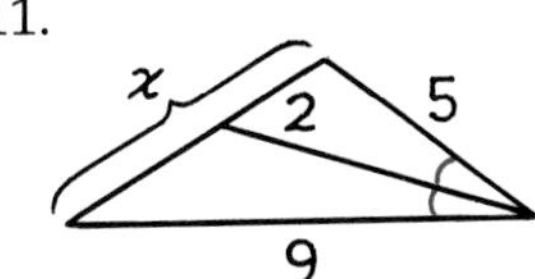

12.

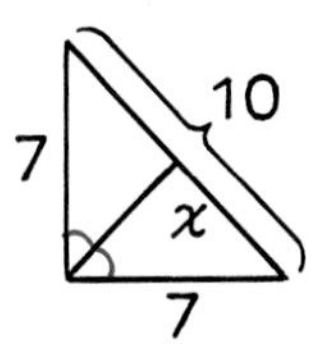

13.

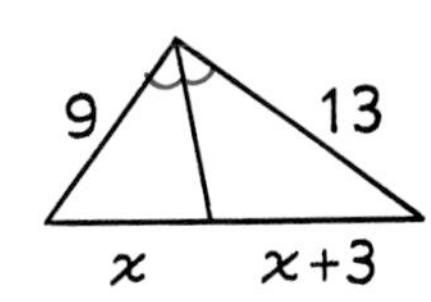

14.

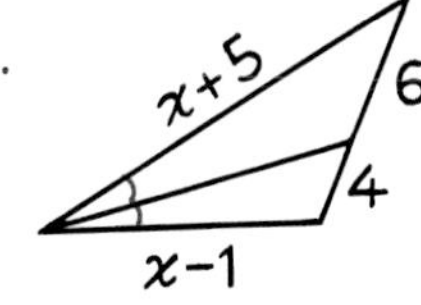

The following exercises are about $\triangle ABC$, in which AB = 8 cm, AC = 10 cm, and BC = 6 cm.

15. Use your straightedge and compass to construct $\triangle ABC$. Bisect $\measuredangle C$ and label the point in which the bisector intersects $\overline{AB}$ D.

16. What kind of triangle is $\triangle ABC$ with respect to its angles?

17. How do you know?

18. Use the Angle Bisector Theorem to find AD and DB. Then, measure $\overline{AD}$ and $\overline{DB}$ on your drawing to check your answers.

In $\triangle OER$, $\overrightarrow{RV}$ bisects $\measuredangle ORE$. The perimeter of $\triangle OER$ is 120 mm, OV = 18 mm, and VE = 27 mm.

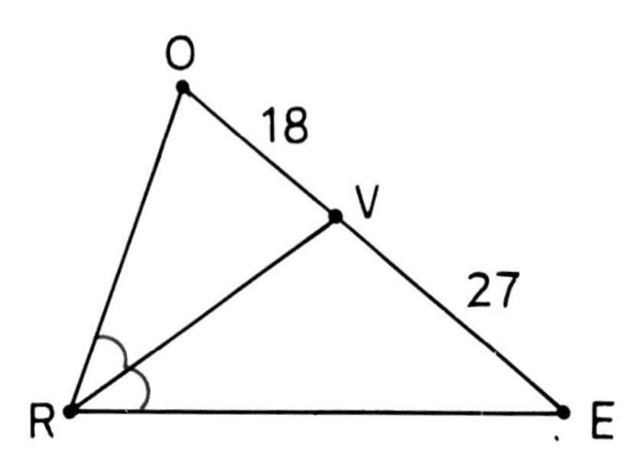

19. Use the Angle Bisector Theorem to find RO and RE. Then, measure $\overline{RO}$ and $\overline{RE}$ to check your answers.

20. What kind of triangle is $\triangle OER$ with respect to its sides?

21. How do you know?

In $\triangle AON$, $\overrightarrow{AU}$ bisects $\measuredangle OAN$ and $\overrightarrow{NR}$ bisects $\measuredangle ONA$. OR = 15 mm, RA = 20 mm, and ON = 30 mm.

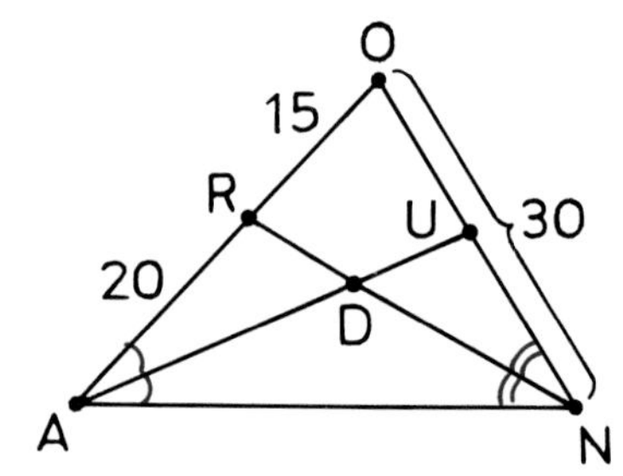

22. Use the Angle Bisector Theorem to find AN.

23. Use the Angle Bisector Theorem to find OU and UN. Then, measure $\overline{OU}$ and $\overline{UN}$ to check your answers.

24. Is $\triangle AOU \sim \triangle NOR$?

Set II

25.

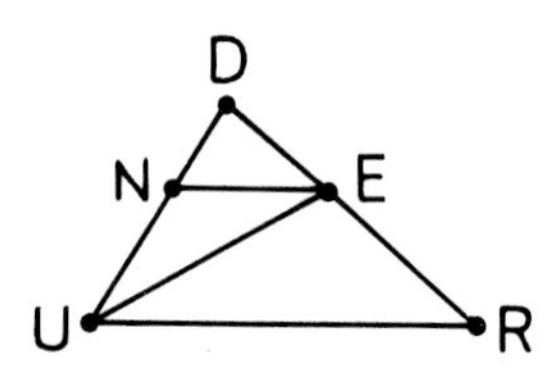

Given: In $\triangle UDR$, $\overrightarrow{UE}$ bisects $\measuredangle DUR$; $\overline{NE} \parallel \overline{UR}$.

Prove: $\frac{DN}{NU} = \frac{DU}{UR}$.

26.

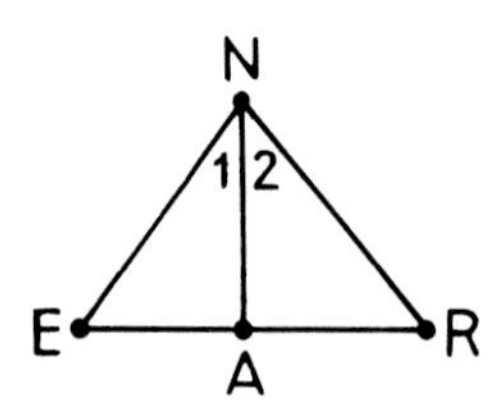

Given: In $\triangle NER$, $\angle 1 = \angle 2$; EA = AR.

Prove: EN = NR.

27.

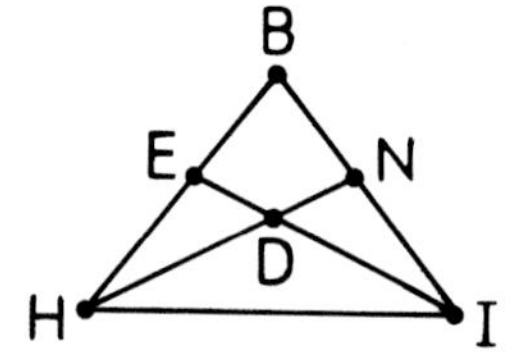

Given: Isosceles $\triangle BHI$ with BH = BI; $\overrightarrow{HN}$ bisects $\measuredangle BHI$ and $\overrightarrow{IE}$ bisects $\measuredangle BIH$.

Prove: $\frac{BE}{EH} = \frac{BN}{NI}$.

28.

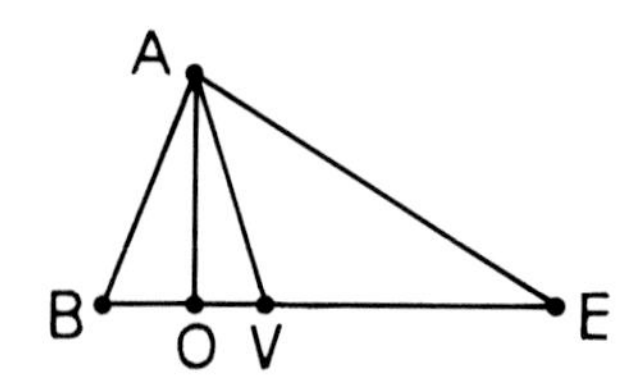

Given: In $\triangle ABE$, $\overrightarrow{AV}$ bisects $\measuredangle BAE$; $\overline{AO} \perp \overline{BE}$.

Prove: $\frac{\alpha\triangle ABV}{\alpha\triangle AVE} = \frac{AB}{AE}$.

Set III

The winning origami entry in the *Scientific American* paper airplane competition is shown in the photograph at the right. It was created by Professor James M. Sakoda of Brown University, Rhode Island.

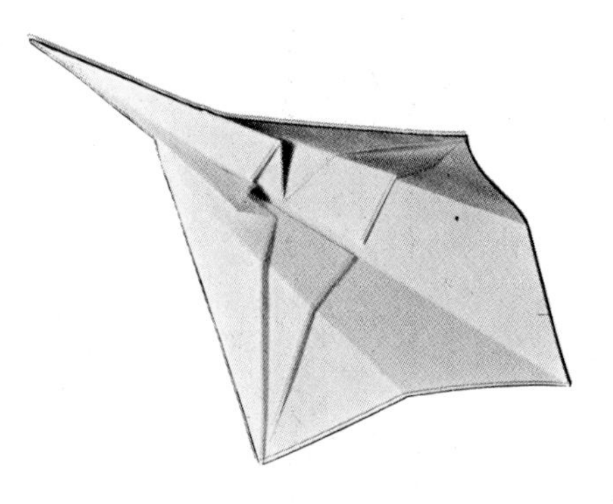

Directions for folding the airplane are given below.

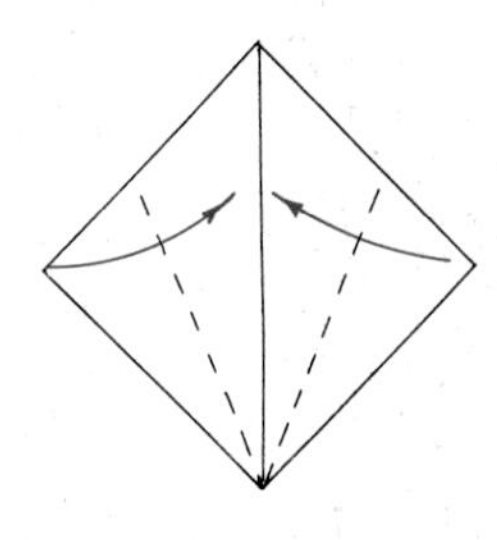

1 Take a square piece of paper. White bond paper $8\frac{1}{2}'' \times 8\frac{1}{2}''$ is suitable. Make a crease along one diagonal and fold two sides to this diagonal line to form the nose and wings of the plane. See next figure for desired result.

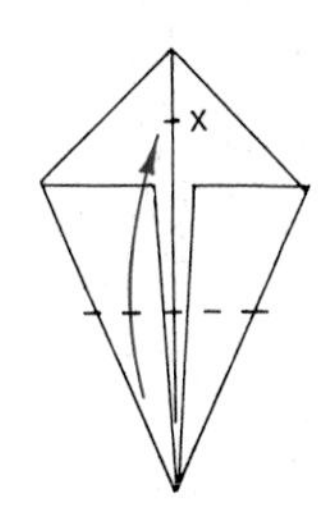

2 Fold the nose of the plane back to the point marked X.

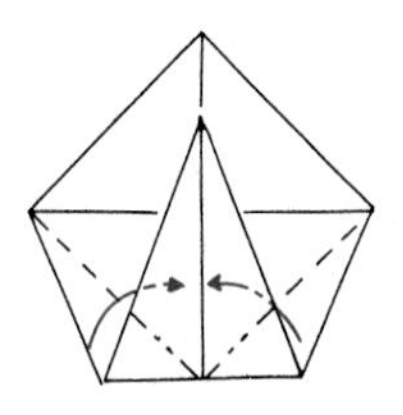

3 Fold over leading edges of the wings; then tuck in between wing and nose.

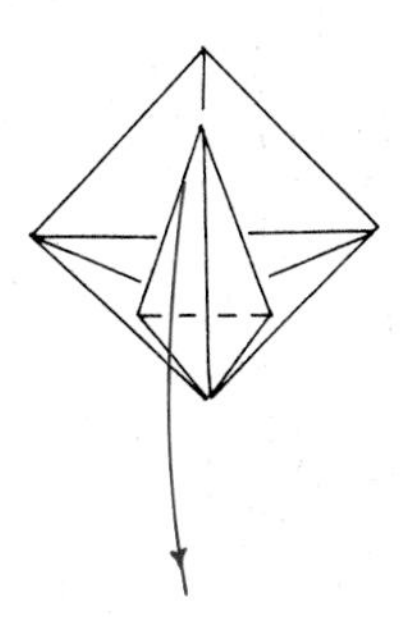

4 Pull nose out forward.

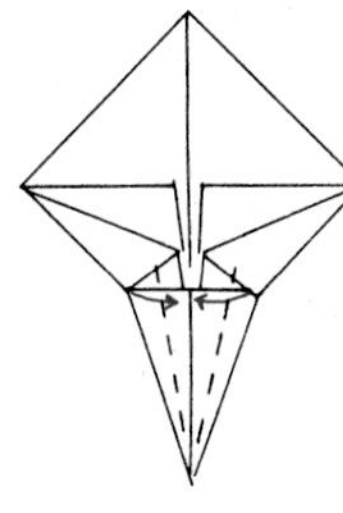

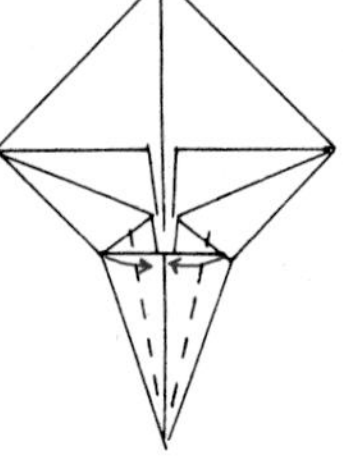

5 Narrow down nose by folding edge to center line.

KEY

– – –	Fold up
– -- –	Fold down

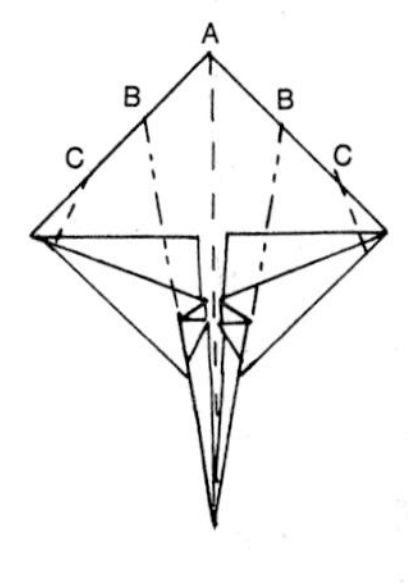

6 Fold up plane in half along center line A. Fold down wings along B, and then spread out horizontally. Turn up trailing edge of wings at slight angle at C. Adjust this angle to provide proper amount of lift. If plane drops rapidly, lift flap up more. If it rises too rapidly and stalls, flatten the flap down more.

National Baseball Library. Cooperstown, NY

Lesson 7

Perimeters and Areas of Similar Polygons

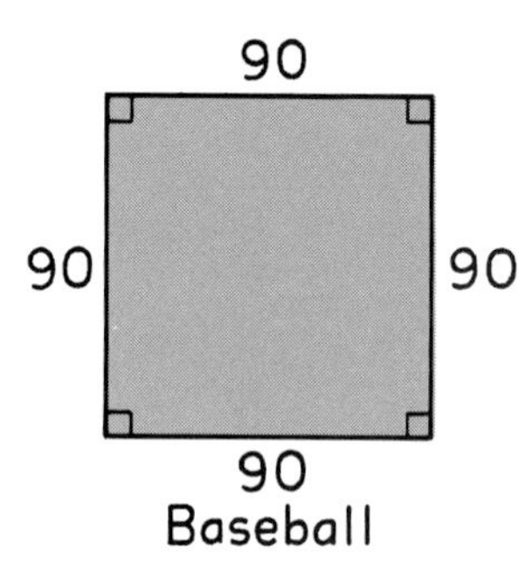

Baseball was first played in the 1830s. Softball, a popular variation of the game, originated approximately fifty years later.

Each side of a baseball diamond is 90 feet long and each side of a softball diamond is 60 feet long. The distance that a batter must run to score a run is the perimeter of the diamond: $4 \times 90 = 360$ feet in the case for baseball and $4 \times 60 = 240$ feet in the case for softball. It is easy to see that the ratio of the perimeters of the two diamonds is equal to the ratio of the lengths of their sides because

$$\frac{360}{240} = \frac{4 \times 90}{4 \times 60} = \frac{90}{60} = \frac{3}{2}.$$

60
60
60
60
Softball

The area enclosed by a baseball diamond is $90^2 = 8{,}100$ square feet and the area enclosed by a softball diamond is $60^2 = 3{,}600$ square feet. It follows from this that the ratio of the areas enclosed by the two diamonds is equal to the *square* of the ratio of the lengths of their sides:

$$\frac{8{,}100}{3{,}600} = \frac{90^2}{60^2} = \left(\frac{90}{60}\right)^2 = \left(\frac{3}{2}\right)^2.$$

These relations of perimeters and areas are true because the diamonds are square in shape and all squares are similar.

We will prove that the same relations are true for any pair of similar polygons.

▶ **Theorem 60**
The ratio of the perimeters of two similar polygons is equal to the ratio of the corresponding sides.

Given: $\triangle ABC \sim \triangle DEF$.*

Prove: $\dfrac{\rho\triangle ABC}{\rho\triangle DEF} = \dfrac{AB}{DE}$.

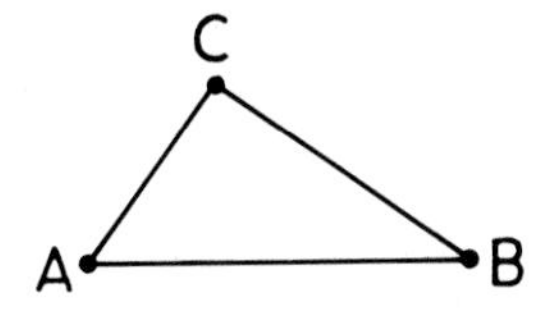

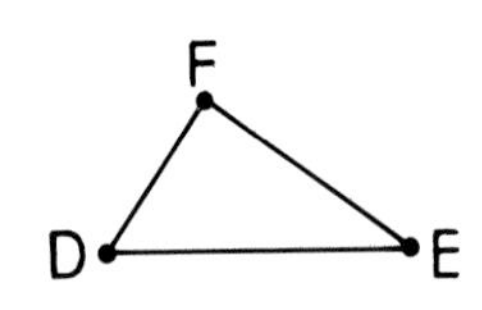

Proof.
Because $\triangle ABC \sim \triangle DEF$, $\dfrac{AB}{DE} = \dfrac{BC}{EF} = \dfrac{CA}{FD}$ (if two polygons are similar, the corresponding sides are proportional). Let $\dfrac{AB}{DE} = r$. Then $\dfrac{BC}{EF} = r$, and $\dfrac{CA}{FD} = r$. It follows that $AB = r \cdot DE$, $BC = r \cdot EF$, and $CA = r \cdot FD$. Adding, we get

$$\begin{aligned} AB + BC + CA &= r \cdot DE + r \cdot EF + r \cdot FD \\ &= r(DE + EF + FD). \end{aligned}$$

But $AB + BC + CA = \rho\triangle ABC$ and $DE + EF + FD = \rho\triangle DEF$ (the perimeter of a polygon is the sum of the lengths of its sides), and so

$$\rho\triangle ABC = r \cdot \rho\triangle DEF \text{ (substitution).}$$

Dividing, we get $\dfrac{\rho\triangle ABC}{\rho\triangle DEF} = r$. Because $\dfrac{AB}{DE} = r$, $\dfrac{\rho\triangle ABC}{\rho\triangle DEF} = \dfrac{AB}{DE}$.

▶ **Theorem 61**
The ratio of the areas of two similar polygons is equal to the square of the ratio of the corresponding sides.

Given: $\triangle ABC \sim \triangle DEF$.

Prove: $\dfrac{\alpha\triangle ABC}{\alpha\triangle DEF} = \left(\dfrac{AB}{DE}\right)^2$.

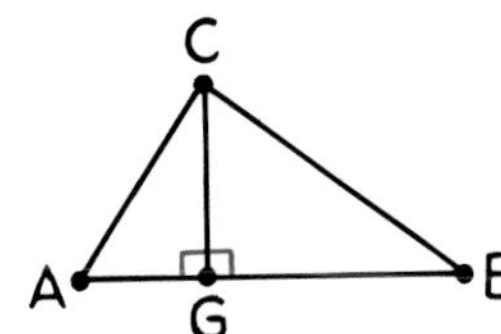

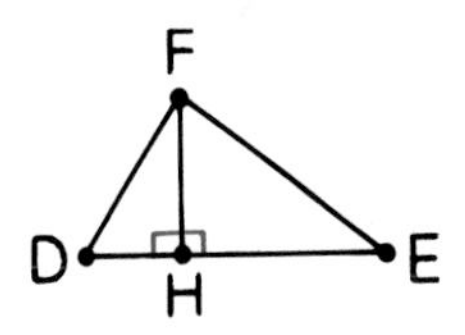

Proof.
Through C, draw $\overline{CG} \perp \overline{AB}$, and through F, draw $\overline{FH} \perp \overline{DE}$ (through a point not on a line, there is exactly one line perpendicular to the line). Because $\overline{CG}$ and $\overline{FH}$ are corresponding altitudes of $\triangle ABC$ and $\triangle DEF$ and $\triangle ABC \sim \triangle DEF$,

$$\frac{CG}{FH} = \frac{AB}{DE} \qquad (1)$$

(corresponding altitudes of similar triangles have the same ratio as the corresponding sides).

* Although the proofs of the theorems in this lesson are written for the case in which the polygons are triangles, comparable proofs can be written for polygons having any number of sides.

Because $\alpha\triangle ABC = \frac{1}{2}AB \cdot CG$ and $\alpha\triangle DEF = \frac{1}{2}DE \cdot FH$,

$$\frac{\alpha\triangle ABC}{\alpha\triangle DEF} = \frac{AB \cdot CG}{DE \cdot FH} \tag{2}$$

by division. From Equations 1 and 2 and substitution, it follows that

$$\frac{\alpha\triangle ABC}{\alpha\triangle DEF} = \frac{AB}{DE} \cdot \frac{AB}{DE} = \left(\frac{AB}{DE}\right)^2.$$

Exercises

Set I

In the figure below, LEAP and FROG are similar rhombuses.

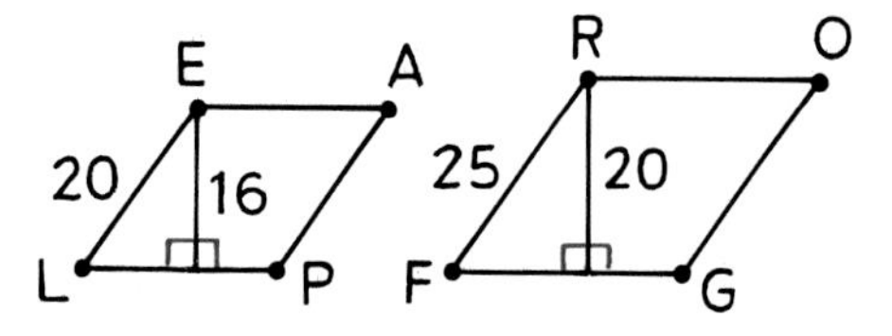

1. Find the value of $\frac{LP}{FG}$ as a common fraction in lowest terms.
2. Find ρLEAP and ρFROG.
3. Use your answers to Exercise 2 to find the value of $\frac{\rho LEAP}{\rho FROG}$ as a common fraction in lowest terms.
4. Find αLEAP and αFROG.
5. Use your answers to Exercise 4 to find the value of $\frac{\alpha LEAP}{\alpha FROG}$ as a common fraction in lowest terms.
6. State the theorem illustrated by your answers to Exercises 1 and 3.
7. State the theorem illustrated by your answers to Exercises 1 and 5.

In the figure below, $\triangle$HOC and $\triangle$KEY are equilateral triangles.

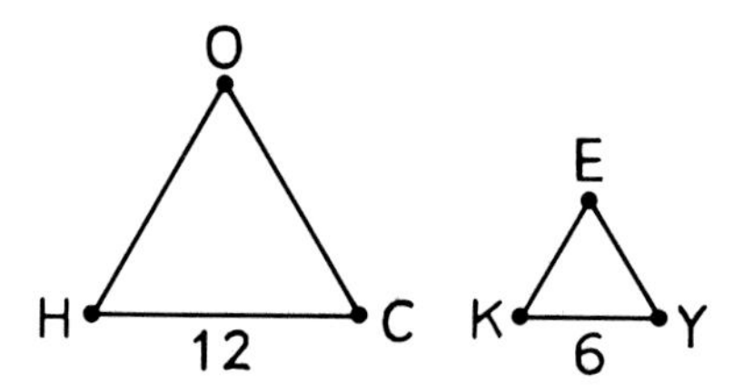

8. Why is $\triangle HOC \sim \triangle KEY$?
9. Find the value of $\frac{HC}{KY}$.
10. Find $\rho\triangle$HOC and $\rho\triangle$KEY.
11. Find the value of $\frac{\rho\triangle HOC}{\rho\triangle KEY}$.
12. Find $\alpha\triangle$HOC and $\alpha\triangle$KEY.
13. Find the value of $\frac{\alpha\triangle HOC}{\alpha\triangle KEY}$.

The pentagons in the figure below are similar.

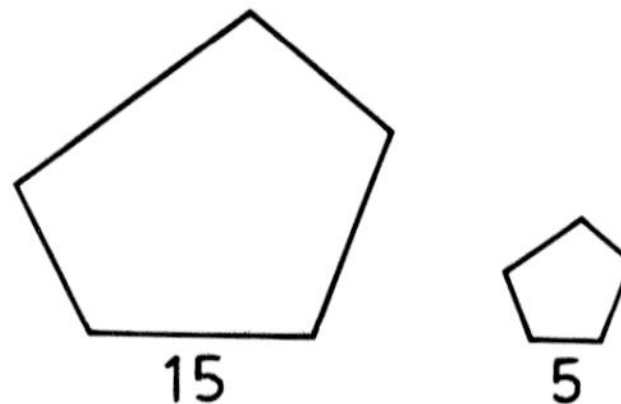

14. Comparing the smaller with the larger, find the ratio of their corresponding sides.

15. What is the ratio of their perimeters?

16. Find the perimeter of the second pentagon if the perimeter of the first is 54.

17. What is the ratio of their areas?

18. Find the area of the second pentagon if the area of the first is 315.

In the figure below,
$\triangle GOF \sim \triangle OLF \sim \triangle GLO$.

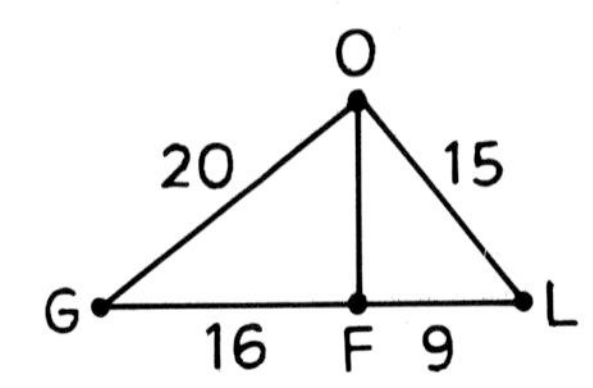

Find the value of each of the following ratios as a common fraction in lowest terms.

19. $\frac{\rho\triangle GOF}{\rho\triangle OLF}$.

20. $\frac{\alpha\triangle GOF}{\alpha\triangle OLF}$.

21. $\frac{\rho\triangle OLF}{\rho\triangle GLO}$.

22. $\frac{\alpha\triangle OLF}{\alpha\triangle GLO}$.

Set II

The perimeter of a trapezoid is 46 and its area is 108.

23. Find the perimeter of a similar trapezoid whose corresponding sides are 2.5 times as long.

24. Find the area of this trapezoid.

25. Find the perimeter of a similar trapezoid whose corresponding sides are one-fifth as long.

26. Find the area of this trapezoid.

In the figure below, HURD and LING are similar rectangles. The area of LING is twice the area of HURD.

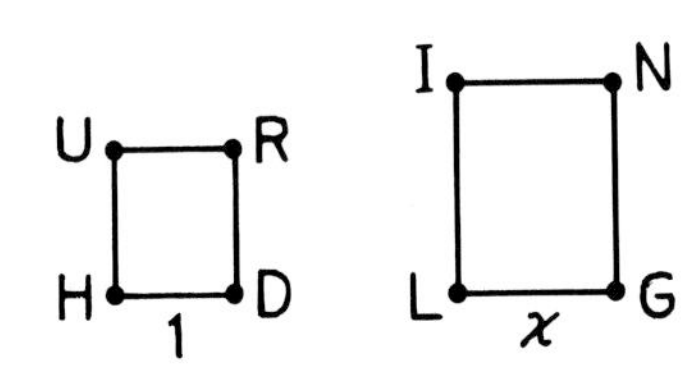

27. Find x.

28. Find the perimeter of HURD if the perimeter of LING is 6.

In the figure below, $\triangle RAC \sim \triangle ING$.

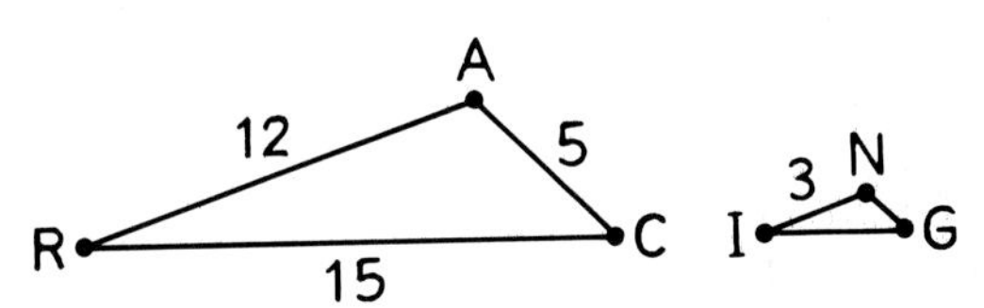

29. Find $\rho\triangle ING$.

30. Find $\alpha\triangle ING$.

The hexagons in the figure below are similar.

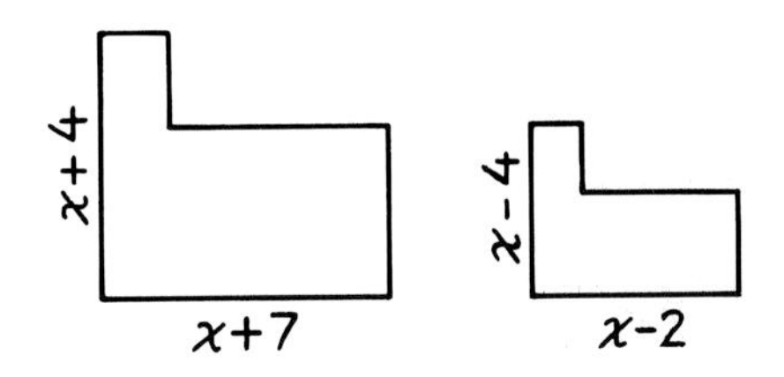

31. Comparing the larger with the smaller, find the ratio of their corresponding sides.

32. Find the area of the smaller, given that the area of the larger is $x^2 - 3x - 7$.

33.

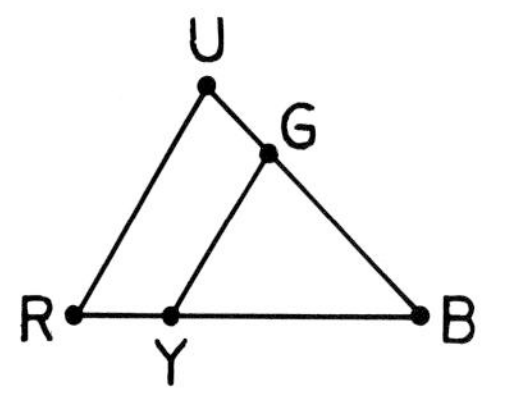

Given: $\triangle RUB$ with $\overline{YG} \parallel \overline{RU}$.

Prove: $\dfrac{\alpha\triangle YGB}{\alpha\triangle RUB} = \left(\dfrac{YB}{RB}\right)^2$.

34.

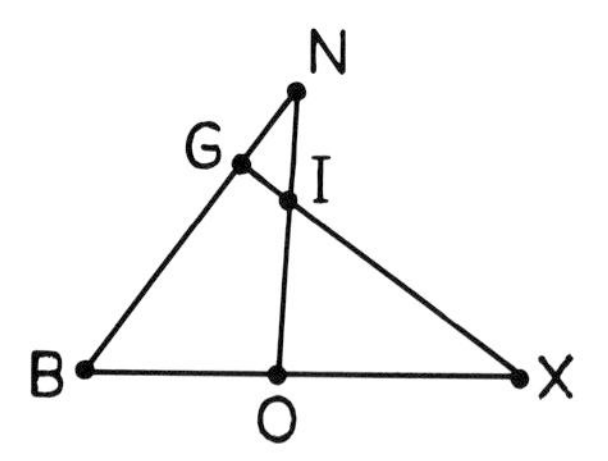

Given: $\triangle BNO$ and $\triangle BXG$ with $\dfrac{BN}{BX} = \dfrac{BO}{BG}$.

Prove: $\dfrac{\rho\triangle BNO}{\rho\triangle BXG} = \dfrac{NO}{XG}$.

Set III

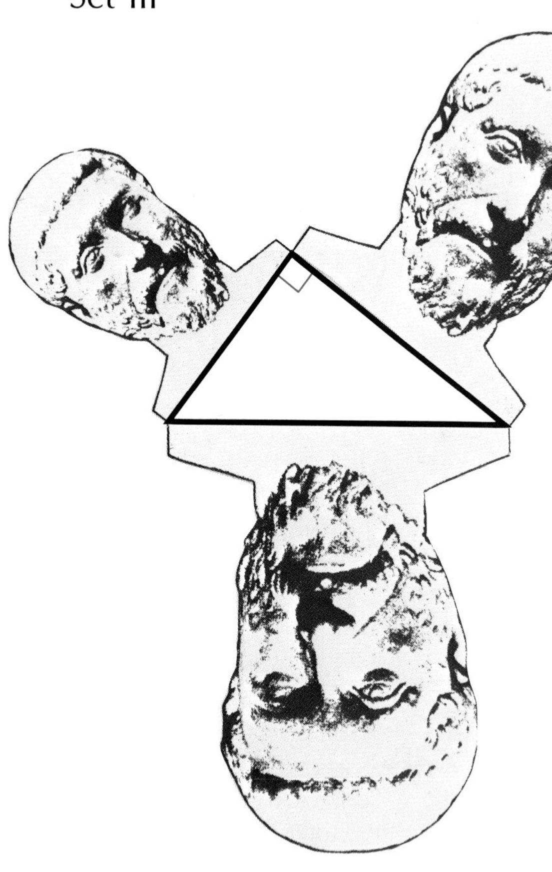

Head of Pythagoras courtesy of Museo Capitolino, Rome

According to Pythagoras, the area of the square on the hypotenuse of a right triangle is equal to the sum of the areas of the squares on the two legs.

Would this still be true if the squares were replaced with equilateral triangles? In other words, is the area of an equilateral triangle on the hypotenuse equal to the sum of the areas of two equilateral triangles on the legs?

What if similar pictures of Pythagoras himself were placed on the three sides of a right triangle. Would the Pythagoras on the hypotenuse be equal to the sum of the Pythagorases on the two legs?

Explain your reasoning.

Chapter 10 / Summary and Review

Basic Ideas

Theorems

55. *The Side-Splitter Theorem.* If a line parallel to one side of a triangle intersects the other two sides in different points, it divides the sides in the same ratio. 354
Corollary. If a line parallel to one side of a triangle intersects the other two sides in different points, it cuts off segments proportional to the sides. 356

56. *The A.A. Similarity Theorem.* If two angles of one triangle are equal to two angles of another triangle, then the triangles are similar. 364
Corollary. Two triangles similar to a third triangle are similar to each other. 365

57. *The S.A.S. Similarity Theorem.* If an angle of one triangle is equal to an angle of another triangle and the sides including these angles are proportional, then the triangles are similar. 369

58. Corresponding altitudes of similar triangles have the same ratio as the corresponding sides. 370

59. *The Angle Bisector Theorem.* An angle bisector in a triangle divides the opposite side into segments that have the same ratio as the other two sides. 376

60. The ratio of the perimeters of two similar polygons is equal to the ratio of the corresponding sides. 381

61. The ratio of the areas of two similar polygons is equal to the square of the ratio of the corresponding sides. 381

Exercises

Set I

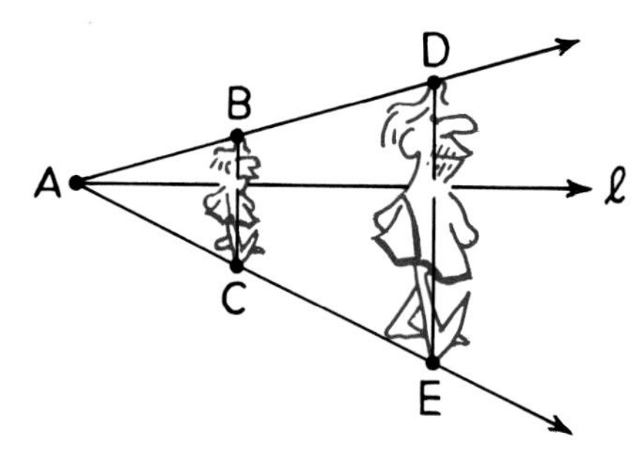

One rule of perspective is illustrated by this figure. In it, points A, B, C, D, and E are coplanar, $\overline{BC} \perp \overline{AE}$, and $\overline{DE} \perp \overline{AE}$.

1. Why is $\overline{BC} \parallel \overline{DE}$?
2. Why is $\frac{AB}{BD} = \frac{AC}{CE}$?
3. Why is $\triangle ABC \sim \triangle ADE$?
4. Why is $\frac{BC}{DE} = \frac{AC}{AE}$?
5. Why is $\frac{\rho\triangle ABC}{\rho\triangle ADE} = \frac{AB}{AD}$?
6. Why is $\frac{\alpha\triangle ABC}{\alpha\triangle ADE} = \left(\frac{AB}{AD}\right)^2$?

Find the length marked x in each of the following figures.

7.

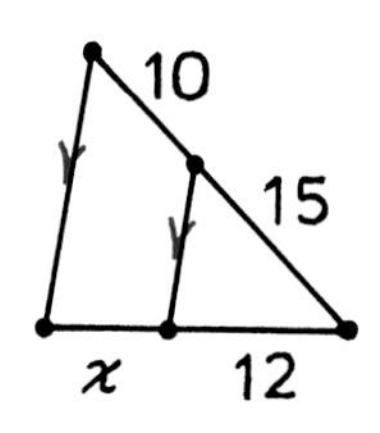

8.

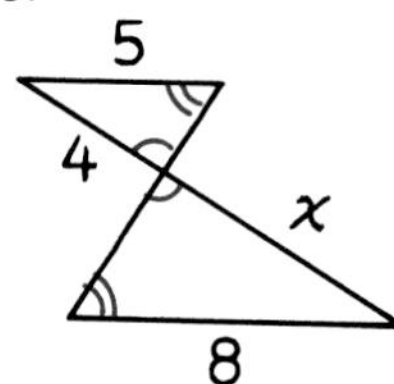

9.

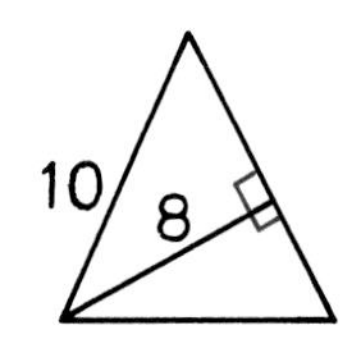

The triangles are similar.

10.

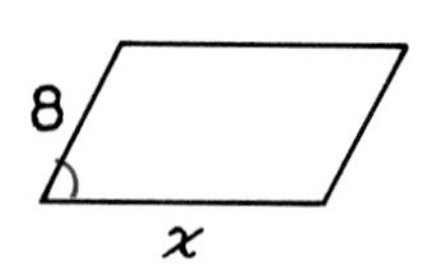

11.

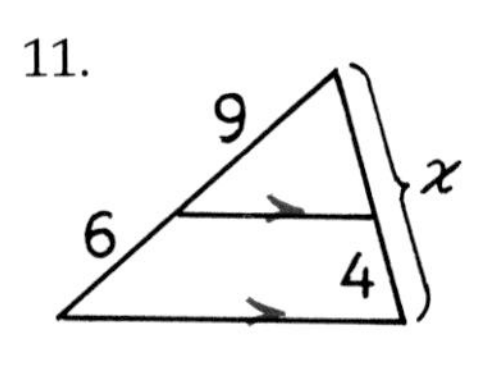

12.

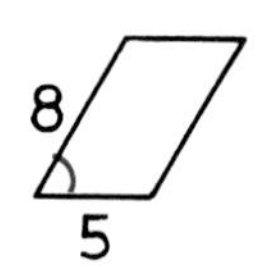

The parallelograms are similar.

The following exercises refer to the figure below. Can you conclude that

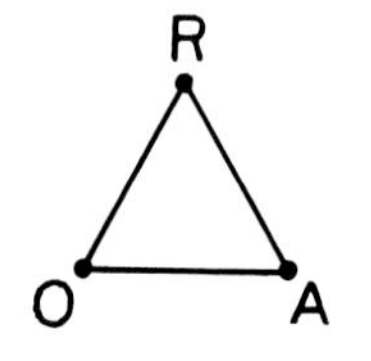

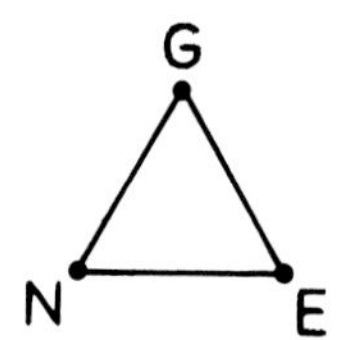

$\triangle ORA \sim \triangle NGE$ if you know that

13. $\triangle ORA$ and $\triangle NGE$ are equilateral?
14. $\angle R = \angle G = 59°$, $\angle O = 60°$, and $\angle N = 61°$?
15. $RO = RA = GN = GE$?

In $\triangle MLN$, $\overline{EO} \parallel \overline{LN}$. Find the value of each of the following ratios in lowest terms.

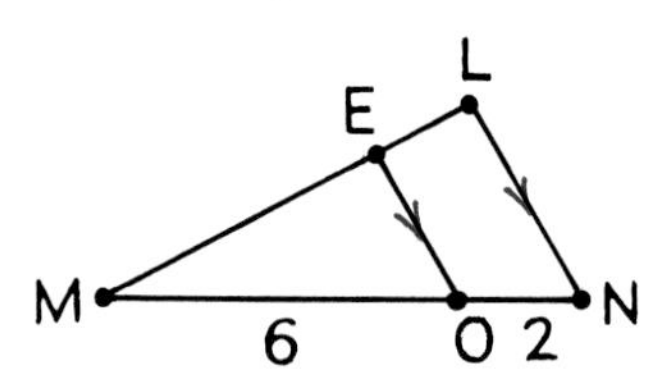

16. $\frac{MO}{MN}$.
17. $\frac{LE}{EM}$.
18. $\frac{EO}{LN}$.
19. $\frac{\rho\triangle LMN}{\rho\triangle EMO}$.
20. $\frac{\alpha\triangle EMO}{\alpha\triangle LMN}$.

In the figure below, $\overrightarrow{LM}$ bisects ∡PLU of △PLU.

21. Find PL.

22. Is △PLU a right triangle?

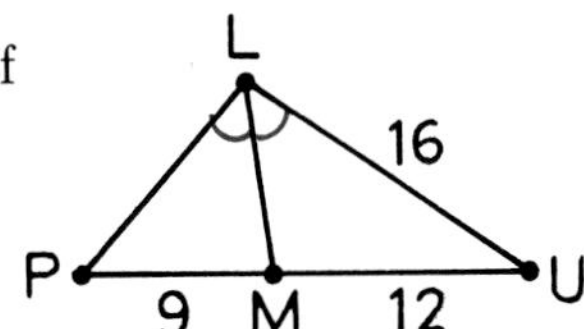

Set II

Photograph by Charles Barr in *Laughing Camera 3,* Hanns Reich Verlag

The dimensions of the photograph above are 36 mm and 28 mm. The width of the frame surrounding it is 6 mm.

23. Find the outside dimensions of the frame.

24. Are the two rectangles in the figure similar?

25. Explain why or why not.

In the figure below, △PEA ~ △ACH.

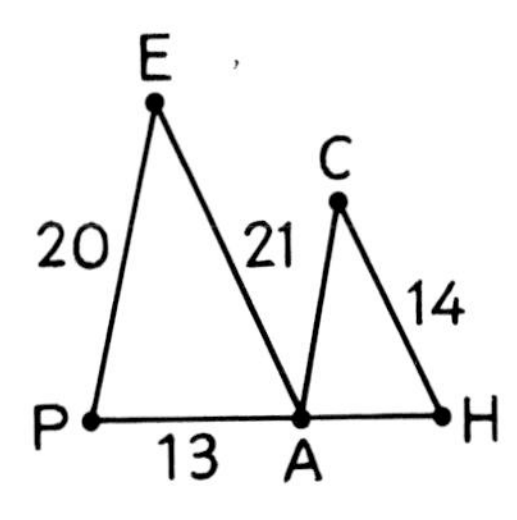

26. Find ρ△PEA.

27. Find α△PEA.

28. Find $\frac{CH}{EA}$.

29. Find ρ△ACH.

30. Find α△ACH.

31.

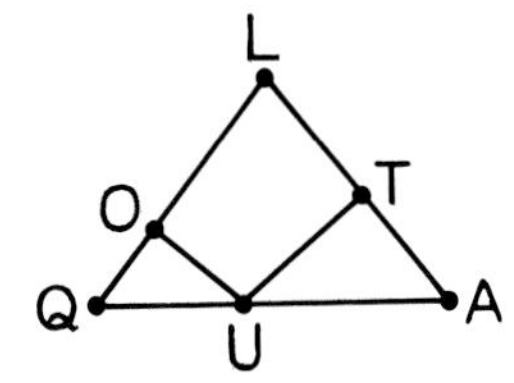

Given: Isosceles △LQA with LQ = LA; $\overline{UO} \perp \overline{LQ}$ and $\overline{UT} \perp \overline{LA}$.

Prove: △QOU ~ △ATU.

32.

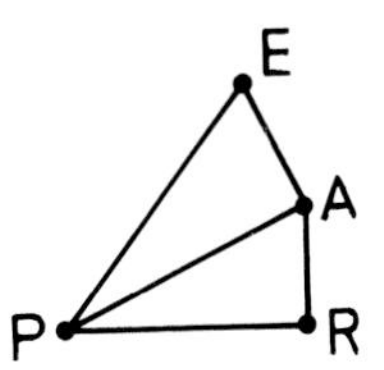

Given: PA is the geometric mean between PE and PR; $\overrightarrow{PA}$ bisects ∡EPR.

Prove: ∠E = ∠PAR.

33.

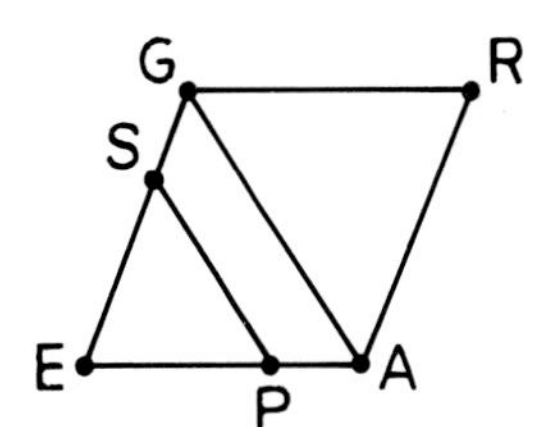

Given: GRAE is a rhombus; ∠ESP = ∠SPE.

Prove: $\overline{SP} \parallel \overline{GA}$.

34.

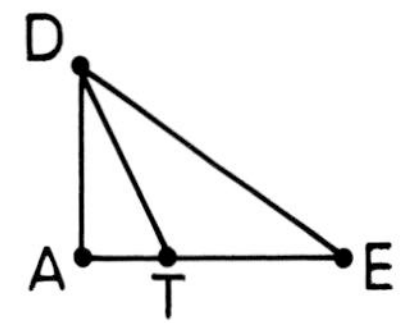

Given: In △DAE, $\overline{DA} \perp \overleftrightarrow{AE}$; $\overrightarrow{DT}$ bisects ∡ADE.

Prove: $\frac{\alpha\triangle DAT}{\alpha\triangle DTE} = \frac{DA}{DE}$.

► **The Pythagorean Theorem**
In a right triangle, the square of the hypotenuse is equal to the sum of the squares of the legs.

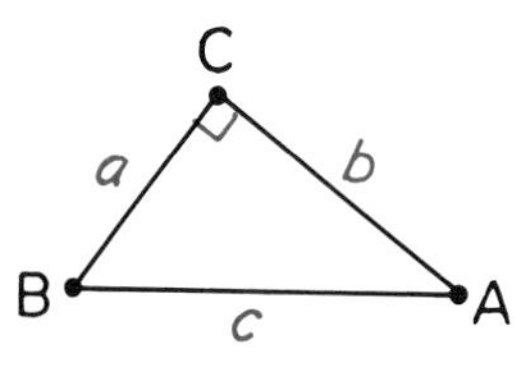

Given: Right $\triangle ABC$ with legs of lengths a and b and hypotenuse of length c.
Prove: $a^2 + b^2 = c^2$.

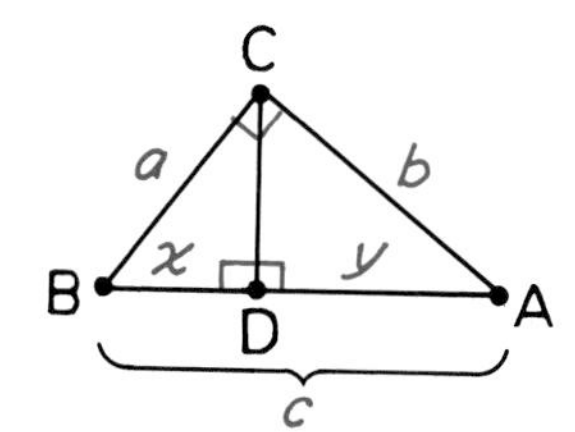

Proof.
Project the legs of $\triangle ABC$ on the hypotenuse by drawing $\overline{CD} \perp \overline{BA}$; let the lengths of the two projections be called x and y.

Because either leg of a right triangle is the geometric mean between the hypotenuse and its projection on the hypotenuse,

$$\frac{c}{a} = \frac{a}{x} \quad \text{and} \quad \frac{c}{b} = \frac{b}{y}.$$

Multiplying, we get

$$a^2 = cx \quad \text{and} \quad b^2 = cy.$$

Adding these equations gives

$$a^2 + b^2 = cx + cy$$
$$= c(x + y).$$

Because $x + y = c$ (betweenness of points: B-D-A, and so BD + DA = BA),

$$a^2 + b^2 = c \cdot c = c^2.$$

Exercises

Set I

Use the Pythagorean Theorem to find the length of the side marked x in each of these triangles. Express your answers in simple radical form.

1. 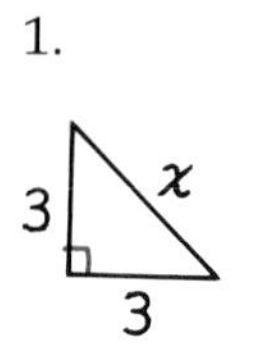

2. 4, x, 4

3. 5, x, 5

4. How does the length of the hypotenuse of an isosceles right triangle seem to be related to the length of each leg?

Find the length of the side marked x in each of these triangles. Express your answers in simple radical form.

5. 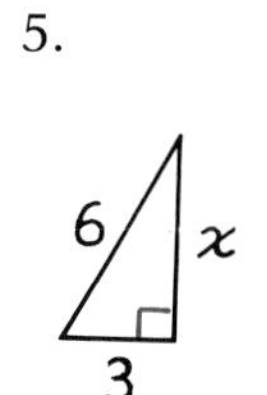

6. 8, x, 4

7. 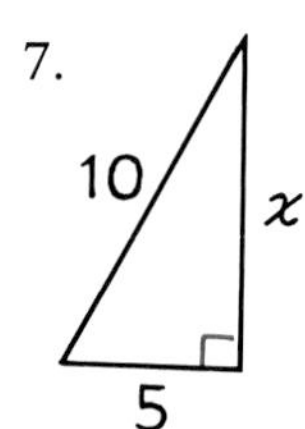

8. If the hypotenuse of a right triangle is twice the length of the shorter leg, how does the length of the longer leg seem

to be related to the length of the shorter leg?

Find the length of the diagonal in each of these rectangles.

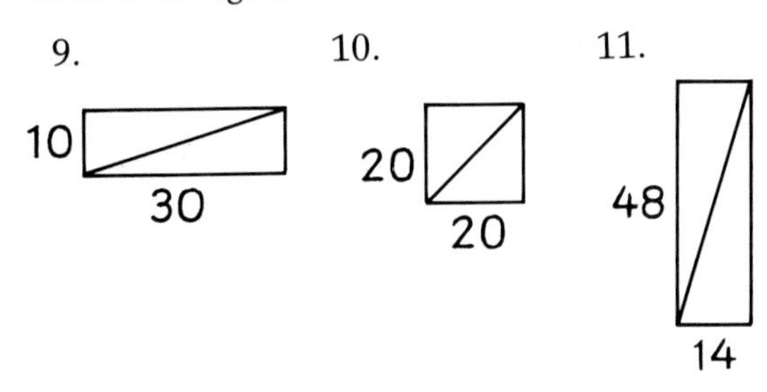

The following exercises are about the two triangles below.

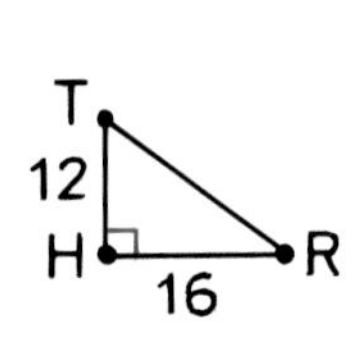

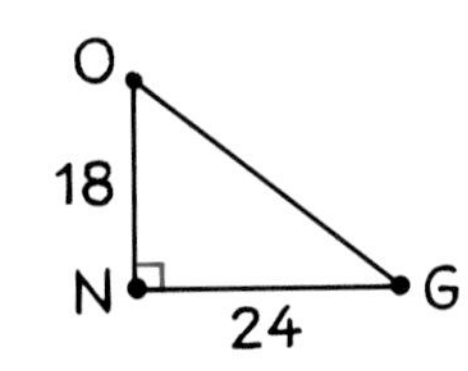

12. Are the triangles similar? Explain why or why not.

13. Find the length of the hypotenuse of each triangle.

14. Is the ratio of the lengths of the hypotenuses of the triangles the number that you would expect? Explain.

15. Find the perimeter of each triangle.

16. Is the ratio of the perimeters of the triangles the number that you would expect? Explain.

17. Find the area of each triangle.

18. Is the ratio of the areas of the triangles the number that you would expect? Explain.

Set II

A set of three integers that can be the lengths of the sides of a right triangle is called a *Pythagorean triple.* We will call a right triangle all of whose sides have lengths that are integers a *Pythagorean triangle.*

The simplest Pythagorean triple is the set "3, 4, 5." These numbers are the lengths of the sides of a "3-4-5" Pythagorean right triangle. The list below contains all of the Pythagorean triples in which no number is more than 50.

3, 4, 5
5, 12, 13
6, 8, 10
7, 24, 25
8, 15, 17
9, 12, 15
9, 40, 41
10, 24, 26
12, 16, 20
12, 35, 37
14, 48, 50
15, 20, 25
15, 36, 39
16, 30, 34
18, 24, 30
20, 21, 29
21, 28, 35
24, 32, 40
27, 36, 45
30, 40, 50

19. Show why the set "7, 24, 25" is a Pythagorean triple.

20. Can all three numbers of a Pythagorean triple be even?

21. Can all three numbers of a Pythagorean triple be odd?

Some of the Pythagorean triples in the list seem to be related to each other. For example, 7, 24, 25 and 14, 48, 50 are related in a special way.

22. Which triple in the list is related to 8, 15, 17 in the same way?

23. Find two triples that are related to 5, 12, 13.

24. Find all of the triples in the list that are related to 3, 4, 5.

Give the missing statements and reasons in the following proof.

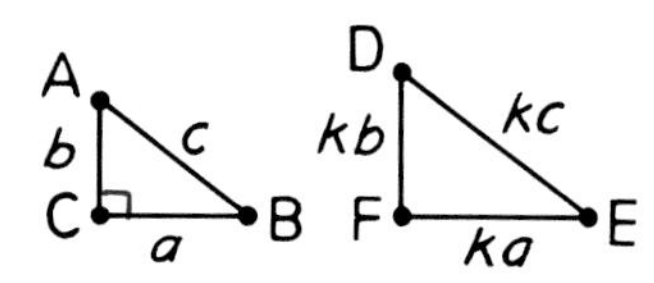

Given: $\triangle ABC$ is a right triangle with sides of lengths a, b, and c; $\triangle DEF$ has sides of lengths ka, kb, and kc.

Prove: $\triangle DEF$ is a right triangle.

25. *Proof.*

Statements	***Reasons***
1. $\triangle ABC$ is a right triangle with sides of lengths a, b, and c; $\triangle DEF$ has sides of lengths ka, kb, and kc.	Given.
2. $a^2 + b^2 = c^2$.	a) ▯▯▯
3. $k^2a^2 + k^2b^2 = k^2c^2$.	b) ▯▯▯
4. $(ka)^2 + (kb)^2 = (kc)^2$.	Substitution.
5. c) ▯▯▯	d) ▯▯▯

In $\triangle CDO$, $\overline{CD} \perp \overline{DO}$ and $\overline{RW} \perp \overline{DO}$.

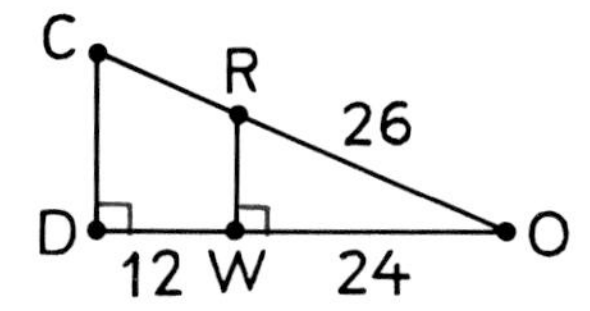

26. Find CR.
27. Find RW.
28. Find CD.

In $\triangle PCK$, $\measuredangle P$ is a right angle and $\overrightarrow{KA}$ bisects $\measuredangle PKC$.

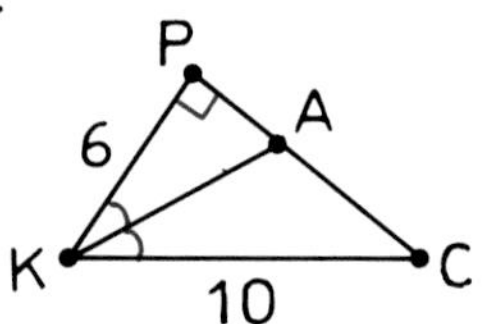

29. Find PC.
30. If PA = x, express AC in terms of x.
31. Find PA and AC.
32. Find KA.
33.

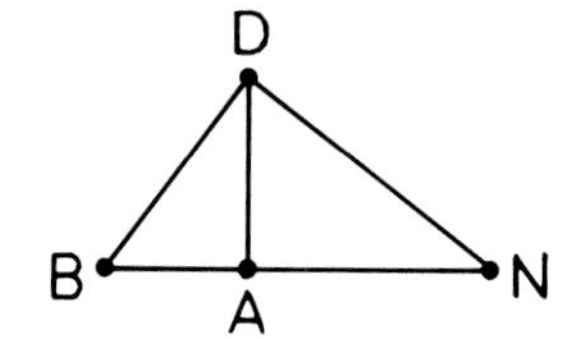

Given: $\triangle DBA$, $\triangle DAN$, and $\triangle DBN$ are right triangles.

Prove: $BA^2 + 2DA^2 + AN^2 = BN^2$.

Set III

By the permission of Johnny Hart and News America Syndicate

An anteater at A wants to walk to a stream ($\overline{BD}$) for a drink and then to an anthill at E for a meal. The shortest path that it can take is the one shown (from A to C to E), in which $\angle 1 = \angle 2$.

What is the length of this path? Show your method.

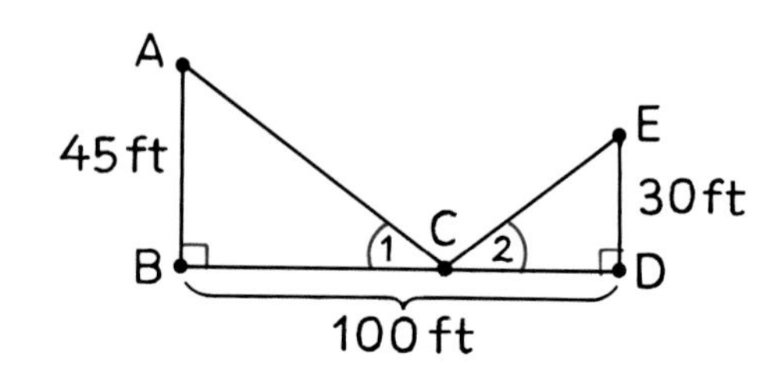

Yale Babylonian Collection

Lesson 3

Isosceles and 30°-60° Right Triangles

Archeologists digging in the land between the Tigris and Euphrates rivers in the late nineteenth century found thousands of clay tablets with writing on them. Some of these tablets, which date back to about 1700 B.C., reveal what the ancient Babylonians knew about mathematics. The one in the photograph above, for example, shows that they knew the relation between the length of a diagonal of a square and the length of one of its sides.* The three wedge-shaped symbols at the upper left indicate that the side of the square is 30 units long. The symbols along the diagonal represent the number 1.41421 and the symbols below them indicate that the diagonal is 42.4263 units long.

These numbers imply that the Babylonians, like the Chinese, knew of the Pythagorean Theorem long before Pythagoras was born. So the theorem was not first discovered by the man for whom it is named.

To understand how the Pythagorean Theorem applies, notice that each diagonal of a square is the hypotenuse of two isosceles right triangles. One of these triangles is shown here. By the Pythagorean Theorem,

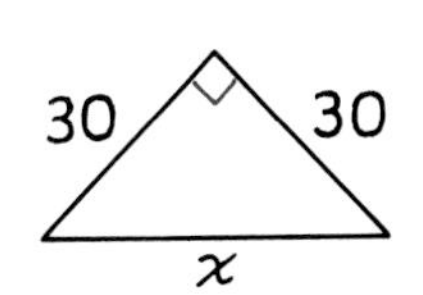

$$x^2 = 30^2 + 30^2 = 2 \cdot 30^2.$$

$$x = \sqrt{2 \cdot 30^2} = 30\sqrt{2}.$$

* *Episodes from the Early History of Mathematics*, by Asger Aaboe (Random House, 1964), pp. 25–27.

The square root of 2 is approximately 1.41421, and so

$$x = 30(1.41421)$$
$$= 42.4263 \text{ (approximately).}$$

The Babylonians recognized that to solve this problem, they could multiply the length of one leg of the triangle by $\sqrt{2}$ to find the length of the hypotenuse. It is this number, in fact, that is written along the diagonal on the tablet.

► **Theorem 63** (The Isosceles Right Triangle Theorem)
In an isosceles right triangle, the hypotenuse is $\sqrt{2}$ times the length of one leg.

Given: An isosceles right triangle with legs of length a and hypotenuse c.

Prove: $c = a\sqrt{2}$.

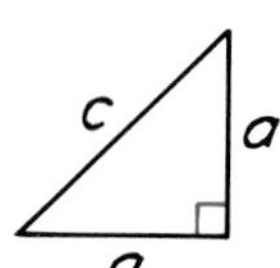

Proof.
By the Pythagorean Theorem,

$$c^2 = a^2 + a^2 = 2a^2.$$
$$c = \sqrt{2a^2} = \sqrt{2} \cdot \sqrt{a^2} = \sqrt{2}a = a\sqrt{2}.$$

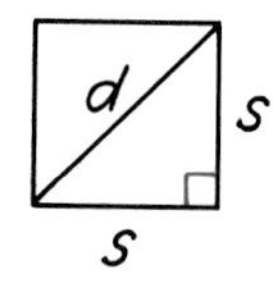

It is convenient to restate this theorem in terms of the diagonal of a square.

► **Corollary**
Each diagonal of a square is $\sqrt{2}$ times the length of one side.

Because each acute angle of an isosceles right triangle has a measure of 45°, we might call it a "45°-45° right triangle." Another important triangle in geometry is the "30°-60° right triangle."

► **Theorem 64** (The 30°-60° Right Triangle Theorem)
In a 30°-60° right triangle, the hypotenuse is twice the length of the shorter leg and the longer leg is $\sqrt{3}$ times the length of the shorter leg.

Given: Right $\triangle ABC$ with right $\measuredangle C$ and $\angle A = 30°$, $\angle B = 60°$; $BC = a$, $AC = b$, and $AB = c$.

Prove: $c = 2a$ and $b = a\sqrt{3}$.

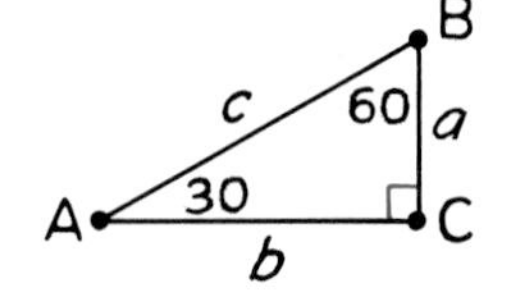

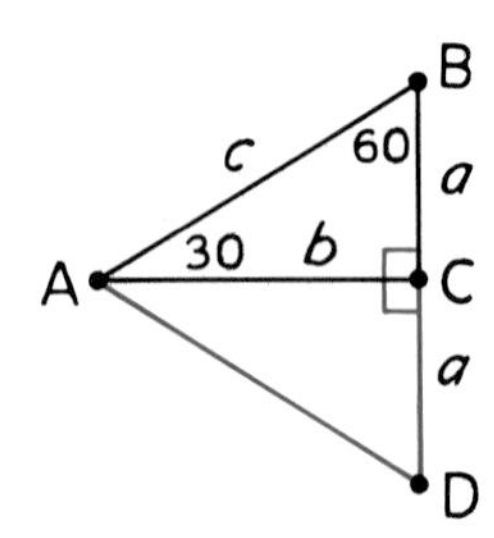

Proof.
Draw $\overleftrightarrow{BC}$ and choose D so that $CD = BC$. Because $\triangle ADC \cong \triangle ABC$ (S.A.S.), $\angle D = \angle B = 60°$ and $\angle CAD = \angle CAB = 30°$; so $\angle BAD = 60°$. Therefore, $\triangle ABD$ is equiangular and, hence, equilateral.

Because $BD = a + a = 2a$, $AB = c$, and $AB = BD$, $c = 2a$.

By the Pythagorean Theorem,

$$a^2 + b^2 = c^2.$$

Substituting, we get

$$a^2 + b^2 = (2a)^2,$$

or

$$a^2 + b^2 = 4a^2.$$
$$b^2 = 3a^2 \text{ (subtraction).}$$

It follows that

$$b = \sqrt{3}a \text{ (square roots property)}$$

or

$$b = a\sqrt{3}.$$

Because $\overline{AC}$ is an altitude of $\triangle ABD$, we can use the theorem we have just proved to derive a formula for the length of an altitude of an equilateral triangle.

► **Corollary**
An altitude of an equilateral triangle is $\frac{\sqrt{3}}{2}$ times the length of one side.

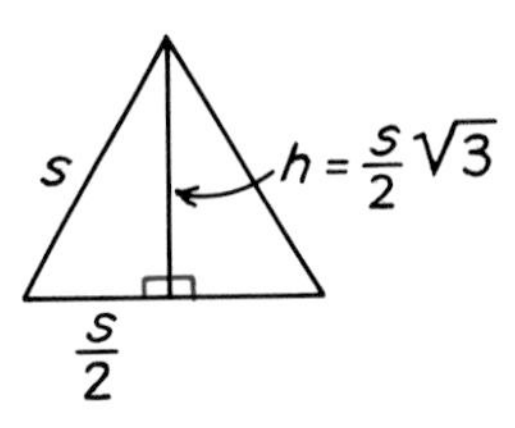

Exercises

Set I

In the figure below, $\triangle ABC$ is an isosceles right triangle and $\triangle DEF$ is a 30°-60° right triangle. Tell how you would find each of the following lengths.

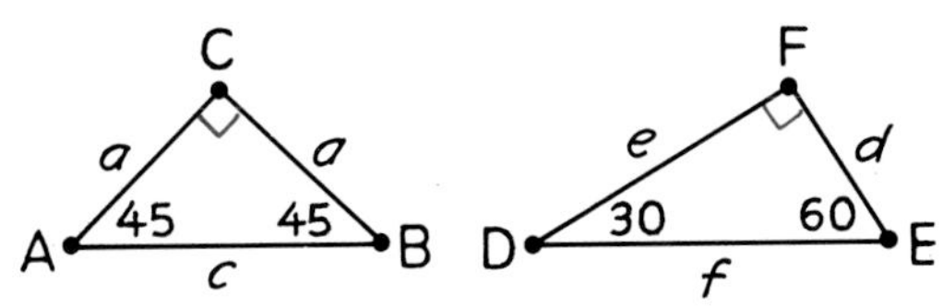

Example: How would you find c if you knew a?

Answer: Multiply a by $\sqrt{2}$.

How would you find

1. a if you knew c?
2. f if you knew d?
3. e if you knew d?
4. d if you knew f?
5. d if you knew e?

Find the length labeled x in each of these isosceles right triangles.

6. 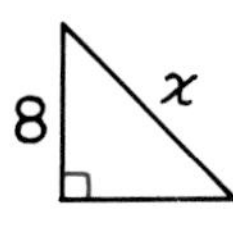

7. 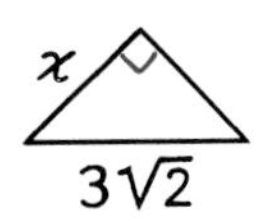

8. 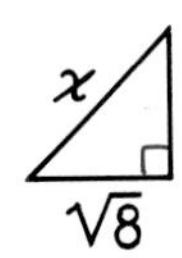

9. 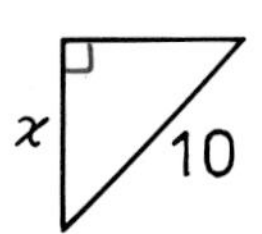

Find the lengths labeled x and y in each of these 30°-60° right triangles.

10.

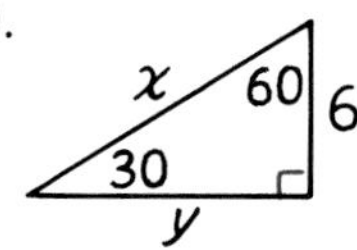

11.

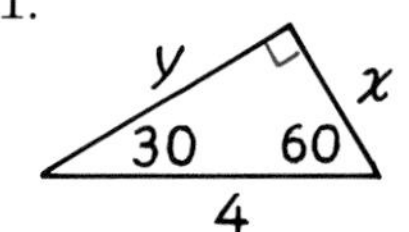

12.

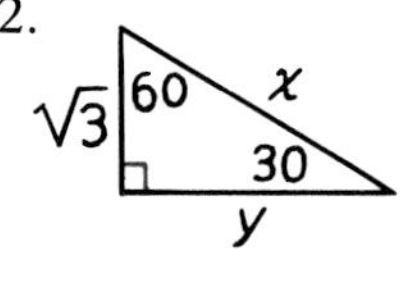

13.

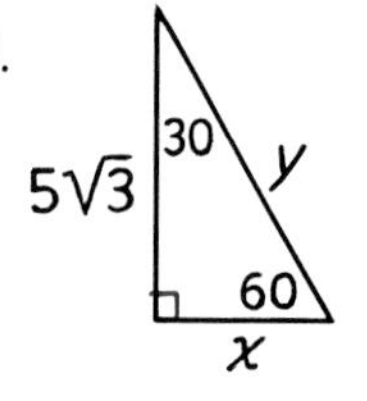

14.

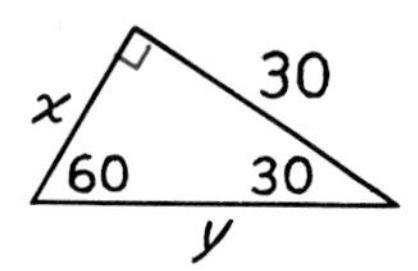

15.

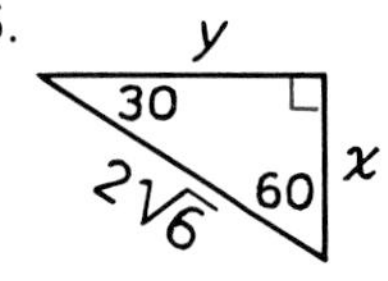

Find the length labeled x in each of these rectangles.

16.

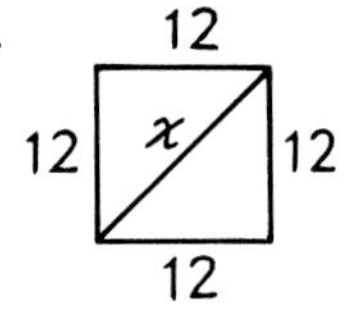

17.

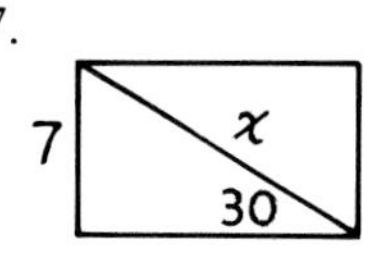

18.

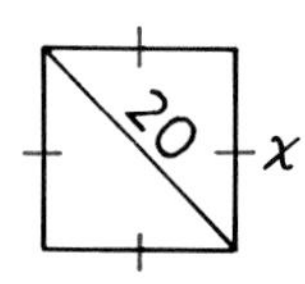

19.

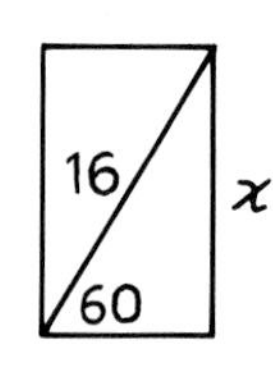

Find the length labeled x in each of these equilateral triangles.

20.

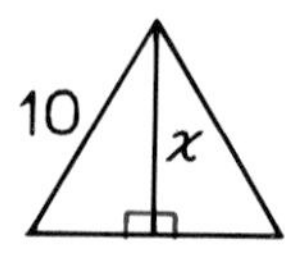

21.

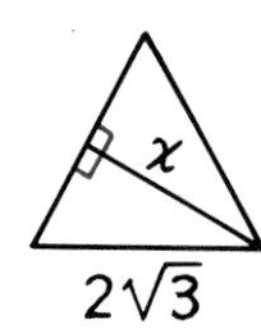

22.

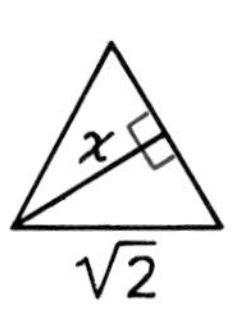

Set II

In the figure below, $\triangle$JAN is a 30°-60° right triangle and $\overline{\text{NO}}$ is the altitude to its hypotenuse.

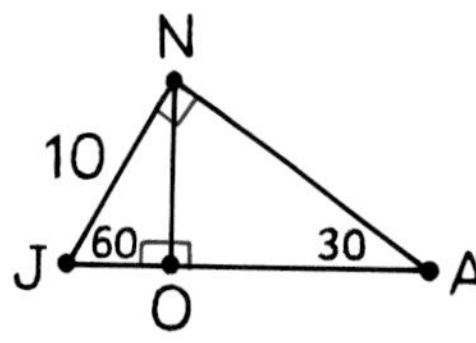

23. Copy the figure and use the 30°-60° Right Triangle Theorem to find the lengths of all of the line segments in it.

24. Use your answers to Exercise 23 to show that $\frac{\text{JA}}{\text{JN}} = \frac{\text{JN}}{\text{JO}}$.

25. What theorem does the proportion in Exercise 24 illustrate?

26. Use your answers to Exercise 23 to show that $\frac{\text{JO}}{\text{NO}} = \frac{\text{NO}}{\text{OA}}$.

27. What theorem does the proportion in Exercise 26 illustrate?

In the figure below, $\triangle$IRA is a 30°-60° right triangle in which $\overrightarrow{\text{IM}}$ bisects $\measuredangle$AIR.

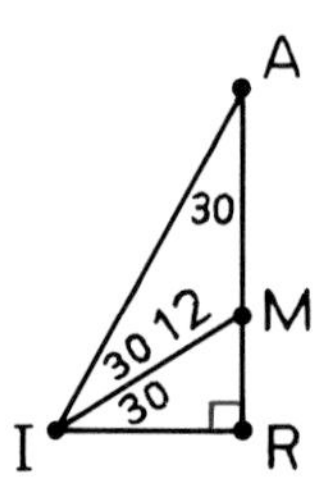

28. Copy the figure and use the 30°-60° Right Triangle Theorem to find the lengths of all of the line segments in it.

29. Use your answers to Exercise 28 to show that $\frac{AM}{MR} = \frac{AI}{IR}$.

30. What theorem does the proportion in Exercise 29 illustrate?

Find the area of each of the following polygons by doing each of the following steps:
a) Copy the figure.
b) Draw an altitude.
c) Find the length of the altitude.
d) Find the area of the polygon.

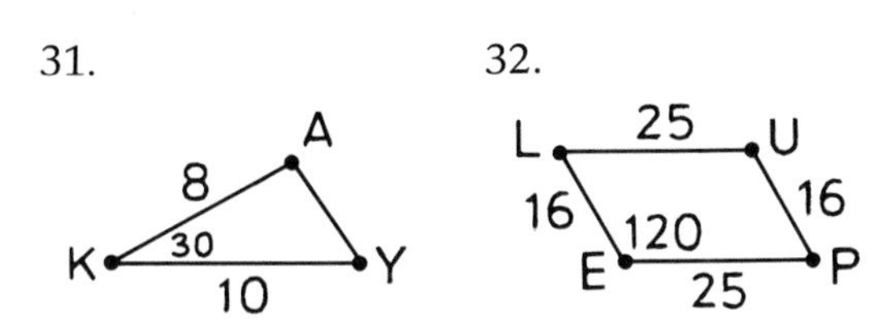

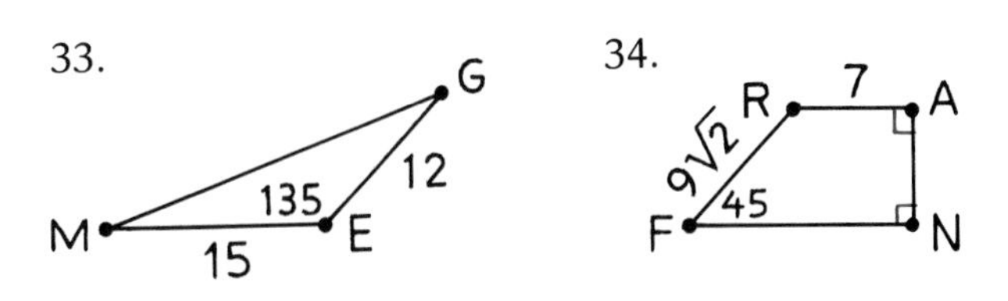

Find the lengths of the other two sides of each of the following triangles.

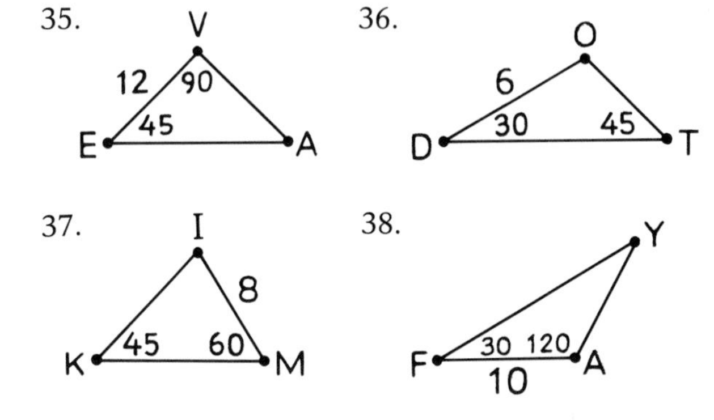

Set III

This article about how to cut down a tree appeared in a newspaper several years ago. You know enough geometry now to be able to recognize that something in the article is incorrect.

1. What is it? Explain.

2. How do you think the article should be corrected? Explain.

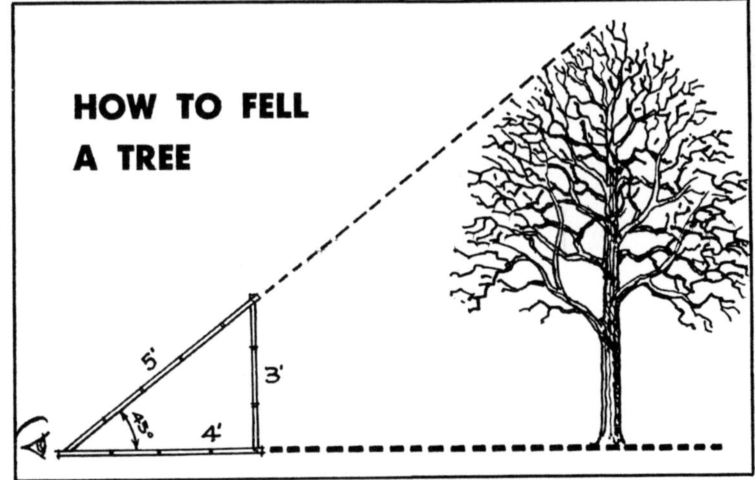

Drawing by Leavitt Dudley

Tree removal used to be a hard, hazardous job, one better left to tree removal companies. But along came the lightweight chain saw and now it's easy to cut a 30-foot-tall tree into neat fireplace-length logs in a few hours. There's just one problem. Felling the tree is still hazardous, although it need not be. If you have analyzed its fall correctly and have made the proper cuts, it will fall exactly where you plan. But if you haven't, no amount of guy wires or ropes can be trusted to guide its crash to the ground.

The first thing you have to do is find how much room it needs when it falls. You can do this by using the Boy Scout method of measuring heights. A triangle with sides of three, four and five feet is placed so that you sight along the five-foot side while you move the triangle to a point where the top of the tree comes into sight, as shown above. That's how far the tree will reach when it's felled.

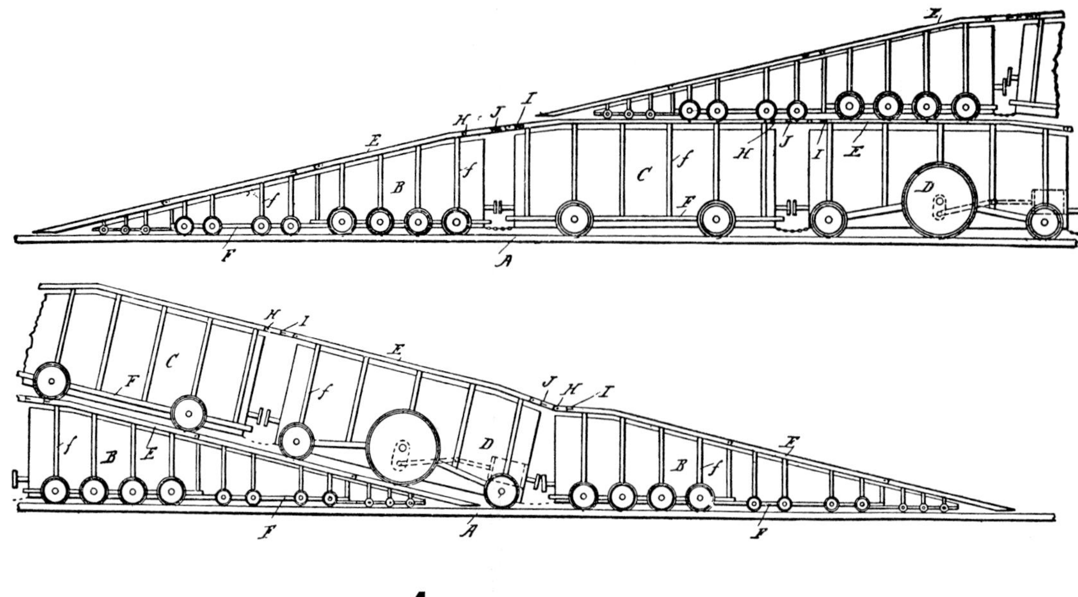

Lesson 4

The Tangent Ratio

Back in 1895, an unusual invention for preventing train collisions was registered in the U.S. Patent Office. It is illustrated in the figures above.* If a train meets another one approaching on the same track, it can run up the slanted rails of a special car, along rails on top of the rest of the cars, and down the other end car. Instead of colliding, one train merely passes over the other! If two trains are traveling in the same direction, one can overtake and pass the other by the same procedure.

The figures reveal that the end cars are approximately right triangular in shape. If the box cars on the train are 14 feet high and the track on one of these end cars rises at a 15° angle with the horizontal, how long must the car be?

If the angle were 30° or 45°, we could use one of the theorems of the last lesson to find the answer. Because it is not, we need to use some other method.

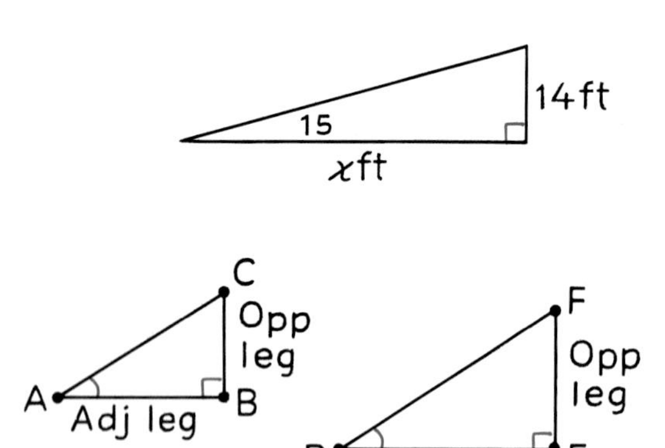

Look at the figure at the left in which $\triangle ABC$ and $\triangle DEF$ are right triangles with $\angle A = \angle D$. Because $\triangle ABC \sim \triangle DEF$,

$$\frac{BC}{EF} = \frac{AB}{DE}.$$

---

*From *Absolutely Mad Inventions,* by A. E. Brown and H. A. Jeffcott, Jr. (Dover).

Multiplying by EF and dividing by AB, we get

$$\frac{BC}{AB} = \frac{EF}{DE}.$$

We have shown that, if two right triangles have an acute angle of one equal to an acute angle of the other, then *the ratio of the length of the leg opposite that angle to the length of the other leg,* called the adjacent leg, *is the same* for both triangles. This ratio, then, depends only on the measure of the acute angle; it does not depend on the size of the triangle. It is called the *tangent* of the angle.

► **Definition**

The ***tangent*** of an acute angle of a right triangle is the ratio of the length of the opposite leg to the length of the adjacent leg.

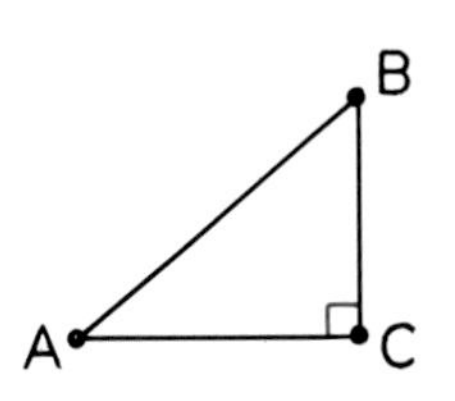

In the figure at the right, the tangent of ∡A is $\frac{BC}{AC}$ and the tangent of ∡B is $\frac{AC}{BC}$. These relations are customarily abbreviated as

$$\tan A = \frac{BC}{AC} \quad \text{and} \quad \tan B = \frac{AC}{BC}.$$

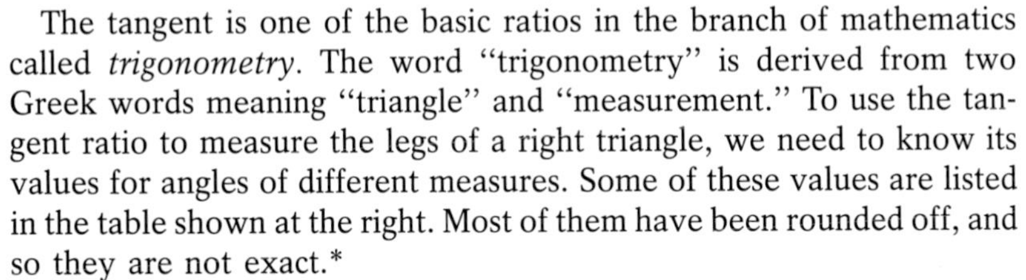

The tangent is one of the basic ratios in the branch of mathematics called *trigonometry.* The word "trigonometry" is derived from two Greek words meaning "triangle" and "measurement." To use the tangent ratio to measure the legs of a right triangle, we need to know its values for angles of different measures. Some of these values are listed in the table shown at the right. Most of them have been rounded off, and so they are not exact.*

Angle	*Tangent*
5°	.087
10°	.176
15°	.268
20°	.364
25°	.466
30°	.577
35°	.700
40°	.839
45°	1.000
50°	1.192
55°	1.428
60°	1.732
65°	2.145
70°	2.747
75°	3.732
80°	5.671
85°	11.430

We can now use the tangent ratio to find the length of the end car if the track on it rises at a 15° angle with the horizontal.

$$\tan 15° = \frac{14}{x}.$$

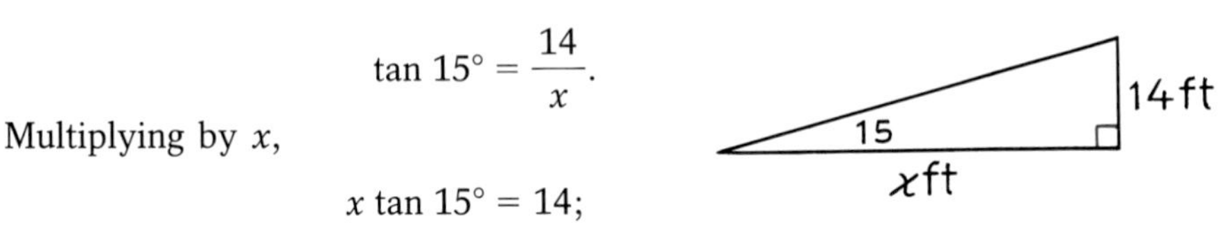

Multiplying by x,

$$x \tan 15° = 14;$$

so

$$x = \frac{14}{\tan 15°}.$$

From the table, $\tan 15° \approx .268$, and so

$$x \approx \frac{14}{.268} \approx 52.2 \text{ ft.}$$

For a rise of 15°, the end car would be about 52 feet long.

---

*Approximations of these values can also be found by using the tangent key of a scientific calculator.

Exercises

Set I

Identify each of the following segments or ratios with respect to right triangle $\triangle TWO$.

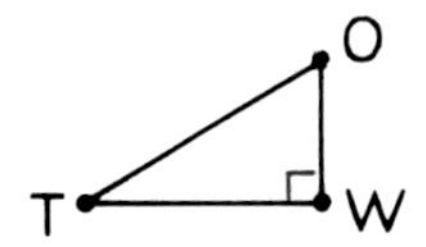

1. The leg opposite $\measuredangle T$.
2. The leg adjacent to $\measuredangle T$.
3. The tangent of $\measuredangle T$.
4. The leg opposite $\measuredangle O$.
5. The leg adjacent to $\measuredangle O$.
6. The tangent of $\measuredangle O$.

Find the tangents of $\measuredangle A$ and $\measuredangle B$ in each figure below.

Example:

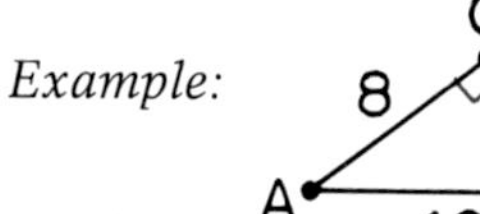

Solution: $\tan A = \frac{6}{8} = \frac{3}{4}$,

$\tan B = \frac{8}{6} = \frac{4}{3}$.

7.

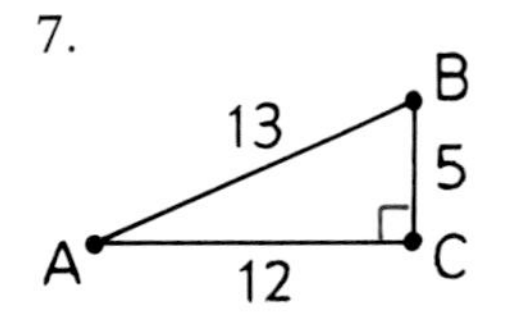

8.

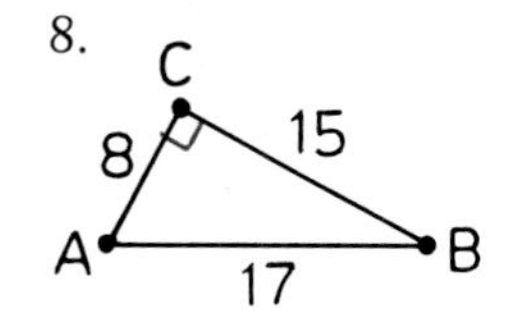

9.

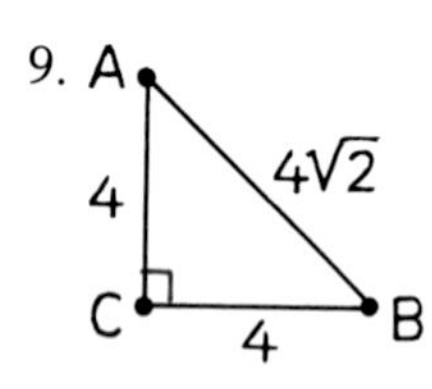

10.

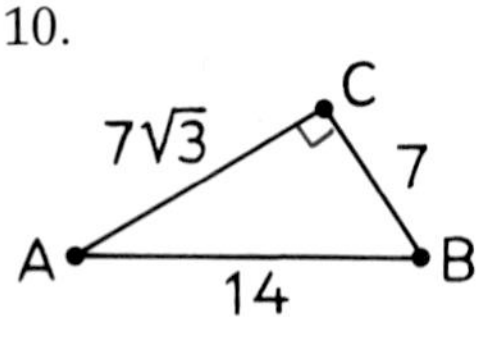

The following exercises refer to the figure below.

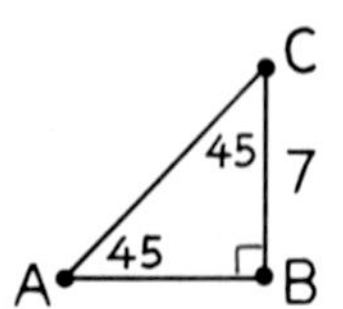

11. Find AB.
12. Use the figure to find the exact value of $\tan 45°$. Check your answer by referring to the table on page 405.

The following exercises refer to the figure below.

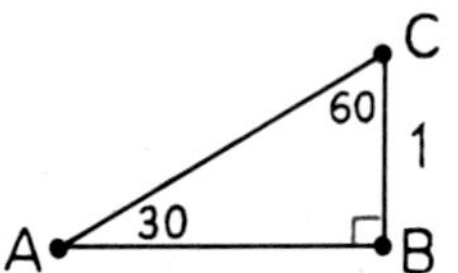

13. Use the 30°-60° Right Triangle Theorem to find AB.
14. Use the information from the figure to find the exact value of $\tan 30°$ in simple radical form.
15. Use your answer to Exercise 14 and the fact that $\sqrt{3} \approx 1.732$ to find $\tan 30°$ to three decimal places.
16. Use the figure to find exact and approximate values of $\tan 60°$. Check your answers to this exercise and Exercise 15 by referring to the table on page 405.

In $\triangle EGH$, $\overline{IT} \perp \overline{EH}$ and $\overline{GH} \perp \overline{EH}$.

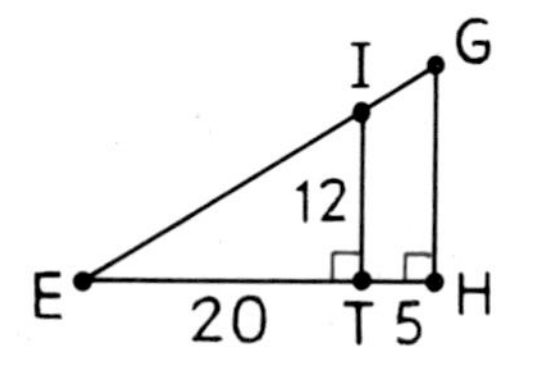

17. Why is $\triangle EIT \sim \triangle EGH$?
18. Find GH.
19. Use $\triangle EIT$ to find the value of $\tan E$ in decimal form.
20. Use $\triangle EGH$ to find the value of $\tan E$ in decimal form.

Solve for the length labeled x in each figure below. Express each length to the nearest tenth. Show your methods.

Example:

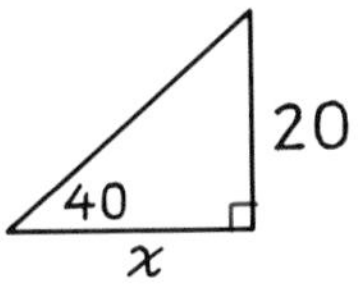

Solution: $\tan 40° = \frac{20}{x}$

$x \tan 40° = 20$

$$x = \frac{20}{\tan 40°} \approx \frac{20}{.839} \approx 23.8$$

21.

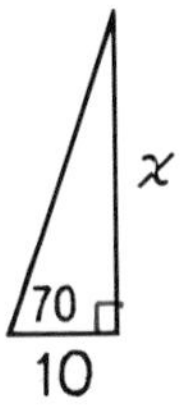

22.

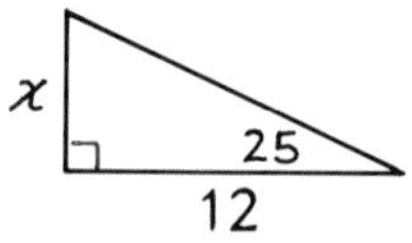

23.

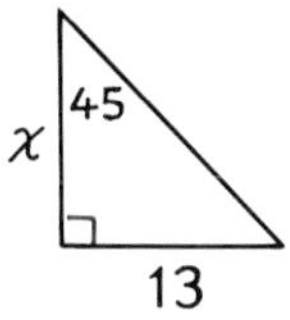

24.

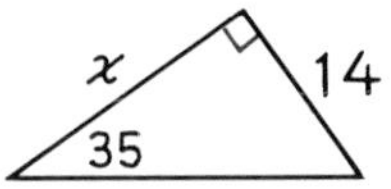

Use the tangent ratio to solve for the angle marked x in each figure below.

Example:

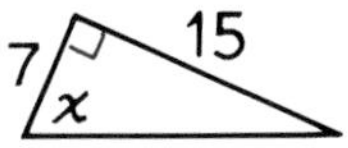

Solution: $\tan x = \frac{15}{7} \approx 2.143$

$x \approx 65°.$

25.

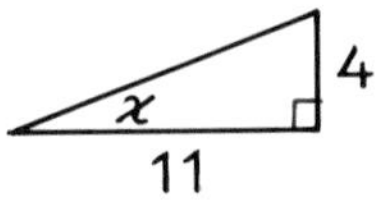

26.

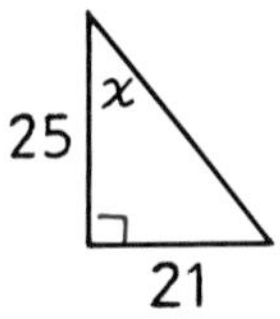

27.

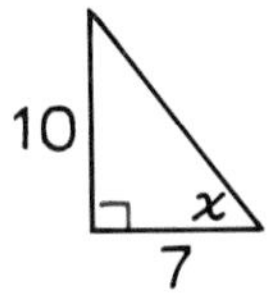

28.

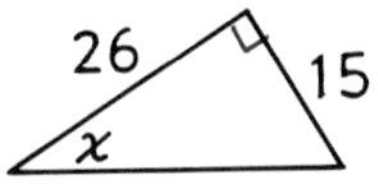

Set II

Write and solve an equation using the tangent ratio for each of the following exercises. Express each length to the nearest integer.

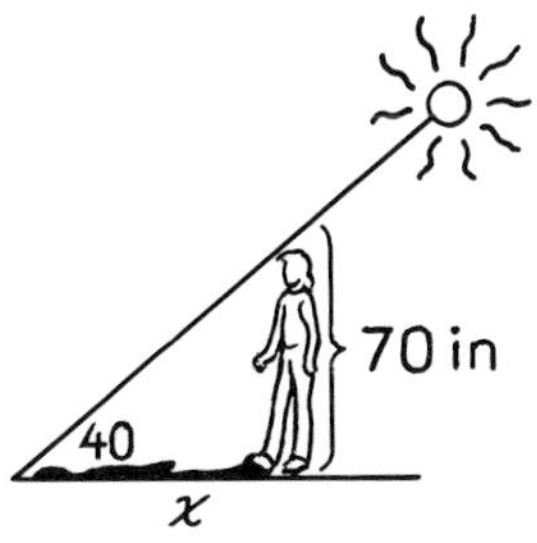

Example: How long is the shadow cast by a person 70 inches tall when the angle of elevation of the sun is 40°?

Solution:

$\tan 40° = \frac{70}{x}$

$x \tan 40° = 70$

$$x = \frac{70}{\tan 40°} \approx \frac{70}{.839} \approx 83.4$$

The shadow is approximately 83 inches long.

29. A street slopes upward at an angle of 15° with the horizontal. How high does it rise over a horizontal distance of 120 meters?

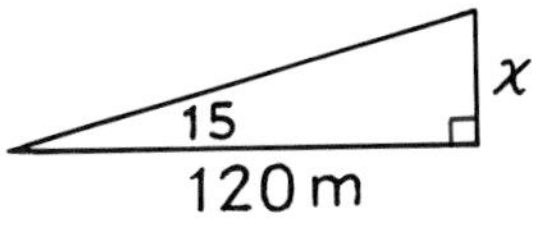

30. Find the distance from the bottom of the ladder to the bottom of the wall given that AB = 16 feet and ∠C = 50°.

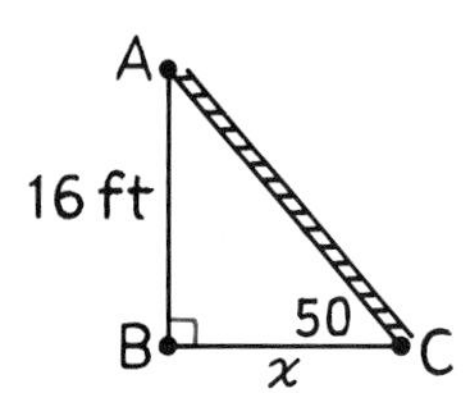

31. Find the distance across the river given that AB = 8 meters and ∠B = 65°.

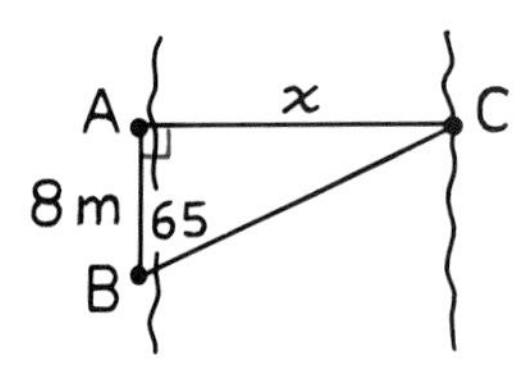

32. Find the angle of elevation of the top of the movie screen as viewed from point C given that AB = 48 feet and BC = 103 feet.

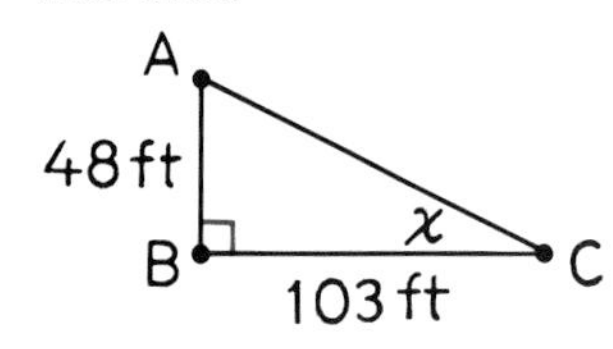

33.

E
T
N

Given: Right △TEN.

Prove: (tan T)(tan E) = 1.

34.

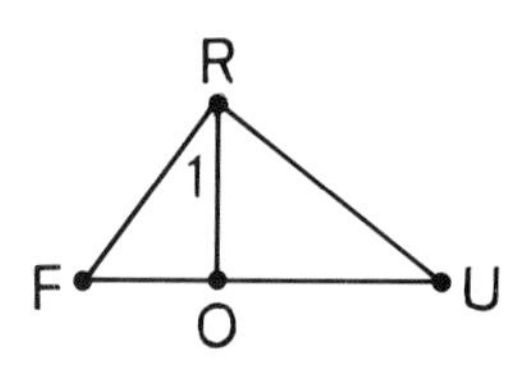

Given: Right △FUR with altitude $\overline{RO}$.

Prove: tan 1 = tan U.

35.

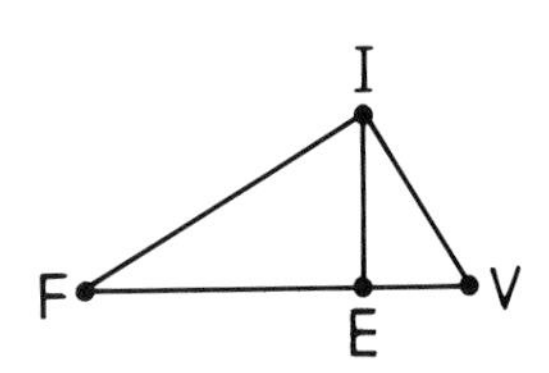

Given: △FIV with $\overline{IE} \perp \overline{FV}$.

Prove: FE tan F = EV tan V.

Set III

A palm tree 21 feet tall casts a horizontal shadow 30 feet long. If the sun rose at 6:00 A.M. and will be directly overhead at noon, what time is it? Show your method.

Drawing by Leo Garel; copyright 1972 by Saturday Review Co.

Frank Siteman/Jeroboam Inc.

Lesson 5

The Sine and Cosine Ratios

How high a kite will fly depends on both the length of its line and the size of its angle of elevation. In kite-flying competitions, performance is often judged by this angle rather than by the altitude of the kite.*

In the figure at the right, $\overline{AB}$ represents the line of a kite and ∡A represents its angle of elevation. Suppose that AB = 500 feet and ∠A = 35°. How high is the kite above the ground?

To figure this out, we need a trigonometric ratio called the *sine* ratio.

► **Definition**

The ***sine*** of an acute angle of a right triangle is the ratio of the length of the opposite leg to the length of the hypotenuse.

In the figure at the right, the sine of ∡A is $\frac{BC}{AB}$ and the sine of ∡B is $\frac{AC}{AB}$.

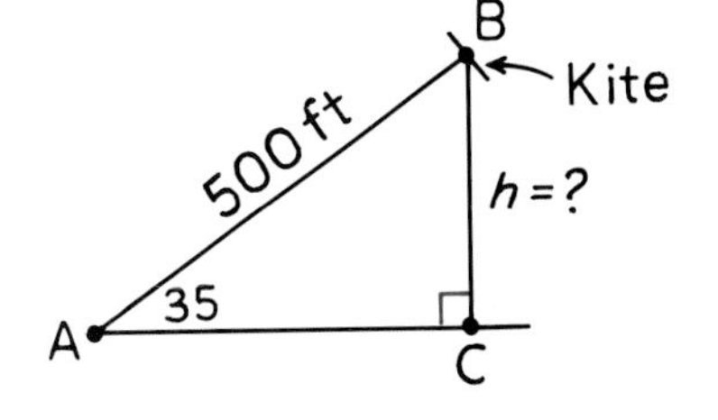

Kites: The Science and the Wonder, by Toshio Ito and Hirotsugu Komura (Japan Publications, Inc., 1983).

These relations are abbreviated as

$$\sin A = \frac{BC}{AB} \quad \text{and} \quad \sin B = \frac{AC}{AB}.$$

The sines of angles of different measures, like their tangents, can be found in tables or by means of a scientific calculator. A table listing these ratios for angles in 1° intervals appears on the next page.

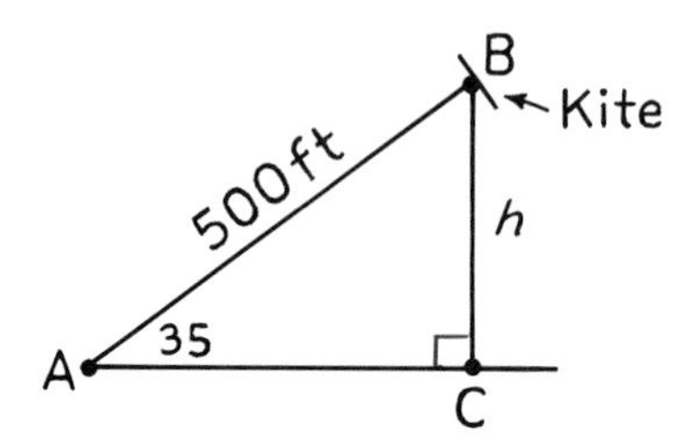

To find the altitude of the kite for which AB = 500 feet and ∠A = 35°, we write the equation

$$\sin 35° = \frac{h}{500}.$$

Multiplying by 500, we get

$$h = 500 \sin 35°.$$

We see from the table that sin 35° ≈ .574, and so

$$h \approx 500(.574)$$
$$\approx 287 \text{ feet.}$$

Another useful trigonometric ratio is the *cosine*. It relates, for a given acute angle in a right triangle, the lengths of the adjacent leg and the hypotenuse.

► **Definition**
The ***cosine*** of an acute angle of a right triangle is the ratio of the length of the adjacent leg to the length of the hypotenuse.

In the figure above, the cosine of ∡A is $\frac{AC}{AB}$ and the cosine of ∡B is $\frac{BC}{AB}$. These relations are abbreviated as

$$\cos A = \frac{AC}{AB} \quad \text{and} \quad \cos B = \frac{BC}{AB}.$$

Table of Trigonometric Ratios

A	sin A	cos A	tan A	A	sin A	cos A	tan A
1°	.017	1.000	.017	46°	.719	.695	1.035
2°	.035	.999	.035	47°	.731	.682	1.072
3°	.052	.999	.052	48°	.743	.669	1.111
4°	.070	.998	.070	49°	.755	.656	1.150
5°	.087	.996	.087	50°	.766	.643	1.192
6°	.105	.995	.105	51°	.777	.629	1.235
7°	.122	.993	.123	52°	.788	.616	1.280
8°	.139	.990	.141	53°	.799	.602	1.327
9°	.156	.988	.158	54°	.809	.588	1.376
10°	.174	.985	.176	55°	.819	.574	1.428
11°	.191	.982	.194	56°	.829	.559	1.483
12°	.208	.978	.213	57°	.839	.545	1.540
13°	.225	.974	.231	58°	.848	.530	1.600
14°	.242	.970	.249	59°	.857	.515	1.664
15°	.259	.966	.268	60°	.866	.500	1.732
16°	.276	.961	.287	61°	.875	.485	1.804
17°	.292	.956	.306	62°	.883	.469	1.881
18°	.309	.951	.325	63°	.891	.454	1.963
19°	.326	.946	.344	64°	.899	.438	2.050
20°	.342	.940	.364	65°	.906	.423	2.145
21°	.358	.934	.384	66°	.914	.407	2.246
22°	.375	.927	.404	67°	.921	.391	2.356
23°	.391	.921	.424	68°	.927	.375	2.475
24°	.407	.914	.445	69°	.934	.358	2.605
25°	.423	.906	.466	70°	.940	.342	2.747
26°	.438	.899	.488	71°	.946	.326	2.904
27°	.454	.891	.510	72°	.951	.309	3.078
28°	.469	.883	.532	73°	.956	.292	3.271
29°	.485	.875	.554	74°	.961	.276	3.487
30°	.500	.866	.577	75°	.966	.259	3.732
31°	.515	.857	.601	76°	.970	.242	4.011
32°	.530	.848	.625	77°	.974	.225	4.331
33°	.545	.839	.649	78°	.978	.208	4.705
34°	.559	.829	.675	79°	.982	.191	5.145
35°	.574	.819	.700	80°	.985	.174	5.671
36°	.588	.809	.727	81°	.988	.156	6.314
37°	.602	.799	.754	82°	.990	.139	7.115
38°	.616	.788	.781	83°	.993	.122	8.144
39°	.629	.777	.810	84°	.995	.105	9.514
40°	.643	.766	.839	85°	.996	.087	11.430
41°	.656	.755	.869	86°	.998	.070	14.301
42°	.669	.743	.900	87°	.999	.052	19.081
43°	.682	.731	.933	88°	.999	.035	28.636
44°	.695	.719	.966	89°	1.000	.017	57.290
45°	.707	.707	1.000				

Exercises

Set I

Express the following ratios for $\triangle GAM$ and $\triangle BLE$ in terms of the lengths of their sides.

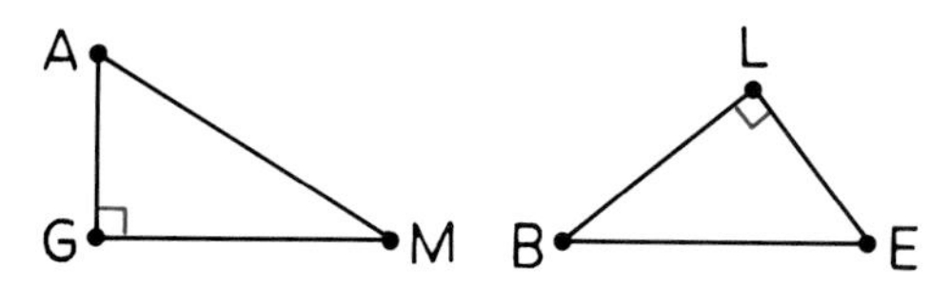

Example: sin M.

Answer: $\sin M = \frac{AG}{AM}$.

1. sin A.
2. cos B.
3. tan M.
4. sin B.
5. cos M.
6. tan E.

The following exercises refer to the figure below.

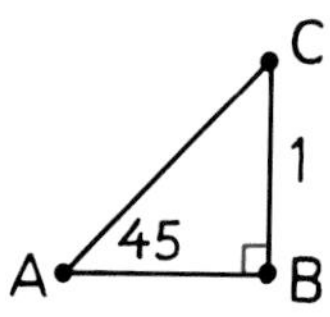

7. Use the Isosceles Right Triangle Theorem to find AC.
8. Use the information from the figure to find the exact value of sin 45° in simple radical form.
9. Use the fact that $\sqrt{2} \approx 1.414$ to find sin 45° to three decimal places. Check your answer by referring to the table on page 411.
10. Use the figure to find cos 45°.

The following exercises refer to the figure below.

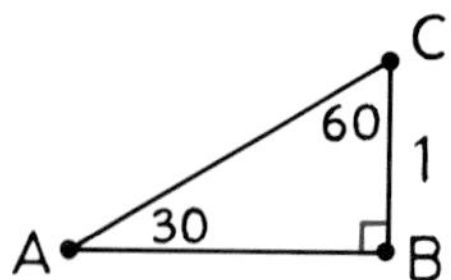

11. Copy the figure and use the 30°-60° Right Triangle Theorem to find the lengths of the other two sides.
12. Use the information from the figure to find the exact value of sin 30° as a common fraction. Check your answer by referring to the table on page 411.
13. Use the figure to find cos 60°.
14. Use the figure to find the exact value of sin 60° in simple radical form.
15. Use the fact that $\sqrt{3} \approx 1.732$ to find sin 60° to three decimal places. Check your answer by referring to the table on page 411.
16. Use the figure to find cos 30°.

Use either the table on page 411 or your calculator to find each of the following angles to the nearest degree. (If the ratio does not appear in the table, choose the angle whose ratio is closest to it.)

17. $\angle C$ if $\sin C = .755$.
18. $\angle H$ if $\cos H = .927$.
19. $\angle A$ if $\tan A = .810$.
20. $\angle N$ if $\sin N = .340$.
21. $\angle C$ if $\tan C = 1.615$.
22. $\angle E$ if $\cos E = .093$.

Solve for x in each figure. Express each length to the nearest tenth. Show your work.

23. 24. 25.

26. 27. 28.

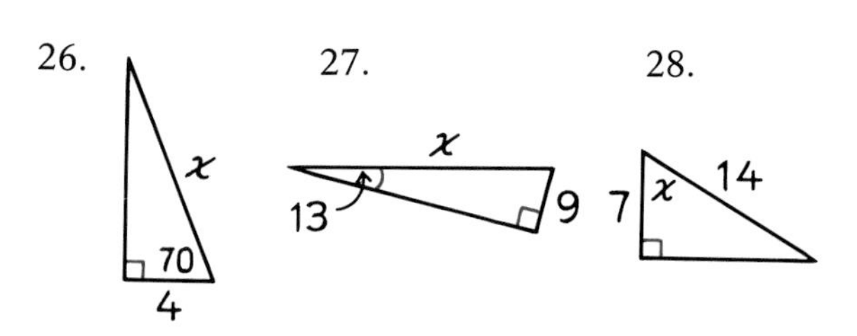

Set II

Write and solve an equation, using an appropriate trigonometric ratio, for each of the following exercises. Express each length to the nearest unit.

29. How tall is a totem pole if the line of sight from a point 100 feet from its base makes an angle of 39° with the horizontal?

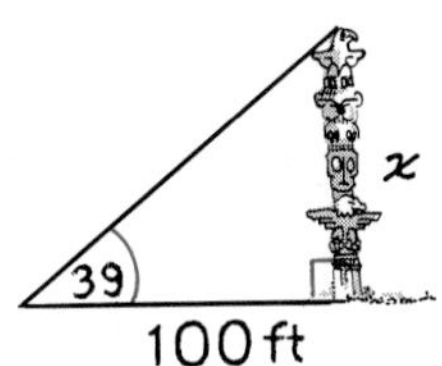

30. A crow sitting on a tree branch 25 feet above the ground sees something to eat on the ground and swoops down on it at an angle of 20° with the horizontal. How long is the crow's flight?

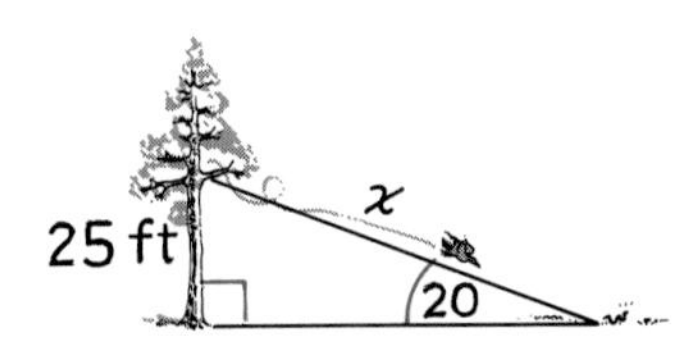

31. In teeing off, a golfer hooks his drive at an angle of 14° from the line to the hole. The hole is 150 yards from the tee. If the ball ends up due left of the hole, how far is it from the hole?

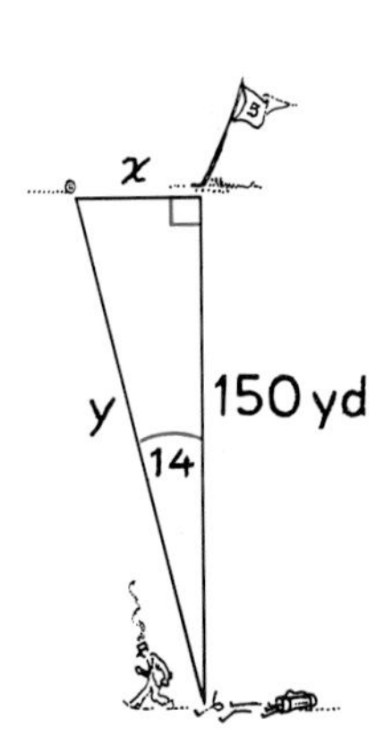

32. How far did the ball go?

33.

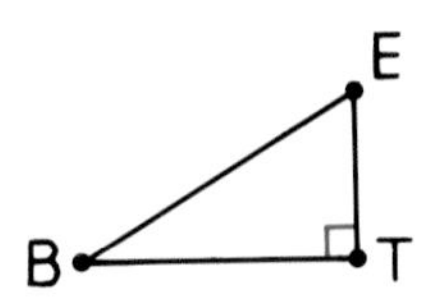

Given: $\triangle BET$ is a right triangle.

Prove: $\sin B = \cos E$.

34.

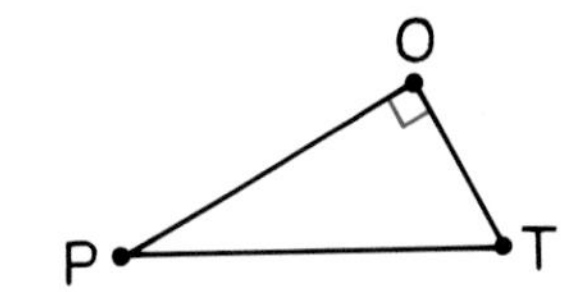

Given: $\triangle POT$ is a right triangle in which $\angle T > \angle P$.

Prove: $\cos P > \cos T$.

35.

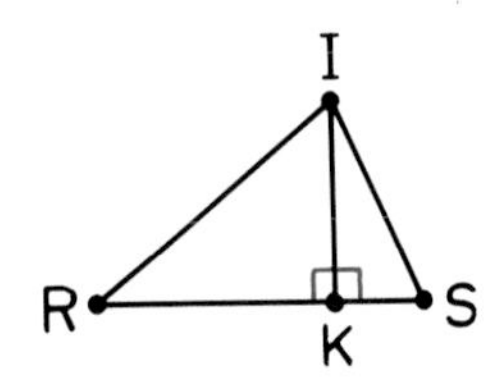

Show that $\alpha\triangle RIS = \frac{1}{2}RI \cdot RS \cdot \sin R$.

Set III

Возьмем произвольный острый угол α и построим прямоугольный треугольник ABC, у которого один из острых углов равен α. Пусть $BC = a$, $CA = b$ и $AB = c$.

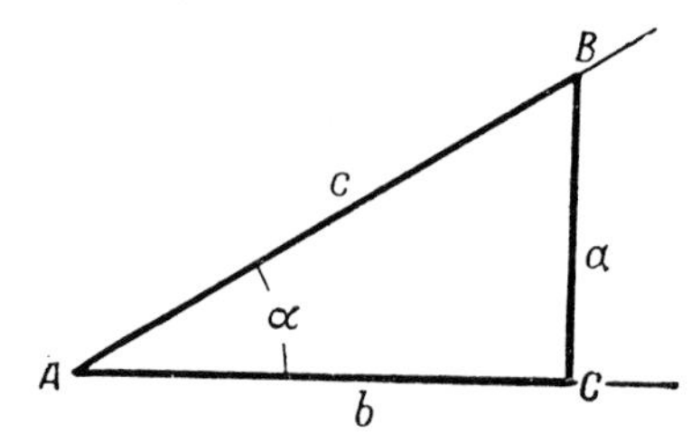

Из прямоугольного треугольника ABC по теореме Пифагора находим:

$$a^2 + b^2 = c^2.$$

Разделив обе части этого равенства на c^2, получим:

$$\left(\frac{a}{c}\right)^2 + \left(\frac{b}{c}\right)^2 = 1,$$

или

$$\sin^2 \alpha + \cos^2 \alpha = 1,$$

так как $\frac{a}{c} = \sin \alpha$ и $\frac{b}{c} = \cos \alpha$.

In this passage from a Russian trigonometry book, a relation is derived that is so useful that it is considered a theorem in the subject.

1. What do you think the theorem says?

2. What do you think the passage says?

Chapter 11 / Summary and Review

Basic Ideas

Cosine 410
Pythagorean triple 397
Sine 409
Tangent 405

Theorems

62. The altitude to the hypotenuse of a right triangle forms two triangles that are similar to each other and to the original triangle. 391
Corollary 1. The altitude to the hypotenuse of a right triangle is the geometric mean between the segments into which it divides the hypotenuse. 391
Corollary 2. Each leg of a right triangle is the geometric mean between the hypotenuse and its projection on the hypotenuse. 391

The Pythagorean Theorem. In a right triangle, the square of the hypotenuse is equal to the sum of the squares of the legs. 396

63. *The Isosceles Right Triangle Theorem.* In an isosceles right triangle, the hypotenuse is $\sqrt{2}$ times the length of one leg. 400
Corollary. Each diagonal of a square is $\sqrt{2}$ times the length of one side. 400

64. *The 30°-60° Right Triangle Theorem.* In a 30°-60° right triangle, the hypotenuse is twice the length of the shorter leg and the longer leg is $\sqrt{3}$ times the length of the shorter leg. 400
Corollary. An altitude of an equilateral triangle is $\frac{\sqrt{3}}{2}$ times the length of one side. 401

Exercises

Set I

By the permission of Johnny Hart and News America Syndicate

In the diagram below, $\overline{AB}$ represents the position of the pole at the beginning of a pole vault.

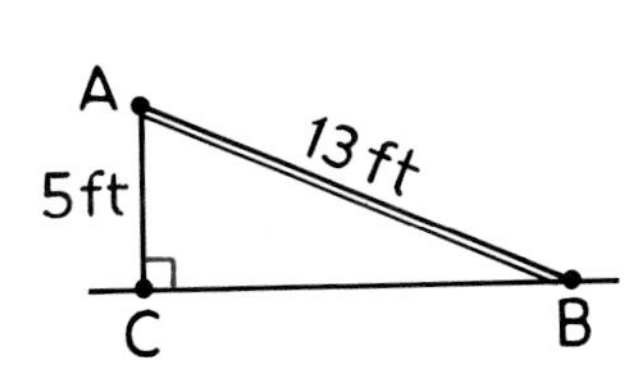

1. Find CB.
2. Find ∠ABC.

Without using trigonometric ratios, find the exact value of x in each of the following figures.

3.

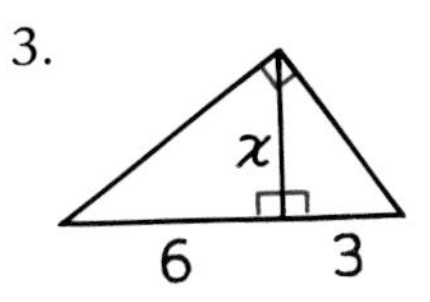

4.

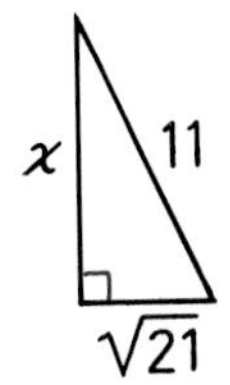

5.

6.

7.

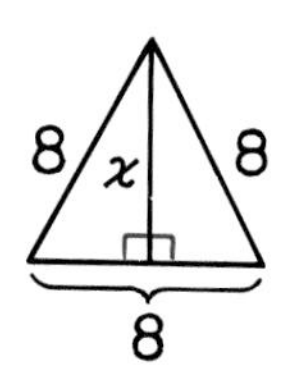

8.

In the figure below, $\triangle CRA$ is a right triangle and $\overline{RN}$ is the altitude to its hypotenuse.

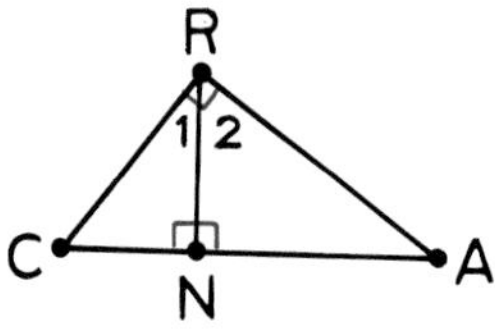

9. Why is $\triangle CRN \sim \triangle RAN$?

10. Why is $\angle 1 = \angle A$ and $\angle 2 = \angle C$?

11. Write the ratio for tan 1 in terms of $\triangle CRN$.

12. Write the ratio for tan A in terms of $\triangle RAN$.

13. Write the proportion that follows from substituting your answers to Exercises 11 and 12 in the equation tan 1 = tan A.

14. What theorem does your answer to Exercise 13 illustrate?

15. Write the ratio for sin 2 in terms of $\triangle RAN$.

16. Write the ratio for sin C in terms of $\triangle CRA$.

17. Write the proportion that follows from substituting your answers to Exercises 15 and 16 in the equation sin 2 = sin C.

18. What theorem does your answer to Exercise 17 illustrate?

The figure below, from a French geometry book published in 1556, illustrates a carpenter's square being used to determine an unknown distance. In it, $\overline{BC} \perp \overline{CF}$ and $\overline{CA} \perp \overline{BF}$. Suppose that AC = 8 feet and AF = 2 feet.

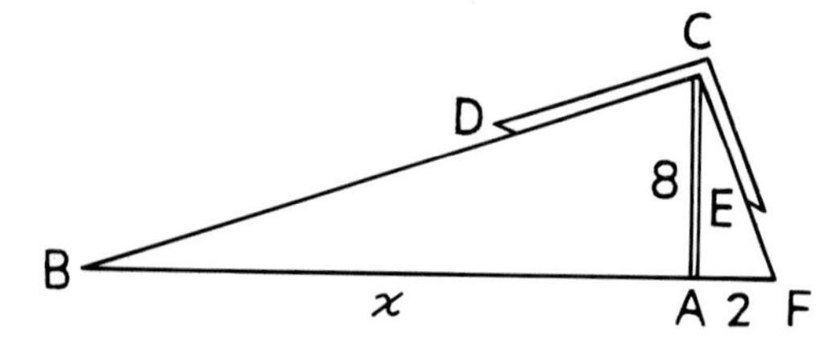

19. Find AB.

20. Find $\angle BCA$.

Solve for x in each figure. Express each length to the nearest tenth. Show your work.

21.

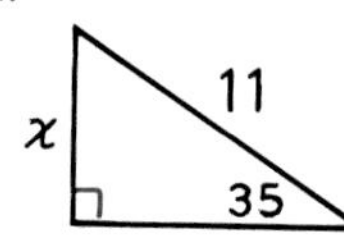

22.

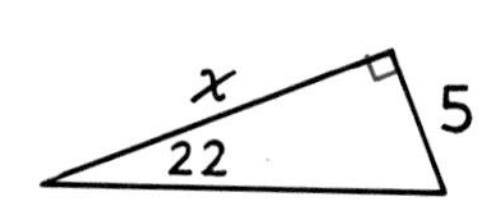

23.

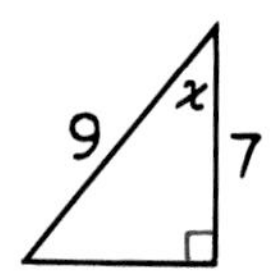

24.

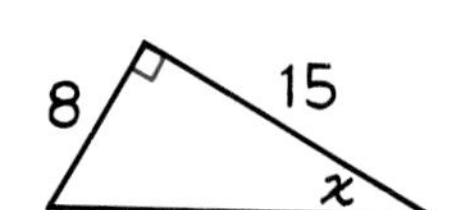

Set II

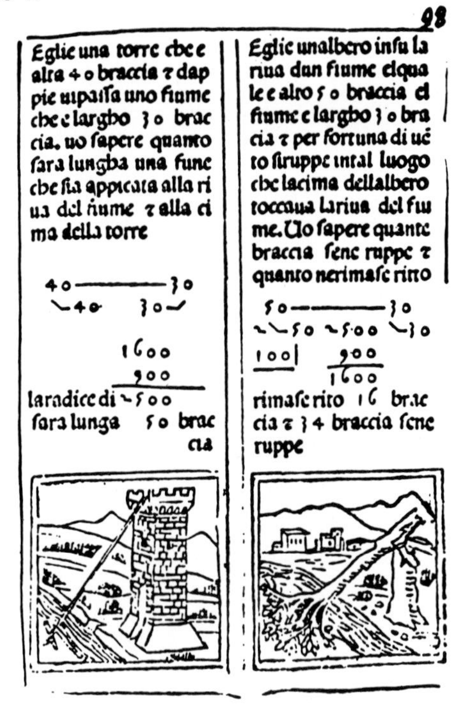

98

Eglie una torre che e alta 40 braccia e dap pie uipassa uno fiume che e largho 30 brac cia. uo sapere quanto sara lungha una fune che sia appicata alla ri ua del fiume e alla ci ma della torre

40 —— 30
40 30
1600
900
la radice di 2500
sara lunga 50 brac cia

Eglie unalbero insu la riua dun fiume elqua le e alto 50 braccia el fiume e largho 30 bra cia e per fortuna di uē to siruppe intal luogo che la cima dellalbero toccaua la riua del fiu me. Uo sapere quante braccia sene ruppe e quanto nerimase ritto

50 —— 30
50 2500 30
100 | 900
1600
rimase ritto 16 brac cia e 34 braccia sene ruppe

This page is from a book printed in Italy in 1491. The first figure shows a ladder leaning against a tower. The tower is 40 feet tall, the bottom of the ladder is 30 feet from the base of the tower, and the ladder just reaches the top of the tower.

25. How long is the ladder?

26. Find the measure of the angle that it makes with the tower.

Draw figures before solving the following exercises.

27. Find the perimeter of a 30°-60° right triangle whose shorter leg is 4.

28. Find the area of an isosceles right triangle whose hypotenuse is 20.

29. Find the perimeter of a rhombus one of whose angles is 60° and whose altitude is 12.

30. Find the area of an isosceles trapezoid one of whose angles is 45°, whose altitude is 7 and whose shorter base is 3.

Write and solve an equation, using an appropriate trigonometric ratio, for each of the following exercises. Express each answer to the nearest unit.

31. How high is an eruption of Old Facefull if the line of sight from a point 90 feet from its base makes an angle of 57° with the horizontal?

32. A 25-foot ladder leans against a burning building. If it just reaches a window 18 feet above the ground, what angle does it make with the horizontal?

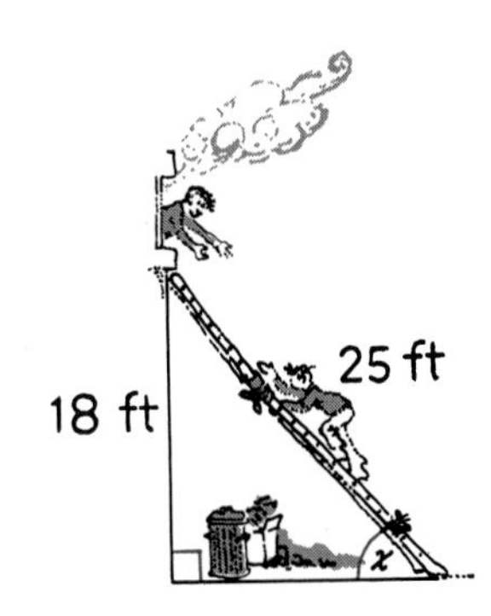

33. What is the angle of elevation of the moon if a 7-foot creature that walks in the night casts a shadow 45 feet long?

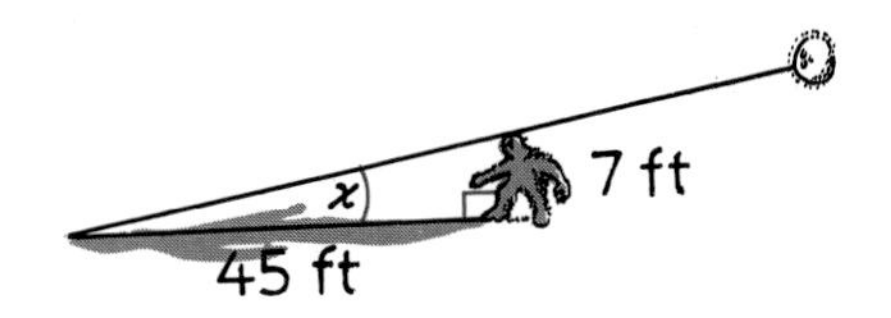

34.

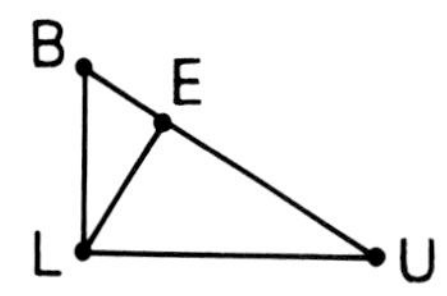

Given: Right $\triangle BLU$ with altitude $\overline{LE}$ to hypotenuse $\overline{BU}$.

Prove: $\dfrac{\alpha\triangle BEL}{\alpha\triangle LEU} = \left(\dfrac{BL}{LU}\right)^2$.

35.

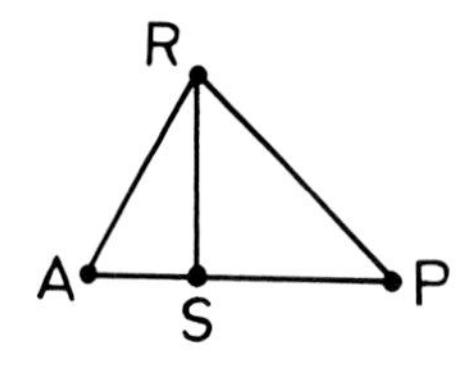

Given: $\triangle RAP$ with altitude $\overline{RS}$.

Prove: $RA^2 - AS^2 = RP^2 - PS^2$.

36.

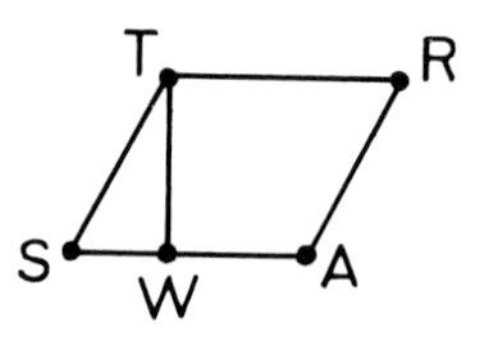

Given: STRA is a parallelogram, $\triangle STW$ is a right triangle and $\sin S = \dfrac{ST}{TR}$.

Prove: ST is the geometric mean between TW and SA.

Chapter 12
CIRCLES

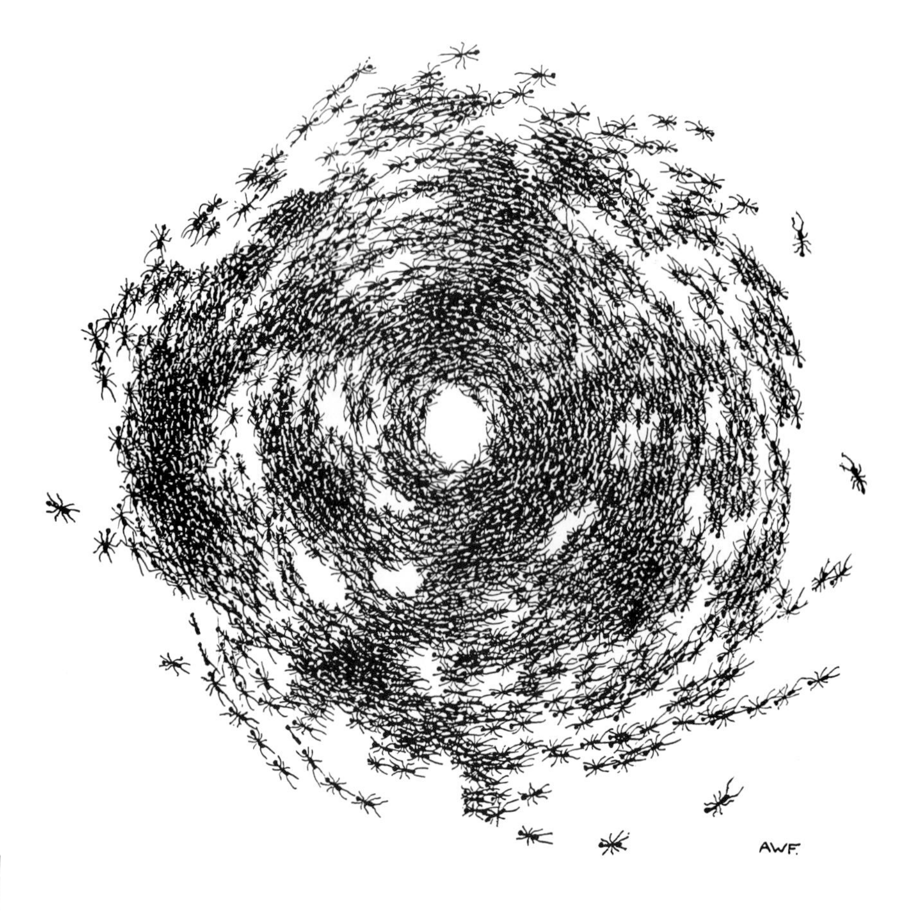

Lesson 1

Circles, Radii, and Chords

Army ants are blind and find their way by following the scent trails left by other ants. When a group of these ants is prevented from traveling along its usual path, the ants sometimes begin milling around in circles. The ants shown in this drawing walked around like this for more than thirty hours, stopping only when all of them were dead.*

The path in which the ants are walking is one of the most familiar of all geometric figures. In this chapter, we will consider the properties of the circle; we begin by reviewing a few familiar terms and introducing some new ones.

► **Definition**

A ***circle*** is the set of all points in a plane that are at a given distance from a given point in the plane.

---

*T. C. Schneirla, *Army Ants,* edited by H. R. Topoff (W. H. Freeman and Company, 1971). Drawing courtesy of the Department Library Services, American Museum of Natural History, neg. 322190.

The given point is called the *center* of the circle and the given distance is called its *radius*.

The ants in the picture seem to be walking in circles that are *concentric*.

► **Definition**
Circles are ***concentric*** iff they lie in the same plane and have the same center.

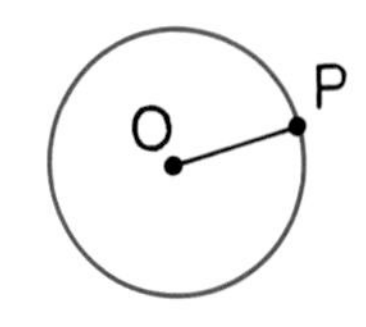

A circle is ordinarily named for its center, and so the figure at the right illustrates circle O. By definition, the radius of circle O is the *distance* between O and P. The word "radius" is also used to name the *line segment*, $\overline{OP}$.*

► **Definition**
A ***radius*** of a circle is a line segment that joins the center of the circle to a point of the circle.

A useful fact that follows directly from the definitions of "circle" and "radius" is:

► **Corollary**
All radii of a circle are equal.

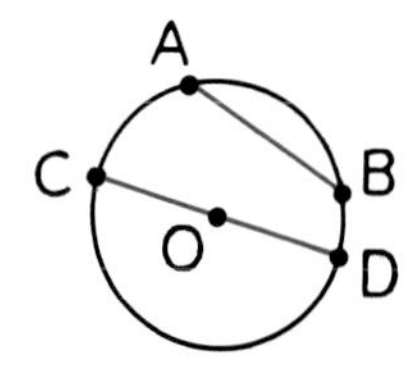

The figure at the right illustrates two more line segments related to circles that are of interest; $\overline{AB}$ is called a *chord* of circle O and $\overline{CD}$ is called a *diameter*.

► **Definitions**
A ***chord*** of a circle is a line segment that joins two points of the circle.

A ***diameter*** of a circle is a chord that contains the center.

The word "diameter," like "radius," is used in two ways. It refers to both a *line segment* and the *number* that is its length. Thinking in terms of numbers, if we represent a circle's diameter by d and its radius by r, then

$$d = 2r.$$

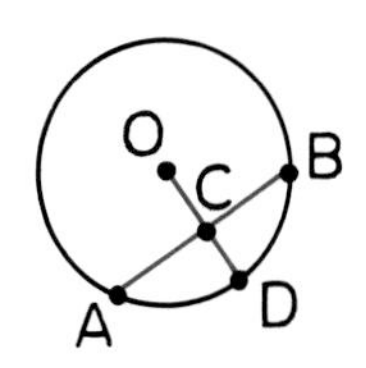

The figure at the right shows a circle in which a radius intersects a chord. If the radius is perpendicular to the chord, we can prove that it bisects it. Conversely, if the radius bisects the chord, we can show that it must be perpendicular to it. These relations, along with a third one, can be proved as theorems.

*It would be better not to use the term "radius" in two different ways, but this is what is ordinarily done. The context in which the word is used, however, should make it clear whether we mean a *line segment* or a *number*.

► **Theorem 65**
If a line through the center of a circle is perpendicular to a chord, it also bisects the chord.

► **Theorem 66**
If a line through the center of a circle bisects a chord that is not a diameter, it is also perpendicular to the chord.

► **Theorem 67**
The perpendicular bisector of a chord (in the plane of a circle) passes through the center of the circle.

Exercises

Set I

In the figure below, one vertex of $\triangle JIG$ is the center of the circle.

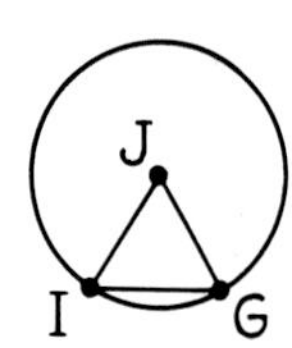

1. What is the name of the circle?
2. How many vertices of the triangle are on the circle?
3. What are $\overline{JI}$ and $\overline{JG}$ called with respect to the circle?
4. What is $\overline{IG}$ called with respect to the circle?
5. What kind of triangle must $\triangle JIG$ be?
6. Why?

When a pebble is dropped in a lake, it forms ripples as shown in this figure.

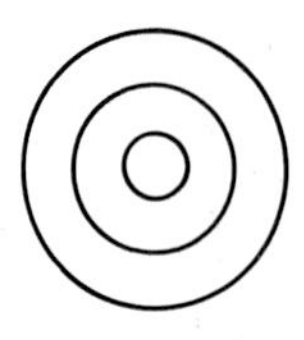

7. What do the circles in the figure seem to have in common?
8. What word names the relation of these circles?

In the figure below, $\overleftrightarrow{AG}$ intersects chord $\overline{TN}$ of circle O at O.

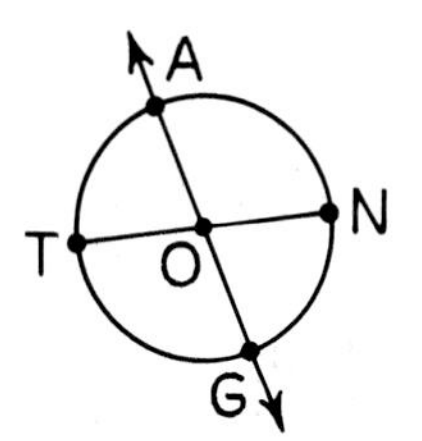

9. What is chord $\overline{TN}$ called with respect to the circle?
10. Does $\overleftrightarrow{AG}$ bisect $\overline{TN}$?
11. Is $\overline{AG} \perp \overline{TN}$?

Give the missing statements and reasons in the following proofs of the theorems in this lesson.

12. *Theorem 65.* If a line through the center of a circle is perpendicular to a chord, it also bisects the chord.

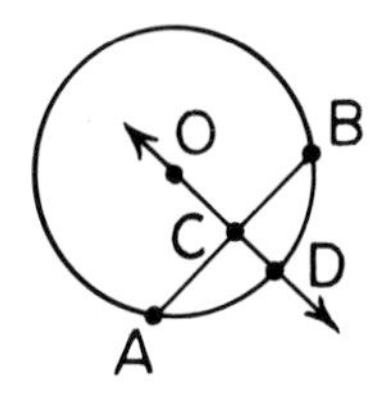

Given: $\overleftrightarrow{OD} \perp \overline{AB}$ in circle O.
Prove: $\overleftrightarrow{OD}$ bisects $\overline{AB}$.

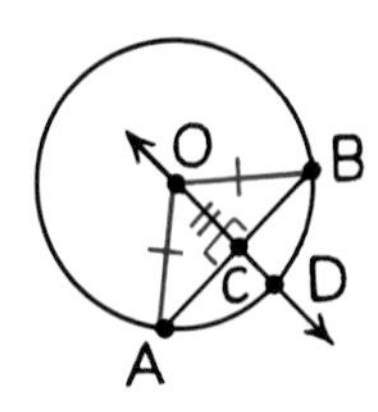

Proof.

Statements	***Reasons***
1. $\overleftrightarrow{OD} \perp \overline{AB}$ in circle O.	Given.
2. a) ▯▯▯▯▯▯	Perpendicular lines form right angles.
3. Draw $\overline{OA}$ and $\overline{OB}$.	b) ▯▯▯▯▯▯
4. △OCA and △OCB are right triangles.	c) ▯▯▯▯▯▯
5. OA = OB.	d) ▯▯▯▯▯▯
6. e) ▯▯▯▯▯▯	Reflexive.
7. △OCA ≅ △OCB.	f) ▯▯▯▯▯▯
8. g) ▯▯▯▯▯▯	If two triangles are congruent, their corresponding sides are equal.
9. h) ▯▯▯▯▯▯	If a line segment is divided into two equal segments, it is bisected.

13. *Theorem 66.* If a line through the center of a circle bisects a chord that is not a diameter, it is also perpendicular to the chord.

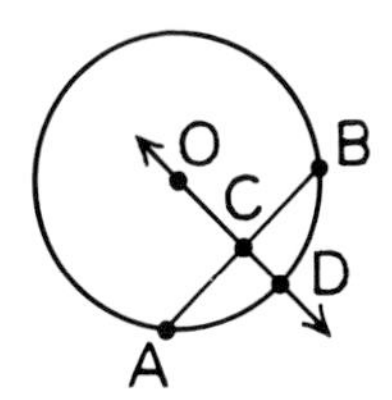

Given: $\overleftrightarrow{OD}$ bisects $\overline{AB}$ in circle O.
Prove: $\overleftrightarrow{OD} \perp \overline{AB}$.

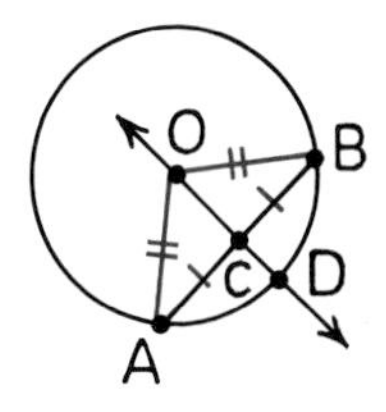

Proof.

Statements	***Reasons***
1. $\overleftrightarrow{OD}$ bisects $\overline{AB}$ in circle O.	Given.
2. a) ▯▯▯▯▯▯	If a line segment is bisected, it is divided into two equal segments.
3. b) ▯▯▯▯▯▯	Two points determine a line.
4. c) ▯▯▯▯▯▯	All radii of a circle are equal.
5. d) ▯▯▯▯▯▯	In a plane, two points each equidistant from the endpoints of a line segment determine the perpendicular bisector of the line segment.

14. *Theorem 67.* The perpendicular bisector of a chord (in the plane of a circle) passes through the center of the circle.

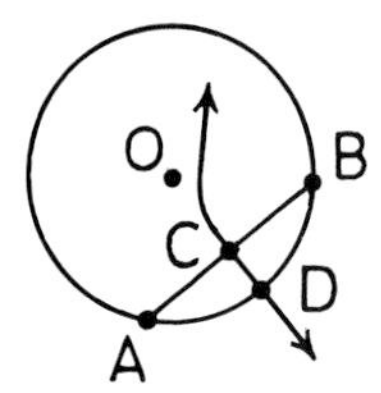

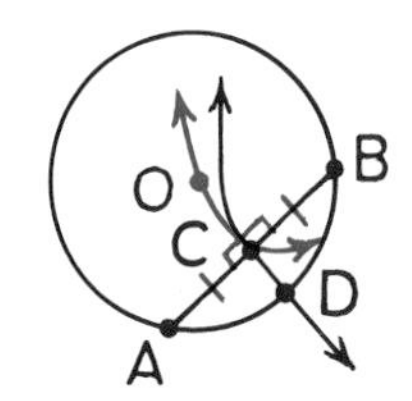

Given: $\overleftrightarrow{CD}$ is the perpendicular bisector of $\overline{AB}$ in circle O.

Prove: $\overleftrightarrow{CD}$ contains point O.

Proof.
Line $\overleftrightarrow{CD}$ is the perpendicular bisector of $\overline{AB}$ in circle O.

Suppose that $\overleftrightarrow{CD}$ does not contain point O. Draw $\overleftrightarrow{OC}$.

a) Why is $\overleftrightarrow{OC} \perp \overline{AB}$?

This means that through point C there are two lines perpendicular to $\overline{AB}$: $\overleftrightarrow{OC}$ and $\overleftrightarrow{CD}$.

b) What theorem does this conclusion contradict?

This tells us that the assumption that $\overleftrightarrow{CD}$ does not contain point O is false. So $\overleftrightarrow{CD}$ contains point O.

Set II

In the figure below, $\overline{TP}$ is a chord of circle A.

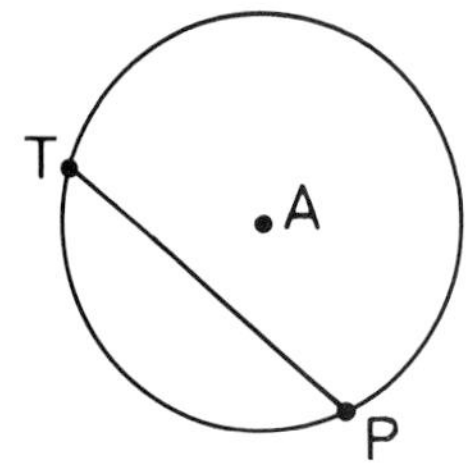

15. Draw the figure. Use your straightedge and compass to construct the perpendicular bisector of $\overline{TP}$.

16. Why does point A lie on the perpendicular bisector of $\overline{TP}$?

Circles A and O intersect in points G and P.

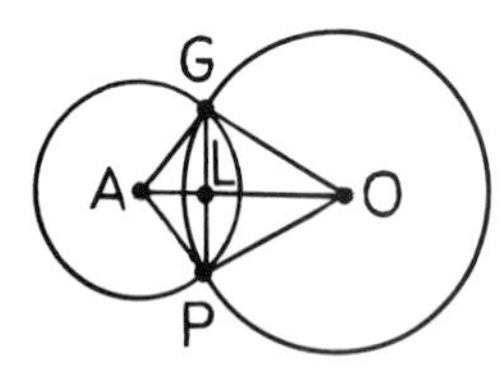

17. Is $\overline{AO}$ the perpendicular bisector of $\overline{GP}$? Explain.

18. Is $\overline{GP}$ the perpendicular bisector of $\overline{AO}$?

In circle H, $\overline{RA}$ bisects $\overline{BU}$; HU = 5 and HM = 3.

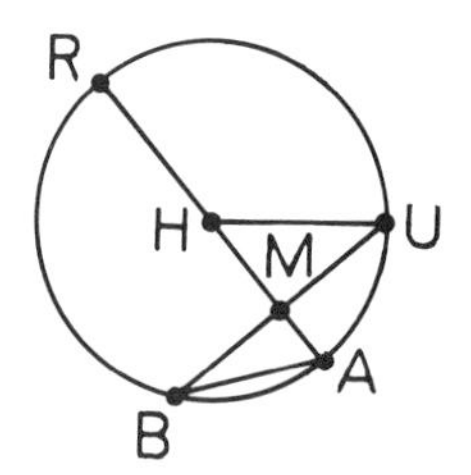

19. Why is $\overline{RA} \perp \overline{BU}$?

20. Draw the figure.

Use your drawing to find each of the following lengths.

21. RA.
22. MU.
23. BU.
24. MA.
25. BA.

In circle N, $\overline{DE} \perp \overline{FL}$ and $\overline{DE} \perp \overline{OK}$; DE = 50, FL = 30, and OK = 48.

26. Why does $\overline{DE}$ bisect $\overline{FL}$ and $\overline{OK}$?

27. Draw the figure.

Use your drawing to find each of the following lengths.

28. NF and NO.

29. NA.

30. NC.

31. If two chords in a circle have different lengths, which chord is closer to the center?

32.

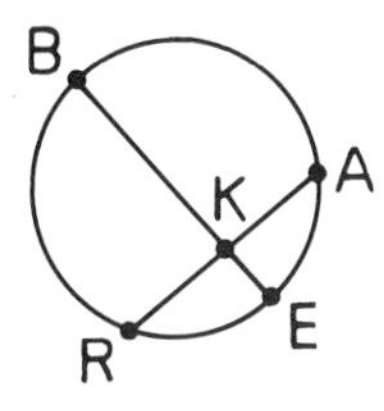

Given: $\overline{BE}$ is the perpendicular bisector of chord $\overline{RA}$.

Prove: $\overline{BE}$ is a diameter of the circle.

33.

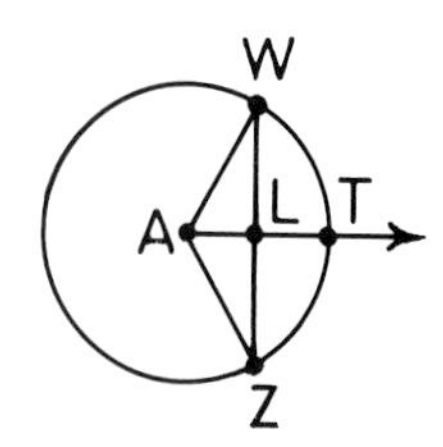

Given: $\overrightarrow{AT}$ bisects $\measuredangle WAZ$ in circle A.

Prove: $\overrightarrow{AT}$ bisects $\overline{WZ}$.

34.

Given: $\overline{LN} \perp \overline{FI}$ in circle G.

Prove: $\triangle FIN$ is isosceles.

35.

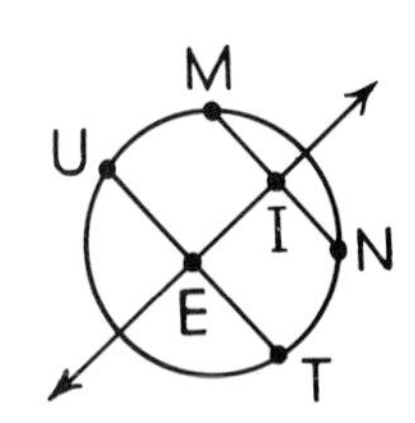

Given: $\overleftrightarrow{IE}$ is the perpendicular bisector of $\overline{MN}$; $\overline{MN} \parallel \overline{UT}$.

Prove: $\overleftrightarrow{IE}$ bisects $\overline{UT}$ *without* adding anything to the figure.

Set III

On a trip to an amusement park, Obtuse Ollie and Acute Alice wondered how tall the ferris wheel was. They asked the man operating the wheel, but he didn't know.

Ollie thought that he could find out by climbing to the top of the wheel, but Alice thought of a simpler way to estimate the height of the wheel that did not require any climbing at all. What do you think it was?

Chicago Historical Society, neg. ICHi-02445. Photograph by W.C.E., 1893.

By the permission of Johnny Hart and News America Syndicate

Lesson 2 Tangents

One way to propel an object with great force into space is to whirl it around a circular path several times before releasing it. The figure below represents a heavy object that is tied to one end of a rope at P and spun around in a circle whose center is the other end of the rope at O. If

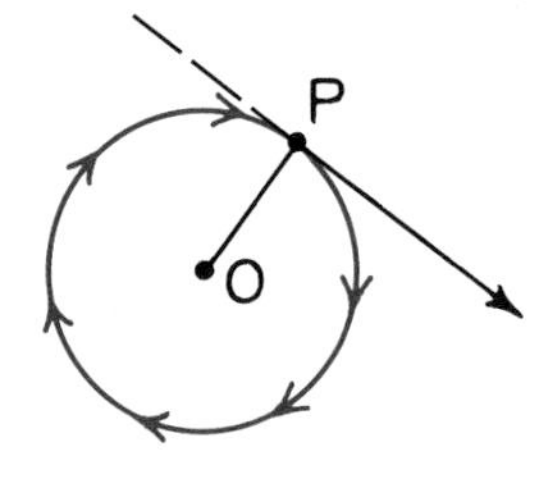

the rope is let go while the object is traveling in this circular path, it will fly off along a path that appears to be a straight line. The figure shows the relation of this line to the circle: it intersects the circle in exactly one point, the point that is the position of the object when the rope is released. Such a line is called a *tangent* to the circle.

► **Definition**
A ***tangent*** to a circle is a line in the plane of the circle that intersects it in exactly one point.

One use of the word "tangent" in everyday conversation refers to someone "going off on a tangent," meaning that they are changing suddenly from one train of thought or course of action to another. Our illustration of the change in path of the whirling object reveals where this expression came from.

The figure above suggests that a tangent to a circle is perpendicular to the radius drawn to the point at which the tangent intersects the circle. We will prove this as our next theorem.

► **Theorem 68**
If a line is tangent to a circle, it is perpendicular to the radius drawn to the point of contact.

Given: $\overleftrightarrow{AB}$ is tangent to circle O at point A.
Prove: $\overleftrightarrow{AB} \perp \overline{OA}$.

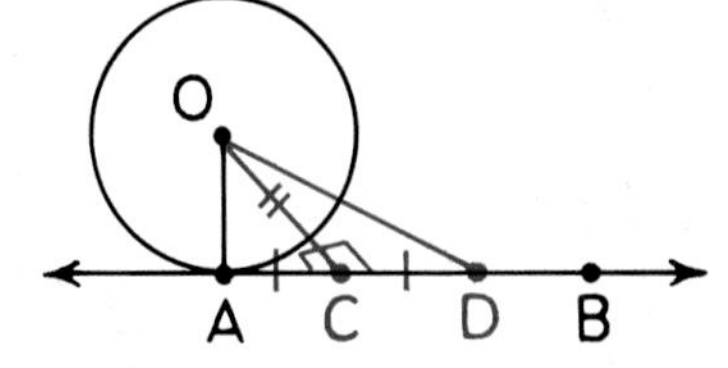

We will prove this theorem by the indirect method.

Proof.
Suppose that $\overleftrightarrow{AB}$ not $\perp \overline{OA}$ (i.e., $\overline{OA}$ not $\perp \overleftrightarrow{AB}$). We can draw $\overline{OC} \perp \overleftrightarrow{AB}$ because, through a point not on a line, there is exactly one perpendicular to the line. Next, we choose point D on $\overleftrightarrow{AB}$ such that CD = CA, and draw $\overline{OD}$.

Now $\triangle OCA \cong \triangle OCD$ (S.A.S.), and so OD = OA. Because OA is the radius of the circle, this means that point D is also at this distance from the center. Therefore, D must be a point on the circle because a circle is the set of all points in the plane at the distance of the radius from the center.

So $\overleftrightarrow{AB}$ intersects the circle in two points: A and D. But this contradicts the fact that $\overleftrightarrow{AB}$ is tangent to the circle, because a tangent to a circle intersects it in exactly one point. This means that our assumption that $\overleftrightarrow{AB}$ not $\perp \overline{OA}$ is wrong, and so $\overleftrightarrow{AB} \perp \overline{OA}$.

The converse of this theorem is also true:

► **Theorem 69**
If a line is perpendicular to a radius at its outer endpoint, then it is tangent to the circle.

Given: Circle O with $\overleftrightarrow{AC} \perp \overline{OA}$.
Prove: $\overleftrightarrow{AC}$ is tangent to circle O.

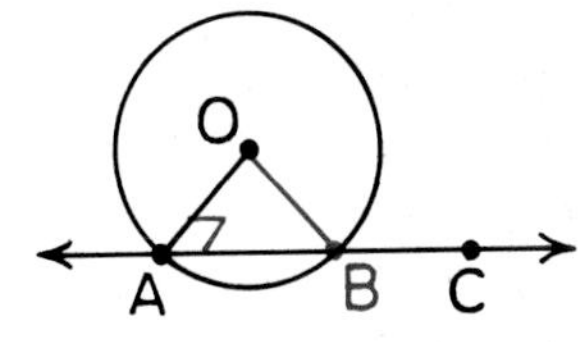

Proof.
Suppose that $\overleftrightarrow{AC}$ is not tangent to circle O. This means that it does not intersect the circle in exactly one point. Suppose that $\overleftrightarrow{AC}$ intersects circle O in point B in addition to point A.

Because $\overleftrightarrow{AC} \perp \overline{OA}$, ∡OAB is a right angle and ∠OAB = 90°. Because OA = OB (all radii of a circle are equal), ∠OBA = ∠OAB. It follows that ∠OBA = 90°, and so ∡OBA is a right angle and $\overleftrightarrow{AC} \perp \overline{OB}$. But $\overleftrightarrow{AC}$ cannot be perpendicular to both $\overline{OB}$ and $\overline{OA}$ because, through a point not on a line, there is exactly one perpendicular to the line. Therefore, our initial assumption is wrong and $\overleftrightarrow{AC}$ is tangent to circle O.

Exercises

Set I

In the figure below, $\overline{BD}$ is tangent to circle N; in $\triangle BND$, BD = 15 and BN = 17.

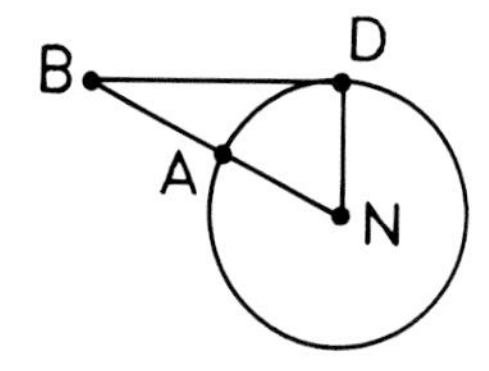

1. What can you conclude about $\overline{BD}$ and $\overline{ND}$?
2. State the theorem that is the basis for your answer.
3. Draw the figure.

Refer to your drawing to find each of the following numbers.

4. DN.
5. BA.
6. $\alpha\triangle BND$.

In the figure below, $\overline{WN}$ is a diameter of circle I, and $\overleftrightarrow{TW}$ and $\overleftrightarrow{NE}$ are tangent to the circle.

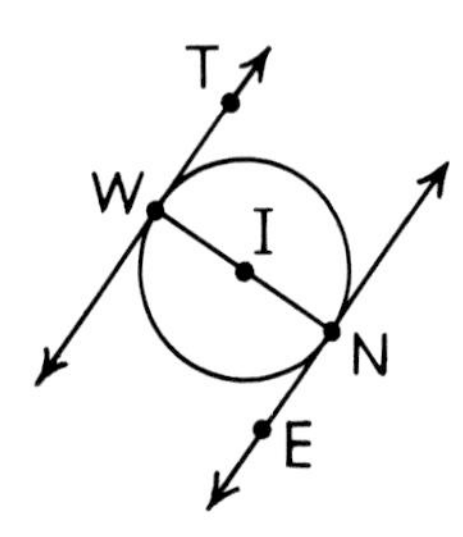

7. What relation does $\overleftrightarrow{TW}$ have to $\overline{WN}$ and $\overleftrightarrow{NE}$ have to $\overline{WN}$?
8. State the theorem that is the basis for your answer.
9. What relation do $\overleftrightarrow{TW}$ and $\overleftrightarrow{NE}$ have to each other?
10. State the theorem that is the basis for your answer.

In the figure below, TRIG is a parallelogram and $\overline{TN}$ bisects $\overline{IG}$ in circle S.

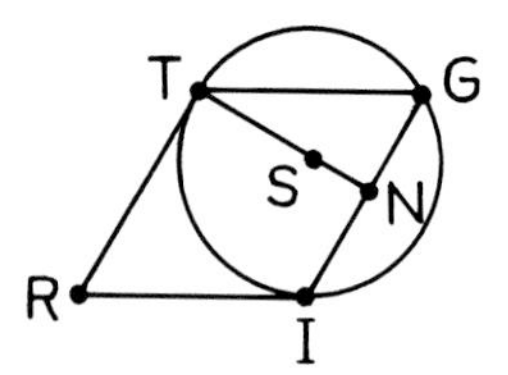

11. Why is $\overline{RT} \parallel \overline{IG}$?
12. Why is $\overline{TN} \perp \overline{IG}$?
13. Why is $\overline{TN} \perp \overline{RT}$?
14. Why is $\overleftrightarrow{RT}$ tangent to circle S?

In the figure below, side $\overline{RE}$ of $\triangle ROE$ is tangent to circle O; $\angle O = 60°$ and OP = 10.

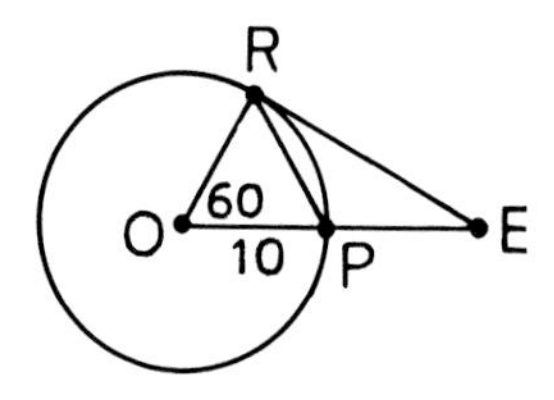

15. What kind of triangle is $\triangle ROP$ with respect to its angles?
16. What kind of triangle is $\triangle ROE$ with respect to its angles?
17. Draw the figure.
18. Find the lengths of the rest of the line segments in the figure.
19. What kind of triangle is $\triangle RPE$ with respect to its sides?
20. Find $\angle PRE$.
21. Find $\angle RPE$.

Set II

In Figure 1 below, the line is tangent to both circles because it intersects each circle in exactly one point.

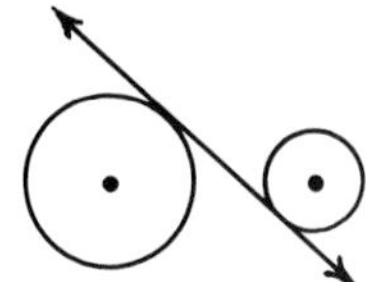

Figure 1

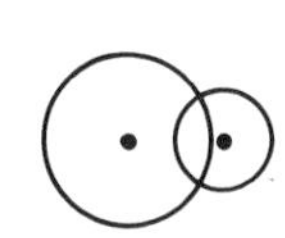

Figure 2

22. Copy Figure 1 and draw three more lines that are also tangent to both circles.

Exactly two lines can be drawn tangent to both of the circles shown in Figure 2.

23. Copy Figure 2 and draw them.
24. Draw two circles for which exactly three lines can be drawn tangent to both circles.
25. Draw two circles for which exactly one line can be drawn tangent to both circles.
26. Draw two circles for which no lines can be drawn tangent to both circles.

In the figure below, side $\overline{BD}$ of $\triangle BID$ is tangent to circle I and $\overline{DA} \perp \overline{BI}$; BI = 25 and ID = 15.

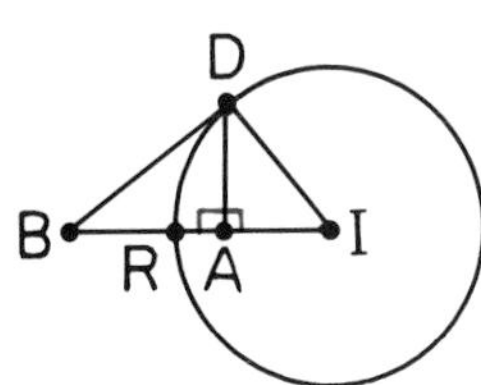

27. Draw the figure.

Find each of the following lengths.

28. BD.
29. BA.
30. DA.
31. RI.
32. BR.

33.

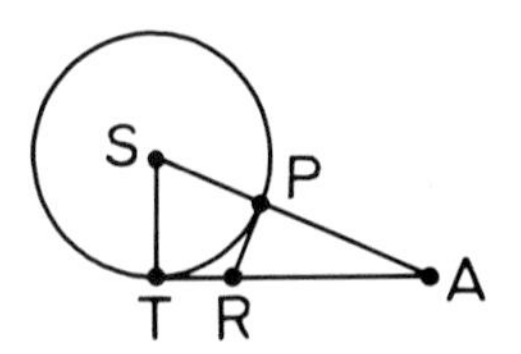

Given: Circle S with tangents $\overline{TA}$ and $\overline{RP}$.
Prove: $\triangle STA \sim \triangle RPA$.

34.

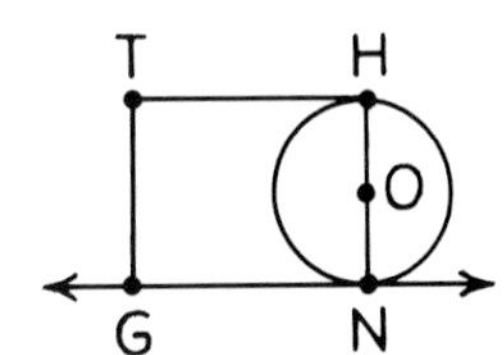

Given: Rectangle THNG and circle O.
Prove: $\overleftrightarrow{GN}$ is tangent to circle O.

35.

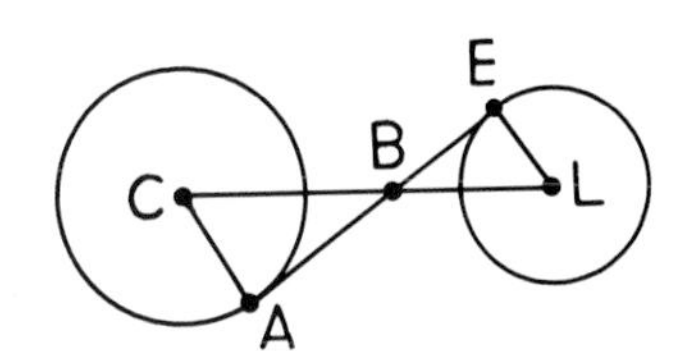

Given: $\overline{AE}$ is tangent to circle C at A and to circle L at E and C-B-L.
Prove: $\angle C = \angle L$.

36.

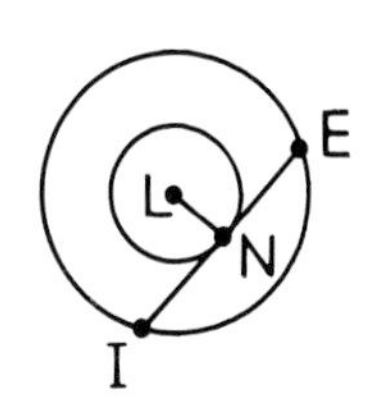

Given: L is the center of both circles and $\overline{IE}$ is tangent to the smaller circle at N.
Prove: IN = NE.

Set III

An Exercise in Op Art

Draw a circle on your paper that has the same radius as the circle shown here. Place your circle over this one, and trace the set of 18 points lettered A through R.

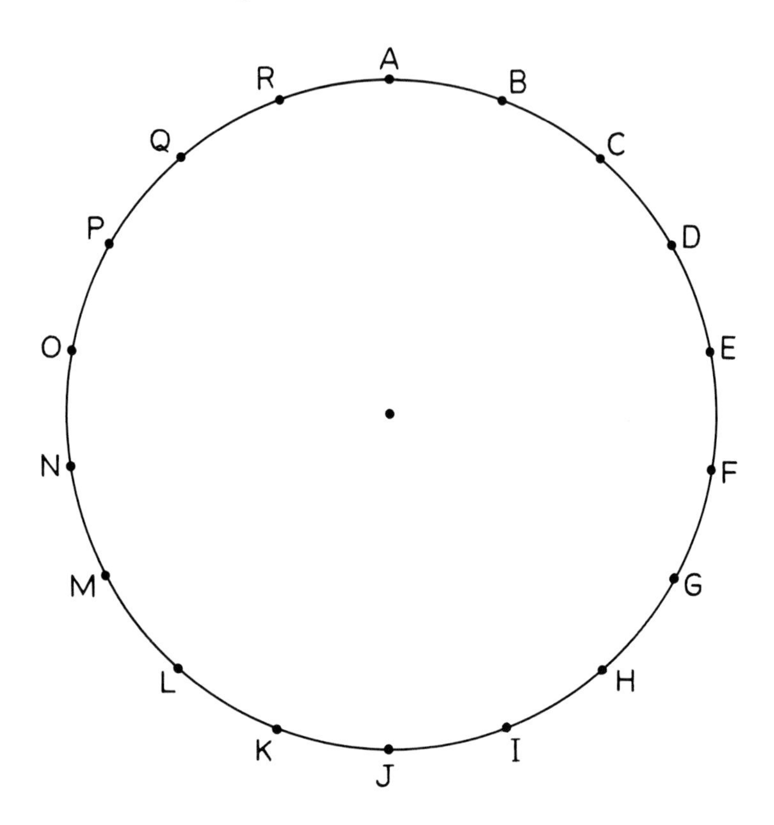

Now, draw the set of chords, $\overline{AI}$, $\overline{BJ}$, $\overline{CK}$, and so forth, around the circle.

Then, draw the following sets of chords, each set in a different color:

$\overline{AH}$, $\overline{BI}$, $\overline{CJ}$, and so forth.
$\overline{AG}$, $\overline{BH}$, $\overline{CI}$, and so forth.
$\overline{AF}$, $\overline{BG}$, $\overline{CH}$, and so forth.
$\overline{AE}$, $\overline{BF}$, $\overline{CG}$, and so forth.

What illusion appears in the finished figure?

Lesson 3

Central Angles and Arcs

A rainbow is produced when sunlight hits a bank of raindrops in either a cloud or falling rain. The drops act as tiny prisms, the apparent color of each drop being determined by the angle at which it is viewed. The rainbow itself consists of a set of circular arcs that have a common

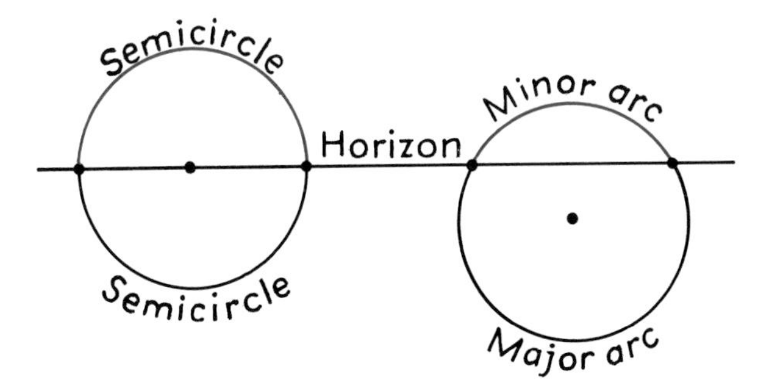

center. If the sun is on the horizon, exactly half of each of these circles, called a *semicircle,* can be seen. For positions of the sun above the horizon, the center of the rainbow is below the horizon. As a result, the visible part of the rainbow is less than a semicircle. It is called a *minor arc.* The part of the rainbow below the horizon is more than a semicircle and is called a *major arc.*

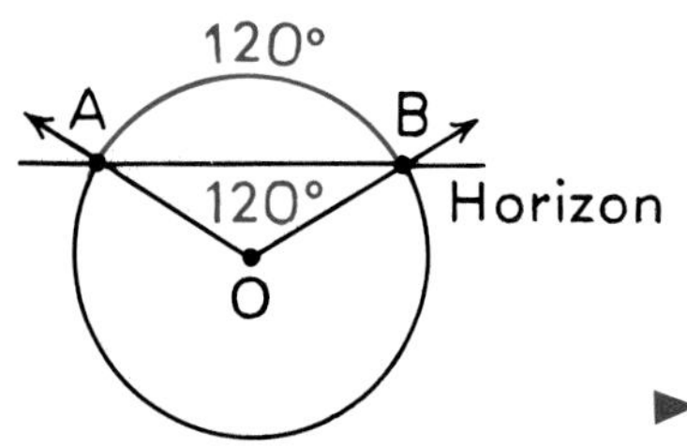

To indicate how much of the rainbow can be seen, it is convenient to measure its arc by means of an angle at its center. For example, if $\measuredangle AOB$ in the diagram at the left has a measure of 120°, then we say that arc AB (written $\overset{\frown}{AB}$) also has a measure of 120°. Angle AOB is called a *central angle* of circle O.

▶ **Definition**
A ***central angle*** of a circle is an angle whose vertex is the center of the circle.

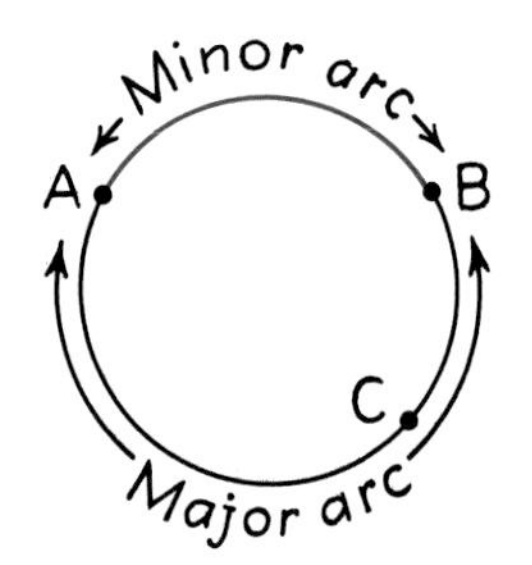

Every pair of points on a circle determines two arcs. A minor arc is usually named by just these two points, called its endpoints. A major arc or a semicircle is named with three letters, the middle letter naming a third point on the arc. In the circle at the left, for example, the symbol $\overset{\frown}{AB}$ refers to the minor arc and the symbol $\overset{\frown}{ACB}$ refers to the major arc.

We have seen that an arc can be measured by the central angle to which it is related; we will call this the "degree measure" of the arc and represent it by the letter m. If $\angle AOB = 120°$, then $m\overset{\frown}{AB} = 120°$ and $m\overset{\frown}{ACB} = 240°$.

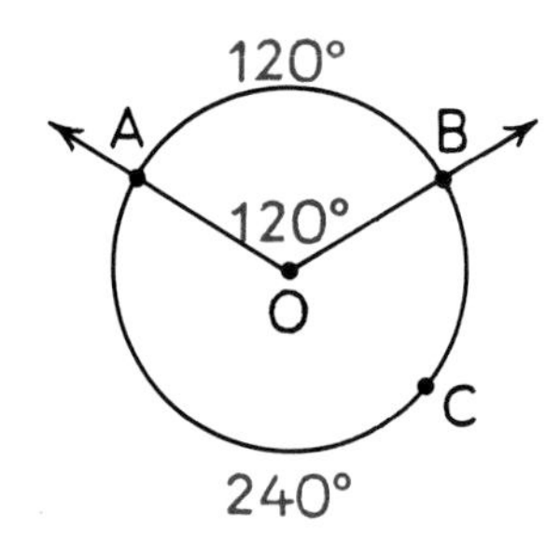

▶ **Definitions**
The ***degree measure*** of a
minor arc is the measure of its central angle;
semicircle is 180°;
major arc is 360° minus the measure of the corresponding minor arc;
circle is 360°.

In the circle above, it appears that $m\overset{\frown}{AC} + m\overset{\frown}{CB} = m\overset{\frown}{ACB}$. Although this equation is similar to the one for betweenness of points, we cannot say that C is between A and B because the three points are not collinear. We will assume that this measure relation between arcs of a circle is true without proving it.

▶ **Postulate 13** (The Arc Addition Postulate)
If C is on $\overset{\frown}{AB}$, then $m\overset{\frown}{AC} + m\overset{\frown}{CB} = m\overset{\frown}{ACB}$.

For simplicity, we will refer to arcs that have equal degree measures as "equal arcs." It is easy to prove that two chords of a circle are equal iff they have equal minor arcs.

► **Theorem 70**
If two chords of a circle are equal, their minor arcs are equal.

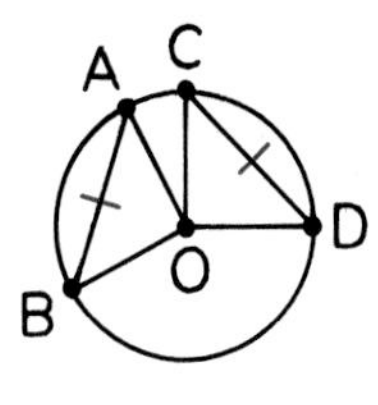

Given: In circle O, AB = CD.
Prove: $m\overset{\frown}{AB} = m\overset{\frown}{CD}$.

Proof.
In circle O, AB = CD. Draw $\overline{OA}$, $\overline{OB}$, $\overline{OC}$, and $\overline{OD}$. Because OA = OC and OB = OD (all radii of a circle are equal), $\triangle AOB \cong \triangle COD$ (S.S.S.).
It follows that $\angle AOB = \angle COD$ and, because $m\overset{\frown}{AB} = \angle AOB$ and $m\overset{\frown}{CD} = \angle COD$ (the degree measure of a minor arc is equal to the measure of its central angle), $m\overset{\frown}{AB} = m\overset{\frown}{CD}$ (substitution).

► **Theorem 71**
If two minor arcs of a circle are equal, their chords are equal.

Given: In circle O, $m\overset{\frown}{AB} = m\overset{\frown}{CD}$.
Prove: AB = CD.

Proof.
In circle O, $m\overset{\frown}{AB} = m\overset{\frown}{CD}$. Draw $\overline{OA}$, $\overline{OB}$, $\overline{OC}$, and $\overline{OD}$. OA = OC and OB = OD. $m\overset{\frown}{AB} = \angle AOB$ and $m\overset{\frown}{CD} = \angle COD$. Because $m\overset{\frown}{AB} = m\overset{\frown}{CD}$, $\angle AOB = \angle COD$. Therefore, $\triangle AOB \cong \triangle COD$ (S.A.S.), and so AB = CD.

Exercises

Set I

In the figure below, $\overline{PS}$ is a diameter of circle O.

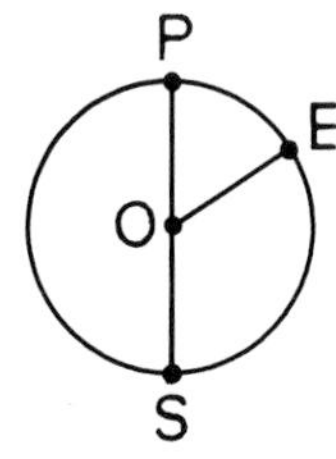

1. What are $\measuredangle$POE, $\measuredangle$EOS, and $\measuredangle$POS called with respect to the circle?
2. Name the minor arc that corresponds to $\measuredangle$POE.
3. Name the major arc that corresponds to $\measuredangle$POE.
4. Name the minor arc that corresponds to $\measuredangle$EOS.
5. Name the major arc that corresponds to $\measuredangle$EOS.
6. What kind of arc is $\overset{\frown}{PES}$?

The center of the circle in this figure is C. Give a reason for each of the following statements.

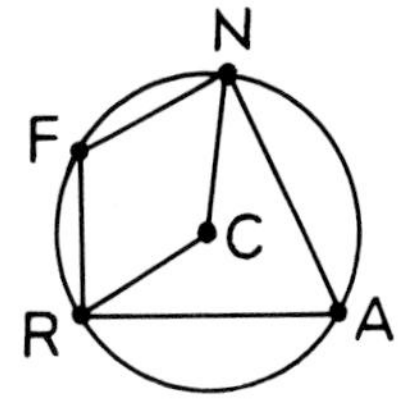

7. CR = CN.
8. $m\overset{\frown}{RN} = \angle C$.
9. $m\overset{\frown}{FR} + m\overset{\frown}{RA} = m\overset{\frown}{FRA}$.
10. If $m\overset{\frown}{NA} = m\overset{\frown}{AR}$, then NA = AR.

In circle D, $\angle RDA = 100°$ and $\angle RDN = 145°$. Find each of the following measures.

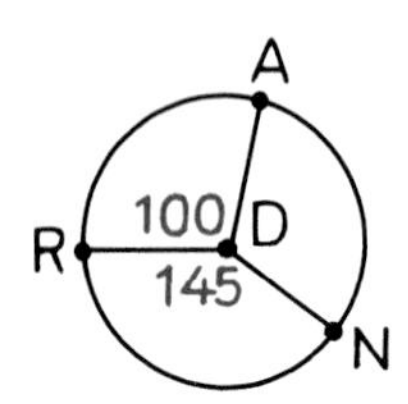

11. $\angle ADN$.
12. $m\overset{\frown}{AR}$ and $m\overset{\frown}{RN}$.
13. $m\overset{\frown}{ARN}$.
14. $m\overset{\frown}{ANR}$.

In the figure below, $\triangle UBL$ and $\triangle RBE$ are equilateral and B is the center of both circles; BU = 4 and UR = 2. Find each of the following measures.

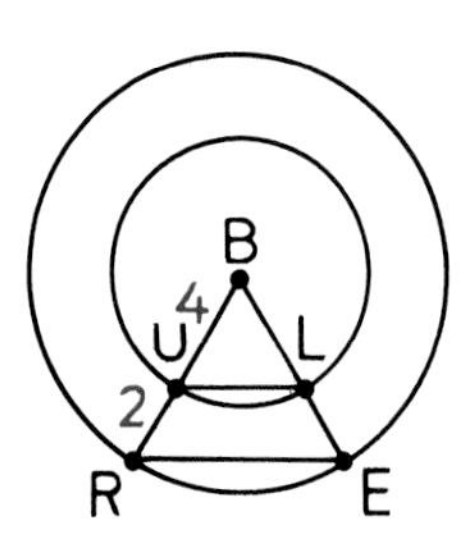

15. UL.
16. RE.
17. $m\overset{\frown}{UL}$.
18. $m\overset{\frown}{RE}$.
19. How do $\overline{UL}$ and $\overline{RE}$ compare in length?
20. How do $\overset{\frown}{UL}$ and $\overset{\frown}{RE}$ compare in measure?

In the figure below, $\triangle LIR$ is a 30°-60° right triangle whose sides are chords of circle A; AR = 10.

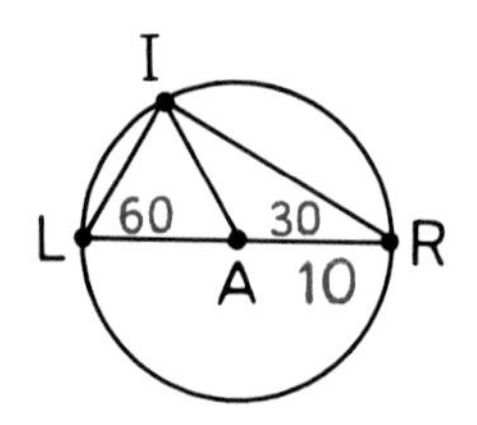

21. Why is $m\overset{\frown}{LI} + m\overset{\frown}{IR} = m\overset{\frown}{LIR}$?
22. Why is LI + IR > LR?

Find each of the following measures.

23. LR.
24. LI.
25. IR. (Use the fact that $\sqrt{3} \approx 1.73$ to express this length to the nearest tenth.)
26. $m\overset{\frown}{LIR}$.
27. $m\overset{\frown}{LI}$.
28. $m\overset{\frown}{IR}$.

Set II

29.

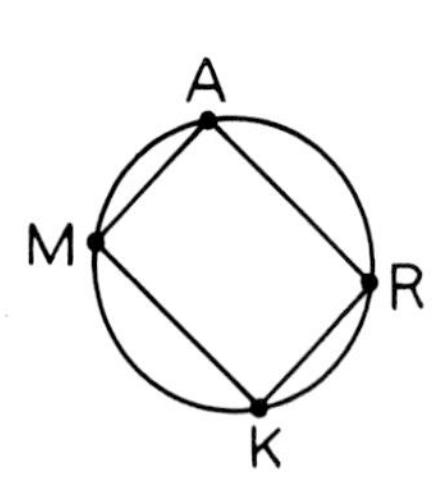

Given: MARK is a rectangle.

Prove: $m\overset{\frown}{MA} = m\overset{\frown}{KR}$.

30.

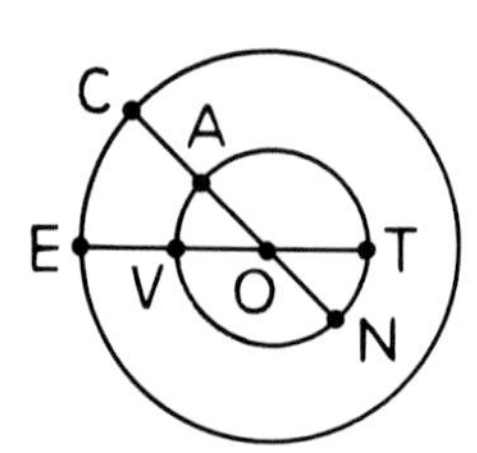

Given: Vertical angles $\measuredangle COE$ and $\measuredangle TON$ are central angles of both circles.

Prove: $m\overset{\frown}{CE} = m\overset{\frown}{TN}$.

31.

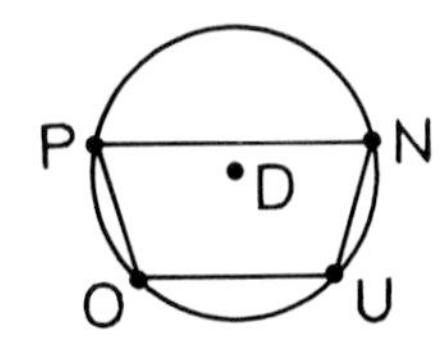

Given: POUN is a trapezoid; $m\widehat{PO} = m\widehat{NU}$.

Prove: $\angle O = \angle U$.

32.

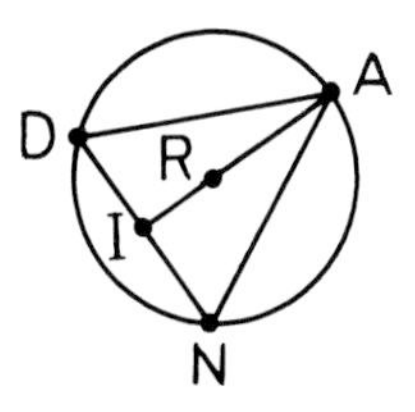

Given: In circle R, $\overline{AI} \perp \overline{DN}$.

Prove: $m\widehat{DA} = m\widehat{NA}$.

33.

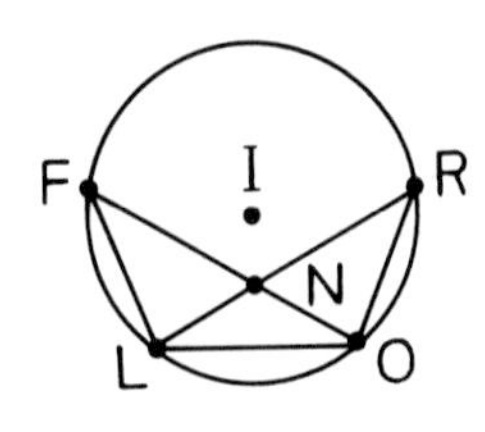

Given: Circle I with FL = RO.

Prove: FO = RL.

34.

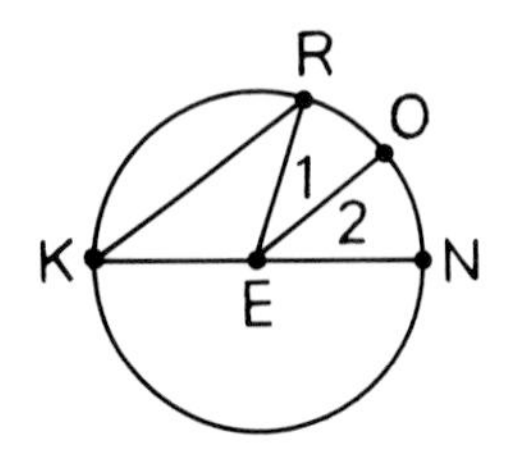

Given: Circle E with $\overline{KR} \parallel \overline{EO}$.

Prove: $m\widehat{RO} = m\widehat{ON}$.

Set III

This photograph of the night sky was taken in Australia and shows stars close to the South Pole. A radar antenna used to study them appears in the foreground.

1. Why do the stars appear as arcs of concentric circles rather than as points of light?
2. How do the arcs of the different stars compare in measure as their distances from the center of the circles (the point above the South Pole) increase? Explain.

Fritz Goro, Life Magazine © 1952 Time Inc.

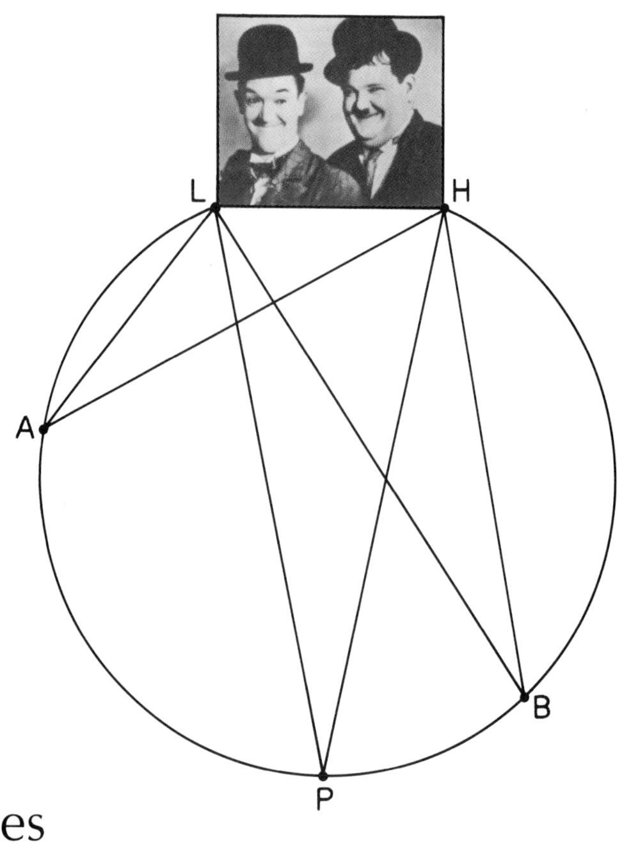

Lesson 4
Inscribed Angles

If you have ever watched a wide-screen movie from a seat near the front of a theater, you know that it is almost impossible to see everything on the screen at once. This is because the angle that the side edges of the screen make with your eyes is very large. At the back of the theater, this angle is much smaller. It is tempting to conclude from this fact that the farther someone is from the screen, the smaller his viewing angle of it will be. But such is not the case.

In the diagram above, $\overline{LH}$ represents the screen and point P represents the center seat of the back row, directly below the projection booth. For a person seated at P, the screen angle to which we have been referring is ∡P. Where else in the theater is the screen angle equal to that at P?

The other locations having the same screen angle are on the circle that contains points L, H, and P. Every angle whose vertex is on this circle and whose sides pass through points L and H is equal to the angle at P. You can verify this for the angles at A and B by measuring both of them and the angle at P with a protractor.

Such angles are called *inscribed angles,* and to prove that they are equal, we will derive a formula for their measure in terms of the measure of the common arc they intercept, $\overset{\frown}{LH}$ ($\overset{\frown}{LH}$ is covered by the screen in the diagram).

► **Definition**
An ***inscribed angle*** is an angle whose vertex is on a circle and each of whose sides intersects it in another point.

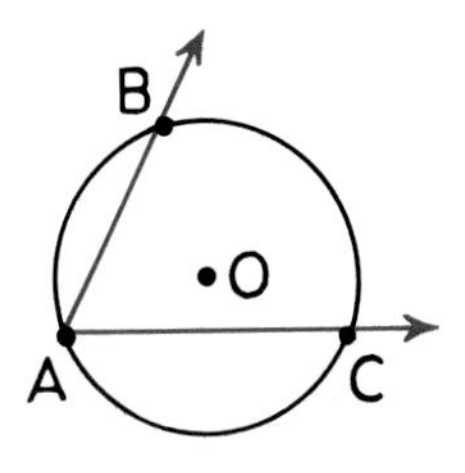

In the figure at the right, ∡A is an inscribed angle of circle O. It is said to *intercept* $\overset{\frown}{BC}$ and to be *inscribed* in $\overset{\frown}{BAC}$. Note that these two arcs make up the entire circle. This is true of every inscribed angle in a circle. It divides the circle into two arcs, one of which it intercepts and the other in which it is inscribed.

By definition, a central angle is equal in measure to the arc of the circle that it intercepts: its minor arc. We will prove that an inscribed angle is equal in measure to *half* its intercepted arc. In each of the figures below, then, $\angle A = \frac{1}{2}m\overset{\frown}{BC}$. To prove this, we have three possibilities to consider: that the center of the circle lies *on a side* of the inscribed angle, that it lies *inside* the angle, or that it lies *outside* the angle.

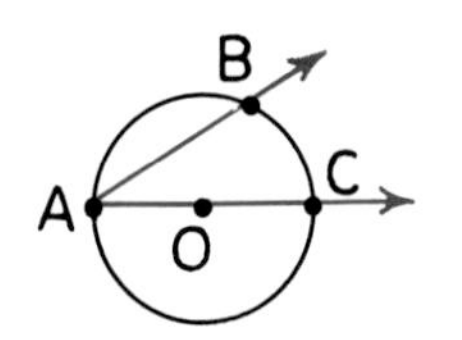

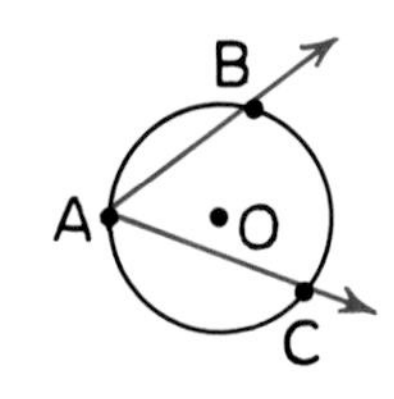

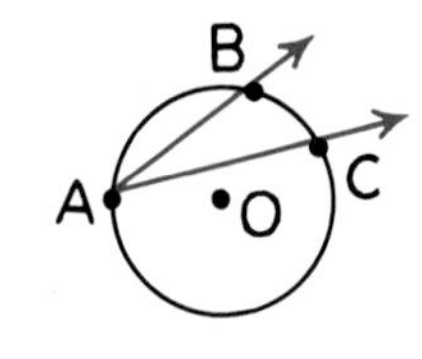

► **Theorem 72**
An inscribed angle is equal in measure to half its intercepted arc.

Given: ∡A is inscribed in circle O.

Prove: $\angle A = \frac{1}{2}m\overset{\frown}{BC}$.

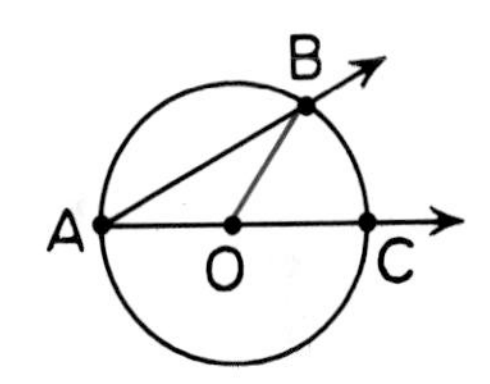

The figure at the right illustrates the first possibility: that O lies on a side of ∡A. To prove the theorem for this case, we draw radius $\overline{OB}$ to form a central angle, ∡BOC, that intercepts the same arc as ∡A: $\overset{\frown}{BC}$. We can then derive an equation relating ∠A and ∠BOC from the fact that they are interior and exterior angles of △AOB.

Proof.
Draw $\overline{OB}$. Because OA = OB (all radii of a circle are equal), ∠ABO = ∠A. Also, ∠BOC = ∠A + ∠ABO (an exterior angle of a triangle is equal in measure to the sum of the measures of the two remote interior angles). Therefore, by substitution, ∠BOC = ∠A + ∠A = 2∠A. Applying the symmetric property to this equation and multiplying both sides by $\frac{1}{2}$, we get $\angle A = \frac{1}{2}\angle BOC$. Because $m\overset{\frown}{BC} = \angle BOC$ (a minor arc is equal in measure to its central angle), we can substitute to get $\angle A = \frac{1}{2}m\overset{\frown}{BC}$.

The other two possibilities, in which O lies inside or outside ∡A, can easily be proved by using this result.

This theorem has a couple of useful corollaries. The first of them justifies the statement that the screen angles at every point on the circle are equal.

► **Corollary 1**
Inscribed angles that intercept the same arc or equal arcs are equal.

► **Corollary 2**
An angle inscribed in a semicircle is a right angle.

Exercises

Set I

In the figure below, R is the center of the circle and $m\widehat{IT} = 140°$.

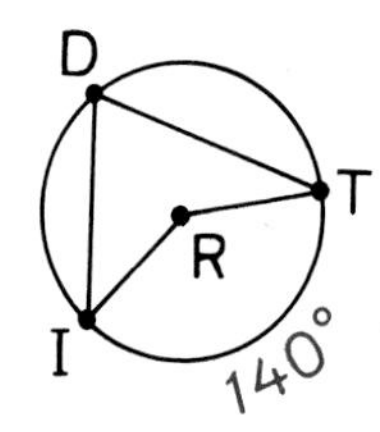

1. What is ∡R called with respect to the circle?
2. What is ∡D called with respect to the circle?
3. What arc do ∡D and ∡R intercept?
4. In what arc is ∡D inscribed?
5. What is the measure of ∡R?
6. What is the measure of ∡D?

In the figure below, $\overline{NL} \perp \overline{KI}$ in circle O.

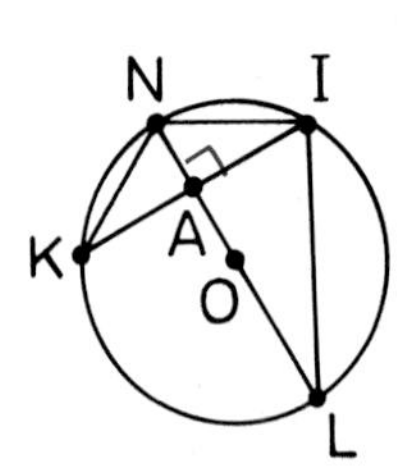

7. Why is $\angle K = \angle L$?
8. Why is ∡NIL a right angle?
9. Why is $\angle KNI = \frac{1}{2} m\widehat{KLI}$?
10. Why is $KA = AI$?
11. Why is $\triangle NAK \cong \triangle NAI$?
12. Why is $NK = NI$?
13. Why is $m\widehat{NK} = m\widehat{NI}$?

In the figure below, $\angle I = 92°$ and $m\widehat{IL} = 124°$.

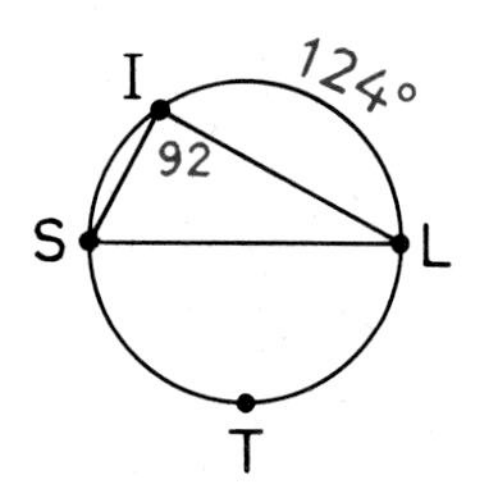

14. Copy the figure.
15. Is $\widehat{SIL}$ a semicircle?
16. Find $\angle S$.
17. Find $m\widehat{STL}$.
18. Find $\angle L$.
19. Find $m\widehat{IS}$.

In the figure below, $\overline{UR}$ is a diameter of circle O, $\angle U = 37°$, and $\angle D = 54°$.

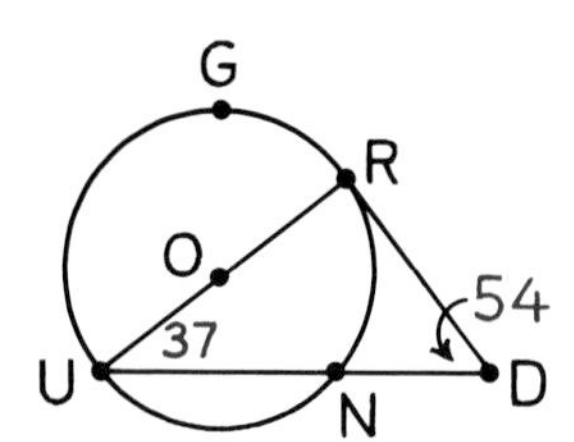

20. Find $\angle R$.
21. Is $\overline{RD}$ tangent to circle O?
22. Is $\overset{\frown}{UGR}$ a semicircle?
23. Find $m\overset{\frown}{RN}$.
24. Find $m\overset{\frown}{UN}$.

Set II

In the figure below, $\overline{SD}$ is a diameter of the circle and $m\overset{\frown}{OD} = 2m\overset{\frown}{SO}$.

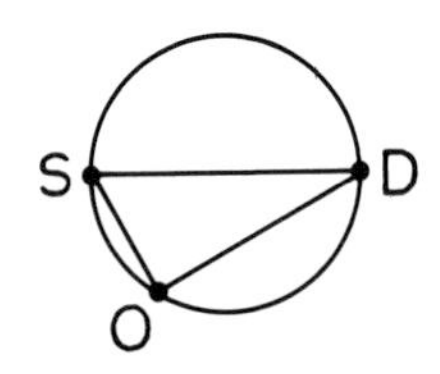

25. Find $m\overset{\frown}{SO}$ and $m\overset{\frown}{OD}$.
26. Find $\angle S$ and $\angle D$.
27. Is OD = 2 SO?

In the figure below, $\overleftrightarrow{GU}$ is tangent to circle O and $\overline{UB}$ is a diameter; $m\overset{\frown}{UM} = x°$. Express each of the following measures in terms of x.

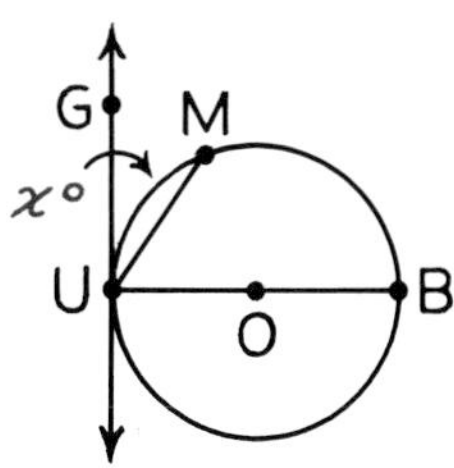

28. $m\overset{\frown}{MB}$.
29. $\angle MUB$.
30. $\angle GUM$.

31.

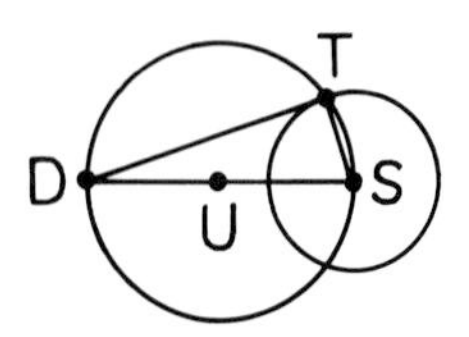

Given: $\overset{\frown}{DTS}$ is a semicircle of circle U and $\overline{ST}$ is a radius of circle S.

Prove: $\overline{DT}$ is tangent to circle S.

32.

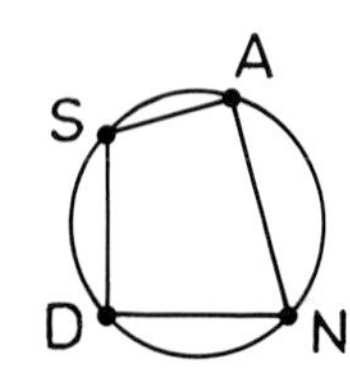

Given: $\overline{SD} \perp \overline{DN}$.

Prove: $m\overset{\frown}{SAN} = 180°$.

33.

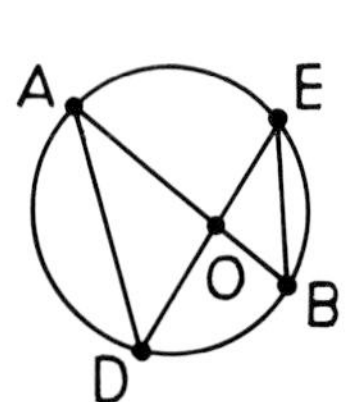

Given: $\overline{AD}$, $\overline{AB}$, $\overline{ED}$, and $\overline{EB}$ are chords of the circle.

Prove: $\triangle ADO \sim \triangle EBO$.

34.

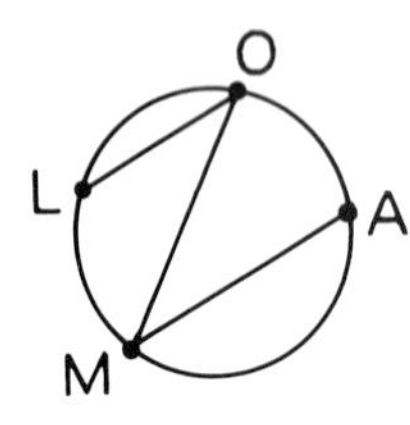

Given: $\overline{LO} \parallel \overline{MA}$ in the circle.
Prove: $m\widehat{LM} = m\widehat{OA}$.

35.

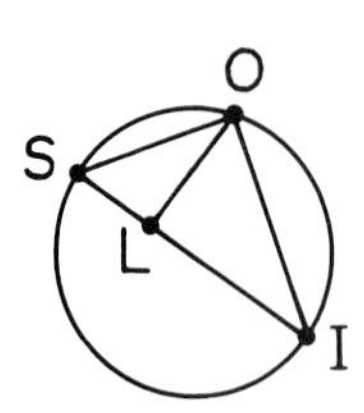

Given: $\widehat{SOI}$ is a semicircle; $\overline{OL} \perp \overline{SI}$.

Prove: $\frac{SL}{OL} = \frac{OL}{LI}$.

36.

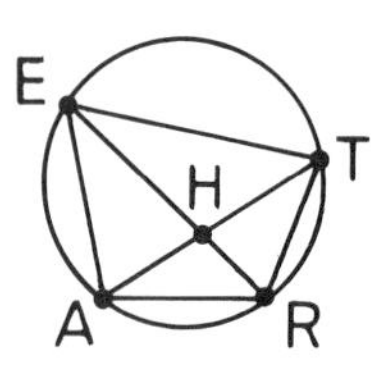

Given: $\overline{EA}$, $\overline{ER}$, $\overline{ET}$, $\overline{AT}$, $\overline{AR}$, and $\overline{RT}$ are chords of the circle and AR = RT.

Prove: $\frac{AH}{HT} = \frac{AE}{ET}$.

Set III

This diagram shows a top view of Acute Alice's new hat after Obtuse Ollie sat on it. On the assumption that $\angle D$ is inscribed in the larger circle, Ollie says that $m\widehat{AG} > m\widehat{BF}$. Alice says that $m\widehat{AG} = m\widehat{BF}$.

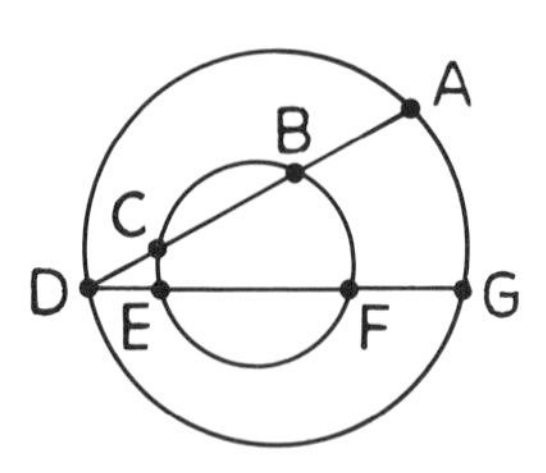

What's your opinion? Explain your reasoning.

Escher Foundation, Haags Gemeentemuseum, The Hague

Lesson 5
Secant Angles

In this circular woodcut by Escher, pictures of angels and devils are fitted together like pieces in a jigsaw puzzle. But, unlike those of a jigsaw puzzle, the parts become progressively smaller and more numerous as they approach the edge. By this device, the artist has managed to convey an impression of the infinite within a finite region.

It is apparent from the intricate and orderly design of the picture that Escher has been strongly influenced by ideas from geometry. For example, if the wing tips of the large devils and angels that meet in the center are joined to it and to each other by line segments, six equilateral triangles are formed.

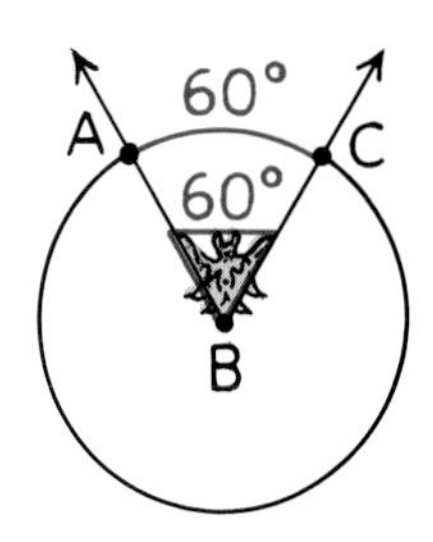

In the first figure at the right, one of these triangles is shown. Because it is equilateral, the angle at the center has a measure of 60°. If we extend its sides so that they intersect the circle, the intercepted arc must also have a measure of 60°.

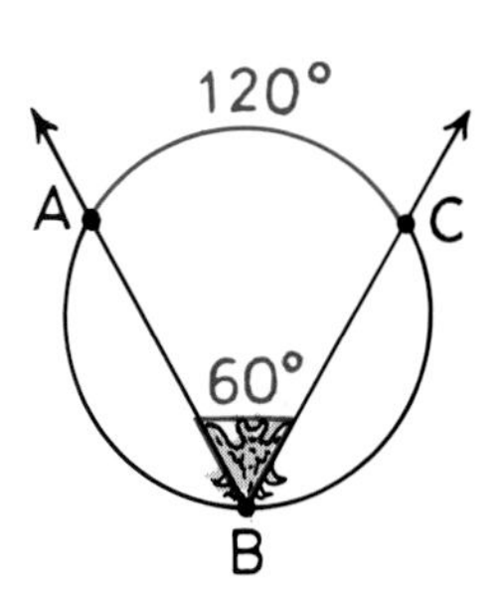

If the vertex of this angle were on the circle rather than at its center, as shown in the second figure, it would be an inscribed angle and would intercept a larger arc with a measure of 120°.

We know the relation between the measures of central and inscribed angles and the arcs that they intercept. What about other angles, such as those shown at the top of the next page?

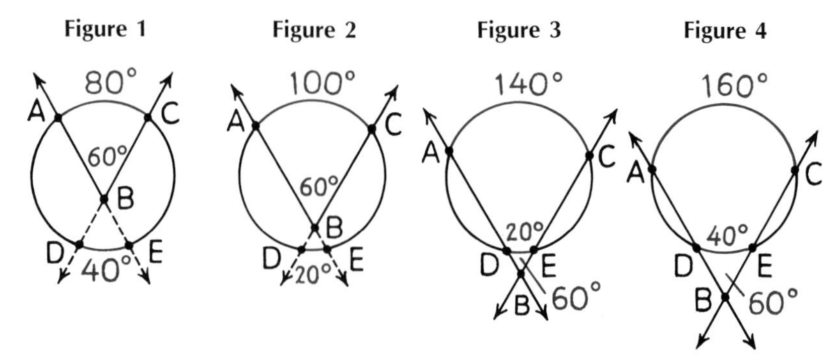

In Figures 1 and 2, the vertex of the angle is inside the circle. Although the size of the arc intercepted by the angle differs in each picture, the *sum of the measures of it and the arc intercepted by its vertical angle does not change.* A comparable result holds for Figures 3 and 4, where the vertex of the angle is outside the circle. Here, the *difference of the measures of the two intercepted arcs does not change.*

The lines that form the angles in these figures are called *secants.*

► **Definition**
A ***secant*** is a line that intersects a circle in two points.

The angles labeled ∡ABC in the figures are called *secant angles.*

► **Definition**
A ***secant angle*** is an angle whose sides are contained in two secants of a circle so that each side intersects the circle in at least one point other than the angle's vertex.

Figures 1 through 4 above suggest the following theorems.

► **Theorem 73**
A secant angle whose vertex is inside a circle is equal in measure to half the sum of the arcs intercepted by it and its vertical angle.

Given: Secant ∡APB with P inside the circle.

Prove: $\angle APB = \frac{1}{2}(m\widehat{AB} + m\widehat{CD})$.

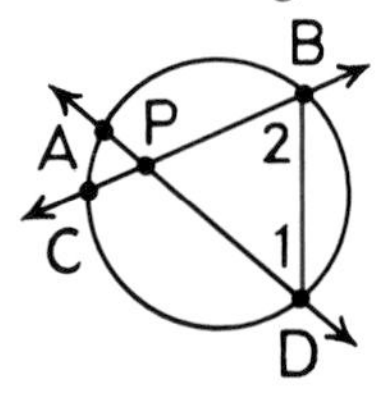

Proof.
Draw $\overline{BD}$. Because $\angle APB = \angle 1 + \angle 2$ (an exterior angle of a triangle is equal in measure to the sum of the measures of the remote interior angles) and $\angle 1 = \frac{1}{2}m\widehat{AB}$ and $\angle 2 = \frac{1}{2}m\widehat{CD}$ (an inscribed angle is equal in measure to half its intercepted arc), it follows by substitution that

$$\angle APB = \frac{1}{2}m\widehat{AB} + \frac{1}{2}m\widehat{CD} = \frac{1}{2}(m\widehat{AB} + m\widehat{CD}).$$

► **Theorem 74**
A secant angle whose vertex is outside a circle is equal in measure to half the positive difference of its intercepted arcs.

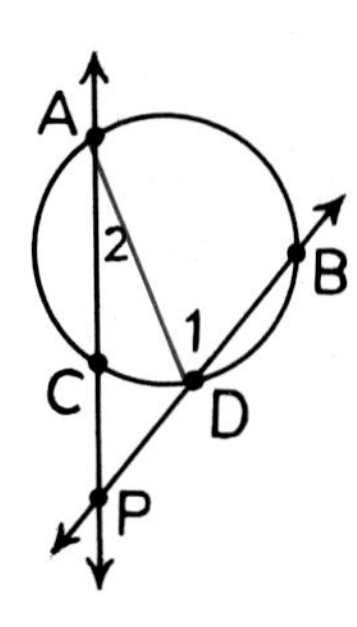

Given: Secant $\measuredangle APB$ with P outside the circle.

Prove: $\angle APB = \frac{1}{2}(m\overset{\frown}{AB} - m\overset{\frown}{CD})$.

Proof.
Draw $\overline{AD}$. Because $\angle 1 = \angle APB + \angle 2$ (an exterior angle of a triangle is equal in measure to the sum of the measures of the remote interior angles), it follows by subtraction that $\angle APB = \angle 1 - \angle 2$. Also, $\angle 1 = \frac{1}{2} m\overset{\frown}{AB}$ and $\angle 2 = \frac{1}{2} m\overset{\frown}{CD}$, so by substitution,

$$\angle APB = \frac{1}{2} m\overset{\frown}{AB} - \frac{1}{2} m\overset{\frown}{CD} = \frac{1}{2}(m\overset{\frown}{AB} - m\overset{\frown}{CD}).$$

Exercises

Set I

In the figure below, $\overline{LH}$ and $\overline{TG}$ are diameters of circle N and $\overline{LH}$ bisects $\overline{TI}$.

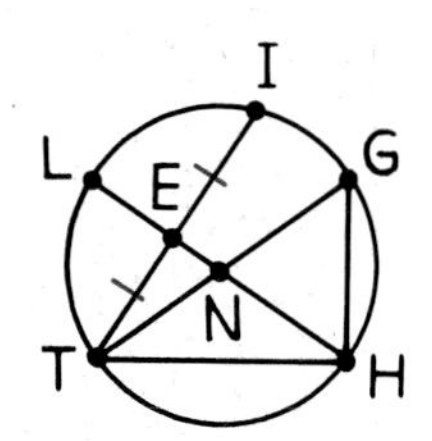

1. Why is $\overline{LH} \perp \overline{TI}$?
2. Why is $m\overset{\frown}{GH} = \angle GNH$?
3. Why is $\angle LEI = \frac{1}{2}(m\overset{\frown}{LI} + m\overset{\frown}{TH})$?
4. Why is LN = NG?
5. Why is $\measuredangle THG$ a right angle?

In the figure below, leg $\overline{NE}$ of right triangle $\triangle SNE$ contains the center of circle L and $m\overset{\frown}{AP} = m\overset{\frown}{PE}$.

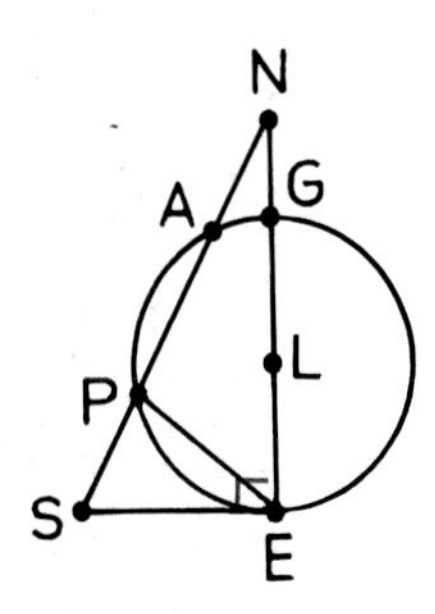

6. Why is $\overline{SE}$ tangent to the circle?
7. Why is $\angle N = \frac{1}{2}(m\overset{\frown}{PE} - m\overset{\frown}{AG})$?
8. Why is AP = PE?
9. Why is $\angle PEG = \frac{1}{2} m\overset{\frown}{PG}$?
10. Why is $m\overset{\frown}{PA} + m\overset{\frown}{AG} = m\overset{\frown}{PG}$?

In the figure below, $\overline{SN}$, $\overline{SI}$, and $\overline{NH}$ are chords of the circle.

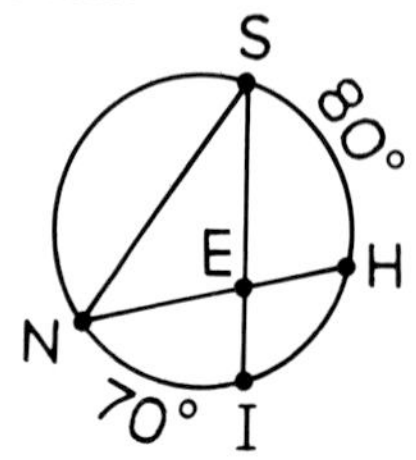

11. Find $\angle S$.
12. Find $\angle N$.
13. Find $\angle SEH$.
14. Why should $\angle SEH = \angle S + \angle N$?

In the figure below, $\measuredangle F$ is a secant angle of the circle.

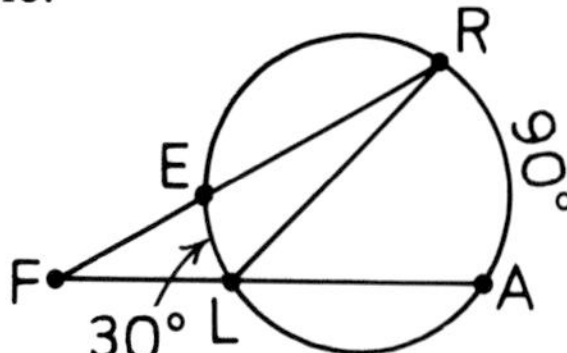

15. Find $\angle R$.
16. Find $\angle F$.
17. Find $\angle RLA$.
18. Why should $\angle RLA = \angle R + \angle F$?

In the figure below, $\overline{BG}$, $\overline{RT}$, and $\overline{IH}$ are chords of the circle and $\overline{RT} \parallel \overline{IH}$.

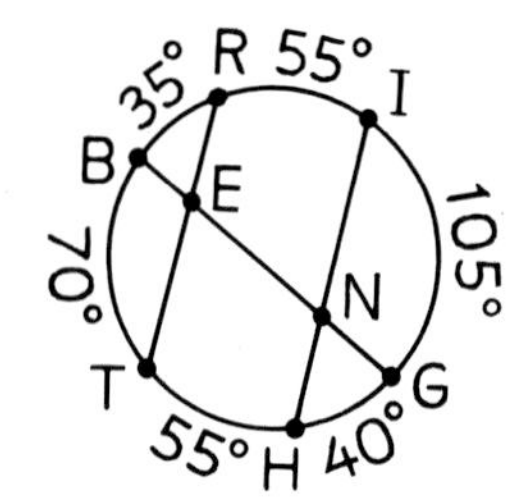

19. Find $\angle BNI$.
20. Find $\angle TEG$.
21. Why should $\angle BNI = \angle TEG$?
22. Find $\angle BNH$.
23. Why should $\angle TEG + \angle BNH = 180°$?

Set II

In the figure below, $\overleftrightarrow{BA}$ and $\overleftrightarrow{LZ}$ intersect at the center of circle E.

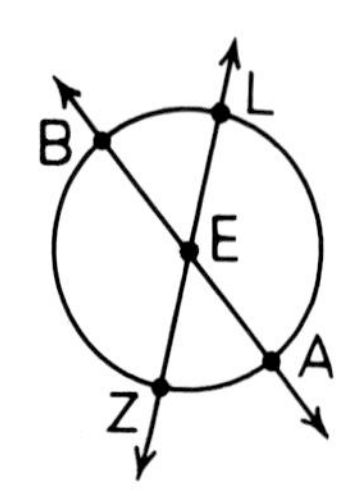

24. Why are $\measuredangle BEL$ and $\measuredangle ZEA$ central angles of circle E?
25. Use the definition of the measure of a minor arc to write equations for the measures of $\measuredangle BEL$ and $\measuredangle ZEA$.
26. Why is $\angle BEL = \angle ZEA$?
27. Why is $m\widehat{BL} = m\widehat{ZA}$?
28. What are $\overleftrightarrow{BA}$ and $\overleftrightarrow{LZ}$ called with respect to the circle?
29. Why is $\measuredangle BEL$ a secant angle of circle E?
30. Use the theorem about a secant angle whose vertex is inside a circle to write an equation for the measure of $\measuredangle BEL$.
31. Show that this equation is equivalent to the equation for the measure of $\measuredangle BEL$ that you wrote in Exercise 25.
32.

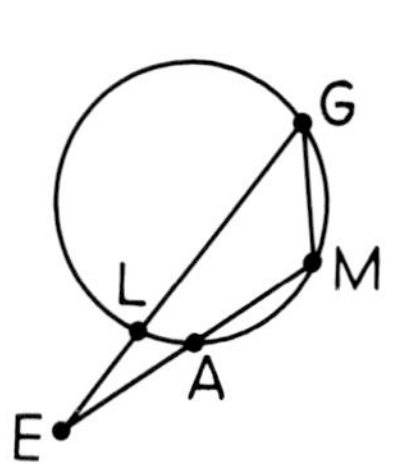

Given: MA = MG.

Prove: $\angle E = \frac{1}{2}(m\widehat{MA} - m\widehat{AL})$.

33.

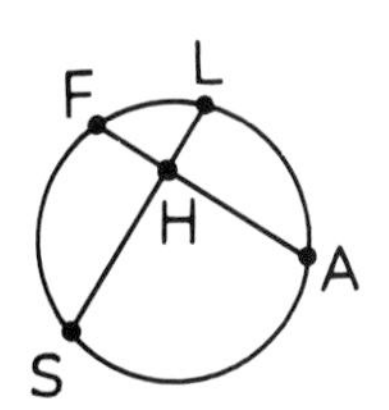

Given: $m\widehat{FL} + m\widehat{SA} = 180°$

Prove: $\overline{LS} \perp \overline{FA}$.

34.

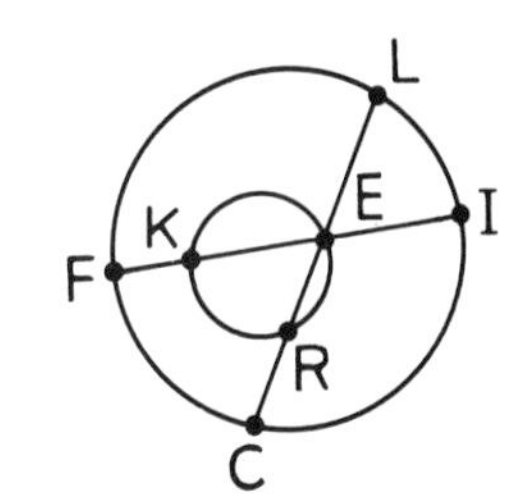

Given: $\overline{FI}$ and $\overline{LC}$ are chords of the larger circle.

Prove: $m\widehat{KR} > m\widehat{LI}$.

35.

Given: $\measuredangle G$ is a secant angle of the circle; $m\widehat{AR} = 2m\widehat{LE}$.

Prove: $\triangle AGE$ is isosceles.

Set III

By the permission of Johnny Hart and News America Syndicate

The King of Id has commissioned a statue of himself for the village square. Unfortunately, what the royal sculptor has turned out so far doesn't seem to be what the king has in mind.

The figure below shows three positions of the king as he walks toward the original statue. The measure of the angle formed by the king's lines of sight to the top and bottom of the statue depends on where the king is standing. Can you describe what happens to it as the king walks from a great distance away to the base of the statue?

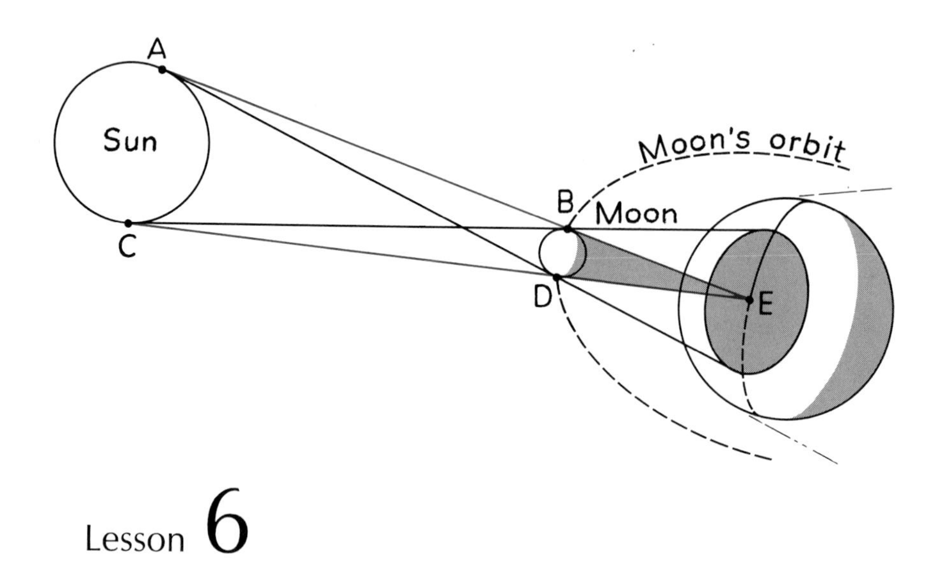

Lesson 6

Tangent Segments

A total eclipse of the sun is an impressive sight. The sky suddenly changes from sunlight to darkness, the stars come out, flowers close up, and birds become confused, thinking that it is night. The darkness lasts for just a few minutes (about seven at the most), and then the light reappears as swiftly as it had gone.

Such an eclipse occurs when the moon passes between the earth and the sun so that its shadow touches the earth. This shadow is a cone. Because of the relative sizes of the moon and the sun and their relative distances from the earth, the tip of the cone often misses the earth. But when the cone touches the earth, a total eclipse can be seen by those in the area of contact. The area of contact moves across the earth's surface as the earth rotates and it is shown as the arc through E in the diagram above.

At the same time, a partial eclipse of the sun can be seen from a much larger region, shown shaded in the diagram. Within this region, the disk of the sun is only partly covered by the moon.

The lines of the shadows in the figure are tangent to the circles that represent the sun and moon. Two of them, $\overleftrightarrow{AB}$ and $\overleftrightarrow{CD}$, meet in point E. The segments $\overline{EA}$ and $\overline{EC}$ are called *tangent segments* from point E to the circle of the sun; segments $\overline{EB}$ and $\overline{ED}$ are tangent segments from point E to the circle of the moon.

Definition

If a line is tangent to a circle, then any segment of the line having the point of tangency as one of its endpoints is a ***tangent segment*** to the circle.

From the diagram, it looks as if one of the two tangent segments from an external point to a circle might be longer than the other. It is easy to prove that, no matter what the "relative positions" of the point and circle, they must be equal.

Theorem 75

The tangent segments to a circle from an external point are equal.

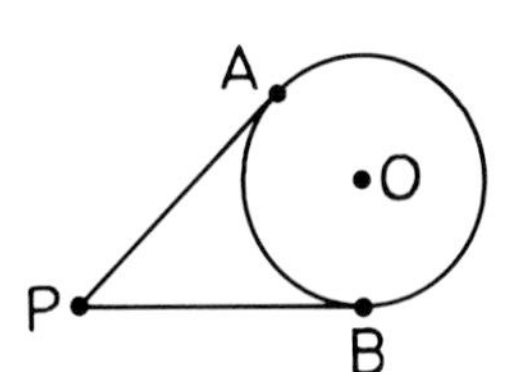

Given: Circle O with tangent segments $\overline{PA}$ and $\overline{PB}$.

Prove: PA = PB.

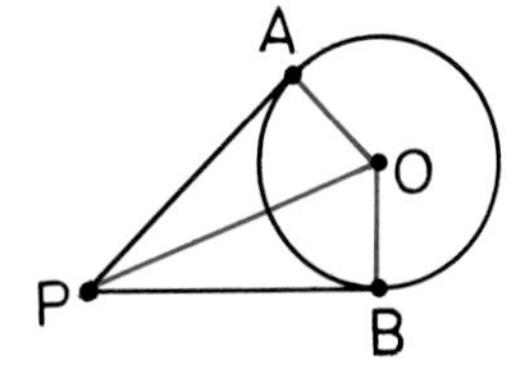

Proof.

Draw $\overline{PO}$, $\overline{OA}$, and $\overline{OB}$. Because $\overline{PA}$ and $\overline{PB}$ are tangent segments to circle O, $\overline{PA} \perp \overline{OA}$ and $\overline{PB} \perp \overline{OB}$ (if a line is tangent to a circle, it is perpendicular to the radius drawn to the point of contact). Therefore, $\measuredangle$PAO and $\measuredangle$PBO are right angles, and so $\triangle$PAO and $\triangle$PBO are right triangles. Because OA = OB (all radii of a circle are equal) and PO = PO, $\triangle PAO \cong \triangle PBO$ (H.L.). Therefore, PA = PB.

We will now consider methods by which tangents to a circle can be constructed.

Construction 9

To construct the tangent to a circle at a given point on the circle.

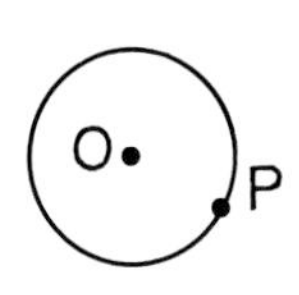

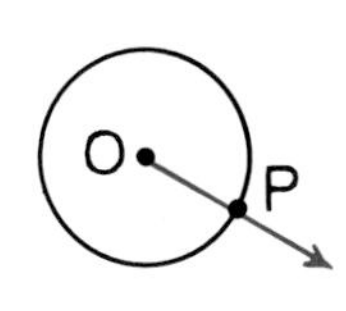

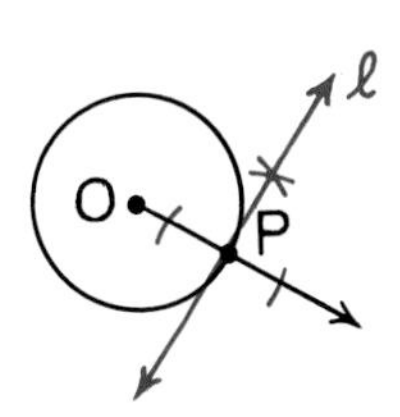

Method.

Let P be the given point on circle O through which a tangent is to be drawn. Draw $\overleftrightarrow{OP}$. Through P, construct line $\ell \perp \overleftrightarrow{OP}$. Line ℓ is the tangent to circle O at P (if a line is perpendicular to a radius at its outer endpoint, it is tangent to the circle).

► **Construction 10**
To construct the tangents to a circle from a given external point.

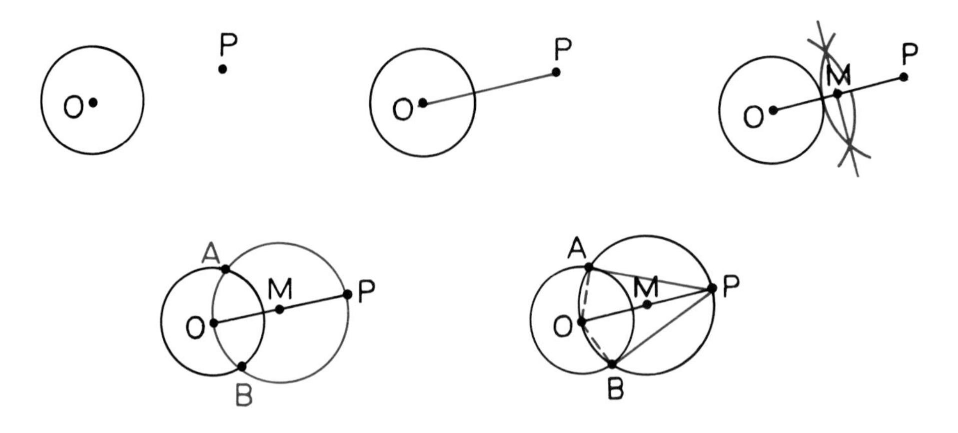

Let P be the given point through which the tangents are to be drawn to circle O. Draw $\overline{OP}$. Bisect $\overline{OP}$ and label its midpoint M. Draw a circle with M as center and MO as radius. Label the two points in which it intersects circle O points A and B. Draw $\overline{PA}$ and $\overline{PB}$. Segments $\overline{PA}$ and $\overline{PB}$ are the tangents from P to circle O.

If $\overline{OA}$ and $\overline{OB}$ are drawn, ∡OAP and ∡OBP are right angles (an angle inscribed in a semicircle is a right angle), and so $\overline{PA} \perp \overline{OA}$ and $\overline{PB} \perp \overline{OB}$. Therefore, $\overline{PA}$ and $\overline{PB}$ are tangent to circle O.

Exercises

Set I

In circle O, $\overrightarrow{EH} \perp \overline{OH}$ and $\overrightarrow{EP} \perp \overline{OP}$.

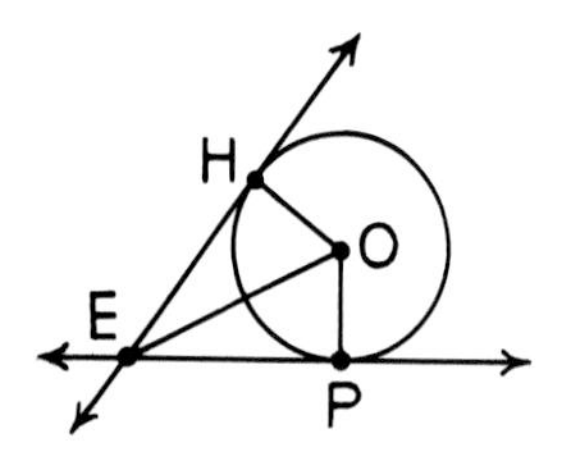

1. Why are $\overrightarrow{EH}$ and $\overrightarrow{EP}$ tangent to circle O?
2. Why is EH = EP?

Tangent segments have been drawn to the circles in the figures below. Find x in each figure.

3.

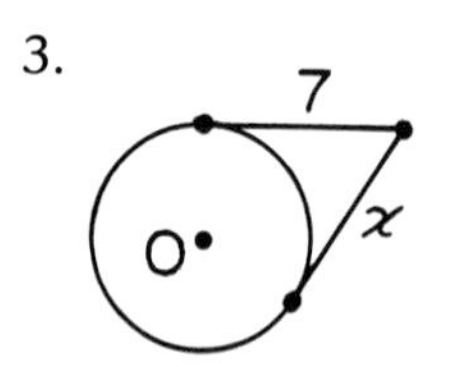

4.

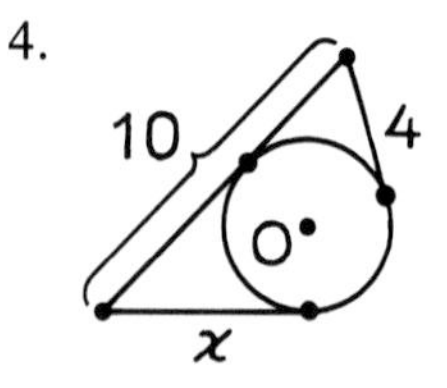

5.

6.

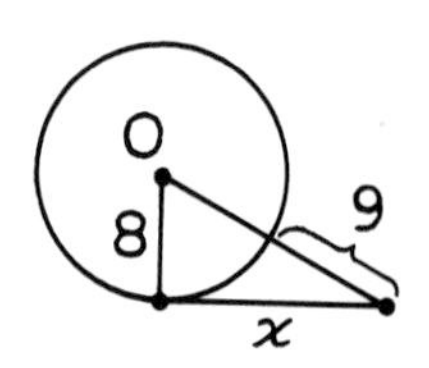

7. 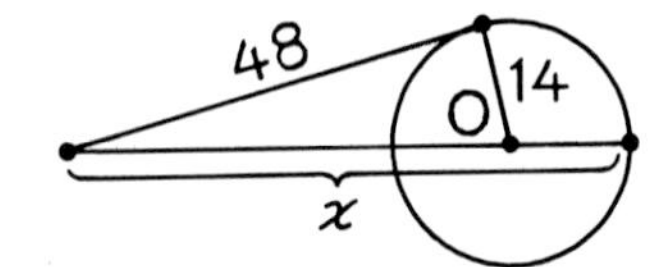

8. 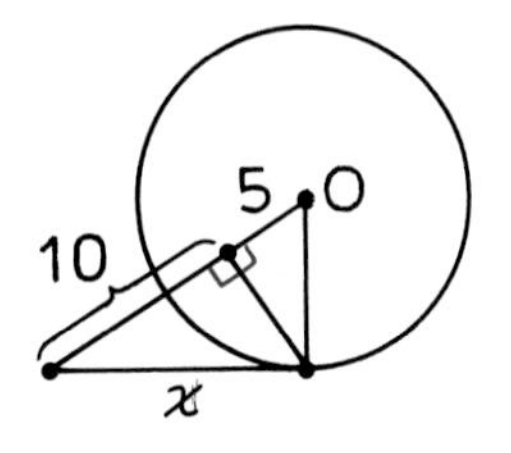

In the figure below, $\overleftrightarrow{BC}$ has been constructed through point B tangent to circle A.

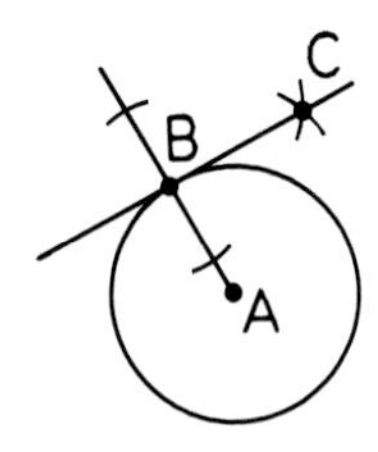

9. Which line was drawn first?
10. What relation does $\overleftrightarrow{BC}$ have to $\overleftrightarrow{AB}$?
11. Why is $\overleftrightarrow{BC}$ tangent to circle A?

In the figure below, $\overleftrightarrow{AC}$ has been constructed from point A tangent to circle B.

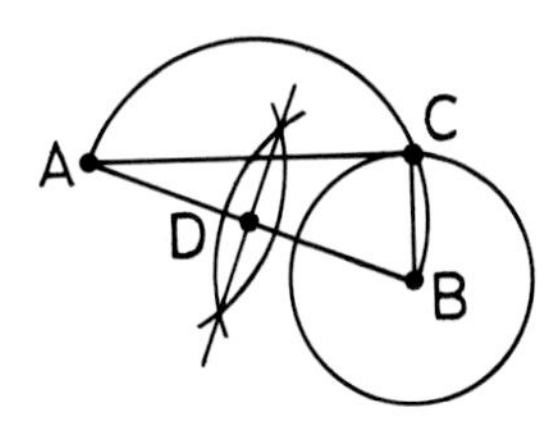

12. Which line was drawn first?

13. What was done to $\overline{AB}$?
14. What kind of arc is $\overset{\frown}{ACB}$?
15. What kind of angle is $\measuredangle C$?
16. Why?
17. Why is $\overline{AC}$ tangent to circle B?

Use your compass to make two copies of the figure shown here.

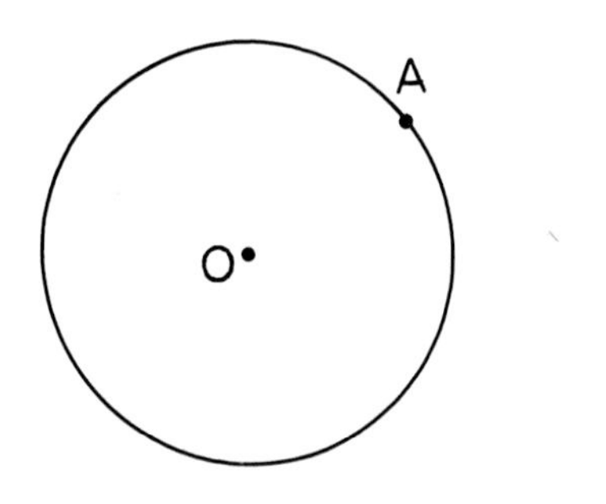

18. Through A in the first figure, construct a tangent to circle O.
19. Through B in the second figure, construct two tangents to circle O.

Copy the figure shown here.

.C

20. With C as its center, construct a circle that is tangent to line ℓ.
21. What is the distance from point C to line ℓ called with respect to the circle?

Set II

In the figure below, $\overline{IK}$ is tangent to circle E and IL = IK.

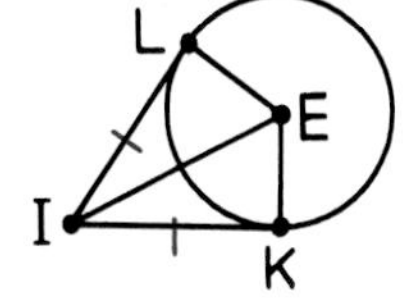

22. Why is $\triangle ILE \cong \triangle IKE$?
23. Why is $\measuredangle K$ a right angle?
24. Why is $\measuredangle L$ a right angle?
25. Why is $\overline{IL}$ tangent to circle E?

The figure below illustrates a connection between the two meanings of the word *tangent* in geometry.

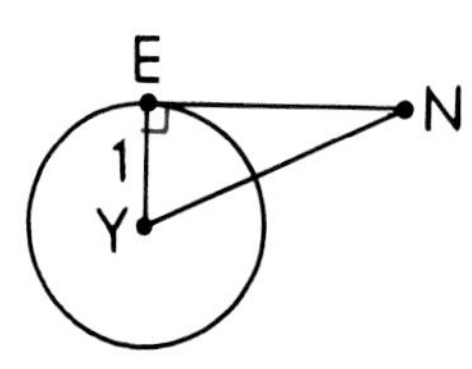

26. What is the definition of *tangent* in terms of a line and a circle?
27. What is the definition of *tangent* in terms of an acute angle of a right triangle?

In the figure above, $\triangle$YEN is a right triangle with right ∡E; Y is the center of the circle, whose radius is 1 unit.

28. Why does it follow from this information that $\overline{EN}$ is tangent to circle Y?
29. Write an equation for the tangent ratio of ∡Y. Use the fact that YE = 1 to simplify it.

30.

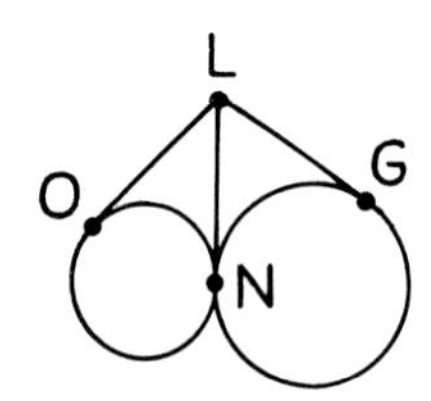

Given: $\overline{LO}$, $\overline{LN}$, and $\overline{LG}$ are tangent segments to the circles.

Prove: LO = LG.

31.

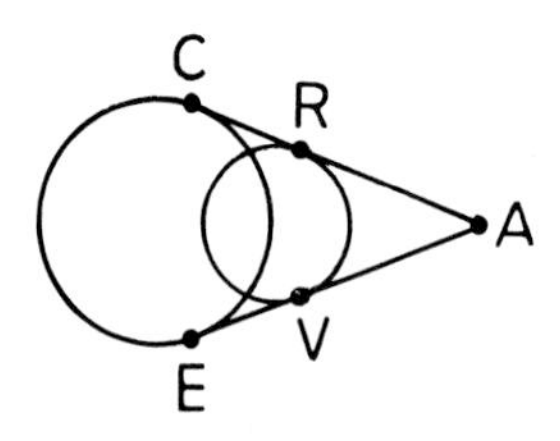

Given: $\overline{CA}$ and $\overline{EA}$ are tangent to both circles.

Prove: CR = EV.

32.

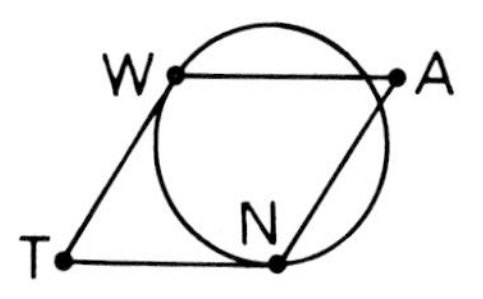

Given: $\overline{TW}$ and $\overline{TN}$ are tangent to the circle; WANT is a parallelogram.

Prove: WANT is a rhombus.

33.

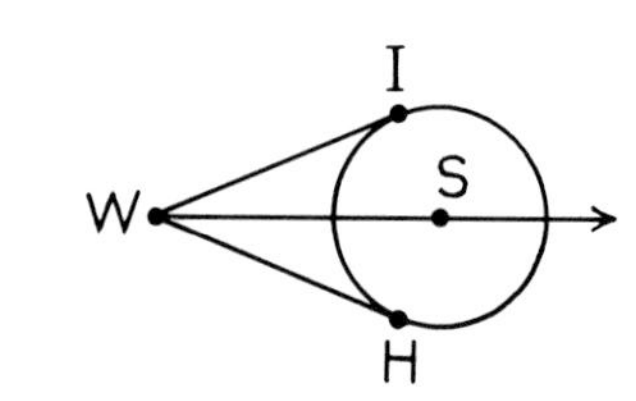

Given: $\overline{WI}$ and $\overline{WH}$ are tangent segments to circle S.

Prove: $\overrightarrow{WS}$ bisects ∡IWH.

34.

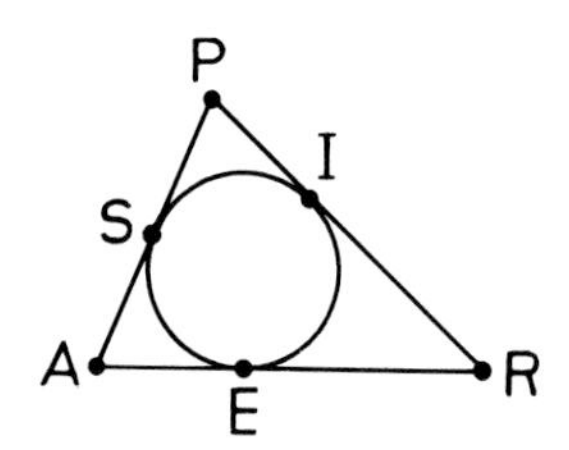

Given: The sides of $\triangle$APR are tangent to the circle; $\angle P = \angle A$.

Prove: PS = SA.

35.

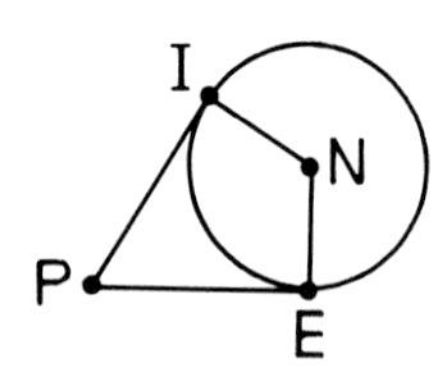

Given: $\overline{PI}$ and $\overline{PE}$ are tangent segments to circle N.

Prove: $\angle P = 180^\circ - m\widehat{IE}$.

Set III

In this figure, $\overline{AD}$ and $\overline{AE}$ are tangent segments to the circle and each has a length of 8. $\overline{BC}$ is tangent to the circle at point F.

What is the perimeter of $\triangle ABC$? Explain your reasoning.

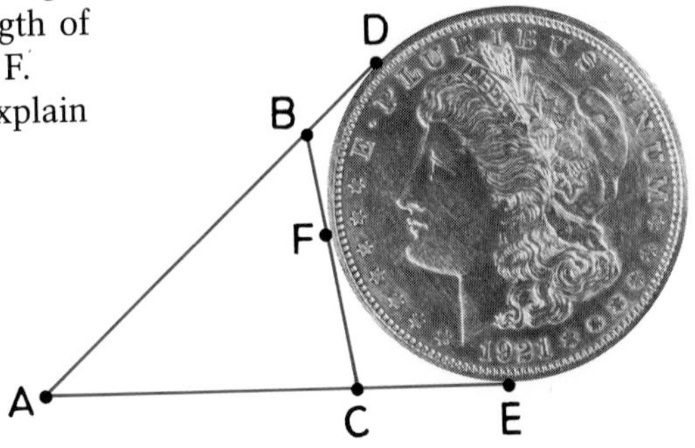

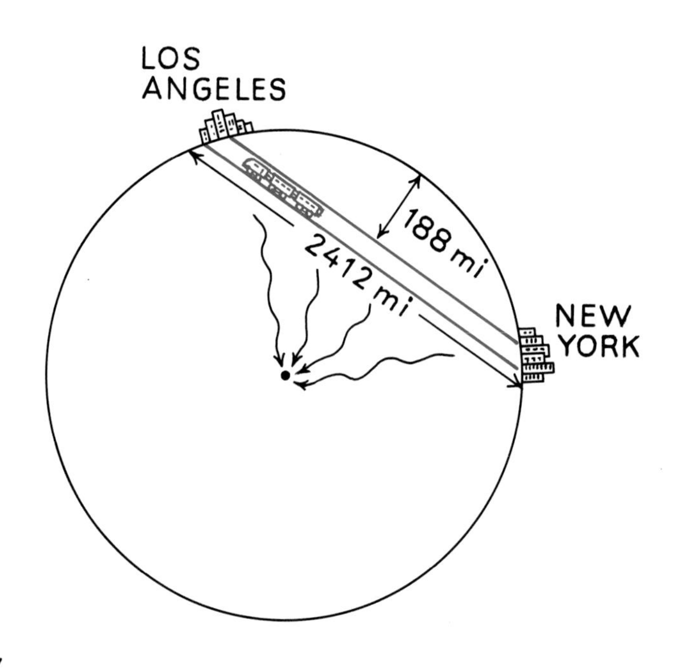

Lesson 7

Chord and Secant Segments

If the principal cities of the world were connected by straight tunnels, trains powered by gravity could travel between them at tremendous speeds. A tunnel from Los Angeles to New York would be 2,412 miles long, dropping at its midpoint to a depth of 188 miles below the earth's surface. A train traveling in this tunnel would be accelerated by the force of gravity during the first half of its journey, gaining just enough speed to coast up to the other end. The trip would take 42.2 minutes. In fact, a trip through a straight tunnel between *any* two points on the earth's surface would take 42.2 minutes. For example, two trains leaving Los Angeles at the same time for New York and London would arrive at their destinations simultaneously: 42.2 minutes later.*

Although this takes some knowledge of physics to prove, there is another remarkable fact about these gravity tunnels that is relatively easy to prove. Suppose that two tunnels are built that cross each other. In the figure at the left, for example, the tunnel from A to B intersects the tunnel from C to D at point P. Then, no matter where point P is, the distances from it to the four points on the earth's surface *always* satisfy

A History of Tunnels, by Patrick Beaver (Citadel Press, 1973).

the following equation:

$$PA \cdot PB = PC \cdot PD.$$

We will refer to this relation as the Intersecting Chords Theorem.

► **Theorem 76** (The Intersecting Chords Theorem)
If two chords intersect in a circle, the product of the lengths of the segments of one chord is equal to the product of the lengths of the segments of the other.

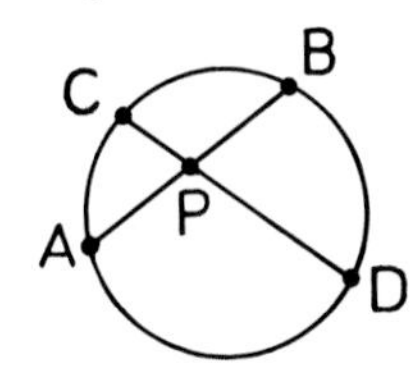

Given: Chords $\overline{AB}$ and $\overline{CD}$ intersect at point P in the circle.

Prove: $PA \cdot PB = PC \cdot PD$.

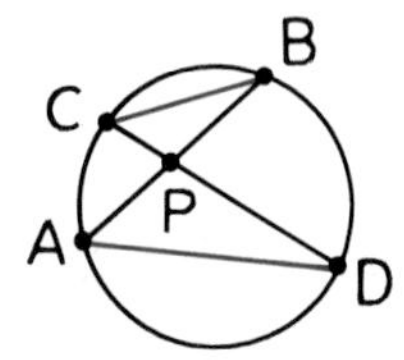

Proof.
Draw $\overline{CB}$ and $\overline{AD}$. Because $\angle A = \angle C$ and $\angle D = \angle B$ (inscribed angles that intercept the same arc are equal), it follows that $\triangle PAD \sim \triangle PCB$. From this we know that

$$\frac{PA}{PC} = \frac{PD}{PB}$$

(corresponding sides of similar triangles are proportional). Multiplying means and extremes, we get

$$PA \cdot PB = PC \cdot PD.$$

It is a remarkable fact that this relation also holds true for chords that "intersect outside" a circle. In the figure below, the lines containing chords $\overline{AB}$ and $\overline{CD}$ have been extended to meet at point P. For any such pair of chords $\overline{AB}$ and $\overline{CD}$, it is true that

$$PA \cdot PB = PC \cdot PD.$$

We will call segments $\overline{PA}$ and $\overline{PC}$ *secant segments* because they intersect the circle in more than one point.

► **Definition**
If a segment intersects a circle in two points, exactly one of which is an endpoint of the segment, then the segment is a ***secant segment*** to the circle.

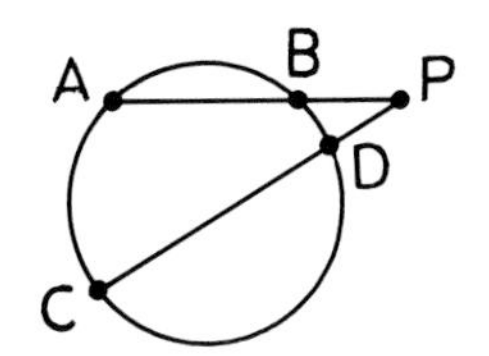

► **Theorem 77** (The Secant Segments Theorem)
If two secant segments are drawn to a circle from an external point, the product of the lengths of one secant segment and its external part is equal to the product of the lengths of the other secant segment and its external part.

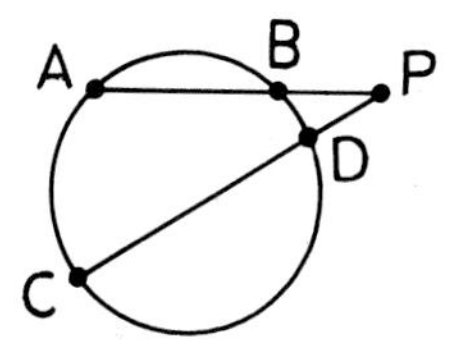

Given: The circle with secant segments $\overline{PA}$ and $\overline{PC}$.

Prove: $PA \cdot PB = PC \cdot PD$.

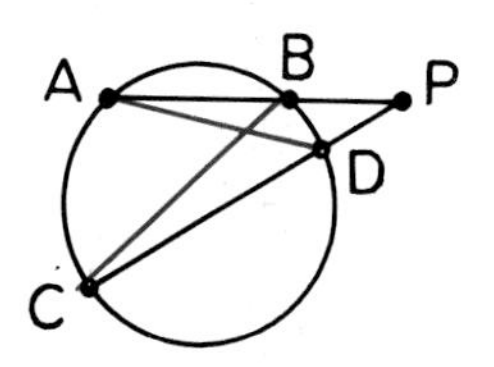

Proof.
Draw $\overline{CB}$ and $\overline{AD}$. Because $\angle A = \angle C$ (inscribed angles that intercept the same arc are equal) and $\angle P = \angle P$, it follows that $\triangle PAD \sim \triangle PCB$. From this we know that

$$\frac{PA}{PC} = \frac{PD}{PB},$$

and so

$$PA \cdot PB = PC \cdot PD.$$

Exercises

Set I

In the figure below, $\overline{JT}$ intersects the circle in points E and T, and $\overline{JS}$ intersects it in point S.

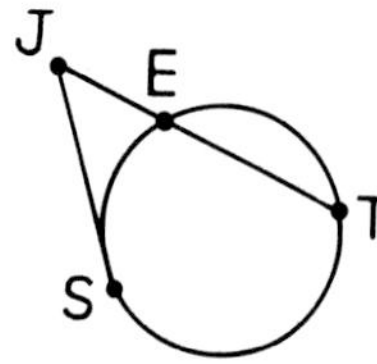

1. What is $\overline{JT}$ called with respect to the circle?
2. What is $\overline{JS}$ called?
3. What is $\overline{ET}$ called?

In the figure below, chords $\overline{CL}$ and $\overline{TO}$ intersect at point S in the circle.

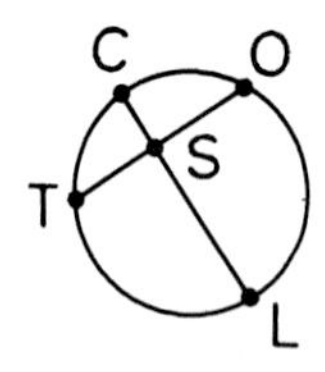

4. Write an equation for this figure that follows from the Intersecting Chords Theorem.
5. Use the fact that $\measuredangle OSL$ is a secant angle of the circle to write an equation for its measure.
6. If $\overline{CL}$ is a diameter of the circle and bisects $\overline{TO}$, what can you conclude?

In the figure below, $\overline{BA}$ and $\overline{BR}$ are secant segments to the circle.

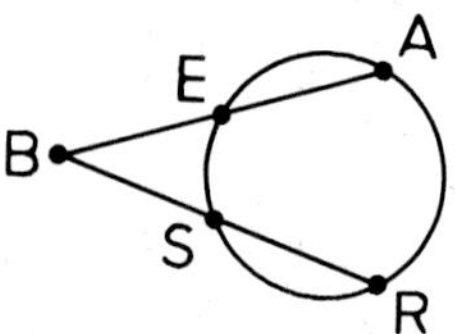

7. Write an equation for this figure that follows from the Secant Segments Theorem.
8. Use the fact that $\measuredangle B$ is a secant angle of the circle to write an equation for its measure.
9. If $m\overset{\frown}{EA} = m\overset{\frown}{SR}$, what can you conclude?

In the figure below, $\overline{CI}$ and $\overline{HE}$ are chords of circle F.

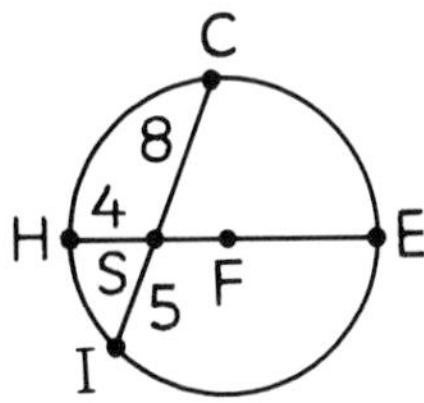

10. Find $\overline{SE}$.

11. Find $\overline{HE}$.

12. Find $\overline{FE}$.

13. Find $\overline{SF}$.

Chords and secant segments have been drawn in the following figures. Find x in each figure.

Example:

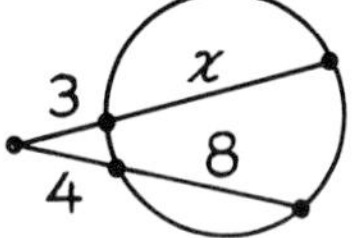

Solution: $(3 + x) \cdot 3 = (4 + 8) \cdot 4$

$9 + 3x = 48$

$3x = 39$

$x = 13$

14.

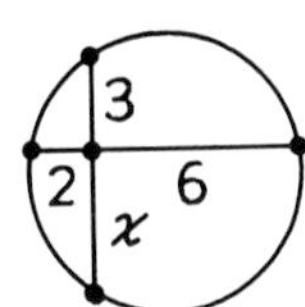

15.

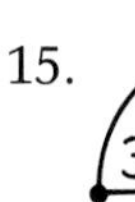
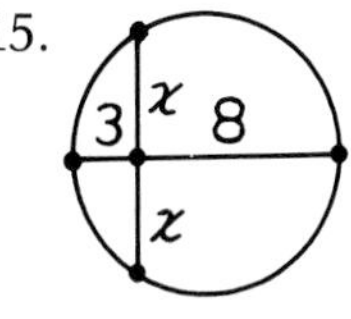

16.

17.

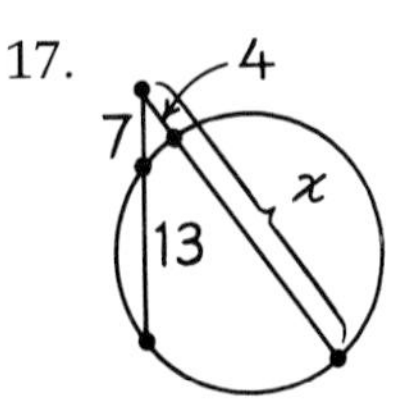

18.

19.

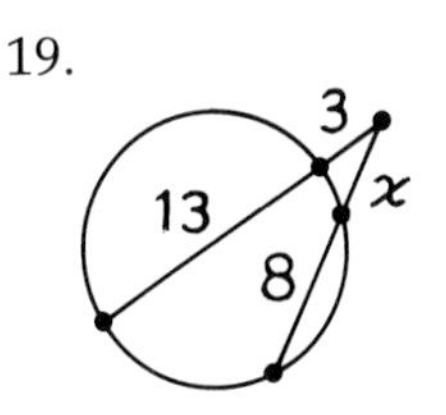

Set II

In the figure below, $\overline{BW}$ is a secant segment and $\overline{BN}$ and $\overline{BS}$ are tangent segments to circle O; BS = 15 and OS = 8. Find each of the following lengths.

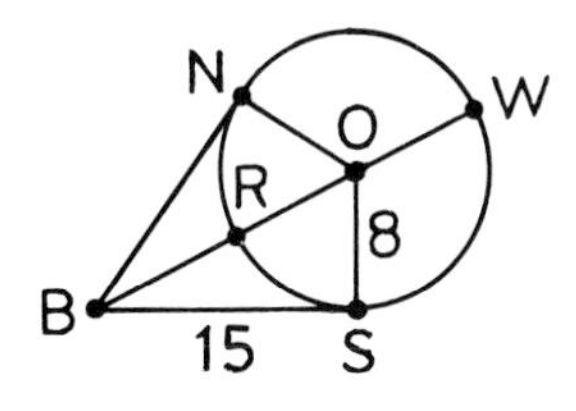

20. BO.

21. BN.

22. BR.

23. BW.

Use an appropriate trigonometric ratio and either the table on page 411 or your calculator to find the measure of each of the following angles and arcs to the nearest degree.

24. ∡BOS.

25. ∡NBS.

26. $\overset{\frown}{WS}$.

27. $\overset{\frown}{NS}$.

In the figure below, $\overline{AO}$ and $\overline{LO}$ are secant segments to the circle and $\overline{LF}$ is a chord; FS = 4, SL = 15, LC = 24, CO = 9, and NO = 11.

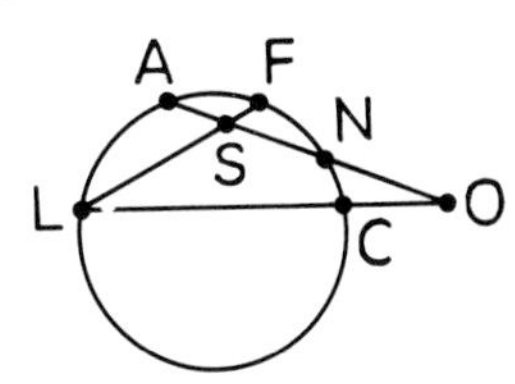

28. Copy the figure and mark these lengths on it.

29. Find AO.

30. Find AN.

31. Find AS.

32.

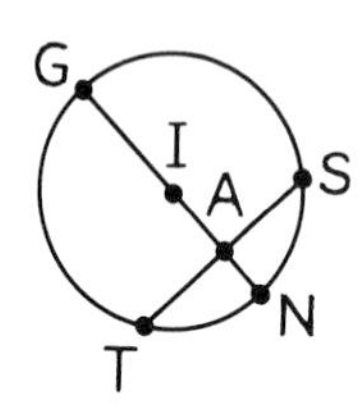

Given: In circle I, $\overline{GN} \perp \overline{TS}$.

Prove: $GA \cdot AN = AS^2$.

33.

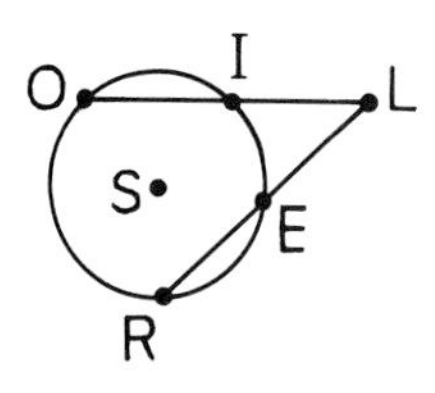

Given: $\overline{LO}$ and $\overline{LR}$ are secant segments to circle S; $LO = LR$.

Prove: $LI = LE$.

34.

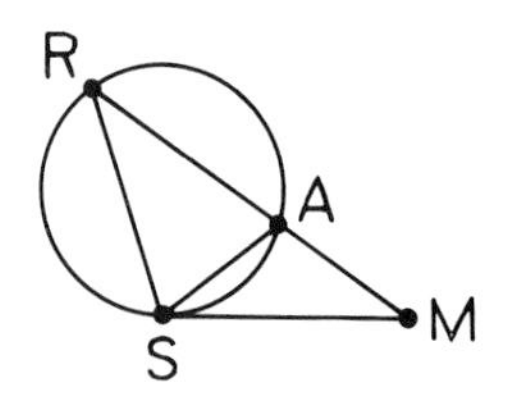

Given: The circle with tangent segment $\overline{MS}$ and secant segment $\overline{MR}$; $\angle R = \angle ASM$.

Prove: $\dfrac{RM}{SM} = \dfrac{SM}{AM}$.

35.

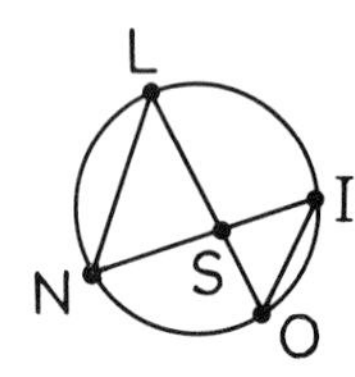

Given: Chords $\overline{LO}$ and $\overline{NI}$ intersect in point S.

Prove: $\dfrac{LN}{IO} = \dfrac{LS}{IS}$.

Set III

This rather challenging problem from a Korean geometry book consists of three consecutive parts. How many of them can you prove? Show your proofs.

△ABC에서 ∠A의 이등분선이 BC와 외접원과의 교점을 각각 D, E라고 하면

① $\triangle ABE \backsim \triangle ADC$

② $AB \cdot AC = AD \cdot AE$

③ $AD^2 = AB \cdot AC - BD \cdot DC$

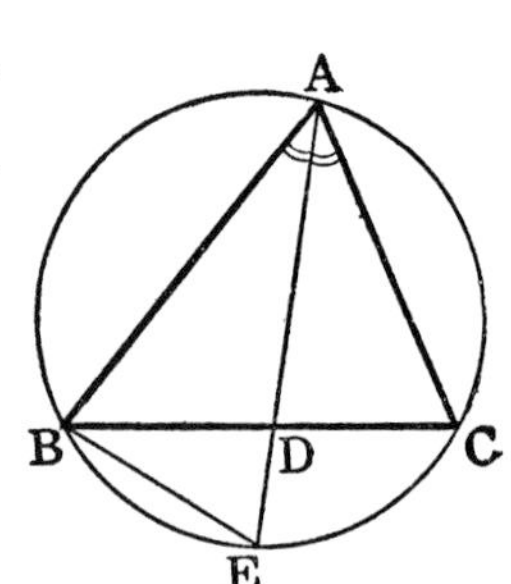

Chapter 12 / Summary and Review

Basic Ideas

Constructions

9. To construct the tangent to a circle at a given point on the circle. 447
10. To construct the tangents to a circle from a given external point. 448

Postulate

13. *The Arc Addition Postulate.* If C is on $\widehat{AB}$, then $m\widehat{AC} + m\widehat{CB} = m\widehat{ACB}$. 432

Theorems

Corollary to the definition of a circle. All radii of a circle are equal. 421

65. If a line through the center of a circle is perpendicular to a chord, it also bisects the chord. 422
66. If a line through the center of a circle bisects a chord that is not a diameter, it is also perpendicular to the chord. 422
67. The perpendicular bisector of a chord (in the plane of a circle) passes through the center of the circle. 422

Courtesy of Elmer Atkins; © 1963 Saturday Review, Inc.

68. If a line is tangent to a circle, it is perpendicular to the radius drawn to the point of contact. 427
69. If a line is perpendicular to a radius at its outer endpoint, then it is tangent to the circle. 427
70. If two chords of a circle are equal, their minor arcs are equal. 433
71. If two minor arcs of a circle are equal, their chords are equal. 433
72. An inscribed angle is equal in measure to half its intercepted arc. 437
    Corollary 1. Inscribed angles that intercept the same arc or equal arcs are equal. 438
    Corollary 2. An angle inscribed in a semicircle is a right angle. 438
73. A secant angle whose vertex is inside a circle is equal in measure to half the sum of the arcs intercepted by it and its vertical angle. 442
74. A secant angle whose vertex is outside a circle is equal in measure to half the positive difference of its intercepted arcs. 443
75. The tangent segments to a circle from an external point are equal. 447

76. *The Intersecting Chords Theorem.* If two chords intersect in a circle, the product of the lengths of the segments of one chord is equal to the product of the lengths of the segments of the other. 453

77. *The Secant Segments Theorem.* If two secant segments are drawn to a circle from an external point, the product of the lengths of one secant segment and its external part is equal to the product of the lengths of the other secant segment and its external part. 454

Exercises

Set I

These bird shapes and many others can be made from the nine pieces of the egg puzzle illustrated below.

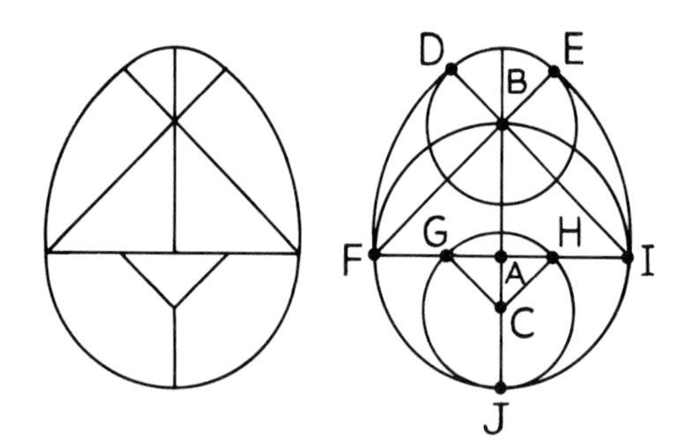

The pieces are shown in the first figure and the pattern for making them is shown in the second figure. The pattern is based on three circles.

With respect to circle A,

1. what is $\overline{FI}$ called?
2. what is $\widehat{FBI}$ called?
3. what is $\measuredangle FBJ$ called?

With respect to circle B,

4. what is $\overline{BE}$ called?
5. what is $\widehat{DE}$ called?
6. what is $\measuredangle DBE$ called?
7. what is $\overline{FE}$ called?

With respect to circle C,

8. what is $\overline{GH}$ called?
9. what is $\widehat{GJH}$ called?
10. what is $\measuredangle HAJ$ called?

Tell whether each of the following statements is true or false.

11. If two circles are concentric, they must lie in the same plane and have the same center.
12. A diameter is a chord.
13. The measure of a minor arc is less than 90°.
14. If a line through the center of a circle bisects a chord, it must also be perpendicular to it.
15. An inscribed angle can intercept a major arc.
16. A secant is a line that intersects a circle in more than one point.
17. If a line is perpendicular to a radius of a circle, it must be tangent to the circle.
18. Two central angles in a circle are equal only if their minor arcs are equal.

Solve for x in each of the following figures.

19.

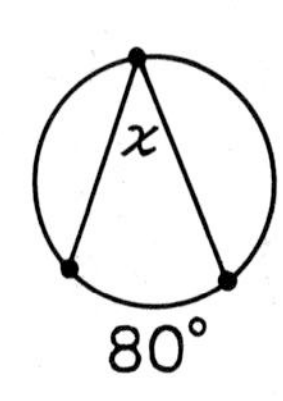

20.

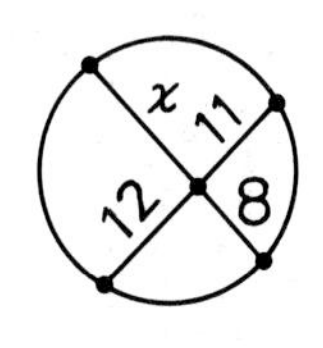

21.

22.

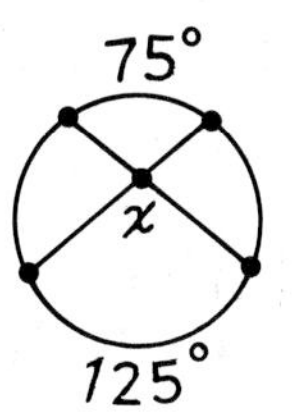

23.

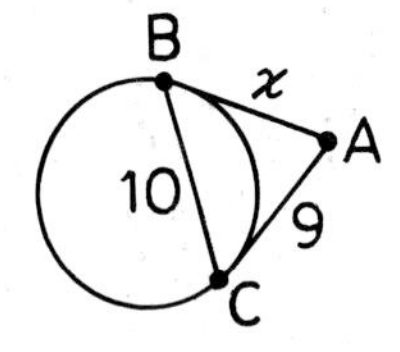

$\overline{AB}$ and $\overline{AC}$ are tangent to the circle.

24.

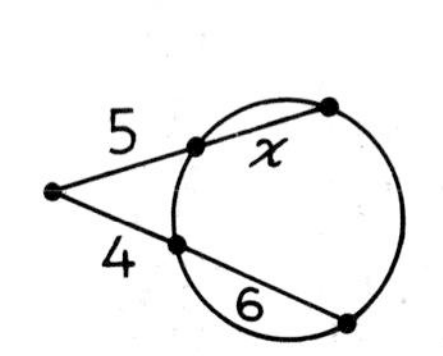

25.

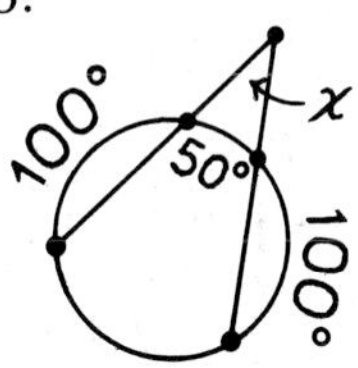

26.

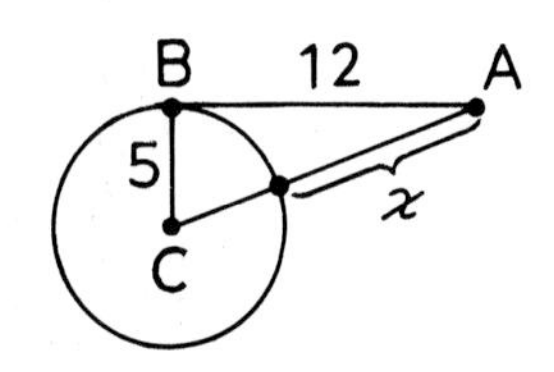

$\overline{AB}$ is tangent to circle C.

Copy the figure below and do the following constructions.

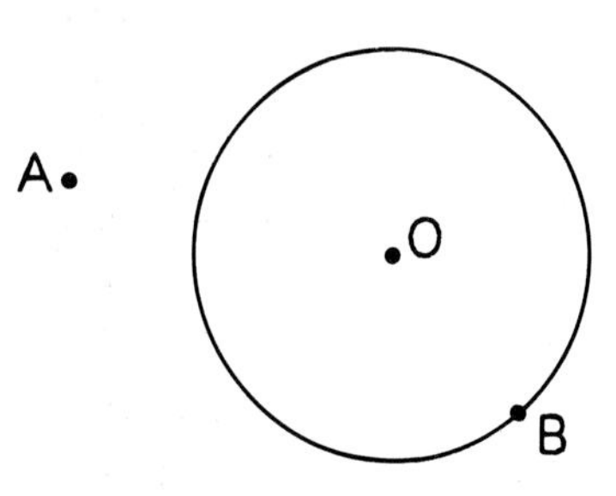

27. Through A, construct two tangents to circle O.

28. Through B, construct a tangent to circle O.

Set II

In the figure below, $\overline{FL}$ is a diameter of circle U and $\overline{FL} \perp \overline{AT}$; FA = 9 and FT = 15.

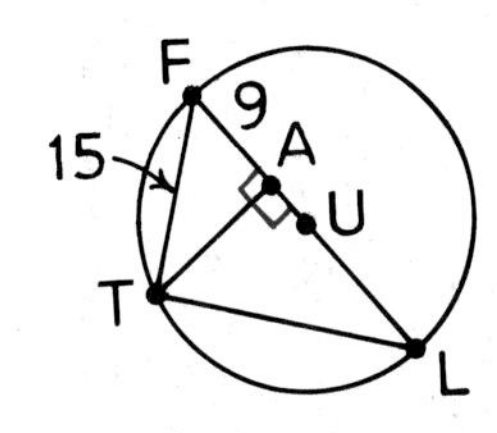

29. Find AL.

30. Find UL.

31. Find $p\triangle$AFT.

32. Find $\alpha\triangle$ALT.

33.

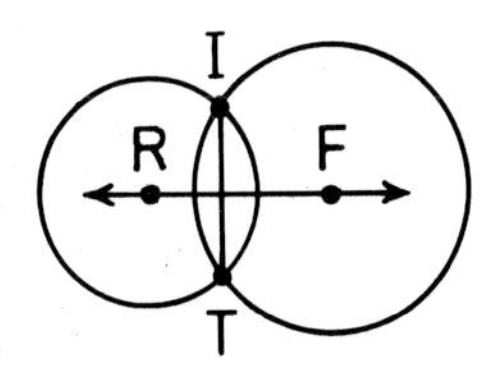

Given: Circles R and F with common chord $\overline{IT}$.

Prove: $\overleftrightarrow{RF}$ is the perpendicular bisector of $\overline{IT}$.

34.

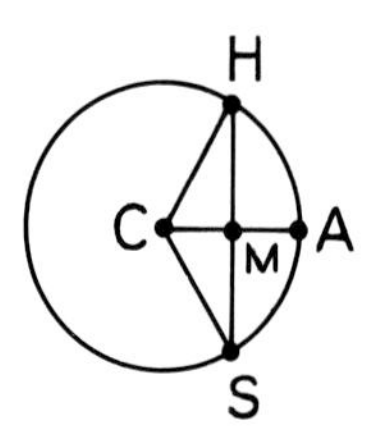

Given: In circle C, $\overline{CA} \perp \overline{HS}$.

Prove: $m\widehat{HA} = m\widehat{AS}$.

35.

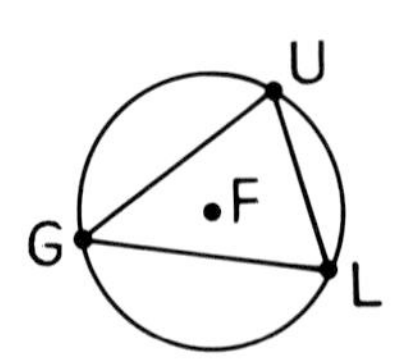

Given: Circle F with $m\widehat{GU} > m\widehat{UL}$.

Prove: $GU > UL$.

36.

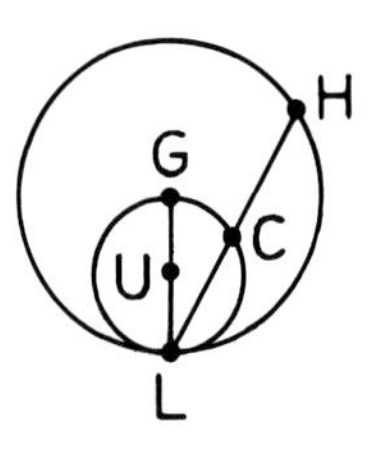

Given: $\overline{GL}$ is a diameter of circle U and $\overline{LH}$ is a chord of circle G.

Prove: $LC = CH$. (Hint: Draw $\overline{GC}$.)

Chapter 13

THE CONCURRENCE THEOREMS

Lesson 1
Concyclic Points

Aerofilms Ltd.

Stonehenge, a great circle of large stones on the Salisbury Plain in England, is thought to have been built between 2000 and 1500 B.C. From the positions of the stones, it seems that Stonehenge was used to determine such things as the beginning of the seasons and eclipses of the sun and moon.

The figure at the left shows the positions of the stones that still stand in the main circle. They illustrate the idea of *concyclic points.*

► **Definition**
Points are ***concyclic*** iff they lie on the same circle.

The center of Stonehenge can be determined from the positions of any three of its stones. To see how, look at the figures below.

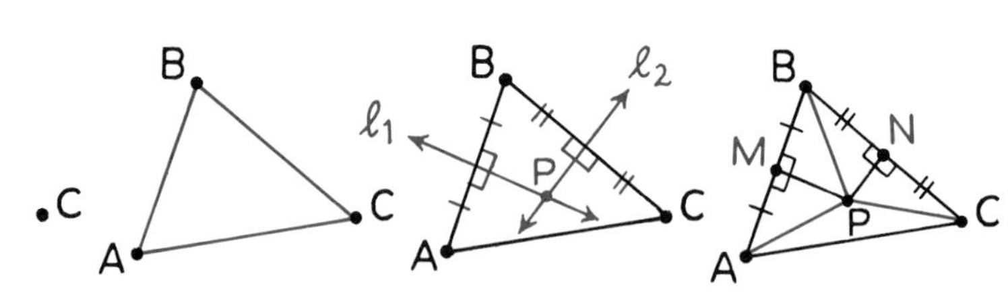

First, the three points representing the positions of the stones can be joined to form a triangle. Second, thinking of the sides of this triangle as chords of a circle, we can use the fact that in a plane the perpendicular bisectors of the chords of a circle pass through its center.

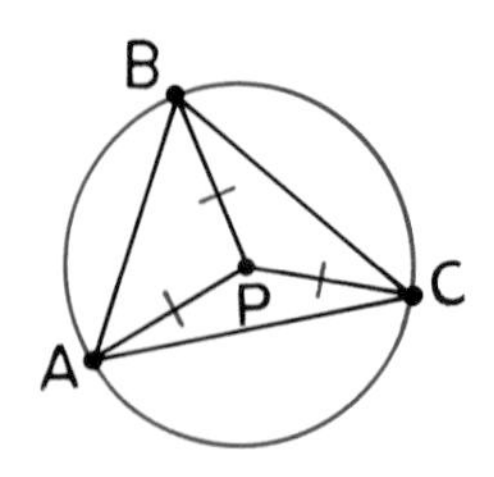

Let ℓ_1 and ℓ_2 be the perpendicular bisectors of $\overline{AB}$ and $\overline{BC}$, respectively, and call their point of intersection P. Draw $\overline{PA}$, $\overline{PB}$, and $\overline{PC}$.

Now, $\triangle PAM \cong \triangle PBM$ (S.A.S.), and so PA = PB; $\triangle PBN \cong \triangle PCN$ (S.A.S.), and so PB = PC. It follows that PA = PB = PC, and so P is equidistant from A, B, and C. Therefore, a circle with P as center and PA as radius contains all three points.

This proof can be applied to all triangles, and so for every triangle there exists a circle that contains all of its vertices.

► **Definition**
A polygon is ***cyclic*** iff there exists a circle that contains all of its vertices.

What we have proved can be stated as the following theorem.

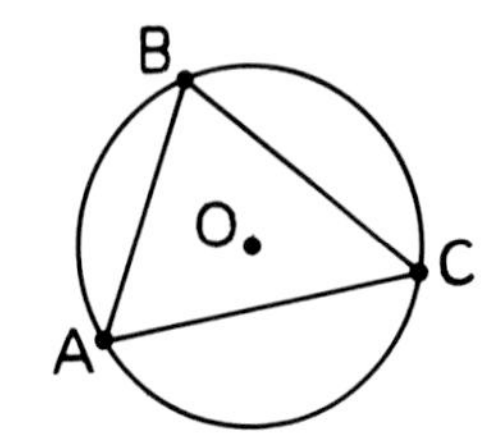

► **Theorem 78**
Every triangle is cyclic.

A polygon is said to be *inscribed* in a circle that contains all of its vertices and the circle is said to be *circumscribed* about the polygon. The circle is called the *circumcircle* of the polygon and its center is called the *circumcenter* of the polygon.

To prove Theorem 78, we drew the perpendicular bisectors of two sides of the triangle and showed that the point in which they intersect is equidistant from the three vertices. It is easy to prove that the perpendicular bisector of the third side passes through this same point. Lines that contain the same point are called *concurrent.*

► **Definition**
Lines are ***concurrent*** iff they contain the same point.

► **Corollary to Theorem 78**
The perpendicular bisectors of the sides of a triangle are concurrent.

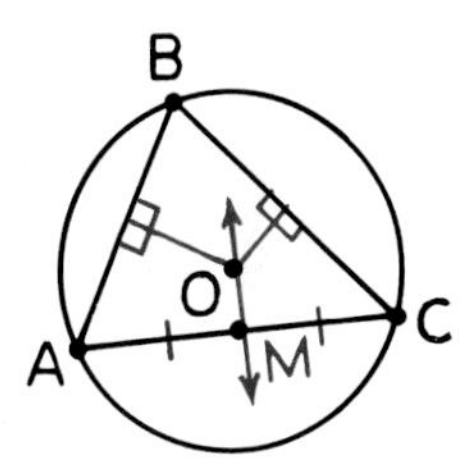

Let O be the center of the circumcircle of $\triangle ABC$ (the point in which the perpendicular bisectors of sides $\overline{AB}$ and $\overline{BC}$ intersect) and let M be the midpoint of side $\overline{AC}$. Draw $\overleftrightarrow{OM}$. Now, $\overline{AC}$ is a chord of circle O and $\overleftrightarrow{OM}$ is a line through the center of the circle that bisects $\overline{AC}$. Hence, $\overleftrightarrow{OM} \perp \overline{AC}$. Therefore, $\overleftrightarrow{OM}$ is the perpendicular bisector of $\overline{AC}$, and so the perpendicular bisectors of the sides of the triangle are concurrent.

Theorem 78 and its corollary suggest that, to construct the circumcircle of a triangle, we must construct the perpendicular bisectors of two of its sides.

► **Construction 11**
To circumscribe a circle about a triangle.

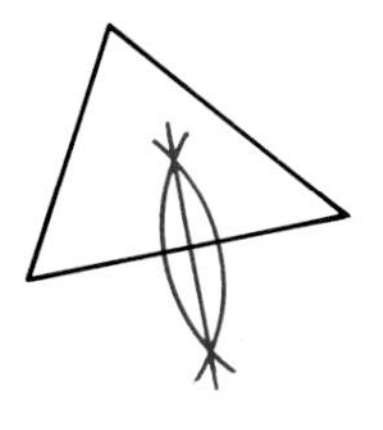

Step 1

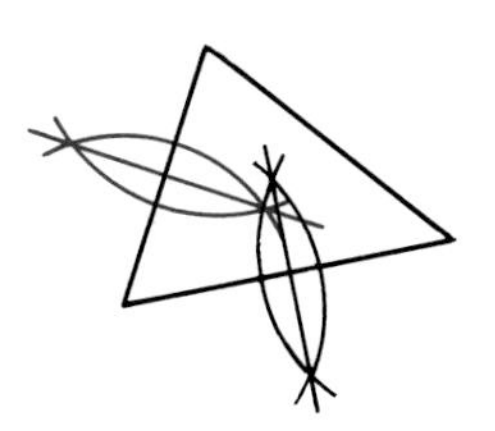

Step 2

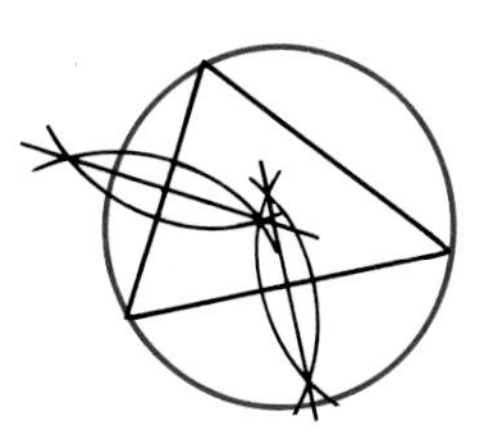

Step 3

Exercises

Set I

In the figure below, the vertices of $\triangle$KEN are on circle O.

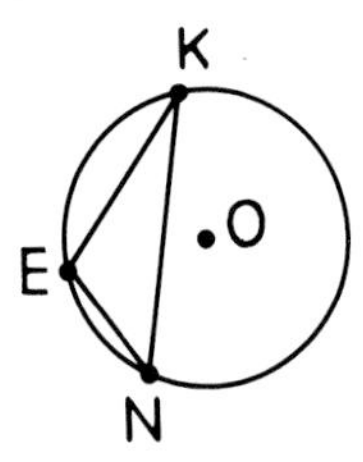

1. What relation do points K, E, and N have to each other?
2. What are $\overline{KE}$, $\overline{EN}$, and $\overline{KN}$ called with respect to the circle?
3. What are $\angle$K, $\angle$E, and $\angle$N called with respect to the circle?
4. What relation do points K, E, and N have to point O?
5. What is circle O called with respect to $\triangle$KEN?
6. What is point O called with respect to the triangle?

Use your straightedge and compass to construct an equilateral triangle each of whose sides is 5 centimeters long.

7. Construct the perpendicular bisectors of all three of its sides.
8. Why would you expect the perpendicular bisectors of the sides to intersect in a common point?
9. What kind of right triangles are formed by this construction?

Trace the following triangles and circumscribe a circle about each.

10.

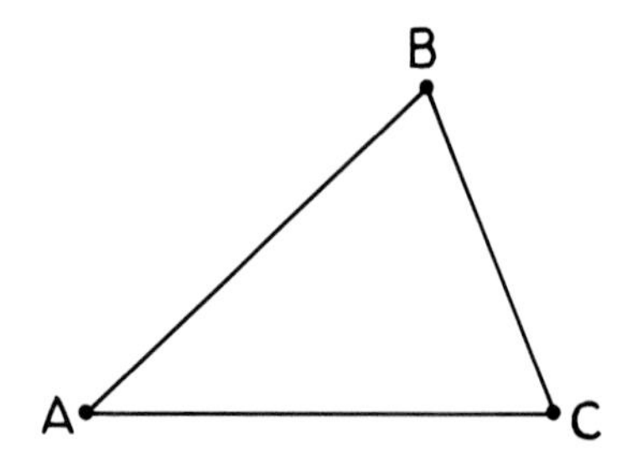

11.

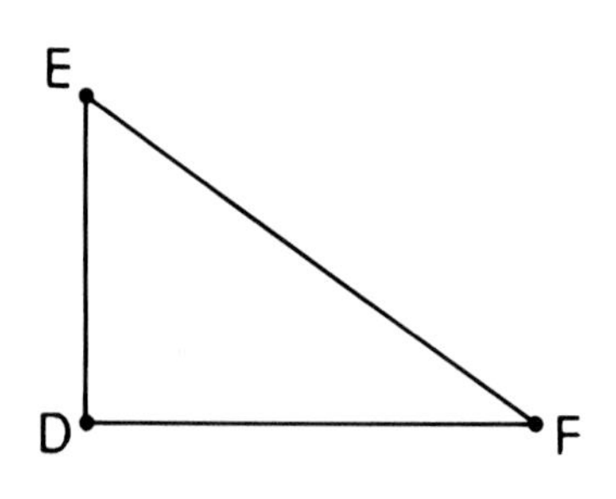

12.

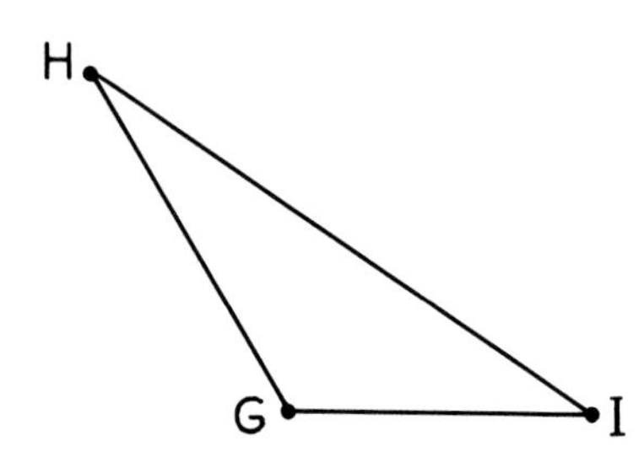

The position of the circumcenter of a triangle is determined by the measures of its angles. For example, Exercise 10 shows that the circumcenter of an acute triangle lies inside it.

13. Where does Exercise 11 seem to indicate that the circumcenter of a right triangle is located?
14. Where does Exercise 12 indicate that the circumcenter of an obtuse triangle is located?

In the figure below, $\triangle$SAT is a 30°-60° right triangle inscribed in circle K; ST = 12. Find each of the following measures.

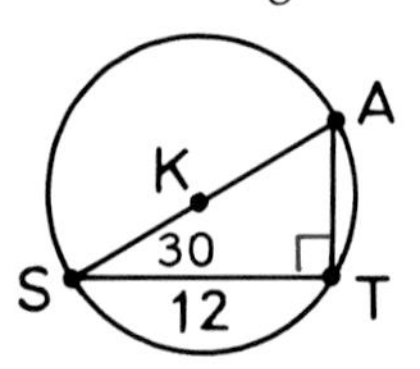

15. $m\widehat{ST}$.
16. AT.
17. KA.

In the figure below, △BAR is an isosceles right triangle inscribed in circle G; $\alpha\triangle BAR = 50$. Find each of the following measures.

18. $m\overset{\frown}{BR}$.
19. BR.
20. BG.

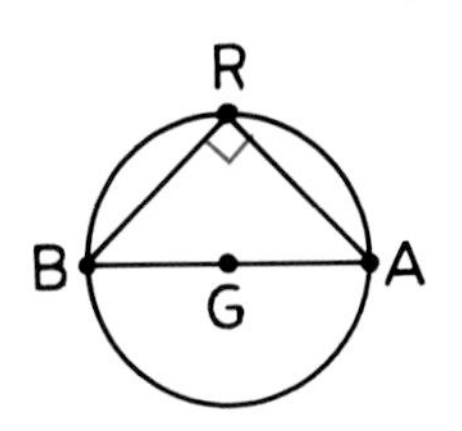

The figures below can be used to prove that the midpoint of the hypotenuse of a right triangle is its circumcenter.

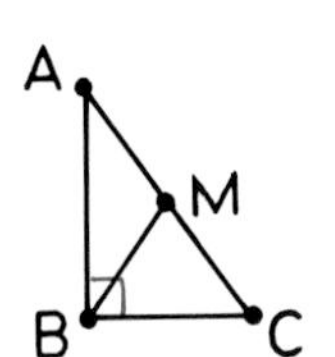

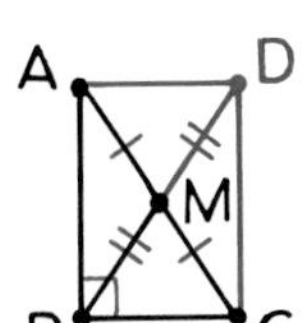

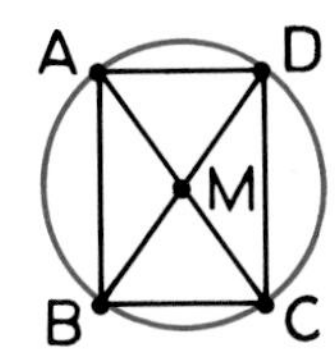

In the first figure, point M is the midpoint of the hypotenuse of right △ABC.

21. Why is MA = MC?

In the second figure, point D has been chosen on $\overleftrightarrow{BM}$ so that MD = BM. $\overline{DA}$ and $\overline{DC}$ have been drawn.

22. Why is ABCD a parallelogram?
23. What special type of parallelogram is ABCD?
24. Why is BD = AC?

Because $MB = \frac{1}{2}BD$ and $MA = MC = \frac{1}{2}AC$, it follows that MB = MA = MC.

25. Why can a circle be drawn with center M and radius MA that goes through points A, B, and C?

Set II

In the figure below, point O is the circumcenter of △PKR and $\overline{OE}$ bisects $\overline{KR}$; KR = 8 and OE = 3.

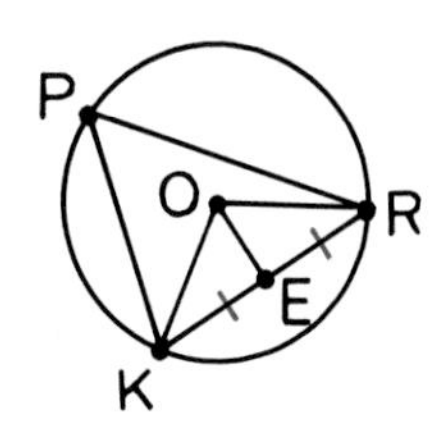

26. Why is $\overline{OE} \perp \overline{KR}$?
27. Find OK.
28. Use the appropriate trigonometric ratio and either the table on page 411 or your calculator to find ∠KOE to the nearest degree.
29. Find ∠KOR.
30. Find $m\overset{\frown}{KR}$.
31. Find ∠P.

32.

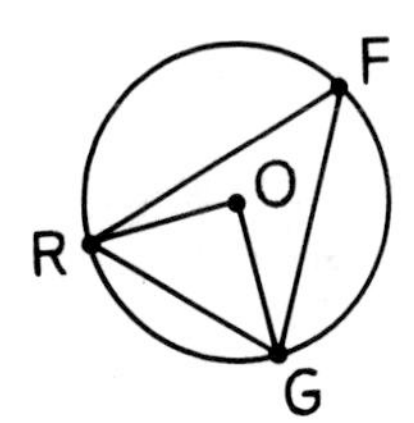

Given: △FRG with circumcircle O.

Prove: △ROG is isosceles.

33.

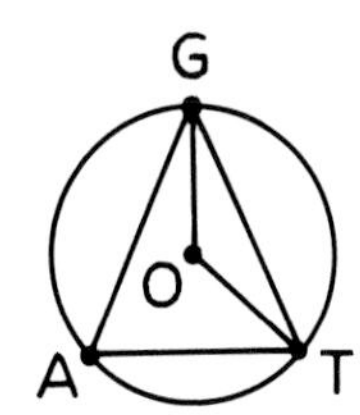

Given: △GAT is inscribed in circle O.

Prove: ∠O = 2∠A.

34.

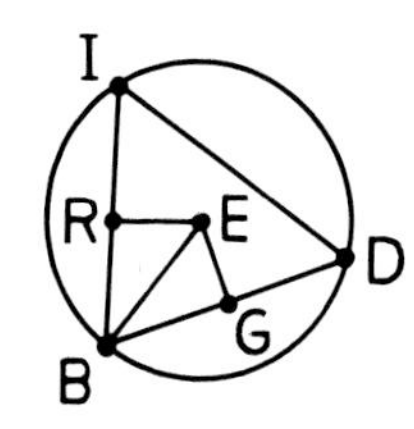

Given: $\triangle$BID with circumcircle E and IR = RB = BG = GD.

Prove: ER = EG.

35.

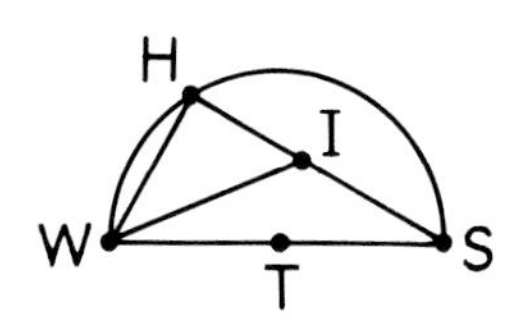

Given: $\triangle$WHS is inscribed in semicircle $\widehat{WHS}$.

Prove: $\triangle$WIS is obtuse.

Set III

Obtuse Ollie put a tall ladder against a wall so that it made 45° angles with the wall and floor. After he had climbed halfway up the ladder, the top end started slipping and slid all the way down the wall. Ollie was too startled to do anything but hold on.

1. Draw two perpendicular line segments to represent the wall and floor. Make an accurate drawing of Ollie's path as the ladder slid down the wall by using one of the shorter edges of a file card to represent the ladder as shown below.

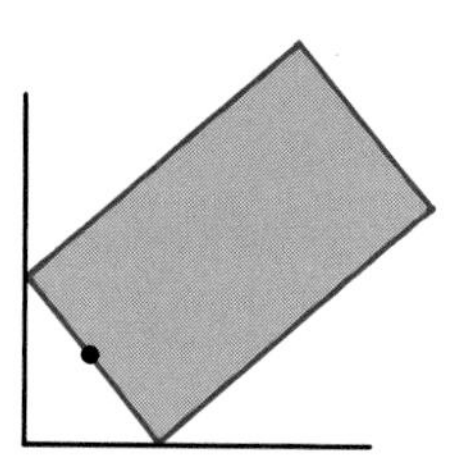

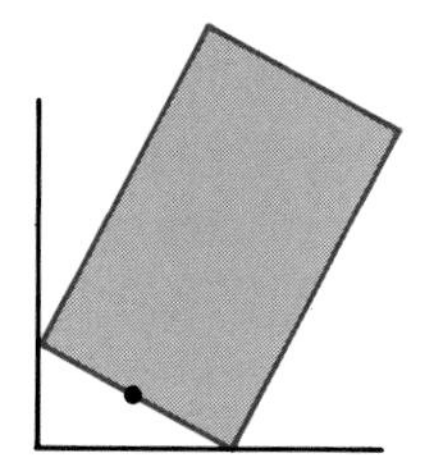

2. What kind of a path do you think Ollie traveled?

3. Use the fact that the midpoint of the hypotenuse of a right triangle is its circumcenter to explain why the path has the shape

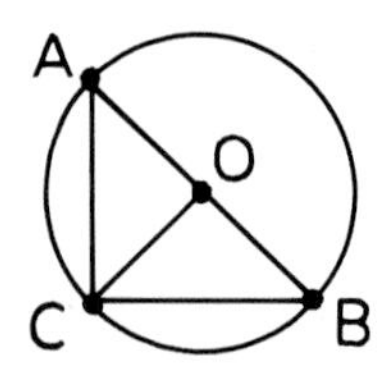

that it does. (Hint: Why, as $\overline{AB}$ moves, does CO always stay the same?)

Lesson 2

Cyclic Quadrilaterals

Leonardo da Vinci applied geometry to his study of the human figure. The illustration above is based on a famous drawing from one of da Vinci's notebooks. One position of the man corresponds to a square and the other position to a circle. If you stood as the man is standing in the second position, would your hands and feet also touch a circle?

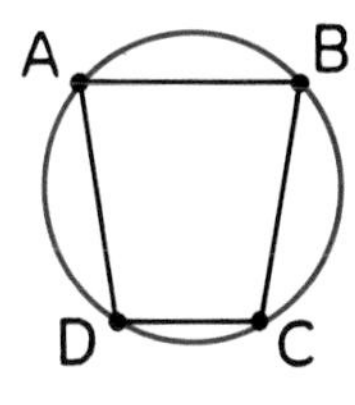

The answer to this question depends on whether the quadrilateral whose vertices correspond to the positions of your hands and feet is cyclic. Unlike triangles, not every quadrilateral is cyclic. Whether or not the vertices of a quadrilateral lie on a circle is determined by the relation of its opposite angles.

► **Theorem 79**
A quadrilateral is cyclic iff a pair of its opposite angles are supplementary.

First, we will prove that, if a quadrilateral is cyclic, its opposite angles are supplementary.

Given: Quadrilateral ABCD is cyclic.
Prove: ∢A and ∢C are supplementary.

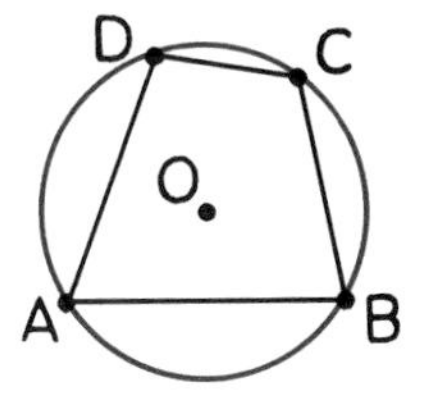

Because quadrilateral ABCD is cyclic, let circle O contain its vertices (if a polygon is cyclic, there exists a circle that contains all of its vertices).

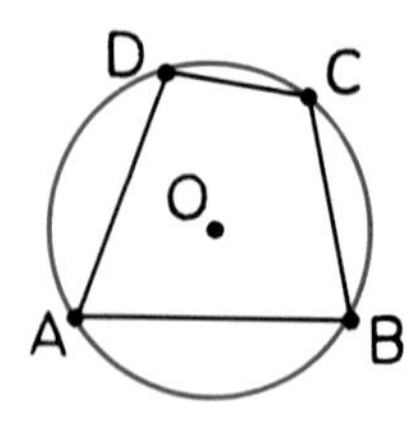

Now, $\angle A = \frac{1}{2} m\overset{\frown}{DCB}$ and $\angle C = \frac{1}{2} m\overset{\frown}{BAD}$ (an inscribed angle is equal in measure to half its intercepted arc). Because $\angle A + \angle C = \frac{1}{2} m\overset{\frown}{DCB} + \frac{1}{2} m\overset{\frown}{BAD}$ (addition), $\angle A + \angle C = \frac{1}{2}(m\overset{\frown}{DCB} + m\overset{\frown}{BAD}) = \frac{1}{2}(360°) = 180°$ (a circle has a degree measure of 360°). Therefore, $\measuredangle A$ and $\measuredangle C$ are supplementary.

Now we will prove the converse: if a pair of opposite angles of a quadrilateral are supplementary, then it is cyclic.

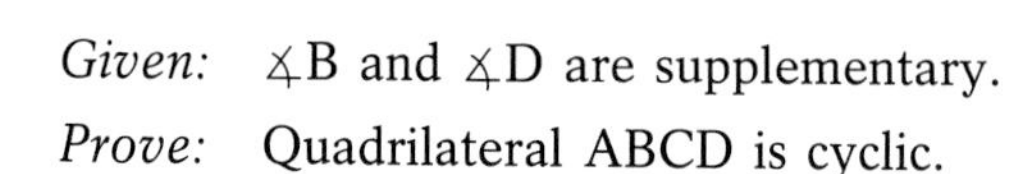

Given: $\measuredangle B$ and $\measuredangle D$ are supplementary.
Prove: Quadrilateral ABCD is cyclic.

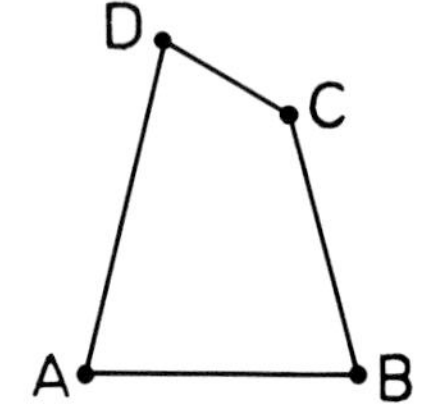

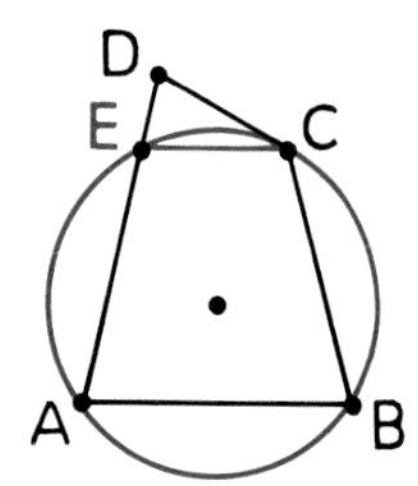

Draw a circle through points A, B, and C (points A, B, and C determine a triangle and every triangle is cyclic). Point D lies either *outside* this circle, *inside* it, or *on* it.

Suppose that D lies outside the circle. Let the second point in which $\overline{AD}$ intersects the circle be called E. Draw $\overline{EC}$.

Now, $\measuredangle B$ and $\measuredangle AEC$ are supplementary (if a quadrilateral is cyclic, its opposite angles are supplementary), and $\measuredangle B$ and $\measuredangle D$ are supplementary (given); so $\angle AEC = \angle D$ (supplements of the same angle are equal). But this contradicts the fact that $\angle AEC > \angle D$ (an exterior angle of a triangle is greater than either remote interior angle), and so our original assumption must be false. In other words, D cannot lie outside the circle.

A similar argument can be used to establish that D cannot lie inside the circle. This means that D must lie on the circle, and so quadrilateral ABCD is cyclic.

Exercises

Set I

The vertices of quadrilateral ZEUS lie on the circle.

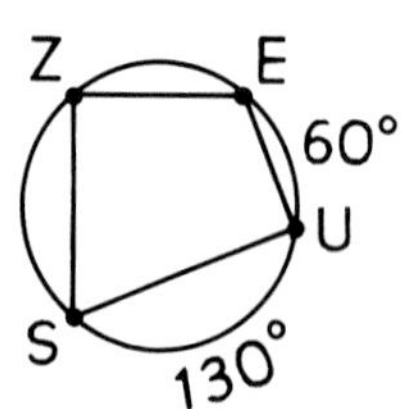

1. What kind of quadrilateral is ZEUS?
2. What relation do $\measuredangle E$ and $\measuredangle S$ have?
3. Given that $m\overset{\frown}{EU} = 60°$ and $m\overset{\frown}{US} = 130°$, find $\angle Z$.
4. Find $\angle U$.

In the figure below, $\overline{EJ}$ and $\overline{EV}$ are tangent to circle O.

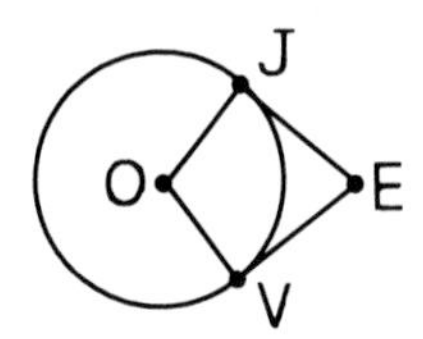

5. What relation do $\overline{EJ}$ and $\overline{EV}$ have to $\overline{OJ}$ and $\overline{OV}$?
6. How do you know?
7. What can you conclude about ∡J and ∡V?
8. Is quadrilateral JOVE cyclic?
9. Explain why or why not.

Quadrilateral ECHO is a rhombus in which ∠E = 112°.

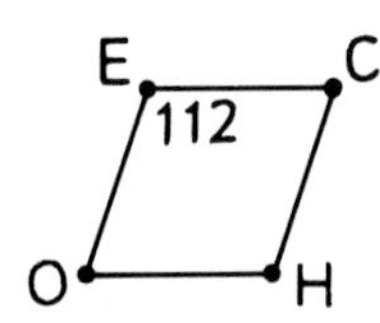

10. Find the measures of ∡C, ∡H, and ∡O.
11. Is quadrilateral ECHO cyclic?
12. Explain why or why not.

In the figure below, side $\overline{VN}$ of quadrilateral VLCN is tangent to circle U; $m\widehat{LC} = 60°$ and $m\widehat{CA} = 40°$.

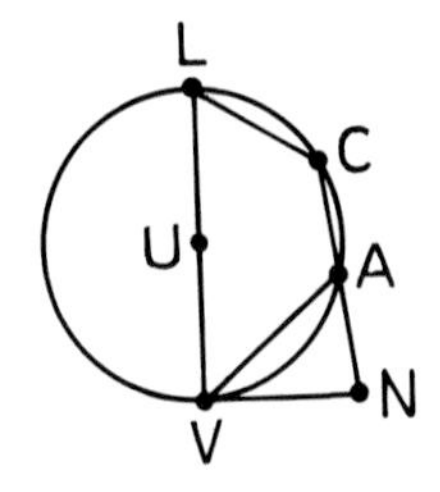

13. Copy the figure and write these measures on it.

Find the measures of the following angles.

14. ∡LVA.
15. ∡C.
16. ∡CAV.
17. ∡L.
18. ∡LVN.
19. ∡N.

20. Give the missing statements and reasons in the following proof that all isosceles trapezoids are cyclic.

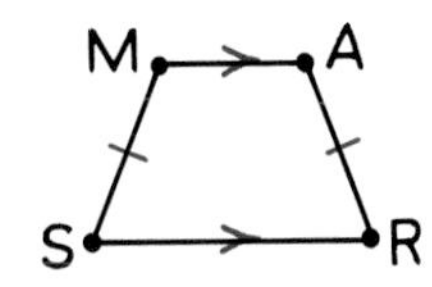

Given: MARS is an isosceles trapezoid with bases $\overline{MA}$ and $\overline{SR}$.

Proof.

Prove: MARS is cyclic.

Statements	***Reasons***
1. MARS is an isosceles trapezoid with bases $\overline{MA}$ and $\overline{SR}$.	Given.
2. $\overline{MA} \parallel \overline{SR}$.	a) ▯▯▯▯▯
3. ∡M and ∡S are supplementary.	b) ▯▯▯▯▯
4. c) ▯▯▯▯▯	If two angles are supplementary, the sum of their measures is 180°.
5. ∠S = ∠R.	d) ▯▯▯▯▯
6. e) ▯▯▯▯▯	Substitution.
7. f) ▯▯▯▯▯	If the sum of the measures of two angles is 180°, the angles are supplementary.
8. g) ▯▯▯▯▯	h) ▯▯▯▯▯

Set II

Trace each of the following quadrilaterals and construct the perpendicular bisectors of its sides. Then try to circumscribe a circle about each quadrilateral.

21.

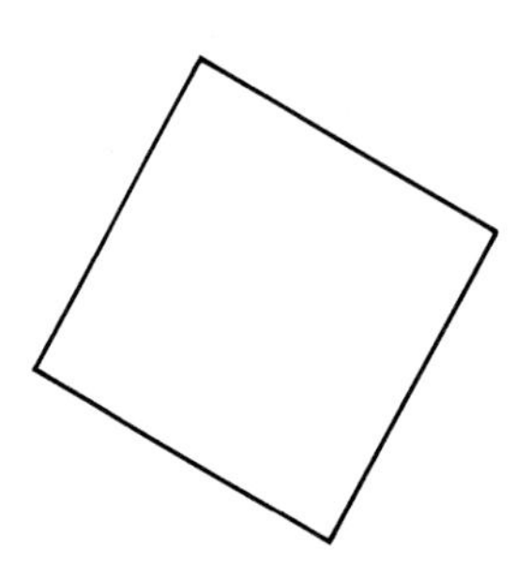

22.

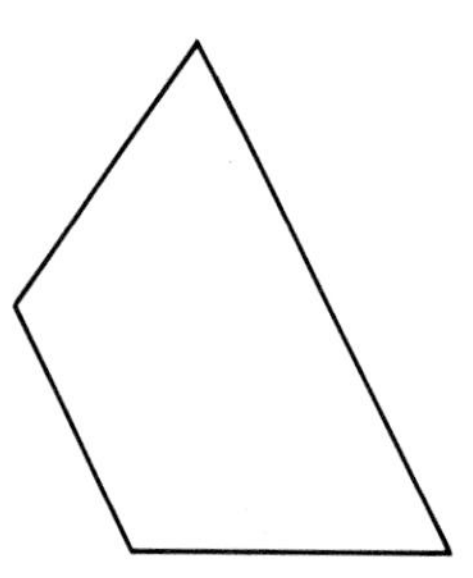

23.

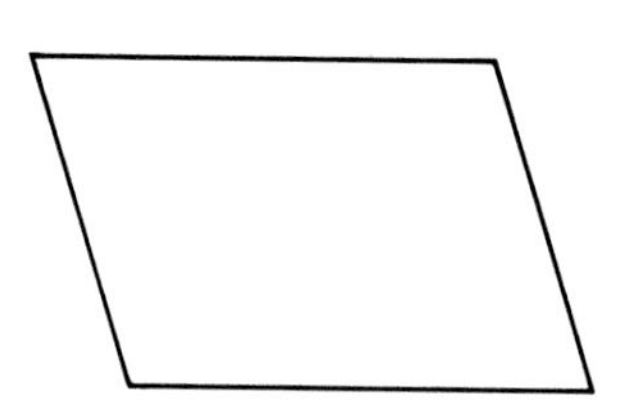

Refer to your drawings for Exercises 21 through 23 to answer the following questions.

24. In order for a quadrilateral to be cyclic, what must be true about the lines that are the perpendicular bisectors of its sides?

25. What relation do the vertices of a cyclic quadrilateral have to the center of its circumscribed circle?

In the figure below, $\overline{LH}$ and $\overline{LS}$ are secant segments to circle O. $\angle HEI = (10x + 3)°$, $\angle EIS = (9x - 2)°$, and $\angle S = (4x + 9)°$.

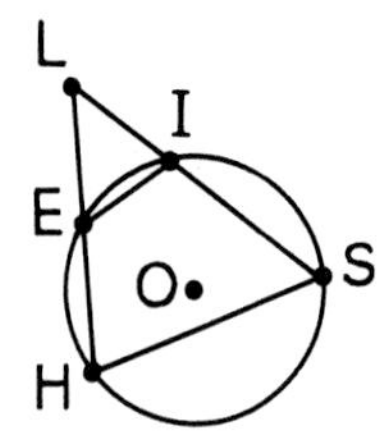

26. Find x.

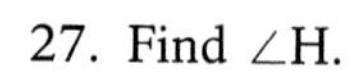

27. Find $\angle H$.

28. Find $\angle L$.

29.

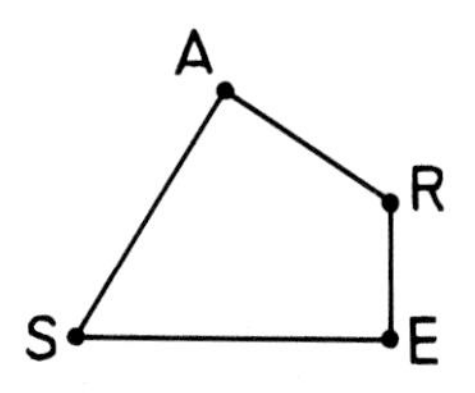

Given: $\overline{AS} \perp \overline{AR}$ and $\overline{ER} \perp \overline{ES}$.

Prove: Quadrilateral ARES is cyclic.

30.

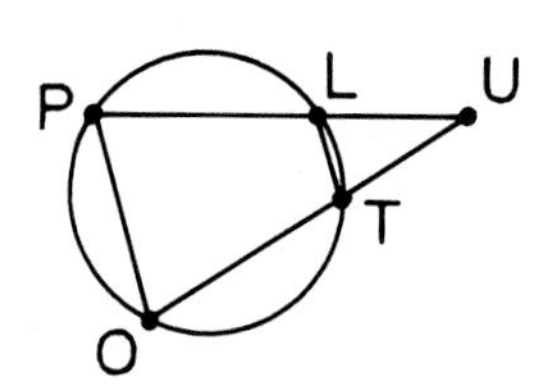

Given: Quadrilateral PLTO is cyclic.

Prove: $\triangle POU \sim \triangle TLU$.

31.

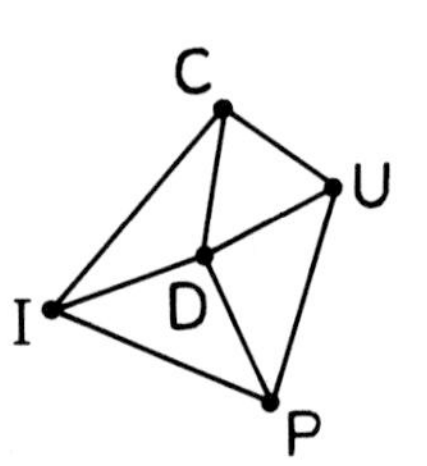

Given: In quadrilateral CUPI, DC = DU = DP = DI.

Prove: $\measuredangle ICU$ and $\measuredangle UPI$ are supplementary.

32.

Given: Quadrilateral VENU with exterior angle $\measuredangle NUS$; $\angle NUS = \angle E$.

Prove: Quadrilateral VENU is cyclic.

Set III

The Indian mathematician Brahmagupta discovered several theorems about cyclic quadrilaterals. You have already seen one of them, an extension of Heron's theorem that can be used to find the area of a cyclic quadrilateral.*

Another theorem of Brahmagupta concerns a cyclic quadrilateral whose diagonals are perpendicular to each other. Draw a large circle and construct two perpendicular chords of unequal lengths in it. Label them $\overline{AB}$ and $\overline{CD}$ and their point of intersection point E. Connect the four points on the circle to form a cyclic quadrilateral ACBD. Find the midpoint of each side of quadrilateral ACBD and draw the four lines determined by point E and these points.

These lines are related to the opposite sides of the quadrilateral in an interesting way. What is it?

*See page 341.

Lesson 3
Incircles

Courtesy of Dr. H. Hashimoto, University of Tokyo

Atoms are so small that it is hard to imagine that they could be photographed. An atom of gold, for example, has a radius of only about one-hundred millionth of a centimeter.

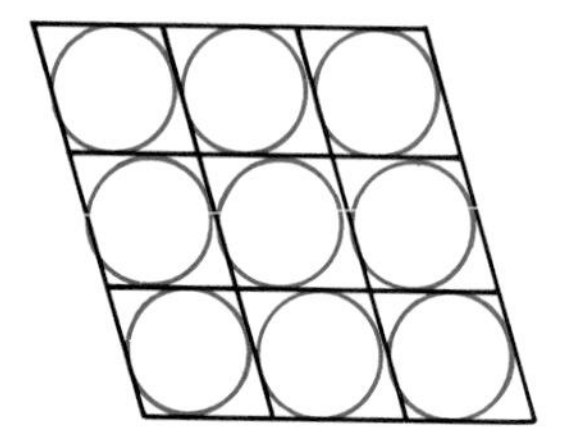

In 1979, Hatsujiro Hashimoto and his research team at Osaka University in Japan used a powerful electron microscope to produce the photograph at the top of this page. It shows a piece of gold film so highly magnified that the individual atoms can be seen as white dots.

The pattern of the atoms is illustrated in the diagram at the left. In this diagram, each atom is represented by a circle. Each circle is *inscribed* in a rhombus and is called its *incircle.*

► **Definition**
A circle is ***inscribed in a polygon*** iff each side of the polygon is tangent to the circle.

As we have just noted, the circle is called the incircle of the polygon. Its center is called the *incenter* of the polygon.

In Lesson 1 of this chapter, we proved that every triangle is cyclic, and so a circle can be circumscribed about every triangle. We will now prove that every triangle also has an incircle. To do this, we will first prove that the angle bisectors of a triangle are concurrent and that the point of concurrency is the incenter of the triangle.

▶ **Theorem 80**
The angle bisectors of a triangle are concurrent.

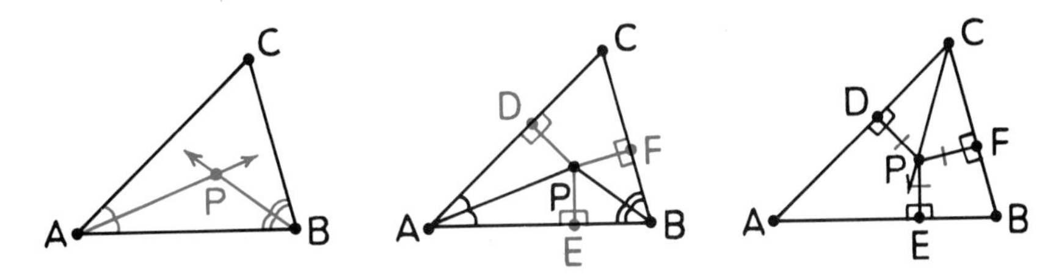

Let $\triangle ABC$ represent any triangle. Let two rays bisect $\measuredangle A$ and $\measuredangle B$ and let their point of intersection be called P.

Through P, draw $\overline{PD} \perp \overline{AC}$, $\overline{PE} \perp \overline{AB}$, and $\overline{PF} \perp \overline{BC}$ (through a point not on a line, there is exactly one perpendicular to the line).

Because $\triangle PAD \cong \triangle PAE$ (A.A.S.), PD = PE; also, because $\triangle PBE \cong \triangle PBF$ (A.A.S.), PE = PF. Hence, PD = PE = PF.

Draw $\overrightarrow{CP}$. Because $\triangle PCD \cong \triangle PCF$ (H.L.), $\angle PCD = \angle PCF$. So $\overrightarrow{CP}$ bisects $\measuredangle ACB$. Therefore, because all three angle bisectors pass through point P, they are concurrent.

▶ **Corollary**
Every triangle has an incircle.

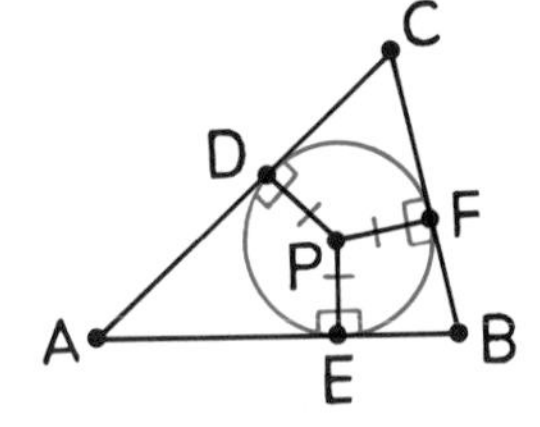

Because PD = PE = PF, a circle can be drawn with P as center and PD as radius that contains all three points, D, E, and F. Because $\overline{AC} \perp \overline{PD}$, $\overline{AB} \perp \overline{PE}$, and $\overline{BC} \perp \overline{PF}$, the sides of the triangle are tangent to the circle (if a line is perpendicular to a radius at its outer endpoint, it is tangent to the circle). Therefore, circle P is inscribed in $\triangle ABC$ and is its incircle by definition.

Theorem 80 and its corollary suggest that, to construct the incircle of a triangle, we must bisect two of its angles and then construct a perpendicular from the point in which the angle bisectors intersect to one of the sides.

▶ **Construction 12**
To inscribe a circle in a triangle.

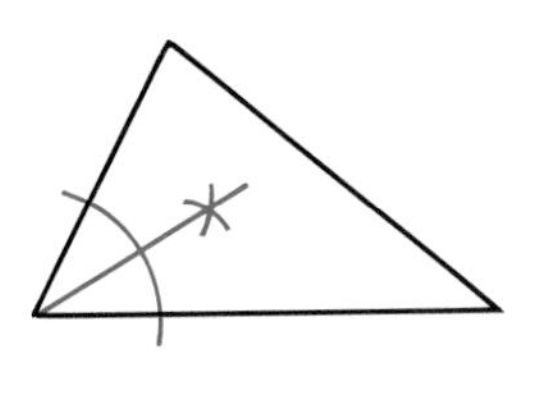

Step 1

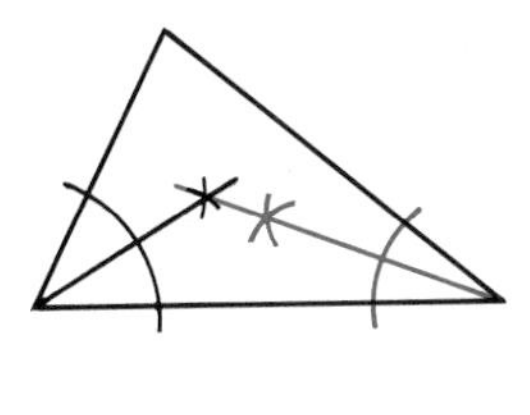

Step 2

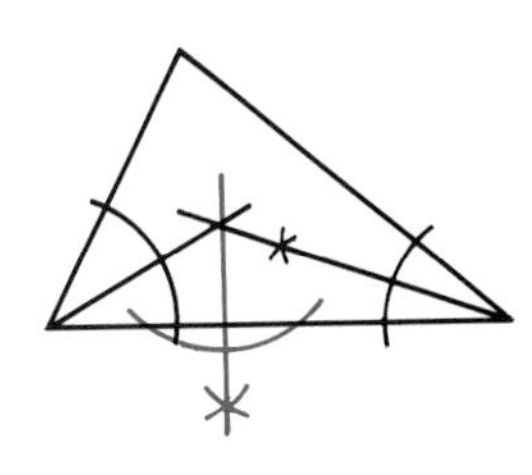

Step 3

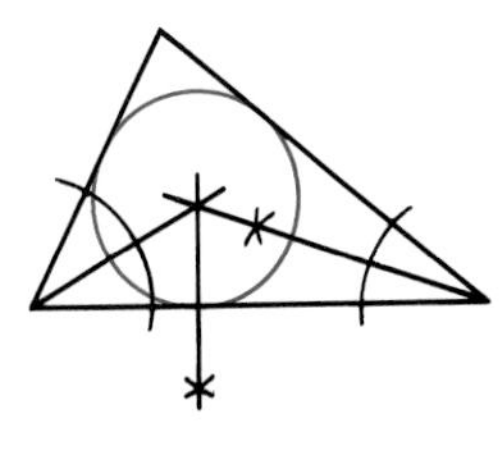

Step 4

Exercises

Set I

Circle T is inscribed in △VRO.

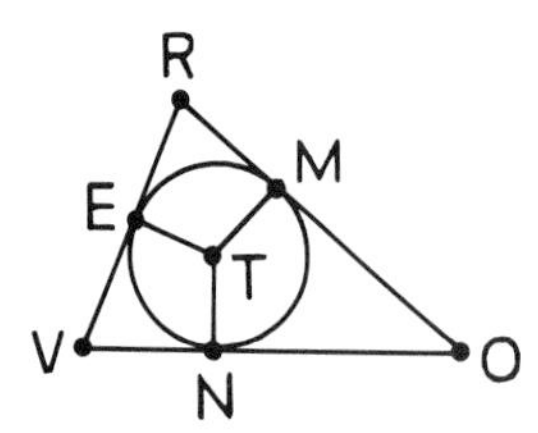

1. What relation do the sides of the triangle have to the circle?
2. What relation do the radii have to the sides?
3. Why?
4. Why is TE = TM = TN?
5. What is circle T called with respect to the triangle?
6. What is point T called with respect to the triangle?

Circle O is inscribed in △NYK.

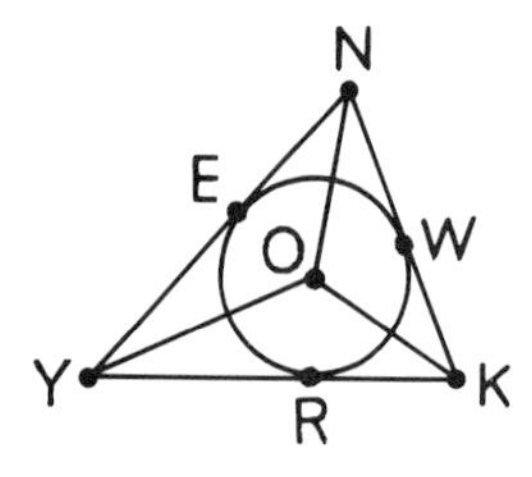

7. Why is NE = NW, YE = YR, and KW = KR?
8. What relation do $\overrightarrow{NO}$, $\overrightarrow{YO}$, and $\overrightarrow{KO}$ have to the angles of △NYK?
9. What relation does point O have to the three sides of the triangle?

Draw large triangles similar to the following triangles and inscribe a circle in each one.

10.

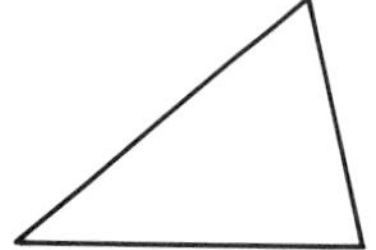

11.

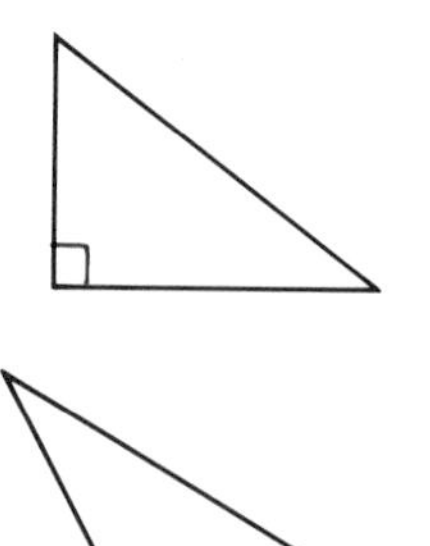

12.

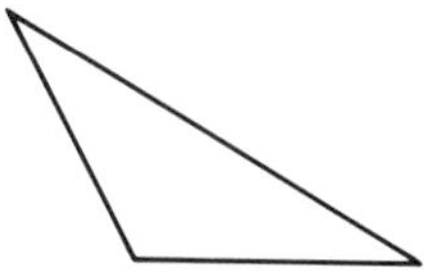

13. Can the incenter of a triangle be outside of the triangle?
14. Can the incenter of a triangle be on a side of the triangle?

Trace each of the following quadrilaterals and construct the bisectors of its angles. Then try to inscribe a circle in each quadrilateral.

15.

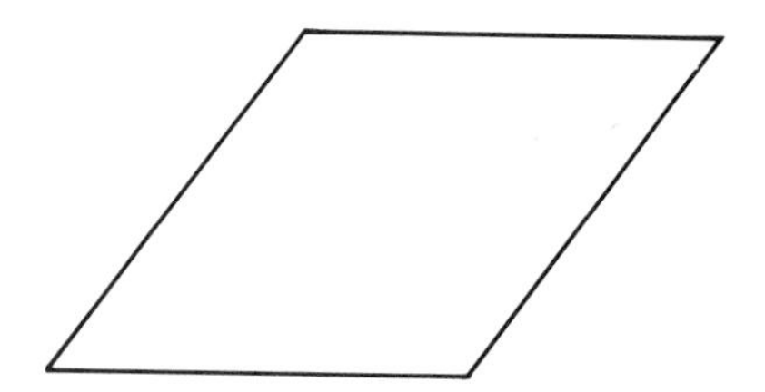

16.

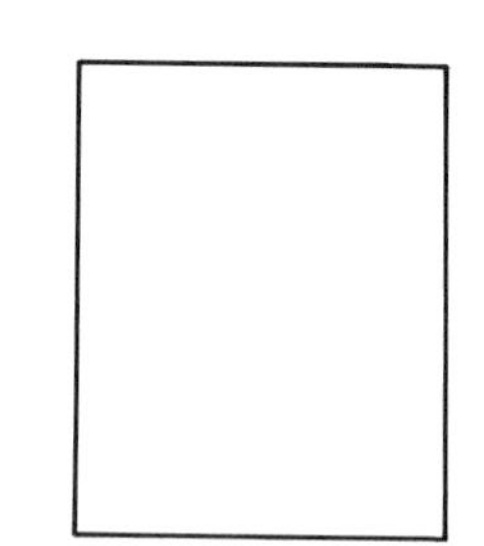

17.

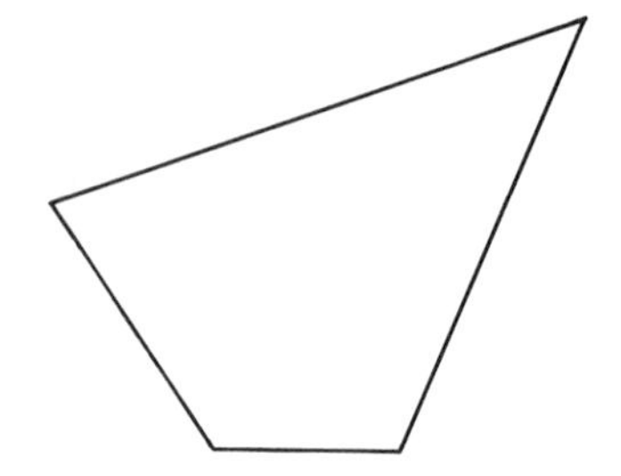

18. 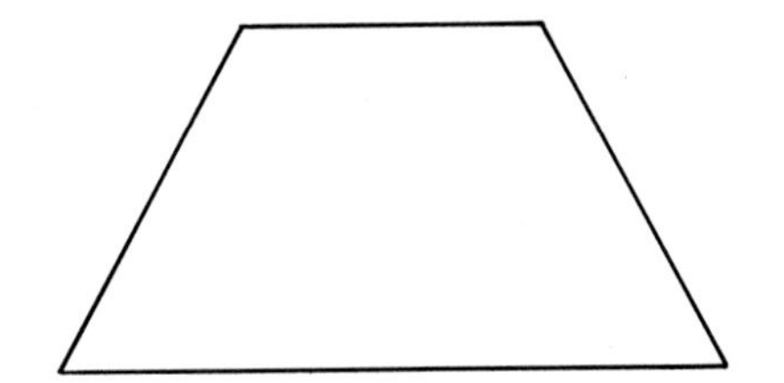

Refer to your drawings for Exercises 15 through 18 to answer the following questions.

19. In order for a quadrilateral to have an incircle, what must be true about the rays that bisect its angles?

20. Do all parallelograms have incircles?

Set II

In the figure below, the sides of $\triangle$AIO are tangent to the circle; AR = 3, IZ = 2, and NO = 5.

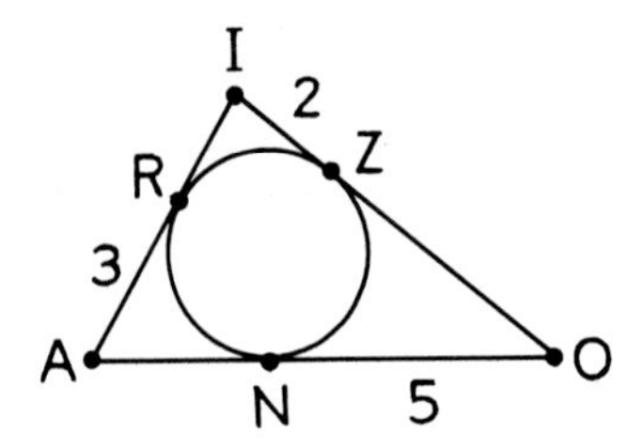

21. Find the lengths of the three sides of $\triangle$AIO.

22. Find $\rho\triangle$AIO.

23. Find $\alpha\triangle$AIO.

In the figure below, circle O is inscribed in square WSIG and circumscribed about square AHNT; OH = 1. Find the following ratios.

24. $\dfrac{\rho \text{WSIG}}{\rho \text{AHNT}}$.

25. $\dfrac{\alpha \text{WSIG}}{\alpha \text{AHNT}}$.

In the figure below, circle Y is the incircle of equilateral $\triangle$KNU; YC = x.

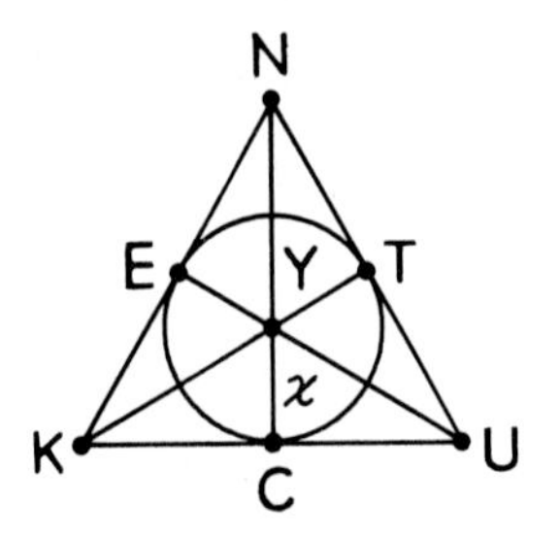

26. What kind of triangles are $\triangle$KNC and $\triangle$KYC?

Find an expression for each of the following in terms of x.

27. NC.

28. KU.

29. $\rho\triangle$KNU.

30. $\alpha\triangle$KNU.

31. 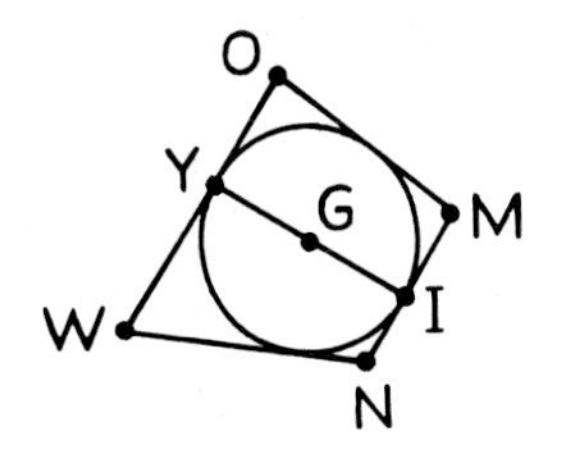

Given: Circle G is inscribed in quadrilateral WOMN and $\overline{YI}$ is a diameter of circle G.

Prove: Quadrilateral WOMN is a trapezoid.

32.

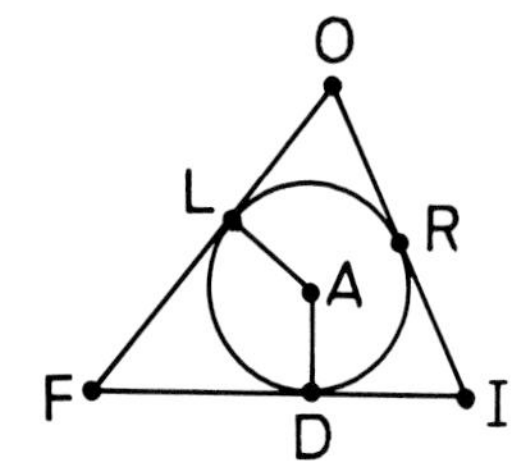

Given: Circle A is inscribed in $\triangle$FOI; $\angle$O = $\angle$I.

Prove: LO = DI.

Set III

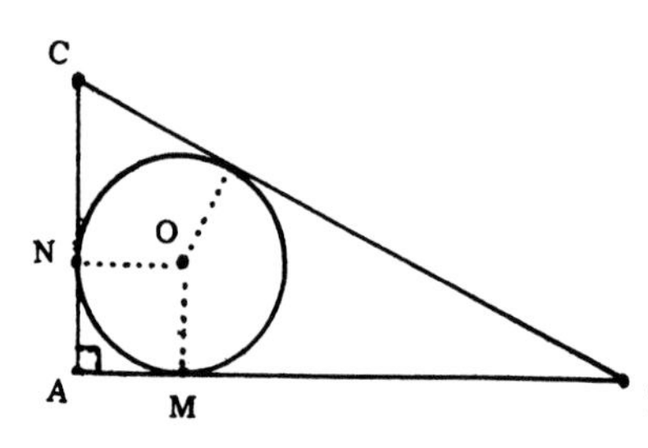

48) במשולש ישר הזווית ABC
חסום מעגל, שמחוגיו ON ו־OM.
הוכח, כי:

$$2OM = AB + AC - BC$$

This problem from a Hebrew geometry book is about a relation between the radius of the incircle of a right triangle and the lengths of the triangle's sides. Can you explain why it is true?

Lesson 4

Ceva's Theorem

Most of the theorems that we have proved in our study of geometry were known to Euclid, who included them in the *Elements*. In fact, after the development of geometry in Greece between 600 and 200 B.C., very few significant additions to the subject were made until the seventeenth century. One of the new theorems that appeared at that time was discovered by an Italian mathematician and engineer named Giovanni Ceva.

Ceva's theorem concerns sets of line segments in a triangle which we will refer to as *cevians*.

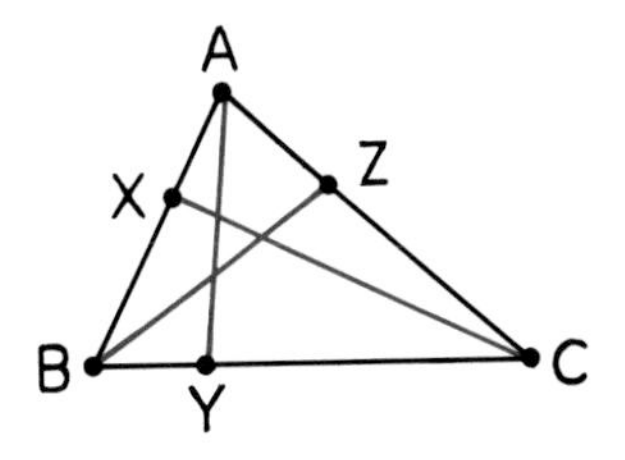

Definition

A ***cevian*** of a triangle is a line segment that joins a vertex of the triangle to a point on the opposite side.

In each of the triangles shown at the right, three cevians have been drawn, one from each vertex. They are $\overline{AY}$, $\overline{BZ}$, and $\overline{CX}$. The cevians divide the sides of each triangle into six segments: $\overline{AX}$, $\overline{XB}$, $\overline{BY}$, $\overline{YC}$, $\overline{CZ}$, and $\overline{ZA}$.

In the second figure, the three cevians are concurrent at point P. Ceva's theorem states that this is true if and only if

$$\frac{AX}{XB} \cdot \frac{BY}{YC} \cdot \frac{CZ}{ZA} = 1.$$

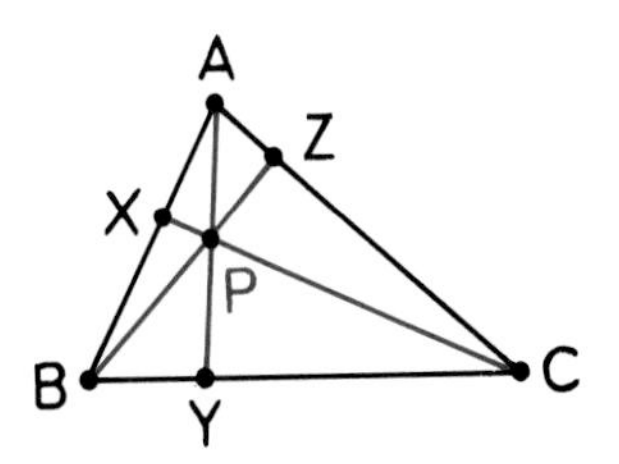

► **Theorem 81** (Ceva's Theorem)
Three cevians $\overline{AY}$, $\overline{BZ}$, and $\overline{CX}$ of $\triangle ABC$ are concurrent iff

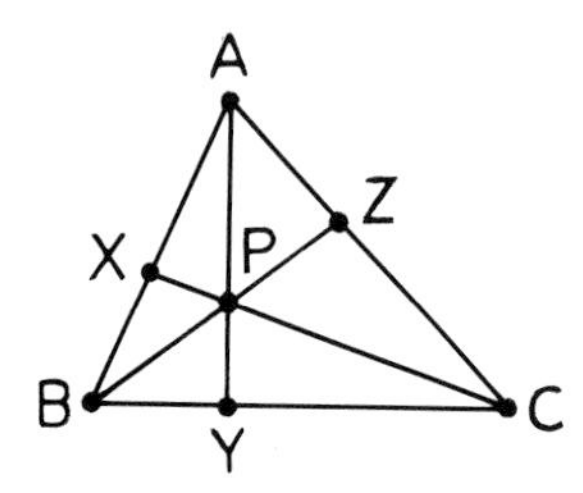

$$\frac{AX}{XB} \cdot \frac{BY}{YC} \cdot \frac{CZ}{ZA} = 1.$$

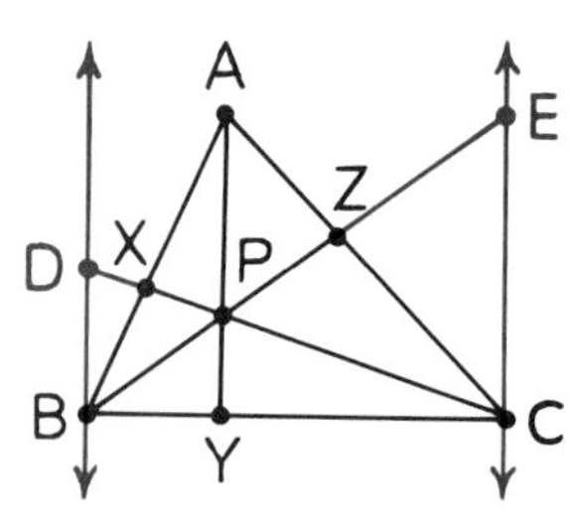

Proof that, if the three cevians are concurrent, then

$$\frac{AX}{XB} \cdot \frac{BY}{YC} \cdot \frac{CZ}{ZA} = 1.$$

Through B and C, draw lines parallel to $\overline{AY}$. Let the points in which they intersect $\overleftrightarrow{CX}$ and $\overleftrightarrow{BZ}$ be called D and E, respectively.

Because of equal vertical angles and equal alternate interior angles $\triangle AXP \sim \triangle BXD$ and $\triangle CZE \sim \triangle AZP$. Therefore,

$$\frac{AX}{BX} = \frac{AP}{BD} \quad \text{and} \quad \frac{CZ}{AZ} = \frac{CE}{AP}.$$

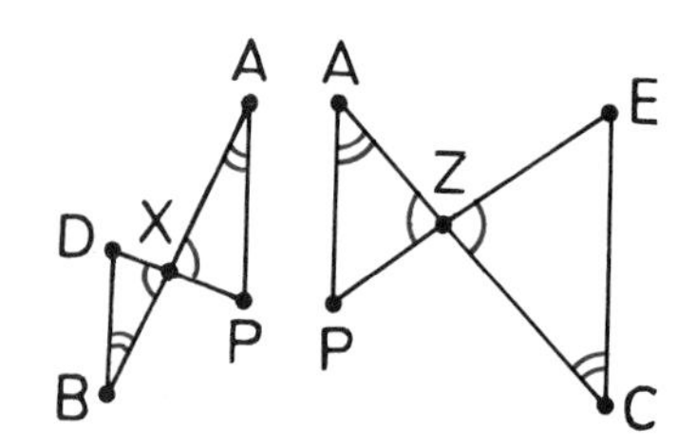

Because of the common angle and equal corresponding angles, $\triangle BYP \sim \triangle BCE$ and $\triangle BCD \sim \triangle YCP$. Therefore,

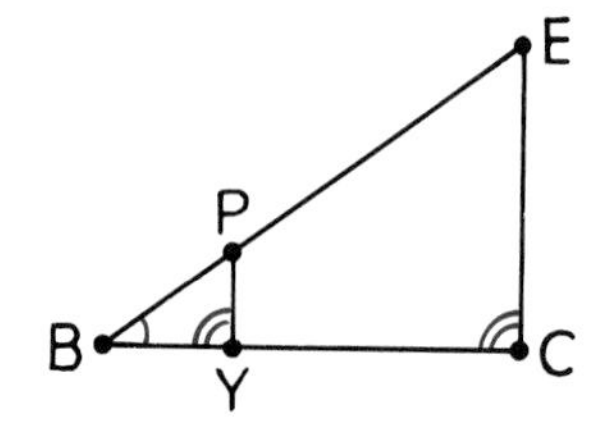

$$\frac{BY}{BC} = \frac{YP}{CE} \quad \text{and} \quad \frac{BC}{YC} = \frac{BD}{YP}.$$

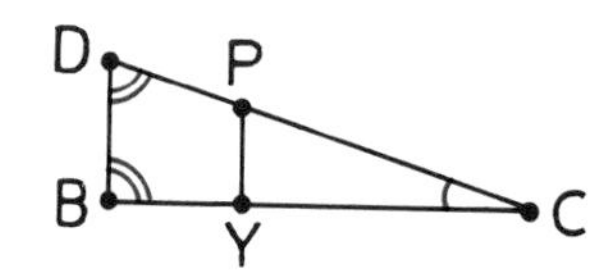

We can now build the conclusion of Ceva's theorem by multiplying the left and right sides of these equations.

$$\frac{AX}{BX} \cdot \frac{BY}{BC} \cdot \frac{BC}{YC} \cdot \frac{CZ}{AZ} = \frac{AP}{BD} \cdot \frac{YP}{CE} \cdot \frac{BD}{YP} \cdot \frac{CE}{AP}.$$

Simplifying,

$$\frac{AX}{BX} \cdot \frac{BY}{\cancel{BC}} \cdot \frac{\cancel{BC}}{YC} \cdot \frac{CZ}{AZ} = \frac{\cancel{AP}}{\cancel{BD}} \cdot \frac{\cancel{YP}}{\cancel{CE}} \cdot \frac{\cancel{BD}}{\cancel{YP}} \cdot \frac{\cancel{CE}}{\cancel{AP}},$$

and so

$$\frac{AX}{BX} \cdot \frac{BY}{YC} \cdot \frac{CZ}{AZ} = 1.$$

Proof of the converse.
If $\frac{AX}{XB} \cdot \frac{BY}{YC} \cdot \frac{CZ}{ZA} = 1$, then the three cevians $\overline{AY}$, $\overline{BZ}$, and $\overline{CX}$ of $\triangle ABC$ are concurrent.

Let $\overline{CX}$ and $\overline{BZ}$ intersect in point P. Draw $\overleftrightarrow{AP}$, and let the point in which it intersects $\overline{BC}$ be called Y′.

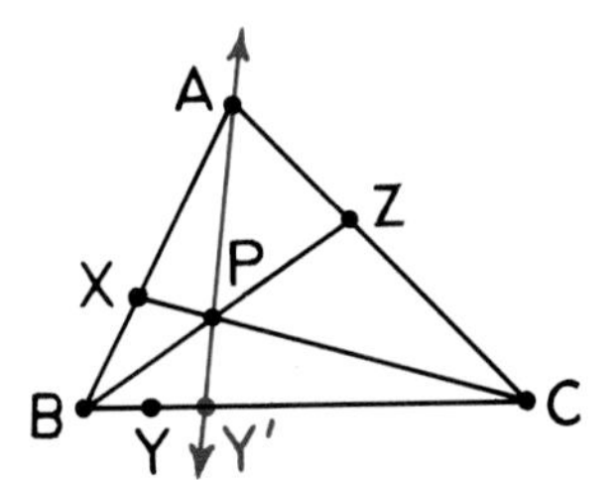

Now

$$\frac{AX}{XB} \cdot \frac{BY'}{Y'C} \cdot \frac{CZ}{ZA} = 1 \quad \text{(proved in the first part)},$$

and

$$\frac{AX}{XB} \cdot \frac{BY}{YC} \cdot \frac{CZ}{ZA} = 1 \quad \text{(by hypothesis)}.$$

So

$$\frac{AX}{XB} \cdot \frac{BY'}{Y'C} \cdot \frac{CZ}{ZA} = \frac{AX}{XB} \cdot \frac{BY}{YC} \cdot \frac{CZ}{ZA},$$

and

$$\frac{BY'}{Y'C} = \frac{BY}{YC} \quad \text{(division)}.$$

Adding 1 to each side of this equation, we get

$$\frac{BY'}{Y'C} + 1 = \frac{BY}{YC} + 1 \quad \text{or}$$

$$\frac{BY' + Y'C}{Y'C} = \frac{BY + YC}{YC}.$$

Because BY′ + Y′C = BC and BY + YC = BC,

$$\frac{BC}{Y'C} = \frac{BC}{YC}.$$

Multiplying means and extremes, we get

$$BC \cdot YC = BC \cdot Y'C.$$

Dividing by BC gives

$$YC = Y'C.$$

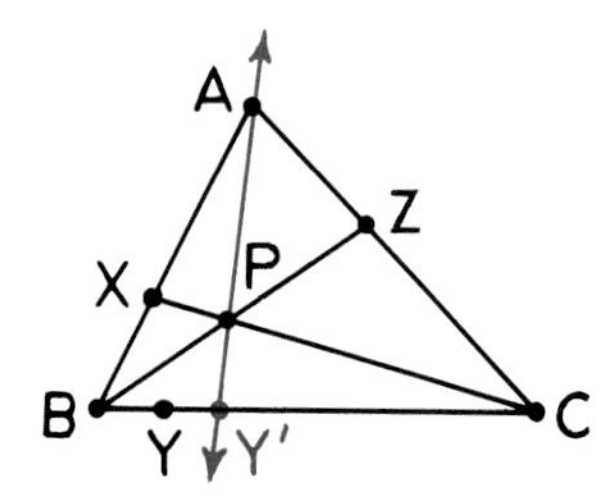

Because CY = CY′, Y and Y′ are the same point (this follows from the Ruler Postulate, which says that the points on a line can be numbered so that to every real number, there corresponds exactly one point). Because Y and Y′ are the same point, $\overleftrightarrow{AY}$ and $\overleftrightarrow{AY'}$ are the same line. Therefore, $\overline{AY}$, $\overline{BZ}$, and $\overline{CX}$ all contain point P and are concurrent.

Exercises

Set I

In the figure below, $\overline{NY}$, $\overline{OE}$, and $\overline{KW}$ are line segments in △NWR.

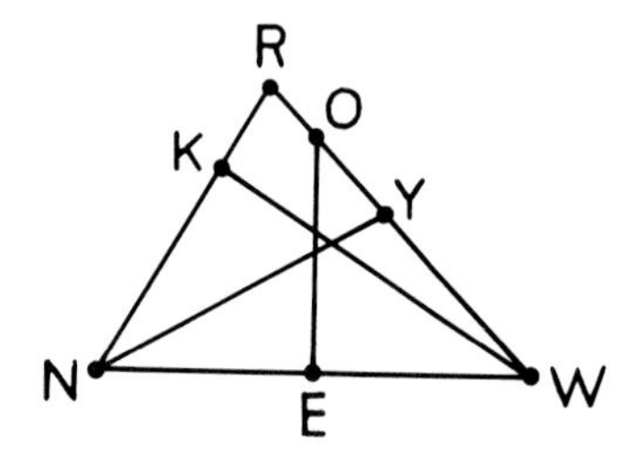

1. Which line segment appears to be an altitude of the triangle?
2. Which line segment appears to be part of an angle bisector of the triangle?
3. Which line segment appears to be a perpendicular bisector of one of the sides of the triangle?
4. Which of these line segments are cevians of △NWR?
5. Do the three line segments appear to be concurrent?

In △PON, $\overline{PE}$, $\overline{OI}$, and $\overline{NH}$ are concurrent at point X. Copy and complete the following equations, which follow from Ceva's Theorem:

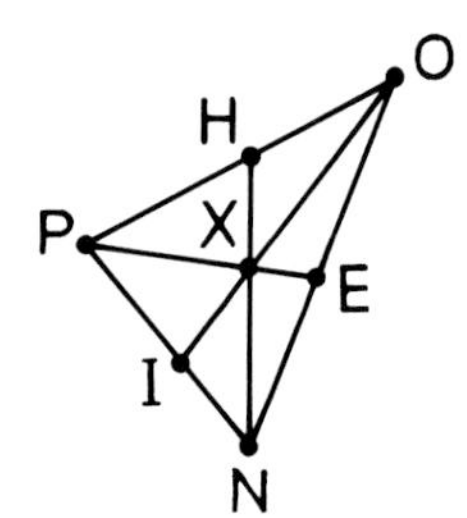

6. $\frac{PH}{HO} \cdot \frac{OE}{EN} \cdot \frac{NI}{IP} = $ ||||||||||.
7. $\frac{PI}{IN} \cdot \frac{NE}{EO} \cdot$ |||||||||| = ||||||||||.
8. $\frac{OE}{EN} \cdot$ |||||||||| · |||||||||| = ||||||||||.

In △ABC, AB = 7 cm, AC = 6 cm, and BC = 5 cm.

9. Construct △ABC. Mark point X on $\overline{AC}$ so that CX = 2 cm and point Y on $\overline{AB}$ so that AY = 4 cm. Draw $\overline{BX}$ and $\overline{CY}$ and label the point in which they intersect P. Draw $\overleftrightarrow{AP}$ and label the point in which it intersects $\overline{BC}$ Z.

Show that your drawing illustrates Ceva's Theorem by finding the following numbers.

10. AX, YB, CZ, and ZB.

11. $\frac{CX}{XA}$.

12. $\frac{AY}{YB}$.

13. $\frac{BZ}{ZC}$.

14. $\frac{CX}{XA} \cdot \frac{AY}{YB} \cdot \frac{BZ}{ZC}$.

Segments $\overline{TE}$, $\overline{PA}$, and $\overline{KO}$ are cevians in $\triangle$TPK. Would they be concurrent if

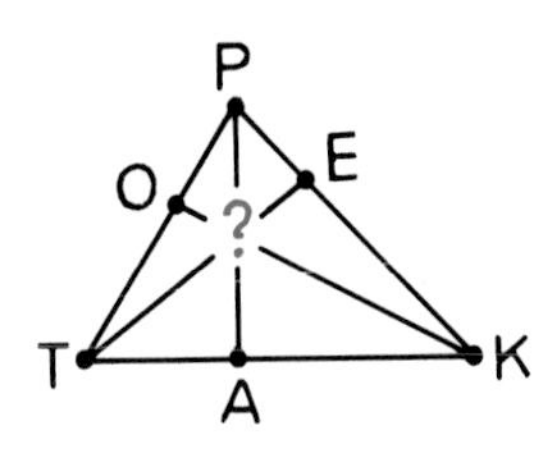

15. $\frac{PO}{OT} = \frac{2}{3}$, $\frac{TA}{AK} = \frac{2}{3}$, and $\frac{KE}{EP} = \frac{9}{4}$?

16. $\frac{TO}{OP} = \frac{PE}{EK} = \frac{KA}{AT}$?

17. PE = EK, KA = AT, and TO = OP?

In $\triangle$ASI, $\overline{AT}$, $\overline{SN}$, and $\overline{IU}$ are concurrent.

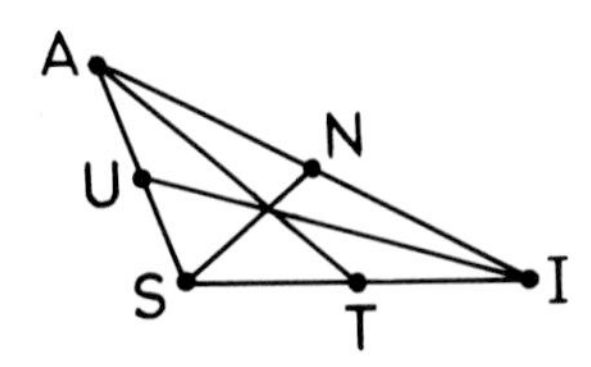

18. If $\frac{ST}{TI} = \frac{4}{3}$ and $\frac{IN}{NA} = \frac{3}{2}$, find $\frac{AU}{US}$.

19. If $\frac{AN}{NI} = \frac{2}{5}$ and IT = TS, find $\frac{SU}{UA}$.

20. If AU = 4, US = 6, ST = 9, TI = 5, and IN = 10, find NA.

Set II

In the figure below, $\overline{SA}$, $\overline{PL}$, and $\overline{UT}$ are concurrent in $\triangle$SPU; SP = 24, PU = 40, SU = 30, ST = 10, and PA = 15.

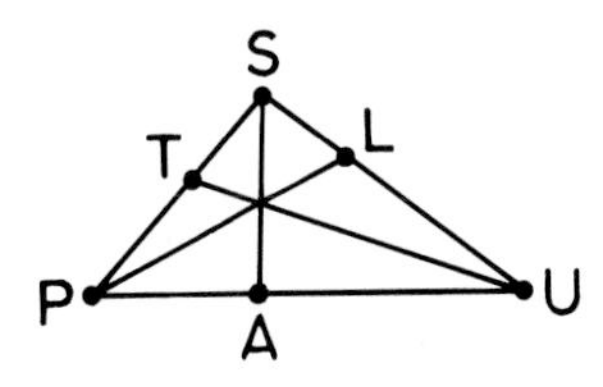

21. Copy the figure and mark the given lengths on it.

22. Find TP and AU.

23. Find SL and LU.

24. Does $\overrightarrow{UT}$ bisect $\measuredangle$SUP? Show why or why not.

In the figure below, $\overline{SK}$, $\overline{ON}$, and $\overline{AP}$ are concurrent in $\triangle$SOA; $\overline{ON} \perp \overline{SA}$, $\overline{SK} \perp \overline{OA}$, OK = SN = 18, and KA = NA = 7.

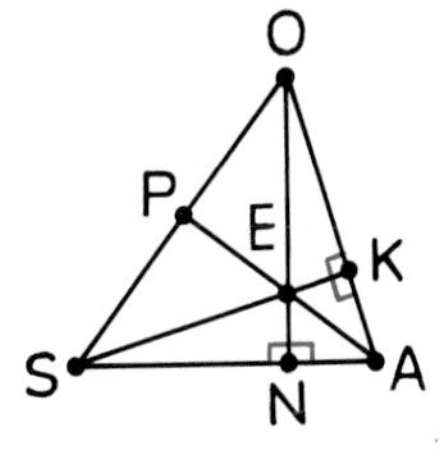

25. Copy the figure and mark the given lengths on it.

26. Find ON and SK.

27. Find SO.

28. Find SP and PO.

29. What can you conclude about △OPA and △SPA?

30. Find PA.

31.

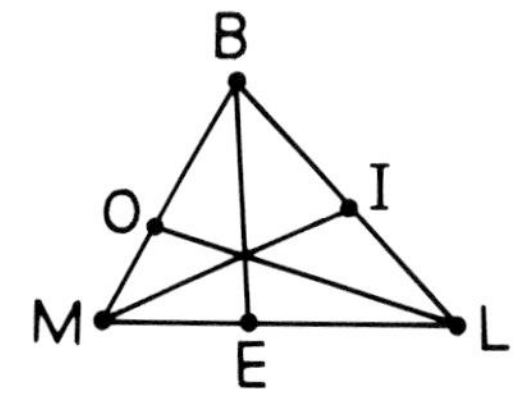

Given: $\overline{MI}$, $\overline{BE}$, and $\overline{LO}$ are concurrent in △MBL.

Prove: MO · BI · LE = OB · IL · EM.

32.

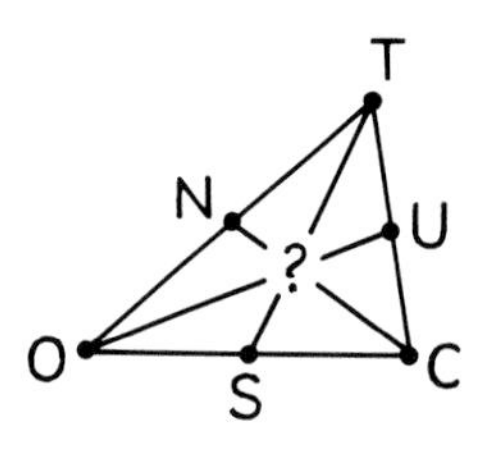

Given: Points N, U, and S are the midpoints of the sides of △TCO.

Prove: $\overline{TS}$, $\overline{UO}$, and $\overline{CN}$ are concurrent.

33.

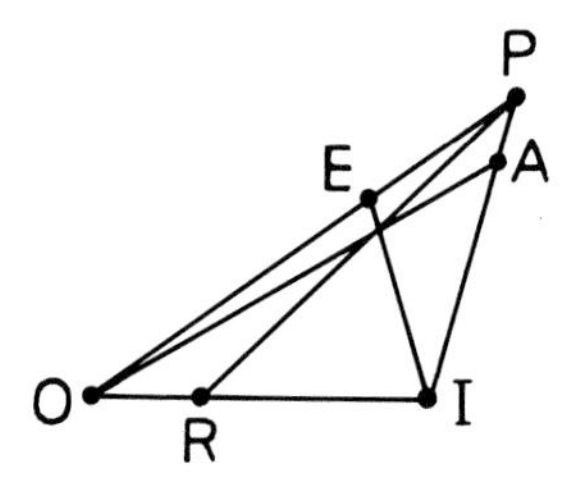

Given: $\overline{PR}$, $\overline{OA}$, and $\overline{IE}$ are concurrent in △POI; EO = 2PE and RI = 3OR.

Prove: IA = 6AP.

34.

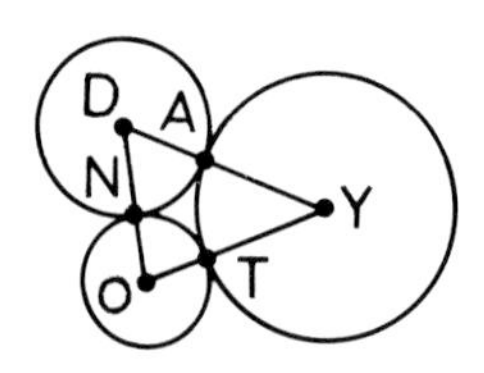

Given: △DYO with circles D, Y, and O.

Prove: $\overleftrightarrow{DT}$, $\overleftrightarrow{YN}$, and $\overleftrightarrow{OA}$ are concurrent.

Set III

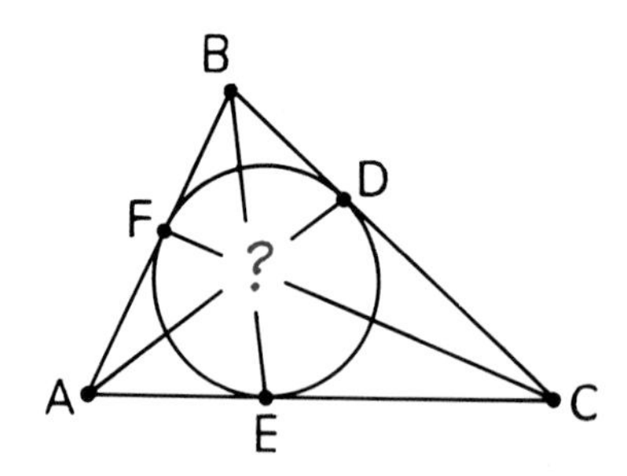

A nineteenth-century French mathematician named Joseph Gergonne proved that, if a circle is inscribed in a triangle, then the cevians joining the vertices of the triangle to the points in which the circle is tangent to the opposite sides of the triangle are concurrent.

How did he do it?

Drawing by O. Soglow; © 1959 The New Yorker Magazine, Inc.

Lesson 5

The Centroid of a Triangle

At what point should a wooden stick be supported so that it is perfectly balanced? If the stick has a uniform thickness and density, it is obvious that this point, called its center of gravity, is its midpoint. Because of this, we might say that the center of gravity of a line segment is its midpoint, even though a line segment has no mass.

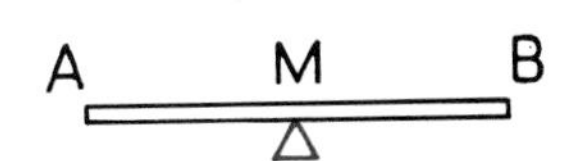

Where is the center of gravity of a triangle? Let us assume that by this we mean, At what point should a triangular board having uniform thickness and density be supported so that it will balance?

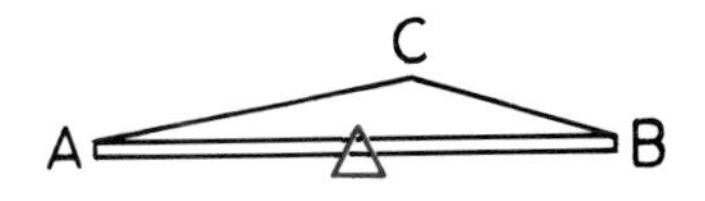

If we try to find this point by trial and error with a large board in the shape of a scalene triangle, we find that it is not the circumcenter, the point in which the perpendicular bisectors of the sides are concurrent. Neither is it the incenter, the point in which the angle bisectors are concurrent. Instead, it is determined by the *medians* of the triangle.

► **Definition**
A ***median*** of a triangle is a cevian joining a vertex to the midpoint of the opposite side.

It is very easy to prove that the three medians of every triangle are concurrent.

► **Theorem 82**
The medians of a triangle are concurrent.

Given: $\overline{AY}$, $\overline{BZ}$, and $\overline{CX}$ are medians of $\triangle ABC$.

Prove: $\overline{AY}$, $\overline{BZ}$, and $\overline{CX}$ are concurrent.

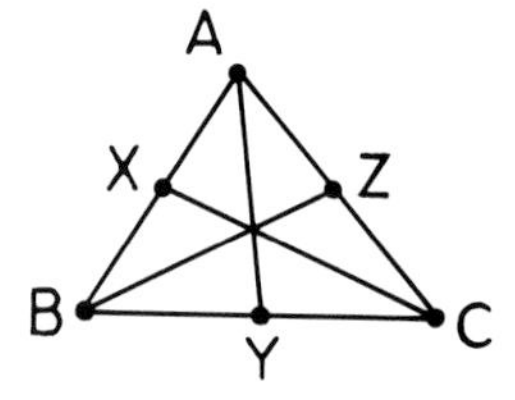

Proof.
Because $\overline{AY}$, $\overline{BZ}$, and $\overline{CX}$ are medians of $\triangle ABC$, it follows that X, Y, and Z are the midpoints of $\overline{AB}$, $\overline{BC}$, and $\overline{CA}$, respectively. Therefore, AX = XB, BY = YC, and CZ = ZA. Hence,

$$\frac{AX}{XB} = 1, \quad \frac{BY}{YC} = 1, \quad \text{and} \quad \frac{CZ}{ZA} = 1 \quad \text{(division)},$$

and so

$$\frac{AX}{XB} \cdot \frac{BY}{YC} \cdot \frac{CZ}{ZA} = 1 \cdot 1 \cdot 1 = 1.$$

It follows from Ceva's Theorem that $\overline{AY}$, $\overline{BZ}$, and $\overline{CX}$ are concurrent.

Archimedes showed that the point in which its medians are concurrent is the center of gravity of a triangular board of uniform thickness and density. Mathematicians refer to it as the *centroid* of the triangle.

► **Definition**
The ***centroid*** of a triangle is the point in which its medians are concurrent.

At this point it will probably come as no surprise to you that the lines containing the *altitudes* of a triangle are also concurrent. A proof of this is included in the exercises.

► **Theorem 83**
The lines containing the altitudes of a triangle are concurrent.

► **Definition**
The ***orthocenter*** of a triangle is the point in which the lines containing its altitudes are concurrent.

In the figure below, the three altitudes of $\triangle ABC$ are concurrent in point O, and so O is the orthocenter of the triangle.

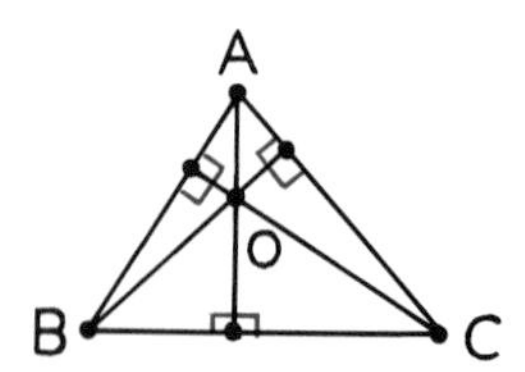

Exercises

Set I

The figures below illustrate four different sets of concurrent lines in a triangle.

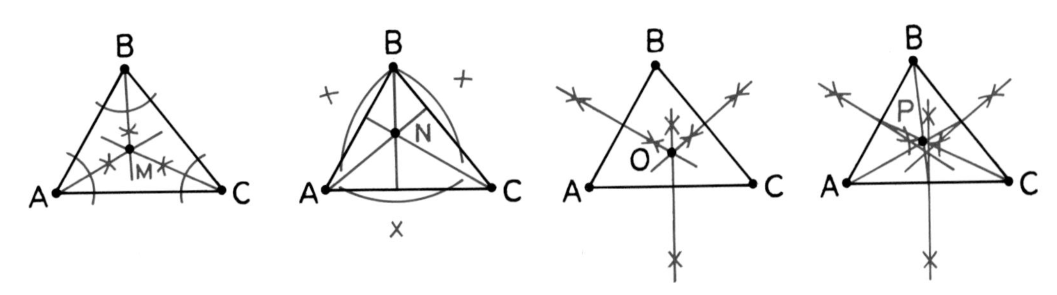

Match the names in the second column with the descriptions in the first column.

1. The lines that meet at M.
2. The line segments that meet at N.
3. The lines that meet at O.
4. The line segments that meet at P.
5. Point M.
6. Point N.
7. Point O.
8. Point P.
9. The point that is equidistant from the vertices of the triangle.
10. The point that is equidistant from the sides of the triangle.
11. The "balancing point" of the triangle.

A. The altitudes.
B. The medians.
C. The angle bisectors.
D. The perpendicular bisectors of the sides.
E. The centroid.
F. The circumcenter.
G. The incenter.
H. The orthocenter.

Make two tracings of △MNO. Then, use your straightedge and compass to locate the following points.

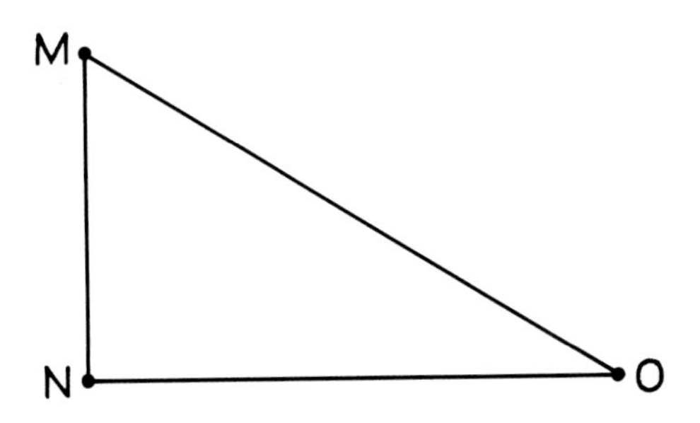

12. In your first drawing, find the centroid of △MNO.
13. In your second drawing, find the orthocenter of △MNO.

Make two tracings of $\triangle PQR$. Then, use your straightedge and compass to locate the following points.

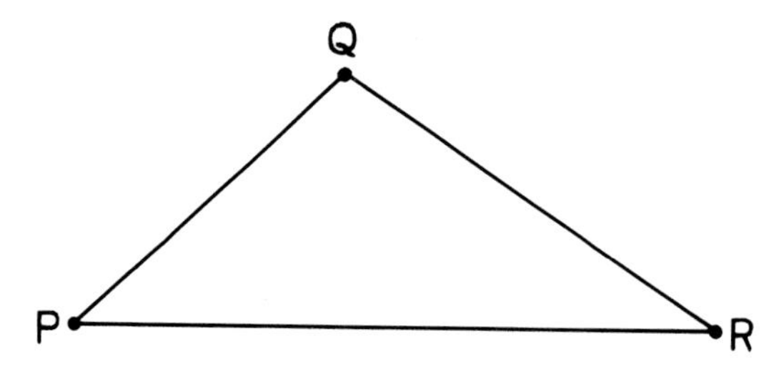

14. In your first drawing, find the centroid of $\triangle PQR$.

15. In your second drawing, find the orthocenter of $\triangle PQR$.

Refer to your drawings for Exercises 12 through 15 to answer the following questions.

16. Where is the orthocenter of a right triangle?

17. Where is the orthocenter of an obtuse triangle?

18. Where is the centroid of a triangle with respect to the triangle?

Set II

In the figure below, $\triangle ABC$ is equilateral and $\overline{CD}$ is one of its medians.

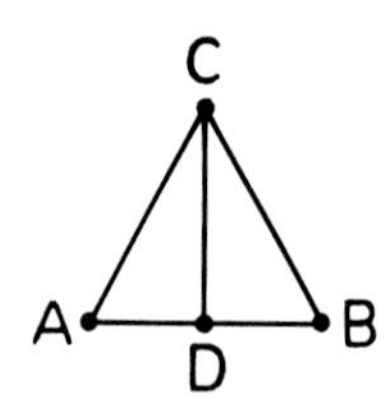

19. What relation do points C and D have to the endpoints of $\overline{AB}$?

20. Why does it follow that $\overleftrightarrow{CD}$ is the perpendicular bisector of $\overline{AB}$?

21. Why is $\overline{CD}$ an altitude of $\triangle ABC$?

22. Why does $\overrightarrow{CD}$ bisect $\measuredangle ACB$?

Exercises 19 through 22 show that the perpendicular bisectors of the sides, the angle bisectors, the altitudes, and the medians of an equilateral triangle are the same.

23. What do these facts prove about the circumcenter, incenter, orthocenter, and centroid of an equilateral triangle?

24. One way to prove that the three altitudes of a triangle are concurrent is shown, in somewhat abridged form, on the following page.

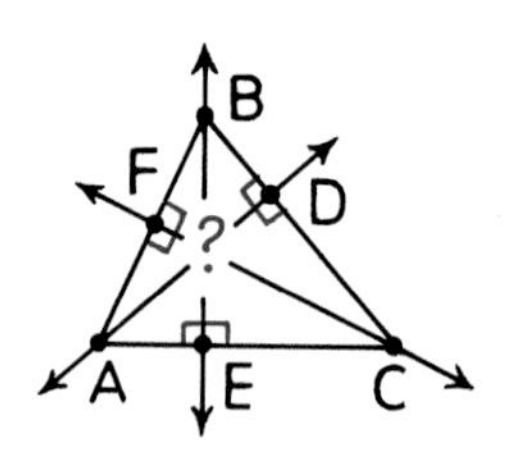

Given: $\overline{AD}$, $\overline{BE}$, and $\overline{CF}$ are altitudes of $\triangle ABC$.

Prove: $\overleftrightarrow{AD}$, $\overleftrightarrow{BE}$, and $\overleftrightarrow{CF}$ are concurrent.

Give the missing statements and reasons.

Proof.

Statements	***Reasons***
1. $\overline{AD}$, $\overline{BE}$, and $\overline{CF}$ are altitudes of $\triangle ABC$.	Given.
2. $\overline{AD} \perp \overline{BC}$, $\overline{BE} \perp \overline{AC}$, and $\overline{CF} \perp \overline{AB}$.	a) ||||||||
3. Through A, draw $\overleftrightarrow{GI} \parallel \overline{BC}$; through B, draw $\overleftrightarrow{GH} \parallel \overline{AC}$; and through C, draw $\overleftrightarrow{IH} \parallel \overline{AB}$.	b) 
4. GBCA and BHCA are parallelograms.	c) 
5. GB = AC and BH = AC.	d) ||||||||
6. e) ||||||||	Substitution.
7. $\overleftrightarrow{BE} \perp \overleftrightarrow{GH}$.	f) ||||||||

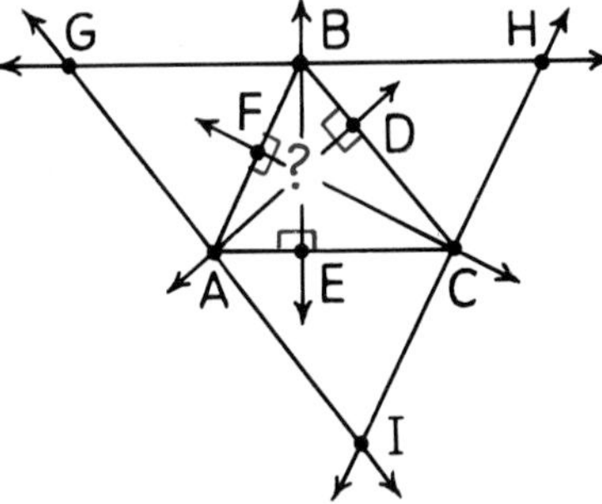

So $\overleftrightarrow{BE}$ is the perpendicular bisector of side $\overline{GH}$ of $\triangle GHI$. In the same way, it can be shown that $\overleftrightarrow{CF}$ is the perpendicular bisector of $\overline{HI}$ and $\overleftrightarrow{AD}$ is the perpendicular bisector of $\overline{GI}$.

8. $\overleftrightarrow{AD}$, $\overleftrightarrow{BE}$, and $\overleftrightarrow{CF}$ are concurrent.	g) ||||||||

Trace $\triangle ABC$.

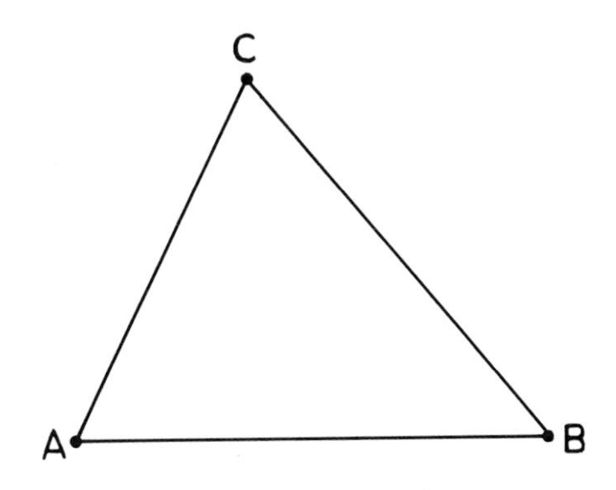

25. Use your straightedge and compass to bisect $\overline{AC}$ and $\overline{BC}$. Label the midpoints D and E respectively. Draw $\overline{AE}$ and $\overline{BD}$. Label the point in which they intersect F. Bisect $\overline{AF}$ and $\overline{BF}$. Label the midpoints G and H respectively.

26. What do points G and F seem to do to $\overline{AE}$?

27. Do points F and H seem to do the same thing to $\overline{DB}$?

Draw $\overline{DE}$, $\overline{EH}$, $\overline{HG}$, and $\overline{GD}$ to form quadrilateral DEHG.

28. Why is $\overline{DE} \parallel \overline{AB}$ and $\overline{GH} \parallel \overline{AB}$?

29. Why is $\overline{DE} \parallel \overline{GH}$?

30. Why is $DE = \frac{1}{2}AB$ and $GH = \frac{1}{2}AB$?

31. Why is DE = GH?

32. What kind of quadrilateral is DEHG?

33. How do you know?

34. Why is GF = FE and DF = FH?

Set III

A board having uniform thickness and density that is triangular in shape can be balanced on its centroid, the point in which the medians of the triangle are concurrent.

If the board were in the shape of a quadrilateral instead, where would its balancing point be? The answer to this question was not discovered until as recently as the last century.

The first step in finding the balancing point is to trisect the sides of the quadrilateral. Trace the figure below, in which this has been done.

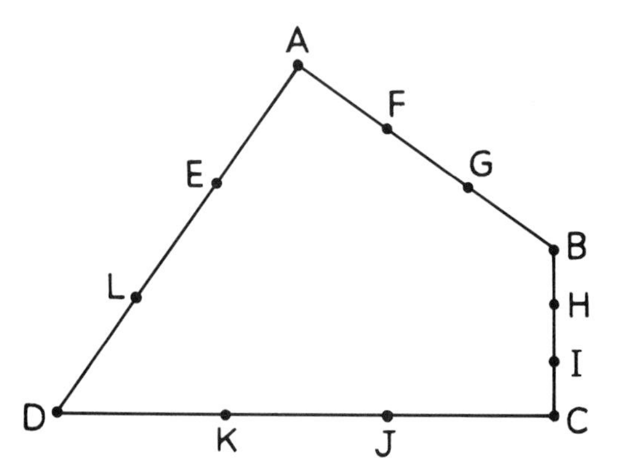

Now, draw $\overline{EF}$, $\overline{GH}$, $\overline{IJ}$, and $\overline{KL}$ and extend them until they meet in four points outside the figure. Label these points M, N, O, and P.

1. What special type of quadrilateral do you suppose MNOP is?

The balancing point of the original quadrilateral ABCD is the same as the balancing point of MNOP.

2. Where do you suppose it is?

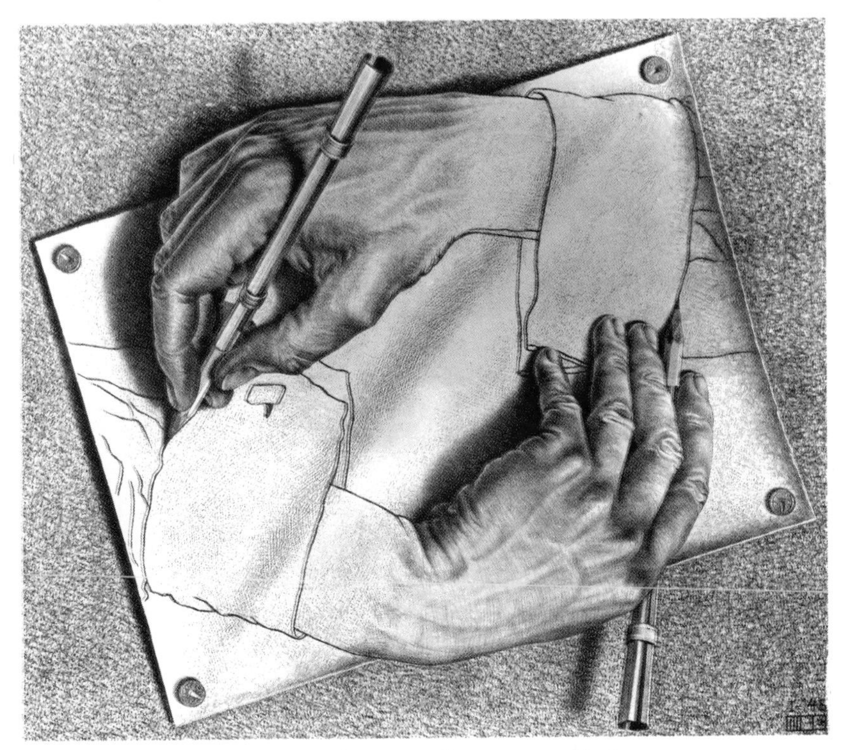

Drawing Hands by M. C. Escher, Escher Foundation, Haags Gemeentemuseum, The Hague

Lesson 6
Constructions

The very first problem presented by Euclid in the *Elements* concerns a method for constructing an equilateral triangle with a straightedge and compass. Mathematicians ever since have been intrigued with determining exactly what constructions are possible: that is, what drawings can be made with these two tools alone. Construction problems can be very challenging and are, for anyone who likes to solve puzzles, a lot of fun. In this lesson we will consider a variety of construction problems concerning triangles and quadrilaterals.

Perhaps the best way to learn how to do original constructions is to study some examples. Our first one will be the construction of a right triangle having c as its hypotenuse and a as one of its legs.

c —————— a ———

Before attempting to do the construction itself, it is a good idea to make a rough sketch of the finished figure in order to get an idea of how to proceed. The completed triangle will look something like the one below.

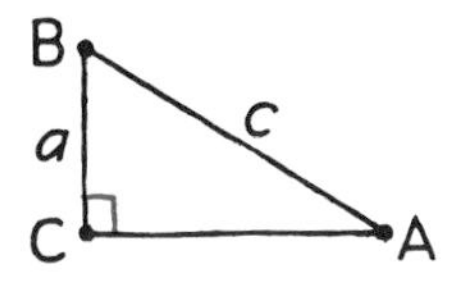

First, we can draw a line ℓ to contain the base and mark a point C on it for the vertex of the right angle. Next, we can construct the line through point C that is perpendicular to line ℓ. This line will contain leg a, and, because we know its length, we can locate point B with a compass. Finally, with B as center, we can draw an arc with radius equal to c intersecting line ℓ at A. Drawing $\overline{BA}$ completes the triangle.

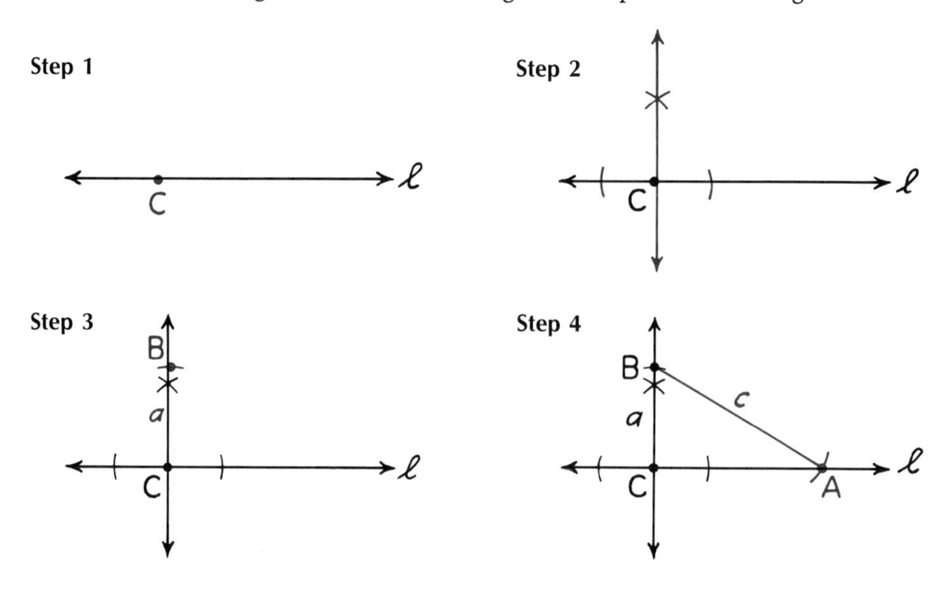

Our second example will be the construction of an isosceles triangle having a as one of its legs and r as the radius of its circumcircle.

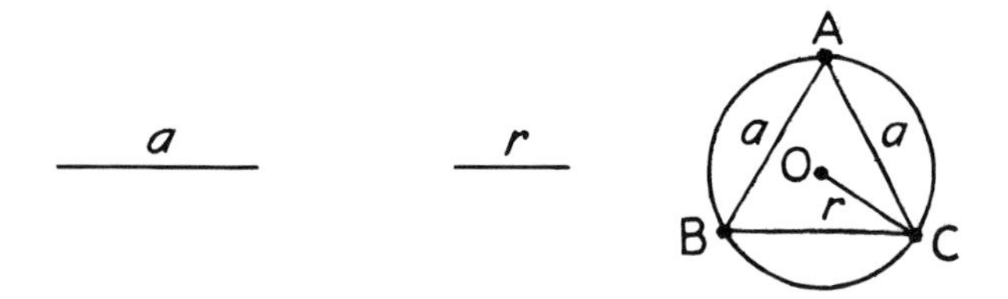

First, we can draw the circumcircle with any point O as center and r as radius. Next, we can choose a point A on the circle for the top vertex of the triangle. With A as center and a as radius, we can draw equal arcs intersecting the circle in points B and C. Finally, we can join A, B, and C with segments to form the triangle.

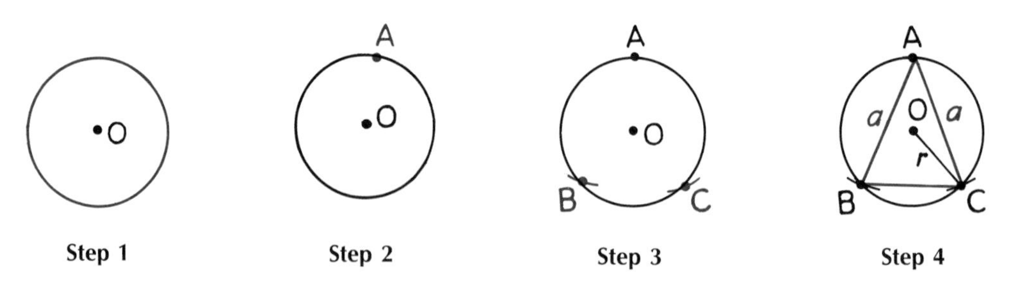

Exercises

Set I

Draw and label segments having the following lengths for use in the exercises of this lesson: a, 1 cm; b, 2 cm; c, 3 cm; d, 4 cm; e, 5 cm; and f, 6 cm. Before doing each construction, make a sketch of the completed figure and then try to develop a plan based on relations that you notice among its parts.

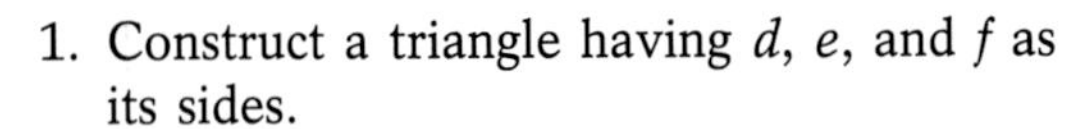

1. Construct a triangle having d, e, and f as its sides.

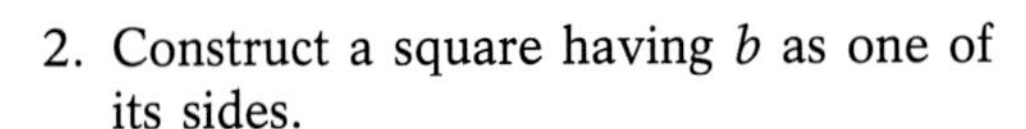

2. Construct a square having b as one of its sides.
3. Construct a right triangle having c and e as its legs.
4. Construct a rectangle having b and e as two of its sides.
5. Construct a square having d as one of its diagonals.
6. Construct a rhombus having c and f as its diagonals.
7. There is often more than one way to do a construction. The first one in this lesson—the right triangle—could also have been constructed as illustrated here.

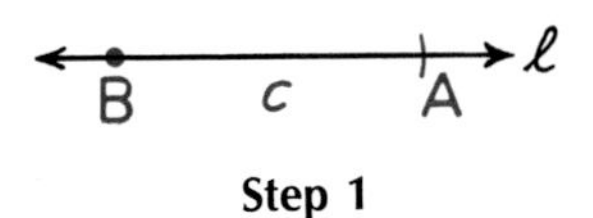

Step 1

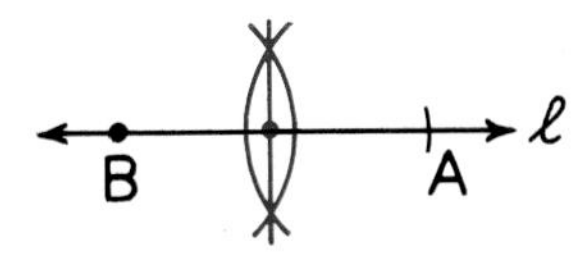

Step 2

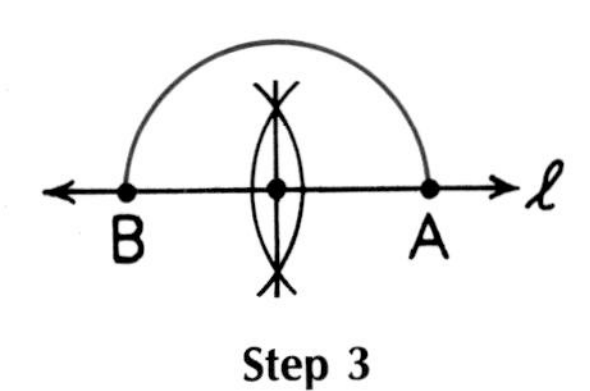

Step 3

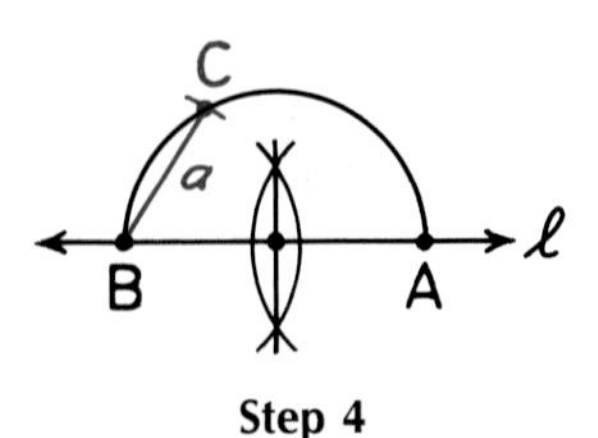

Step 4

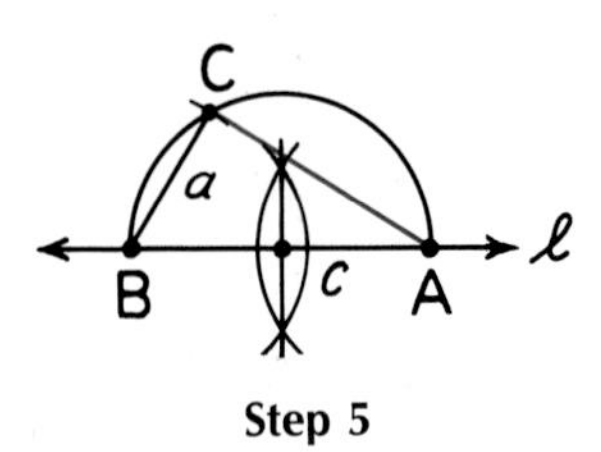

Step 5

a) What was done in step 2?
b) What was done in step 3?
c) When $\overline{BC}$ and $\overline{CA}$ are drawn in step 5, how do we know that $\angle C$ is a right angle?

8. Use any method you wish to construct a right triangle having b as one leg and f as its hypotenuse.
9. Construct a triangle having d and e as two of its sides and c as the radius of its circumcircle.
10. Construct a parallelogram having an angle of 60° and c and d as two of its sides. (Hint: Construct an equilateral triangle to form the 60° angle.)
11. Construct a rhombus having an angle of 45° and c as one of its sides.
12. Construct an isosceles right triangle having e as its hypotenuse.
13. Construct an isosceles trapezoid having f as its longer base, base angles of 60°, and b as each leg.
14. Construct a circle having b as its radius and inscribe a rectangle having b as one of its sides in the circle.

Set II

In some of the following exercises, two figures of different shapes can be constructed using the given parts. For example, in constructing a triangle having x and y as two of its sides and z as the altitude to the third side, we have a choice at the end.

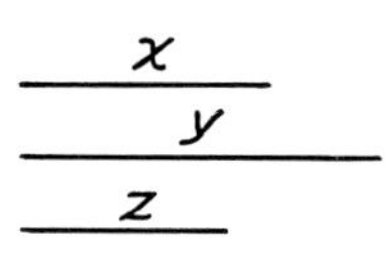

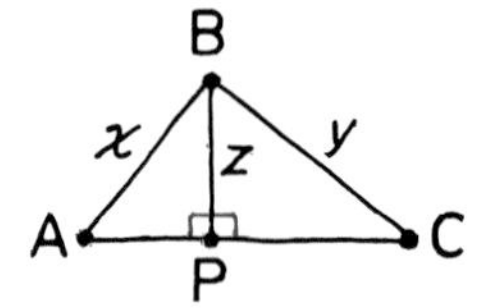

Step 1

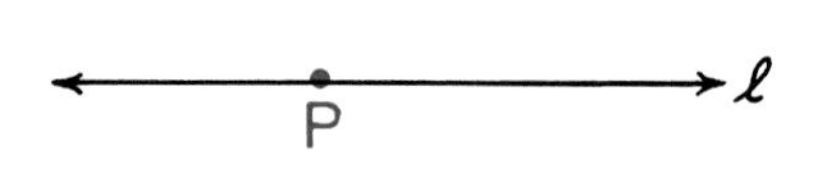

Step 2

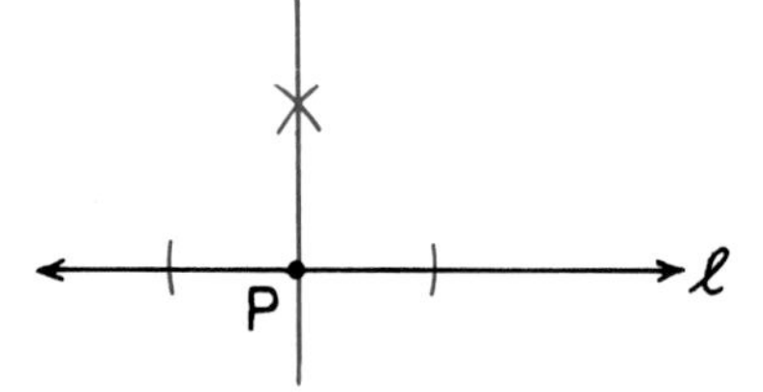

Step 3

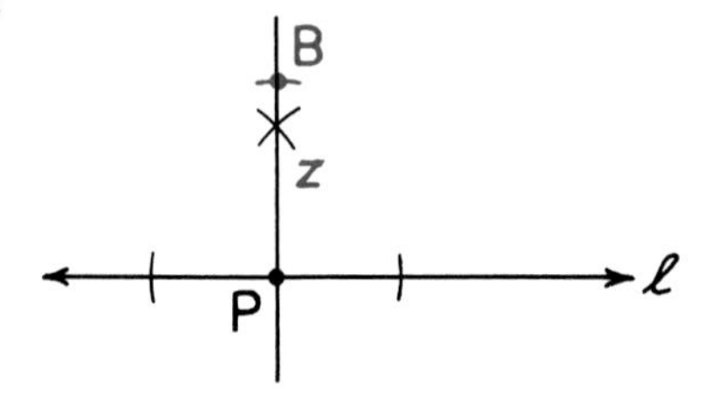

Step 4

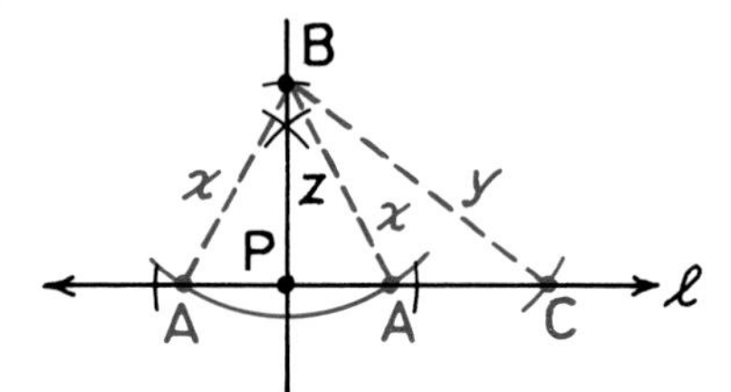

Step 5a

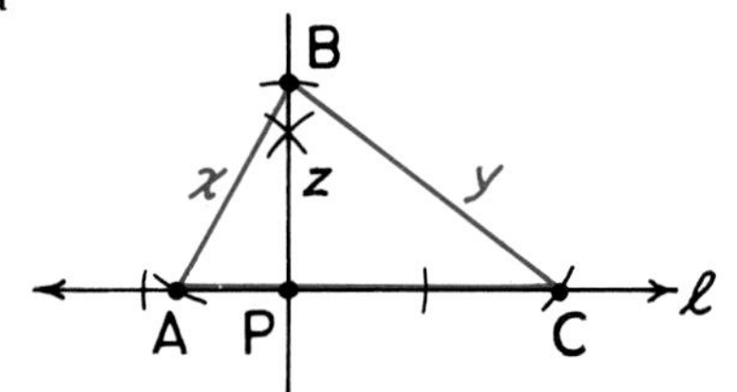

Step 5b

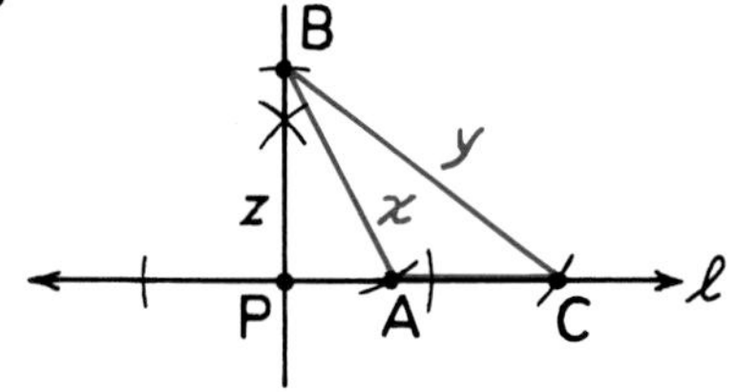

15. Construct an isosceles triangle having c as the radius of its circumcircle and e as its base. (There are two possibilities.)

16. Construct a rectangle having b as the radius of its circumcircle and a as one of its sides.

17. Construct a 30°-60° right triangle having c as its shorter leg.

18. Construct a 30°-60° right triangle having e as its longer leg.

19. Construct a triangle having b and f as two of its sides and d as the median to side f.

20. Construct a right triangle having d as one of its legs and b as the altitude to its hypotenuse.

21. Construct a triangle having d and f as two of its sides and c as the altitude to side f. (There are two possibilities.)

22. Construct a right triangle having c as one of its legs and a as the radius of its incircle.

23. Construct a triangle having f as one of its sides and d and e as the altitudes to the other two sides.

24. Construct a trapezoid having b and f as its bases and c and d as its legs.

Set III

The French general and emperor Napoleon was also an amateur mathematician. He was especially interested in geometry and may have been the first to discover the following rather remarkable construction.

Draw a large scalene triangle of any shape. Construct three equilateral triangles, each sharing one side with the scalene triangle and each facing outward from it. Find the centers of the equilateral triangles and join them to form a fifth triangle.

What do you notice?

Chapter 13 / Summary and Review

Basic Ideas

Centroid 484
Cevian 477
Circumcenter 463
Circumscribed circle 463
Concurrent lines 463
Concyclic points 462
Cyclic polygon 463
Incenter 472
Inscribed circle 472
Median 483
Orthocenter 484

Constructions

11. To circumscribe a circle about a triangle. 463
12. To inscribe a circle in a triangle. 473

Theorems

78. Every triangle is cyclic. 463
    Corollary. The perpendicular bisectors of the sides of a triangle are concurrent. 463
79. A quadrilateral is cyclic iff a pair of its opposite angles are supplementary. 467
80. The angle bisectors of a triangle are concurrent. 473
    Corollary. Every triangle has an incircle. 473
81. *Ceva's Theorem.* Three cevians $\overline{AY}$, $\overline{BZ}$, and $\overline{CX}$ of $\triangle ABC$ are concurrent iff $\frac{AX}{XB} \cdot \frac{BY}{YC} \cdot \frac{CZ}{ZA} = 1.$ 478
82. The medians of a triangle are concurrent. 484
83. The lines containing the altitudes of a triangle are concurrent. 484

Exercises

Set I

Copy $\triangle ABC$ and make each of the following constructions.

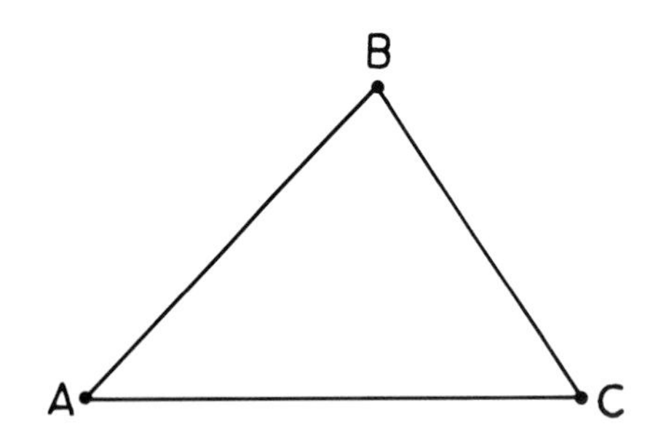

1. Circumscribe a circle about it.
2. Inscribe a circle in it.

Copy $\triangle DEF$ and make each of the following constructions.

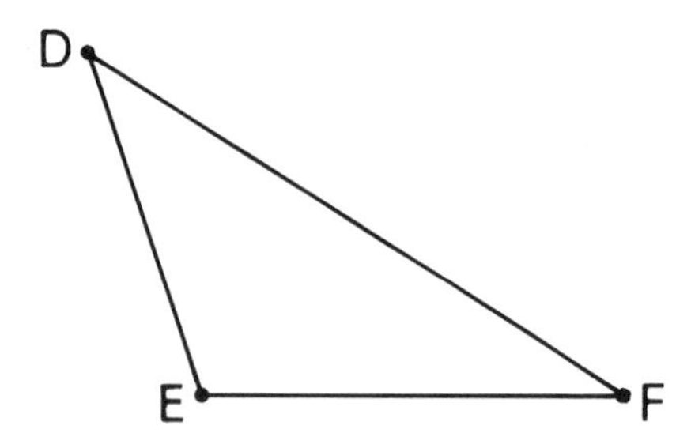

3. Construct the median to side $\overline{DE}$.
4. Construct the altitude to side $\overline{EF}$.

Lines $\overleftrightarrow{RE}$, $\overleftrightarrow{HK}$, and $\overleftrightarrow{IO}$ are the perpendicular bisectors of the sides of $\triangle ATC$.

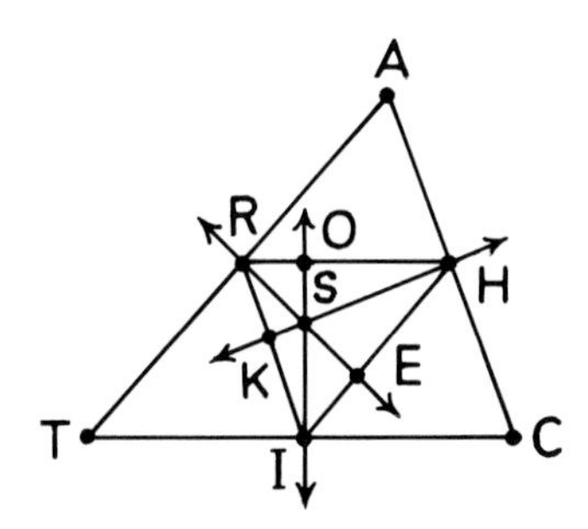

5. What is point S called with respect to $\triangle ATC$?
6. What are $\overline{RH}$, $\overline{HI}$, and $\overline{IR}$ called with respect to $\triangle ATC$?
7. Why is each side of $\triangle RHI$ parallel to one of the sides of $\triangle ATC$?
8. Why is $\overline{RE} \perp \overline{IH}$, $\overline{HK} \perp \overline{RI}$, and $\overline{IO} \perp \overline{RH}$?
9. What are $\overline{RE}$, $\overline{HK}$, and $\overline{IO}$ called with respect to $\triangle RHI$?
10. What is point S called with respect to $\triangle RHI$?

Copy circle O and point P as shown here.

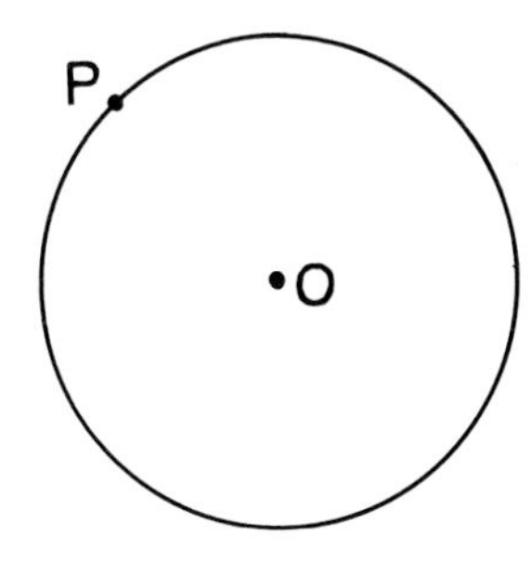

11. Draw at least ten chords of the circle that have P as one endpoint. Without constructing them, mark the midpoint of each of these chords.
12. What relation do these midpoints seem to have to each other?

Line segments $\overline{TN}$, $\overline{RP}$, and $\overline{IU}$ are cevians of $\triangle TRI$.

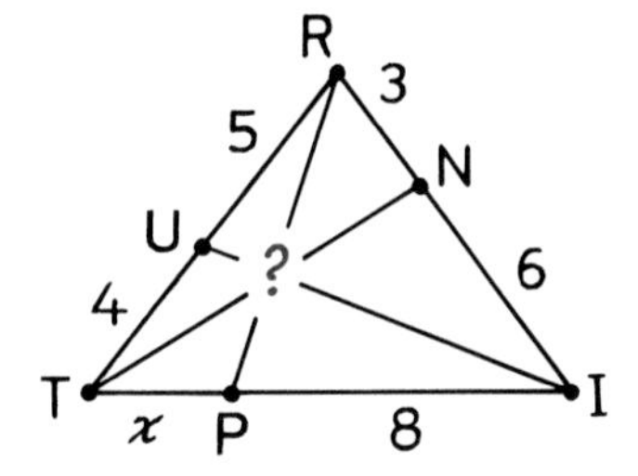

13. Define cevian.
14. Copy and complete the following statement of Ceva's Theorem: Three cevians $\overline{TN}$, $\overline{RP}$, and $\overline{IU}$ of $\triangle TRI$ are ▒▒▒ iff ▒▒▒.
15. Given that Ceva's Theorem applies to $\triangle TRI$, find x.

Copy $\triangle SIA$.

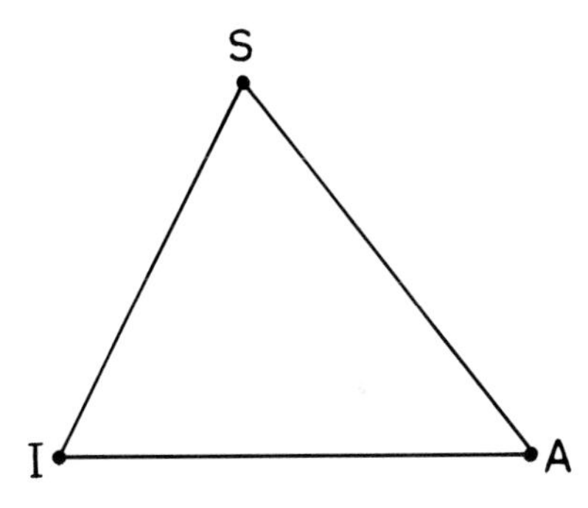

16. Construct its altitudes; label them $\overline{SN}$, $\overline{IC}$, and $\overline{AP}$ and the point in which they intersect H. Draw $\triangle PCN$.
17. What is point H called with respect to $\triangle SIA$?
18. It is possible to prove that point H is the incenter of $\triangle PCN$. What does this imply about the relation of $\overleftrightarrow{SN}$, $\overleftrightarrow{IC}$, and $\overleftrightarrow{AP}$ to $\triangle PCN$?
19. Which lines in the figure are equidistant from point H?

Segments $\overline{RD}$ and $\overline{AS}$ are altitudes of $\triangle$RAI. Use this information to tell whether each of the following statements *must be true, may be true,* or *appears to be false.*

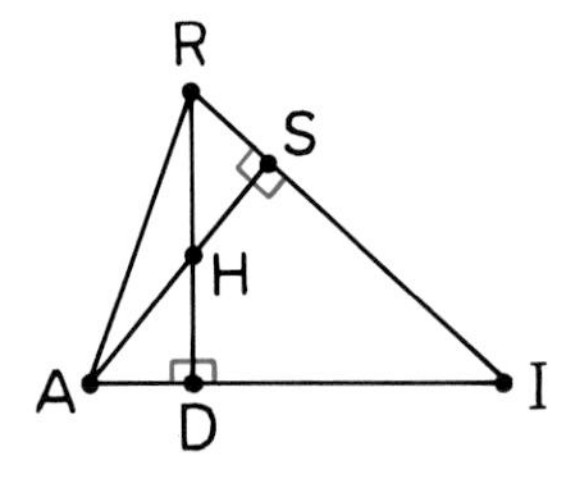

20. Point H is the orthocenter of $\triangle$RAI.

21. RD = AS.

22. The three segments $\overline{RA}$, $\overline{AI}$, and $\overline{IR}$ are concurrent.

23. The four points D, H, S, and I are concyclic.

24. $\alpha\triangle RAI = \frac{1}{2}RI \cdot AS$.

25. $\triangle RDI \sim \triangle ASI$.

Set II

Draw and label segments having the following lengths: a, 1 cm; b, 2 cm; c, 3 cm; d, 4 cm; e, 5 cm; and f, 6 cm. Use them to make the following constructions.

26. Construct an isosceles triangle having c as its base and d as the corresponding altitude.

27. Construct a rectangle having d as one of its sides and e as one of its diagonals.

28. Construct a 30°-60° right triangle having e as its hypotenuse.

29. Construct a parallelogram having an angle of 45° and c and e as two of its sides.

30. Construct an equilateral triangle having b as the radius of its circumcircle.

31.

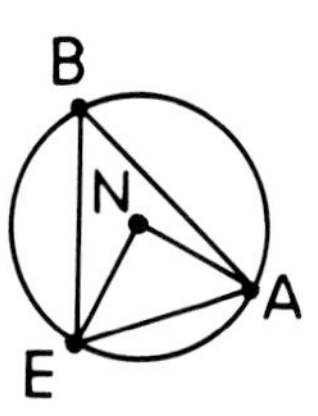

Given: $\triangle$BEA with circumcircle N.

Prove: $\angle EBA = \frac{1}{2}\angle ENA$.

32.

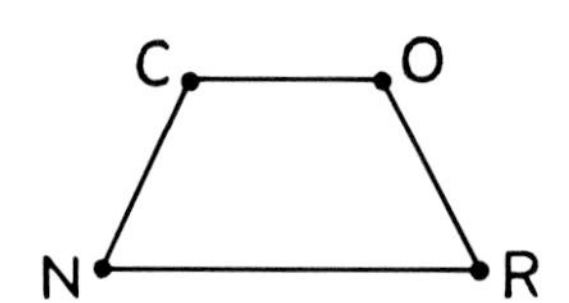

Given: CORN is an isosceles trapezoid with bases $\overline{CO}$ and $\overline{NR}$.

Prove: CORN is cyclic.

Chapter 14

REGULAR POLYGONS AND THE CIRCLE

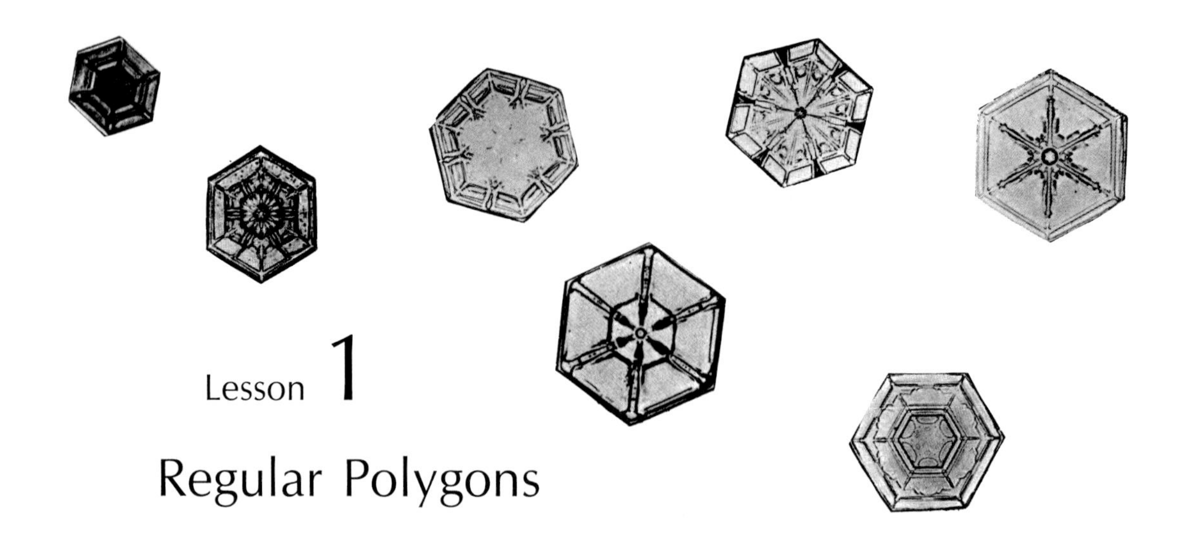

Lesson 1

Regular Polygons

Snowflakes are a beautiful example of geometry in nature. A farmer-meteorologist in Vermont spent many winters taking photographs of thousands of them through a microscope; some of his pictures are shown here.

Although it has been said that no two snowflakes are alike, those illustrated here have several basic properties in common. They are all six-sided and hence hexagonal in shape. Furthermore, each is convex and has equal sides and equal angles; such polygons are called *regular.*

► **Definition**
A ***regular polygon*** is a convex polygon that is both equilateral and equiangular.

You know that one of these properties cannot exist without the other in polygons having three sides. We have proved that all equilateral triangles are equiangular and conversely.

By the permission of Johnny Hart and News America Syndicate

If a polygon has more than three sides, however, having one property does not imply the other. Consider quadrilaterals, for example. A rhombus is equilateral but not necessarily equiangular, whereas a rectangle is equiangular but not necessarily equilateral. Only squares are both.

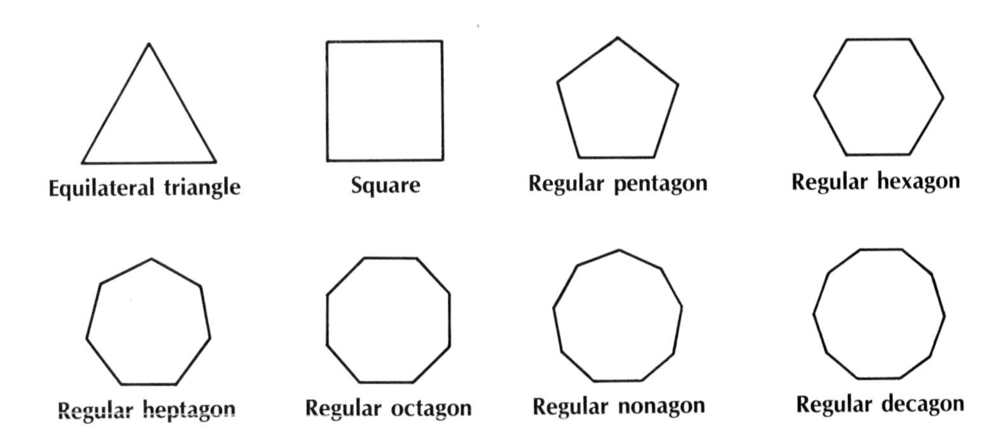

Figures representing regular polygons having from three through ten sides are shown above. It is evident that, as the number of sides of a regular polygon increases, it looks more and more like a circle. This is a consequence of the fact that every regular polygon is cyclic.

► **Theorem 84**
Every regular polygon is cyclic.

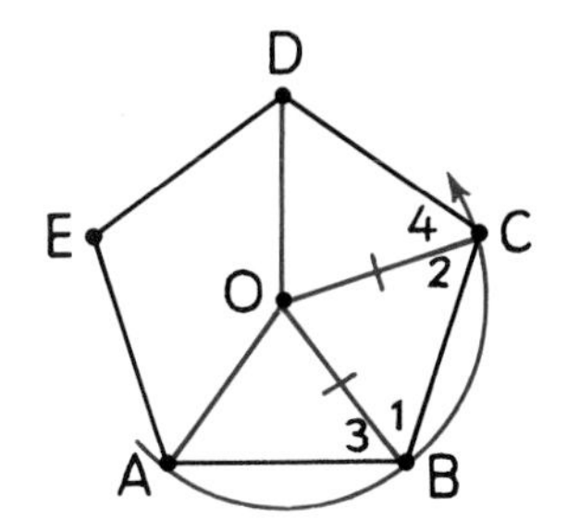

Given: ABCDE is a regular polygon.
Prove: ABCDE is cyclic.

To illustrate our proof, we will use a regular pentagon. The proof, however, applies to all regular polygons because it does not depend on the number of sides of the pentagon.

The idea is to draw a circle through three vertices of the polygon and then prove, by means of congruent triangles, that the distance from each of the remaining vertices to the center of this circle is equal to its radius. From this it follows that they also lie on the circle.

Proof.
Draw the circle that contains points A, B, and C and let its center be called O. Draw $\overline{OA}$, $\overline{OB}$, $\overline{OC}$, and $\overline{OD}$.

We know that OB = OC (all radii of a circle are equal), and so $\angle 1 = \angle 2$ (if two sides of a triangle are equal, the angles opposite them are equal). Now $\angle ABC = \angle BCD$ (a regular polygon is equiangular), and so $\angle 3 = \angle 4$ (subtraction). Also, AB = CD (a regular polygon is equilateral), and so $\triangle OBA \cong \triangle OCD$ (S.A.S.). Therefore, OD = OA. Because

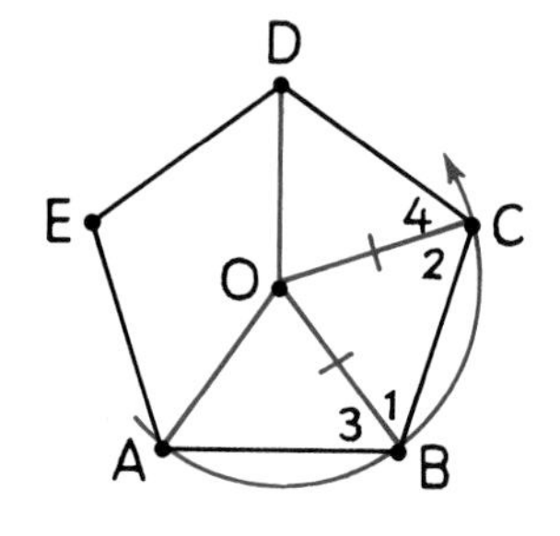

OA is the radius of the circle, it follows that OD is also. Hence, D lies on circle O, because a circle is the set of all points in a plane at a distance of one radius from its center.

To prove that E also lies on circle O, $\overline{OE}$ can be drawn and $\triangle ODE$ proved congruent to $\triangle OCB$ in the same way.

Recall that a polygon whose vertices lie on a circle is *inscribed* in the circle and that the circle is *circumscribed* about the polygon. Other terms that are used in referring to regular polygons are illustrated and defined below.

Definitions

The ***center*** of a regular polygon is the center of its circumscribed circle.

A ***radius*** of a regular polygon is a line segment that joins its center to a vertex.

An ***apothem*** of a regular polygon is a perpendicular line segment from its center to one of its sides.

A ***central angle*** of a regular polygon is an angle formed by radii drawn to two consecutive vertices.

Radius

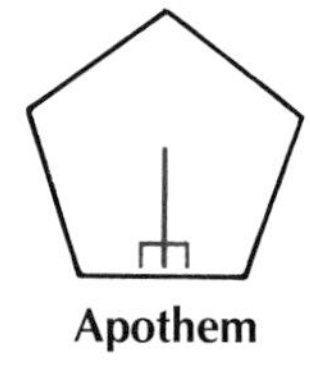

Apothem

Central angle

Exercises

Set I

Tell whether each of the following statements is true or false.

1. Every triangle is cyclic.
2. If a triangle is equilateral, it must be regular.
3. If a quadrilateral is equilateral, it must be regular.
4. If a quadrilateral is equiangular, it must be regular.
5. If a quadrilateral is equiangular, it must be cyclic.
6. If a polygon is regular, it must be convex.
7. If a polygon is regular, it must be cyclic.
8. If a polygon is cyclic, it must be regular.
9. If a polygon is regular, it must be equilateral.
10. If a polygon is equilateral, it must be equiangular.

Circle O is circumscribed about regular pentagon NITRE; $\overline{OG} \perp \overline{RE}$.

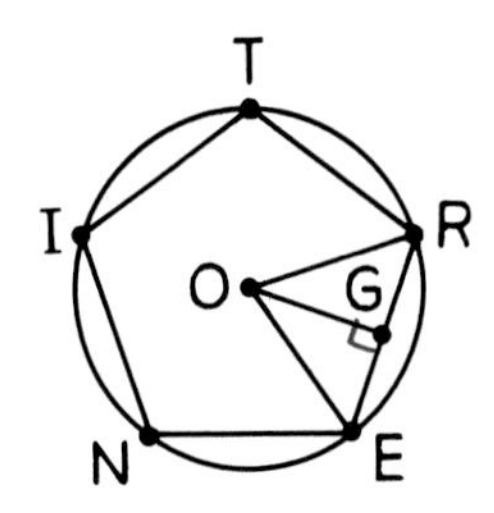

11. What is point O called with respect to the pentagon?
12. What is $\overline{OG}$ called?
13. How do we know that $\overline{OG}$ bisects $\overline{RE}$?
14. What are $\overline{OR}$ and $\overline{OE}$ called with respect to NITRE?
15. What kind of triangle is $\triangle ORE$?
16. What is $\angle ROE$ called with respect to NITRE?
17. Does $\overrightarrow{OG}$ bisect $\angle ROE$?

The figures below illustrate the central angles of some regular polygons.

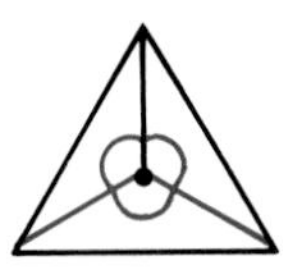
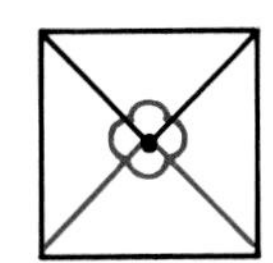
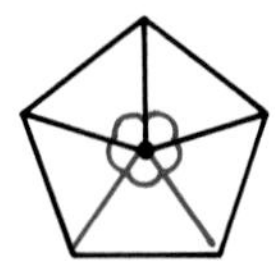
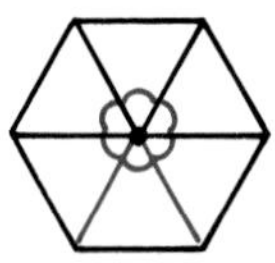

18. As the number of sides of a regular polygon increases, how does the measure of one of its central angles change?
19. Find the measure of a central angle of each figure shown.
20. Find the measure of a central angle of a regular decagon.
21. How would you express the measure of a central angle of a regular polygon that has n sides in terms of n?

Set II

The figure below suggests a way to construct a regular hexagon.

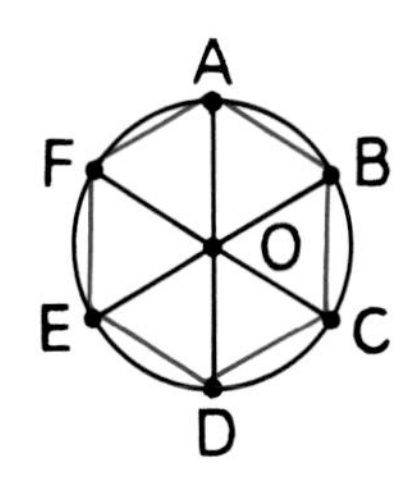

22. What kind of triangles surround point O?
23. How do the sides of the hexagon compare in length to the radius of the circle?
24. Use your straightedge and compass to construct a regular hexagon by inscribing it in a circle.

The figure below suggests a way to construct a square.

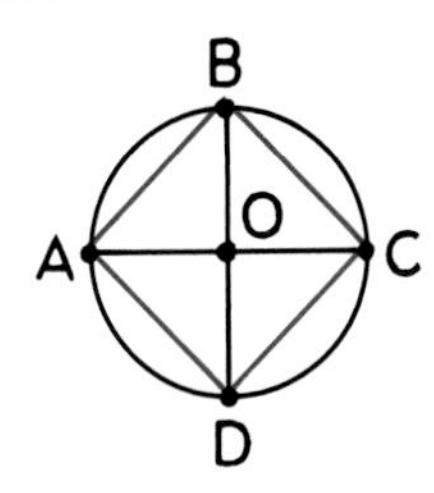

25. What relation do $\overline{AC}$ and $\overline{BD}$ have to each other?
26. Use your straightedge and compass to construct a square by inscribing it in a circle.
27. Construct a regular octagon. (Hint: Begin by inscribing a square in a circle.)
28. Construct a regular dodecagon. (Hint: Begin by inscribing a regular hexagon in a circle.)

In the figure below, $\overline{OT}$ is an apothem of regular pentagon KRYPN; $\angle O = 36°$ and ON = 10. Use the appropriate trigonometric

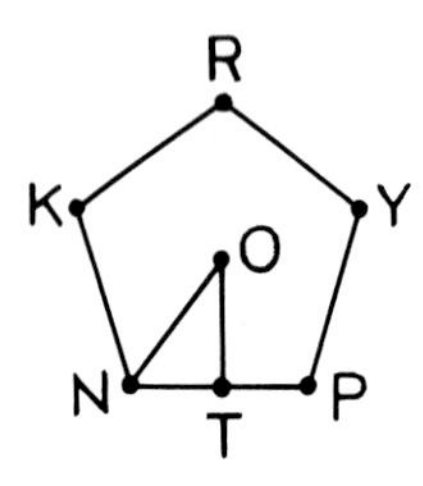

ratio and either the table on page 411 or your calculator to find each of the following lengths to the nearest tenth.

29. OT.

30. NT.

31.

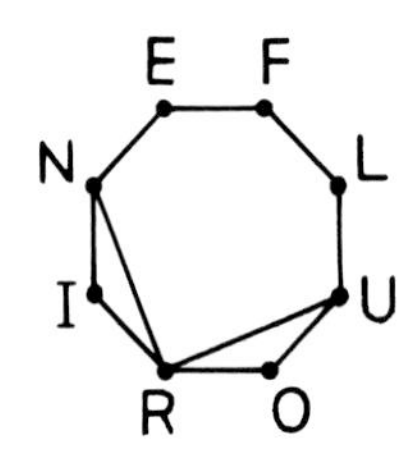

Given: FLUORINE is a regular polygon with diagonals $\overline{NR}$ and $\overline{RU}$.

Prove: NR = RU.

32.

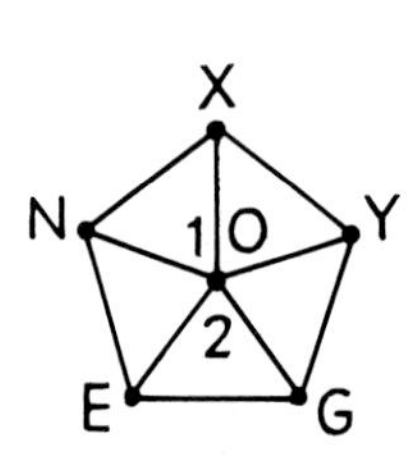

Given: XYGEN is a regular polygon with central angles 1 and 2.

Prove: $\angle 1 = \angle 2$. (Hint: circumscribe circle O about XYGEN.)

33.

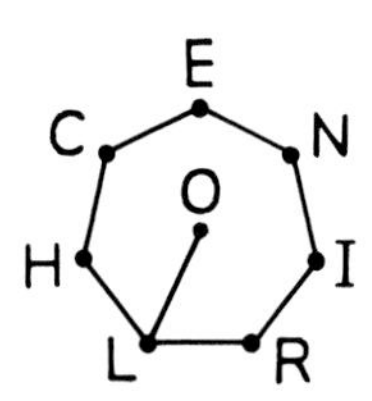

Given: CHLRINE is a regular polygon with center O.

Prove: $\overrightarrow{LO}$ bisects ∡HLR.

34.

Given: HELIUM is a regular hexagon with diagonals $\overline{HL}$ and $\overline{MI}$.

Prove: $\overline{HL} \parallel \overline{MI}$ without adding anything to the figure.

Set III

Three Dissection Puzzles.*

Carefully trace the figures shown here and cut each one apart.

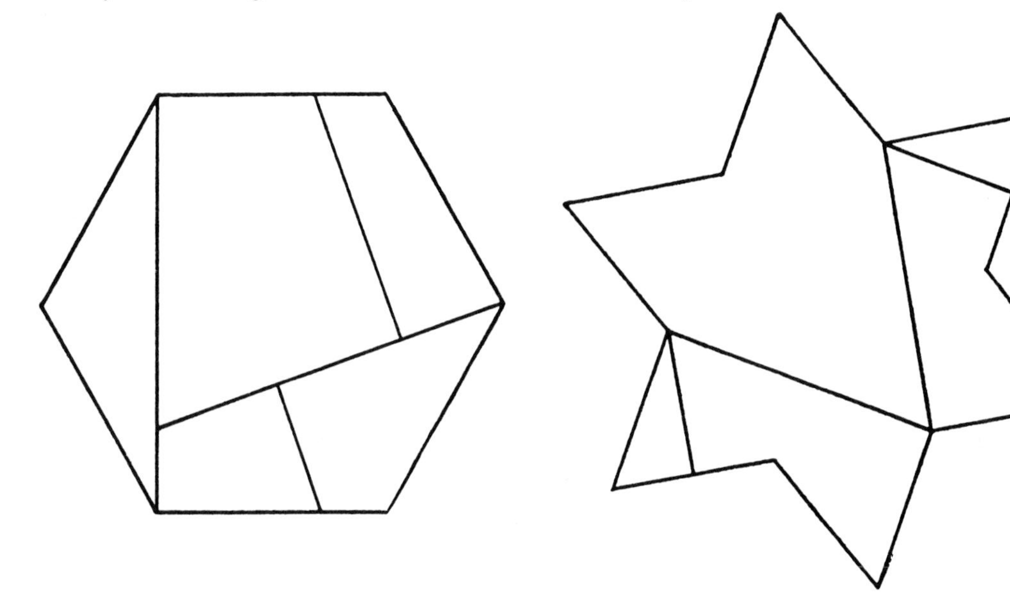

Can you rearrange the pieces of

1. the hexagon to form a square?
2. the star to form an equilateral triangle?
3. the cross to form a regular dodecagon?

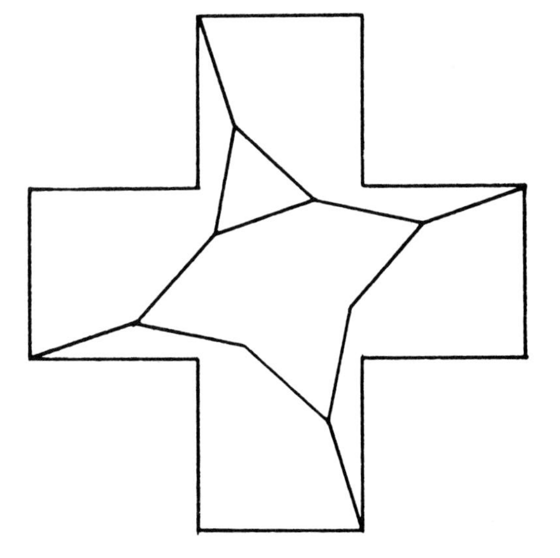

*From "Geometric Dissections," in *The Unexpected Hanging* by Martin Gardner (Simon & Schuster, 1969), Chapter 4. Originally published in "Mathematical Games" by Martin Gardner. Copyright © 1961 by Scientific American, Inc.

General Turtle

Lesson 2

The Perimeter of a Regular Polygon

General Turtle is a company that makes a small device known as a "turtle." If one of these turtles is linked to a computer, it can be ordered to do such things as move in a given direction, turn through a given angle, and trace the path it follows with a pen. Its purpose is to help elementary school children learn mathematics.*

Among the tricks that the turtle can perform when given the correct directions is the drawing of regular polygons. For example, if it is told to move 10 cm and turn right 72°, and then to repeat these two actions until it returns to its starting point, it will produce a regular pentagon.

How far does the turtle travel in drawing a regular polygon? The answer to this question depends on two things: the number and length of the polygon's sides. The total distance covered is the perimeter of the polygon and can be expressed by the equation

$$\rho = ns,$$

where ρ is the perimeter of the polygon, n is the number of sides, and s is the length of one side. For example, the turtle would move 50 cm in drawing a regular pentagon that has 10-cm sides.

---

* *Turtle Geometry: The Computer as a Medium for Exploring Mathematics*, by Harold Abelson and Andrea A. diSessa (MIT Press, 1981).

The perimeter of a regular polygon can also be expressed in terms of the length of its radius. In the figure at the right, a regular pentagon with sides of length s has been inscribed in a circle with radius r. Segment $\overline{OM}$ is an apothem of the pentagon, and so $\overline{OM}$ is perpendicular to $\overline{AB}$ and hence bisects it. (If a line through the center of a circle is perpendicular to a chord, it also bisects it.) So $AM = \frac{1}{2}AB = \frac{1}{2}s$.

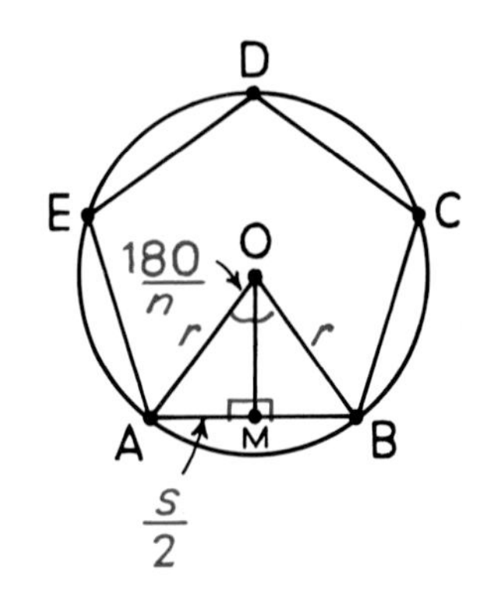

It is easy to see that $\triangle AOM \cong \triangle BOM$, and therefore $\angle AOM = \angle BOM$. This means that $\overrightarrow{OM}$ bisects central $\measuredangle AOB$. Because a central angle of a regular n-gon has a measure of $\frac{360°}{n}$, $\angle AOM = \frac{1}{2}\left(\frac{360}{n}\right) = \frac{180°}{n}$.

Now, to find a relation between r and s we can apply the sine ratio to right $\triangle AOM$.

$$\sin\frac{180}{n} = \frac{\frac{s}{2}}{r}.$$

Multiplying both sides of the equation by r,

$$r\sin\frac{180}{n} = \frac{s}{2},$$

and by 2,

$$s = 2r\sin\frac{180}{n}.$$

Because $\rho = ns$, we get, by substitution,

$$\rho = n\left(2r\sin\frac{180}{n}\right) = 2\left(n\sin\frac{180}{n}\right)r.$$

The product $n\sin\frac{180}{n}$ depends only on n, the number of sides of the polygon. We will use a capital N to represent this product so that

$$N = n\sin\frac{180}{n}.$$

Substituting into the equation above, we get

$$\rho = 2Nr.$$

We have developed a formula for the perimeter of a regular polygon having a given number of sides in terms of its radius. Because this formula does not depend on the fact that we used a polygon with five sides to derive it, we will state it as a general theorem.

► **Theorem 85**

The perimeter of a regular polygon having n sides is $2Nr$, in which r is the length of its radius and $N = n\sin\frac{180}{n}$.

The following problem illustrates how to use this theorem. Suppose that the turtle has been told to draw a regular pentagon with a 10-cm radius. How far would it travel?

$$\begin{aligned}
p &= 2Nr = 2\left(n \sin \frac{180}{n}\right)r \\
&= 2\left(5 \sin \frac{180}{5}\right)(10) \\
&= 100 \sin 36^\circ \\
&\approx 100(.5878) \\
&\approx 58.78
\end{aligned}$$

The turtle would move approximately 58.78 cm.

Exercises

Use either your calculator or the table below to do the exercises in this lesson.

A Short Table of Sines

A	*sin A*	*A*	*sin A*	*A*	*sin A*	*A*	*sin A*
1°	.0175	5°	.0872	12°	.2079	30°	.5000
2°	.0349	6°	.1045	15°	.2588	36°	.5878
3°	.0523	9°	.1564	18°	.3090	45°	.7071
4°	.0698	10°	.1736	20°	.3420	60°	.8660

Set I

CRAWFIS is a regular polygon with center H.

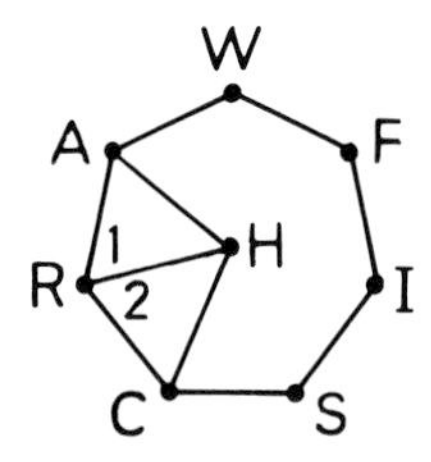

1. What are $\overline{HA}$, $\overline{HR}$, and $\overline{HC}$ called with respect to CRAWFIS?
2. Why is $\triangle RAH \cong \triangle RCH$?
3. Why is $\angle 1 = \angle 2$?
4. Do your answers to Exercises 1 through 3 depend on the fact that CRAWFIS has seven sides?
5. What does a radius of a regular polygon do to the angle of the polygon to whose vertex it is drawn?

LBSTER is a regular hexagon with center O.

6. Find $\angle 1$.
7. Why is OL = OB?
8. Why is $\angle 2 = \angle 3$?
9. Find $\angle 2$ and $\angle 3$.
10. Why is $\triangle$OLB equilateral?

Suppose that OL = 7.

11. Find LB.
12. Find ρLBSTER.
13. Use Theorem 85 to find ρLBSTER.

SHRI is a square with radius $\overline{PS}$ and apothem $\overline{PM}$.

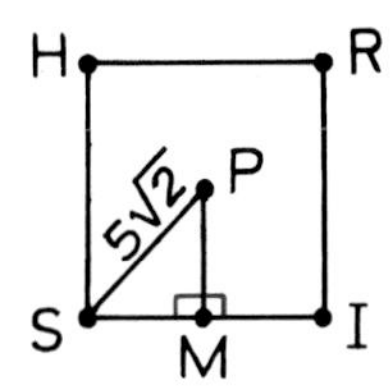

14. What kind of triangle is $\triangle$SMP?

Suppose that PS = $5\sqrt{2}$.

15. Find SM.
16. Find SI.
17. Use your answer to Exercise 16 to find ρSHRI.
18. Use Theorem 85 to find ρSHRI. Round your answer to the nearest integer. ($\sqrt{2} \approx 1.414$.)

CLM is an equilateral triangle with radius $\overline{SL}$ and apothem $\overline{SA}$.

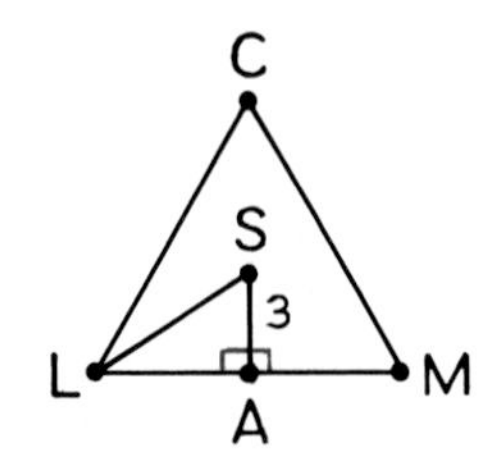

19. What kind of triangle is $\triangle$LAS?

Suppose that SA = 3.

20. Find SL.
21. Find LA.
22. Find LM.
23. Use your answer to Exercise 22 to find ρCLM.
24. Change your answer to Exercise 23 to decimal form. ($\sqrt{3} \approx 1.732$.)
25. Use Theorem 85 to find $\rho\triangle$CLM.

Set II

Each of the four regular polygons shown here is inscribed in a circle with a radius of 10.

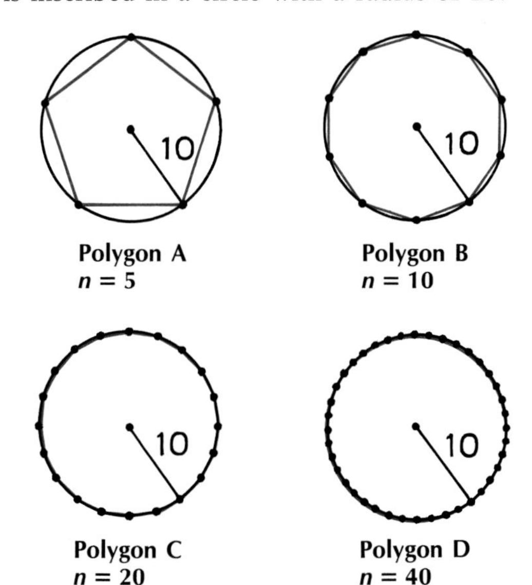

Polygon A $n = 5$

Polygon B $n = 10$

Polygon C $n = 20$

Polygon D $n = 40$

26. Copy and complete the following table.

Polygon	n	$\frac{180}{n}$	$\sin\frac{180}{n}$	$n\sin\frac{180}{n}$ *	$2Nr$
A				2.94	
B					
C					
D			.0785		

*Round each number in this column to the nearest hundredth.

Refer to your table for Exercise 26 to tell what happens to each of the following quantities as n increases.

Example: $\frac{180}{n}$.

Answer: $\frac{180}{n}$ decreases.

27. $\sin \frac{180}{n}$.

28. $n \sin \frac{180}{n}$.

29. $2Nr$.

Which one of the following conclusions about the perimeter of a regular polygon is correct?

30. As the number of sides of a regular polygon inscribed in a circle is repeatedly doubled,
    a) the perimeter is also repeatedly doubled.
    b) the perimeter increases by equal amounts.
    c) the perimeter increases by successively smaller amounts.

Use Theorem 85 to find the perimeters of the following polygons. *Round each answer to the nearest integer.*

31. A regular nonagon with radius 6.

32. A regular nonagon with radius 12.

33. A regular polygon having 90 sides and radius 6.

Refer to your answers to Exercises 31 through 33 to answer the following questions.

34. Which seems to have a greater effect on the perimeter of a regular polygon: increasing its radius or increasing the number of its sides?

35. Which one of the following regular polygons do you think has the greatest perimeter?
    a) A polygon with 50 sides and radius 4.
    b) A polygon with 200 sides and radius 4.
    c) A polygon with 50 sides and radius 12.

Set III

Suppose that you took a walk in which you went ten feet forward, turned right one degree, walked ten feet, turned right another degree, and so on until you returned to your starting point.

1. What regular polygon would you walk along?

2. Approximately how much time do you think the walk would take you? Explain the basis for your answer.

3. Estimate your maximum distance from your starting point. Explain the basis for your answer. ($\sin 0.5° \approx 0.0087$.)

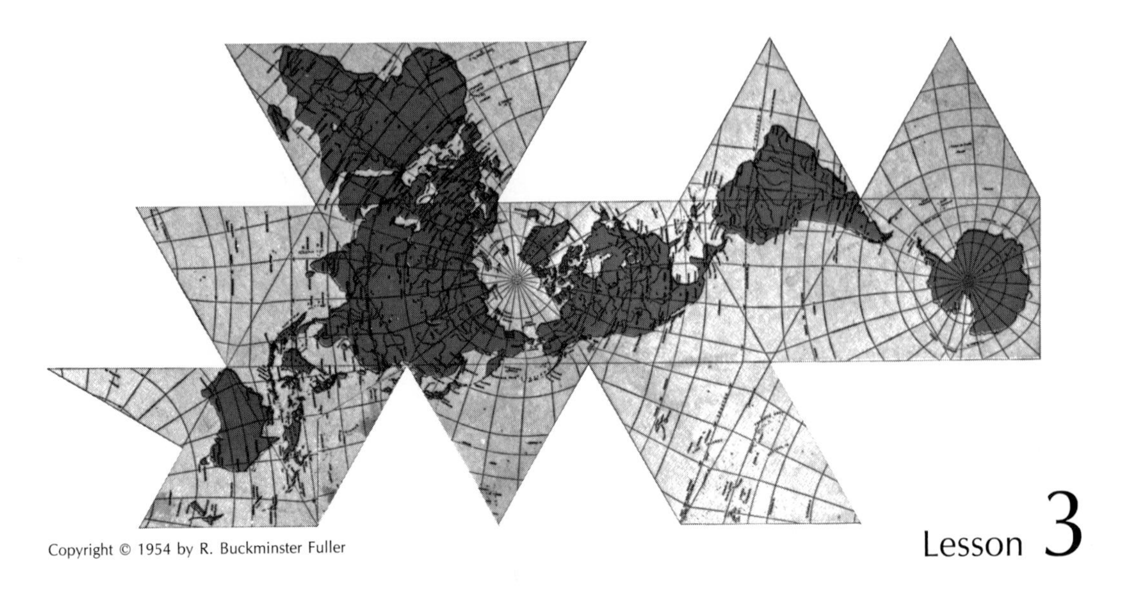

Lesson 3
The Area of a Regular Polygon

Perhaps the cleverest map of the earth ever devised is one created by Buckminster Fuller, the inventor of the geodesic dome. It consists of twenty equilateral triangles, each of which contains an equal amount of the earth's surface. The map can be folded along the sides of these triangles into a three-dimensional "globe" called an *icosahedron.*

If a copy of Mr. Fuller's icosahedron map were made in which each edge was 10 centimeters long, what would its area be? We can easily answer this question by using the formula for the area of an equilateral triangle,

$$\alpha = \frac{s^2}{4}\sqrt{3},$$

in which s is the length of a side. The map consists of twenty equilateral triangles, and so its area is

$$20\left(\frac{10^2}{4}\sqrt{3}\right) = 20(25\sqrt{3}) = 500\sqrt{3} \approx 500(1.732) = 866 \text{ cm}^2.$$

The solution of this problem required a formula for the area of a regular polygon, the equilateral triangle:

$$\alpha = \frac{s^2}{4}\sqrt{3}.$$

The only other regular polygon for which we have such a formula is the square:

$$\alpha = s^2.$$

To find the areas of other regular polygons, it would be useful to have a more general formula. We will derive such a formula, not in terms of the length of a *side*, but rather in terms of the length of the *radius* of the polygon.

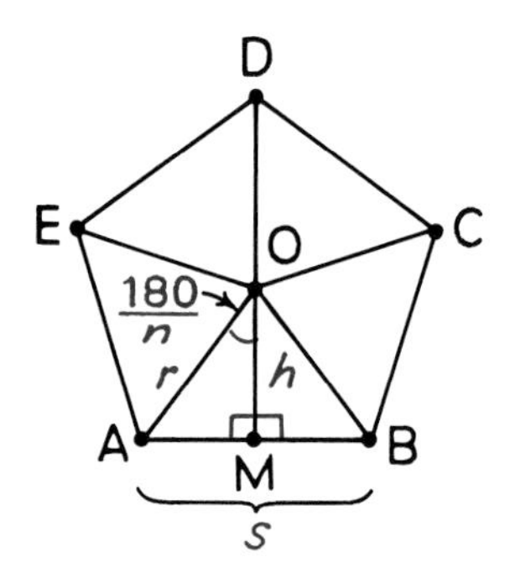

We will use a regular pentagon to develop the formula, but our argument will not depend on the fact that it has five sides.

If all of the radii of a regular polygon having n sides are drawn, they divide it into n congruent isosceles triangles. (In this pentagon, there are five of these triangles.)

The area of one of these triangles, such as $\triangle AOB$, is $\frac{1}{2}sh$ because apothem $\overline{OM}$ is the altitude to base $\overline{AB}$. It follows that the total area of the n triangles in the figure is $n\left(\frac{1}{2}sh\right)$, so that

$$\alpha = \frac{1}{2}nsh.$$

Because the perimeter of a regular polygon having n sides is ns, we can also write

$$\alpha = \frac{1}{2}\rho h. \tag{1}$$

We now have a formula for the area of a regular polygon having n sides in terms of its perimeter and apothem. We will now restate it in terms of the polygon's radius.

By Theorem 85,

$$\rho = 2\mathrm{N}r = 2n \sin \frac{180}{n} r. \tag{2}$$

Applying the cosine ratio to right $\triangle AOM$ gives

$$\cos \frac{180}{n} = \frac{h}{r}.$$

Multiplying both sides of this equation by r gives

$$h = r \cos \frac{180}{n}. \tag{3}$$

Substituting (2) and (3) into (1), we get

$$\alpha = \frac{1}{2}\left(2n \sin \frac{180}{n} r\right)\left(r \cos \frac{180}{n}\right)$$

$$= \left(n \sin \frac{180}{n} \cos \frac{180}{n}\right) r^2.$$

We will let a capital M represent the product

$$n \sin \frac{180}{n} \cos \frac{180}{n}.$$

This reduces our area formula to

$$\alpha = Mr^2.$$

▶ **Theorem 86**
The area of a regular polygon having n sides is Mr^2, in which r is the length of its radius and $M = n \sin \frac{180}{n} \cos \frac{180}{n}$.

To illustrate how this theorem is used, we will find the area of the regular pentagon with a 10-centimeter radius drawn by the turtle that is described on page 504.

$$\begin{aligned}\alpha = Mr^2 &= \left(n \sin \frac{180}{n} \cos \frac{180}{n}\right) r^2\\ &= \left(5 \sin \frac{180}{5} \cos \frac{180}{5}\right)(10)^2\\ &= 500 \sin 36° \cos 36°\\ &\approx 500(.588)(.809) \approx 238.\end{aligned}$$

The area of the regular pentagon would be about 238 square centimeters.

Exercises

Use either your calculator or the table below to do the exercises in this lesson.

A Table of Sine-Cosine Products

A	*sin A cos A*	*A*	*sin A cos A*	*A*	*sin A cos A*	*A*	*sin A cos A*
1°	.01745	5°	.08682	12°	.2034	30°	.4330
2°	.03488	6°	.1040	15°	.2500	36°	.4755
3°	.05226	9°	.1545	18°	.2939	45°	.5000
4°	.06959	10°	.1710	20°	.3214	60°	.4330

Set I

VAMPI is a regular pentagon with center R; $\overline{RE} \perp \overline{VI}$.

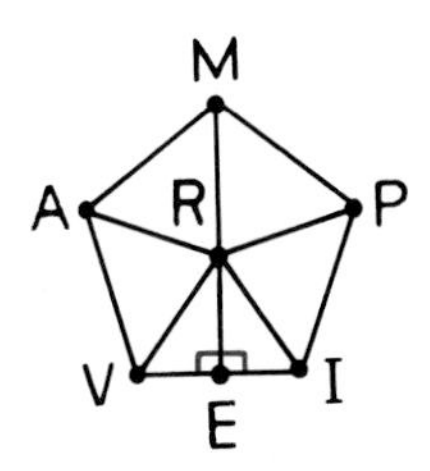

1. What is $\overline{RE}$ called with respect to VAMPI?
2. What is $\overline{RE}$ called with respect to $\triangle$VRI?
3. Why is $\triangle VRI \cong \triangle IRP \cong \triangle PRM \cong \triangle MRA \cong \triangle ARV$?
4. Why is $\alpha\triangle VRI = \alpha\triangle IRP = \alpha\triangle PRM = \alpha\triangle MRA = \alpha\triangle ARV$?
5. How does αVAMPI compare to $\alpha\triangle$VRI?
6. Write an expression for $\alpha\triangle$VRI in terms of VI and RE.
7. Write an expression for αVAMPI in terms of VI and RE.

WRAT is a square with center H; RA = 6.

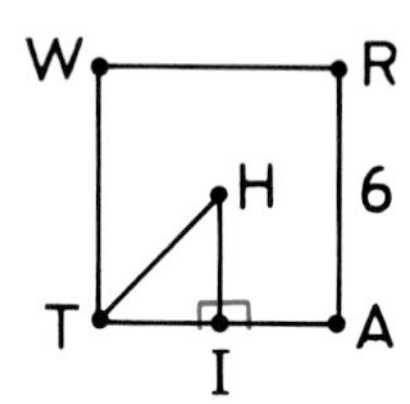

8. Use the formula for the area of a square to find αWRAT.
9. Find TI.
10. Find HT.
11. Use Theorem 86 to find αWRAT.

GHT is an equilateral triangle with center O; GT = $4\sqrt{3}$.

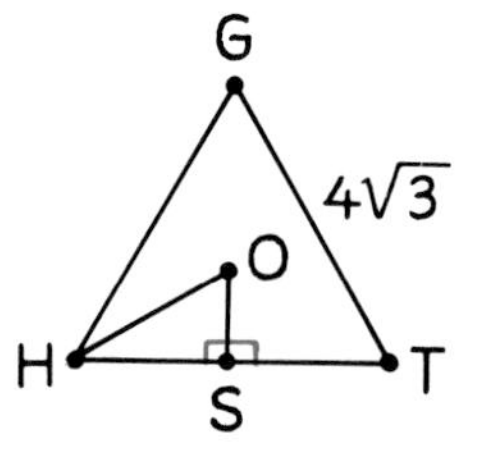

12. Use the formula for the area of an equilateral triangle to find $\alpha\triangle$GHT.
13. Change your answer to Exercise 12 to decimal form. ($\sqrt{3} \approx 1.732$.)
14. Find HS. 15. Find OS. 16. Find OH.
17. Use Theorem 86 to find $\alpha\triangle$GHT.

Set II

Each of the four regular polygons shown here is inscribed in a circle with a radius of 10.

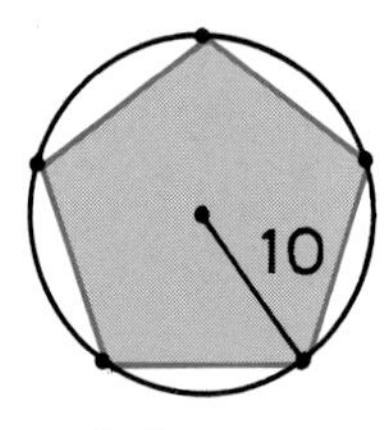

Polygon A
$n = 5$

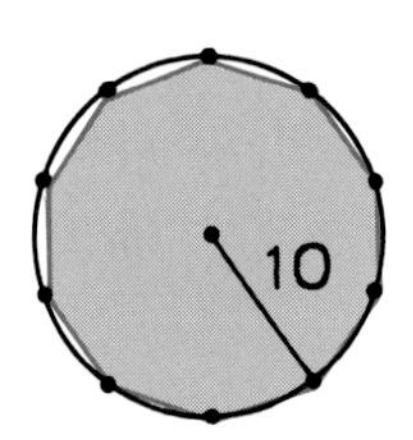

Polygon B
$n = 10$

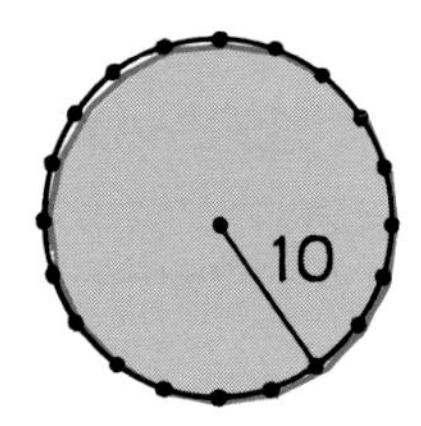

Polygon C
$n = 20$

Polygon D
$n = 40$

18. Copy and complete the following table.

Polygon	n	$\frac{180}{n}$	$\sin\frac{180}{n}\cos\frac{180}{n}$	$n\sin\frac{180}{n}\cos\frac{180}{n}$ *	Mr^2
A	5			2.38	
B					
C					
D			.0782		

*Round each number in this column to the nearest hundredth.

Refer to your table for Exercise 18 to tell what happens to each of the following quantities as n increases.

Example: $\frac{180}{n}$.

Answer: $\frac{180}{n}$ decreases.

19. $\sin\frac{180}{n}\cos\frac{180}{n}$.

20. $n\sin\frac{180}{n}\cos\frac{180}{n}$.

21. Mr^2.

Which one of the following statements about the area of a regular polygon is correct?

22. As the number of sides of a regular polygon inscribed in a circle is repeatedly doubled,
    a) the area is also repeatedly doubled.
    b) the area increases by equal amounts.
    c) the area increases by successively smaller amounts.

Use Theorem 86 to find the areas of the following polygons. *Round each answer to the nearest integer.*

23. A regular polygon having 18 sides with radius 5.

24. A regular polygon having 18 sides with radius 10.

25. A regular polygon having 180 sides with radius 5.

Refer to your answers to Exercises 23 through 25 to answer the following questions.

26. Which seems to have a greater effect on the area of a regular polygon: increasing its radius or increasing the number of its sides?

27. Which one of the following regular polygons do you think has the greatest area?
    a) A polygon with 100 sides and radius 7.
    b) A polygon with 100 sides and radius 21.
    c) A polygon with 300 sides and radius 7.

In the figure below, PHANTM is a regular hexagon with center O; OH = 8. Find each

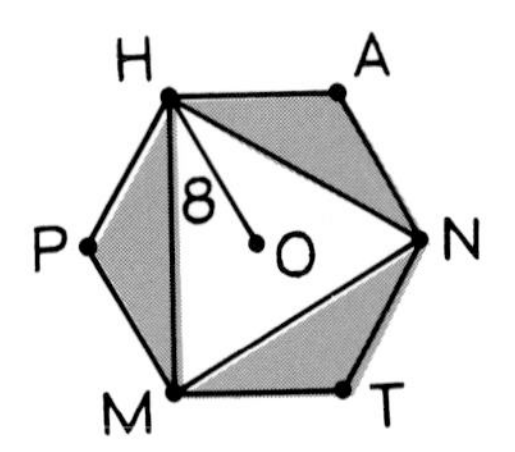

of the following quantities. Round each answer to the nearest tenth.

28. $\rho\triangle$HNM.

29. ρPHANTM.

30. $\rho\triangle$HAN.

31. $\alpha\triangle$HNM.

32. αPHANTM.

33. $\alpha\triangle$HAN.

Set III

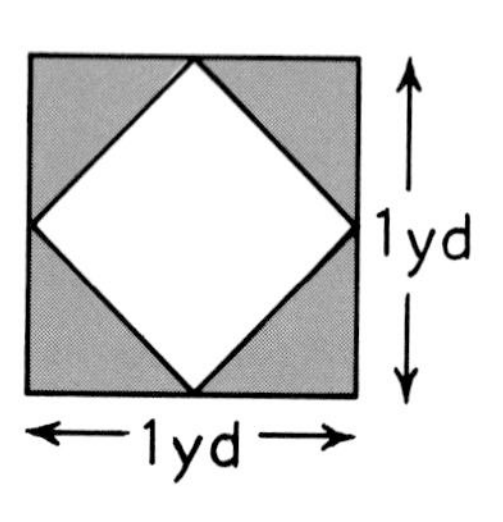

There is an old puzzle about a man who had a window that was a yard square. He decided that it let in too much light, so he boarded up half of it and still had a square window a yard high and a yard wide. The puzzle is to figure out how he did it and the answer is shown here. As you can see, the man formed a smaller square by joining the midpoints of the sides of the original square.

Now suppose that the original window was in the shape of a regular hexagon and that it was boarded up in the same way to form a smaller hexagon. Can you figure out how the amount of light let in by the smaller of the hexagonal windows would compare with that let in by the larger? (Hint: Draw the figure. Draw a radius of the large hexagon and a radius of the small hexagon so that a 30°-60° right triangle is formed.)

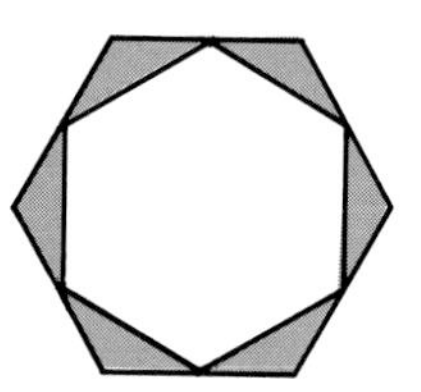

Photograph from the exhibition Mathematica designed by Charles and Ray Eames for IBM, reprinted with permission

Lesson 4
Limits

This photograph of a boy holding a mirror was taken by pointing the camera toward another mirror facing him. As a result, his image is reflected back and forth between the two mirrors a seemingly infinite number of times.

In the picture, each image is half the height of the previous one so that, if the first image is 1 unit tall, the successive heights of the images are

$$1, \frac{1}{2}, \frac{1}{4}, \frac{1}{8}, \frac{1}{16}, \text{ and so on.}$$

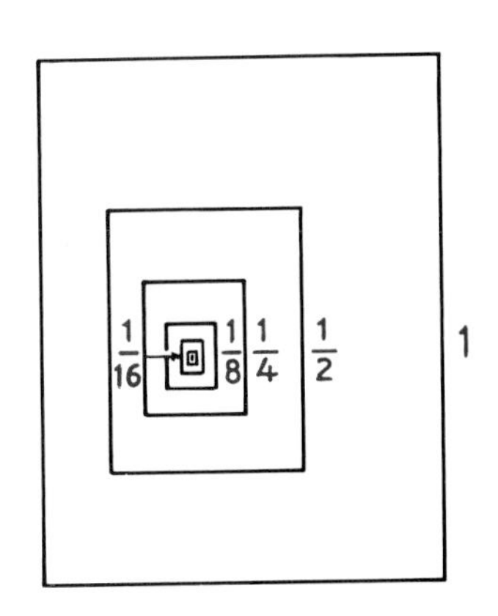

The drawing represents the images as a sequence of similar rectangles. After a certain point, it is difficult to draw any more because the rectangles become so small. The further along we go, the closer and closer their heights approach zero. The tenth rectangle, for example, has a height of $\frac{1}{512}$ or about 0.002 unit. The hundredth rectangle has a height of about

$$0.000000000000000000000000000002 \text{ unit.}$$

The numbers in a sequence are called its *terms*. The first five terms in the sequence of heights of the rectangles are

$$1, \frac{1}{2}, \frac{1}{4}, \frac{1}{8}, \frac{1}{16}.$$

If each successive term in this sequence is half the previous term, we will never get to the number zero. However, if we go far enough, we can find a term that is as *close to zero* as we wish. Because of this, we say that the terms in this sequence approach *zero as a limit.*

If we represent the height of the nth rectangle in the sequence as h_n, then as n, the number of the rectangle, gets larger and larger, h_n approaches zero. In symbols, this is written

$$\lim_{n \to \infty} h_n = 0.$$

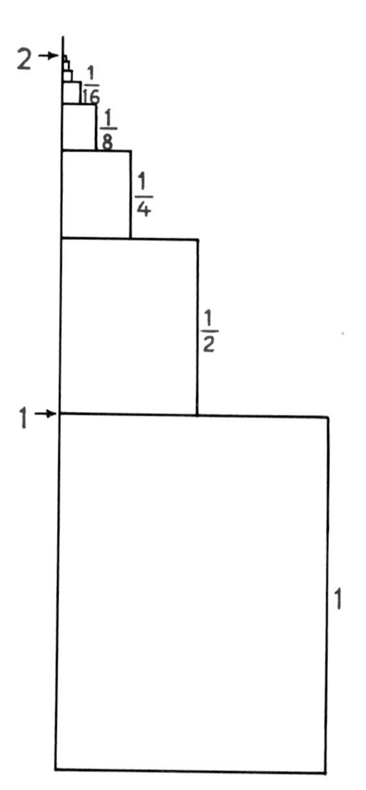

Another example of a limit consists of arranging the rectangles in a vertical stack as shown here. If "all" of the rectangles were included, how tall would the stack be? From the figure, the answer seems to be 2, and the list below seems to confirm this.

The height of the stack of

the first rectangle $= 1$

the first two rectangles $= 1 + \frac{1}{2} = 1\frac{1}{2}$

the first three rectangles $= 1 + \frac{1}{2} + \frac{1}{4} = 1\frac{3}{4}$

the first four rectangles $= 1 + \frac{1}{2} + \frac{1}{4} + \frac{1}{8} = 1\frac{7}{8}$

the first five rectangles $= 1 + \frac{1}{2} + \frac{1}{4} + \frac{1}{8} + \frac{1}{16} = 1\frac{15}{16}.$

A stack of the first ten rectangles would be about 1.998 units high, whereas a stack of the first hundred rectangles would be about 1.9999999999999999999999999998 units high!

The first five terms in the sequence of sums of the heights of the rectangles are

$$1,\ 1\frac{1}{2},\ 1\frac{3}{4},\ 1\frac{7}{8},\ 1\frac{15}{16}.$$

The height of a finite stack of rectangles will never actually be 2, but if we include enough of them in the stack, we can make its height *as close to* 2 as we wish. So the height of the stack *approaches 2 as a limit.* If the height of the stack of n rectangles is represented as H_n, we have

$$\lim_{n \to \infty} H_n = 2.$$

Because a mathematically precise definition of the word *limit* would be quite difficult to understand, an informal explanation is given instead. The two examples that you have just seen should help you understand it.

An Informal Explanation of "Limit"

Let $a_1, a_2, a_3, \ldots, a_n, \ldots$ be a sequence of numbers. Then the number L is the ***limit*** of this sequence if, by letting n get sufficiently large, the successive differences between L and a_n can be made as small as we wish.

Exercises

Set I

Consider the sequence $\frac{1}{1}, \frac{1}{2}, \frac{1}{3}, \frac{1}{4}, \frac{1}{5}, \ldots$

1. What is the sixth term of this sequence?
2. What is the nth term of this sequence?
3. As n gets larger and larger, what happens to the terms of this sequence?
4. What number do you think is the limit of this sequence?

Consider the sequence 3, 6, 9, 12, 15, . . .

5. What is the sixth term of this sequence?
6. What is the nth term of this sequence?
7. As n gets larger and larger, what happens to the terms of this sequence?
8. Do you think that this sequence has a limit?

Consider the sequence 0.3, 0.33, 0.333, . . .

9. What is the seventh term of this sequence?
10. What would you have to write to express the nth term of this sequence?
11. As n gets larger and larger, what happens to the terms of this sequence?
12. What number do you think is the limit of this sequence?

Write each of the following statements in symbols.

Example: The limit of the sequence whose nth term is $\frac{1}{n}$ is 0.

Answer: $\lim_{n \to \infty} \frac{1}{n} = 0.$

13. The limit of the sequence whose nth term is $\frac{n+5}{2n}$ is $\frac{1}{2}$.
14. The limit of the sequence whose nth term is $\frac{1-n}{n+4}$ is -1.

Consider the sequence whose nth term is $\frac{3n}{n+1}$. Express each of the following terms of this sequence both as a common fraction and in decimal form.

Example: The third term.

Answer: Letting $n = 3$, we get

$$\frac{3(3)}{3+1} = \frac{9}{4} = 2.25.$$

15. The fourth term.
16. The ninth term.
17. The 99th term.
18. The 999th term.
19. What number do you think is equal to $\lim_{n \to \infty} \frac{3n}{n+1}$?

Consider the sequence whose nth term is $\frac{5-n}{5n}$. Express each of the following terms of this sequence both as a common fraction and in decimal form.

20. The first term.
21. The fifth term.
22. The tenth term.

23. The 100th term.

24. The 1,000,000th term.

25. What number do you think is equal to $\lim_{n\to\infty} \frac{5 - n}{5n}$?

Consider the sequence whose first term is 2 and whose other terms are given by the expression $\frac{n^2 - 1}{n - 1}$. Find each of the following terms of this sequence.

26. The second term.

27. The third term.

28. The tenth term.

29. The 101st term.

30. Do you think that this sequence has a limit? If so, what is it?

Set II

In the formula for the perimeter of a regular polygon having n sides, N represents the product $n \sin \frac{180}{n}$.

31. Copy and complete the table below.

n	$\frac{180}{n}$	$\sin \frac{180}{n}$	$n \sin \frac{180}{n}$
20	▯▯▯	0.15643	▯▯▯
60	▯▯▯	0.052336	▯▯▯
180	▯▯▯	0.017452	▯▯▯
540	▯▯▯	0.0058177	▯▯▯

The table suggests that, as n increases, the product $n \sin \frac{180}{n}$ approaches a limit.

32. What seems to be the value of this limit to the nearest hundredth?

33. Although $\frac{22}{7}$ is not equal to this limit, this fraction is sometimes used to approximate it. Express $\frac{22}{7}$ as a decimal to the nearest hundredth to show why.

34. The number that is the value of $\lim_{n\to\infty} n \sin \frac{180}{n}$ has a name. The name is also the name of a letter of the Greek alphabet. What do you think it is?

In the formula for the area of a regular polygon having n sides, M represents the product $n \sin \frac{180}{n} \cos \frac{180}{n}$.

n	$n \sin \frac{180}{n}$	$\cos \frac{180}{n}$	$n \sin \frac{180}{n} \cos \frac{180}{n}$
90	3.1409546 . . .	0.9993908 . . .	3.1390413 . . .
180	3.1414331 . . .	0.9998476 . . .	3.1409547 . . .
360	3.1415527 . . .	0.9999619 . . .	3.1414331 . . .
720	3.1415826 . . .	0.9999904 . . .	3.1415527 . . .
1440	3.1415901 . . .	0.9999976 . . .	3.1415826 . . .
2880	3.1415919 . . .	0.9999994 . . .	3.1415901 . . .

35. What seems to be the value of $\lim_{n \to \infty} \cos \frac{180}{n}$?

36. What seems to be the value of $\lim_{n \to \infty} n \sin \frac{180}{n} \cos \frac{180}{n}$?

Set III

This photograph seems to show an infinite sequence of elephants. Suppose that the biggest elephant is 10 feet tall and that each succeeding elephant is half as tall as the one before. If, starting with the 10-foot elephant, there were an infinite number of such smaller elephants and each stood on the back of the next larger one, do you think that the stack of elephants would be infinitely tall or would it have a finite height?

Explain your reasoning. (If you think the height of the stack has a limit, what do you think it would be?)

Photograph by Seidenstücker in *Laughing Camera for Children*, Hanns Reich Verlag

By the permission of Johnny Hart and News America Syndicate

Lesson 5

The Circumference and Area of a Circle

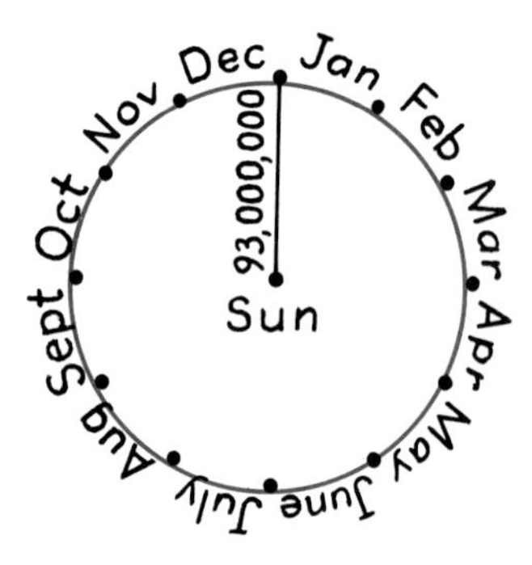

Earth's orbit

The earth travels around the sun in an orbit that is very close to being circular. The radius of the orbit is 93,000,000 miles and, as the diagram at the left illustrates, it takes one year for the earth to move once around it. How far does the earth travel during this time?

To answer this question, we need a way to measure the "distance around a circle." This distance is called the *circumference* of the circle. The sequence of figures below suggests a way of defining it.

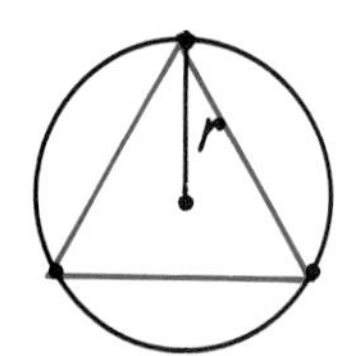

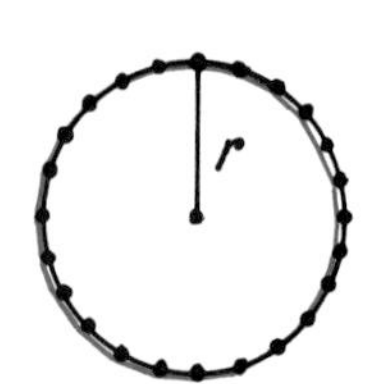

As the number of sides of a regular polygon inscribed in a circle increases, the polygon looks more and more like the circle itself. The perimeter of the polygon apparently becomes a better and better approximation of the circumference of the circle.

► **Definition**
The ***circumference*** of a circle is the limit of the perimeters of the inscribed regular polygons.

A table for the perimeters of the polygons on the preceding page is given here.

n	$p = 2Nr$
3	$2\left(3 \sin \frac{180}{3}\right)r = 2(2.59807\ldots)r$
6	$2\left(6 \sin \frac{180}{6}\right)r = 2(3.00000\ldots)r$
12	$2\left(12 \sin \frac{180}{12}\right)r = 2(3.10582\ldots)r$
24	$2\left(24 \sin \frac{180}{24}\right)r = 2(3.13262\ldots)r$

Continuing this table with some larger values of n suggests that, as n increases, N approaches some number as its limit.

n	$p = 2Nr$
100	$2\left(100 \sin \frac{180}{100}\right)r = 2(3.14107\ldots)r$
1000	$2\left(1000 \sin \frac{180}{1000}\right)r = 2(3.14158\ldots)r$
10000	$2\left(10000 \sin \frac{180}{10000}\right)r = 2(3.14159\ldots)r$

It can be proved that this is true and that the limit is the number π:

$$3.1415926536\ldots.$$

Because the circumference of a circle is the limit of the perimeters of the inscribed regular polygons, we get the following theorem.

► **Theorem 87**
The circumference of a circle is $2\pi r$, in which r is the length of its radius.

A formula for the area of a circle can be obtained in a similar way. The figures below suggest that, as the number of sides of a regular polygon inscribed in a circle increases, the area of the polygon becomes a better and better approximation of the area of the circle.

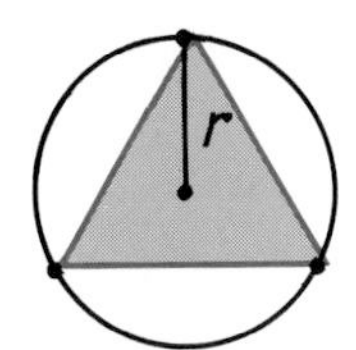

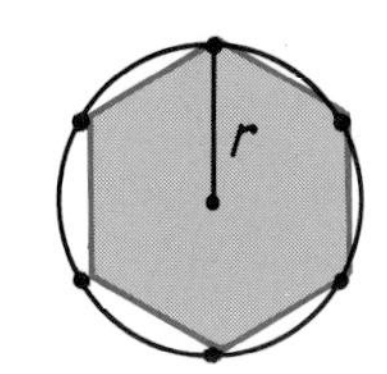

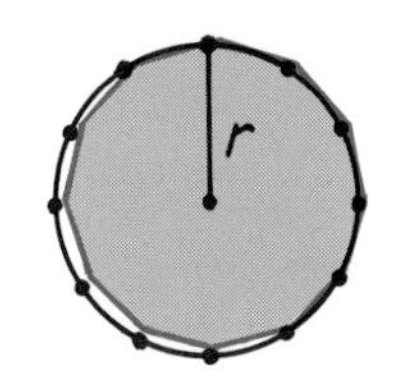

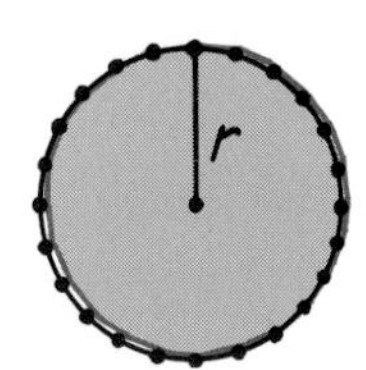

► **Definition**
The ***area*** of a circle is the limit of the areas of the inscribed regular polygons.

A table for the areas of the polygons at the bottom of page 521 is given here.

n	$\alpha = Mr^2$
3	$\left(3 \sin \frac{180}{3} \cos \frac{180}{3}\right)r^2 = (1.29903\ldots)r^2$
6	$\left(6 \sin \frac{180}{6} \cos \frac{180}{6}\right)r^2 = (2.59807\ldots)r^2$
12	$\left(12 \sin \frac{180}{12} \cos \frac{180}{12}\right)r^2 = (3.00000\ldots)r^2$
24	$\left(24 \sin \frac{180}{24} \cos \frac{180}{24}\right)r^2 = (3.10582\ldots)r^2$

Continuing this table with some larger values of n suggests that as n increases, M approaches some number as its limit.

n	$\alpha = Mr^2$
100	$\left(100 \sin \frac{180}{100} \cos \frac{180}{100}\right)r^2 = (3.13952\ldots)r^2$
1000	$\left(1000 \sin \frac{180}{1000} \cos \frac{180}{1000}\right)r^2 = (3.14157\ldots)r^2$
10000	$\left(10000 \sin \frac{180}{10000} \cos \frac{180}{10000}\right)r^2 = (3.14159\ldots)r^2$

It can be proved that this is true and that the limit is again the number π. Because the area of a circle is the limit of the areas of the inscribed regular polygons, we get the following theorem.

► **Theorem 88**
The area of a circle is πr^2, in which r is the length of its radius.

The number π, which relates both the circumference and the area of a circle to its radius, is irrational, and so its decimal form neither ends nor repeats. Its value has been calculated on modern computers to more than 29,000,000 decimal places! Rounded to 10 places, it is:

$$3.1415926536.$$

Exercises

Set I

As the number of sides of a regular polygon inscribed in a circle increases,

1. what measurement of the circle do the perimeters of the polygons approach as a limit?
2. what measurement of the circle do the areas of the polygons approach as a limit?

Use the indicated lengths to find the circumference and area of each of the following circles. Leave your answers in terms of π.

Example:

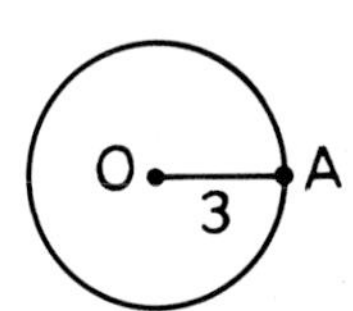

$\overline{OA}$ is a radius.

Answer: $c = 2\pi r$, and so $c = 2\pi(3) = 6\pi$; $a = \pi r^2$, and so $a = \pi(3)^2 = 9\pi$.

3.

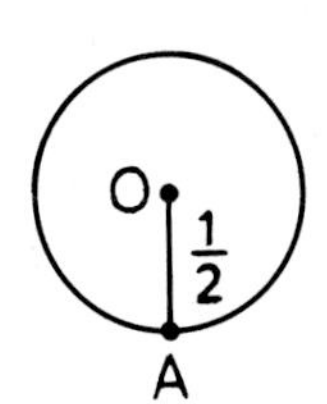

$\overline{OA}$ is a radius.

4.

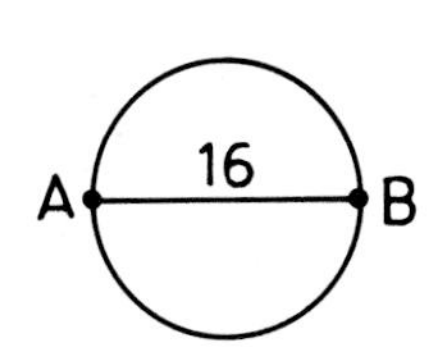

$\overline{AB}$ is a diameter.

The following description from the Bible of a circular pool in Solomon's temple suggests a very simple approximation of π.

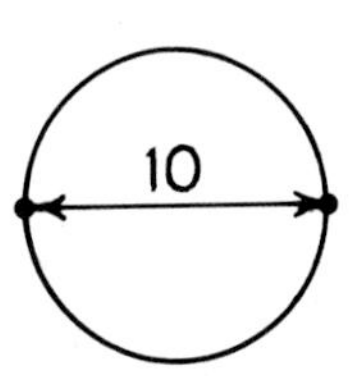

"Also he made a molten sea of ten cubits from brim to brim . . . and a line of thirty cubits did compass it round about."*

5. What was the circumference of this pool?
6. What was its diameter?
7. What was its radius?
8. What approximation of π follows from these numbers?

The formula $c = 2\pi r$ relates the circumference of a circle to its radius.

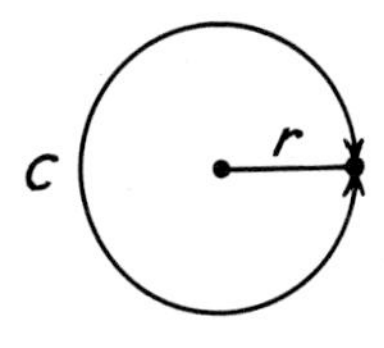

9. Write the equation that results from dividing both sides of $c = 2\pi r$ by $2r$.
10. What does the equation $d = 2r$ mean?
11. To what number is the ratio of the circumference of a circle to its diameter equal?

*II Chronicles 4:2.

The circles below have radii a and b.

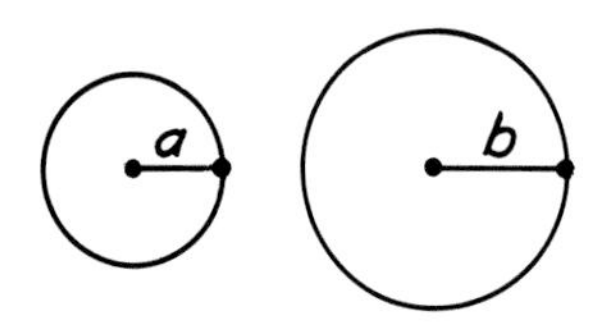

12. Write an expression for the circumference of each circle.
13. Show that the ratio of the circumferences of two circles is equal to the ratio of their radii.
14. Write an expression for the area of each circle.
15. Show that the ratio of the areas of two circles is equal to the square of the ratio of their radii.
16. If the radii of two circles are 6 and 8, find the ratio of their circumferences.
17. Find the ratio of their areas.

Set II

Numbers that are sometimes used as approximations of π are 3.14 and $3\frac{1}{7}$. To four decimal places, π is 3.1416.

18. Express $3\frac{1}{7}$ in decimal form, rounding your answer to four decimal places.
19. Which approximation of π is better: 3.14 or $3\frac{1}{7}$?
20. An especially good approximation of π is the fraction $\frac{355}{113}$. To how many decimal places does it give the correct value?

In the figure below, $\overline{IU}$ and $\overline{ME}$ are chords of circle T.

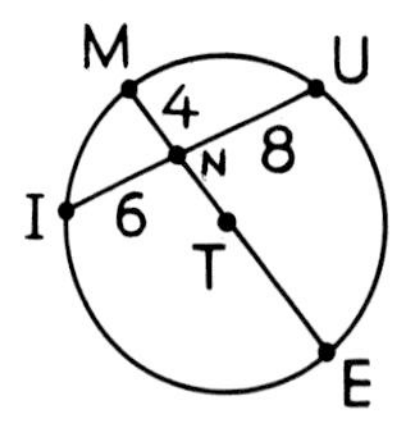

21. Find ME.
22. Find the area of the circle.

In the figure below, $\overline{SC}$ and $\overline{SD}$ are secant segments to circle N.

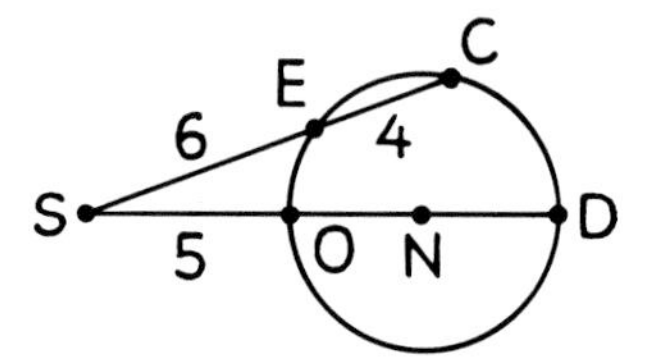

23. Find ND.
24. Find the circumference of the circle.

Find the *exact* area of the shaded region in each of the following figures. (This means leaving your answers in terms of π.)

25.

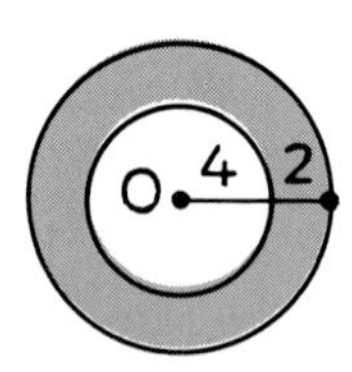

The two circles are concentric with center O.

26.

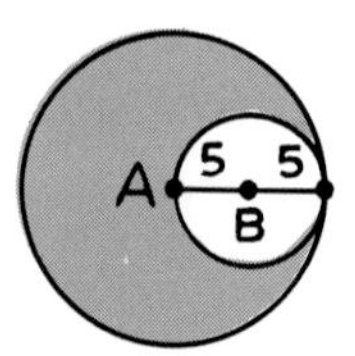

The centers of the circles are A and B.

27.

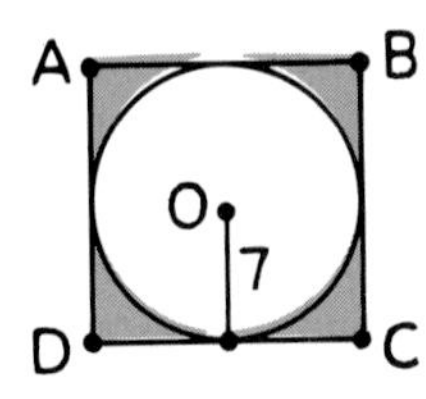

Circle O is inscribed in square ABCD.

28.

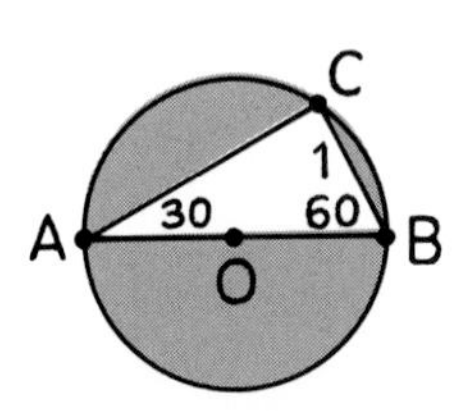

Right $\triangle ABC$ is inscribed in circle O.

29.

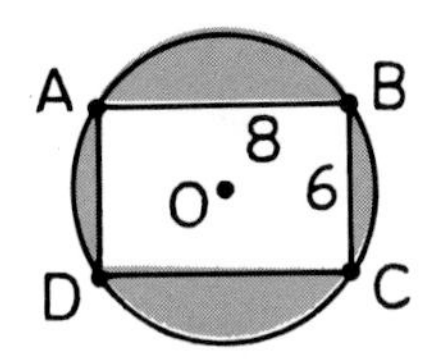

Rectangle ABCD is inscribed in circle O.

30.

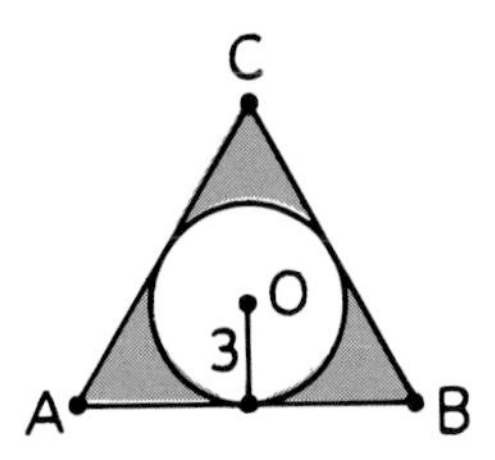

Circle O is inscribed in equilateral $\triangle ABC$.

The ancient Egyptians found the area of a circle by multiplying the square of its diameter by $\frac{64}{81}$.

31. What answer did they get when they used this method to find the area of a circle whose diameter was 18?
32. Find the correct area of the circle. Round your answer to the nearest integer.
33. Find the correct formula for the area of a circle in terms of its diameter.

The earth is about 93 million miles from the sun.

34. On the assumption that the earth's orbit is circular, find the approximate distance that the earth moves during the course of a year.
35. Approximately how far does the earth move each day in traveling around the sun?

Set III

Obtuse Ollie measured his circumference and found that it was 47 inches.

1. On the assumption that he is perfectly round in the middle, what is Ollie's approximate radius?
2. On the assumption that you are perfectly round in the middle, what is your approximate radius? Explain your reasoning.

Lesson 6

Sectors and Arcs

Drawing by Tom Henderson; Parade Magazine

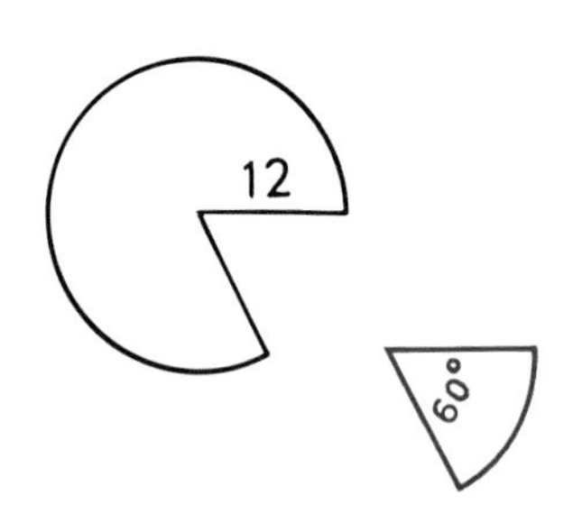

Suppose that a slice is cut from a large pizza as shown in the figure at the left. If the central angle of the slice is 60°, how does it compare in size with the pizza before it was cut?

Because 60° is one-sixth of 360°, the area of the slice would be one-sixth the area of the entire pizza. If the radius of the pizza is 12 inches, its area would be $\pi(12)^2$ or 144π square inches and the area of the slice would be $\frac{1}{6}(144\pi)$ or 24π square inches.

The slice is an example of a *sector* of a circle.

► **Definition**

A ***sector*** of a circle is a region bounded by an arc of the circle and the two radii to the endpoints of the arc.

The method that we used to find the area of the slice illustrates the following theorem, which is stated without proof.

▶ Theorem 89

The area of a sector whose arc has a measure of $m°$ is $\frac{m}{360}\pi r^2$, in which r is the radius of the circle.

As another example of how this theorem is used, consider the problem of finding the area of the rest of the pizza. It is also a sector, and its arc has a measure of 300°. Its area, then, is

$$\frac{300}{360}\pi(12)^2 = \frac{5}{6}(144\pi) = 120\pi \text{ in}^2.$$

Now that we have considered how the *area* of a sector can be determined, we will consider the problem of finding the *length of its arc*. For the pizza slice, we know that the *measure* of its arc is the same as the measure of its central angle: 60°. As we have already noted, this is one-sixth of 360°, the measure of the entire circle. Thus, it is reasonable that the *length* of the arc would be one-sixth of the length of the circle—in other words, one-sixth of its circumference. Because the circumference of the circle is

$$2\pi(12) \quad \text{or} \quad 24\pi \text{ in},$$

the length of the arc would be

$$\frac{1}{6}(24\pi) \quad \text{or} \quad 4\pi \text{ in}.$$

Again our method is stated as a theorem without proof.

▶ Theorem 90

The length of an arc whose measure is $m°$ is $\frac{m}{360}2\pi r$, in which r is the radius of the circle.

We will use the letter ℓ to denote the length of an arc in the same way that we use the letter m to denote its measure.

Exercises

Set I

In circle O, OG = 6 and $\angle$O = 30°.

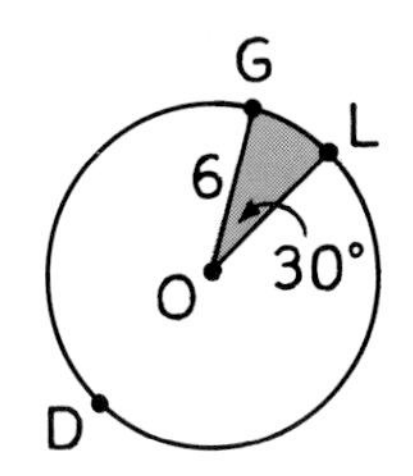

1. Find the circumference of the circle.
2. Find $m\overset{\frown}{GL}$.
3. Find $\ell\overset{\frown}{GL}$.
4. Find $m\overset{\frown}{GDL}$.
5. Find $\ell\overset{\frown}{GDL}$.
6. Find the area of the circle.
7. Find the area of the shaded sector.

In circle E, EA = 4 and $\angle E = 135°$.

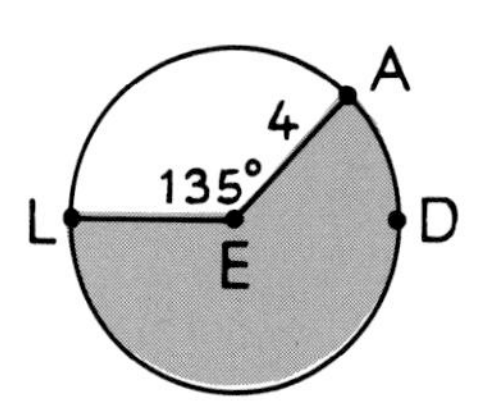

8. Find the circumference of the circle.
9. Find $m\widehat{LDA}$.
10. Find $\ell\widehat{LDA}$.
11. Find the area of the circle.
12. Find the area of the shaded sector.

The radii and measures of these arcs are given in the table below.

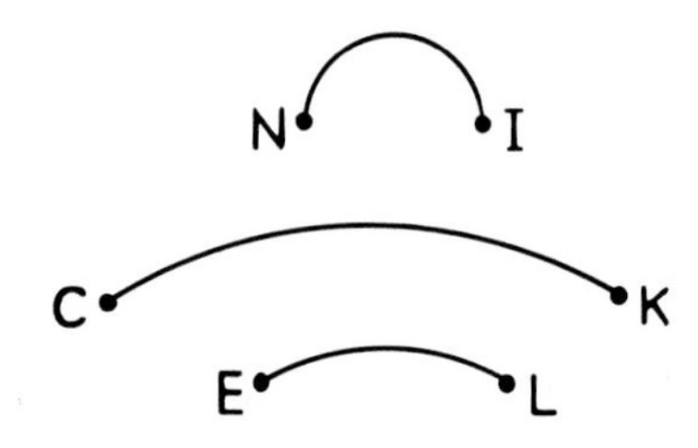

Arc	*Radius of arc*	*Measure of arc*
$\widehat{NI}$	1	180°
$\widehat{CK}$	6	60°
$\widehat{EL}$	3	60°

13. Two of the arcs have equal lengths. Which do they appear to be?
14. Find the lengths of all three arcs.
15. If two arcs have equal measures, does it follow that they have equal lengths?
16. If two arcs have equal lengths, does it follow that they have equal measures?

Obtuse Ollie baked Acute Alice a pie for her birthday. Alice ate the shaded sector shown in the diagram and Ollie ate the rest.

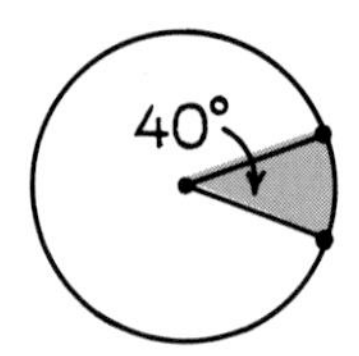

The diameter of the pie was 10 inches.

17. Find the area of the part Alice ate. Express your answer to the nearest square inch.
18. Find the area of the part Ollie ate. Express your answer to the nearest square inch.

Set II

Semicircles have been drawn with the midpoints of two sides of rectangle COAL as their centers; CO = 15 and CL = 8.

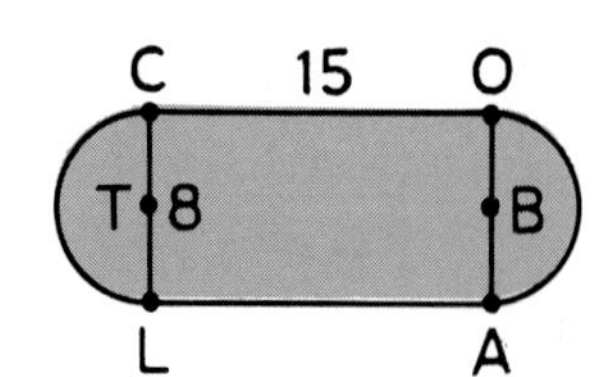

19. Find the exact length of the outer border of the figure.
20. Find the approximate length of the outer border. Express your answer to the nearest integer.
21. Find the exact area of the figure.
22. Find the approximate area of the figure. Express your answer to the nearest integer.

The shaded region in the figure below is called a *segment* of the circle. Its area can be

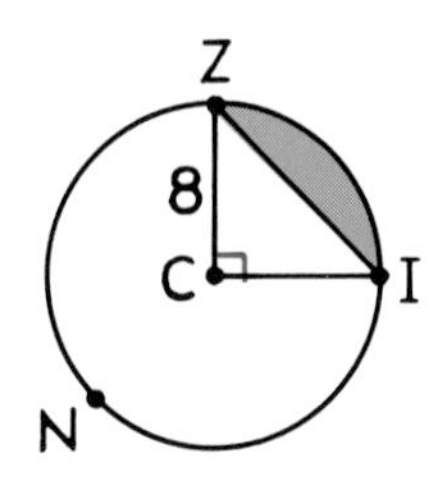

found by subtracting the area of $\triangle ZCI$ from the area of sector ZCI.

23. Find the exact area of the circle.
24. Find the exact area of sector ZCI.
25. Find the exact area of $\triangle ZCI$.
26. Find the exact area of the shaded region.
27. Find the exact area of the region bounded by $\overline{ZI}$ and $\widehat{ZNI}$.

Find the area of the shaded region in each of the following figures.

28.

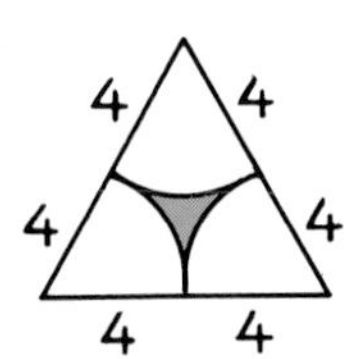

The arcs have their centers at the corners of the triangle.

29.

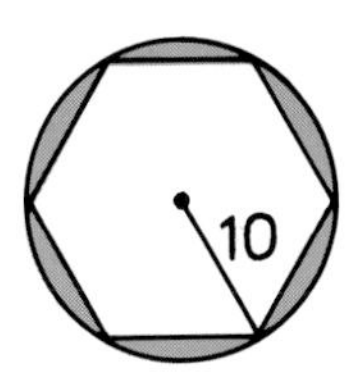

A regular hexagon is inscribed in the circle.

30.

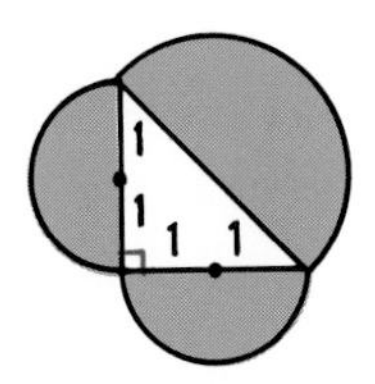

Semicircles are drawn on the sides of an isosceles right triangle.

31.

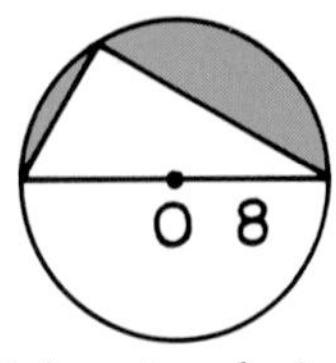

A 30°-60° right triangle is inscribed in the circle.

32.

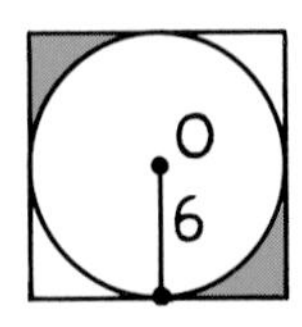

The circle is inscribed in the square.

33.

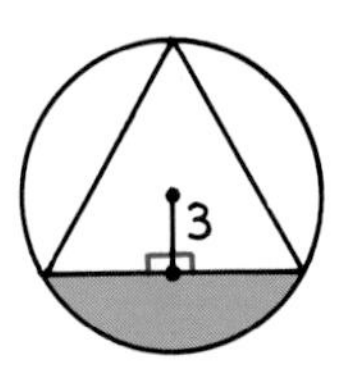

An equilateral triangle is inscribed in the circle.

34.

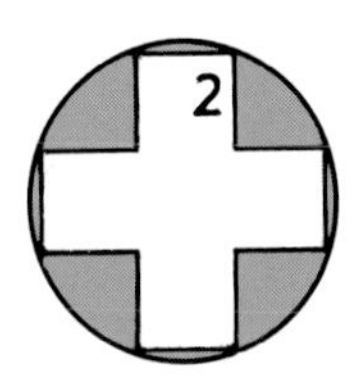

The cross is equilateral.

35.

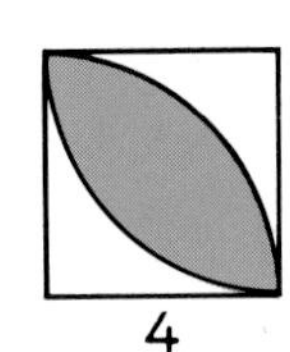

The arcs have their centers at the opposite corners of the square.

Set III

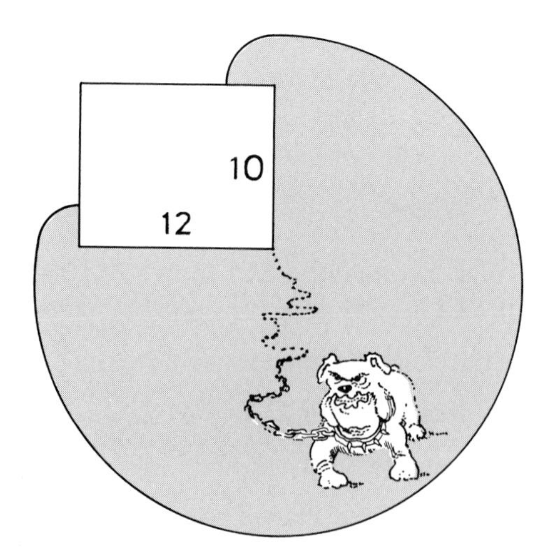

Dilcue tied his pet bulldog with a rope to one corner of a shed 12 feet long and 10 feet wide. If the rope is 15 feet long, can you figure out the approximate area within biting distance? Express your answer to the nearest square foot.

Chapter 14 / Summary and Review

Basic Ideas

Theorems

84. Every regular polygon is cyclic. 499

85. The perimeter of a regular polygon having n sides is $2Nr$, in which r is the length of its radius and

$$N = n \sin \frac{180}{n}. \quad 505$$

86. The area of a regular polygon having n sides is Mr^2, in which r is the length of its radius and

$$M = n \sin \frac{180}{n} \cos \frac{180}{n}. \quad 511$$

87. The circumference of a circle is $2\pi r$, in which r is the length of its radius. 521

88. The area of a circle is πr^2, in which r is the length of its radius. 522

89. The area of a sector whose arc has a measure of $m°$ is $\frac{m}{360}\pi r^2$, in which r is the radius of the circle. 527

90. The length of an arc whose measure is $m°$ is $\frac{m}{360}2\pi r$, in which r is the radius of the circle. 527

Exercises

Set I

Tell whether each of the following statements is true or false.

1. An apothem of a regular polygon bisects one of its sides.
2. A polygon is cyclic iff it is regular.
3. If the number of sides of a regular polygon inscribed in a circle is doubled, the perimeter of the polygon is also doubled.
4. Each central angle of a regular polygon having n sides has a measure of $\frac{360°}{n}$.
5. The ratio of the circumference to the diameter of a circle does not depend on the circle's size.
6. The length of a semicircle with radius r is πr.
7. The areas of two sectors of a circle have the same ratio as the measures of their arcs.
8. The area of a circle is $\frac{1}{4}\pi d^2$, in which d is the length of its diameter.

Polygon CHARLESTN is a regular nonagon with center O. Find the measure of each of the following angles.

9. $\angle$O.
10. $\angle$ONT.
11. $\angle$CNT.

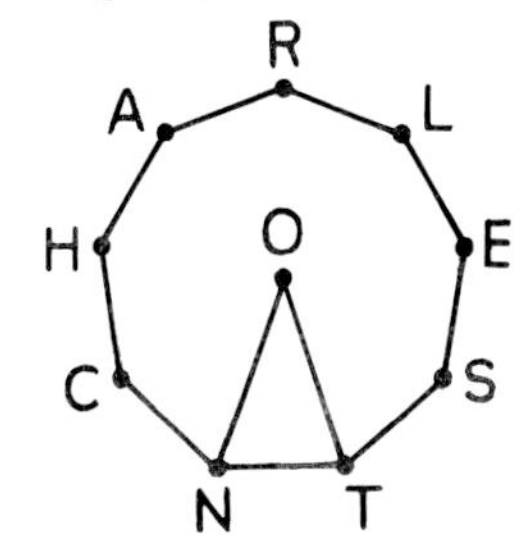

The nth term of a sequence is $\frac{2n + 1}{n}$.
Express each of the following terms of this sequence both as a common fraction and in decimal form.

12. The fifth term.
13. The tenth term.
14. The 1000th term.
15. What do you think is the value of $\lim_{n \to \infty} \frac{2n + 1}{n}$?

Each of these regular polygons is inscribed in a circle with radius 10.

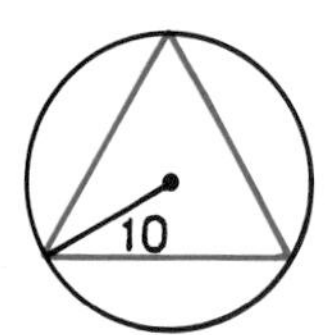

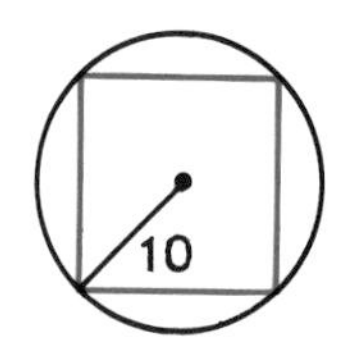

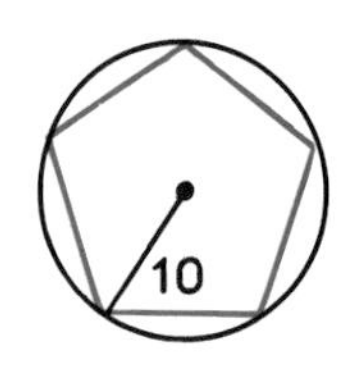

Use either your calculator or the table below to find each of the following numbers.

A	*sin A*	*sin A cos A*
36°	.5878	.4755
45°	.7071	.5000
60°	.8660	.4330

16. Find the perimeter of each polygon. Round each answer to the nearest integer.
17. Find the area of each polygon. Round each answer to the nearest integer.
18. What measurement of the circles do the perimeters of the polygons approach as a limit?
19. Find the exact value of this limit and its approximate value, rounded to the nearest integer.
20. What measurement of the circles do the areas of the polygons approach as a limit?
21. Find the exact value of this limit and its approximate value, rounded to the nearest integer.

Set II

A Japanese mathematician of the seventeenth century found the area of a circle by dividing it into rectangles as shown here. If the entire circle were divided into rectangles in this way, however, the rectangles would not fill the circle.

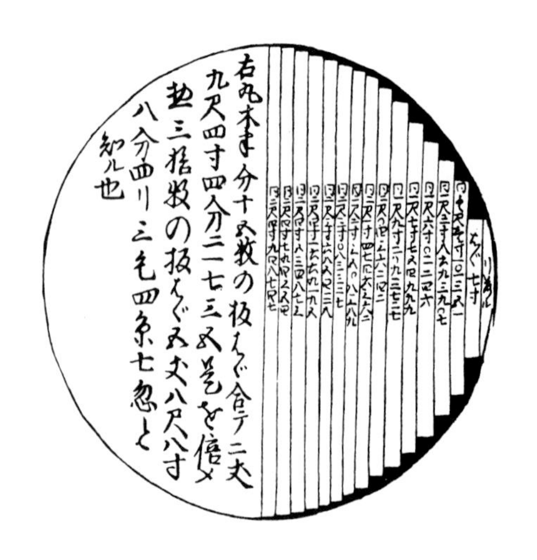

22. If n represents the number of rectangles and all the rectangles have the same width, what happens to this width as n increases?
23. How does the sum of the areas of the rectangles change as n increases?
24. If S_n is the sum of the areas of the rectangles, what does $\lim_{n \to \infty} S_n$ seem to be?

In the Olympics shot-put, the shot is thrown from a circle with a radius of 2.1 meters. The shot is thrown into a sector with a central angle of 45° and a radius of 23 meters.

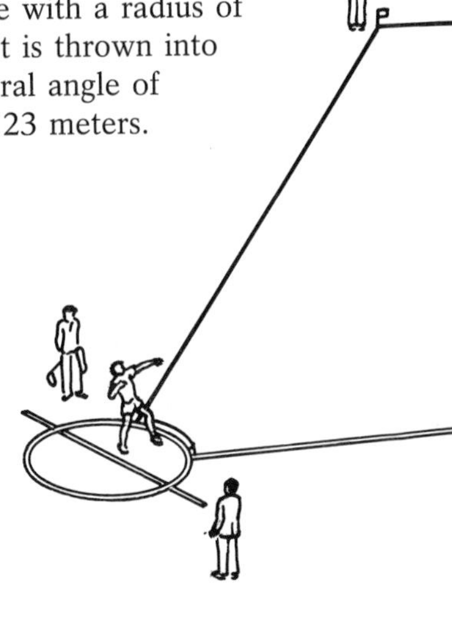

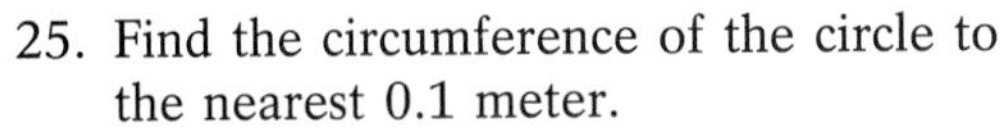

Richard Mackson/SPORTS ILLUSTRATED

25. Find the circumference of the circle to the nearest 0.1 meter.
26. Find the area of the circle to the nearest 0.1 square meter.
27. Find the length of the arc of the sector to the nearest 0.1 meter.
28. Find the area of the sector to the nearest 0.1 square meter.

Find the area of the shaded region in each of these figures.

29.

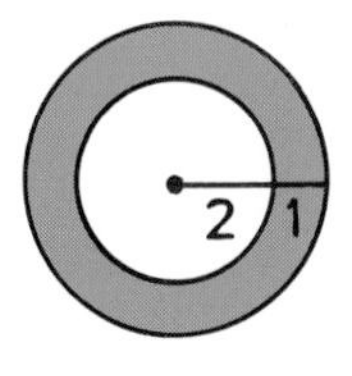

The circles are concentric.

30.

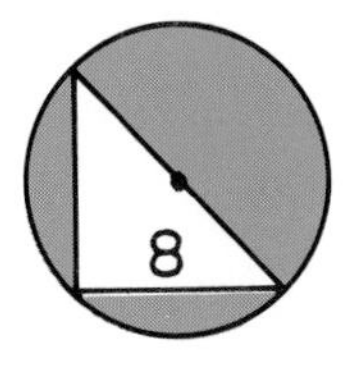

An isosceles right triangle is inscribed in the circle.

31.

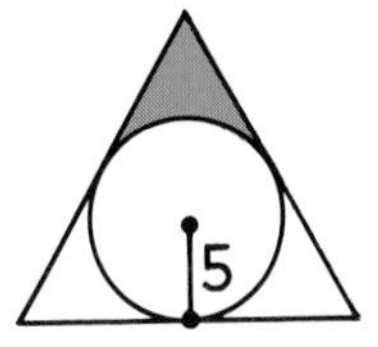

The circle is inscribed in an equilateral triangle.

32.

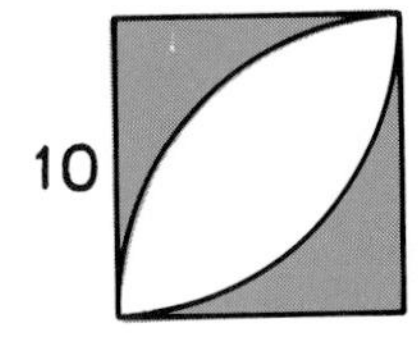

The arcs have their centers at the opposite corners of the square.

In the figure below, the radii of each circle are tangent segments to the other circle; $\angle ONE = 120°$ and $NO = 6$.

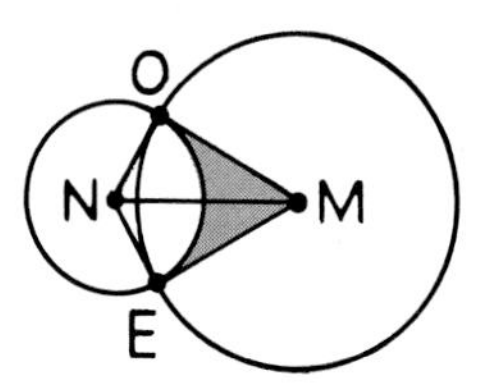

33. Find MO.

34. Find αNOME.

35. Find the area of the shaded region.

36. 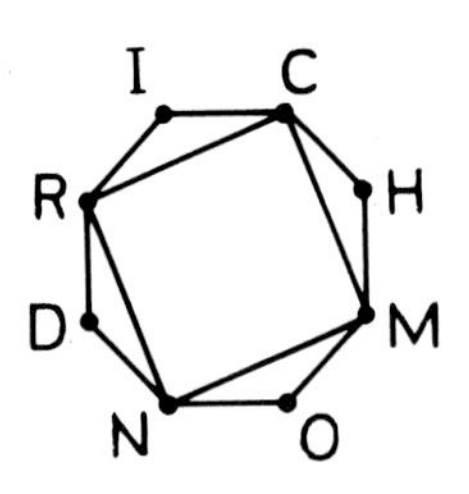

Given: RICHMOND is a regular octagon.

Prove: RCMN is a rhombus.

37. 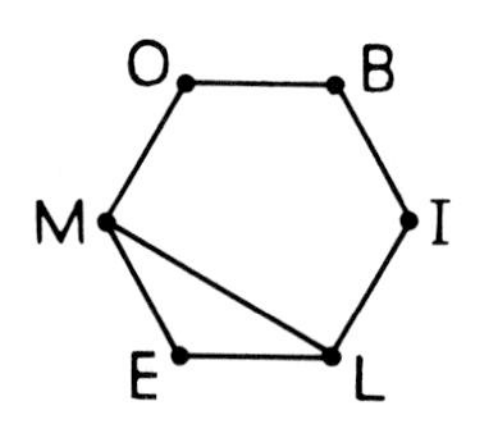

Given: Regular hexagon MOBILE with diagonal $\overline{ML}$.

Prove: $ML > ME$.

Chapter 15

GEOMETRIC SOLIDS

Lesson 1
Lines and Planes in Space

Escher Foundation, Haags Gemeentemuseum, The Hague

The staircase on the roof of the building in this picture is one that can exist only in the imagination. The picture, by Maurits Escher, is titled *Ascending and Descending* and the people walking the stairs are doing just that. But those on the outside are *always* ascending, whereas those on the inside are *always* descending! Two of them, one watching from a lower level and the other sitting on the stairs at the bottom, see the futility of it all and refuse to join the rest.

Escher's picture is a complex perspective drawing that illustrates a number of special line and plane relations in space. In this lesson, we will study some of these relations.

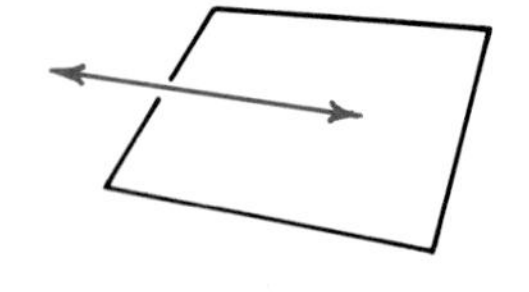

First, look at the line of the railing that the person at the lower left is leaning against. It is *parallel* to the plane of the ground. The diagram at the left shows this line-plane relation. Remember that a plane, being infinite in extent, cannot be represented in its entirety in a picture. Instead, we draw a rectangle (usually a perspective view of one) that lies in the plane.

► **Definition**
A ***line and a plane are parallel*** iff they do not intersect.

The lines containing the vertical edges of the building, on the other hand, are *perpendicular* to the plane of the ground.

We cannot say that a line is perpendicular to a plane if they form right angles, because, as the first figure below shows, there *are* no angles. However, the line *does* form right angles with lines in the plane that

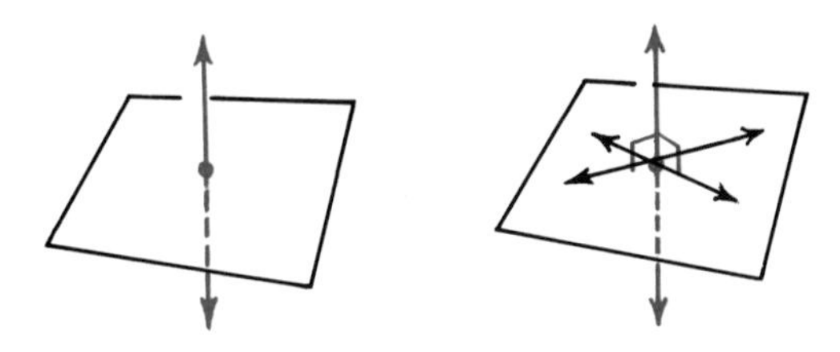

pass through the point of intersection, and we can base our definition on this fact.

► **Definition**
A ***line and a plane are perpendicular*** iff they intersect and the line is perpendicular to every line in the plane that passes through the point of intersection.

The walls of the building that face us lie in *parallel planes*. The figure below illustrates two of these planes.

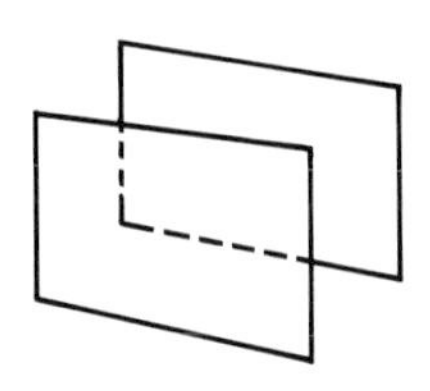

► **Definition**
Two ***planes are parallel*** iff they do not intersect.

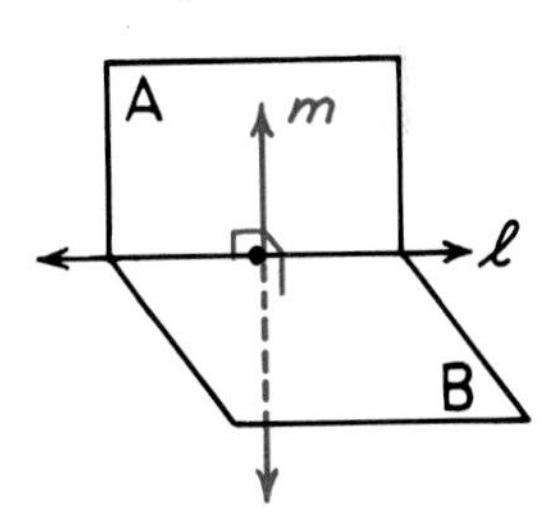

We will assume that, if two planes do intersect, their intersection is a line. The floor and one of the walls of Escher's building lie in intersecting planes that are *perpendicular* to each other. In the figure at the right, these planes are represented as A and B and the line in which they intersect as ℓ. If a line m is drawn in plane A so that it is perpendicular to line ℓ, it will also be perpendicular to plane B. It is on this basis that we define *perpendicular planes*.

► **Definition**
Two ***planes are perpendicular*** iff one plane contains a line that is perpendicular to the other plane.

We will use the term "oblique" to refer to a line and a plane, or two planes, that intersect without being perpendicular to each other.

► **Definition**
A ***line and a plane (or two planes) are oblique*** iff they are neither parallel nor perpendicular.

Exercises

Set I

Some of the planes and lines containing parts of the "impossible staircase" shown below have been named with capital and small letters, respectively. Refer to them in answering the following questions.

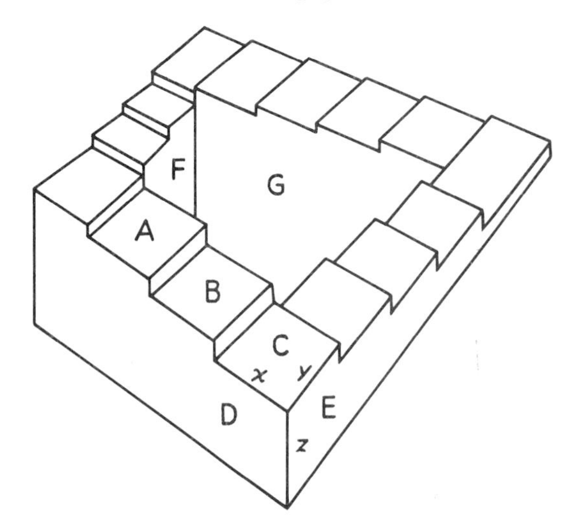

What relation does each of the following pairs of planes appear to have?

1. A and B.
2. C and E.
3. D and G.
4. D and F.

What relation do each of the following lines and planes appear to have?

5. x and E.
6. y and B.
7. z and F.
8. z and G.
9. y and G.

In the figure below, lines ℓ and m intersect plane A in point P. Lines n and o lie in plane A.

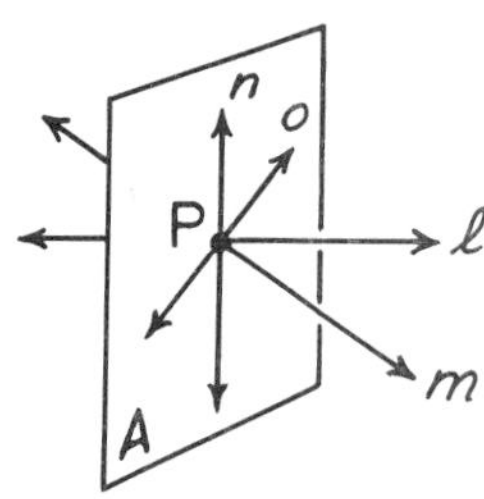

10. If $\ell \perp n$, does it follow that $\ell \perp$ A?
11. If $\ell \perp$ A, does it follow that $\ell \perp o$?
12. Can n and o both be perpendicular to ℓ?
13. If m is not perpendicular to A, can m be perpendicular to o?
14. If m is not perpendicular to A, what word names their relation?

In the figure below, $\overleftrightarrow{GE} \perp$ B and $\overrightarrow{UR}$, $\overrightarrow{UL}$, and $\overrightarrow{UM}$ lie in the plane.

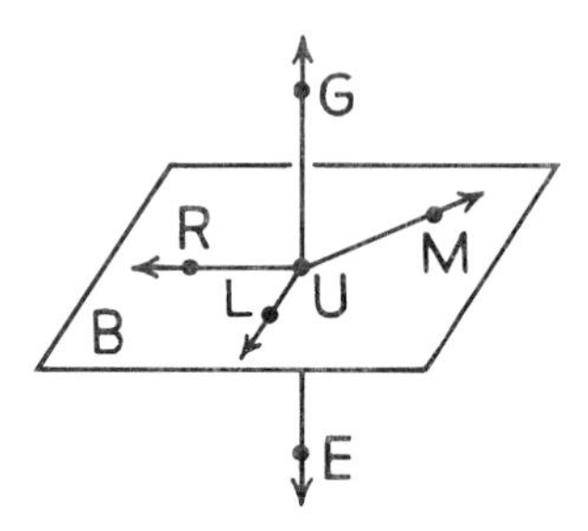

15. What can you conclude about $\overleftrightarrow{GE}$ and $\overrightarrow{UR}$, $\overleftrightarrow{GE}$ and $\overrightarrow{UL}$, and $\overleftrightarrow{GE}$ and $\overrightarrow{UM}$?
16. Name every angle in the figure that you know for certain to be a right angle.

In the figure below, plane O contains $\overleftrightarrow{WL}$ and $\overleftrightarrow{WL} \perp$ G. Plane G contains $\overrightarrow{WR}$.

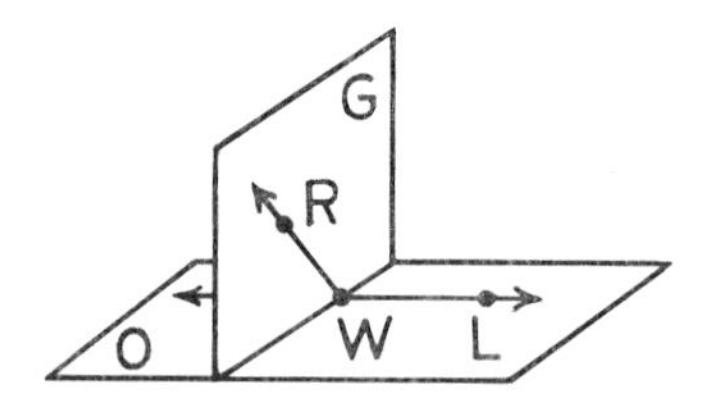

17. What relation do planes G and O have to each other?
18. How do you know?
19. What relation do $\overleftrightarrow{WL}$ and $\overrightarrow{WR}$ have to each other?
20. How do you know?

Set II

Tell whether each of the following statements is true or false. If you think that a statement is false, draw a diagram to illustrate why.

21. If two planes intersect, their intersection is a line.
22. If three planes intersect, their intersection is a single point.
23. If a line intersects a plane that does not contain it, then the line and plane intersect in exactly one point.
24. If a line does not intersect a plane, it is parallel to the plane.
25. If one plane contains a line that is perpendicular to another plane, then the planes are perpendicular.
26. If two lines are perpendicular to the same plane, the lines are parallel to each other.
27. If two planes are perpendicular to a third plane, the planes are parallel to each other.
28. If two planes are parallel to a third plane, they are parallel to each other.
29. If two lines lie in parallel planes, the lines are parallel to each other.
30. If a line is perpendicular to one of two parallel planes, it is perpendicular to the other.

31.

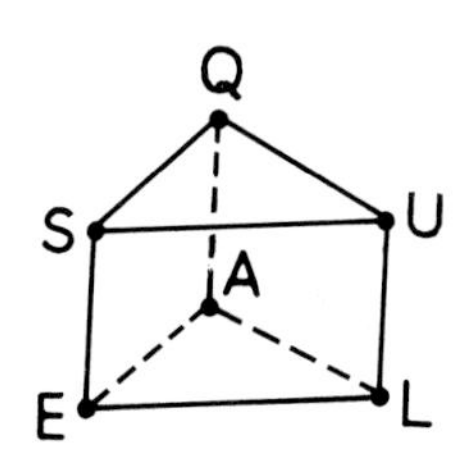

Given: SQAE, QULA, and SULE are parallelograms.

Prove: $\triangle SQU \cong \triangle EAL$.

32.

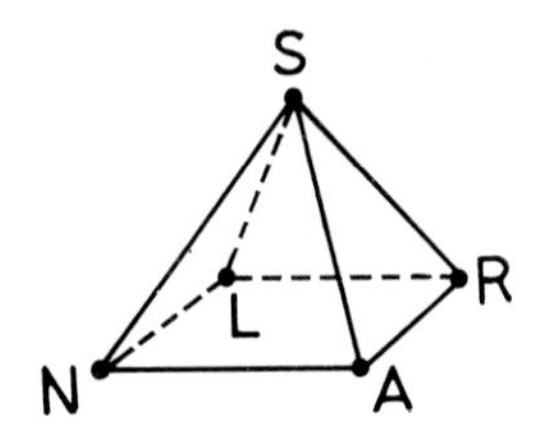

Given: $\triangle SNA \cong \triangle SAR \cong \triangle SRL \cong \triangle SLN$.

Prove: $\overline{LR} \parallel \overline{NA}$.

33.

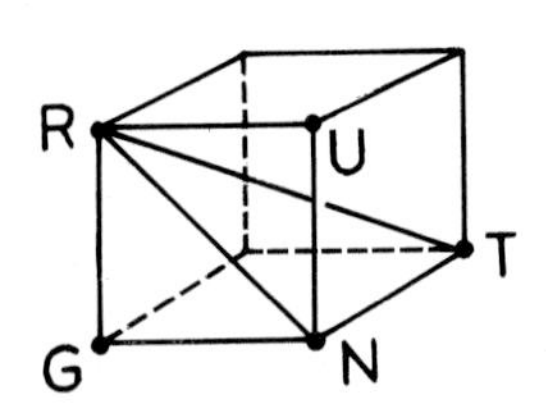

Given: $\overline{TN} \perp$ plane GRUN.

Prove: $\triangle RNT$ is a right triangle.

34.

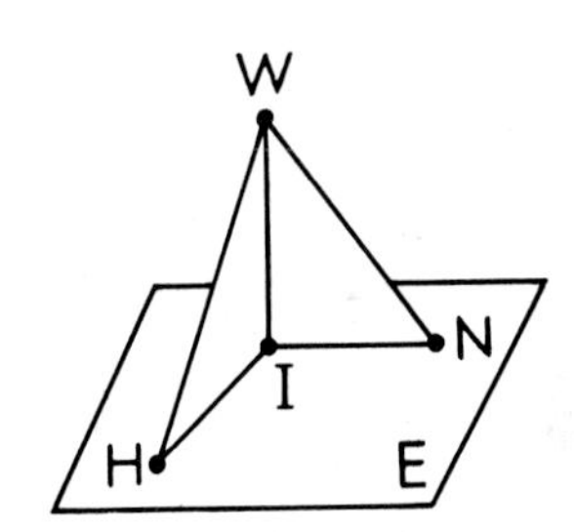

Given: $\overline{WI} \perp E$;
$WH = WN$.

Prove: $\angle H = \angle N$.

Set III

This picture by Escher, titled *Belvedere,* illustrates another impossible building. Can you explain why?

Lesson 2

Rectangular Solids

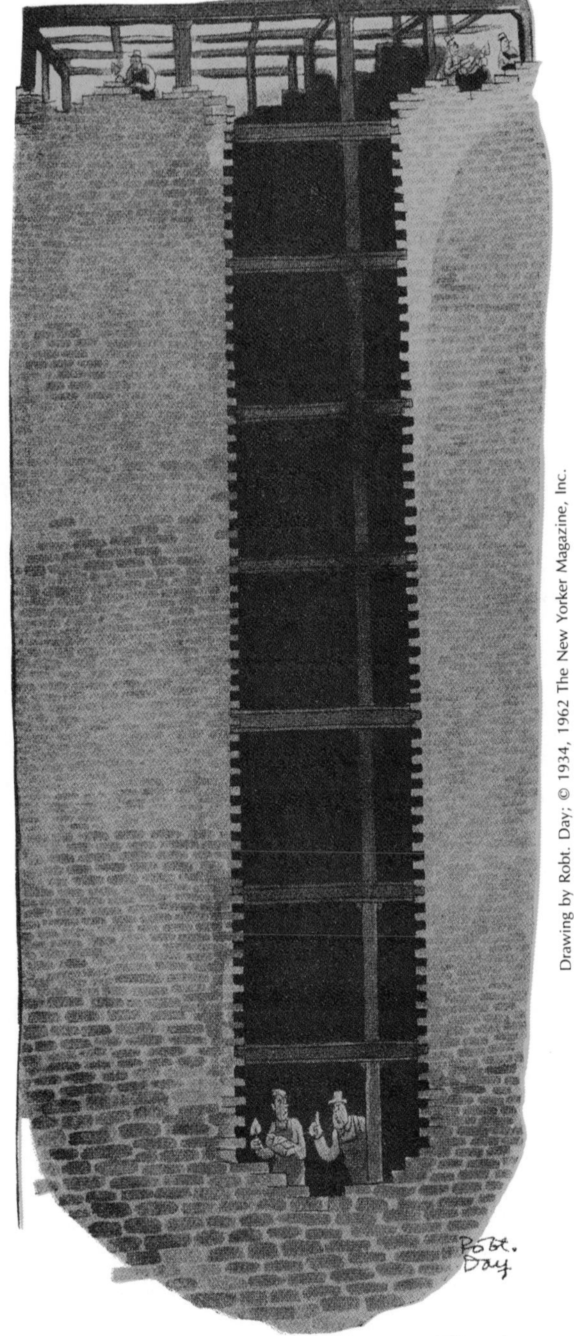

"See here, Pritchard, you're falling behind."

Years ago, students spent a full year learning plane geometry before taking a course in solid geometry. In this chapter, we will study in an informal way some of the basic topics included in that second course. Instead of continuing to define every term precisely as we did in the preceding lesson, we will take for granted the meanings of words in some of the definitions throughout the rest of this chapter. Furthermore, we will not attempt to prove every theorem but will merely present informal arguments to make them seem reasonable. This approach will enable us to explore a wider variety of topics than would otherwise be possible.

We begin our study with the geometric solid illustrated in this cartoon. A brick is a good model of a *polyhedron* because its faces are flat and polygonal in shape.

► **Definition**

A ***polyhedron*** is a solid bounded by parts of intersecting planes.

The intersecting planes form polygonal regions that are called the *faces* of the polyhedron. Their sides are called the *edges* of the polyhedron and their vertices are called its *vertices*.

A brick illustrates a special type of polyhedron called a *rectangular solid.*

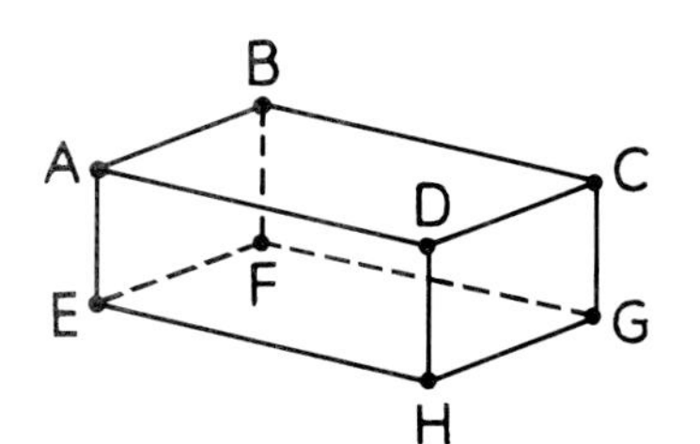

► **Definition**
A ***rectangular solid*** is a polyhedron that has six rectangular faces.

It is evident from the diagram at the left that the intersecting faces of a rectangular solid lie in perpendicular planes and that its opposite faces lie in parallel planes. It is also apparent that a rectangular solid has eight vertices. Two vertices of the solid that are not vertices of the same face are called *opposite vertices.* For example, in the diagram, one pair of opposite vertices is A and G and another pair is B and H.

A line segment that joins two opposite vertices of a rectangular solid is called a *diagonal* of the solid. Every rectangular solid has four diagonals and it is easy to prove that they have equal lengths.

The lengths of the three edges of a rectangular solid that meet at one of its vertices are the *dimensions* of the solid and are usually called its *length, width,* and *height.*

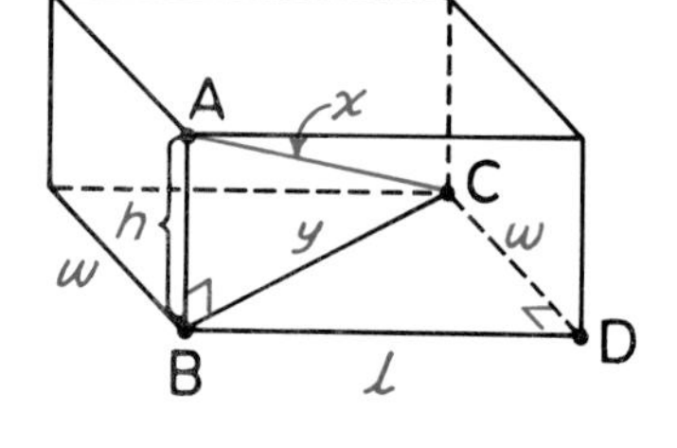

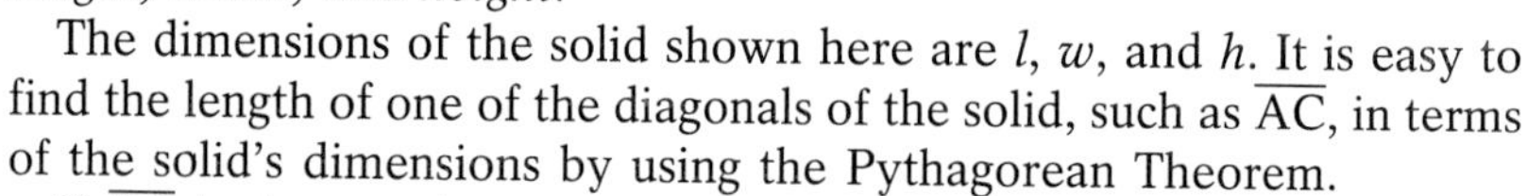

The dimensions of the solid shown here are l, w, and h. It is easy to find the length of one of the diagonals of the solid, such as $\overline{AC}$, in terms of the solid's dimensions by using the Pythagorean Theorem.

If $\overline{BC}$ is drawn, then $\triangle ABC$ and $\triangle BCD$ are right triangles ($\overline{AB}$ is perpendicular to the plane of the base of the solid, and so it must be perpendicular to $\overline{BC}$). Now in right $\triangle ABC$,

$$x^2 = y^2 + h^2,$$

and in right $\triangle BCD$,

$$y^2 = l^2 + w^2.$$

Substituting, we get

$$x^2 = l^2 + w^2 + h^2,$$

and taking square roots,

$$x = \sqrt{l^2 + w^2 + h^2}.$$

► **Theorem 91**
The length of a diagonal of a rectangular solid is $\sqrt{l^2 + w^2 + h^2}$, in which l, w, and h are its dimensions.

If all three dimensions of a rectangular solid are equal, it is a *cube.* If we let e represent the length of one edge of a cube, it follows that the length of one of its diagonals is

$$\sqrt{e^2 + e^2 + e^2} = \sqrt{3e^2} = e\sqrt{3}.$$

► **Corollary**
The length of a diagonal of a cube is $e\sqrt{3}$, in which e is the length of one of its edges.

Exercises

Set I

The figure below represents a cube.

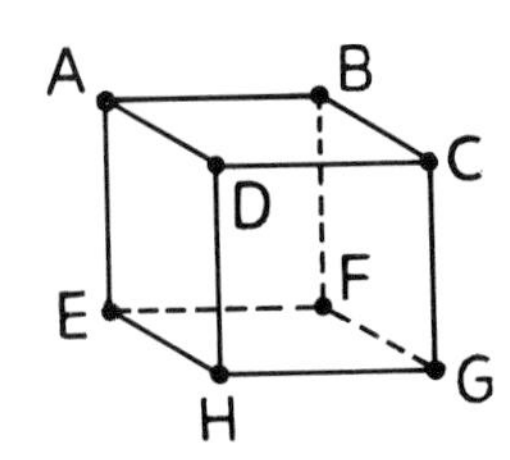

1. How many faces does it have?
2. How many vertices does it have?
3. How many edges does it have?
4. Name the vertex of the cube that is opposite E.
5. Name the edges that are parallel to $\overline{AD}$.
6. Name all of the edges that are perpendicular to $\overline{DC}$.

The figure below represents a rectangular solid.

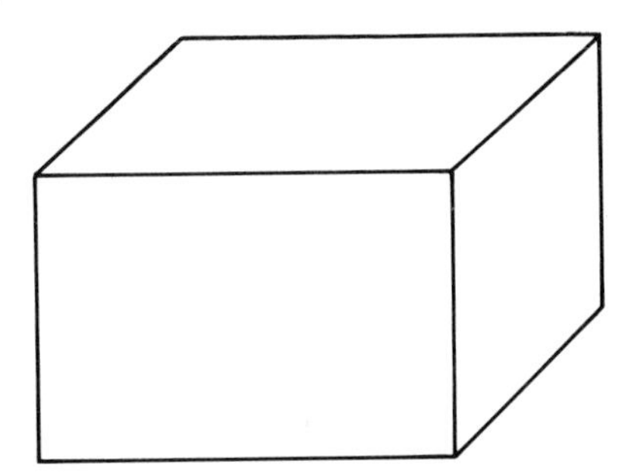

7. Trace it and then add the three hidden edges to your drawing as dotted line segments. Also draw all of the diagonals of the solid.
8. How many diagonals does a rectangular solid have?

Tell whether you think the following statements about the diagonals of a rectangular solid are true or false.

9. They have equal lengths.
10. They bisect each other.
11. They are perpendicular to each other.
12. They are concurrent.

The figure below represents a cube.

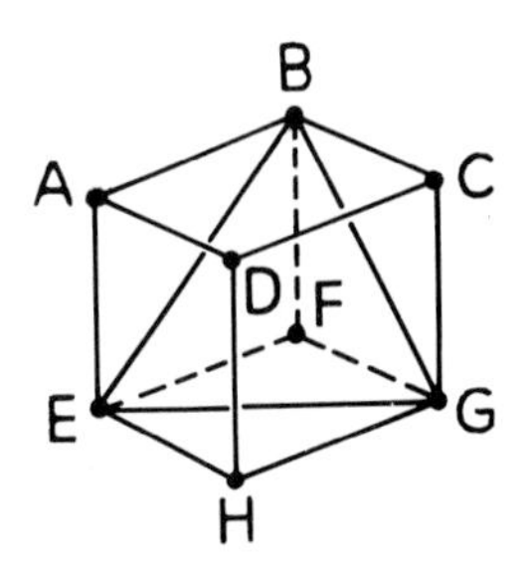

13. Find $\angle$AEB.
14. Find $\angle$BEG.
15. Find $\angle$AEG.
16. Is $\angle AEB + \angle BEG = \angle AEG$?
17. Is $\overrightarrow{EA}$-$\overrightarrow{EB}$-$\overrightarrow{EG}$?

The figure below represents a cube, each of whose edges has a length of 5.

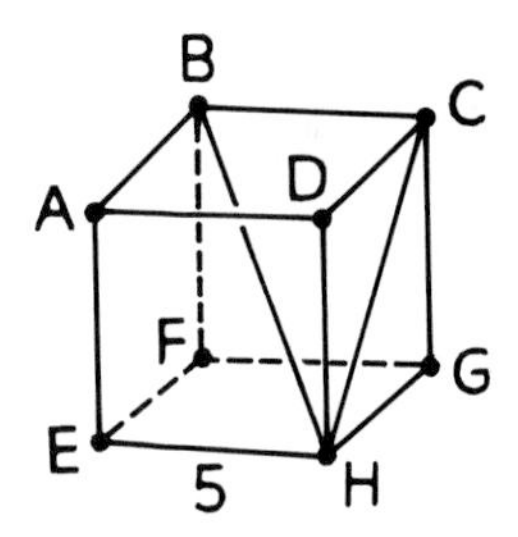

18. Which line segment is a diagonal of the cube?
19. Find CH.
20. Find BH.
21. What kind of triangle is $\triangle$BCH with respect to its sides?
22. What kind of triangle is $\triangle$BCH with respect to its angles?

Set II

Find the length marked x in each of the following rectangular solids. Express irrational answers in simple radical form.

Example:

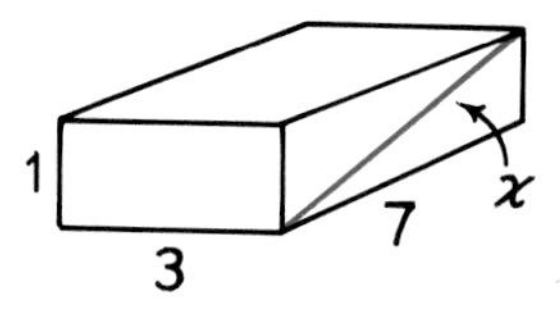

Solution: x is the length of a diagonal of a rectangular face with dimensions 7 and 1;

$$x^2 = 7^2 + 1^2$$
$$= 49 + 1 = 50$$
$$x = \sqrt{50} = \sqrt{25 \cdot 2} = 5\sqrt{2}.$$

23.

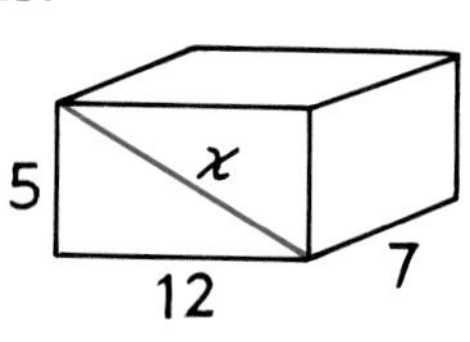

24.

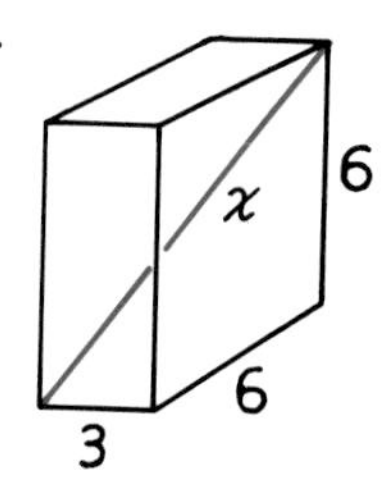

25.

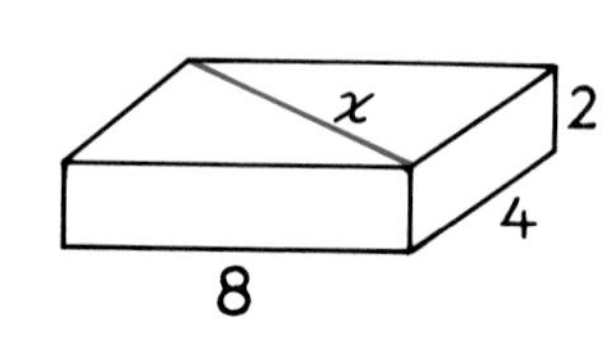

26.

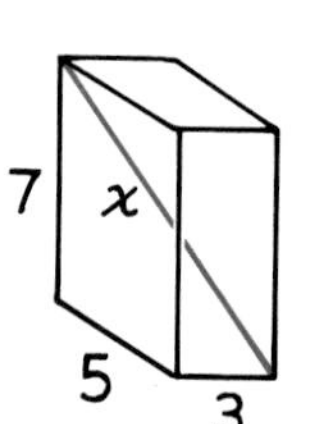

27.

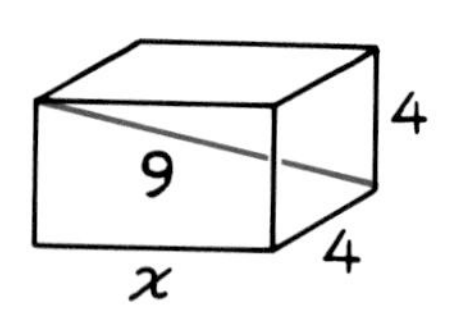

28.

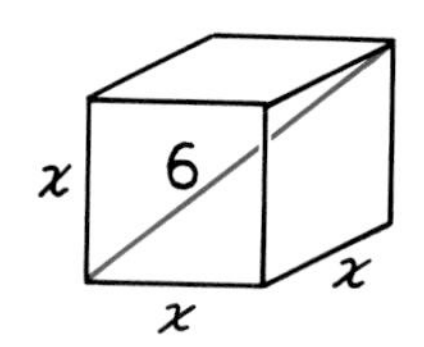

29.

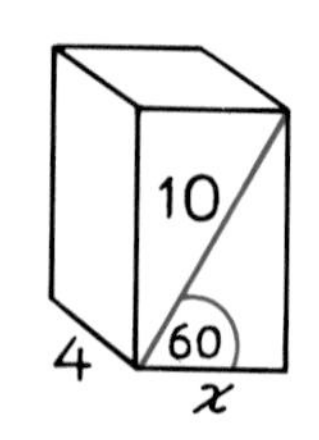

30.

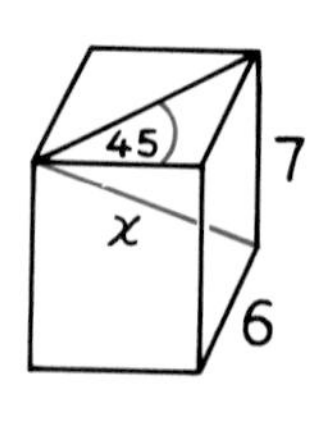

Dilcue bought a pair of skis that are 6 feet 10 inches long. He wanted to keep them in a broom closet that is 2 feet wide, 3 feet deep, and 6 feet high, but they are too long to fit as shown.

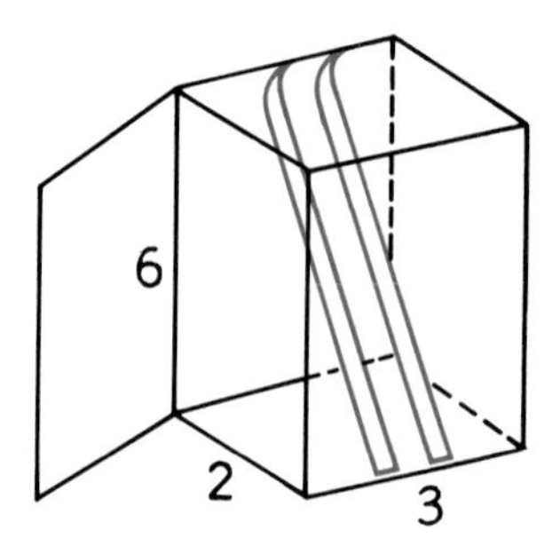

31. Explain why. ($\sqrt{5} \approx 2.24$ and $\sqrt{10} \approx 3.16$.)

32. Are there any other ways he could put the skis in the closet so that they might fit?

The intersections of a geometric solid and various planes are called its cross sections. Three different cross sections of a cube, all regular polygons, are shown in the following figures. Find the area of each cross section, given that each edge of the cubes is 8. Express your answers both in exact form and to the nearest integer.

33.

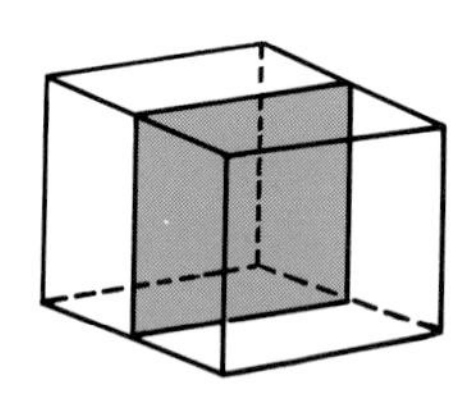

The corners of the square are midpoints of the edges of the cube.

34.

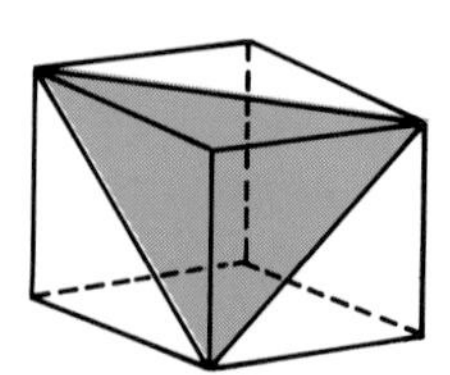

The corners of the triangle are vertices of the cube.

35.

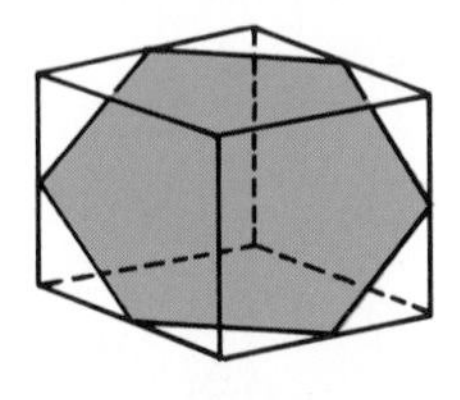

The corners of the hexagon are midpoints of the edges of the cube.

Set III

Some people use the word "square" when they really mean "cube." This mistake is understandable because a cube seems to be the three-dimensional equivalent of the two-dimensional square. Because

A square

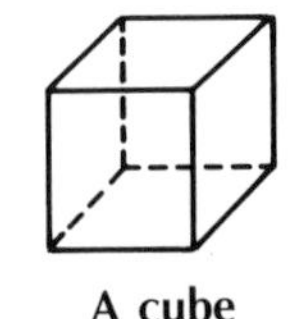

A cube

it is easy to picture figures in both two and three dimensions, it is natural to wonder whether there is such a thing as a four-dimensional space and, if so, what a figure in such a space would look like. For example, what is the four-dimensional equivalent of a cube?

Many mathematicians have thought about this; in fact, a name has been invented for such a figure: it is called a *hypercube*.*

One way to make a picture of a hypercube is based on the fact that, if a cube whose edges are made of sticks is viewed from a close distance, it looks like a small square inside a large square. In three dimensions, a hypercube viewed from a close distance might look like a small cube inside a large cube.

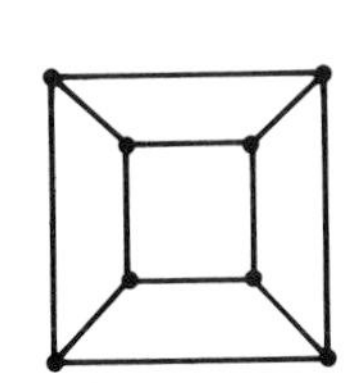

A cube seen as a square in a square in two dimensions.

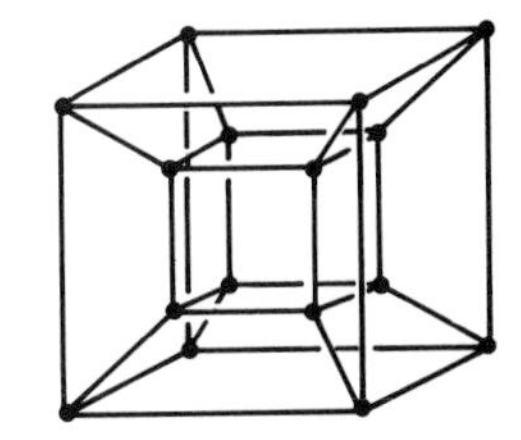

A hypercube seen as a cube in a cube in three dimensions.

The picture of the cube shows that it has 8 vertices, 12 edges, and 6 faces. (Although all 6 faces are square, 4 of them are distorted in the picture.)

Use the picture of the hypercube to determine how many vertices, edges, square faces, and cubical solids a hypercube has.

---

*See *The Fourth Dimension,* by Rudy Rucker (Houghton Mifflin, 1984), pp. 25–37.

National Park Service

Lesson 3
Prisms

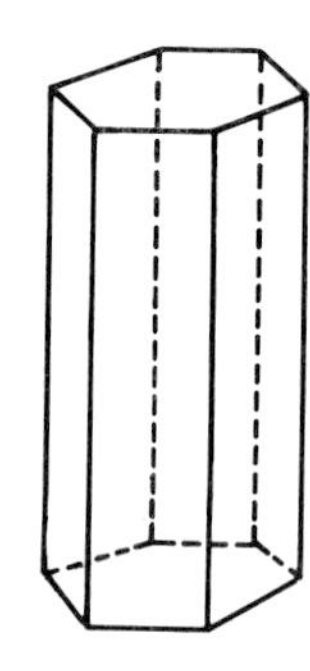

The Devil's Post Pile is a set of tall columns of rock in the Sierra Nevada mountains of California. These columns, some of which are 60 feet high, are an impressive display of geometric polyhedra in nature. Most of them are hexagonal in shape as shown in the diagram.

This polyhedron is an example of a *prism.* Notice that its top and bottom faces lie in parallel planes and that the edges joining the vertices of one face to those of the other are parallel to each other.

► **Definition**
Suppose that A and B are two parallel planes, R is a polygonal region in one plane, and ℓ is a line that intersects both planes but not R. The set of all segments parallel to line ℓ that join a point of region R to a point of the other plane form a ***prism.***

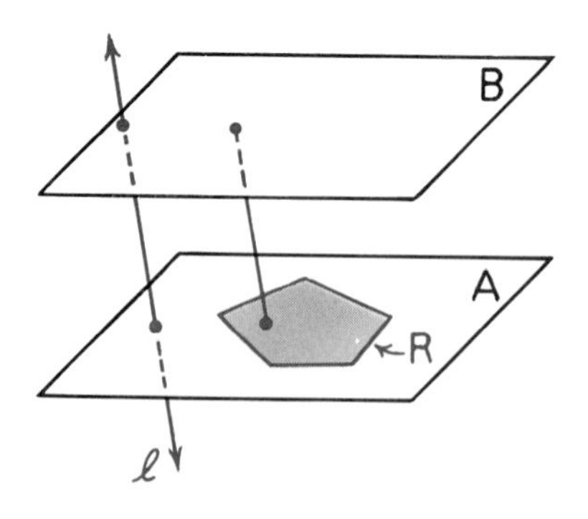

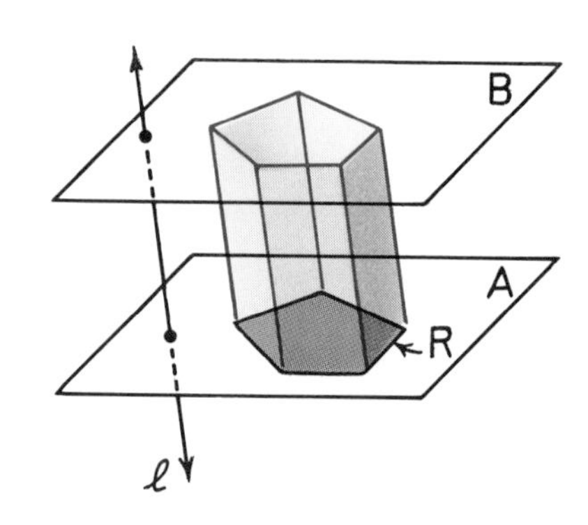

The two faces of the prism that lie in these parallel planes are called its *bases.* It can be shown that the bases of a prism are always congruent to each other. The rest of the faces of the prism, called its *lateral faces,* are parallelograms. The edges of the prism in which the lateral faces intersect are called its *lateral edges.*

Prisms are classified according to two properties: the relation of their lateral edges to the planes containing their bases, and the shape of their bases. The prism shown between planes A and B in the figure on the facing page is an *oblique pentagonal* prism: its lateral edges are oblique to its bases and its bases are pentagons. The prism representing one of the columns of the Devil's Post Pile is a *right hexagonal* prism.

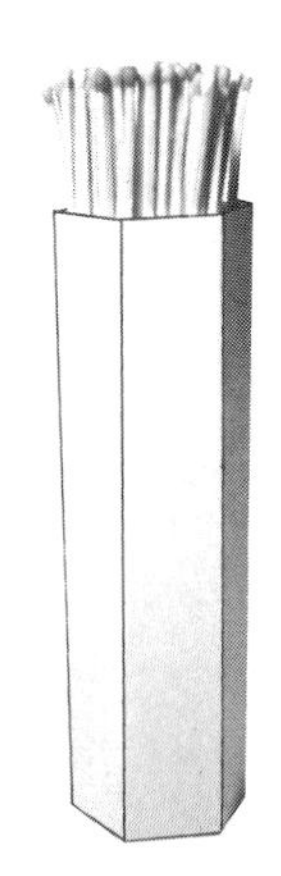

Long matches for lighting fireplaces are often stored in boxes that are in the shape of right hexagonal prisms. If a label is wrapped around the six lateral faces of the box, its area is called the *lateral area* of the prism. The *total area* of the prism, on the other hand, includes the areas of its two bases.

► **Definitions**

The ***lateral area*** of a prism is the sum of the areas of its lateral faces.

The ***total area*** of a prism is the sum of its lateral area and the areas of its bases.

Exercises

Set I

The figure below represents an oblique triangular prism.

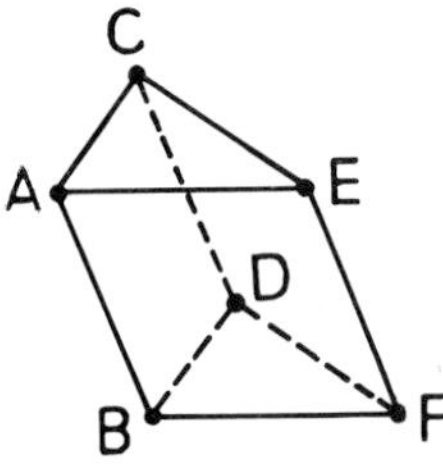

1. Name the bases of this prism.
2. Name its lateral edges.
3. Are its lateral edges parallel to each other?
4. Are its lateral edges perpendicular to the bases?

Tell whether each of the following figures appears to be a prism. If you think that a figure is a prism, identify the shape of its bases.

Example: 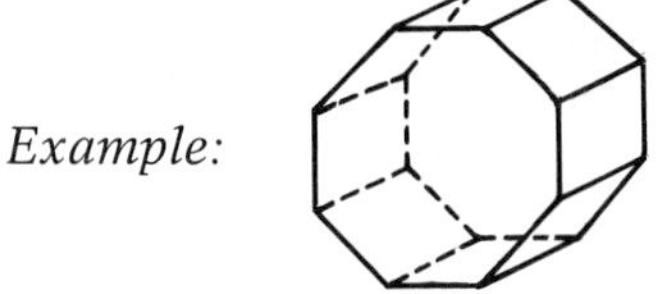

Answer: Yes; octagons.

5. 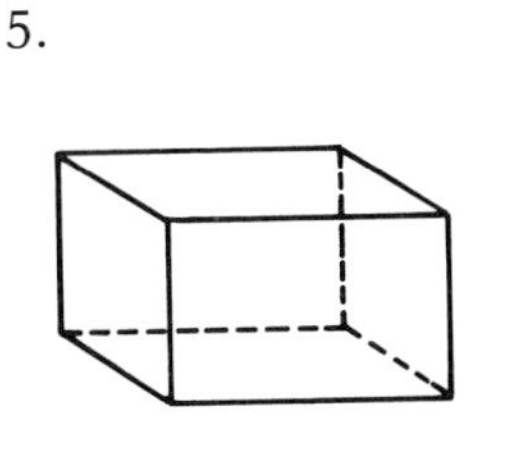

6. 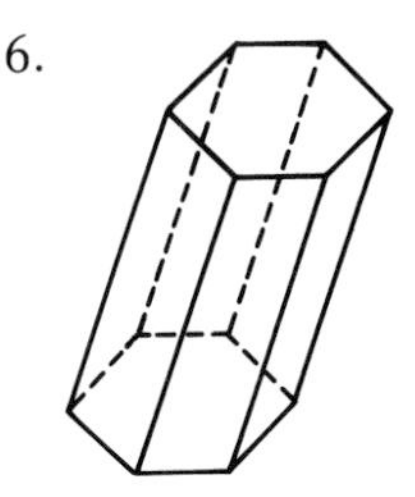

7.

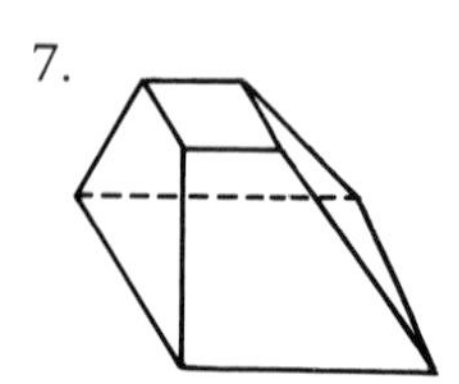

8.

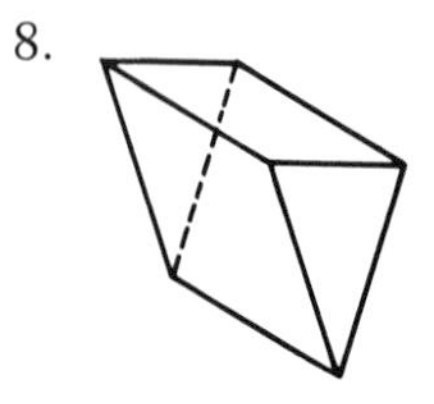

9.

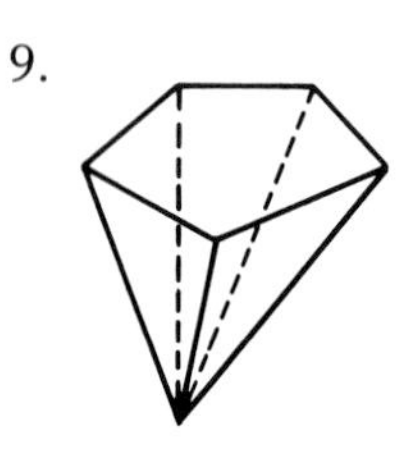

10.

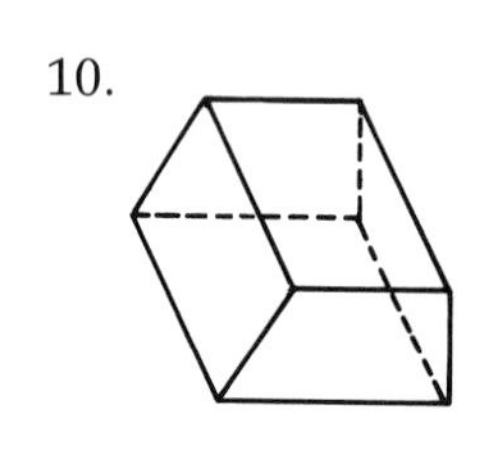

The figure below represents a face and edge of a prism.

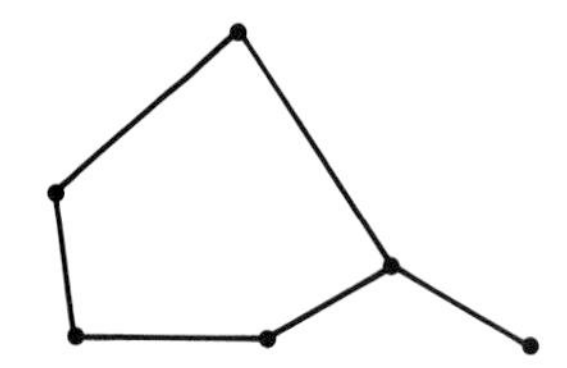

11. Trace the figure and complete it.
12. What kind of prism is it?
13. How many vertices does the prism have?
14. How many edges does it have?
15. How many faces does it have altogether?

Obtuse Ollie has decided that, because the least number of sides that a polygon can have is *three*, the least number of faces that a prism can have is *four*. Acute Alice disagrees, saying that four walls do *not* a prism make.

16. What's your opinion?
17. What is the least number of *edges* that a prism can have?
18. What is the least number of *vertices* that a prism can have?

Set II

A cube is a prism.

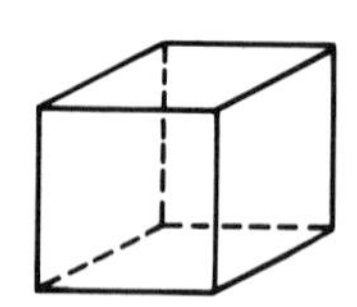

19. What kind of prism is it?

Suppose that each edge of a cube has a length of 5. Find each of the following areas.

20. The area of one face.
21. The lateral area of the cube.
22. The total area of the cube.

Suppose that each edge of a cube has a length of e. Write an expression for each of the following areas.

23. The area of one face.
24. The lateral area of the cube.
25. The total area of the cube.

A rectangular solid is a prism. The dimensions of the prism below are 3, 7, and 6.

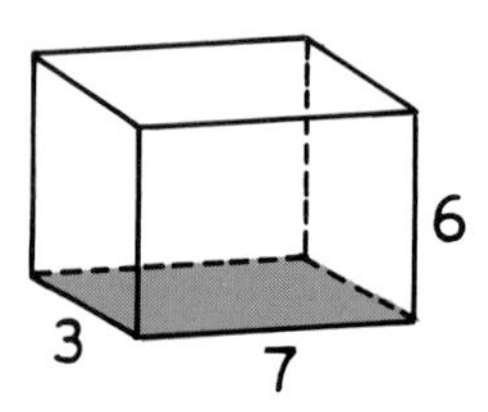

26. Find the area of the bottom of this prism.
27. Find the total area of the prism.

The dimensions of the prism below are a, b, and c.

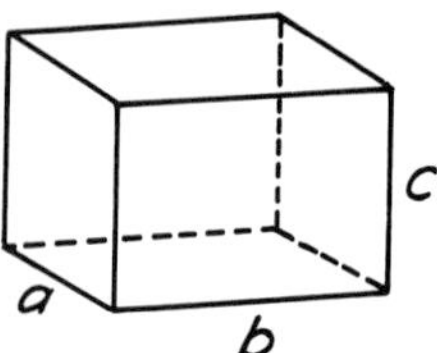

28. Write an expression for the total area of this prism.

The bases of the prism below are right triangles and its lateral faces are rectangles.

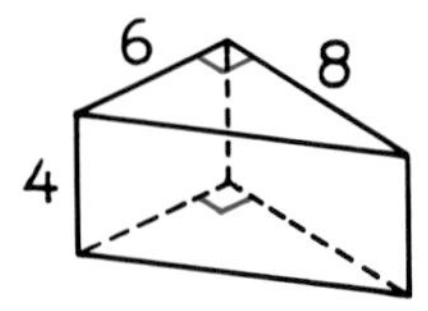

29. Find the area of one of its bases.
30. Find its lateral area.
31. Find its total area.

The pattern in the next column consists of six squares and two regular hexagons. If it were cut out, it could be folded to form a polyhedron.

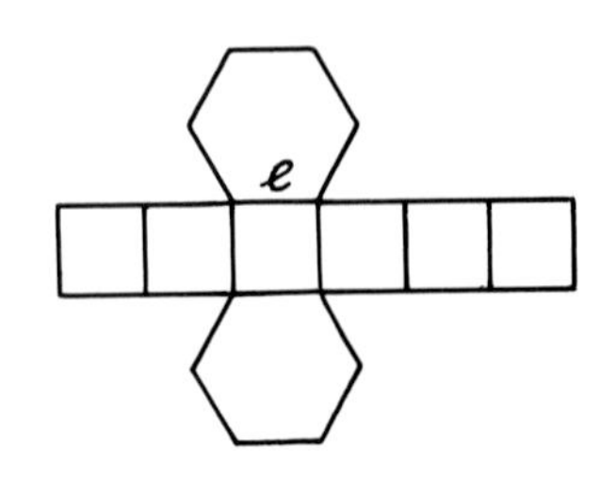

32. Which polygons would be its bases?
33. What kind of polyhedron would it be?
34. Write an expression for its lateral area.
35. Use the fact that a regular hexagon can be divided into six equilateral triangles to write an expression for the area of one of the hexagons.
36. Write an expression for the total area of the polyhedron.

Set III

This figure represents a "transparent" cube in which every edge can be seen.

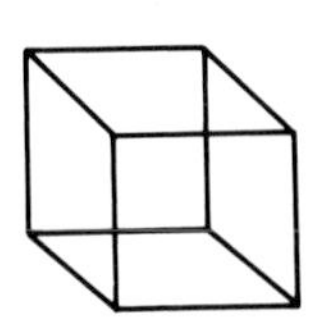

Which edges cannot be seen if the cube were solid depends on your point of view, as the two figures at the right show.

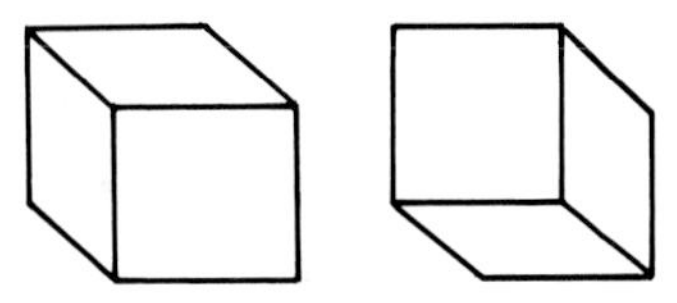

The figure below represents a transparent view of three intersecting prisms.

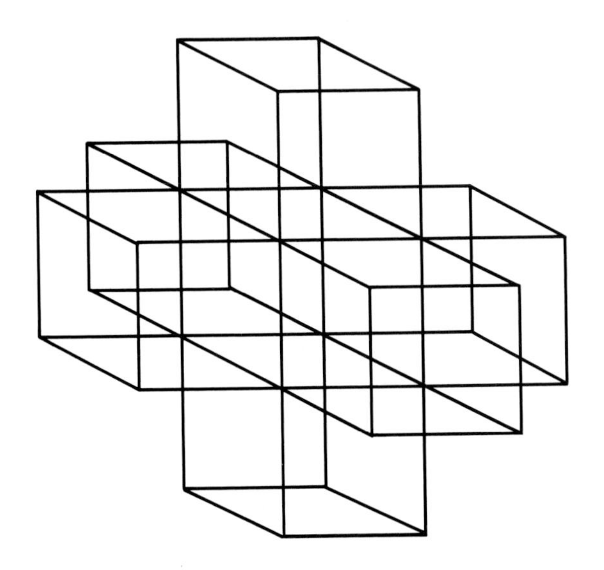

Like the cube in the example above, some of the line segments in this figure can be removed in two different ways to show two different views of the figure as a solid.

Can you draw both views? (Put your paper over the figure and trace part of it to make each drawing.)

Joe Munroe, Life Magazine © 1959 Time Inc.

Lesson 4

The Volume of a Prism

A fad among college students in the late fifties was cramming. Not cramming for exams, but cramming as many people as possible into a small space. This photograph shows twenty-two people who have managed to crowd themselves into a telephone booth. This certainly doesn't leave very much space for each one! Just exactly how much would it be?

The measure of a region of space is called its *volume.* We will assume that the volume of a three-dimensional region, like the area of a two-dimensional region, is a positive number. The volume of the telephone booth can be expressed in terms of the number of cubes measuring 1 foot along each edge that it can contain.

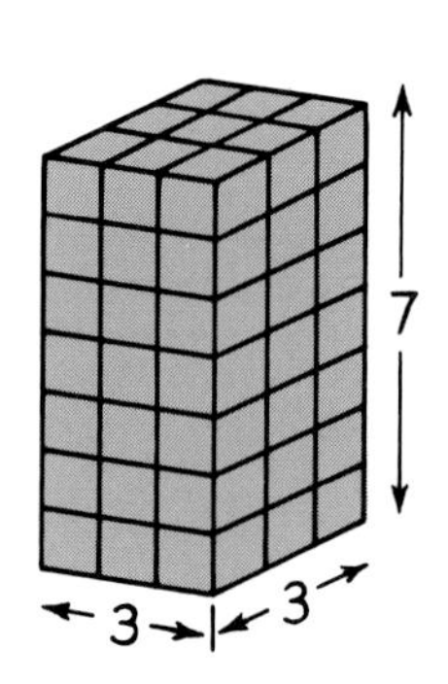

Suppose that the base of the booth is a square 3 feet on a side and that the booth is 7 feet high. A layer of $3 \cdot 3 = 9$ cubes could be put on the floor of the booth and 7 such layers would reach the ceiling. The volume of the telephone booth would be

$$3 \cdot 3 \cdot 7 = 63 \text{ cubic feet.}$$

If all twenty-two people were completely inside the booth, each would occupy, on the average, a space of less than 3 cubic feet!

It is not possible to divide every rectangular solid into a whole number of unit cubes as we have just done with the telephone booth. Nevertheless, it seems reasonable that the volume of such a solid could still be found by multiplying its three dimensions. If the dimensions of a rectangular solid are l, w, and h, and its volume is V, then we will assume that

$$V = lwh,$$

or, because $B = lw$, in which B is the area of one of the bases,

$$V = Bh.$$

Because a rectangular solid is a right rectangular prism, we now have a formula for finding the volume of one type of prism. Before considering how the volumes of other prisms can be found, we need to define the *altitude* of a prism.

► **Definition**
An ***altitude*** of a prism is a line segment that connects the planes of its bases and is perpendicular to both of them.

To find the volume of an oblique rectangular prism, we might imagine that it is made from a deck of playing cards. The volume of the prism is equal to the sum of the volumes of all of the cards. Now imagine pushing the cards so that they form a right rectangular prism. Its base has the same area as before, and its altitude remains unchanged. Furthermore,

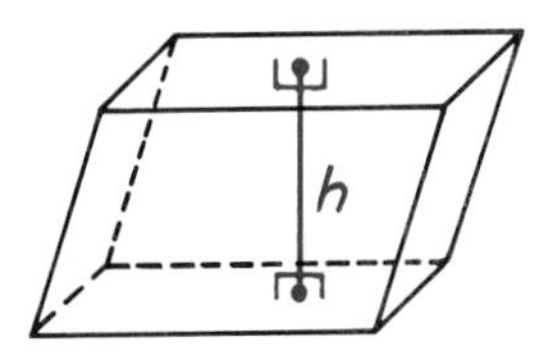

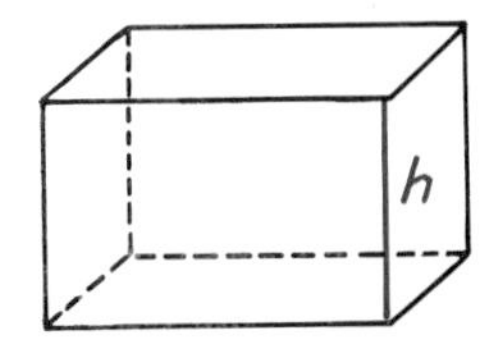

the cards have the same volume that they had before, so that the volume of the prism *also* remains *unchanged.* This means that the volume of an oblique rectangular prism also can be found by the formula

$$V = Bh.$$

So far we have considered how to determine the volumes of prisms whose bases are *rectangles.* To find the volumes of prisms whose bases are other shapes, we will compare their *cross sections.*

► **Definition**
A ***cross section*** of a geometric solid is the intersection of a plane and the solid.

For example, one of the cross sections of the triangular prism shown below is $\triangle ABC$.

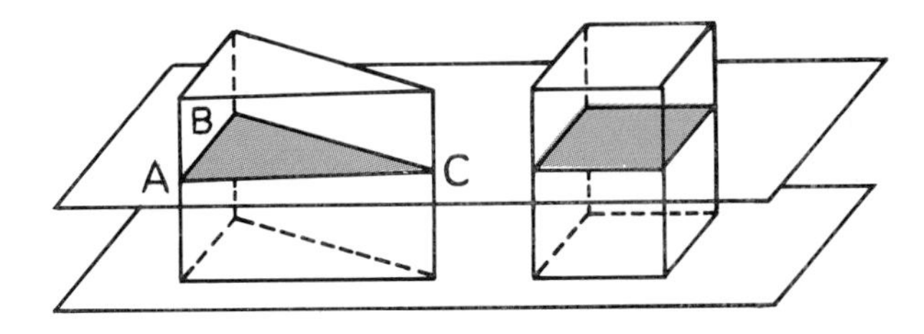

Now suppose that the bases of two prisms lie in the same plane and that every plane parallel to this plane that intersects both prisms cuts off cross sections with the same area. It seems reasonable to conclude from this fact that the two prisms have the same volume. Because we cannot prove this, we will state it, in more general terms, as a postulate. It was first stated in this form by an Italian mathematician, Cavalieri, who lived in the seventeenth century and was a pupil of Galileo.

► **Postulate 14** (Cavalieri's Principle)
Consider two geometric solids and a plane. If every plane parallel to this plane that intersects one of the solids also intersects the other so that the resulting cross sections have the same area, then the two solids have the same volume.

It can be proved that all cross sections of a prism are congruent to its bases and hence have the same area that they have. From this and Cavalieri's Principle, it follows that two prisms that have bases of equal area and altitudes that are equal must have equal volumes. Hence, the volume of *every prism* can be found by the same formula. We will now state this conclusion as a general postulate.

► **Postulate 15**
The volume of any prism is Bh, in which B is the area of one of its bases and h is the length of its altitude.

Formulas for the volumes of a rectangular solid and cube now follow as corollaries to this postulate.

► **Corollary 1**
The volume of a rectangular solid is lwh, in which l, w, and h are its length, width, and height.

► **Corollary 2**
The volume of a cube is e^3, in which e is the length of one of its edges.

Exercises

Set I

One yard contains 3 feet.

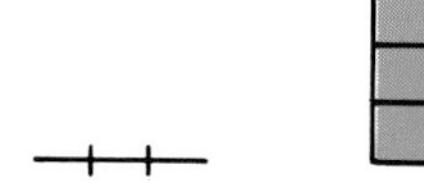

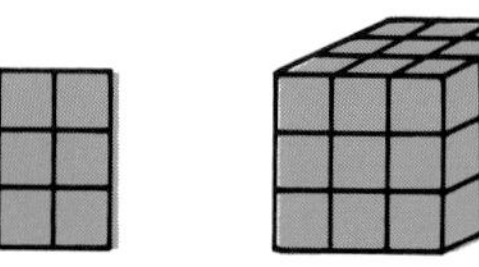

1. How many square feet does one square yard contain?
2. How many cubic feet does one cubic yard contain?

One foot contains 12 inches.

3. How many square inches does one square foot contain?
4. How many cubic inches does one cubic foot contain?

Find the volume of each of the following prisms.

5.

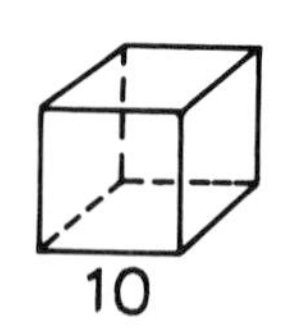

This is a cube.

6.

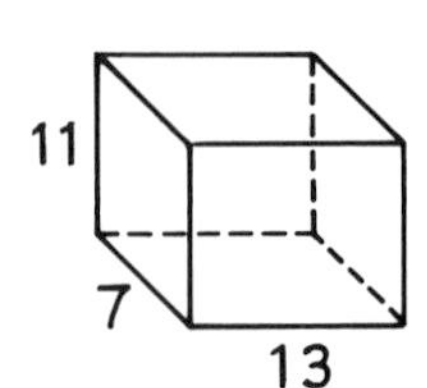

This is a rectangular solid.

7.

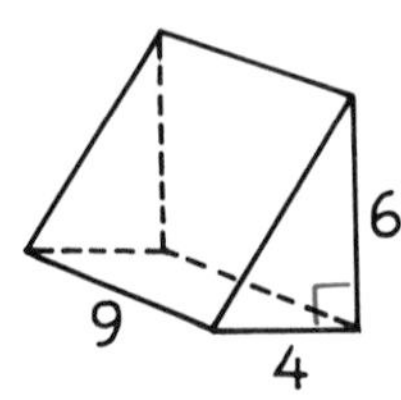

This is a right prism whose bases are right triangles.

8.

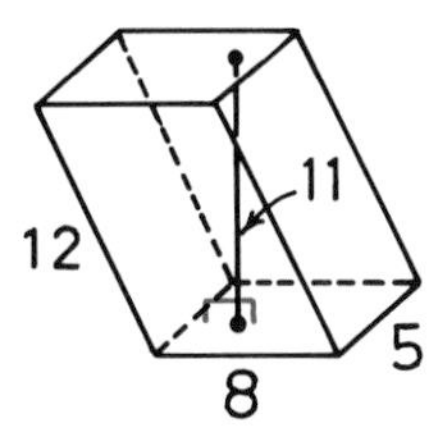

This is an oblique prism whose bases are rectangles.

9.

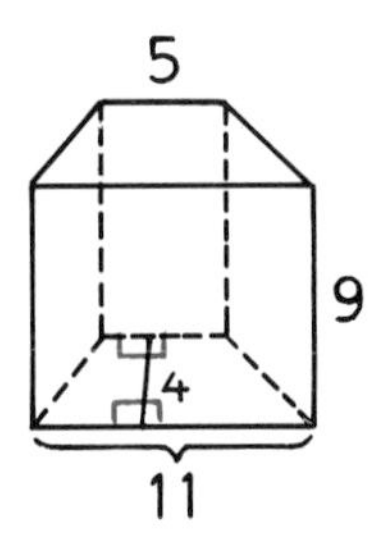

This is a right prism whose bases are trapezoids.

10.

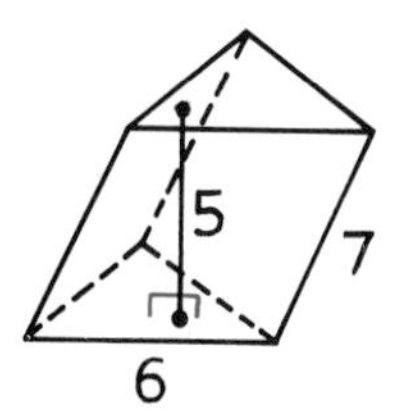

This is an oblique prism whose bases are equilateral triangles.

The bases of the two prisms shown in the figure below lie in the same plane.

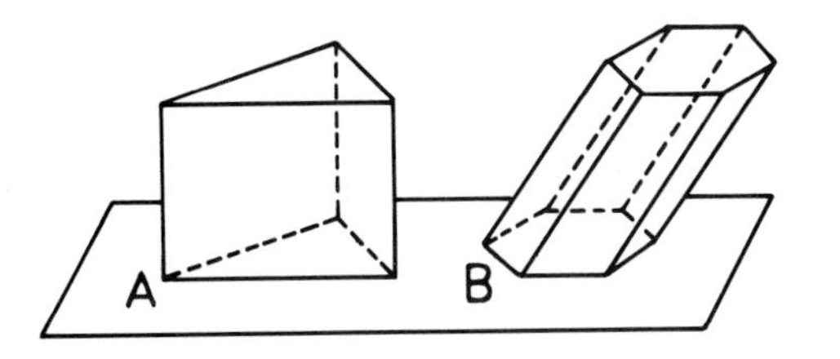

11. What kind of prism does each solid appear to be?

12. What can you conclude about the prisms if every plane that intersects them and is parallel to the plane containing their bases cuts off cross sections of equal areas?

The figure below represents a box in the shape of a rectangular solid with dimensions 10 cm, 20 cm, and 15 cm.

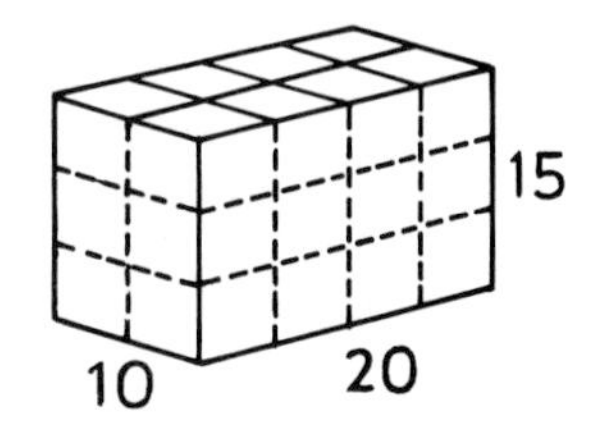

13. Find its total surface area.

14. Find its volume.

15. How many cubes measuring 5 cm on each edge could the box contain?

Three cubes have edges of lengths 1, 2, and 3 units respectively.

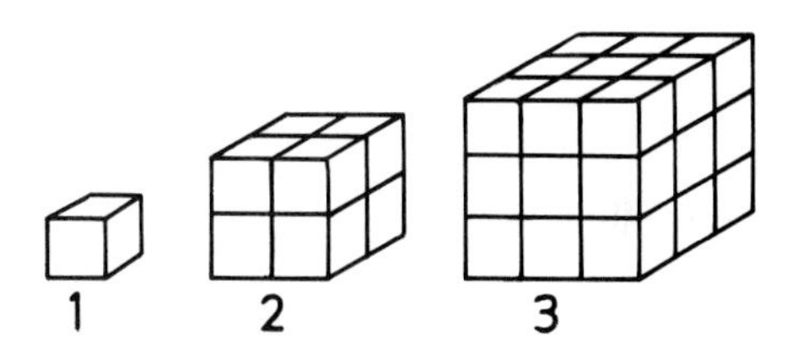

16. Find the total area of each cube.

What happens to the total area of a cube

17. if the length of one edge is doubled?

18. if the length of one edge is tripled?

19. Find the volume of each cube.

What happens to the volume of a cube

20. if the length of one edge is doubled?

21. if the length of one edge is tripled?

Set II

The figure below represents a prism having an altitude of 15.

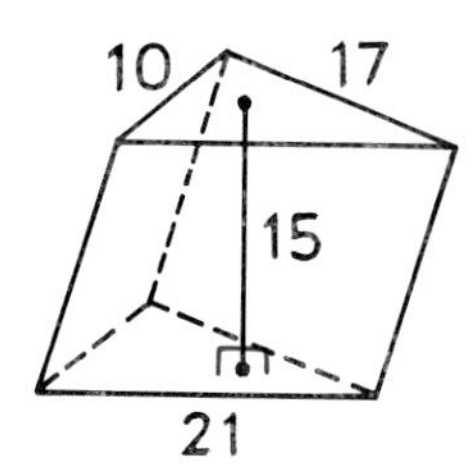

22. What kind of prism does it appear to be?
23. Find the area of one of its bases.
24. Find its volume.

The figure below represents a garage.

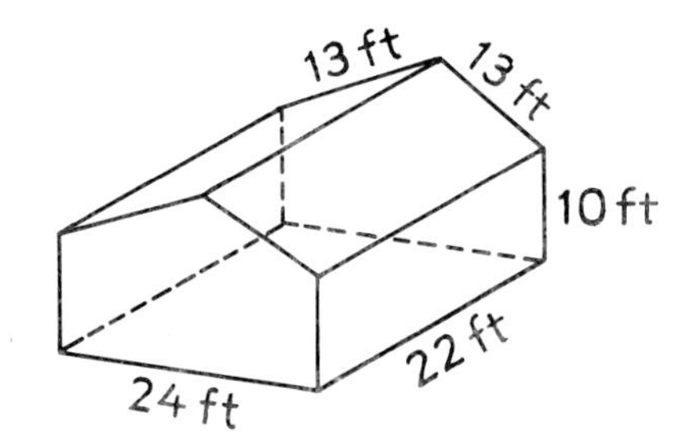

25. What kind of prism is it?
26. Find the area of the floor.
27. Find the area of the roof.
28. Find the total area of the four walls.
29. Find the volume of the garage.

The figure below represents a swimming pool.

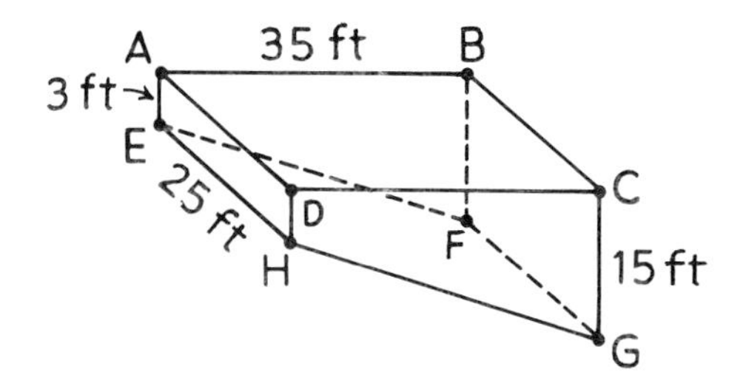

30. What kind of prism is it?
31. Find the area of the top of the pool.
32. Find the area of the bottom.
33. Find the area of one of the two congruent side walls.
34. Find the volume of the pool.

Several steps of a concrete staircase are shown here. Each step is a right triangular prism and has the dimensions indicated.

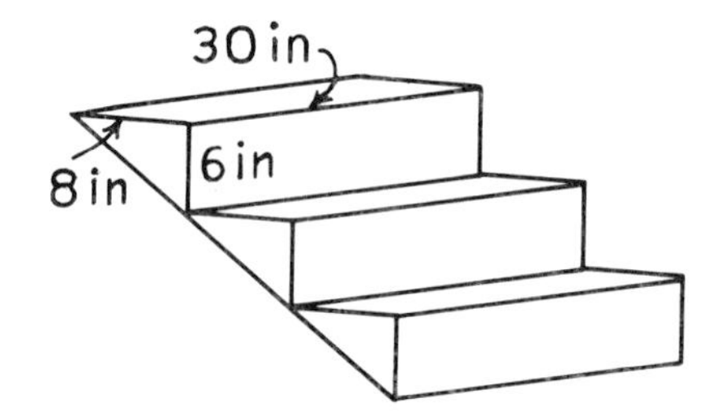

35. Find the volume of one of the steps in cubic inches.
36. If the staircase contains 24 steps in all, find the total volume of concrete in the steps. Express your answer in cubic feet.

Set III

According to U.S. postal regulations, the maximum size that a package sent in the mail can have is "100 inches in combined length and girth."

If you wanted to send a package in the shape of a cube in the mail, what is the largest volume that it could have? Show your method.

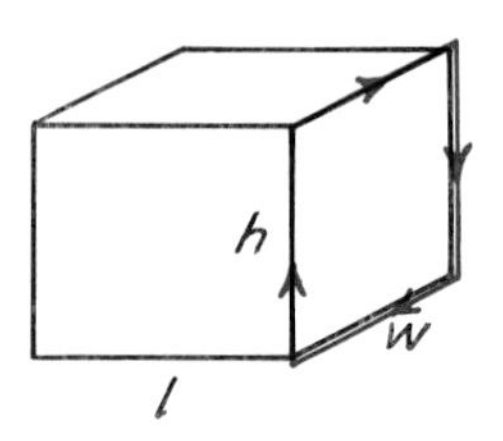

Girth = $2w + 2h$

Eliot Elisofon, Life Magazine © Time Inc.

Lesson 5
Pyramids

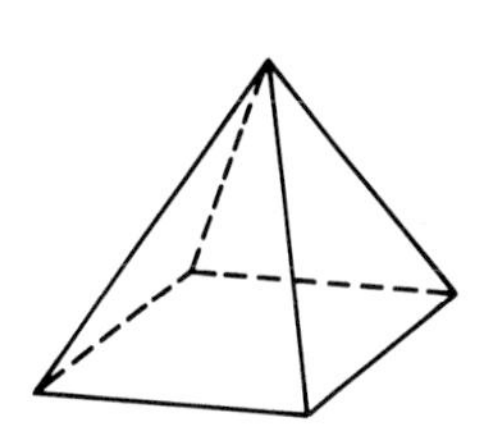

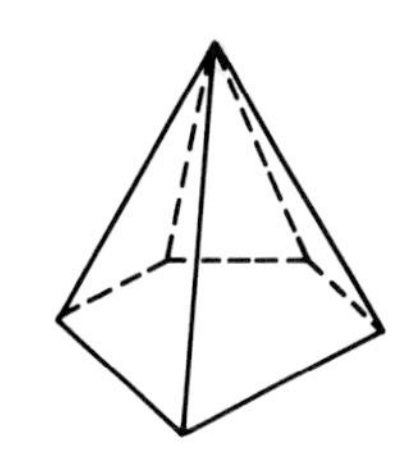

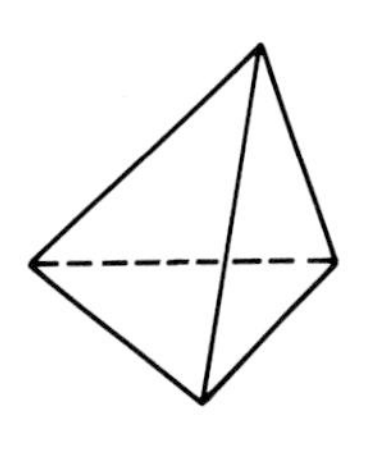

The largest of all man-made geometric solids was built more than four thousand years ago. It is the Great Pyramid in Egypt, the only one of the "seven wonders of the world" still in existence. This pyramid, one of about eighty such structures built by the ancient Egyptians, is comparable in height to a forty-story building and covers an area of more than 13 acres. It was put together from more than two million stone blocks, weighing between 2 and 150 tons each!

The figures at the right represent pyramids of several different types. Although the Egyptians consistently chose the square for the shape of the bases of their pyramids, other polygons can also be used.

► **Definition**
Suppose that A is a plane, R is a polygonal region in plane A, and P is a point not in plane A. The set of all segments that join P to a point of region R form a ***pyramid.***

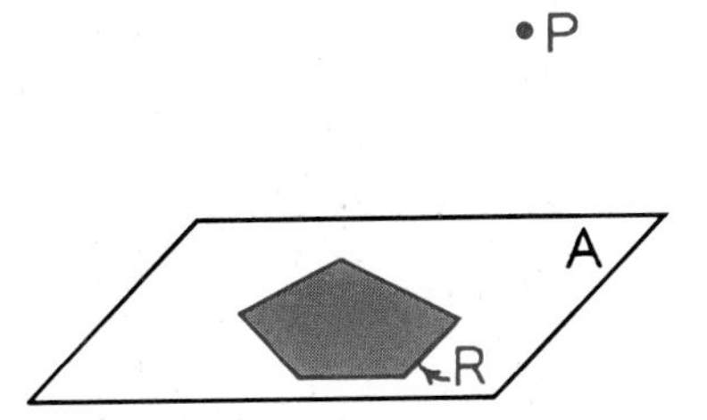

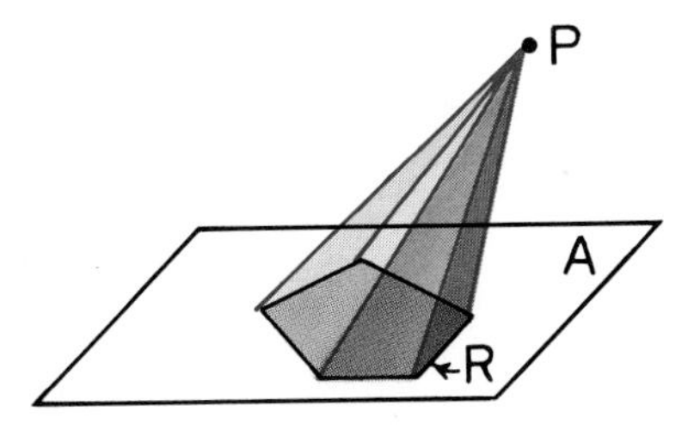

The face of the pyramid that lies in this plane is called its *base*. The rest of its faces are called *lateral faces* and the edges in which they intersect each other are called its *lateral edges*. The lateral edges meet at the *vertex* of the pyramid. Although the base of a pyramid can be any polygonal region, its lateral faces are always triangular.

The height of a pyramid is measured by the length of its *altitude*.

► **Definition**

The ***altitude*** of a pyramid is the perpendicular line segment joining its vertex to the plane of its base.

The Great Pyramid is an example of a *regular* pyramid.

► **Definition**

A ***regular pyramid*** is a pyramid whose base is a regular polygon and whose lateral edges are equal.

It can be proved that the altitude of such a pyramid joins its vertex to the center of its base.

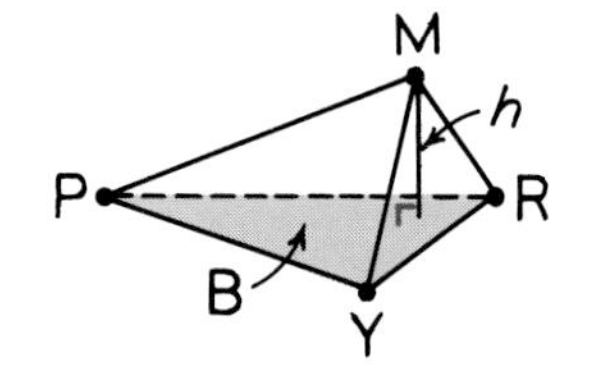

The amount of space occupied by the Great Pyramid is enormous: it contains more than 91,000,000 cubic feet of rock. How can such a volume be determined? To develop a formula for the volume of a pyramid, we can imagine building a prism having the same base and altitude as the pyramid in the following way.

Consider pyramid PYRM with base of area B and altitude of length h. Construct $\overline{PI}$ and $\overline{RD}$ so that they are both parallel to $\overline{YM}$ and equal to it in length. Join points M, I, and D to determine the upper base of the prism.

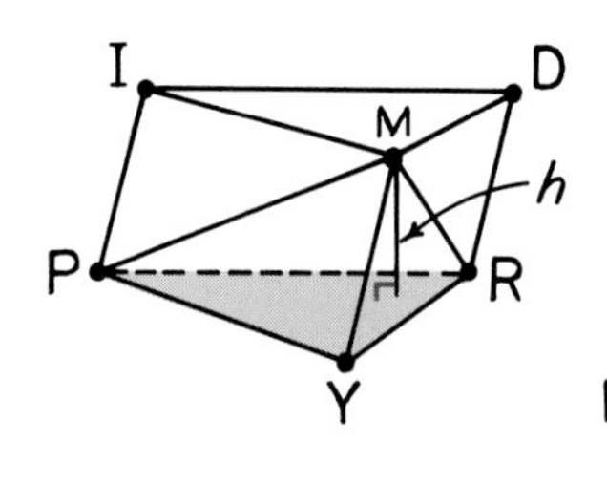

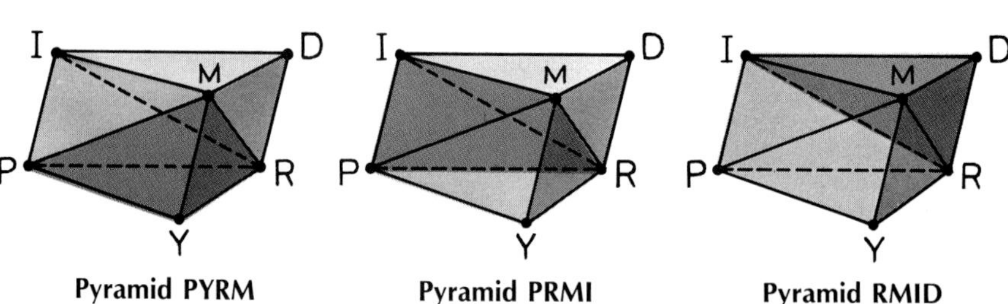

It is possible to cut the prism that we have formed into three pyramids, one of which is the original pyramid. Each one is shaded in one of the figures above. Furthermore, it can be shown that these pyramids have equal volumes. This means that the volume of each is one-third the volume of the prism, or $\frac{1}{3}Bh$. This result is true of all pyramids. We will state it as a theorem, even though we have not proved it.

► **Theorem 92**

The volume of any pyramid is $\frac{1}{3}Bh$, in which B is the area of its base and h is the length of its altitude.

Exercises

Set I

The solid illustrated in the figure below is a regular pentagonal pyramid; $\overline{FG} \perp P$.

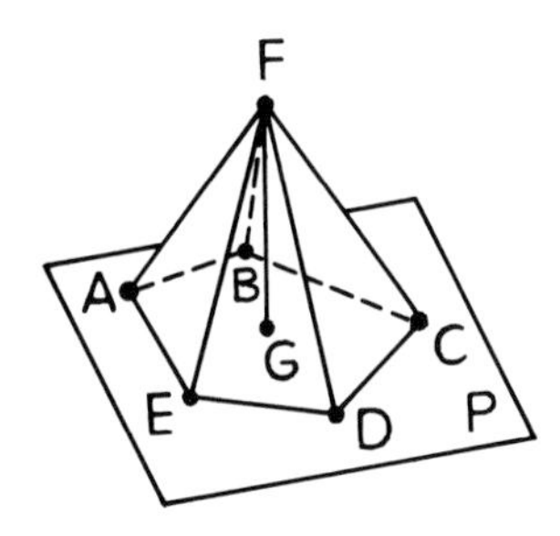

1. Name the vertex of the pyramid.
2. Name its altitude.
3. Name one of its lateral edges.
4. Are its lateral edges equal?
5. What kind of base does the pyramid have?
6. What kind of lateral faces does it have?
7. What is point G called with respect to ABCDE?

The figure below represents a face and an edge of a pyramid.

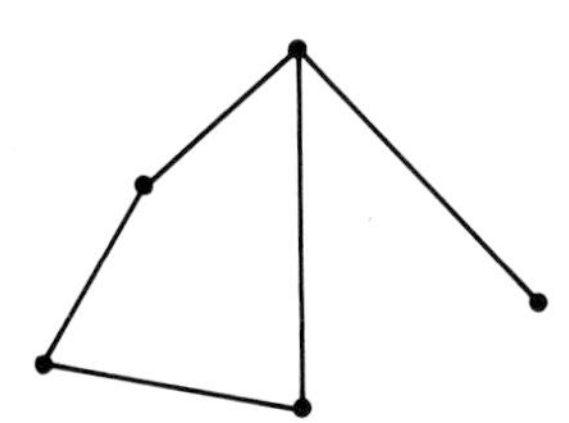

8. Trace the figure and complete it.
9. How many vertices does the pyramid have altogether?
10. How many edges does it have?
11. How many faces does it have?

Find the volume of each of the following pyramids.

12. 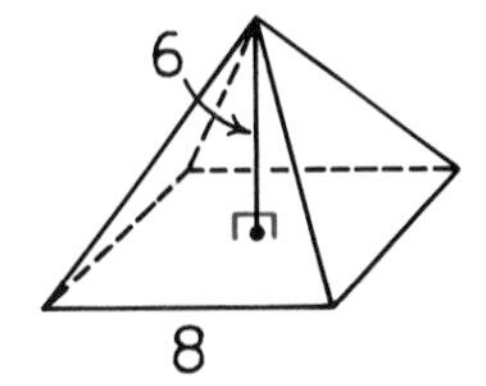

This is a regular pyramid.

13. 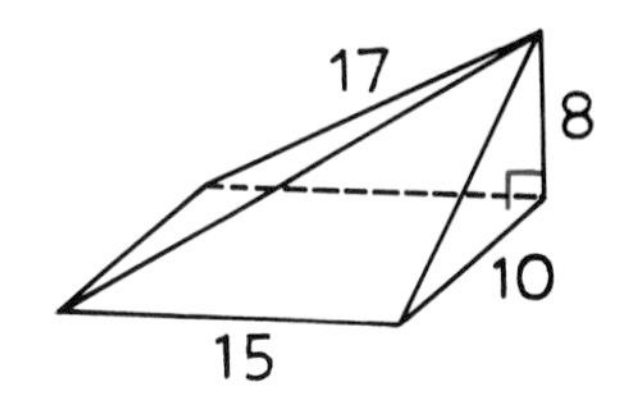

This is a pyramid whose base is a rectangle.

14. 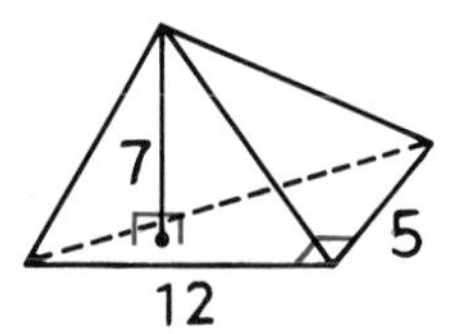

This is a pyramid whose base is a right triangle.

15. 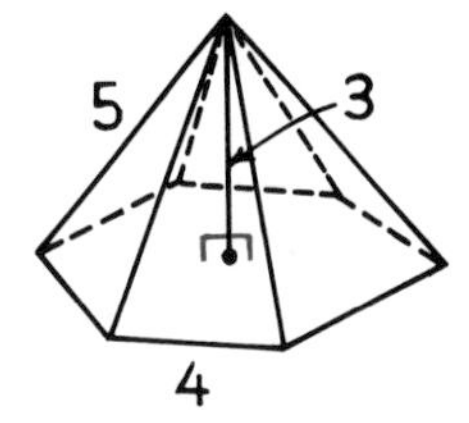

This is a regular pyramid.

16. 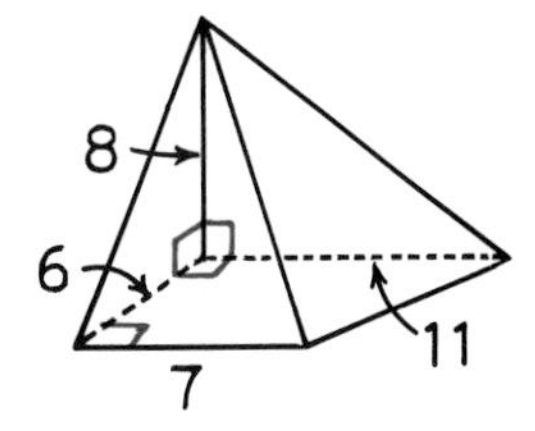

This is a pyramid whose base is a trapezoid.

The figures below represent three rectangular pyramids and their altitudes.

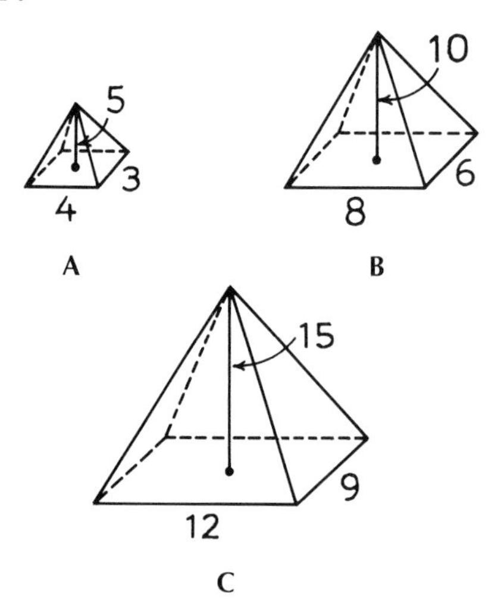

17. Find the volume of each pyramid.
18. How many times the dimensions of pyramid A are the corresponding dimensions of pyramid B?
19. How many times the volume of pyramid A is the volume of pyramid B?
20. How many times the dimensions of pyramid A are the corresponding dimensions of pyramid C?
21. How many times the volume of pyramid A is the volume of pyramid C?

The bases of the two pyramids shown in the figure below lie in the same plane. If

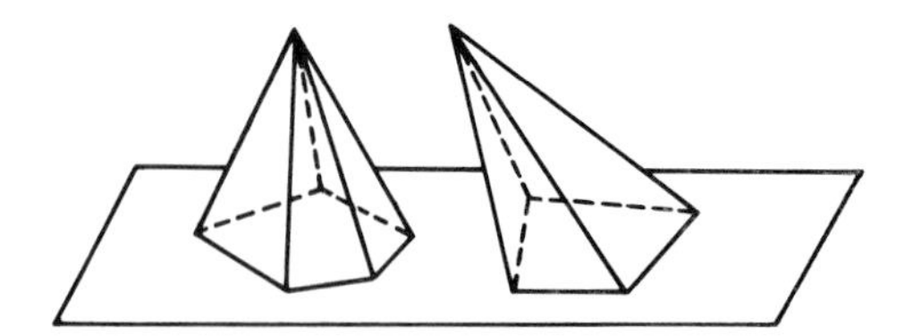

every plane that intersects them and is parallel to the plane containing their bases cuts off cross sections of equal areas,

22. does it follow that the pyramids have equal areas?
23. does it follow that the pyramids have equal volumes?

Set II

The figure below represents a pyramid contained inside a prism. The vertex of the pyramid lies on the upper base of the prism.

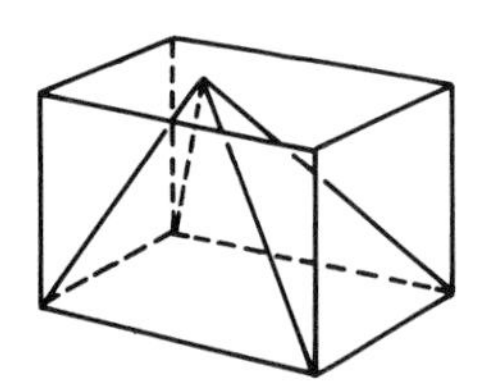

24. How does the volume of the pyramid compare with the volume of the prism?
25. How does the volume of the space between the walls of the pyramid and prism compare with the volume of the pyramid?

The cube below consists of six identical pyramids, each having one of the faces of the cube for its base. Each edge of the cube is 6 units long.

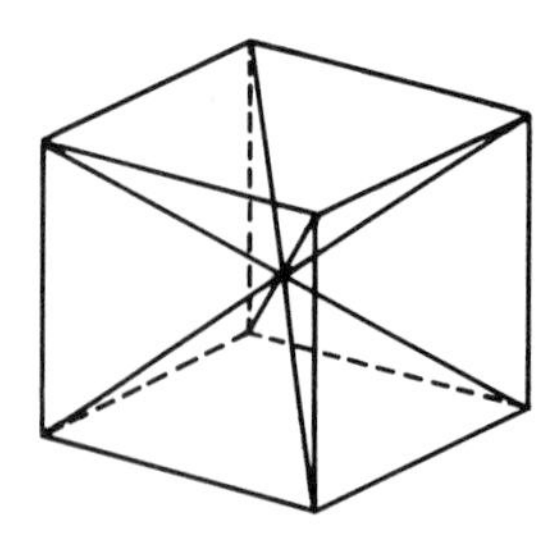

26. Find the volume of one of the pyramids by using the formula for the volume of a pyramid.
27. Find the volume of one of the pyramids *without* using the formula for the volume of a pyramid.

The figure below represents a regular pyramid. Each edge of its base has a length of 16 and its altitude, $\overline{PO}$, has a length of 15. Point E is the midpoint of $\overline{DC}$.

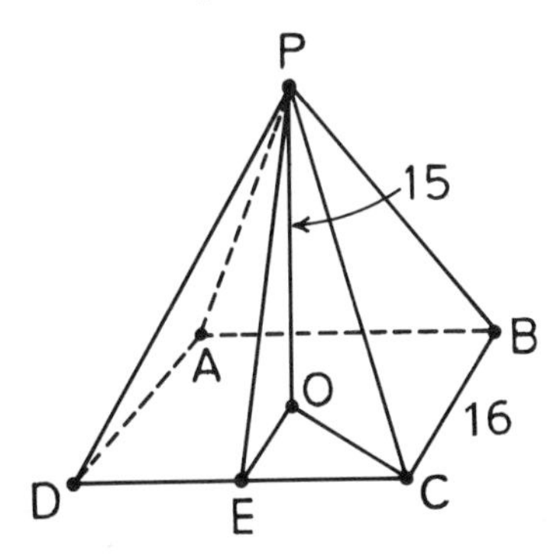

28. Find PE.

29. Find the lateral area of the pyramid.

30. Find the volume of the pyramid.

A *frustrum* of a pyramid is the part of the pyramid included between its base and a plane parallel to its base. The figure in the next column represents a triangular pyramid whose altitude, $\overline{GC}$, is one of its lateral edges. The shaded part of the figure is a frustrum of the pyramid.

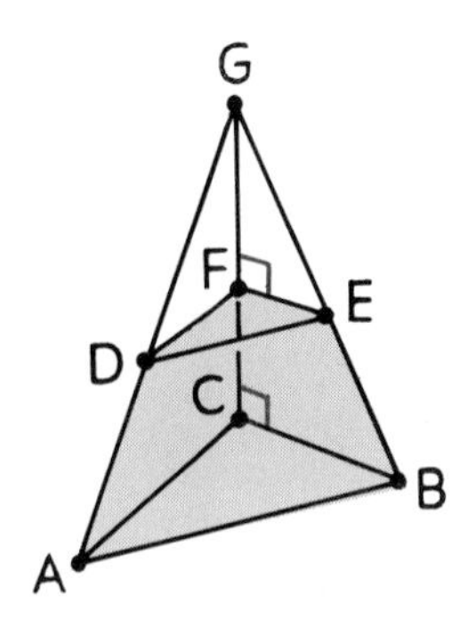

31. ABED is one of the lateral faces of the frustrum. What kind of quadrilateral is it?

32. $\triangle ABC$ and $\triangle DEF$ are the bases of the frustrum. What relation do they appear to have to each other?

Suppose that GF = 6, FC = 4, and $\alpha\triangle ABC = 30$.

33. Find the volume of the pyramid whose base is $\triangle ABC$.

34. Find $\alpha\triangle DEF$.

35. Find the volume of the pyramid whose base is $\triangle DEF$.

36. Find the volume of the frustrum.

Set III

The following question appeared on the PSAT test given in October 1980.

In pyramids ABCD and EFGHI shown here, all faces except base FGHI are equilateral triangles of equal size. If face ABC were placed on face EFG so that the vertices of the triangles coincide, how many exposed faces would the resulting solid have?

(A) Five
(B) Six
(C) Seven
(D) Eight
(E) Nine

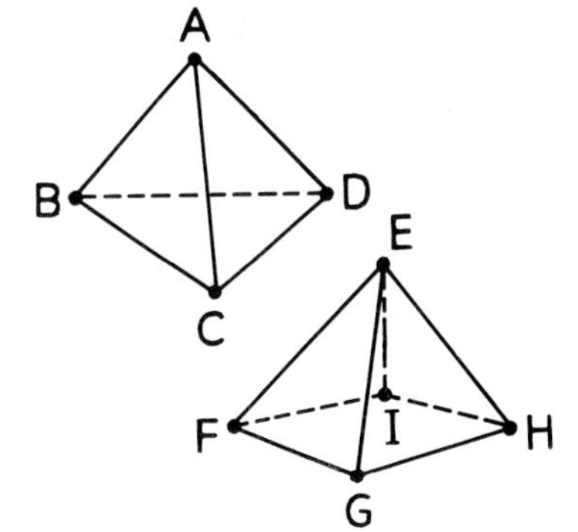

Daniel Lowen, a student at Cocoa Beach High School in Florida, showed that the "expected" answer to the question was wrong.

1. What do you think the "expected" answer was? Explain your reasoning.

Make models of the two pyramids and put them together as described in the question.

2. What do you think the correct answer to the question is? Explain.

Lesson 6
Cylinders and Cones

The Minneapolis Institute of Art

This picture, part of a painting titled *Euclidean Walks* by René Magritte, contains several deliberate visual tricks. One of them concerns the roof of the tower and the street extending out to the horizon. They are almost identical in appearance, and yet the tower roof is shaped like a *cone,* a three-dimensional geometric solid, whereas the surface of the street is a two-dimensional figure bounded by parallel lines.

A cone is very much like a pyramid. Its base, however, is bounded by a circle rather than a polygon. And, instead of having a set of flat triangular faces, it has a single curved surface called its lateral surface.

We will use the term *circular region* to mean the union of a circle and its interior. With this agreement, it is easy to define the term *cone.* We simply replace the word "polygonal" in the definition of a pyramid with the word "circular."

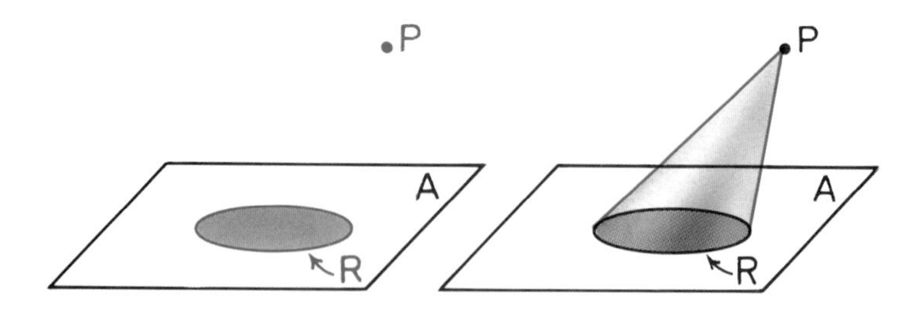

► **Definition**
Suppose that A is a plane, R is a circular region in plane A, and P is a point not in plane A. The set of all segments that join P to a point of region R form a ***cone.***

The words "base," "vertex," and "altitude" are used in the same sense with respect to cones that they are with pyramids. The line segment joining the vertex of a cone to the center of its base is called its *axis*. A cone is either *right* or *oblique*, depending on whether its axis is perpendicular or oblique to its base.

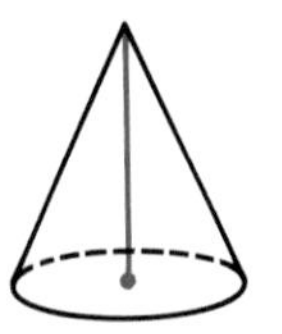

A right cone

Just as cones are the circular counterparts of *pyramids*, so are cylinders the circular counterparts of *prisms*. Again, by changing the word "polygonal" to "circular," the definition of a prism becomes that of a cylinder.

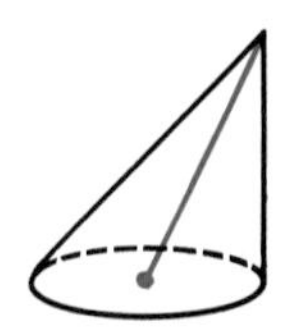

An oblique cone

Definition
Suppose that A and B are two parallel planes, R is a circular region in one plane, and ℓ is a line that intersects both planes but not R. The set of all segments parallel to line ℓ that join a point of region R to a point of the other plane form a ***cylinder.***

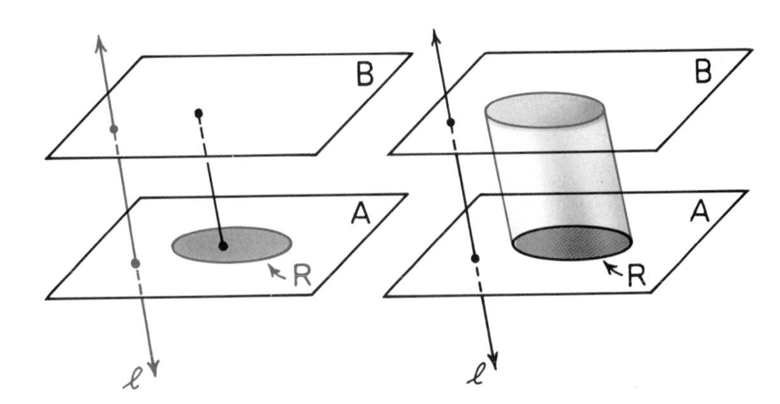

Every cylinder has three surfaces: two flat ones, which are its bases, and a curved one, which is its lateral surface. The *axis* of a cylinder is the line segment joining the centers of its bases. Cylinders, like cones, are classified as *right* or *oblique*, depending on the direction of their axes with respect to their bases.

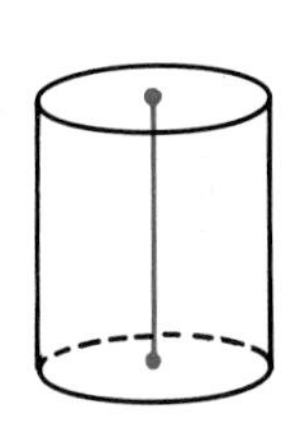

A right cylinder

Because a cylinder can be closely approximated by a prism and a cone approximated by a pyramid, the formulas for the volumes of prisms and pyramids can be used to find the volumes of cylinders and cones as well. They are restated as the following theorems without proof.

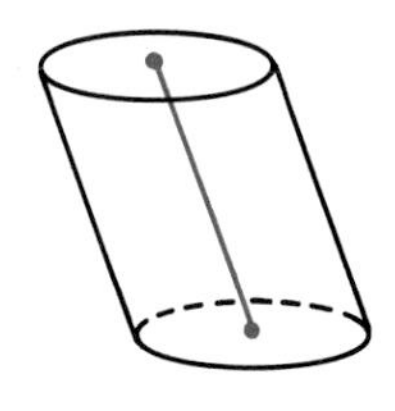

An oblique cylinder

Theorem 93
The volume of a cylinder is $\pi r^2 h$, in which r is the radius of its bases and h is the length of its altitude.

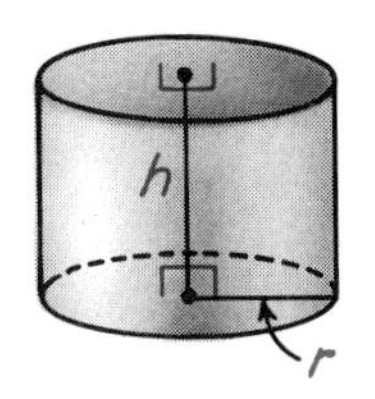

▶ **Theorem 94**

The volume of a cone is $\frac{1}{3}\pi r^2 h$, in which r is the radius of its base and h is the length of its altitude.

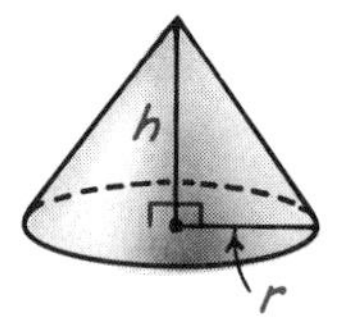

Exercises

Set I

The figure below represents a cone. B is the center of its circular end.

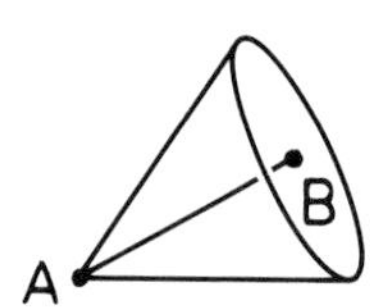

1. What is point A called with respect to the cone?
2. What is the circular region with center B called with respect to the cone?
3. What is $\overline{AB}$ called?
4. What is the curved surface of the cone called?
5. What must be true if the figure is a right cone?

Find the exact volume of each of these cylinders and cones.

6.

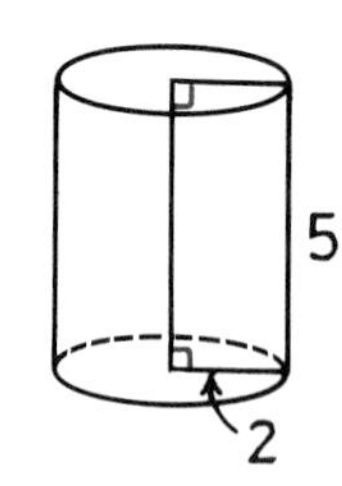

7.

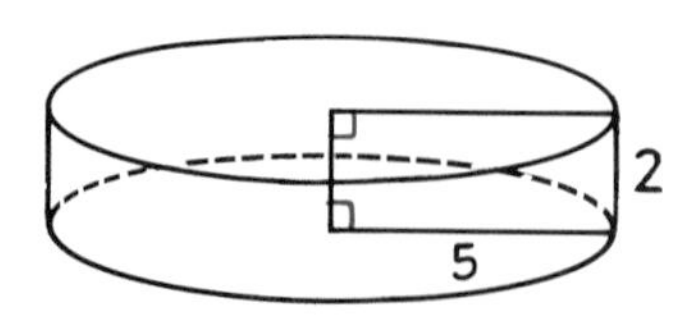

8.

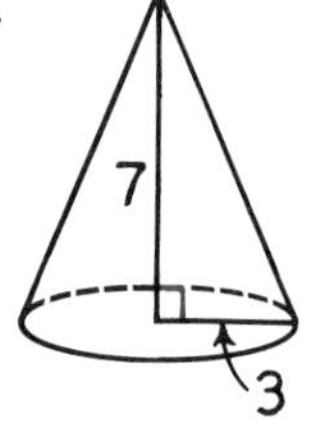

9.

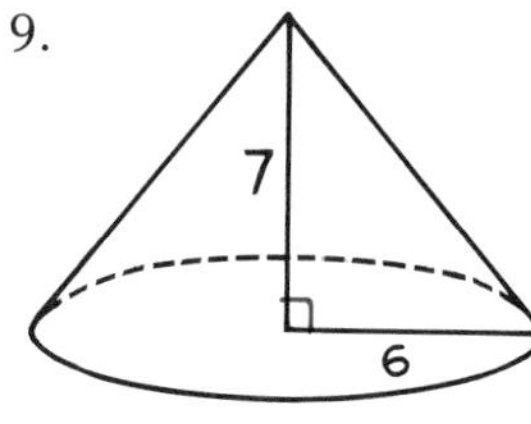

10.

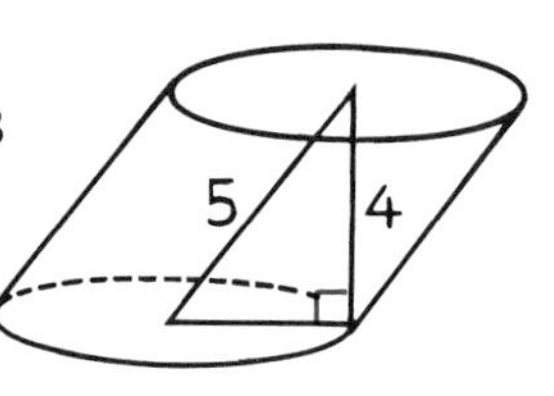

11.

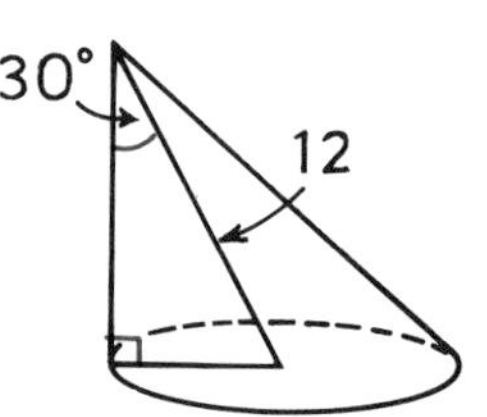

Every plane that intersects solids A and B and that is parallel to plane C cuts off cross sections of equal areas. Solid A is a cylinder whose altitude is 7 and whose base has a radius of 3.

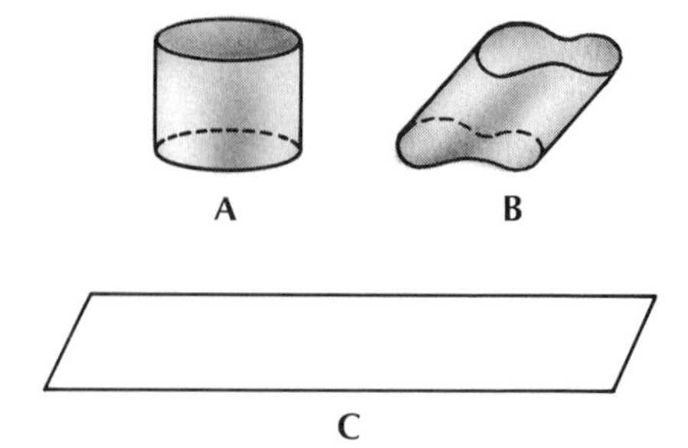

12. Find the volume of solid B.
13. What is the basis for your answer to Exercise 12?

The figures below represent right cones.

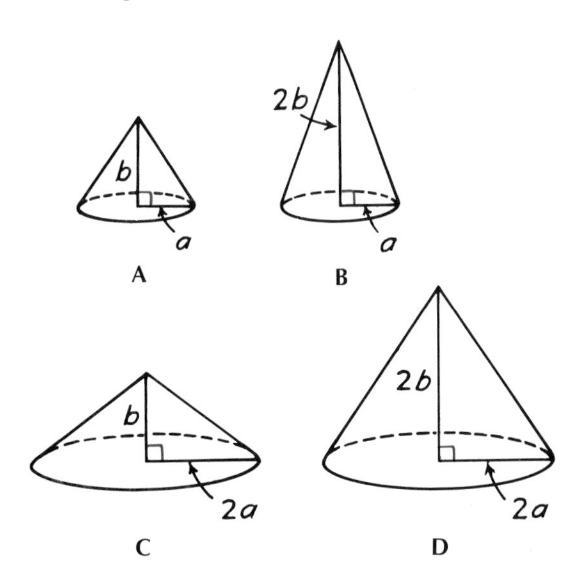

Write an expression for the volume of

14. cone A.
15. cone B.
16. cone C.
17. cone D.

Use your answers to Exercises 14 through 17 to tell what happens to the volume of a cone

18. if its altitude is doubled and the radius of its base remains unchanged.
19. if the radius of its base is doubled and its altitude remains unchanged.
20. if both its altitude and the radius of its base are doubled.

Set II

The figure below shows what the surface of a right cylinder would look like if it were flattened out.

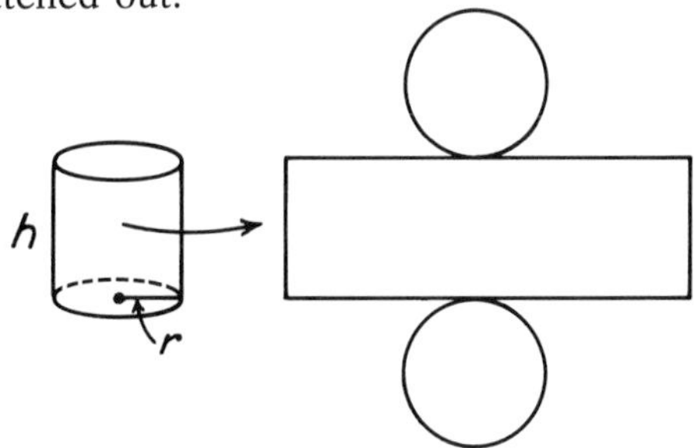

21. The lateral area of a cylinder is the area of its curved surface. Write an expression for the lateral area in terms of r and h.
22. Write an expression for the total area in terms of r and h.

The figure below represents a box that contains six right cylinders.

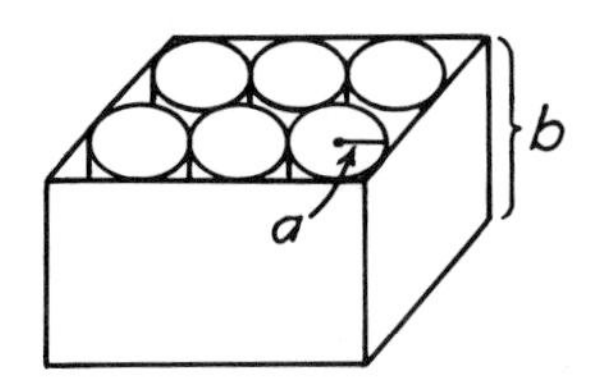

23. Write an expression for the total volume of the cylinders.
24. Write an expression for the volume of the box.
25. What percentage of the volume of the box is filled by the cylinders?

This "hourglass" consists of two identical cones contained in a right cylinder. The cylinder is 6 inches tall and the radius of its base is 2 inches long.

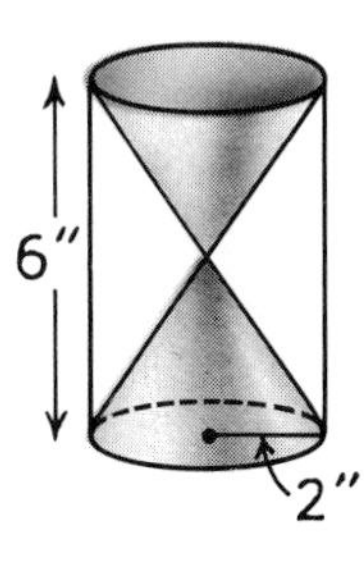

26. Find the exact volume of the shaded region of the figure.
27. Find the exact volume of the space between the cones and the cylinder in the same figure.

Photograph by Syndication International in *Laughing Camera 3,* Hanns Reich Verlag

This cat (or cats?) is walking through a cylindrical pipe.

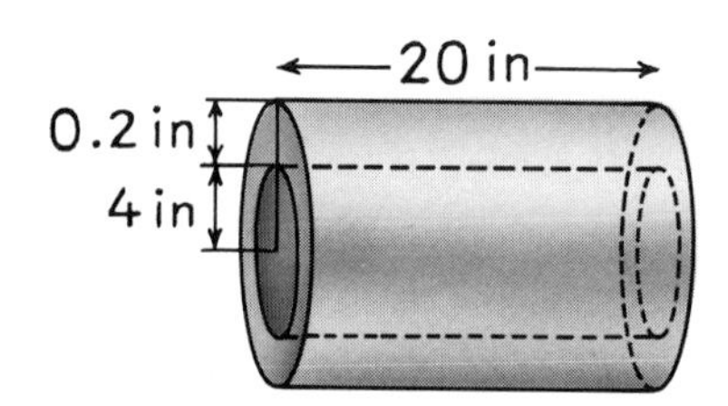

28. If the dimensions of the pipe are as shown in the figure above, find the volume of material that it contains. Express your answer to the nearest integer.

A frustrum of a cone is the part of the cone included between its base and a plane parallel to its base. The figure below represents a right cone whose axis is $\overline{AD}$; AB = 6, BC = 4, and BD = 3. The shaded part of the figure is a frustrum of the cone.

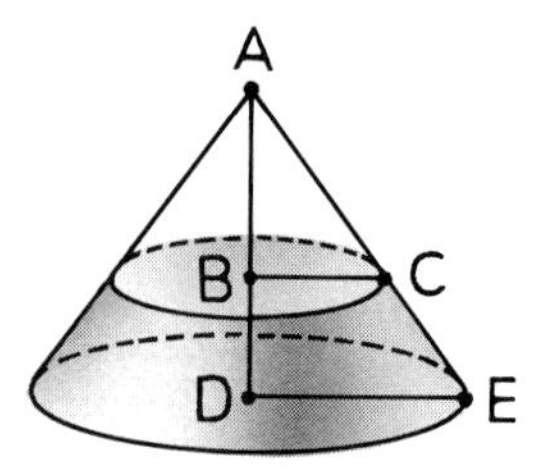

29. Find DE.
30. Find the volume of the frustrum.

The lateral area of a cone is the area of its curved surface. If the curved surface of a right cone is flattened out, it has the shape of a sector of a circle.

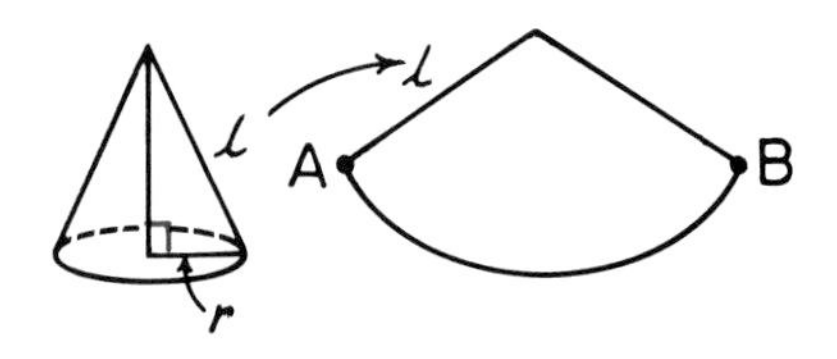

31. Express the length of $\overset{\frown}{AB}$ in terms of r.
32. Express the lateral area of the cone in terms of r and l.

Set III

In the blood disease multiple myeloma, blood cells stick together like piles of checker pieces. This causes the area of the exposed surfaces of the blood cells to be reduced.

On the assumption that a blood cell is a right cylinder with a radius of 3.6 micrometers and an altitude of 2.2 micrometers, find each of the following quantities to the nearest integer.

1. The total area of three separate blood cells.
2. The exposed area of three blood cells that are stuck together as shown in the picture at the left.
3. What percentage of the area of three separate blood cells remains exposed after they become stuck together?

Wide World Photos, Inc.

Lesson 7
Spheres

At the age of 81, Mr. Luke Roberts decided to start collecting string. This photograph shows him standing with the result of his unusual hobby, a ball of string three feet in diameter! How heavy would a ball of string this size be? To answer this, it would be helpful to know how to find the volume of a sphere.

► **Definition**
A ***sphere*** is the set of all points in space that are at a given distance from a given point.

The given distance is called the *radius* of the sphere and the given point is called its *center*. By the volume of a sphere, we mean the volume of the solid consisting of the sphere and its interior. The volume of a sphere is determined by its radius just as is the area of a circle.

To find a formula for the volume of a sphere, we will use Cavalieri's Principle. We begin by imagining a sphere of radius r sliced by a plane at a distance d from its center. The cross section of the sphere is a circle; we will let its radius be x.

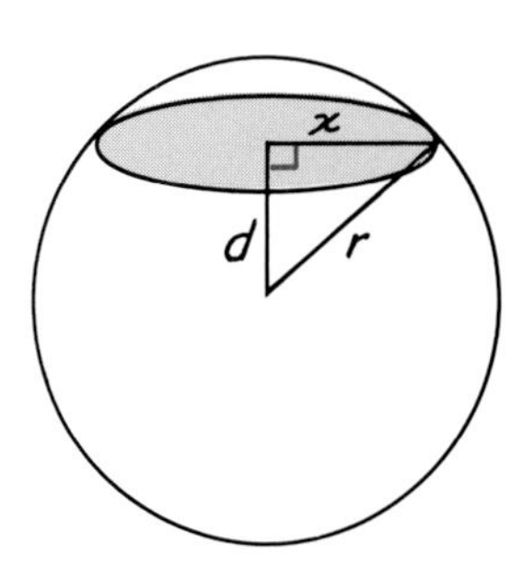

By the Pythagorean Theorem,

$$x^2 + d^2 = r^2, \quad \text{and so} \quad x^2 = r^2 - d^2.$$

Because the cross section is a circle, its area is πx^2. Substituting for x^2, we have

$$\alpha_{\text{cross section}} = \pi(r^2 - d^2) = \pi r^2 - \pi d^2.$$

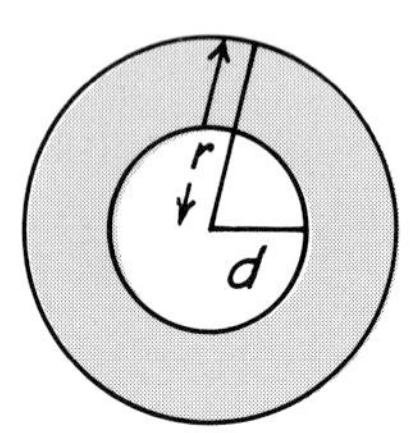

This result can be interpreted as the difference between the areas of two circles having radii r and d. In the adjoining figure, this is the area of the shaded region between the two circles.

Now to apply Cavalieri's Principle to finding the volume of the sphere, we will construct a geometric solid that has this kind of cross section. It consists of a right cylinder from which two identical cones have been removed, as shown in the figure below. The cylinder has the

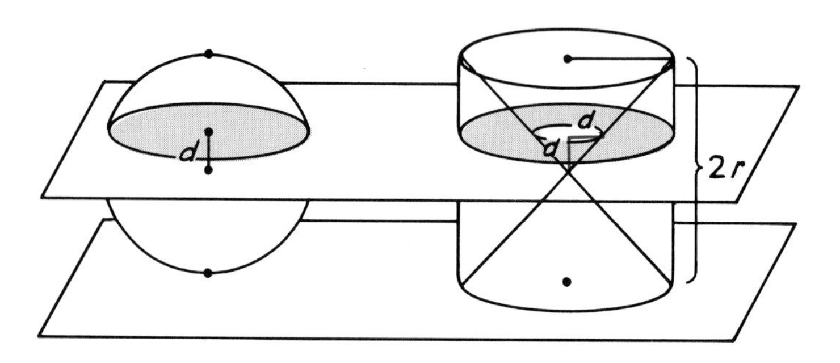

same radius, r, and the same height, $2r$, as the sphere and is situated so that the two solids rest on the same plane. The two hollowed-out cones meet at the center of the cylinder and each shares one of its bases.

Consider a cross section of this solid at a distance d from its center. It is bounded by two circles with radii of r and d and so its area is

$$\pi r^2 - \pi d^2.$$

Because this is the same as the area of the corresponding cross section of the sphere, it follows from Cavalieri's Principle that the volumes of the two solids are the same.

We can easily find the volume of the cylindrical solid. It is equal to

$$\begin{aligned} V_{\text{cylinder}} - V_{\text{two cones}} &= \\ \pi r^2(2r) - 2\left[\frac{1}{3}\pi r^2(r)\right] &= \\ 2\pi r^3 - \frac{2}{3}\pi r^3 &= \frac{4}{3}\pi r^3. \end{aligned}$$

It follows that this is also the volume of the sphere.

► **Theorem 95**

The volume of a sphere is $\frac{4}{3}\pi r^3$, in which r is its radius.

Next we will use an intuitive argument to derive a formula for the surface area of a sphere. To do this, we will divide the sphere into a large number of small "pyramids." Imagine that the surface of the sphere is separated into a large number of tiny "polygons." They are not actually polygons because there are no straight line segments on the surface of a sphere. However, the shorter the "line segments", the closer they come to forming polygons.

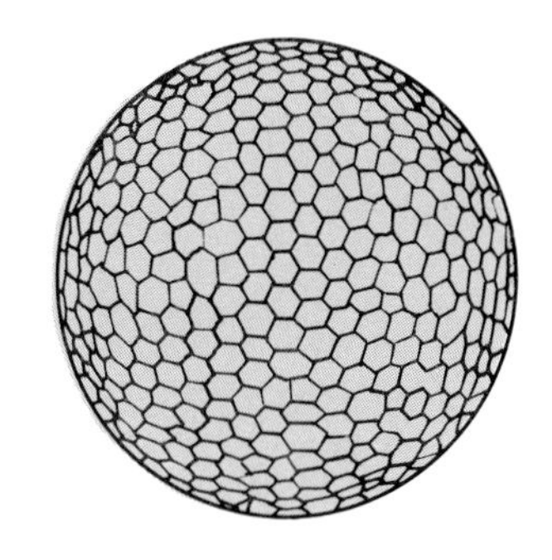

Now imagine joining the corners of all of these "polygons" to the center of the sphere so that they become the bases of a set of "pyramids" all with a common vertex, the center of the sphere. All of the pyramids, then, have altitudes equal to the radius of the sphere.

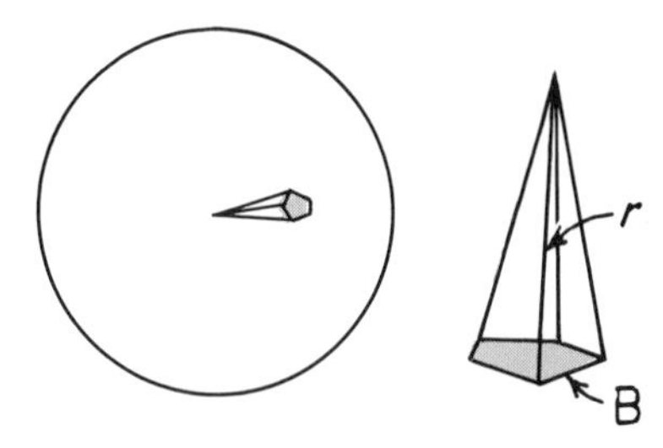

The volume of one of these pyramids is $\frac{1}{3}Br$, in which B is the area of its base and r is the length of its altitude. The volume of the sphere is the sum of the volumes of all of the pyramids. If the areas of their bases are B_1, B_2, B_3, and so on, then their volumes are $\frac{1}{3}B_1r$, $\frac{1}{3}B_2r$, $\frac{1}{3}B_3r$, and so on, and

$$\begin{aligned} V_{\text{sphere}} &= \frac{1}{3}B_1r + \frac{1}{3}B_2r + \frac{1}{3}B_3r + \cdots \\ &= \frac{1}{3}r(B_1 + B_2 + B_3 + \cdots) \end{aligned}$$

Now $\alpha_{\text{sphere}} = B_1 + B_2 + B_3 + \cdots$, and so

$$V_{\text{sphere}} = \frac{1}{3}r\alpha_{\text{sphere}}.$$

Solving this equation for α_{sphere}, we get

$$\alpha_{\text{sphere}} = \frac{3V_{\text{sphere}}}{r}.$$

But $V_{\text{sphere}} = \frac{4}{3}\pi r^3$, and so

$$\alpha_{\text{sphere}} = \frac{3\left(\frac{4}{3}\pi r^3\right)}{r} = 4\pi r^2.$$

As already noted, the argument in developing this formula has been an intuitive one. Although polygons and pyramids have been talked about, no such figures can really exist. So we cannot consider our argument a proof. It is possible by means of the calculus, however, to derive the *same result* without making any approximations such as we have made.

▶ **Theorem 96**
The surface area of a sphere is $4\pi r^2$, in which r is its radius.

Exercises

Set I

Find the exact surface area and volume of each of the following spheres.

Example:

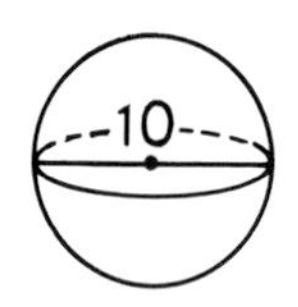

Solution: The radius of this sphere is 5, and so

$$\alpha_{\text{sphere}} = 4\pi 5^2 = 4\pi(25) = 100\pi;$$
$$V_{\text{sphere}} = \frac{4}{3}\pi 5^3 = \frac{4}{3}\pi(125) = \frac{500}{3}\pi.$$

1.

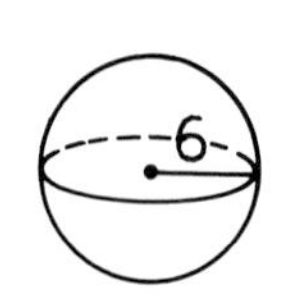

2.

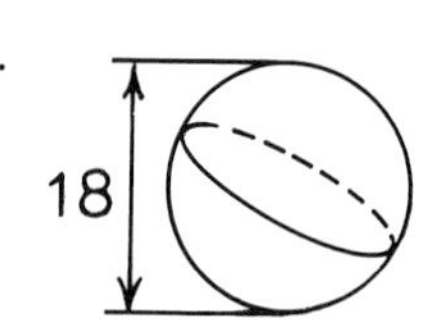

3.

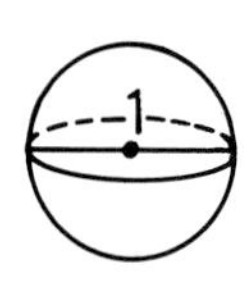

4.

Two spheres have radii of x and $2x$ units, respectively.

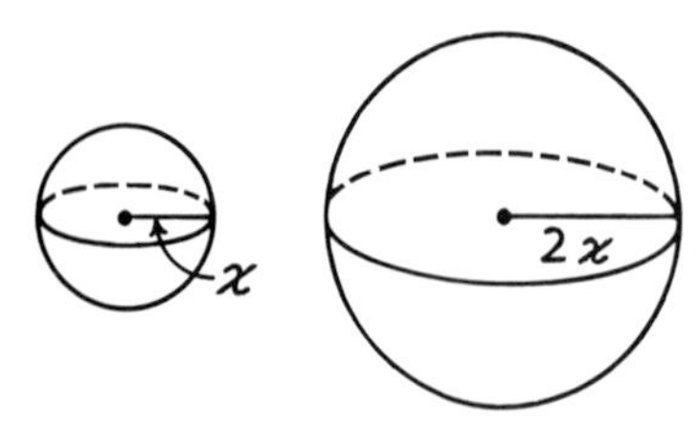

5. Write expressions for the surface areas of the spheres.

6. Write expressions for the volumes of the spheres.

Refer to your answers to Exercises 5 and 6 to answer the following questions.

7. What happens to the surface area of a sphere if its radius is doubled?

8. What happens to the volume of a sphere if its radius is doubled?

If a sphere is sliced through its center into two identical parts, each part is called a *hemisphere.* Suppose that a hemisphere has radius r. Write an expression for each of the following quantities.

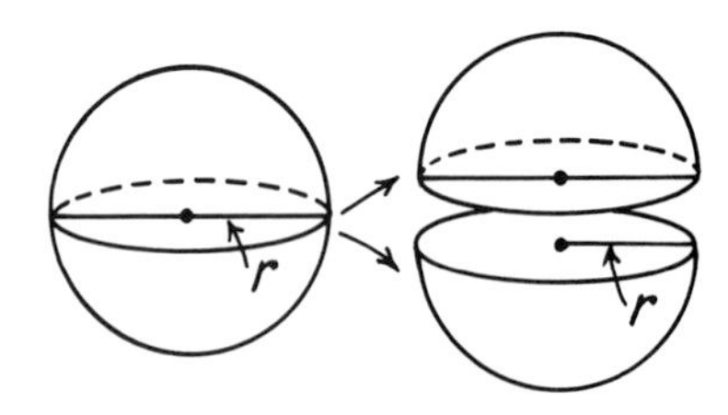

9. The volume of the hemisphere.

10. The area of its curved surface.

11. Its total surface area.

The figure below represents sphere O intersected by plane P in more than one point. Points A and B represent any two points in the intersection and $\overline{OM}$ is perpendicular to plane P.

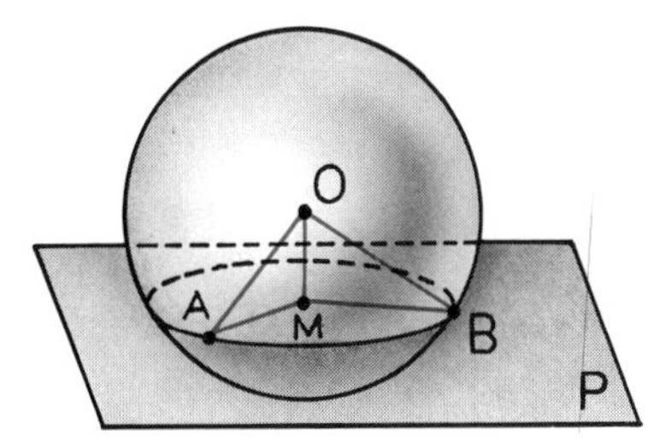

12. Why is OA = OB?
13. Why is $\overline{OM} \perp \overline{MA}$ and $\overline{OM} \perp \overline{MB}$?
14. Why is $\triangle OMA \cong \triangle OMB$?
15. Why is MA = MB?
16. What does the fact that MA = MB imply about the curve that is the intersection of sphere O and plane P?

The earth as seen from Apollo 17.

The radius of the earth is approximately 4,000 miles.

17. Find the length of the earth's equator. Round your answer to the nearest thousand miles.
18. Find the area of the earth. Round your answer to the nearest million square miles.
19. Find the volume of the earth. Round your answer to the nearest billion cubic miles.

In the figure below, the radius of the base of the right cone is equal to the radius of the sphere. The two solids have the same height.

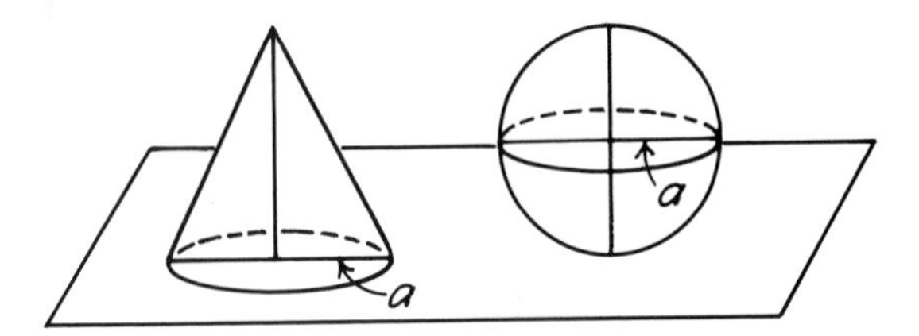

20. Write an expression for the exact volume of the cone.
21. Write an expression for the exact volume of the sphere.
22. Exactly how does the volume of the cone compare with that of the sphere?

The figure below represents a right cylinder from which two identical cones have been removed. Write an expression for each of the following volumes.

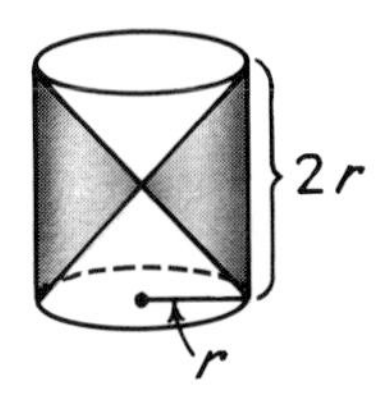

23. The volume of the right cylinder.
24. The volume of one of the two hollowed-out cones.
25. The volume of the solid.

The figure below represents a sphere inscribed in a cube. This means that the sphere touches each face of the cube in a point. Each edge of the cube has a length of e.

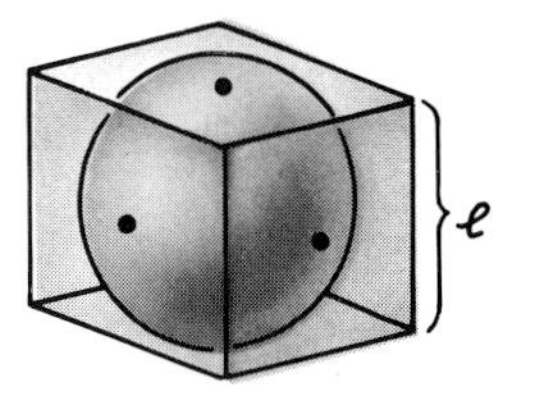

26. Write an expression for the area of the sphere.

27. Write an expression for the area of the cube.

28. Find the ratio of the areas of the sphere and cube.

29. Write an expression for the volume of the sphere.

30. Write an expression for the volume of the cube.

31. Find the ratio of the volumes of the sphere and cube.

Mr. Roberts's ball of string (pictured at the beginning of this lesson) had a diameter of 3 feet.

32. Find its volume in cubic feet.

33. Assuming that one cubic inch of string weighs 0.03 pound and that the ball was solid string, find how much the ball weighed.

Set III

The following problem appears in an old book of puzzles by Sam Loyd, a man who invented hundreds of clever puzzles in the late nineteenth and early twentieth centuries.*

A balloon for a trip to the moon is attached to a ball of wire in which the wire is one hundredth of an inch thick. If the ball of wire was originally two feet in diameter and was wound so solidly that there was no air space, what was the total length of the wire?

In his book, Mr. Loyd claimed that the problem could be solved without being concerned about the value of π. Can you do it?

*Originally in Sam Loyd's *Cyclopedia of Puzzles,* privately published in 1914. Included in *Mathematical Puzzles of Sam Loyd,* Volume 2, selected and edited by Martin Gardner (Dover, 1960).

"And one more thing. Don't cross your eyes while I'm speaking with you."

Lesson 8
Similar Solids

In *Gulliver's Travels* by Jonathan Swift, Gulliver's first voyage took him to Lilliput, a land of people "not six inches high." The emperor of Lilliput issued a decree that, among other things, specified the amount of food to which Gulliver was entitled. As Gulliver describes it:

"The Reader may please to observe that . . . the Emperor stipulates to allow me a Quantity of Meat and Drink sufficient for the Support of 1728 Lilliputians. Some time after, asking a Friend at Court how they came to fix on that determinate Number, he told me that his Majesty's Mathematicians, having taken the Height of my Body . . . and finding it to exceed theirs in the Proportion of Twelve to One, they concluded from the Similarity of their Bodies that mine must contain at least 1728 of theirs, and consequently would require as much Food as was necessary to support that Number of Lilliputians."

In making their calculations, the emperor's mathematicians assumed that Gulliver's body was *similar* to that of a Lilliputian. Two geometric solids are similar if they have the *same shape*. For example, all spheres are similar but all cylinders are not.

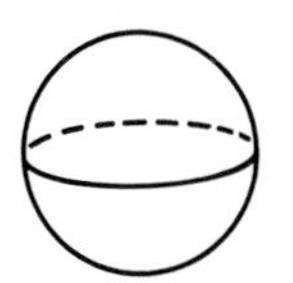
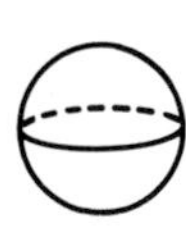
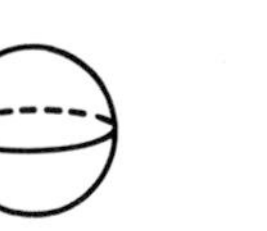

These spheres are similar. **These cylinders are not similar.**

If the solids are polyhedra, their corresponding faces are similar. This means that all of their corresponding dimensions are proportional. For example, the two rectangular solids shown here are similar if

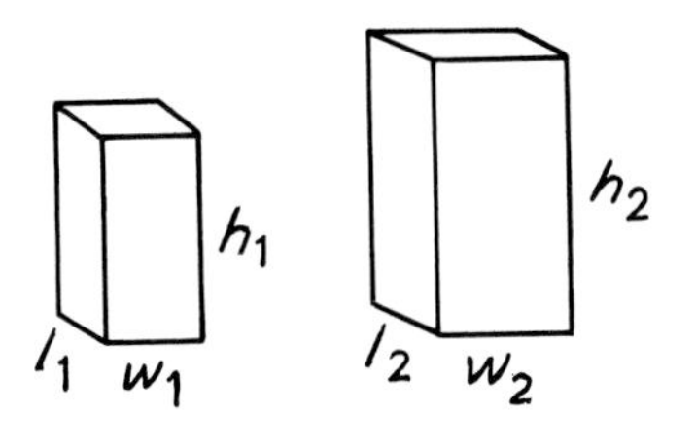

$$\frac{l_1}{l_2} = \frac{w_1}{w_2} = \frac{h_1}{h_2}.$$

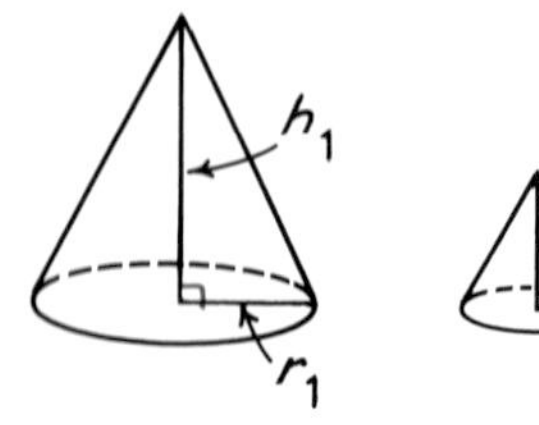

Two right cylinders or cones are similar if their altitudes have the same ratio as the radii of their bases. For example, the two right cones shown here are similar if

$$\frac{h_1}{h_2} = \frac{r_1}{r_2}.$$

How do two similar solids compare in volume? To find out, we will consider a pair of similar rectangular solids, A and B, in which the dimensions of one solid are twice those of the other. The volume of solid

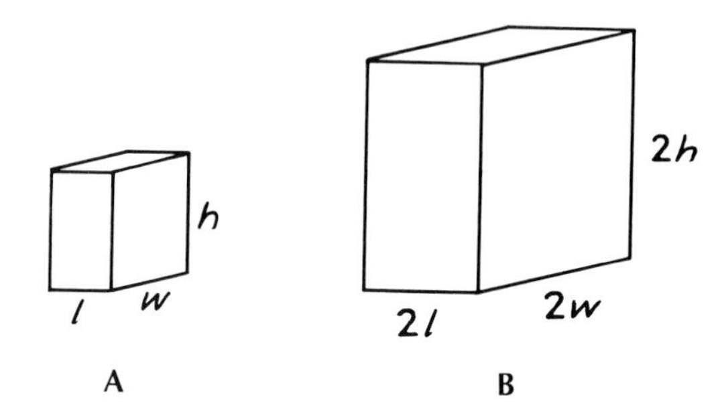

A is lwh and the volume of solid B is $(2l)(2w)(2h) = 8lwh$. So the volume of solid B is 8, or 2^3, times that of solid A.

This relation is also true of other similar solids. Because Gulliver was 12 times as tall as one of the Lilliputians, the emperor's mathematicians concluded that his volume was $12^3 = 1728$ times one of theirs.

In another passage of Swift's book, the emperor's tailors made Gulliver a suit of clothes. If the suit was made of the same fabric used to make their own clothing, how much material did the tailors need?

To answer this, we need to know how the surface areas of two similar solids compare. Consider the rectangular solids A and B above. The surface area of solid A is $2lw + 2wh + 2lh = 2(lh + wh + lh)$ and the surface area of solid B is $2(2l)(2w) + 2(2w)(2h) + 2(2l)(2h) = 8(lw + wh + lh)$. The surface area of solid B is 4, or 2^2, times that of solid A.

Again this relation is true for all similar solids. The emperor's tailors needed $12^2 = 144$ times as much material to make a suit for Gulliver as for themselves.

► **Theorem 97**

The ratio of the surface areas of two similar solids is equal to the square of the ratio of any pair of corresponding dimensions.

► **Theorem 98**

The ratio of the volumes of two similar solids is equal to the cube of the ratio of any pair of corresponding dimensions.

Exercises

Set I

The two cylinders in the figure below are similar. The ratio of their altitudes is 3.

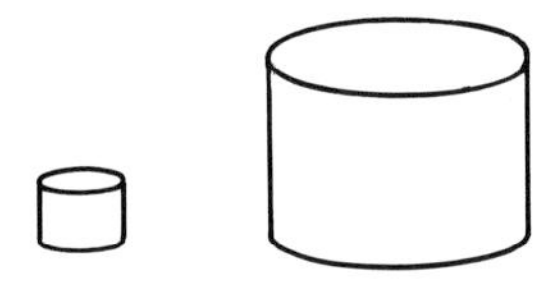

1. What is the ratio of their surface areas?
2. What is the ratio of their volumes?

The two solids in the figure below are similar. Each dimension of the second solid is 4 times the corresponding dimension of the first solid.

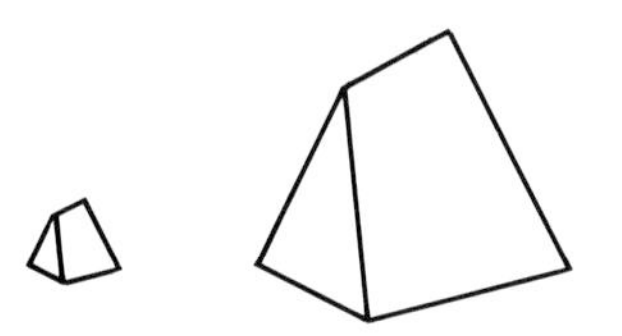

3. How many times the surface area of the first solid is the surface area of the second solid?
4. How many times the volume of the first solid is the volume of the second solid?

The two solids in the figure below are similar. Express each of the following ratios as a common fraction in simplest terms, comparing the first solid to the second.

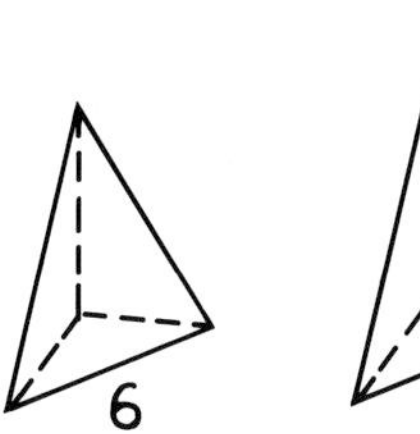

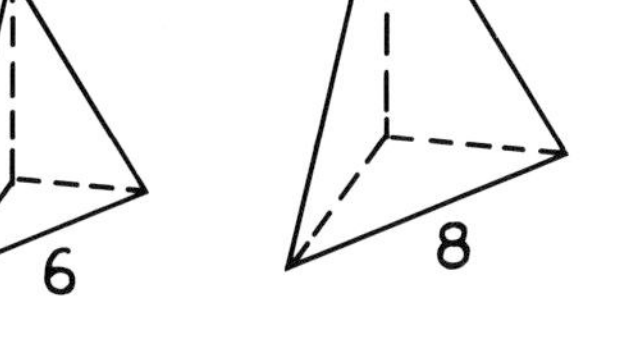

5. What is the ratio of the corresponding dimensions of the solids?
6. What is the ratio of their areas?
7. What is the ratio of their volumes?

Any two cubes are similar.

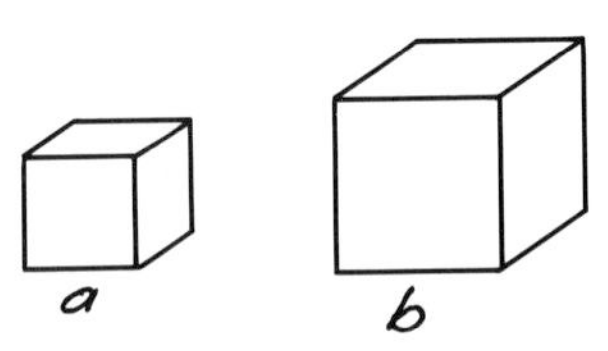

8. Write an expression for the ratio of the corresponding dimensions of the two cubes shown in the figure.
9. Write an expression for the surface area of each cube.
10. Write an expression for the ratio of the surface areas of the cubes.
11. Write an expression for the volume of each cube.
12. Write an expression for the ratio of the volumes of the cubes.

The two rectangular pyramids in the figure below are similar. In finding each of the following ratios, compare the first pyramid with

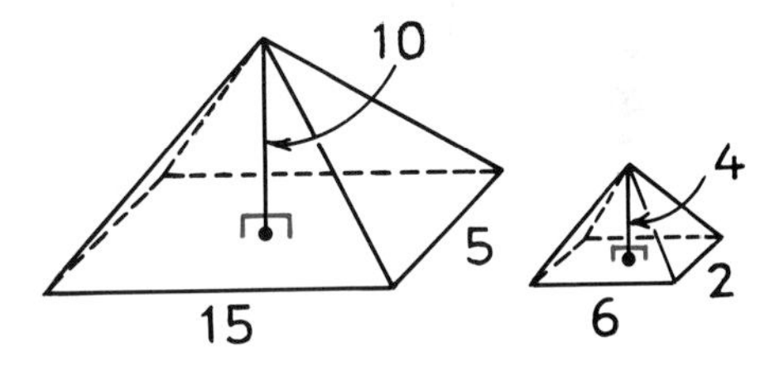

the second. Express each ratio as a common fraction in simplest terms.

13. Find the ratio of their corresponding dimensions.
14. Find the perimeter of the base of each pyramid.
15. Find the ratio of the perimeters of the bases.
16. How is this ratio related to the ratio of the corresponding dimensions?

17. Find the area of the base of each pyramid.
18. Find the ratio of the areas of the bases.
19. How is this ratio related to the ratio of the corresponding dimensions?
20. Find the volume of each pyramid.
21. Find the ratio of the volumes of the pyramids.
22. How is this ratio related to the ratio of the corresponding dimensions?

Set II

The two rectangular solids in the figure below are similar. Find each of the following quantities.

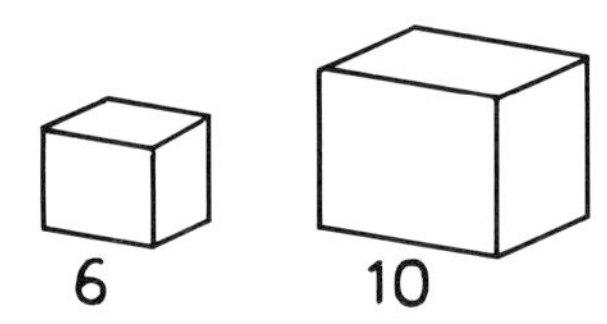

Example: The area of the second solid if the area of the first solid is 135.

Solution: $\frac{135}{x} = \left(\frac{6}{10}\right)^2$ (the ratio of the areas is equal to the square of the ratio of the corresponding dimensions).

$\frac{135}{x} = \left(\frac{3}{5}\right)^2$, $\frac{135}{x} = \frac{9}{25}$,

$9x = 3375$, $x = 375$.

23. The volume of the second solid if the volume of the first solid is 108.
24. A diagonal of the second solid if a diagonal of the first solid is 9.

The two right cones in the figure below are similar. Find each of the following quantities.

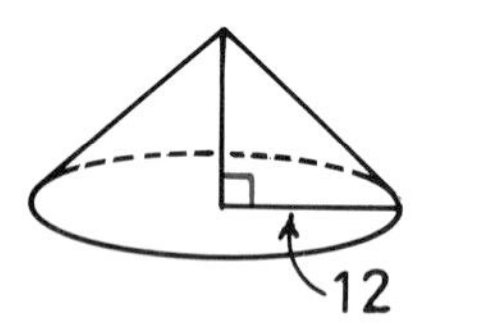

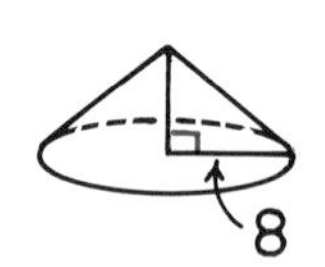

25. The altitude of the second cone if the altitude of the first cone is 3.
26. The area of the second cone if the area of the first cone is 720.
27. The volume of the second cone if the volume of the first cone is 324.

The two bones shown in the figure below are similar, and the second is three times as long as the first. The strengths of two similar bones are proportional to their cross-sectional areas.

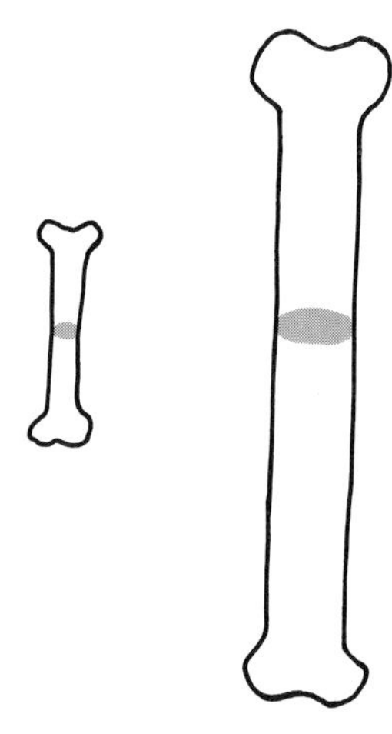

28. How much stronger than the first bone do you think the second one is?

The bones are from two animals that have similar shapes but different sizes. The weights of two similar animals are proportional to their volumes.

29. How many times heavier than the first animal would the second one be?
30. Which animal has a skeleton more capable of supporting its own weight?

The Rin-Tin-Tin Company sells dog food in tin cans of two sizes: "large" and "colossal." The cans are 8 and 12 inches tall, respectively, and are similar in shape.

31. If the large can sells for 24¢ and the colossal can sells for 72¢, which is a better buy?

Suppose that the walls of the two cans have the same thickness.

32. If the large can contains 4¢ worth of metal, what is the value of the metal in the colossal can?

33. If it costs the company 8¢ to produce the dog food in the large can, how much does it cost to produce the dog food in the colossal can?

34. How does the selling price of each can compare with the cost of the materials (the metal and the dog food) in it?

A person's weight is approximately proportional to his or her volume. Suppose that someone is five feet tall and weighs 100 pounds. How much would a person similar in shape weigh who was

35. six feet tall?

36. seven feet tall?

Set III

By the permission of Johnny Hart and News America Syndicate

Clams range in size from as little as a pinhead to more than four feet in length! If a clam 2.4 inches long weighs one ounce, how much would you expect a clam having the same shape and a length of four feet to weigh? Explain your reasoning.

Chapter 15 / Summary and Review

Basic Ideas

Postulates

14. *Cavalieri's Principle.* Consider two geometric solids and a plane. If every plane parallel to this plane that intersects one of the solids also intersects the other so that the resulting cross sections have the same area, then the two solids have the same volume. 552

15. The volume of any prism is Bh, where B is the area of one of its bases and h is the length of its altitude. 552

Theorems

91. The length of a diagonal of a rectangular solid is $\sqrt{l^2 + w^2 + h^2}$, in which l, w, and h are its dimensions. 542
    Corollary. The length of a diagonal of a cube is $e\sqrt{3}$, in which e is the length of one of its edges. 542

    Corollary 1 to Postulate 15. The volume of a rectangular solid is lwh, in which l, w, and h are its length, width, and height. 552
    Corollary 2 to Postulate 15. The volume of a cube is e^3, in which e is the length of one of its edges. 552

92. The volume of any pyramid is $\frac{1}{3}Bh$, in which B is the area of its base and h is the length of its altitude. 556

93. The volume of a cylinder is $\pi r^2 h$, in which r is the radius of its bases and h is the length of its altitude. 561

94. The volume of a cone is $\frac{1}{3}\pi r^2 h$, in which r is the radius of its base and h is the length of its altitude. 562

95. The volume of a sphere is $\frac{4}{3}\pi r^3$, in which r is its radius. 566

96. The surface area of a sphere is $4\pi r^2$, in which r is its radius. 568

97. The ratio of the surface areas of two similar solids is equal to the square of the ratio of any pair of corresponding segments. 572

98. The ratio of the volumes of two similar solids is equal to the cube of the ratio of any pair of corresponding segments. 572

Exercises

Set I

Tell whether each of the following statements is true or false.

1. If a line does not intersect a plane, it must be parallel to it.
2. All rectangular solids are prisms.
3. If two cones have equal bases and equal altitudes, they must also have equal volumes.
4. Two planes are perpendicular if one plane contains a line that is perpendicular to the other plane.
5. The lateral faces of every pyramid are isosceles triangles.
6. The diagonals of a cube are perpendicular to each other.
7. All square pyramids are similar.
8. If the radius of a sphere is doubled, its surface area is also doubled.

Find the exact volume of each of these solids.

9.

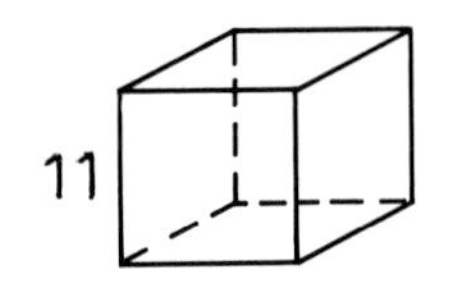

A cube.

10.

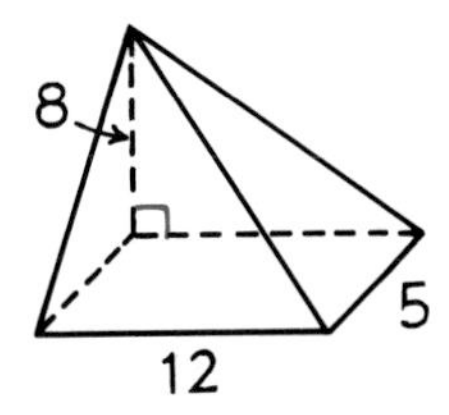

A rectangular pyramid.

11.

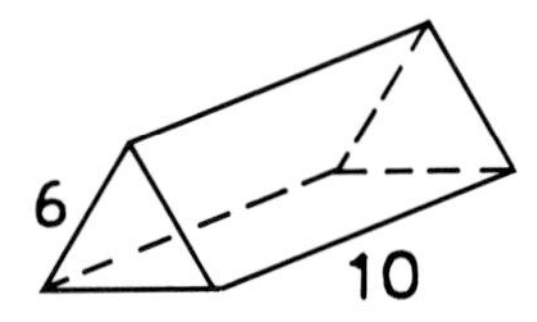

A right prism whose bases are equilateral triangles.

12.

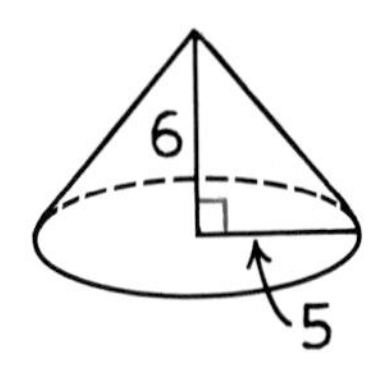

A right cone.

The two right cylinders shown below are similar. In finding each of the following ratios, compare the first cylinder with the second.

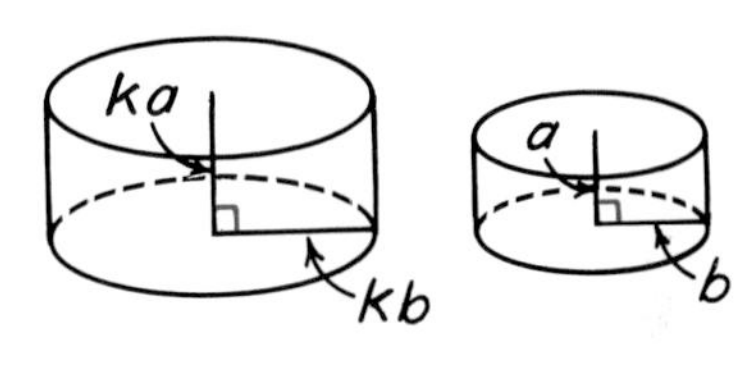

13. Find the ratio of their altitudes.
14. Write an expression for the volume of each cylinder.
15. Find the ratio of their volumes.
16. Write an expression for the lateral area of each cylinder.
17. Find the ratio of their lateral areas.

The two figures shown below represent rectangular solids.

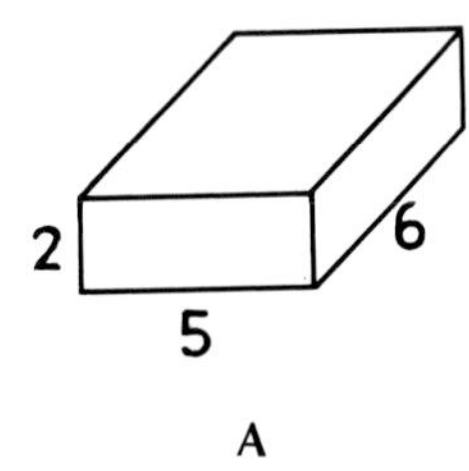

A

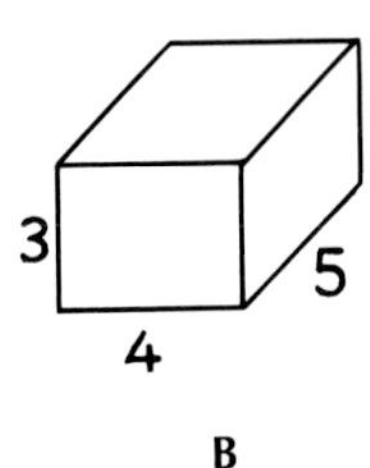

B

18. Find the volume of each solid.
19. Find the surface area of each solid.
20. Find the length of a diagonal of each solid.

Set II

The base of the Great Pyramid is bounded by a square each of whose sides is 230 meters long.

Suppose that the small pyramid built by this Egyptian child is similar to it and that each edge of its base is 1 meter long.

By the permission of Ed Fisher

21. How do the two pyramids compare in height?
22. How do the two pyramids compare in surface area?
23. Find the area of the Great Pyramid, given that the area of the small pyramid is 2.6 square meters. Give your answer to the nearest thousand square meters.
24. How do the two pyramids compare in volume?
25. Find the volume of the Great Pyramid, given that the volume of the small pyramid is 0.21 cubic meter. Give your answer to the nearest 100,000 cubic meters.

The figures below represent a right cylinder and a sphere of equal height. The radius of the bases of the cylinder is equal to the radius of the sphere.

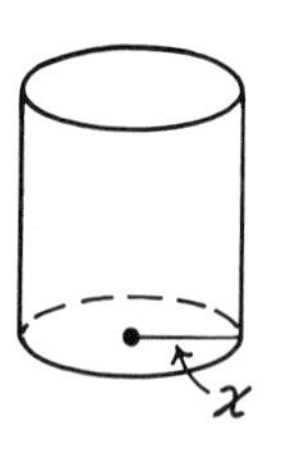

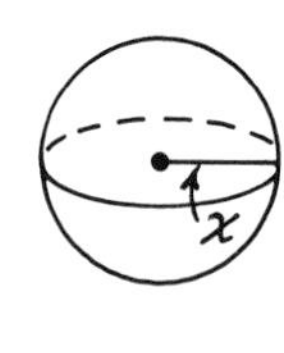

26. Write an expression for the volume of each solid.
27. What fraction of the volume of the cylinder is the volume of the sphere?
28. Write an expression for the lateral area of the cylinder and an expression for the area of the sphere.
29. How do the two areas compare?

The formulas for the surface area and volume of a sphere can be expressed in terms of the diameter, d, instead of the radius, r.

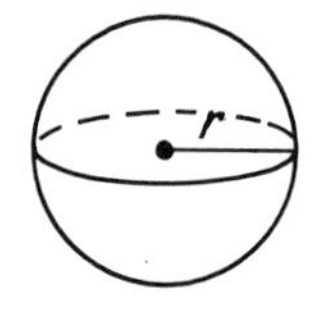

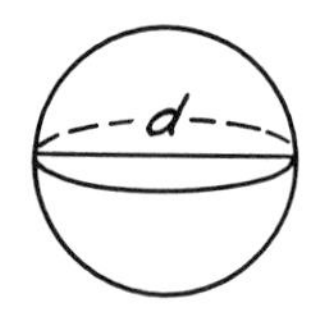

30. Write an equation for r in terms of d.
31. Find an equation for the surface area in terms of d.
32. Find an equation for the volume in terms of d.

Chapter 16

NON-EUCLIDEAN GEOMETRIES

By the permission of Johnny Hart and News America Syndicate

Lesson 1 Geometry on a Sphere

Euclid defined parallel lines as "lines that, being in the same plane and being produced indefinitely in both directions, do not meet one another in either direction." In this cartoon, Peter has attempted to demonstrate the existence of parallel lines by tracing them on the surface of the earth. Unfortunately, his two-pronged stick for drawing the lines was worn down to a nub in the course of the long journey, and so Peter's friends remain skeptical on seeing that his "parallel lines" have come together to meet in a common point at the end of the trip.

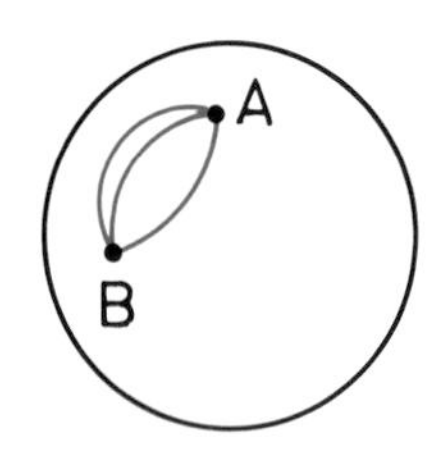

The cartoon raises a number of questions about the geometry that we have been studying and its relation to the earth on which we live. Consider, for example, the distance between two points. In a plane, it is measured along the line determined by the points. On the surface of a sphere, it must be measured along a curved path. Although there are many curved paths between two such points, one of them is the shortest. It is along the *great circle* through the two points.

► **Definition**

A ***great circle*** of a sphere is a set of points that is the intersection of the sphere and a plane containing its center.

Examples of great circles on the earth are the equator and the meridians (the circles that pass through the north and south poles). Because distances on the surface of the earth are measured along great circles, we might consider them "lines" in our development of geometry on a sphere.

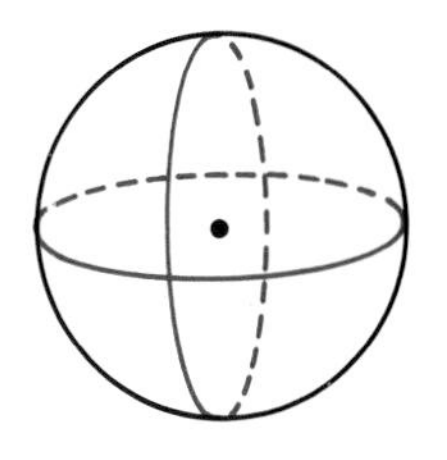
"Lines" on a sphere

If we do, however, then we need to reconsider our postulates about points and lines. Do two points determine a line on a sphere? It depends on where the points are. In the first figure below, there is exactly one

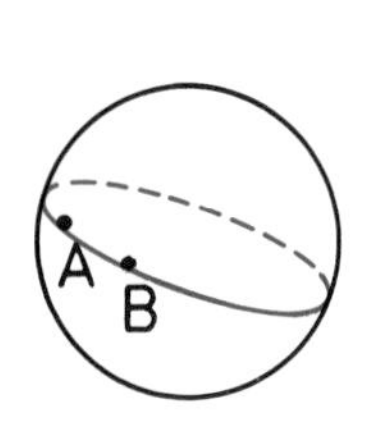

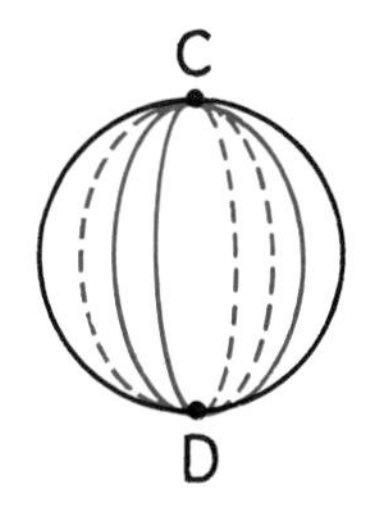

line through points A and B. In the second figure, however, there is more than one line through points C and D. This is due to the fact that points C and D are a pair of *polar points.*

► **Definition**
Polar points are the points of intersection of a line through the center of the sphere with the sphere.

To eliminate the difficulty of having more than one line through a pair of polar points, we might consider them to count as just one point. By agreeing that, when we refer to a given *pair* of points it is understood that they are *not polar,* we have rescued our postulate that two points determine a line. For a given *pair* of points, there is exactly one line that contains them.

With our new ideas about points and lines comes a surprising result. The figure shown here represents a point, P, and a line, ℓ, that does not contain it. Through P, how many lines can be drawn parallel to line ℓ?

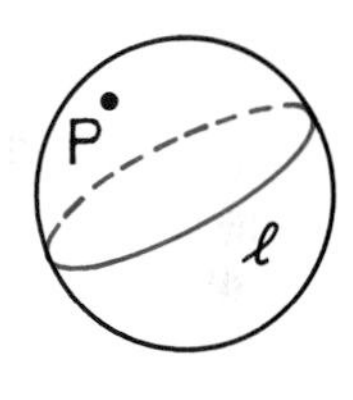

Because our definition of parallel lines states that they lie in the same plane and do not intersect, we will evidently need to consider the sphere to be a "plane." Through P, how many lines can be drawn on the sphere that do not intersect line ℓ? The answer is *none. Every* great circle of a sphere intersects all other great circles of the sphere!

What does all this mean? It means that we have the beginning of a geometry in which the Parallel Postulate that we have used in the past no longer applies, for, through a point not on a line, there are now *no lines parallel to the line.*

With other changes in our postulates concerning distance and betweenness and this new assumption that there are no parallel lines, we can develop a new geometry with all sorts of unexpected theorems. Although these theorems flatly contradict others that we have proved in our study of geometry, they make sense in terms of the new parallel postulate and in terms of each other. They are part of a *non-Euclidean* geometry. In this chapter, we will become acquainted with the two main non-Euclidean geometries.

Exercises

In the following exercises, we will refer to the geometry that we have been studying throughout the course as *Euclidean* geometry and to our new geometry on a sphere as *sphere* geometry. We will restrict our comparison of the two geometries to their two-dimensional versions only; this means that in each case we will assume that all "points" and "lines" lie in the same "plane."

Set I

The terms *point, line,* and *plane* are pictured differently in Euclidean and sphere geometry.

Which of these descriptions, A, B, or C, fits each of the following terms in Euclidean geometry?

A. Is straight and infinite in extent.

B. Is flat and infinite in extent.

C. Has a position but no extent.

1. Point.
2. Line.
3. Plane.

Which of these descriptions, A, B, or C, fits each of the following terms in sphere geometry?

A. A pair of polar points.

B. The surface of the sphere.

C. A great circle of a sphere.

4. Point.
5. Line.
6. Plane.

The figures below represent a line and a plane that contains it in Euclidean geometry and in sphere geometry.

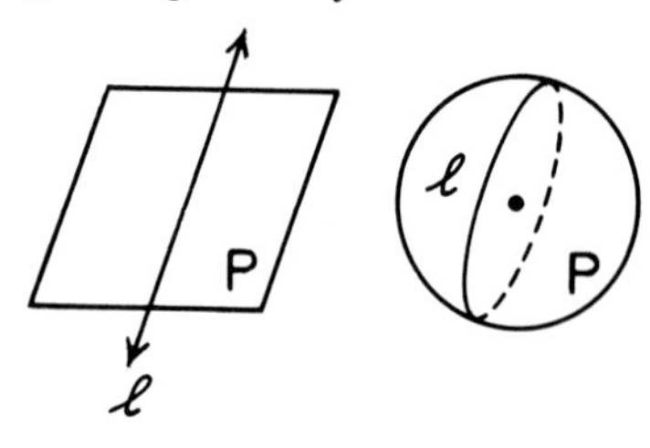

Does a line have endpoints

7. in Euclidean geometry?
8. in sphere geometry?

Does a line have a finite length

9. in Euclidean geometry?
10. in sphere geometry?

Does a line divide a plane that contains it into two separate regions

11. in Euclidean geometry?
12. in sphere geometry?

The figures below represent three points and a line that contains them in Euclidean geometry and in sphere geometry.

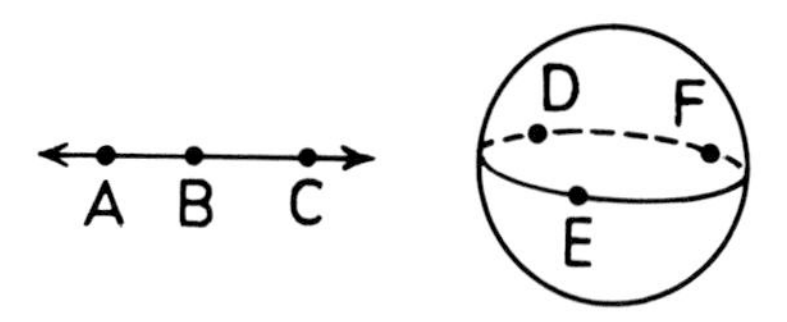

13. Are points A, B, and C collinear?
14. Are points D, E, and F collinear?
15. If three points in Euclidean geometry are collinear, does it follow that exactly one of them is between the other two?
16. If three points in sphere geometry are collinear, does it follow that exactly one of them is between the other two?

The figures below represent two lines that intersect in Euclidean geometry and in sphere geometry.

If two lines intersect, is it true that they intersect in no more than one point

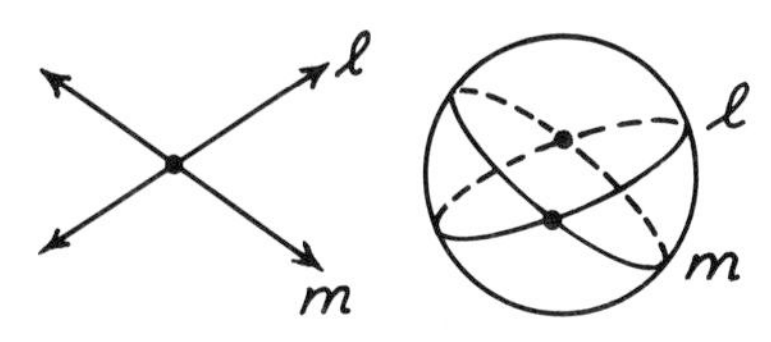

17. in Euclidean geometry?
18. in sphere geometry?

The figure below represents two points and a line that contains them in Euclidean geometry.

19. Make a drawing that represents two points and a line that contains them in sphere geometry.

Do two points determine a line

20. in Euclidean geometry?
21. in sphere geometry?

Set II

The figure below illustrates two lines, ℓ and m, that intersect to form right angles.*

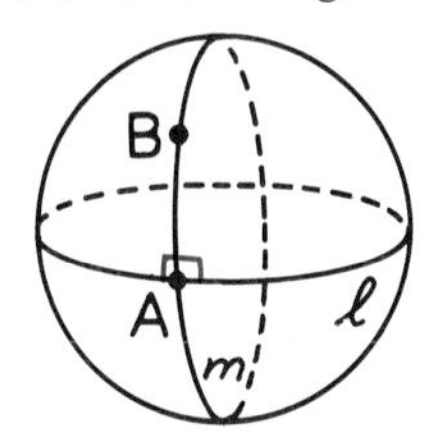

22. What word names the relation between lines ℓ and m?
23. Do you think that any lines other than m can be drawn through point A that also have this relation to line ℓ?
24. Do you think that any lines other than m can be drawn through point B that have the same relation to line ℓ?

In the figure below, both lines m and n are perpendicular to line ℓ.*

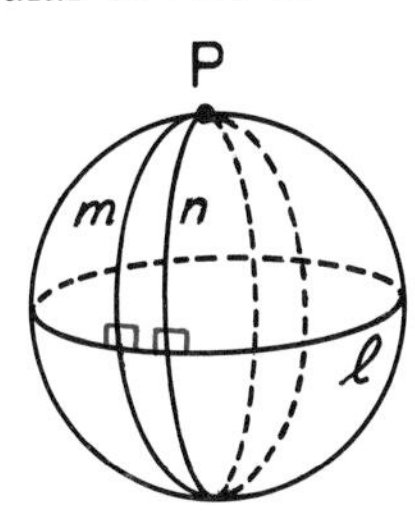

*We will take the liberty of using such words as "angle" and "perpendicular" in sphere geometry without attempting to define them.

25. In Euclidean geometry, what is the relation between two lines that form equal corresponding angles with a transversal?
26. Is this also true in sphere geometry?
27. In Euclidean geometry, what relation do two lines have that lie in a plane and are perpendicular to a third line?
28. Is this also true in sphere geometry?

The figure below seems to illustrate two lines in sphere geometry that are parallel. In fact, they are what Peter was trying to draw in the cartoon strip.

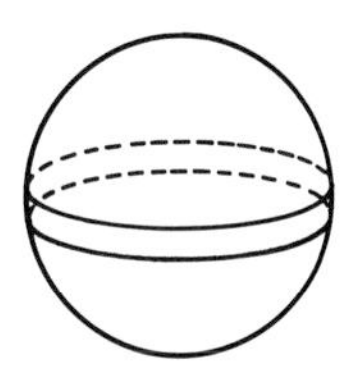

29. Why do the two curves seem to be parallel?
30. Because parallel lines do not exist on a sphere, something is wrong with our thinking about this figure. What is it?

In the figure below, point P is the pole of line ℓ; ABP is a triangle with exterior ∡PBX.

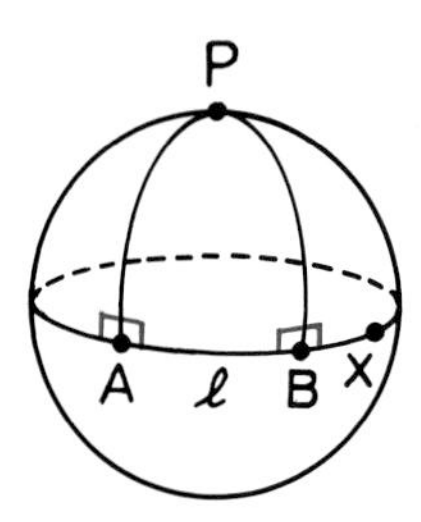

31. In Euclidean geometry, what relations does an exterior angle of a triangle have to the remote interior angles?
32. Are these relations true in sphere geometry?
33. In Euclidean geometry, what is the sum of the measures of the angles of a triangle?
34. What can you conclude about the sum of the measures of the angles of a triangle in sphere geometry?
35. How many right angles can a triangle have in Euclidean geometry?
36. How many right angles do you think a triangle can have in sphere geometry?

The figure below represents a rectangle in sphere geometry.

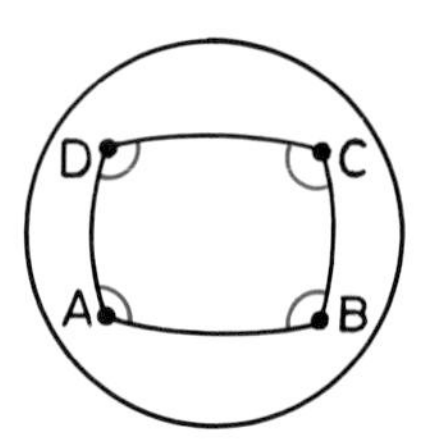

37. What is a rectangle?
38. What kind of angles do you think this rectangle has?
39. Are rectangles in sphere geometry parallelograms? Explain why or why not.
40. Do you think that a rectangle in sphere geometry could be a square?

Set III

According to one model of the universe, called Einstein's spherical universe, the geometrical properties of space are comparable to those of the surface of a sphere.*

1. If this model is correct, what conclusion follows about the volume of the space of the universe? Explain your reasoning.
2. If an astronaut were to travel far enough along a straight line, what would this model predict? Explain your reasoning.

---

*See, for example, Chapters 16 and 17 in *Einstein's Universe*, by Nigel Calder (Viking, 1979).

Lesson 2

The Saccheri Quadrilateral

The first book to contain some non-Euclidean geometry was published in 1733, about two thousand years after Euclid wrote the *Elements.* It was written by an Italian priest named Girolamo Saccheri and, oddly enough, was intended by its author to prove that Euclidean geometry is the only logically consistent geometry possible. In fact, he named the book *Euclid Freed of Every Flaw.*

Saccheri planned to reach his goal by making what he thought were some false assumptions about a special quadrilateral. By reasoning indirectly from these assumptions, he felt certain that contradictions would eventually develop that could, in turn, be used to prove the Parallel Postulate. Instead of this happening, however, Saccheri ended up creating part of a new geometry without realizing it. Unfortunately for him, other mathematicians at the time paid little attention to his book and it was soon forgotten, not to be rediscovered until 1889, more than a century later. By that time, other mathematicians had independently recreated what Saccheri had developed; and it is their names that are now closely associated with non-Euclidean geometry—especially those of Nicholas Lobachevsky, a Russian, and Bernhard Riemann, a German.

Saccheri began his work with a quadrilateral that has a pair of sides perpendicular to a third side. We will call such a quadrilateral "biperpendicular."

▶ **Definition**
A ***biperpendicular quadrilateral*** is a quadrilateral that has a pair of sides both of which are perpendicular to a third side.

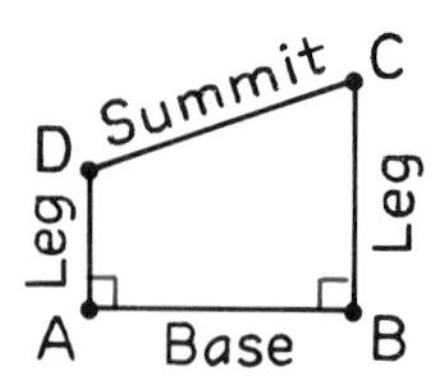

As you can see from the labeling of this figure, the two sides perpendicular to the same side are called the *legs,* the side to which they are perpendicular is called the *base,* and the side opposite the base is called the *summit.* The angles at A and B are called *base angles* of the quadrilateral and the angles at C and D are called *summit angles.* Furthermore, we will refer to angle C as the summit angle opposite leg $\overline{AD}$ and to angle D as the summit angle opposite leg $\overline{BC}$.

A biperpendicular quadrilateral whose legs are equal might be called "isosceles," but in honor of Saccheri it is usually called a *Saccheri quadrilateral.*

▶ **Definition**
A ***Saccheri quadrilateral*** is a biperpendicular quadrilateral whose legs are equal.

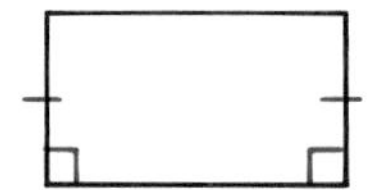

A Saccheri quadrilateral

A Saccheri quadrilateral looks very much like a rectangle; that is, the summit angles look as if they must also be right angles, so that the figure is equiangular. It is easy to prove this in Euclidean geometry. In a geometry that is non-Euclidean, however, a Saccheri quadrilateral is *not* a rectangle because, with a different postulate about parallel lines, it can be proved that its summit angles are *not* right angles.

Before we look further into this, we will state some theorems about biperpendicular quadrilaterals that are true in both Euclidean and the non-Euclidean geometries. Each of them can be proved without using the Parallel Postulate, and outlines of the proofs are in the exercises. Although these theorems are not very interesting in themselves, we will use them in the next lesson to derive some of the strange results that Saccheri obtained.

▶ **Theorem 99**
The summit angles of a Saccheri quadrilateral are equal.

► **Theorem 100**
The line segment joining the midpoints of the base and summit of a Saccheri quadrilateral is perpendicular to both of them.

The next two theorems are comparable to a pair of theorems about inequalities in triangles that you already know.

► **Theorem 101**
If the legs of a biperpendicular quadrilateral are unequal, then the summit angles are unequal and the larger angle is opposite the longer leg.

► **Theorem 102**
If the summit angles of a biperpendicular quadrilateral are unequal, then the legs are unequal and the longer leg is opposite the larger angle.

Exercises

Set I

The following questions refer to these quadrilaterals.

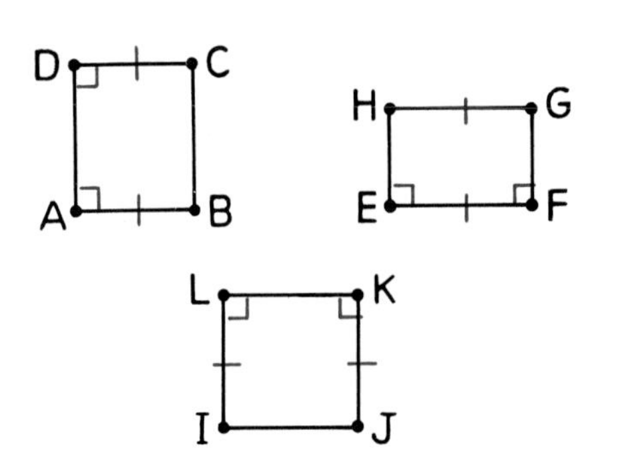

1. Which of them are Saccheri quadrilaterals?
2. What is $\overline{AD}$ called with respect to quadrilateral ABCD?
3. What are ∡A and ∡D called?
4. What are $\overline{IL}$ and $\overline{JK}$ called with respect to quadrilateral IJKL?
5. What is $\overline{IJ}$ called?
6. What are ∡I and ∡J called?

Answer the questions about the proofs of the following theorems.

Theorem 99 The summit angles of a Saccheri quadrilateral are equal.

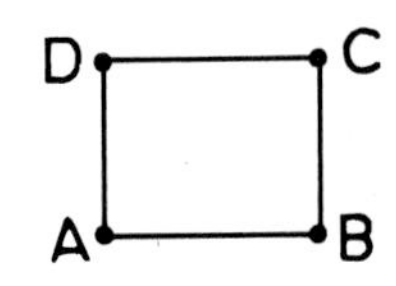

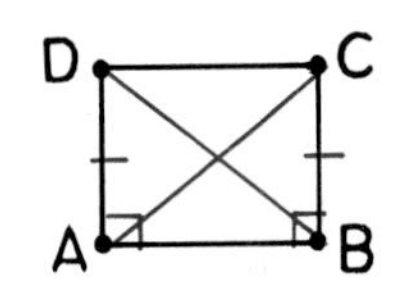

Given: Saccheri quadrilateral ABCD with base $\overline{AB}$.

Prove: $\angle D = \angle C$.

To prove this theorem, we can draw $\overline{AC}$ and $\overline{DB}$ and show that two pairs of triangles are congruent.

7. Why is $AD = BC$?
8. Why is $\angle DAB = \angle CBA$?
9. Why is $\triangle DAB \cong \triangle CBA$?

It follows that $AC = DB$.

10. Why is $\triangle DAC \cong \triangle CBD$?
11. Why is $\angle ADC = \angle BCD$?

Theorem 101 If the legs of a biperpendicular quadrilateral are unequal, then the summit angles are unequal and the larger angle is opposite the longer leg.

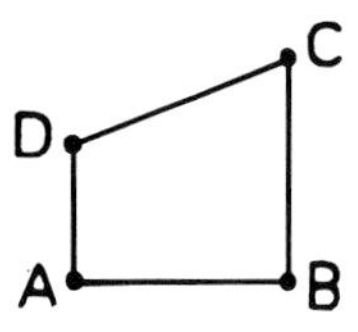

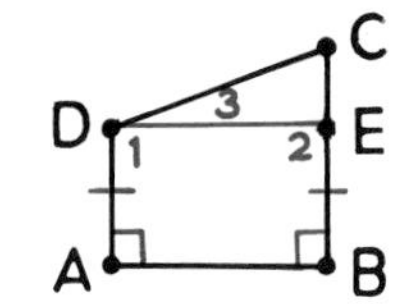

Given: Biperpendicular quadrilateral ABCD with base $\overline{AB}$; CB > DA.

Prove: $\angle D > \angle C$.

Because ABCD is a biperpendicular quadrilateral with base $\overline{AB}$, we know that $\overline{DA} \perp \overline{AB}$ and $\overline{CB} \perp \overline{AB}$. Also, CB > DA.

Choose point E on $\overline{BC}$ so that BE = AD and draw $\overline{DE}$.

12. Why is ABED a Saccheri quadrilateral?
13. Why is $\angle 1 = \angle 2$?
14. Because $\angle ADC = \angle 1 + \angle 3$, $\angle ADC > \angle 1$. Why?
15. It follows that $\angle ADC > \angle 2$. Why?
16. Why is $\angle 2 > \angle C$?
17. Why is $\angle ADC > \angle C$?

Theorem 102 If the summit angles of a biperpendicular quadrilateral are unequal, then the legs are unequal and the longer leg is opposite the larger angle.

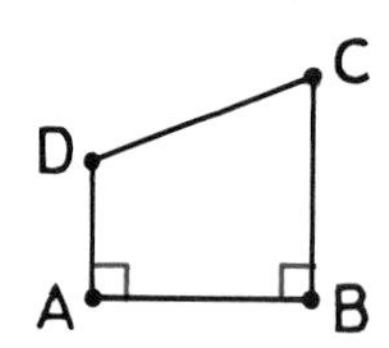

Given: Biperpendicular quadrilateral ABCD with base $\overline{AB}$; $\angle D > \angle C$.

Prove: CB > DA.

We will prove this theorem by the indirect method.

18. Either CB < DA, CB = DA, or CB > DA. Why?

Suppose that CB < DA.

19. Then it follows that $\angle D < \angle C$. Why? This contradicts the fact that $\angle D > \angle C$.

Suppose that CB = DA.

20. What kind of quadrilateral must ABCD be if this is true?
21. It follows that $\angle D = \angle C$. Why? This also contradicts the fact that $\angle D > \angle C$.

The only possibility remaining is that CB > DA.

Theorem 100 The line segment joining the midpoints of the base and summit of a Saccheri quadrilateral is perpendicular to both of them.

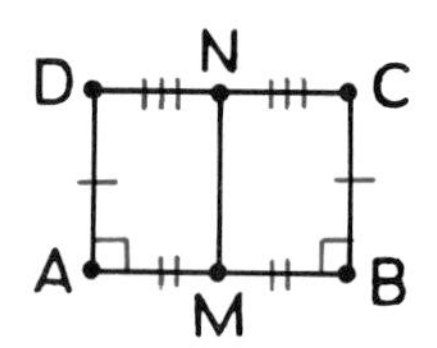

Given: Saccheri quadrilateral ABCD with $\overline{MN}$ joining the midpoints of $\overline{AB}$ and $\overline{CD}$ respectively.

Prove: $\overline{MN} \perp \overline{AB}$ and $\overline{MN} \perp \overline{DC}$.

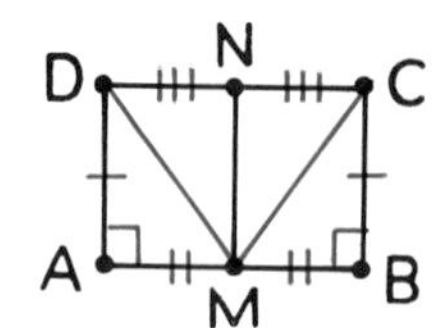

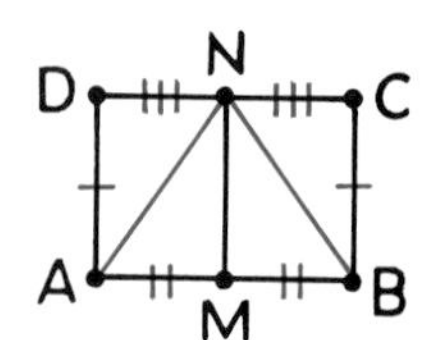

This theorem can be proved without using any facts about parallel lines. Answer the following questions related to how this could be done. (Each answer may include several ideas.)

First, we can draw $\overline{DM}$ and $\overline{CM}$.

22. How can it be shown that DM = CM?

Because N is the midpoint of $\overline{DC}$, we know that DN = NC.

23. How does it follow that $\overline{MN} \perp \overline{DC}$?

Next, we draw $\overline{NA}$ and $\overline{NB}$.

24. How can it be shown that $\triangle DNA \cong \triangle CNB$?

25. Explain, without proving any more triangles congruent, why $\overline{MN} \perp \overline{AB}$.

Set II

Refer to the definitions and theorems in this lesson to answer the questions about the figures.

In the figure below, DEIG is a Saccheri quadrilateral with $\overline{DE} \perp \overline{EI}$ and $\overline{GI} \perp \overline{EI}$; $\overline{NS}$ joins the midpoints of $\overline{DG}$ and $\overline{EI}$.

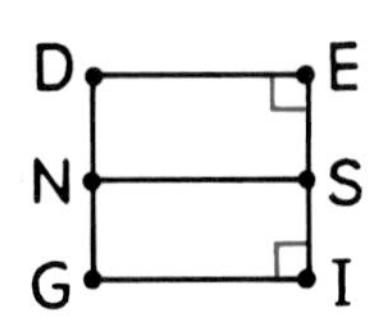

26. Why is DE = GI?

27. Why is $\angle D = \angle G$?

28. Why is $\overline{NS} \perp \overline{DG}$ and $\overline{NS} \perp \overline{EI}$?

In the figure below, MD > ET and $\angle O > \angle H$.

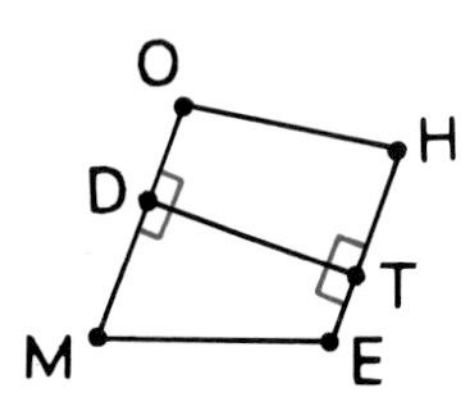

29. Why is $\angle E > \angle M$?

30. Why is HT > OD?

In Euclidean geometry, it is easy to prove that a Saccheri quadrilateral is a parallelogram.

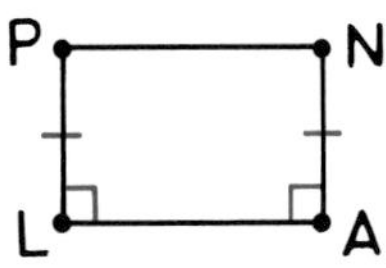

31. Use this figure to explain how this could be done.

32. Why is the summit of a Saccheri quadrilateral in Euclidean geometry equal to its base?

33. Explain why, in Euclidean geometry, the summit angles of a Saccheri quadrilateral are right angles.

34.

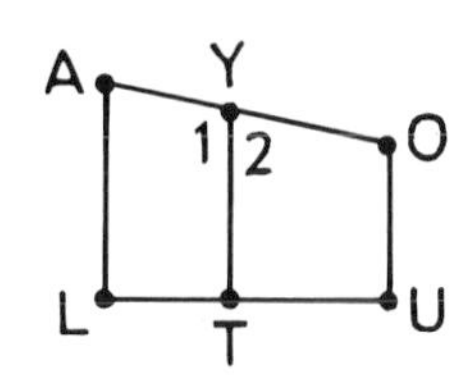

Given: In quadrilateral LAOU, $\overline{AL} \perp \overline{LU}$, $\overline{YT} \perp \overline{LU}$, and $\overline{OU} \perp \overline{LU}$; $\angle 1 > \angle A$ and $\angle O > \angle 2$.

Prove: AL > OU.

35.

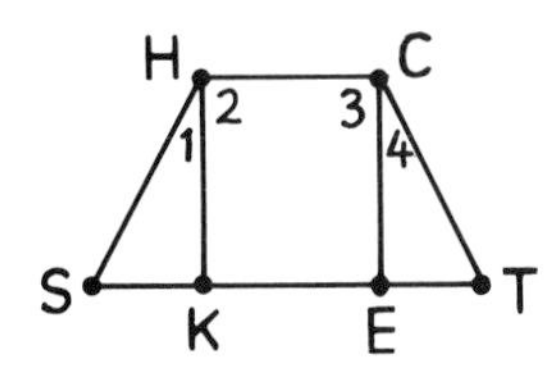

Given: In quadrilateral STCH, $\overline{HK} \perp \overline{ST}$, $\overline{CE} \perp \overline{ST}$, HK = CE, and $\angle 1 = \angle 4$.

Prove: $\angle SHC = \angle HCT$.

Set III

Among the many people who have tried to prove Euclid's Parallel Postulate was a Persian mathematician who was the court astronomer of the grandson of the famous Genghis Khan. His name was Nasir Eddin and he lived in the thirteenth century.

Nasir Eddin began by supposing that ℓ and m are two lines such that perpendiculars to m from points on ℓ make unequal angles with ℓ.

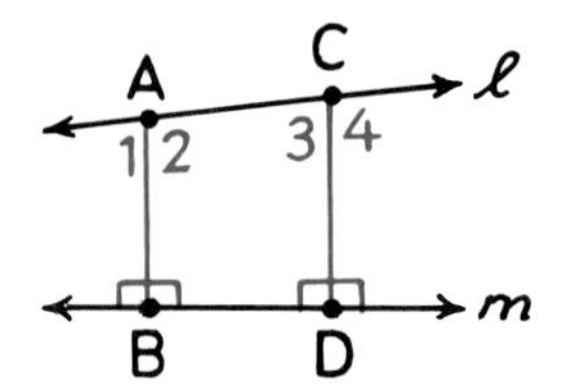

1. If $\measuredangle 1$ and $\measuredangle 3$ are acute and $\measuredangle 2$ and $\measuredangle 4$ are obtuse, which segment must be longer: $\overline{AB}$ or $\overline{CD}$?
2. Why?

Nicholas Lobachevsky

Bernhard Riemann

Lesson 3
The Geometries of Lobachevsky and Riemann

After proving that the summit angles of a Saccheri quadrilateral are equal, Saccheri realized that, if he could prove them to be right angles, he could use this fact to prove the Parallel Postulate. This he planned to accomplish by reasoning indirectly. He would show that the assumptions that the summit angles were either acute or obtuse would lead to contradictions.

Saccheri did, in fact, eventually arrive at a contradiction from the hypothesis that the summit angles were obtuse. But to eliminate the possibility that they were acute proved to be far more difficult. In fact, instead of causing contradictions, the "acute angle" hypothesis became the beginning of one of the non-Euclidean geometries. Saccheri and his contemporaries did not really comprehend the significance of what he had started, and so it was not until many years later that the idea that a non-Euclidean geometry could make sense was accepted. Three men are given credit for independently realizing this: the great German mathematician Carl Friedrich Gauss; Janos Bolyai, a Hungarian; and Nicholas Lobachevsky. Because the non-Euclidean geometry based on the "acute angle" hypothesis has often simply been called *Lobachevskian,* we will refer to it by that name.

► **The Lobachevskian Postulate**
The summit angles of a Saccheri quadrilateral are acute.

By means of this postulate, the following theorems can be proved.

► **Lobachevskian Theorem 1**
In Lobachevskian geometry, the summit of a Saccheri quadrilateral is longer than its base.

► **Lobachevskian Theorem 2**
In Lobachevskian geometry, a midsegment of a triangle is less than half as long as the third side.

Of course, these theorems contradict theorems in Euclidean geometry, *but only those based on the Parallel Postulate.* Because the Parallel Postulate states that through a point not on a line there is exactly one parallel to the line, these new theorems are consistent with the idea that through a point not on a line, there is *not* exactly one parallel to the line!

You will recall that in sphere geometry, there are *no* parallels to a line through a point not on it. However, this result contradicts other Euclidean theorems proved *before* the Parallel Postulate. In fact, it is equivalent to assuming that the summit angles of a Saccheri quadrilateral are *obtuse,* and this is why Saccheri was able to eliminate this possibility.

Nevertheless, if we are willing to change some other postulates related to distance and betweenness as well as the Parallel Postulate, a logically consistent non-Euclidean geometry can be developed in which there are no parallels at all. The German mathematician Bernhard Riemann was the first to understand this and we will refer to the geometry that he created as *Riemannian geometry.*

The table below summarizes the basic differences between Euclidean geometry and these two non-Euclidean geometries. In each case, either statement can be proved to be a logical consequence of the other.

Statement	*Euclid*	*Lobachevsky*	*Riemann*
Through a point not on a line, there is	exactly one parallel to the line.	more than one parallel to the line.	no parallel to the line.
The summit angles of a Saccheri quadrilateral are	right.	acute.	obtuse.

Exercises

In the exercises of this lesson and the following one, we will restrict our proofs to Lobachevskian geometry because in this geometry only the Parallel Postulate is changed. This means that we can use any idea considered before Chapter 6 in this book.

Set I

Answer the following questions about the proofs of the theorems in this lesson.

Lobachevskian Theorem 1 In Lobachevskian geometry, the summit of a Saccheri quadrilateral is longer than its base.

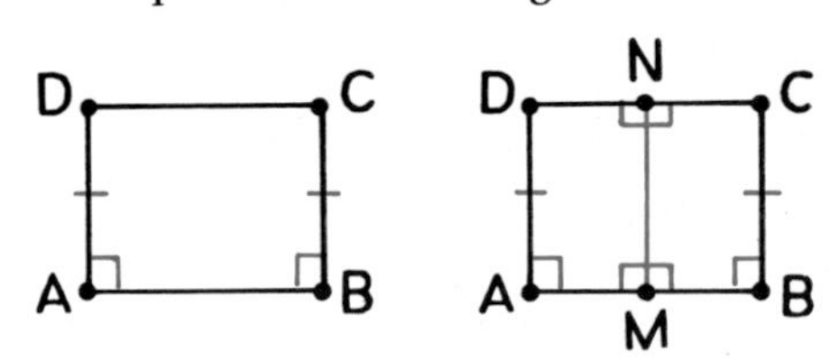

Given: Saccheri quadrilateral ABCD with base $\overline{AB}$.

Prove: DC > AB.

Because ABCD is a Saccheri quadrilateral with base $\overline{AB}$, $\overline{CB} \perp \overline{AB}$. Therefore, ∡B is a right angle, and hence $\angle B = 90°$.

1. Why is ∡C acute?
2. Why is $\angle C < 90°$?
3. Why is $\angle C < \angle B$?

Next, we let M and N be the midpoints of $\overline{AB}$ and $\overline{DC}$, respectively, as shown in the second figure.

4. What permits us to do this?
5. After drawing $\overline{MN}$, how do we know that $\overline{MN} \perp \overline{AB}$ and $\overline{MN} \perp \overline{DC}$?
6. Why is CNMB a biperpendicular quadrilateral?

Because we have already proved that $\angle C < \angle B$, we know that $\angle B > \angle C$.

7. How does it follow that NC > MB?
8. Why is $NC = \frac{1}{2}DC$ and $MB = \frac{1}{2}AB$?
9. Why is $\frac{1}{2}DC > \frac{1}{2}AB$?
10. Why is DC > AB?

Lobachevskian Theorem 2 In Lobachevskian geometry, a midsegment of a triangle is less than half as long as the third side.

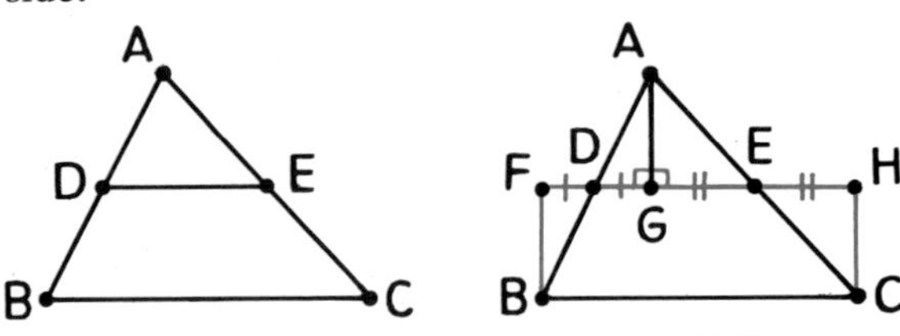

Given: △ABC with midsegment $\overline{DE}$.

Prove: $DE < \frac{1}{2}BC$.

Through A, draw $\overline{AG} \perp \overline{DE}$. Draw $\overleftrightarrow{DE}$ and choose points F and H on $\overleftrightarrow{DE}$ so that DF = DG and EH = EG. Draw $\overline{BF}$ and $\overline{CH}$.

11. Explain why △ADG ≅ △BDF and △AEG ≅ △CEH.

Because the triangles are congruent, BF = AG and AG = CH.

12. Why does it follow that BF = CH?

Angles F and H are right angles because they are equal to the right angles at G, and so $\overline{BF} \perp \overline{FH}$ and $\overline{CH} \perp \overline{FH}$.

13. Why is BFHC a Saccheri quadrilateral?
14. Why is BC > FH?

Because FH = FD + DG + GE + EH, FD = DG, and GE = EH, it follows that FH = DG + DG + GE + GE = 2DG + 2GE = 2(DG + GE).

15. Why is FH = 2DE?
16. Why is BC > 2DE?
17. Why is $\frac{1}{2}BC > DE$ so that $DE < \frac{1}{2}BC$?

Set II

The figure below represents a triangle in Lobachevskian geometry. Points E, L, and N

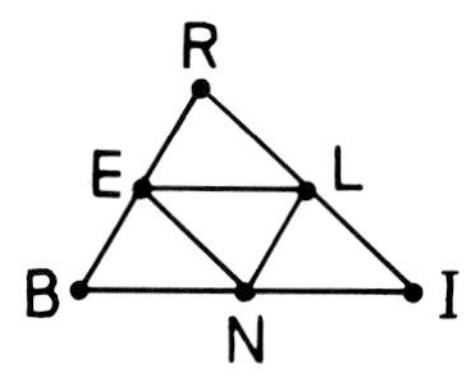

are the midpoints of its sides; BR = 8, RI = 10, and BI = 12.

18. What can you conclude about the lengths of $\overline{LN}$, $\overline{NE}$, and $\overline{EL}$?
19. What can you conclude about the perimeter of $\triangle$LNE?

In the figure below, FRIL represents a Saccheri quadrilateral in Lobachevskian geometry; $\overline{FL}$ is its base. Use this information to

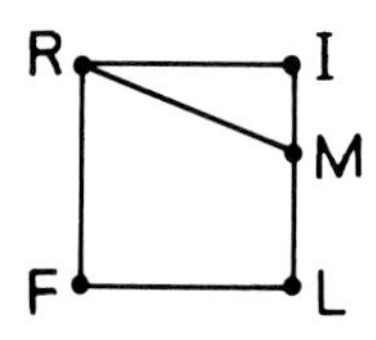

tell whether each of the following statements *must be true, may be true,* or *is false.*

20. FR = LI.
21. FR > LM.
22. $\angle$RML > $\angle$FRM.
23. RI = FL.
24. $\measuredangle$L is a right angle.
25. $\measuredangle$I is acute.
26. $\measuredangle$RML is obtuse.

The figure below represents a quadrilateral in Lobachevskian geometry. In it, $\overline{CD} \perp \overline{DA}$, $\overline{WA} \perp \overline{DA}$, CO = OW, DR = RA and CD = WA.

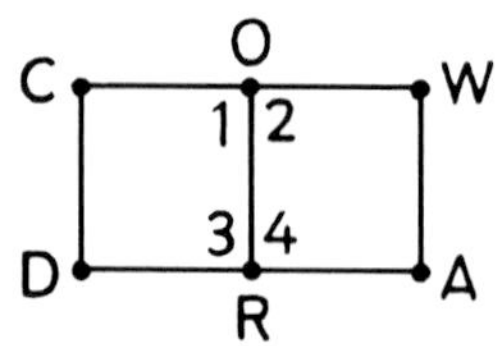

27. Copy the figure and mark all of this information on it.
28. What kind of quadrilateral is DAWC?
29. What kind of angles are $\measuredangle$C and $\measuredangle$W?
30. What kind of angles are $\measuredangle$1, $\measuredangle$2, $\measuredangle$3, and $\measuredangle$4?
31. What kind of quadrilaterals are DROC and RAWO?
32. Why is RO < DC and RO < AW?
33. State a theorem about the length of the line segment that joins the midpoints of the summit and base of a Saccheri quadrilateral in Lobachevskian geometry.

The sphere geometry that we studied in Lesson 1 makes a good model for understanding Riemannian geometry because there are no parallel lines in either. The figure below shows a Saccheri quadrilateral on a sphere.

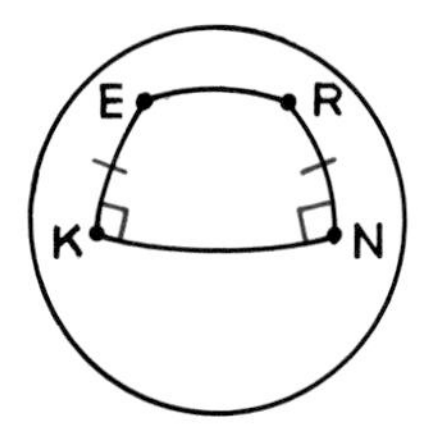

34. What kind of angles do $\measuredangle$E and $\measuredangle$R seem to be?
35. Does your answer agree with the table on page 592?
36. How does the length of $\overline{ER}$ seem to compare to the length of $\overline{KN}$?
37. State a theorem in Riemannian geometry that you think corresponds to Lobachevskian Theorem 1.

The figure below shows a triangle on a sphere; $\overline{RN}$ is one of its midsegments.

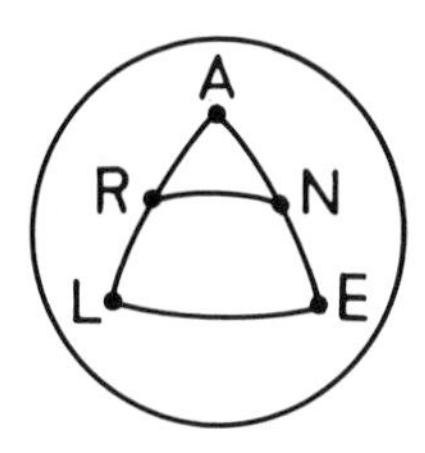

38. State a theorem in Riemannian geometry that you think corresponds to Lobachevskian Theorem 2.

39. Do you think that a midsegment of a triangle in Riemannian geometry is parallel to the third side of the triangle? Explain why or why not.

Set III

"He keeps his place clean as a whistle."

Each of the homes in this cartoon that doesn't seem to be as clean as a whistle is supported by four posts. These posts are not perpendicular to the ground but are slanted toward each other so that their tops are closer together than their bottoms. The figure below illustrates the relation of the posts to one another.

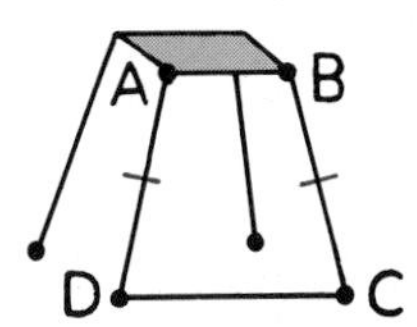

Suppose that AD = BC and that AB < DC. Is it possible for the posts to be perpendicular to the floor of the home they support? Specifically, can it be that $\overline{AD} \perp \overline{AB}$ and $\overline{BC} \perp \overline{AB}$? Explain.

Hale Observatories

Lesson 4

The Triangle Angle Sum Theorem Revisited

Photography by Robert M. Gottschalk

Albert Einstein

We live in a mysterious universe. Before Einstein developed his theory of relativity, physical space was assumed to be Euclidean. Although mathematicians in the nineteenth century had come to accept the idea that the geometry of Euclid is not the only one possible, it is still hard to believe that a non-Euclidean geometry can give a better description of the universe. Yet Einstein's theory of relativity suggests that this may be the case.*

At about the time that Lobachevsky and Bolyai were developing the non-Euclidean geometry called Lobachevskian, Karl Friedrich Gauss carried out an experiment concerning the angles of a triangle. You know that in Euclidean geometry the sum of the measures of the angles of a triangle is 180°. According to the geometry of Lobachevsky, it is *less* than 180°; in Riemannian geometry, it is *more* than 180°.

---

*For more information about the possible geometries of space, see *Black Holes and Warped Spacetime,* by William J. Kaufmann, III (W. H. Freeman and Company, 1979), pp. 179–197.

In an attempt to discover which of these geometries is the one that fits physical space, Gauss measured the angles of a giant triangle. The sides of this triangle were formed by light rays sent between three mountain tops in Germany. The sum of the measures of the angles of the triangle turned out to be within just a few seconds of 180°. Even though Gauss's measurements were as accurate as was possible at the time, the fact that the sum was slightly more than 180° could have been due to experimental error. Furthermore, we now know that the mountain-top triangle was much too small to determine if physical space is non-Euclidean, even though the lengths of its sides ranged from 60 to 110 kilometers. According to both Lobachevskian and Riemannian geometry, the smaller the triangle, the closer the sum of the measures of its angles is to 180°. If space can be described by either of these non-Euclidean geometries, then a triangle with three stars for its vertices, rather than three mountain tops, would be required to determine which one is correct.

In the non-Euclidean geometries, the sum of the measures of the angles of a triangle is not 180°. As a consequence of this fact, it can be proved that there are no scale models in these geometries. In other words, two figures cannot be similar without being congruent.

It is no wonder that, with such surprising results as these, it took a long time for the non-Euclidean geometries to be accepted. Yet we already know enough about Lobachevskian geometry to be able to prove the results as theorems.

► **Lobachevskian Theorem 3**
In Lobachevskian geometry, the sum of the measures of the angles of a triangle is less than 180°.

► **Corollary**
In Lobachevskian geometry, the sum of the measures of the angles of a quadrilateral is less than 360°.

► **Lobachevskian Theorem 4**
In Lobachevskian geometry, if two triangles are similar, they must also be congruent.

At the beginning of this course, we compared geometry to a game and said that the nature of a game depends on the rules by which it is played. With your brief introduction to the non-Euclidean geometries, you may be able to more fully understand the meaning of this.

In the past century, many other geometries have been developed in addition to the three that we have studied. Human curiosity will undoubtedly lead to the invention of still others in the future. In fact, one twentieth-century mathematician has said, "When a man stops wondering and asking and playing, he is through."*

---

*In *Mathematics and the Imagination,* by Edward Kasner and James R. Newman (Simon & Schuster, 1940).

Exercises

Set I

Answer the following questions about the proofs of the theorems in this lesson.

Lobachevskian Theorem 3 In Lobachevskian geometry, the sum of the measures of the angles of a triangle is less than 180°.

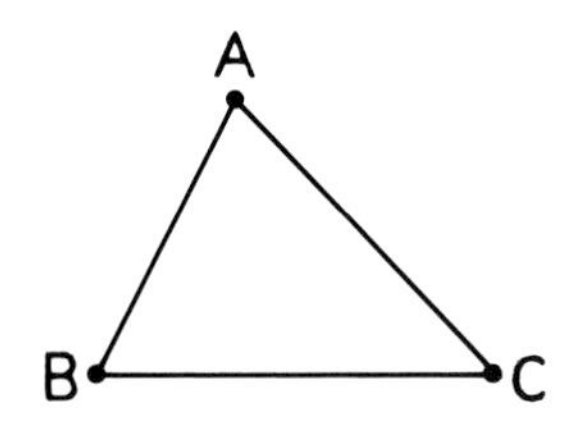

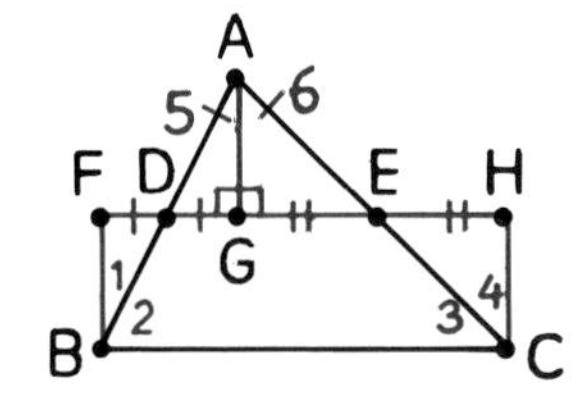

Given: △ABC.

Prove: $\angle A + \angle B + \angle C < 180°$.

The proof begins with the addition of several points and lines to the triangle to form a figure like the one that we used to prove Lobachevskian Theorem 2.

Let D and E be the midpoints of sides $\overline{AB}$ and $\overline{AC}$ of the triangle and draw $\overleftrightarrow{DE}$. Through A, draw $\overline{AG} \perp \overleftrightarrow{DE}$. Choose points F and H on $\overleftrightarrow{DE}$ so that DF = DG and EH = EG. Draw $\overline{BF}$ and $\overline{CH}$.

1. Why does it follow that △AGD ≅ △BFD and △AGE ≅ △CHE?
2. Which angles of quadrilateral BFHC are right angles?
3. Which sides of quadrilateral BFHC are equal?
4. What kind of quadrilateral is BFHC?
5. Why are ∡FBC and ∡HCB acute?
6. What can be concluded about ∠FBC + ∠HCB from the fact that ∠FBC < 90° and ∠HCB < 90°?

Because ∠FBC = ∠1 + ∠2 and ∠HCB = ∠3 + ∠4, it follows that ∠1 + ∠2 + ∠3 + ∠4 < 180°.

7. To which angles in the figure are ∡1 and ∡4 equal?
8. Why is ∠5 + ∠2 + ∠3 + ∠6 < 180°?
9. Why does it follow from this that ∠BAC + ∠ABC + ∠ACB < 180°?

Corollary In Lobachevskian geometry, the sum of the measures of the angles of a quadrilateral is less than 360°.

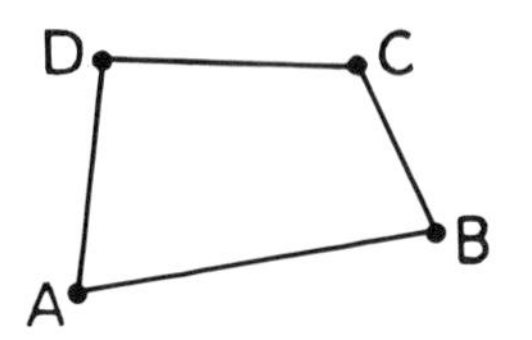

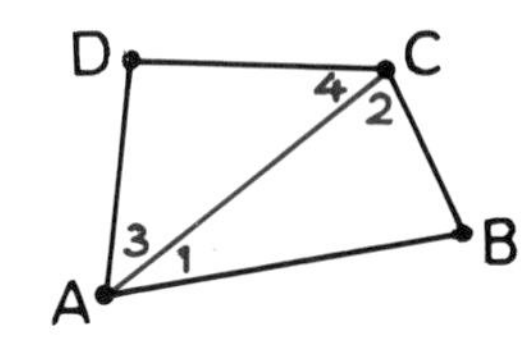

Given: Quadrilateral ABCD.

Prove: $\angle A + \angle B + \angle C + \angle D < 360°$.

Draw $\overline{AC}$.

10. Why is ∠1 + ∠B + ∠2 < 180° and ∠3 + ∠D + ∠4 < 180°?
11. Why is ∠1 + ∠B + ∠2 + ∠3 + ∠D + ∠4 < 360°?
12. Why is ∠1 + ∠3 = ∠DAB and ∠2 + ∠4 = ∠DCB?
13. Why is ∠DAB + ∠B + ∠DCB + ∠D < 360°?

Lobachevskian Theorem 4 In Lobachevskian geometry, if two triangles are similar, they must also be congruent.

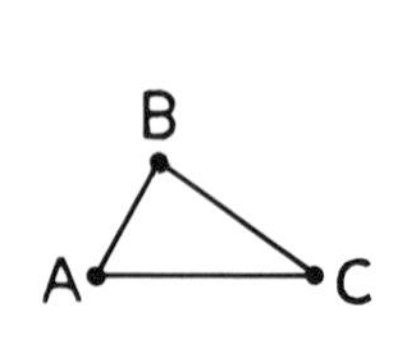

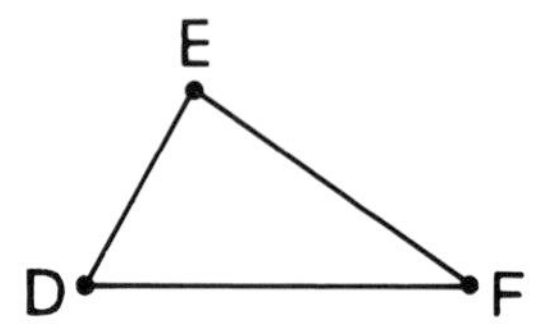

Given: △ABC ~ △DEF.

Prove: △ABC ≅ △DEF.

By hypothesis, $\triangle ABC \sim \triangle DEF$.

14. Why is $\angle D = \angle A$, $\angle E = \angle B$, and $\angle F = \angle C$?

Suppose that $\triangle ABC$ and $\triangle DEF$ are not congruent. If this is the case, then $BA \neq ED$ and $BC \neq EF$, because if either pair of these sides were equal, then the triangles would be congruent by A.S.A.

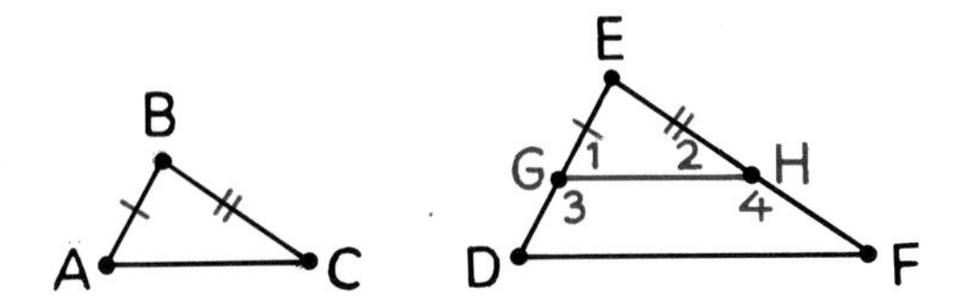

We will now copy the smaller triangle on the larger one as shown above. Choose point G on $\overline{ED}$ so that $EG = BA$ and point H on $\overline{EF}$ so that $EH = BC$. Draw $\overline{GH}$.

15. Why is $\triangle GEH \cong \triangle ABC$?

It follows that $\angle 1 = \angle A$ and $\angle 2 = \angle C$ because they are corresponding parts of these triangles.

16. Why is $\angle D = \angle 1$ and $\angle F = \angle 2$?

17. Why is $\angle 1 + \angle 3 = 180°$ and $\angle 2 + \angle 4 = 180°$?

Adding these equations, we get $\angle 1 + \angle 2 + \angle 3 + \angle 4 = 360°$.

18. Why is $\angle D + \angle F + \angle 3 + \angle 4 = 360°$?

19. If we assume from the figure that DGHF is a quadrilateral, the equation in Exercise 18 is impossible. Why?

Because we have arrived at a contradiction, our initial assumption that $\triangle ABC$ and $\triangle DEF$ are not congruent is false! So $\triangle ABC \cong \triangle DEF$.

Set II

In Lobachevskian geometry, the sum of the measures of the angles of a triangle is less than 180°. Use this fact to decide on answers to the following questions about triangles in Lobachevskian geometry.

20. If two angles of one triangle are equal to two angles of another triangle, does it follow that the third pair of angles must be equal?

21. Are the acute angles of a right triangle complementary?

22. What can you conclude about the measure of each angle of an equilateral triangle?

In the figure below, $\triangle EDA$ represents a triangle in Lobachevskian geometry and $\measuredangle 2$ is one of its exterior angles. Copy and complete the following statements.

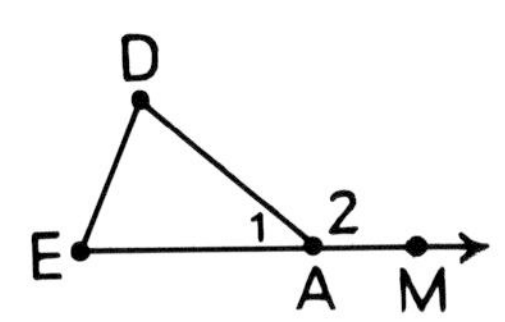

23. $\angle E + \angle D + \angle 1$ ▒▒▒ 180°.

24. $\angle 1 + \angle 2$ ▒▒▒ 180°.

25. Use your answers to Exercises 23 and 24 to derive a relation between $\angle E + \angle D$ and $\angle 2$.

26. State a theorem in Lobachevskian geometry suggested by your answer to Exercise 25.

The sum of the measures of the angles of a triangle in Riemannian geometry is more than 180°.

27. Draw a figure to represent a fairly large triangle on a sphere to show why this is plausible.

28. Do you think that any number more than 180° can be the sum of the angles of a triangle in Riemannian geometry?

In quadrilateral BRIE, $\measuredangle B$, $\measuredangle R$, and $\measuredangle I$ are right angles.

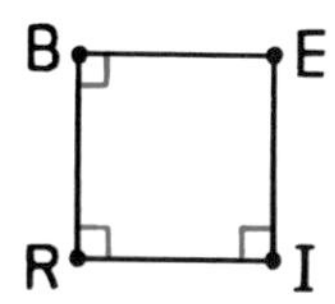

29. In Euclidean geometry, what can you conclude about $\measuredangle E$?

30. What kind of quadrilateral must BRIE be in Euclidean geometry?

31. In Lobachevskian geometry, what can you conclude about $\measuredangle E$?

32. Can BRIE be a Saccheri quadrilateral in Lobachevskian geometry?

Tell whether each of the following statements is true or false.

33. In Euclidean geometry, if two triangles are congruent, they are also similar.

34. In Lobachevskian geometry, if two triangles are congruent, they are also similar.

35. In Euclidean geometry, if two triangles are similar, they are also congruent.

36. In Lobachevskian geometry, if two triangles are similar, they are also congruent.

In $\triangle GUA$, $\angle G = \angle 2$.

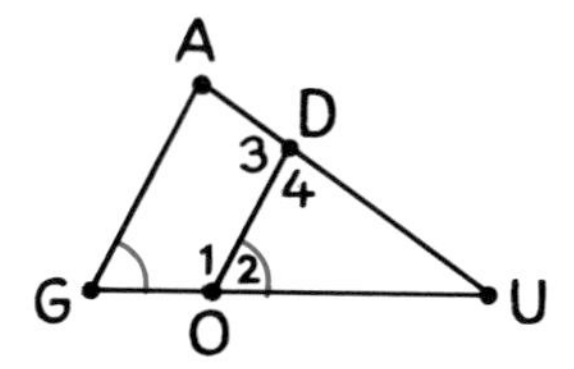

37. In Euclidean geometry, what can you conclude about $\triangle OUD$ and $\triangle GUA$?

38. What kind of quadrilateral must DOGA be in Euclidean geometry?

39. In Lobachevskian geometry, can you conclude that $\triangle OUD \sim \triangle GUA$?

40. Explain why or why not.

Set III

Obtuse Ollie is upset by all of these strange theorems from non-Euclidean geometry. A theorem of Euclidean geometry that he especially likes is the one stating that an angle inscribed in a semicircle is a right angle. His intuition tells him that the theorem is probably true in the other geometries as well.

By drawing radius $\overline{UL}$ and beginning with the four numbered angles shown here, it is possible to arrive at a definite conclusion about this. Can you do so? Explain your reasoning.

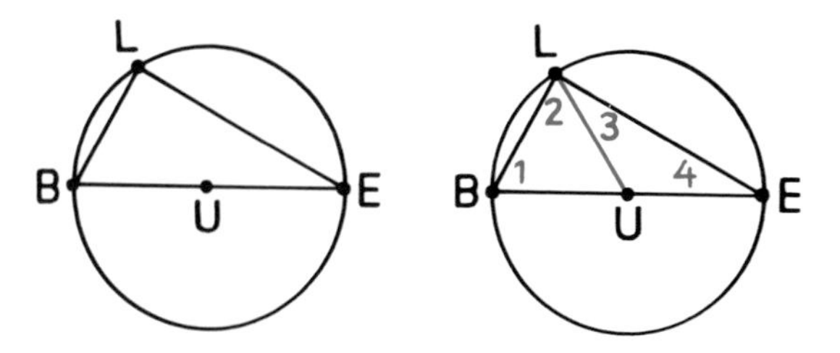

Chapter 16 / Summary and Review

Circle Limit III by M. C. Escher. Escher Foundation, Haags Gemeentemuseum, The Hague

Basic Ideas

Biperpendicular quadrilateral 586
Great circle 580
Lobachevskian geometry 591
Polar points 581
Riemannian geometry 592
Saccheri quadrilateral 586

Postulate

The Lobachevskian Postulate. The summit angles of a Saccheri quadrilateral are acute. 592

Theorems

99. The summit angles of a Saccheri quadrilateral are equal. 586

100. The line segment joining the midpoints of the base and summit of a Saccheri quadrilateral is perpendicular to both of them. 587

101. If the legs of a biperpendicular quadrilateral are unequal, then the summit angles are unequal and the larger angle is opposite the longer leg. 587

102. If the summit angles of a biperpendicular quadrilateral are unequal, then the legs are unequal and the longer leg is opposite the larger angle. 587

L.1. In Lobachevskian geometry, the summit of a Saccheri quadrilateral is longer than its base. 592

L.2. In Lobachevskian geometry, a midsegment of a triangle is less than half as long as the third side. 592

L.3. In Lobachevskian geometry, the sum of the measures of the angles of a triangle is less than 180°. 597
Corollary. In Lobachevskian geometry, the sum of the measures of the angles of a quadrilateral is less than 360°. 597

L.4. In Lobachevskian geometry, if two triangles are similar, they must also be congruent. 597

Statement	*Euclid*	*Lobachevsky*	*Riemann*
Through a point not on a line, there is	exactly one parallel to the line.	more than one parallel to the line.	no parallel to the line.
The summit angles of a Saccheri quadrilateral are	right.	acute.	obtuse.
The sum of the measures of the angles of a triangle is	180°.	less than 180°.	more than 180°.

Exercises

Set I

Tell whether each of the following statements about lines and angles in sphere geometry is true or false.

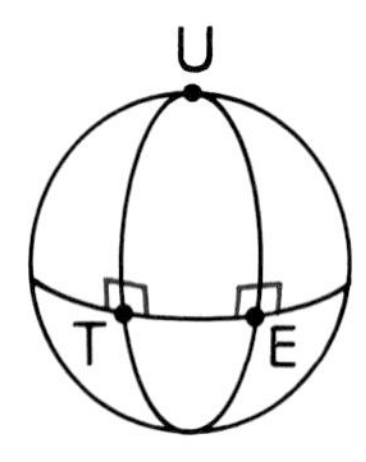

1. Through a point not on a line, there is exactly one line perpendicular to the line.
2. In a plane, two lines perpendicular to a third line are parallel to each other.
3. If two lines intersect, they intersect in no more than one point.
4. The sum of the measures of the angles of a triangle is 180°.
5. An exterior angle of a triangle is greater than either remote interior angle.

In the figure below, INCA is a quadrilateral in Lobachevskian geometry; ∡I and ∡N are right angles and AI = IN = NC.

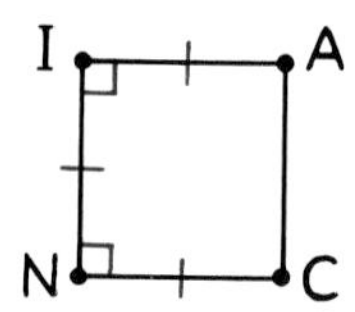

6. Is INCA a Saccheri quadrilateral?
7. What can you conclude about the length of $\overline{AC}$?
8. Can INCA be a rhombus?
9. What kind of angles are ∡A and ∡C?
10. Can INCA be a rectangle?

In the figure below, $\ell_1 \parallel \ell_2$, $\overline{MK} \perp \ell_2$, $\overline{OW} \perp \ell_2$, and $\overline{HA} \perp \ell_2$.

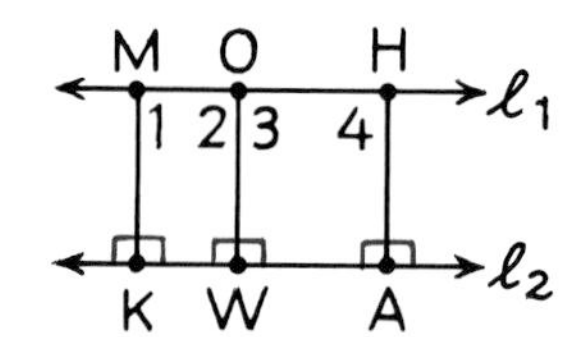

11. What can you conclude about the lengths of $\overline{MK}$, $\overline{OW}$, and $\overline{HA}$ in Euclidean geometry?

We will show, by an indirect proof, that this conclusion is false in Lobachevskian geometry. Suppose that MK = OW = HA.

12. What kind of quadrilaterals are MKWO and OWAH?

13. What kind of angles are $\measuredangle 1$, $\measuredangle 2$, $\measuredangle 3$, and $\measuredangle 4$? (Remember that we are thinking in terms of Lobachevskian geometry.)

14. How does your answer to the preceding question contradict the fact that $\measuredangle 2$ and $\measuredangle 3$ are a linear pair?

This contradiction shows that the assumption that MK = OW = HA is false.

15. Can two parallel lines in Lobachevskian geometry be everywhere equidistant?

In the figure below, $\triangle$ATE is a right triangle in Euclidean geometry and $\overline{ZC}$ is one of its midsegments.

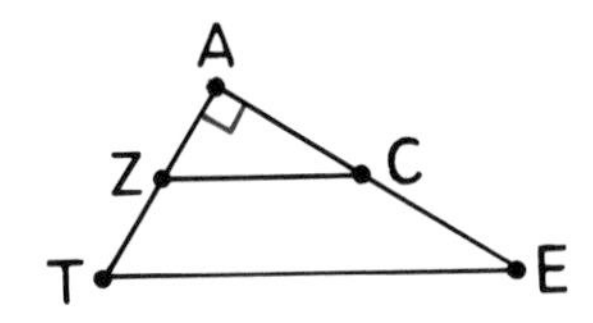

16. Why is $AT^2 + AE^2 = TE^2$?

Because Z and C are the midpoints of $\overline{AT}$ and $\overline{AE}$,

$$AZ = \frac{1}{2}AT \quad \text{and} \quad AC = \frac{1}{2}AE.$$

17. Why does it follow that 2AZ = AT and 2AC = AE?

18. Why is $(2AZ)^2 + (2AC)^2 = TE^2$?

It follows that

$$4AZ^2 + 4AC^2 = TE^2,$$

so that

$$4(AZ^2 + AC^2) = TE^2.$$

Because

$$AZ^2 + AC^2 = ZC^2,$$
$$4ZC^2 = TE^2.$$

19. Why does it follow from this that 2ZC = TE?

20. Why does it follow that $ZC = \frac{1}{2}TE$?

21. Is it true that $ZC = \frac{1}{2}TE$ in Lobachevskian geometry?

22. What does this imply about the Pythagorean Theorem in Lobachevskian geometry?

Set II

The artist Maurits Escher said of his work: "By keenly confronting the enigmas that surround us, and by considering and analyzing the observations that I had made, I ended up in the domain of mathematics. . . . I often seem to have more in common with mathematicians than with my fellow artists."*

His print at the beginning of this review illustrates a model devised by a French mathematician, Henri Poincaré, for visualizing the theorems of Lobachevskian geometry. To understand this model, it is necessary to know what *orthogonal circles* are.

*M. C. Escher, *The Graphic Work of M. C. Escher*, rev. ed. (Meredith Press, 1967), p. 10.

The figure below represents two orthogonal circles O and I intersecting at points H and P. The tangents to the circles at these points have been drawn.

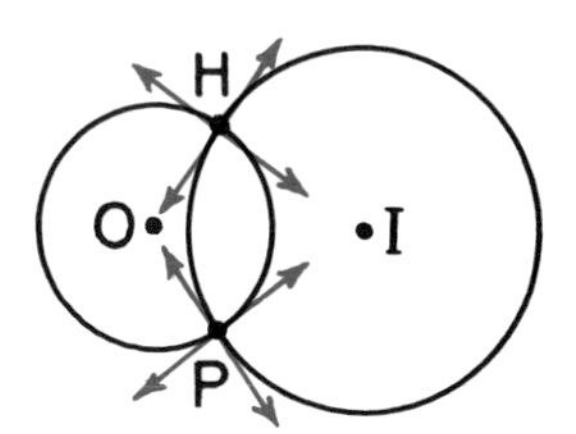

23. On the basis of this figure, define *orthogonal circles.*

In Poincaré's model of Lobachevskian geometry, points of the plane are represented by points in the interior of a circle and lines by both the diameters of the circle and the arcs of circles orthogonal to it. Examples of some lines in this model are shown in the figure below. The white arcs through the backbones of the fish in Escher's print are also "lines."

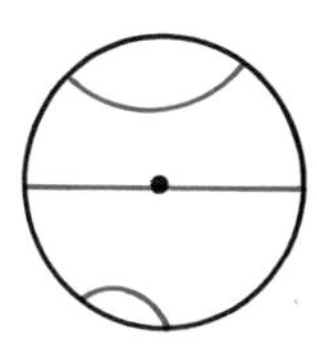

24. Through two points in a circle, there is exactly one arc of a circle orthogonal to it. What postulate does this illustrate?

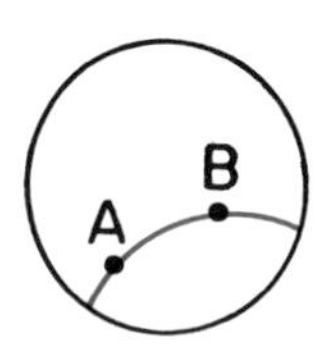

25. The figure below shows an "orthogonal arc" and several such arcs through point P that do not intersect it. What idea in Lobachevskian geometry does this illustrate?

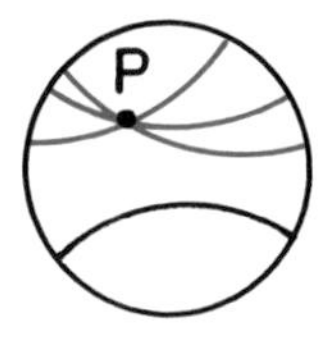

State a theorem or postulate in Lobachevskian geometry suggested by each of the following figures.

26.

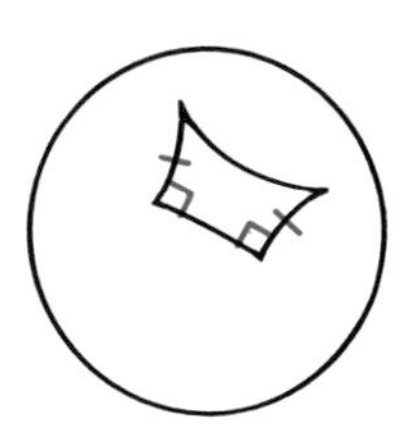

27.

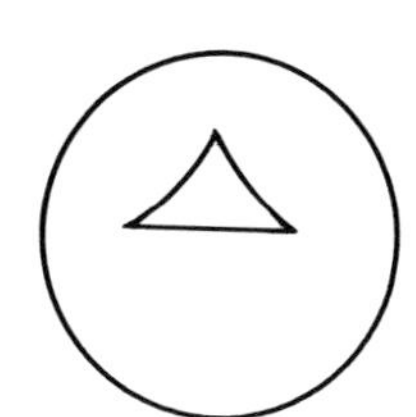

28.

Chapter 17

COORDINATE GEOMETRY

Drawing by Chas. Addams; © 1957, 1985 The New Yorker Magazine, Inc.

Lesson 1

Coordinate Systems

One way to locate a position on a map is with coordinates. Books on reading and making maps usually begin with an explanation of coordinate systems.* Coordinates are also the basis for an important branch of mathematics called *coordinate,* or *analytic, geometry.*

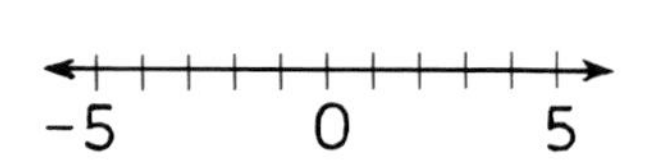

According to the Ruler Postulate, the points on a line can be numbered so that to every point there corresponds a real number called its coordinate and to every real number there corresponds a point. Furthermore, the distance between two points is the absolute value of the difference of their coordinates. A line numbered in this way is a *one-dimensional coordinate system.*

If two lines numbered according to the Ruler Postulate are perpendicular to each other, as shown in this figure, the lines become the *axes* of

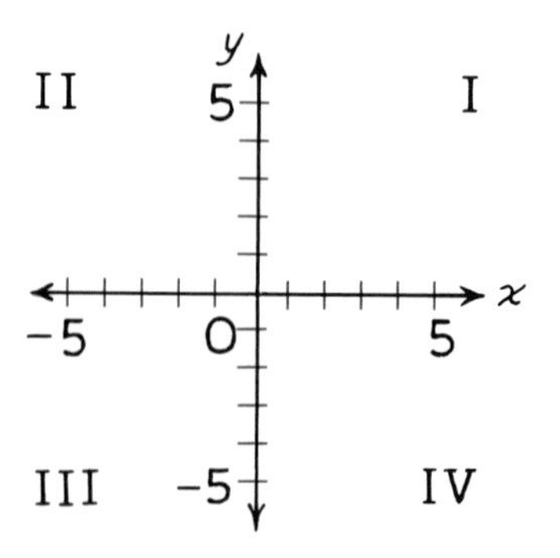

*See, for example, *Mapping,* by David Greenhood (University of Chicago Press, 1964), whose first chapter is titled "How To Find Places: Coordinates."

a *two-dimensional coordinate system*. The axes are labeled x and y, and the point in which they intersect, zero on each axis, is labeled with a capital O and is the *origin* of the coordinate system. The axes separate the plane that contains them into four regions called *quadrants*. The quadrants are identified by numbers, shown as Roman numerals in the figure.

Two numbers are needed to locate a point in a two-dimensional coordinate system. These numbers are found by drawing perpendiculars from the point to the axes as shown in these figures.

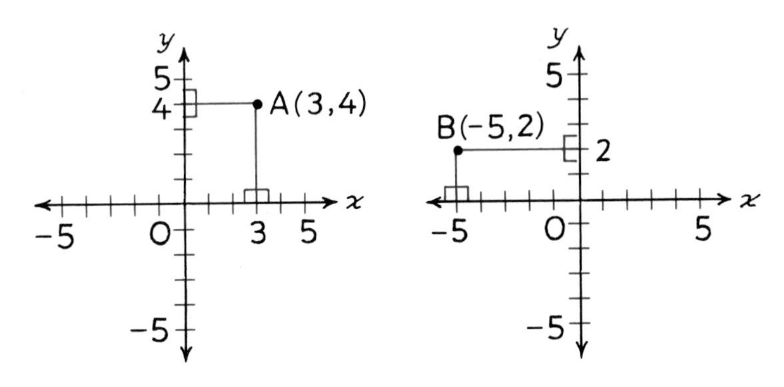

► **Definitions**

The ***x*-coordinate** of a point is the coordinate of the foot of the perpendicular from the point to the x-axis.

The ***y*-coordinate** of the point is the coordinate of the foot of the perpendicular from the point to the y-axis.

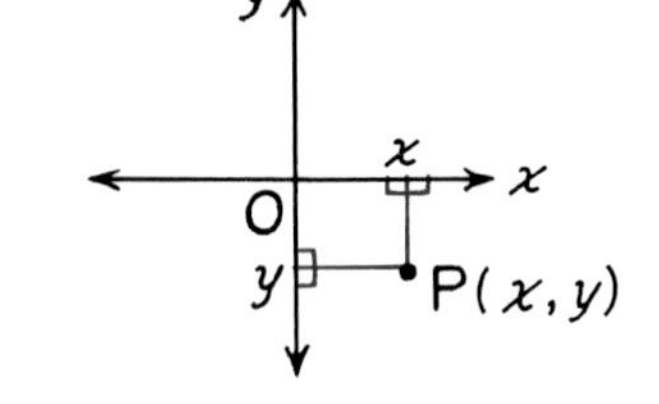

The coordinates of a point are written in parentheses and separated by a comma, the x-coordinate always being listed first: (x, y).

Exercises

Set I

The following exercises refer to the figure below.

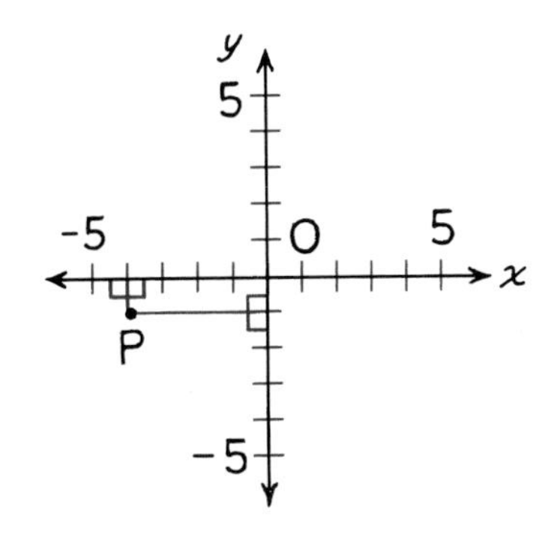

1. What are the lines labeled x and y called?
2. What is the point labeled O called?
3. Which quadrant is point P in?
4. What is the x-coordinate of point P?
5. What is the y-coordinate of point P?

Write the coordinates of the indicated points in this graph.

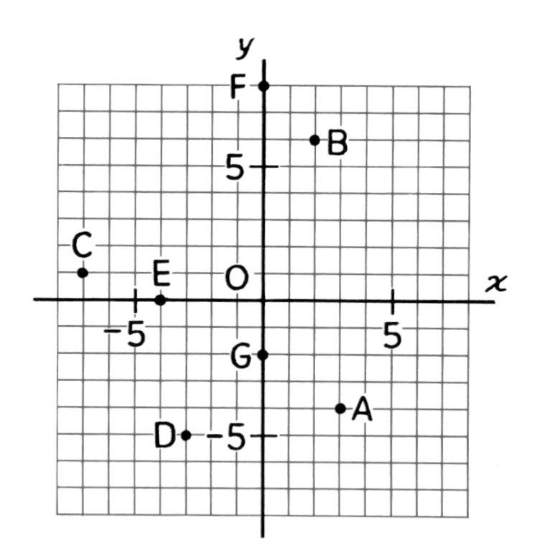

Example: Point A.
Answer: (3, -4).

6. Point B.
7. Point C.
8. Point D.
9. Point E.
10. Point F.
11. Point G.

On graph paper, draw a pair of axes extending 6 units in each direction from the origin.

12. Plot the following points: A(2, 4), B(3, 6), C(0, 0), D(-1, -2), E(-3, -6).
13. What relation do the five points have to each other?
14. In what way is the y-coordinate of each of these points related to the x-coordinate?

The following exercises refer to the figure below.

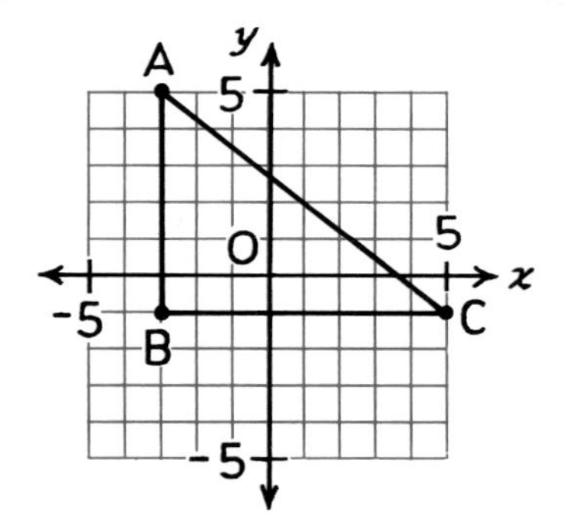

15. Which vertex of △ABC has the largest x-coordinate?
16. Which vertex has the largest y-coordinate?
17. Find the length of the shortest side of △ABC.
18. Find the length of the hypotenuse.

All of the line segments in the figure below are either horizontal or vertical. Find

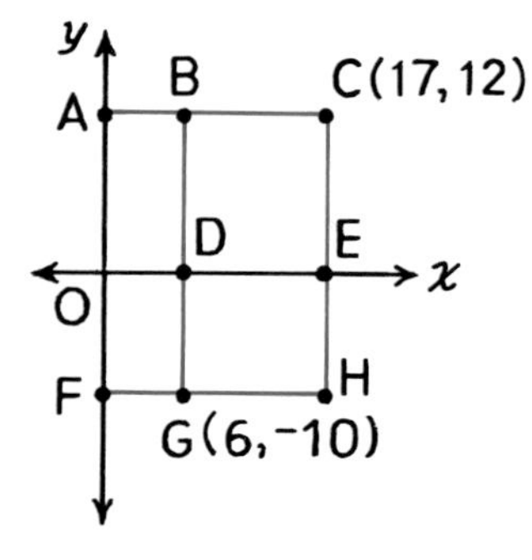

the coordinates of each of the following points.

19. Point E.
20. Point D.
21. Point A.
22. Point F.
23. Point B.
24. Point H.

Set II

On graph paper, draw a pair of axes extending 5 units in each direction from the origin.

25. Plot the following points and join them to form quadrilateral ABCD: A(5, 0), B(5, -2), C(-1, -2), D(-1, 0).
26. What kind of quadrilateral is ABCD?
27. Find its perimeter.
28. Find its area.
29. Find the length of one of its diagonals.

Tell where each of the following points is located.

Example: A point whose x-coordinate is 0.
Answer: On the y-axis.

30. A point both of whose coordinates are positive.
31. A point both of whose coordinates are negative.
32. A point whose y-coordinate is 0.
33. A point both of whose coordinates are 0.

A three-dimensional coordinate system can be formed by placing a third line, called the z-axis, perpendicular to the x- and y-axes at their point of intersection. The axes are usually drawn in the perspective shown in the figure below, so that the y- and z-axes are in the plane of the page and the x-axis is perpendicular to the plane of the page.

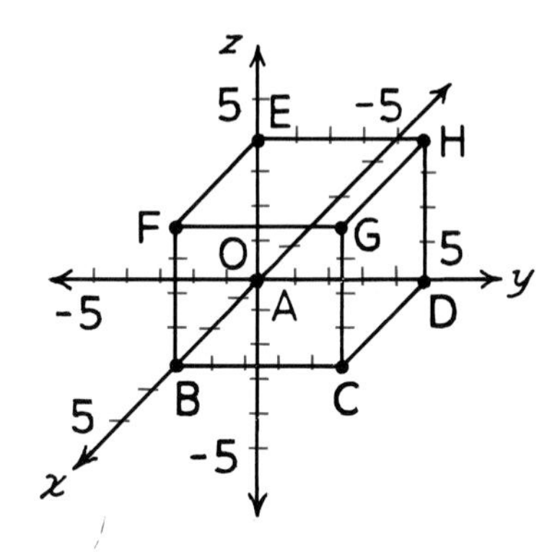

The figure shows a rectangular solid, one of whose vertices, A, is at the origin of the coordinate system. The coordinates of A are (0, 0, 0) and the coordinates of B are (3, 0, 0).

34. What are the coordinates of the other six vertices of the solid?
35. Which vertices of the solid are on the y-axis?
36. Which coordinates of these vertices are equal to zero?
37. Which vertices of the solid are in the plane that contains the x- and z-axes?
38. Which coordinate of these vertices is equal to zero?

Set III

Here is an exercise that requires patience but has an amusing result. Draw a pair of axes extending from -10 to +10 on the x-axis and from -10 to +15 on the y-axis. Connect the points in each set with straight line segments in the order given. After you have connected the points in one set, start all over again with the next. In other words, *do not connect* the last point in each set to the first point in the next one.

Set 1. (4, 1), (5, 1), (6, -2), (5, -5), (4, -5), (3, -2), (4, 1).

Set 2. (1, 11), (0, 12), (-1, 11), (0, 10), (1, 10), (-1, 8), (1, 8).

Set 3. (2, 13), (5, 13), (6, 12), (6, 9).

Set 4. (4, 8), (3, 6), (8, 5), (8, 4), (7, 4), (4, 3), (3, 2), (2, 2), (3, 4), (6, 5).

Set 5. (-3, -2), (-3, 0), (2, -1), (2, 0), (-1, 1), (0, 5), (-2, 1), (-4, 2), (-1, 5), (-2, 5), (-2, 4).

Set 6. (1, -1), (1, -5), (-3, -4), (-3, -7), (-4, -8), (-1, -8), (-1, -5).

Set 7. (-1, -1), (-1, -3), (-4, -2), (-4, -8), (-7, -8), (-6, -7), (-4, -7).

Set 8. (-1, 8), (0, 6), (-4, 6), (-3, 3), (-4, 2), (-5, 2), (-3, 0).

Set 9. (5, -8), (7, -5), (7, 1), (6, 3), (4, 3), (3, 2), (2, 0), (2, -5), (3, -8), (7, -8), (9, -5), (9, 1), (8, 3), (7, 4).

Set 10. (5, 8), (5, 11).

Set 11. (3, 10), (4, 10), (4, 9), (3, 8).

Set 12. (3, 12), (2, 11), (3, 11).

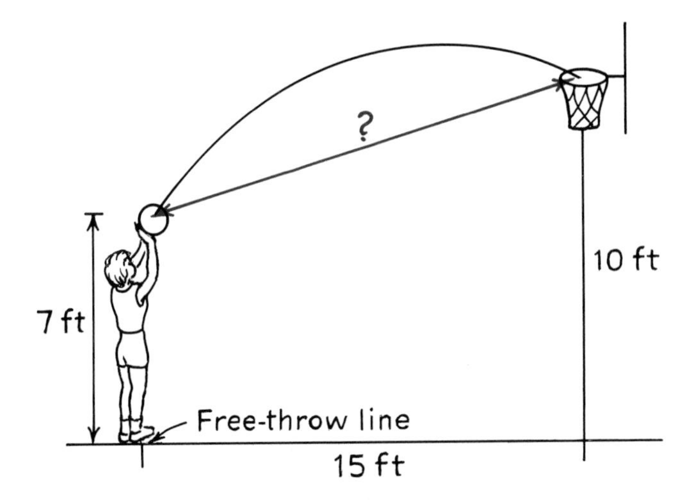

Lesson 2
The Distance Formula

When a basketball player makes a free throw, how far is the ball from the basket when it is released? The distance depends on the height of the player. In the figure above, the ball is thrown from a height of 7 feet.

One way to find the distance is to use a coordinate system. The distance between two points in such a system can be calculated from the coordinates of the points. The method is based on the Pythagorean Theorem.

In this figure, points P_1 and P_2 have coordinates (x_1, y_1) and (x_2, y_2).

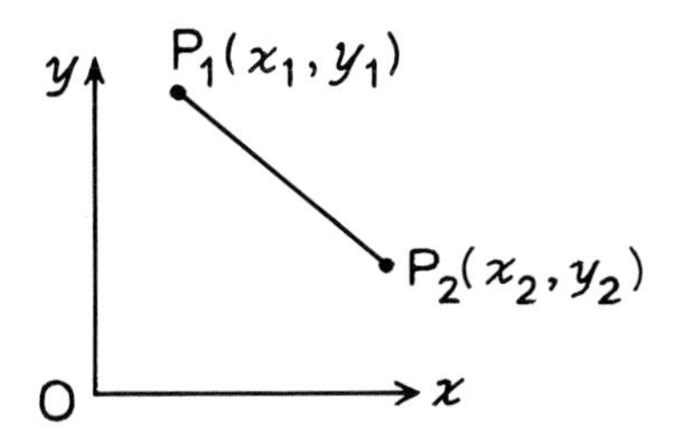

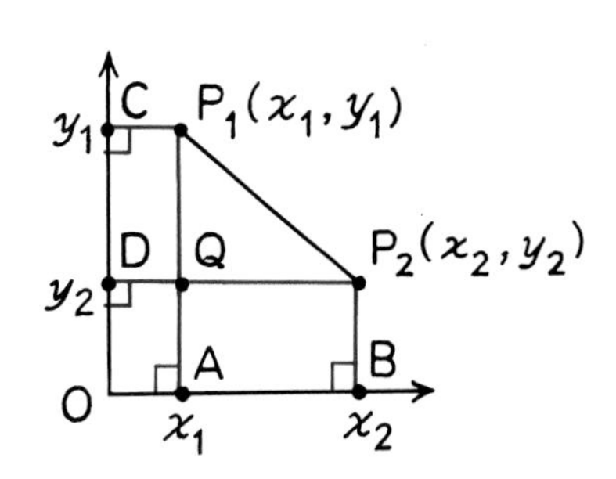

First, we draw lines through P_1 and P_2 perpendicular to the x- and y-axes. These lines form right $\triangle P_1QP_2$ and parallelograms QP_2BA and P_1QDC. By the Pythagorean Theorem,

$$P_1P_2^2 = QP_2^2 + P_1Q^2. \qquad (1)$$

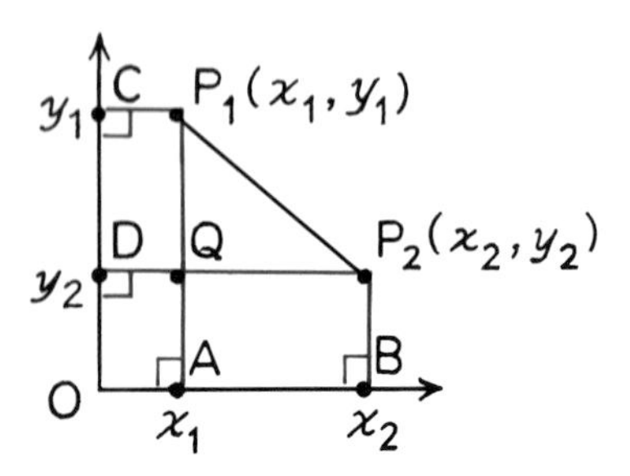

Because AB = QP_2 and CD = P_1Q (the opposite sides of a parallelogram are equal), we can substitute into equation 1 to get

$$P_1P_2^{\,2} = AB^2 + CD^2. \tag{2}$$

By the Ruler Postulate,

$$AB = |x_2 - x_1| \quad \text{and} \quad CD = |y_2 - y_1|.$$

Therefore,

$$AB^2 = (x_2 - x_1)^2 \quad \text{and} \quad CD^2 = (y_2 - y_1)^2.$$

Substituting into equation 2,

$$P_1P_2^{\,2} = (x_2 - x_1)^2 + (y_2 - y_1)^2,$$

and taking square roots, we get

$$P_1P_2 = \sqrt{(x_2 - x_1)^2 + (y_2 - y_1)^2}.$$

This equation expresses the distance between points P_1 and P_2 in terms of their coordinates. We will call it *the distance formula.*

► **Theorem 103** (The Distance Formula)
The distance between the points $P_1(x_1, y_1)$ and $P_2(x_2, y_2)$ is

$$\sqrt{(x_2 - x_1)^2 + (y_2 - y_1)^2}.$$

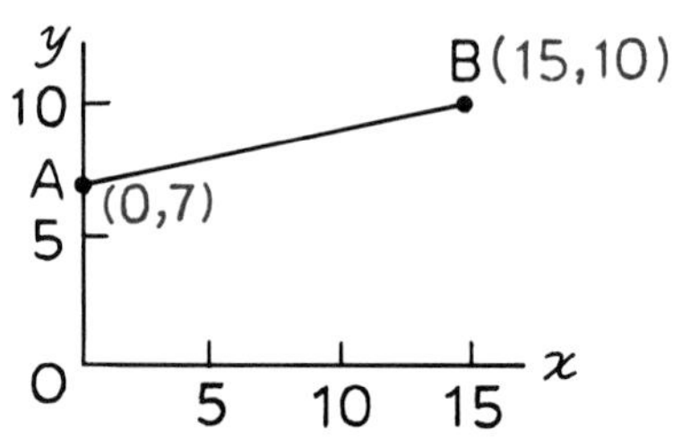

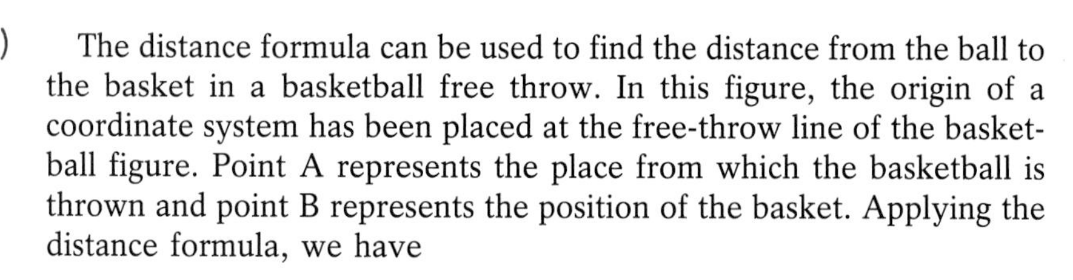

The distance formula can be used to find the distance from the ball to the basket in a basketball free throw. In this figure, the origin of a coordinate system has been placed at the free-throw line of the basketball figure. Point A represents the place from which the basketball is thrown and point B represents the position of the basket. Applying the distance formula, we have

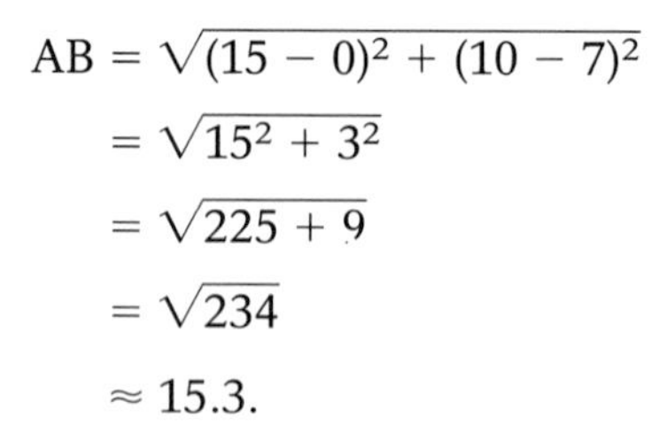

$$\begin{aligned} AB &= \sqrt{(15 - 0)^2 + (10 - 7)^2} \\ &= \sqrt{15^2 + 3^2} \\ &= \sqrt{225 + 9} \\ &= \sqrt{234} \\ &\approx 15.3. \end{aligned}$$

The distance from the ball to the basket is approximately 15.3 feet.

Exercises

Set I

Use the distance formula to find the distances between the following pairs of points. Express irrational answers in simple radical form.

1. (11, 0) and (5, 8).
2. (4, 7) and (7, 4).
3. (0, 0) and (12, -5).
4. (3, -1) and (1, -8).
5. (-2, 11) and (6, 15).
6. (4, 40) and (-3, 16).

Triangle ABC has vertices A(0, 0), B(4, 3), and C(0, 6).

7. Plot the points and draw the triangle.
8. Find the length of each of its sides.
9. What kind of triangle is $\triangle$ABC?
10. Find its perimeter.

Points A, B, and C have coordinates A(-2, 5), B(1, 3), and C(7, -1).

11. Find AB, BC, and AC.
12. Do A, B, and C determine a triangle?
13. Explain why or why not.
14. Plot the three points.

A circle has center A(2, 0) and goes through point B(6, 5).

15. Find the radius of the circle.
16. Does the circle go through point C(-3, 4)?
17. Does the circle go through point D(5, -6)?
18. Plot points A, B, C, and D and draw the circle.

Set II

Rectangle ABCD has vertices A(-3, -2), B(0, -8), C(12, -2) and D(9, 4).

19. Find AB and BC.
20. Find the perimeter of the rectangle.
21. Find its area.
22. Find the length of each of its diagonals.
23. Plot the points and draw the rectangle.

Quadrilateral ABCD has vertices A(0, 2), B(7, 1), C(2, -4), and D(-5, -3).

24. Find the length of each of its sides.
25. What kind of quadrilateral is ABCD?
26. Plot the points and draw the quadrilateral.
27. Find the length of each of its diagonals.
28. Find its area.

Write expressions for the distances between the following pairs of points.

29. $(0, 0)$ and (a, b).
30. $(2a, 0)$ and $(0, 2b)$.
31. (a, a) and $(a + b, a + b)$.
32. $(a, 0)$ and $\left(\frac{1}{2}a, \frac{1}{2}a\sqrt{3}\right)$.

Set III

The following problem is from a textbook on coordinate geometry published in 1892.*

> One endpoint of a line segment whose length is 13 is the point (-4, 8); the y-coordinate of the other endpoint is 3. What is its x-coordinate?

Show how to solve the problem by using either the distance formula or a graph.

---

* *Elements of Analytic Geometry*, by G. A. Wentworth (Ginn and Company, 1892).

Ames/NASA

Lesson 3
Slope

The airplane in this photograph can take off from very short runways because of its ability to climb at an unusually large angle. Called a VSTOL (Vertical Short Take Off and Landing) plane, it can climb vertically 12 meters for every 15 meters that it moves horizontally forward. The ratio of these distances,

$$\frac{12}{15} \quad \text{or} \quad \frac{4}{5},$$

is the *slope* of the path of the plane.

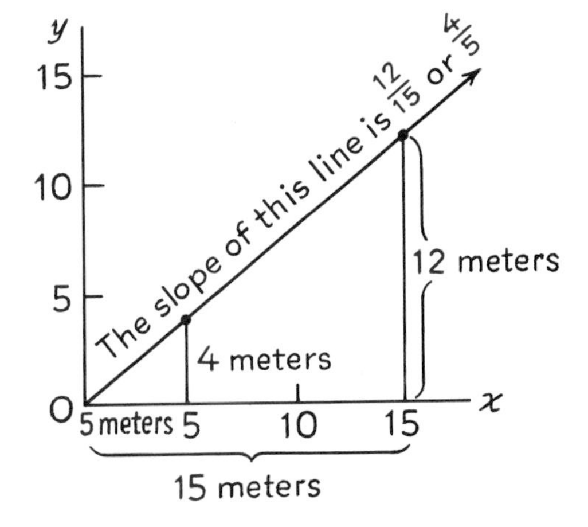

The direction of a line in the coordinate plane is also given by a number called its slope. To find the slope of a line, we choose two points on it. As we go from the left point to the right point, we move a certain distance to the right, called the *run*, and a certain distance vertically, called the *rise*.

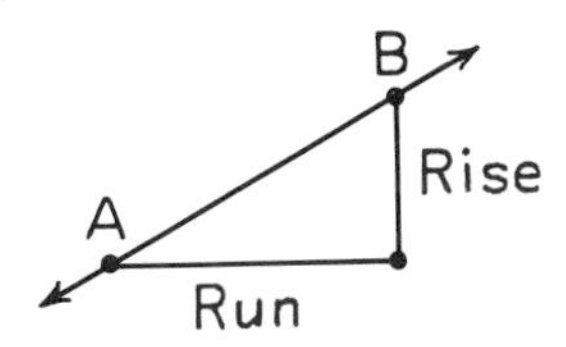

The slope of the line is found by dividing the rise by the run:

$$\text{slope} = \frac{\text{rise}}{\text{run}}.$$

In the figures below, the slope of $\overleftrightarrow{AB}$ is $\frac{3}{4}$ and the slope of $\overleftrightarrow{CD}$ is $-\frac{2}{5}$.

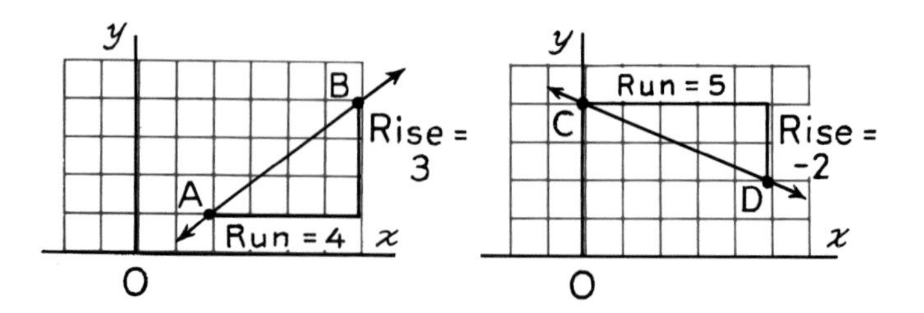

The slope of a line that is not vertical can be found from the coordinates of two points on the line because they can be used to determine the rise and run.

► **Definition**

The ***slope m*** of a nonvertical line that contains the points $P_1(x_1, y_1)$ and $P_2(x_2, y_2)$ is

$$m = \frac{y_2 - y_1}{x_2 - x_1}.$$

The slope of a line can be a positive number or a negative number as the examples above illustrate. It can also be zero. For example, in the first figure below, $\overleftrightarrow{EF}$ is horizontal. The slope of $\overleftrightarrow{EF}$ is zero because

$$m = \frac{3 - 3}{6 - 2} = \frac{0}{4} = 0.$$

The same type of calculation can be used to show that *the slope of every horizontal line is zero.*

If a line is vertical, its slope is undefined because division by zero is undefined. For example, in the second figure, $\overleftrightarrow{GH}$ is vertical. We cannot find the slope of $\overleftrightarrow{GH}$ because

$$m = \frac{5 - 2}{4 - 4} = \frac{3}{0} = ?$$

It is for this reason that *the slope of every vertical line is undefined.*

Exercises

Set I

The following questions refer to the figure below.

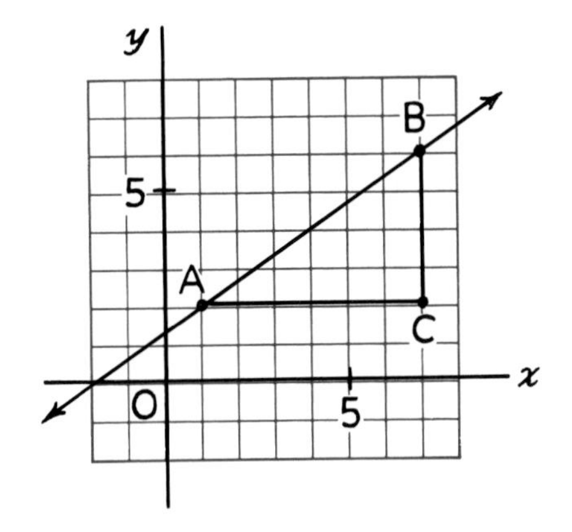

1. What are the coordinates of point A?
2. What are the coordinates of point B?
3. What is BC?
4. What is AC?
5. What is the slope of $\overleftrightarrow{AB}$?

The following questions refer to the figure below.

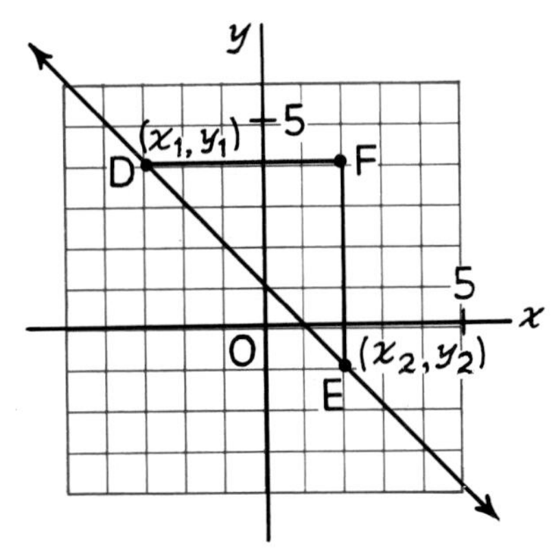

6. To what numbers are x_1 and y_1 equal?
7. To what numbers are x_2 and y_2 equal?
8. To what number is $y_2 - y_1$ equal?
9. To what number is $x_2 - x_1$ equal?
10. What is the slope of $\overleftrightarrow{DE}$?

If possible, find the slope of the line through each of the following pairs of points.

Example: (2, 1) and (10, -3).

Solution: $m = \frac{-3 - 1}{10 - 2} = \frac{-4}{8} = -\frac{1}{2}$.

11. (0, 0) and (5, 10).
12. (7, 4) and (1, 6).
13. (-8, -7) and (12, 9).
14. (-5, 3) and (4, 3).
15. (1, -10) and (0, 15).
16. (11, -2) and (11, 13).

In the figure below, the coordinates of point A are (c, d).

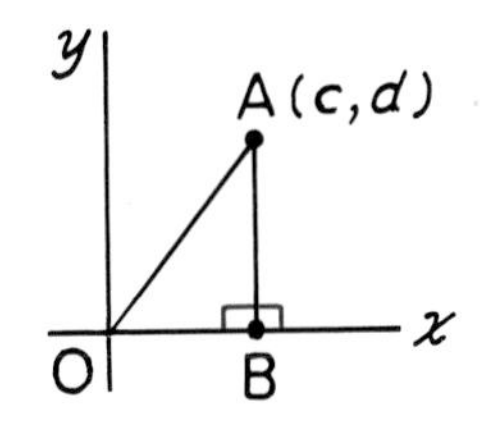

17. Which segment in the figure has length c?
18. Which segment has length d?
19. What is $\frac{d}{c}$ called with respect to $\overleftrightarrow{OA}$?
20. What is $\frac{d}{c}$ called with respect to $\measuredangle$AOB?

Set II

Triangle ABC has vertices A(8, 4), B(-6, 2), and C(-4, -2).

21. Find the length of each side.
22. Show that △ABC is a right triangle.
23. Which two sides of △ABC are perpendicular?
24. Find the slope of each side of the triangle.
25. In what way are the slopes of the perpendicular sides of △ABC related to each other?

Quadrilateral ABCD has vertices A(16, 0), B(6, -5), C(-5, -7), and D(5, -2).

26. Find the length of each of its sides.
27. What kind of quadrilateral is ABCD?
28. Find the slope of each side.
29. In what way are the slopes of the parallel sides of quadrilateral ABCD related to each other?
30. Find the slope of each diagonal.
31. In what way are the slopes of the diagonals of ABCD related to each other?

Write expressions for the slopes of the lines through the following pairs of points.

32. $(0, 0)$ and (a, b).
33. (a, a) and $(a + b, a + b)$.
34. $(a, 0)$ and (b, c).
35. (a, b) and (b, a).

Set III

This figure represents right △ABC but it is so tall that it will not fit on this page.

1. Find the slope of $\overline{AC}$, given that the slope of $\overline{AB}$ is 0.
2. How tall is △ABC? Express your answer to the nearest inch.

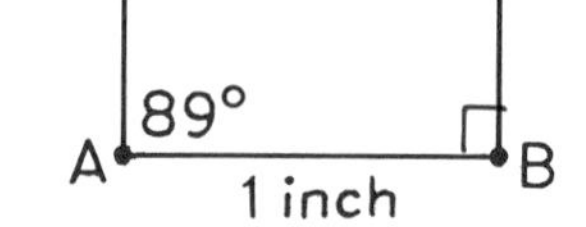

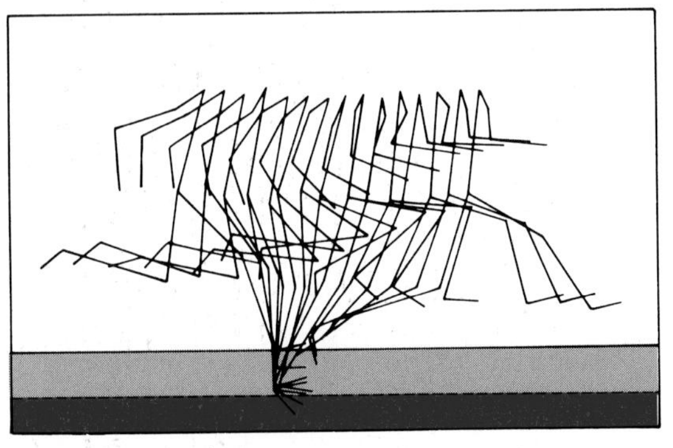
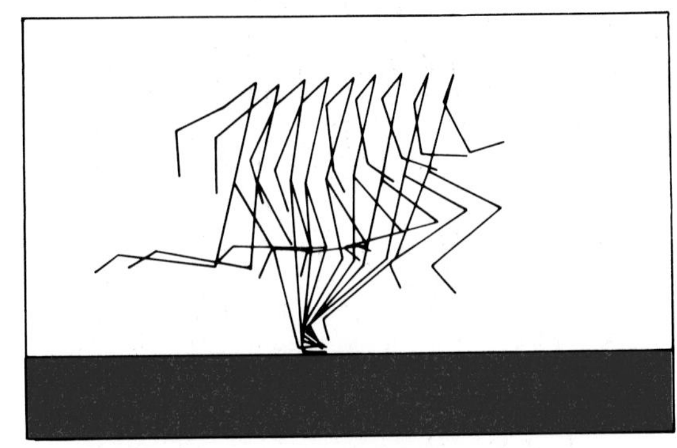

Courtesy of Tom McMahon

Lesson 4

Parallel and Perpendicular Lines

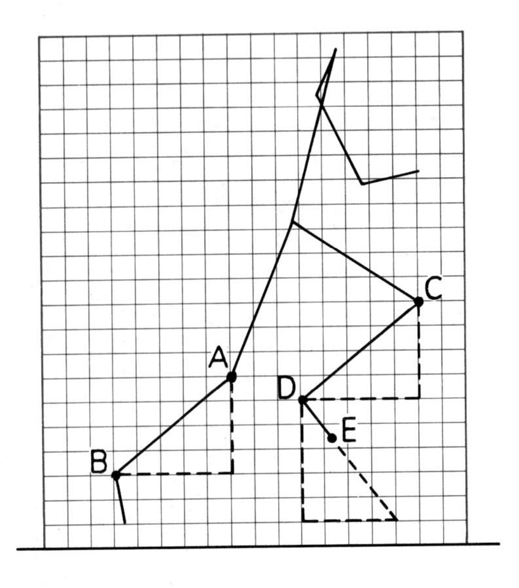

When Harvard University built a new indoor track several years ago, a computer was used to help in determining the ideal surface. High-speed pictures taken of runners were reduced by the computer to stick figures such as the two shown above. The drawings compare the movement of a runner on a very soft surface (left) and on a very hard surface (right). These drawings, together with other data, revealed that the best surface is much springier than the track designers had originally thought.*

The line segments in the computer drawings have many different slopes. The slopes of the segments that are parallel or perpendicular, however, are related in very simple ways. The last position of the runner in the second drawing, shown here, suggests what these relations are. Segments $\overline{AB}$ and $\overline{CD}$ appear to be parallel and both have a slope of $\frac{4}{5}$. Segments $\overline{CD}$ and $\overline{DE}$ appear to be perpendicular; their slopes are $\frac{4}{5}$ and $-\frac{5}{4}$.

* "Tuning the Track," by Anthony Chase in *Newton at the Bat: The Science in Sports,* edited by Eric W. Schrier and William F. Allman (Scribners, 1984), pp. 59–62.

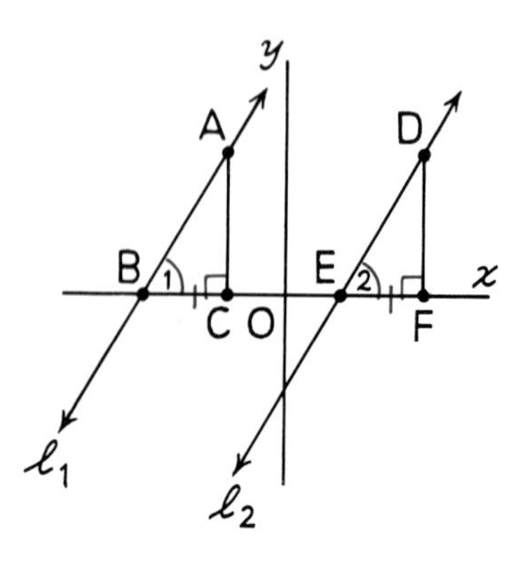

Because the direction of a line in the coordinate plane is given by its slope, it is not surprising that nonvertical* parallel lines have the same slope. In this figure, $\ell_1 \parallel \ell_2$. It follows that $\angle 1 = \angle 2$ because parallel lines form equal corresponding angles with a transversal.

Triangles ABC and DEF have been drawn so that BC = EF, $\overline{AC} \perp \overline{BC}$, and $\overline{DF} \perp \overline{EF}$. It follows that $\triangle ABC \cong \triangle DEF$ (A.S.A.), and so AC = DF. Dividing this equation by BC = EF, we get

$$\frac{AC}{BC} = \frac{DF}{EF}.$$

But $\frac{AC}{BC}$ is the slope of ℓ_1 and $\frac{DF}{EF}$ is the slope of ℓ_2. Representing these slopes as m_1 and m_2, we have

$$m_1 = m_2.$$

It can also be shown that lines that have the same slope are parallel. These facts suggest the following theorem.

► **Theorem 104**
Two nonvertical lines are parallel iff their slopes are equal.

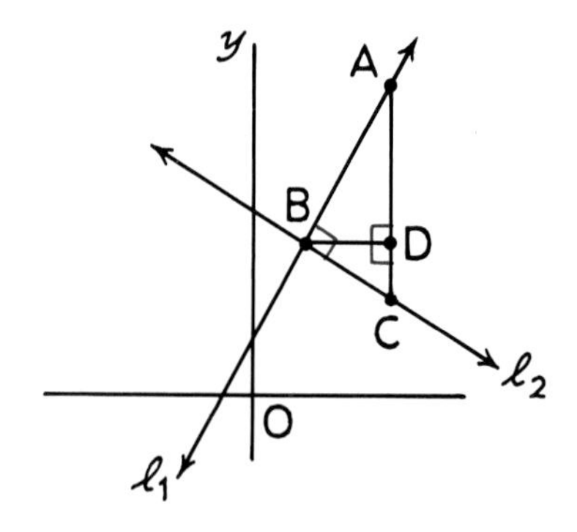

We can prove that the slopes of perpendicular lines are also algebraically related. In this figure, $\ell_1 \perp \ell_2$. Triangle ABC has been drawn so that $\overline{AC}$ is parallel to the y-axis and $\overline{BD}$ is parallel to the x-axis. Because $\triangle ABC$ is a right triangle in which $\overline{BD}$ is the altitude to the hypotenuse $\overline{AC}$, it follows that

$$\frac{AD}{BD} = \frac{BD}{DC}$$

(the altitude to the hypotenuse of a right triangle is the geometric mean between the segments of the hypotenuse).

But $\frac{AD}{BD}$ is the slope of ℓ_1 and $\frac{-DC}{BD}$ is the slope of ℓ_2. Representing these slopes as m_1 and m_2, we have

$$m_1 = -\frac{1}{m_2}, \quad \text{or} \quad m_1 m_2 = -1.$$

It can be shown that the converse of this is also true.

► **Theorem 105**
Two nonvertical lines are perpendicular iff the product of their slopes is -1.

---

*Recall that the slope of a vertical line is undefined.

Exercises

Set I

The following exercises refer to the figure below.

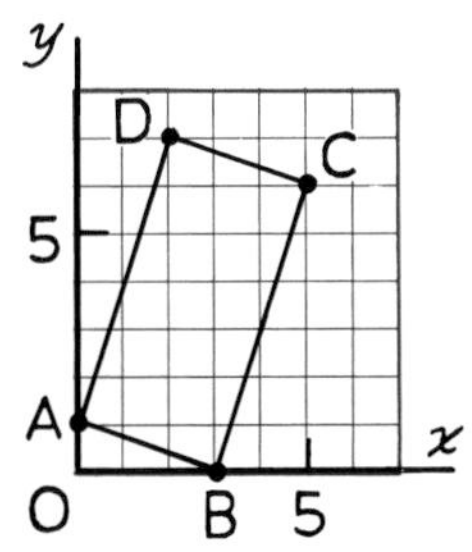

1. Which sides of quadrilateral ABCD have a slope of 3?
2. What can you conclude about these sides?
3. State the theorem that is the basis for your answer.
4. Which sides of the quadrilateral have a slope of $-\frac{1}{3}$?
5. What can you conclude about these sides?
6. What can you conclude about each pair of consecutive sides of the quadrilateral?
7. State the theorem that is the basis for your answer.
8. What kind of quadrilateral is ABCD?

Line ℓ_1 has a slope of $\frac{1}{4}$.

9. What is the slope of line ℓ_2 if $\ell_1 \perp \ell_2$?
10. What is the slope of line ℓ_3 if $\ell_2 \parallel \ell_3$?
11. What can you conclude from your answers to Exercises 9 and 10 about lines ℓ_1 and ℓ_3?
12. What theorem does this conclusion illustrate?

Find the slopes of the following lines.

Example: A line perpendicular to the line through (-2, 3) and (4, 6).

Solution: The slope of the line through (-2, 3) and (4, 6) is

$$\frac{6 - 3}{4 - -2} = \frac{3}{6} = \frac{1}{2},$$

and so the slope of a line perpendicular to it is -2.

13. A line parallel to the line through the origin and (12, 9).
14. A line perpendicular to the line through (4, 8) and (-1, 10).
15. A line parallel to the x-axis.
16. A line perpendicular to the x-axis.
17. A line containing one side of a parallelogram if the endpoints of the opposite side are (-9, 2) and (-1, -5).
18. A line containing one diagonal of a rhombus if the endpoints of the other diagonal are (0, 8) and (2, -4).

Set II

Line segments are parallel iff they lie in parallel lines. Consider the points A(3, 1), B(6, 7), C(0, -5), and D(2, -1).

19. Find the slope of $\overline{AB}$.
20. Find the slope of $\overline{CD}$.
21. Plot the four points and draw $\overline{AB}$ and $\overline{CD}$.
22. Does $\overline{AB}$ appear to be parallel to $\overline{CD}$?

Line segments are perpendicular iff they lie in perpendicular lines. Consider the points A(-4, 6), B(-2, 0), C(2, -3), and D(5, -2).

23. Find the slope of $\overline{AB}$.

24. Find the slope of $\overline{CD}$.

25. Plot the four points and draw $\overline{AB}$ and $\overline{CD}$.

26. Is $\overline{AB} \perp \overline{CD}$?

Triangle ABC has vertices A(0, 0), B(5, 2), and C(7, -3).

27. Find the slope of each side of the triangle.

28. Show that △ABC is a right triangle.

29. Plot the points and draw △ABC.

30. Show that △ABC is isosceles.

Consider the points A(0, 1), B(5, 4), C(3, -2), and D(18, y).

31. Find the value of y for which $\overline{AB} \parallel \overline{CD}$.

32. Find the value of y for which $\overline{AB} \perp \overline{CD}$.

Set III

In calculus, the slope of a curve at a point on the curve is defined to be the slope of the line that is tangent to the curve at that point.

In this figure, for example, the slope of the circle at P_1 is the slope of tangent line ℓ.

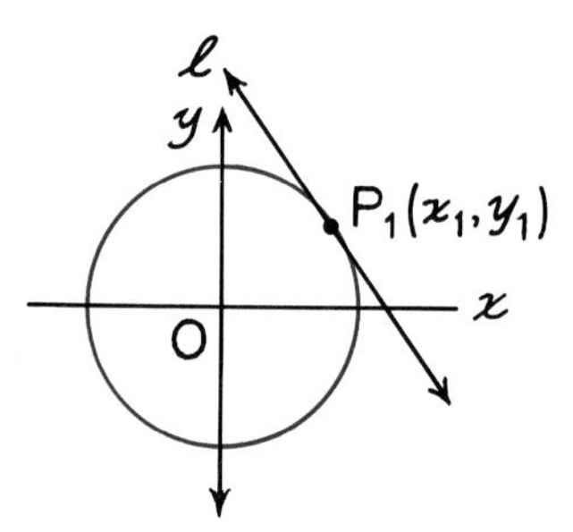

1. Find an expression for the slope of the circle at P_1 in terms of x_1 and y_1. Show your method.

2. Where is the slope of the circle in this figure positive?

3. Does the circle have a slope at every point on it? Explain why or why not.

Franklin Berger © 1977

Lesson 5
The Midpoint Formula

Steve McPeak holds several records for tightrope walking. One of his amazing feats was to walk up a wire rising at an angle of 40° to a height of 45 feet in 71 seconds.

The line in the figure below represents the wire and Steve's positions when he was midway up the wire and at the top. The origin of a coordinate system has been placed at A, the bottom end of the wire. The coordinates of B indicate that Steve had traveled 54 feet horizontally and 45 feet vertically when he made it to the top.

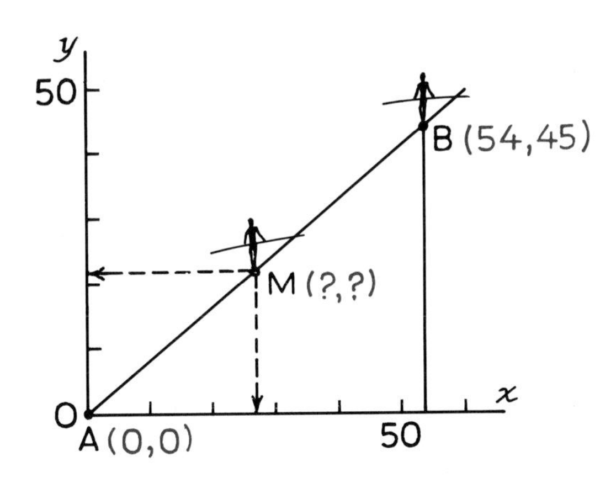

What were his coordinates at the midpoint of the trip? It seems reasonable to guess that they must have been half of those at the end: (27, 22.5). This is easy to prove.

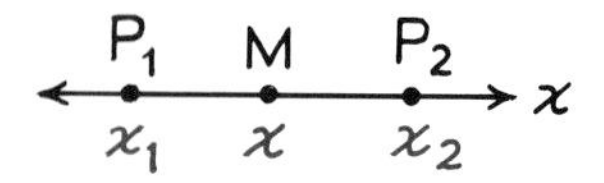

First, we will find the coordinate of the midpoint of a line segment in a *one*-dimensional coordinate system. Suppose that $\overline{P_1P_2}$ is a line segment on the x-axis and that M is its midpoint. We represent the coordinates of the three points as x_1, x, and x_2 as shown and assume that $x_2 > x > x_1$.

Because M is the midpoint of $\overline{P_1P_2}$, we know that

$$P_1M = MP_2. \tag{1}$$

By the Ruler Postulate,

$$P_1M = |x - x_1| = x - x_1$$

and

$$MP_2 = |x_2 - x| = x_2 - x.$$

Substituting into equation 1 gives

$$x - x_1 = x_2 - x.$$

Solving this equation for x in terms of x_1 and x_2, we get

$$2x = x_1 + x_2$$

$$x = \frac{x_1 + x_2}{2}.$$

This result turns out the same if $x_2 < x < x_1$. Therefore, the coordinate of the midpoint is the average of the coordinates of the endpoints of the line segment.

This fact can now be applied to finding the coordinates of the midpoint of a line segment in a two-dimensional coordinate system. In this figure, M is the midpoint of $\overline{P_1P_2}$. We begin by drawing lines through P_1,

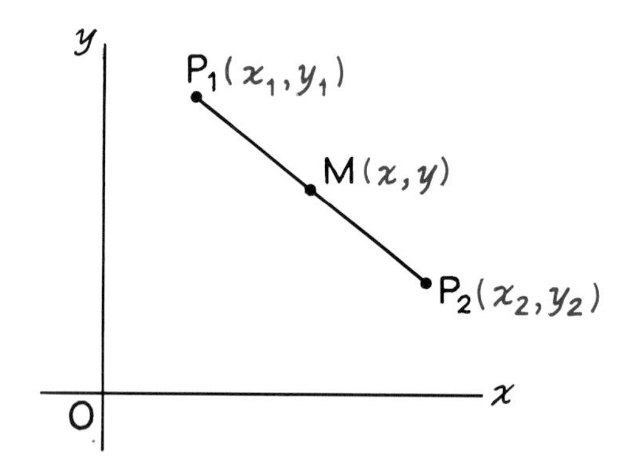

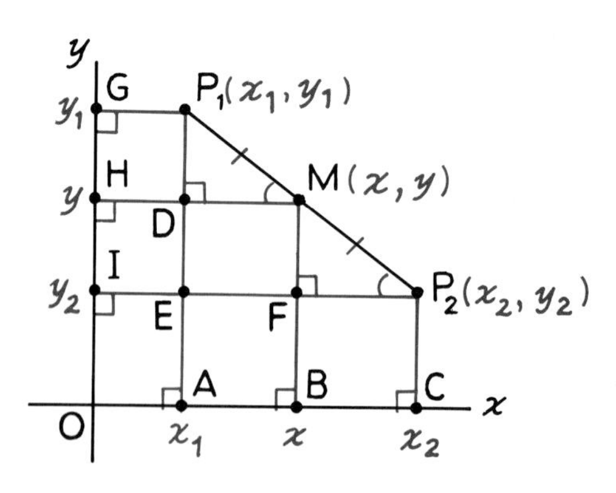

M, and P_2 perpendicular to the x- and y-axes. These lines form right triangles $\triangle P_1DM$ and $\triangle MFP_2$ and parallelograms DMBA and FP_2CB. The triangles are congruent by A.A.S., and so

$$DM = FP_2.$$

Also, AB = DM and BC = FP_2 (the opposite sides of a parallelogram are equal), and so

$$AB = BC.$$

This means that B is the midpoint of $\overline{AC}$, and so

$$x = \frac{x_1 + x_2}{2}.$$

The same reasoning can be used to show that

$$y = \frac{y_1 + y_2}{2}.$$

These results can be stated as the following theorem, which we will refer to as *the midpoint formula.*

► **Theorem 106** (The Midpoint Formula)
The midpoint of the line segment whose endpoints are $P_1(x_1, y_1)$ and $P_2(x_2, y_2)$ has the coordinates

$$\left(\frac{x_1 + x_2}{2}, \frac{y_1 + y_2}{2}\right).$$

Exercises

Set I

Write an expression in terms of the coordinates of the endpoints of $\overline{P_1P_2}$ for each of the following quantities.

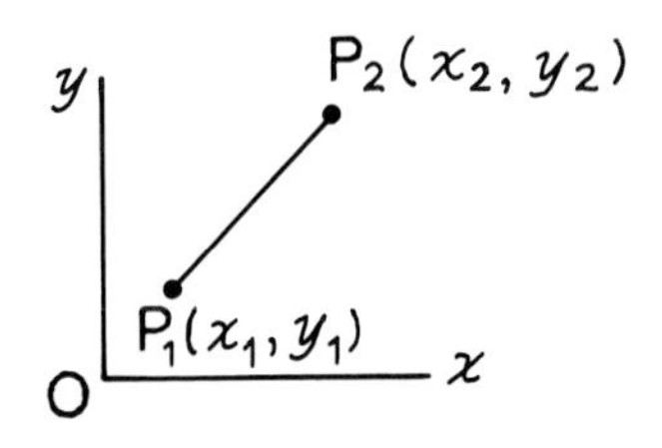

1. The length of $\overline{P_1P_2}$.
2. The slope of $\overline{P_1P_2}$.
3. The coordinates of the midpoint of $\overline{P_1P_2}$.

Use the midpoint formula to find the coordinates of the midpoints of the following line segments in the figure below. Then refer to the figure to see if your answers seem reasonable.

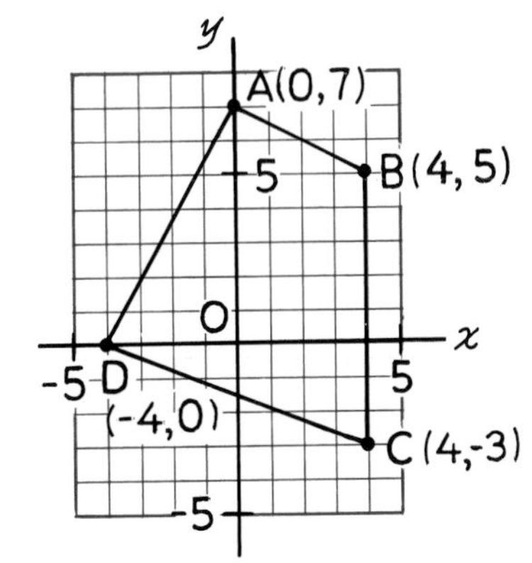

4. $\overline{AB}$.
5. $\overline{BC}$.
6. $\overline{AD}$.
7. $\overline{DC}$.

Use the midpoint formula to find the coordinates of the point midway between each of the following pairs of points.

8. (0, 0) and (12, -20).
9. (5, -8) and (-5, 8).
10. (3, -6) and (4, -7).
11. (-11, 9) and (1, -13).

Quadrilateral ABCD has vertices A(4, 9), B(8, -3), C(2, -7), and D(-2, 5).

12. Find the coordinates of the midpoint of diagonal $\overline{AC}$.
13. Find the coordinates of the midpoint of diagonal $\overline{BD}$.
14. What can you conclude about diagonals $\overline{AC}$ and $\overline{BD}$?
15. What kind of quadrilateral is ABCD?
16. How do you know?
17. Plot points A, B, C, and D; draw quadrilateral ABCD and its diagonals.

Set II

Triangle ABC has vertices A(6, 7), B(-4, 9), and C(0, 1).

18. Find the coordinates of D, the midpoint of $\overline{AB}$.
19. Find the coordinates of E, the midpoint of $\overline{AC}$.
20. Find the slope of $\overline{DE}$.
21. Find the slope of $\overline{BC}$.
22. What can you conclude about $\overline{DE}$ and $\overline{BC}$?
23. Plot points A, B, C, D, and E; draw △ABC and $\overline{DE}$.

Points A(3, 2) and B(-1, -6) are endpoints of a diameter of circle C.

24. Find the coordinates of point C.
25. Find the length of the radius of the circle.
26. Point D has coordinates (4, 1). Where is D with respect to the circle?
27. One endpoint of diameter $\overline{EF}$ of the circle is E(5, -4). Find the coordinates of point F.
28. Plot points A, B, C, D, E, and F; draw the circle.

Triangle ABC has vertices A(0, -3), B(1, 5), and C(7, 1).

29. Find the length of each of its sides.
30. Find the coordinates of the midpoint of its shortest side.
31. Find the slope of the median to its shortest side.
32. Find the slope of its shortest side.
33. What can you conclude about the relation between the median to the shortest side and the shortest side?
34. Plot points A, B, and C; draw △ABC and the median to the shortest side.

Set III

The coordinates of the centroid of a triangle are related to the coordinates of its vertices in a simple way.

1. Plot the points A(0, 9), B(2, 1), and C(10, 11), and draw the triangle that they determine. Find the midpoints of the sides, and draw the medians of the triangle.

2. Refer to your figure to tell what the coordinates of the centroid of the triangle seem to be.

3. In general, how do you think the coordinates of the centroid of a triangle, (x, y), are related to the coordinates of the vertices of the triangle, (x_1, y_1), (x_2, y_2), and (x_3, y_3)?

Portrait of Descartes courtesy of the New York Public Library, Astor, Lenox, and Tilden Foundations

L A

GEOMETRIE.

LIVRE PREMIER.

Des problefmes qu'on peut conftruire fans y employer que des cercles & des lignes droites.

Ous les Problefmes de Geometrie fe peuuent facilement reduire a tels termes, qu'il n'eft befoin par aprés que de connoiftre la longeur de quelques lignes droites, pour les conftruire.

Et comme toute l'Arithmetique n'eft compofée, que de quatre ou cinq operations, qui font l'Addition, la Souftraction, la Multiplication, la Diuifion, & l'Extraction des racines, qu'on peut prendre pour vne efpece de Diuifion : Ainfi n'at'on autre chofe a faire en Geometrie touchant les lignes qu'on cherche, pour les preparer a eftre connuës, que leur en adioufter d'autres, ou en ofter, Oubien en ayant vne, que ie nommeray l'vnité pour la rapporter d'autant mieux aux nombres, & qui peut ordinairement eftre prife a difcretion, puis en ayant encore deux autres, en trouuer vne quatriefme, qui foit à l'vne de ces deux, comme l'autre eft a l'vnité, ce qui eft le mefme que la Multiplication; oubien en trouuer vne quatriefme, qui foit a l'vne de ces deux, comme l'vnité

Commét le calcul d'Arithmetique fe rapporte aux operations de Geometrie.

Pp eft

Lesson 6 Coordinate Proofs

The branch of mathematics known as *coordinate* or *analytic geometry* was created in the seventeenth century by the famous French mathematician and philosopher René Descartes. Descartes decided that mathematics, to quote his words, "is a more powerful instrument of knowledge than any other that has been bequeathed to us by human agency." This belief led him to apply deductive reasoning, the method used by Euclid in the *Elements,* to other areas of study. Descartes's creation of coordinate geometry combined geometry and algebra to form a single subject more powerful than either by itself had been before. His idea was to use coordinates and algebra to discover facts that only ingenuity could find using the methods we studied previously. The portrait of Descartes at the top of this page was copied from a painting by Frans Hals that hangs in the Louvre in Paris. The first page of his work on coordinate geometry, published in 1637, is also shown.

The methods of coordinate geometry can be used to prove some of the theorems of plane geometry. The procedure is to place a coordinate system on the figure illustrating the proof and then to use algebra to establish the relation to be proved.

Here is an example of a proof by the methods of coordinates.

Theorem.
The diagonals of a rectangle are equal.

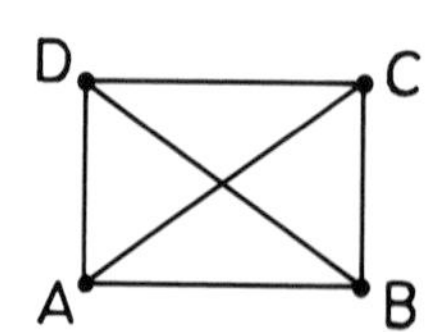

Given: Rectangle ABCD with diagonals $\overline{AC}$ and $\overline{DB}$.

Prove: AC = DB.

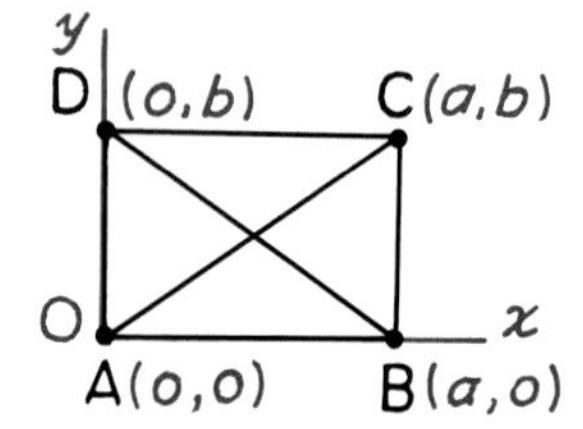

Proof.
Place a coordinate system on rectangle ABCD as shown on the preceding page. By the distance formula,

$$AC = \sqrt{(a-0)^2 + (b-0)^2} = \sqrt{a^2 + b^2}$$

and

$$DB = \sqrt{(a-0)^2 + (0-b)^2} = \sqrt{a^2 + b^2}$$

and so AC = DB (substitution).

This proof is as simple as it is because the placement of the coordinate system on the rectangle makes the coordinates of its vertices easy to work with. It is usually a good idea to place the origin of the coordinate system at one vertex of the polygon and the x-axis along one side.

Here is another example of a proof by coordinates.

Theorem.
A midsegment of a triangle is parallel to the third side.

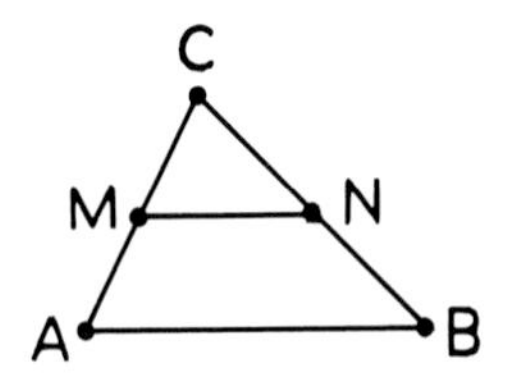

Given: $\triangle$ABC with midsegment $\overline{MN}$.

Prove: $\overline{MN} \parallel \overline{AB}$.

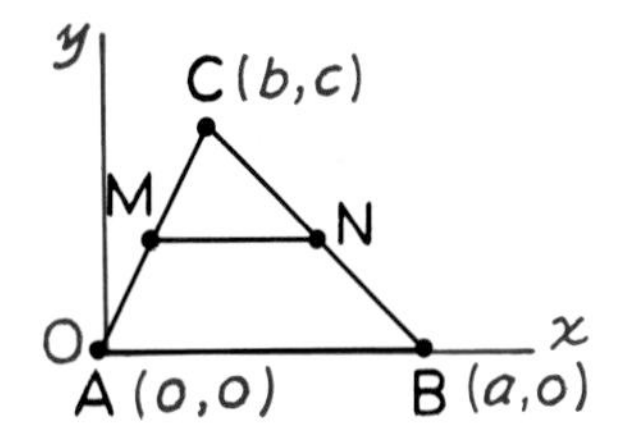

Proof.
Place a coordinate system on $\triangle$ABC as shown above. By the midpoint formula, points M and N have coordinates

$$\left(\frac{b}{2}, \frac{c}{2}\right) \text{ and } \left(\frac{a+b}{2}, \frac{c}{2}\right).$$

From the definition of slope, it follows that the slope of $\overline{MN}$ is

$$\frac{\frac{c}{2} - \frac{c}{2}}{\frac{a+b}{2} - \frac{b}{2}} = \frac{0}{\frac{a+b}{2} - \frac{b}{2}} = 0.$$

The slope of $\overline{AB}$ is

$$\frac{0-0}{a-0} = \frac{0}{a} = 0^*,$$

and so $\overline{MN} \parallel \overline{AB}$ (two nonvertical lines are parallel iff they have the same slope).

*This is to be expected because we placed the x-axis on $\overline{AB}$.

Exercises

Set I

In the figure below, quadrilateral ABCD is a trapezoid with bases $\overline{DC}$ and $\overline{AB}$ and diagonals $\overline{AC}$ and $\overline{DB}$. Write an expression in

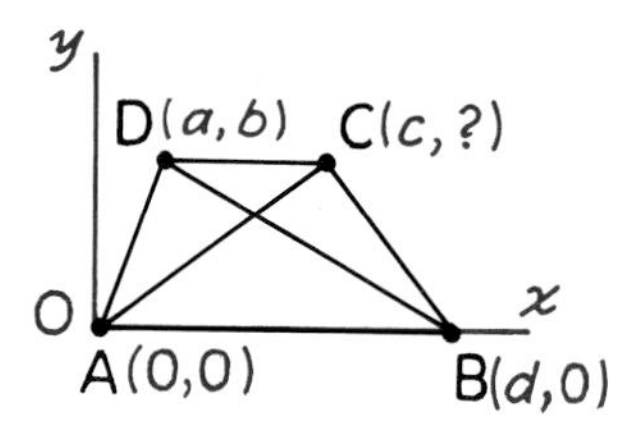

terms of the given coordinates for each of the following quantities.

1. The y-coordinate of point C.
2. The slope of $\overline{AC}$.
3. The coordinates of the midpoint of $\overline{DB}$.
4. The length of $\overline{DC}$.
5. The length of $\overline{AD}$.

In the figure below, $\triangle$ABC is an isosceles right triangle with hypotenuse $\overline{BC}$. Write an

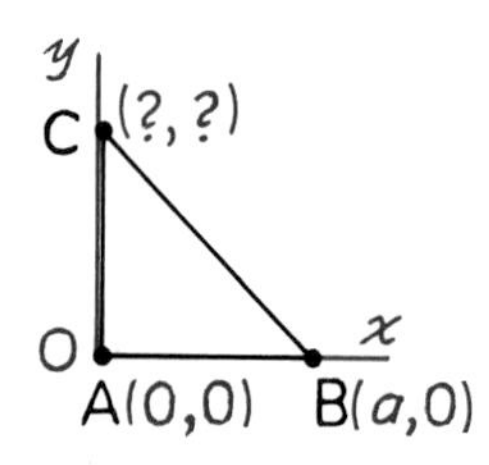

expression in terms of the given coordinates for each of the following quantities.

6. The coordinates of point C.
7. The length of $\overline{CB}$.
8. The coordinates of the midpoint of $\overline{CB}$.

In the figure below, ABCD is a quadrilateral with vertices as shown.

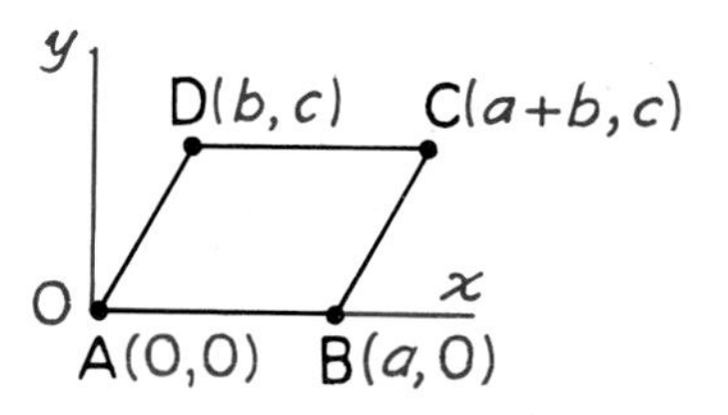

9. Find the slope of $\overline{AB}$ and the slope of $\overline{DC}$.
10. Why is $\overline{AB} \parallel \overline{DC}$?
11. Find the length of $\overline{AB}$ and the length of $\overline{DC}$.
12. What kind of quadrilateral is ABCD?
13. How do you know?

Exercises 9 through 13 show that a quadrilateral whose vertices can be labeled as shown in the figure above is a parallelogram.

14. Copy the figure and draw diagonals $\overline{AC}$ and $\overline{DB}$.
15. Write expressions for the coordinates of the midpoints of $\overline{AC}$ and $\overline{DB}$.
16. What theorem about the diagonals of a parallelogram does the fact that these expressions are equal prove?

Set II

The figure below can be used to prove that the median to the hypotenuse of a right triangle is half as long as the hypotenuse.

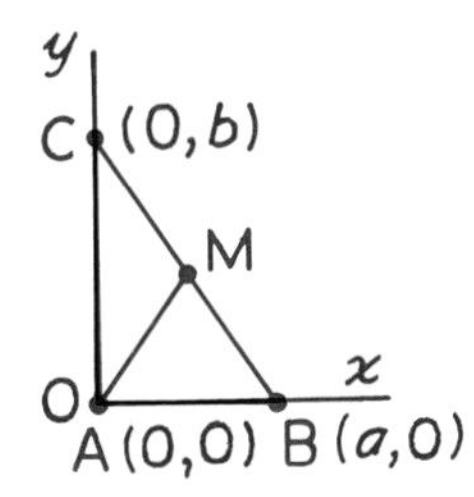

17. What are the coordinates of point M?
18. Write an expression for AM and simplify it.
19. Write an expression for BC.

The figure below can be used to prove that, if the diagonals of a parallelogram are equal, then it is a rectangle.

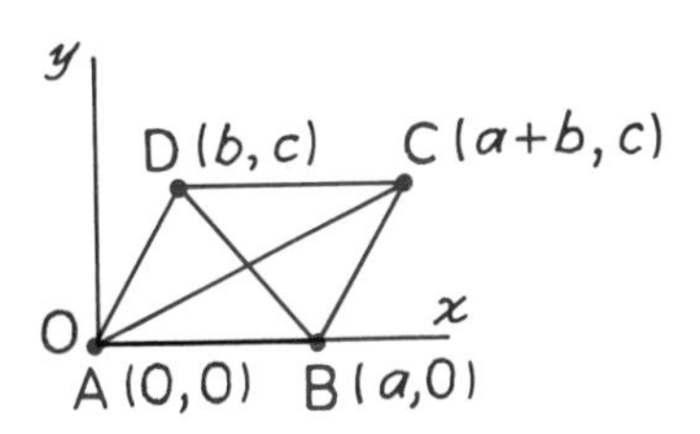

20. Write an expression for AC.
21. Write an expression for DB.
22. Why is

$$\sqrt{(a + b)^2 + c^2} = \sqrt{(a - b)^2 + c^2}?$$

Squaring both sides of this equation, we get

$$(a + b)^2 + c^2 = (a - b)^2 + c^2,$$

and so $a^2 + 2ab + b^2 + c^2 = a^2 - 2ab + b^2 + c^2$.

23. Why does it follow that $2ab = -2ab$?
24. Why does it follow that $4ab = 0$?
25. Why does it follow that $b = 0$?
26. If $b = 0$, what can you conclude about the position of point D?
27. What can then be concluded about $\measuredangle$DAB?

Because the consecutive angles of a parallelogram are supplementary and the opposite angles are equal, we can conclude that $\measuredangle$ADC, $\measuredangle$DCB, and $\measuredangle$ABC are right angles.

28. Why is $\angle DAB = \angle ADC = \angle DCB = \angle ABC$?
29. Why does it follow that ABCD is a rectangle?

The figure below can be used to prove that the line segments joining the midpoints of the consecutive sides of a quadrilateral form a parallelogram.

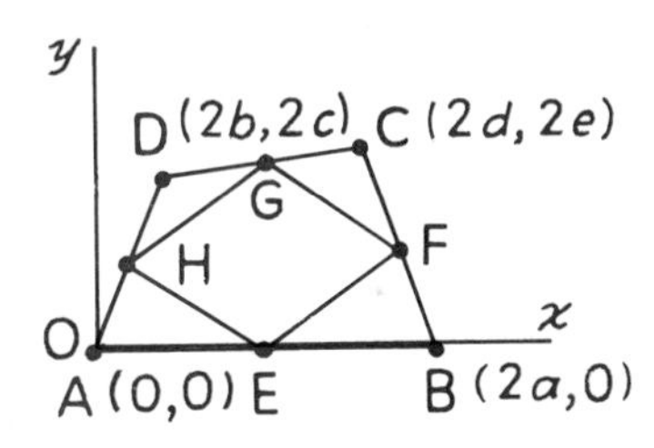

30. Copy the figure and write in the coordinates of the midpoints of the sides in terms of the coordinates given. (The coordinates of the vertices of ABCD have been written as multiples of 2 to make the work simpler.)
31. Show that HG = EF.
32. Show that $\overline{HG} \parallel \overline{EF}$.
33. How can you conclude from Exercises 30 through 32 that EFGH is a parallelogram?

The figure below can be used to prove that the diagonals of a rhombus are perpendicular.

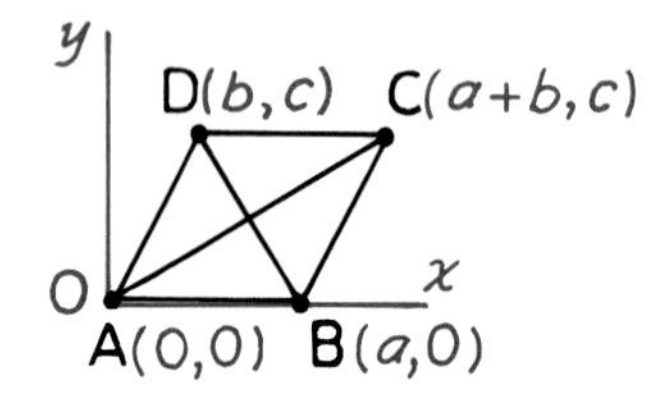

34. Write a proof. (Hint: Use the fact that AD = AB to write an equation relating a^2, b^2, and c^2.)

Set III

All of the exercises in this lesson have been about polygons. Here is one about a circle.

In this figure, a coordinate system has been placed on circle O so that the origin of the system is at the center of the circle.

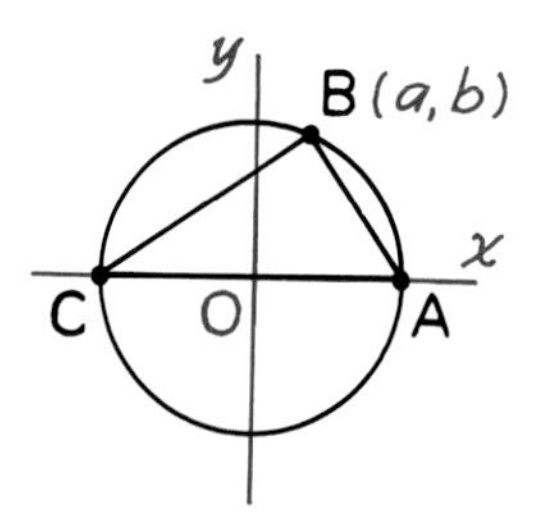

1. Find the coordinates of points A and C in terms of the coordinates of point B.
2. Use this information to prove the theorem that an angle inscribed in a semicircle is a right angle.

Chapter 17 / Summary and Review

CALCOMP, California Computer Products, Inc.

A drawing produced by a computer plotter using coordinate geometry.

Basic Ideas

Theorems

103. *The Distance Formula.* The distance between the points $P_1(x_1, y_1)$ and $P_2(x_2, y_2)$ is $\sqrt{(x_2 - x_1)^2 + (y_2 - y_1)^2}$. 612

104. Two nonvertical lines are parallel iff their slopes are equal. 620

105. Two nonvertical lines are perpendicular iff the product of their slopes is -1. 620

106. *The Midpoint Formula.* The midpoint of the line segment whose endpoints are $P_1(x_1, y_1)$ and $P_2(x_2, y_2)$ has the coordinates $\left(\frac{x_1 + x_2}{2}, \frac{y_1 + y_2}{2}\right)$. 625

Exercises

Set I

Name each of the following parts of the figure below.

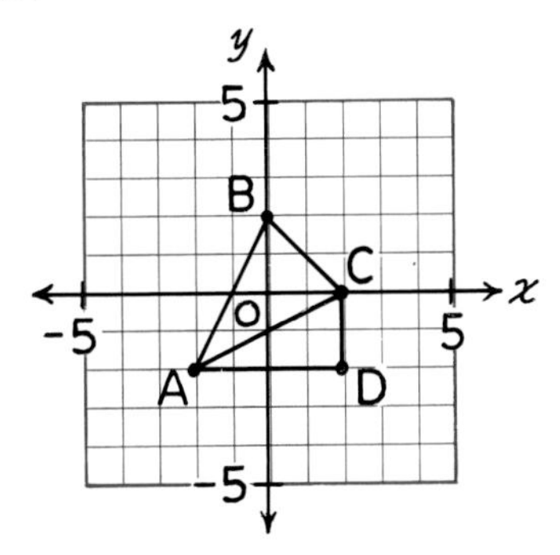

1. The point whose coordinates are (0, 2).
2. The line segment whose length is 2.
3. The line segment whose slope is 0.
4. The line segment whose slope is 2.
5. The line segment whose slope is undefined.
6. The line segment whose slope is negative.

Points A and B have coordinates (0, -7) and (-12, 9) respectively. Find each of the following quantities.

7. The distance between A and B.
8. The coordinates of the midpoint of $\overline{AB}$.
9. The slope of $\overline{AB}$.
10. The slope of the perpendicular bisector of $\overline{AB}$.

Quadrilateral ABCD has vertices A(-4, 3), B(2, 1), C(-1, -3), and D(-4, -2).

11. Find the slope of each side of the quadrilateral.
12. What relation does $\overline{AB}$ have to $\overline{CD}$?
13. Do $\overline{BC}$ and $\overline{DA}$ have the same relation?
14. Find BC and DA.
15. What kind of quadrilateral is ABCD?
16. Plot the points and draw quadrilateral ABCD.

Set II

Draw a pair of axes extending 8 units in each direction from the origin.

17. Use your compass to construct a circle with center at the origin and a radius of 5 units. Plot the points A(1, 7) and B(4, 3).
18. Draw $\overline{OB}$ and find its slope.
19. Draw $\overline{AB}$ and find its slope.
20. What relation do the slopes of $\overline{OB}$ and $\overline{AB}$ have?
21. What relation do $\overline{OB}$ and $\overline{AB}$ have?
22. What does this prove about $\overline{AB}$ with respect to circle O?

On graph paper, draw a pair of axes extending 5 units in each direction from the origin.

23. Plot the following points and connect them to form quadrilateral ABCD: A(3, 2), B(1, -2), C(-3, 0), and D(-1, 4).
24. What kind of quadrilateral is ABCD?
25. Find its perimeter.
26. Find its area.

In the figure below, $\overline{MN}$ is a midsegment of $\triangle$ABC. Write expressions in terms of the given coordinates for each of the following quantities.

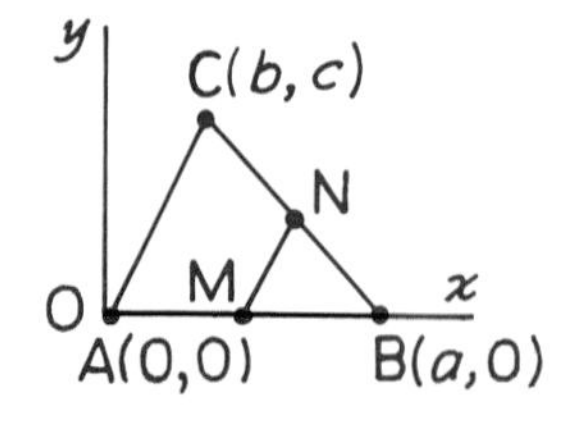

27. The coordinates of points M and N.
28. The length of $\overline{AC}$.
29. The length of $\overline{MN}$.
30. What theorem do the results of Exercises 28 and 29 prove?

The figure below can be used to prove that, if the diagonals of a quadrilateral bisect each other, then it is a parallelogram.

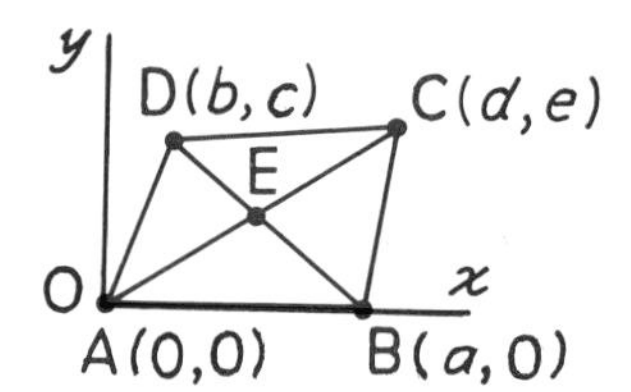

If $\overline{AC}$ and $\overline{DB}$ bisect each other, then E is the midpoint of both segments.

31. Write an expression for the coordinates of E as the midpoint of $\overline{AC}$.
32. Write an expression for the coordinates of E as the midpoint of $\overline{DB}$.

It follows from Exercises 31 and 32 that

$$\frac{d}{2} = \frac{a+b}{2} \quad \text{and} \quad \frac{e}{2} = \frac{c}{2}.$$

33. Why does it follow from these equations that

$$d = a + b \quad \text{and} \quad e = c?$$

34. Use this information to show that $\overline{DC} \parallel \overline{AB}$ and DC = AB.
35. Why is ABCD a parallelogram?

FINAL REVIEW

Set I

Write the letter of the correct answer.

1. The point that is equidistant from the vertices of a triangle is its
   a) circumcenter.
   b) orthocenter.
   c) incenter.
   d) centroid.
2. Which of the following sets of numbers can be the lengths of the sides of a right triangle?
   a) 4, 8, 9.
   b) 9, 16, 25.
   c) 20, 21, 29.
   d) 12, 12, 17.
3. Which one of the following statements about an apothem of a regular polygon is *not necessarily true?*
   a) It is shorter than the radius of the polygon.
   b) It is equal in length to one of the sides.
   c) It bisects one of the sides.
   d) It is perpendicular to one of the sides.
4. Which of the following polygons are similar?
   a) All isosceles triangles.
   b) All squares.
   c) All rectangles.
   d) All pentagons.
5. Heron's Theorem is used to find the
   a) semiperimeter of a triangle.
   b) perimeter of a cyclic quadrilateral.
   c) area of a right triangle given the lengths of its legs.
   d) area of a triangle given the lengths of its sides.
6. A central angle of a regular decagon has a measure of
   a) 18°.
   b) 30°.
   c) 36°.
   d) 72°.
7. Which one of the following statements about a regular pyramid is *not necessarily true?*
   a) Its base is a regular polygon.
   b) Its lateral edges are equal.
   c) Its lateral faces are equilateral triangles.
   d) Its altitude joins its vertex to the center of its base.
8. The limit of the sequence whose nth term is $\frac{n+2}{n}$ is
   a) 0.
   b) 1.
   c) 2.
   d) ∞.
9. Which one of the following polygons is not always cyclic?
   a) A scalene triangle.
   b) A rectangle.
   c) A regular hexagon.
   d) A parallelogram.
10. Which one of the following statements is *not necessarily true?*
    a) All angles that intercept the same arc or equal arcs of a circle are equal.
    b) The tangent segments to a circle from an external point are equal.
    c) If a diameter of a circle is perpendicular to a chord, it also bisects it.
    d) If a line is tangent to a circle, it is perpendicular to the diameter drawn to the point of contact.

Read the following statements carefully. If a statement is always *true,* write true. If it is not, *do not write false.* Instead, write a word or words that could replace the underlined word to make the statement true. Some of the questions in this section may have more than one correct answer; however, do not make a change in any statement that is already true.

11. A <u>median</u> of a triangle is any line segment that joins a vertex to a point on the opposite side.
12. If b is the geometric mean between a and c, then $b = \underline{\sqrt{a^2 + c^2}}$.
13. An angle inscribed in a major arc is <u>acute.</u>
14. The area of a <u>parallelogram</u> is the product of the lengths of a base and corresponding altitude.
15. The ratio of the circumference to the radius of a circle is $\underline{\pi}$.
16. If e is the length of one of the edges of a cube, the <u>volume</u> of the cube is $6e^2$.
17. A <u>tangent</u> to a circle intersects the circle in two points.
18. Corresponding <u>angles</u> of similar triangles have the same ratio as the corresponding sides.
19. The perpendicular bisectors of the chords of a circle (in the plane of the circle) are <u>collinear.</u>
20. The sine of an acute angle of a right triangle is the ratio of the length of the opposite leg to the length of the <u>adjacent leg</u>.

In the figure below, ∡B and ∡C are right angles.

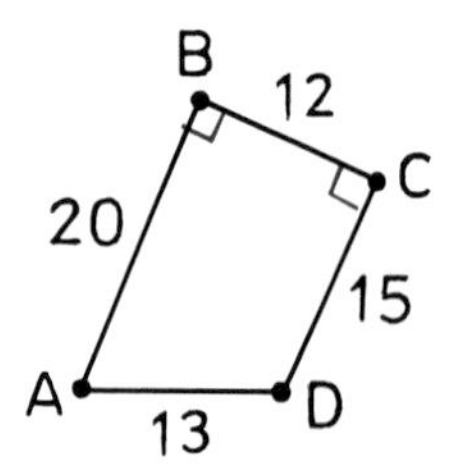

21. What kind of quadrilateral is ABCD?
22. Find its area.

In circle O, $m\widehat{AB} = 140°$ and $m\widehat{BC} = 120°$.

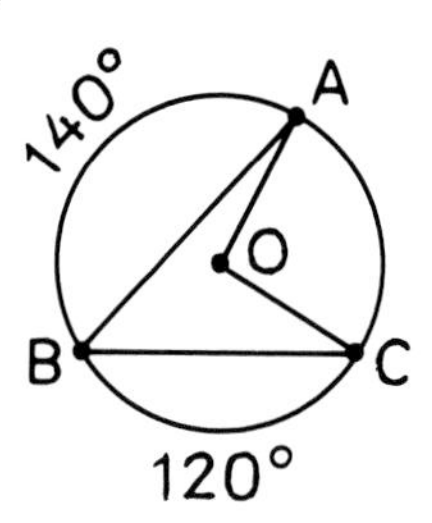

23. Find ∠AOC.
24. Find ∠B.

In the figure below, $\overline{AB} \perp \overline{BC}$, ACDE is a square, AB = 15, and BC = 8.

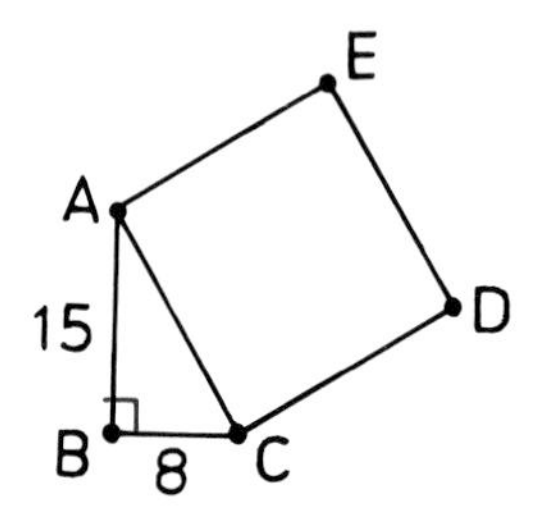

25. Find αACDE.
26. Find $\alpha\triangle$ABC.

In △ACE, $\overline{BD} \parallel \overline{AE}$ and $\overline{FD} \parallel \overline{AC}$. Use this information to tell whether each of the following statements *must be true, may be true,* or *appears to be false.*

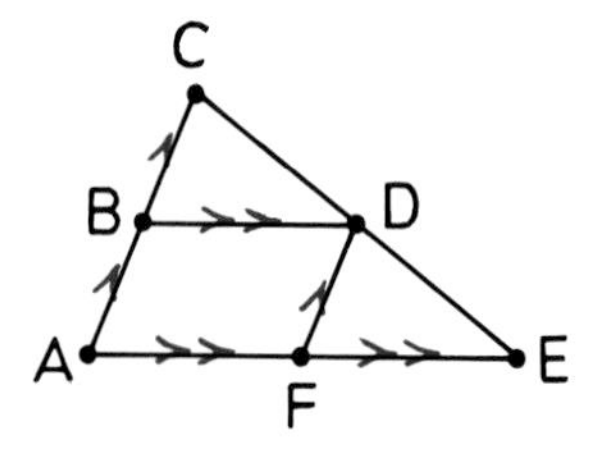

27. △CBD ~ △DFE.
28. ∠A = ∠BDF.
29. $\overline{BD}$ is a midsegment of △ACE.
30. $\dfrac{AF}{AE} = \dfrac{CD}{CE}$.
31. $AC^2 + CE^2 = AE^2$.

Isosceles trapezoid ABCD is inscribed in circle O; $m\widehat{DC} = 80°$.

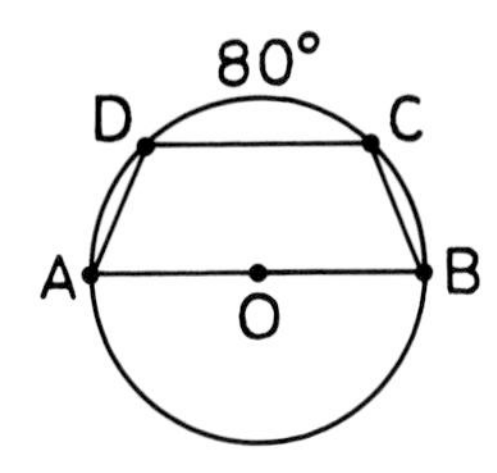

32. Find $m\widehat{CB}$.

33. Find ∠D.

In right △ABC, $\overline{BD} \perp \overline{AC}$, AD = 9, and DC = 16.

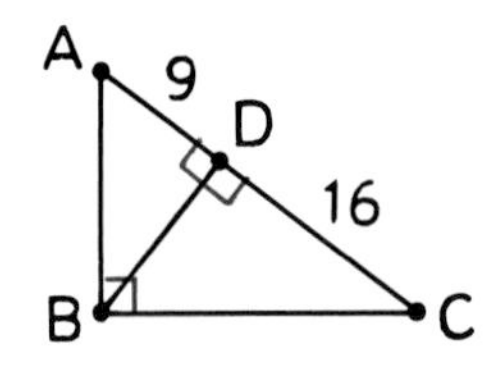

34. Find BD.

35. Find α△ABC.

The figure below represents a cube; AB = 4.

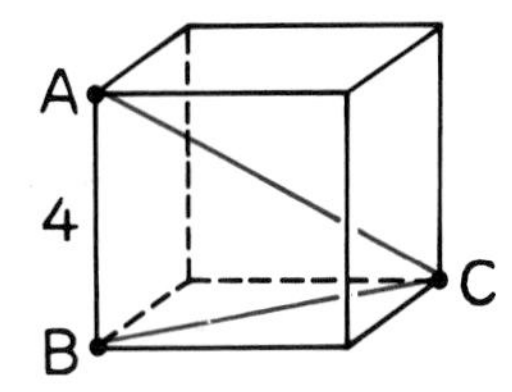

36. Find AC.

37. Find BC.

△ABC is a 30°-60° right triangle in which AC = 10.

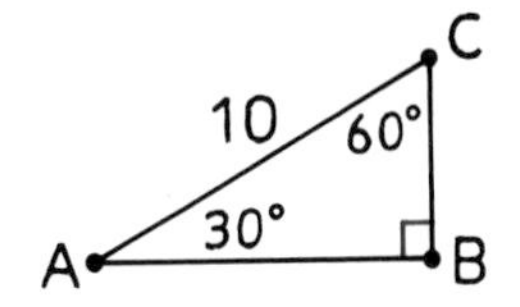

38. Find BC and AB.

39. Use the figure to find the value of sin 30°.

40. Use the figure to find the exact value of tan 60°.

Set II

In the figure below, $\overline{AC}$ and $\overline{AD}$ are secant segments to the circle and $\overline{BD}$ and $\overline{EC}$ are chords; $m\widehat{BE} = 50°$ and $m\widehat{CD} = 110°$.

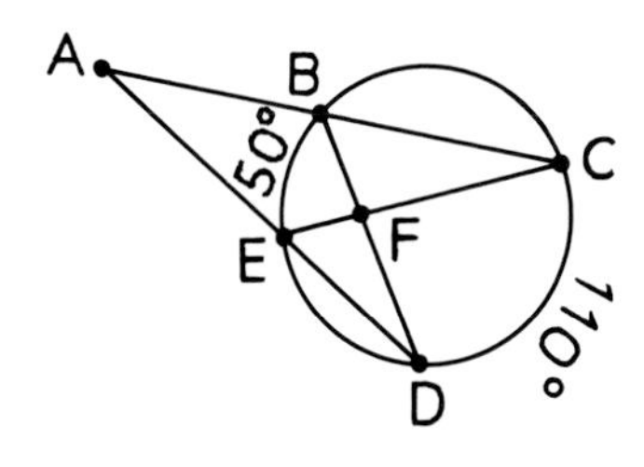

41. Find ∠A.

42. Find ∠CED.

43. Find ∠CFD.

In the figure below, ABCDE ~ AFGHI; AB = 6 and BF = 4. Find each of the following ratios.

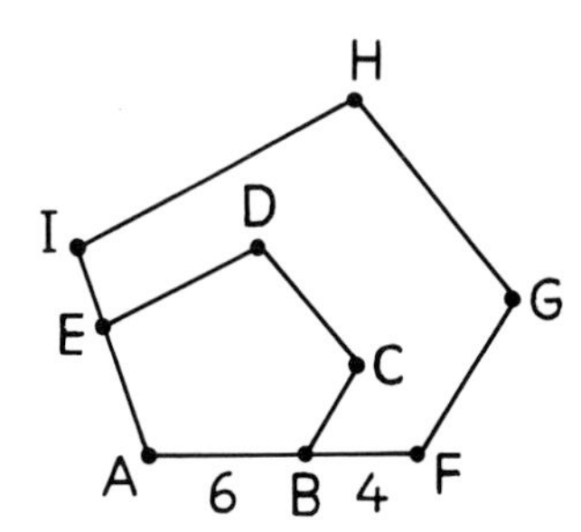

44. $\dfrac{\rho ABCDE}{\rho AFGHI}$.

45. $\dfrac{\alpha ABCDE}{\alpha AFGHI}$.

In the figure below, $\overparen{ADC}$ is a semicircle and $\overline{DB} \perp \overline{AC}$.

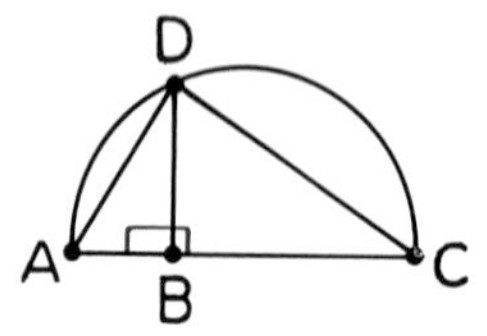

46. Why is $\measuredangle ADC$ a right angle?

47. Why does it follow that $\triangle ABD \sim \triangle DBC \sim \triangle ADC$?

The dimensions of a rectangular solid are 2, 5, and 14.

48. Find its total area.

49. Find its volume.

50. Find the length of one of its diagonals.

In ▱ABCD, $\overline{BE} \perp \overline{AD}$, $\angle A = 60°$, AB = 8, and BC = 10.

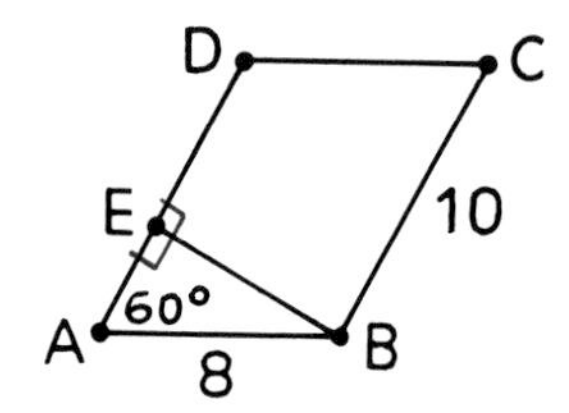

51. Find DE.

52. Find BE.

53. Find α▱ABCD.

Chords $\overline{AC}$ and $\overline{BD}$ intersect at point E in circle O; AE = 30, BE = 9, and CE = 6.

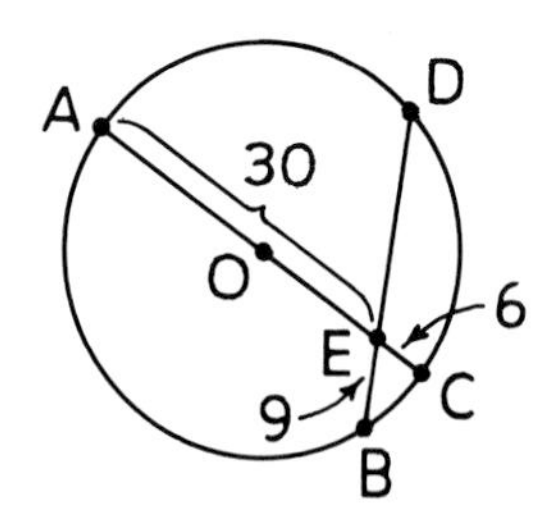

54. Find DE.

55. Find OC.

56. Find the circumference of circle O.

57. Find the area of circle O.

In $\triangle ABC$, $\overrightarrow{AD}$ bisects $\measuredangle CAB$ and $\overline{DE} \parallel \overline{CA}$; AC = 8, CD = 6, and AB = 12.

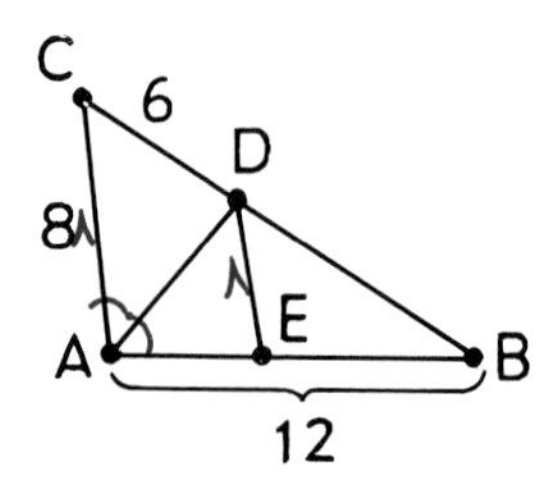

58. Find DB.

59. Find AE.

60. Find DE.

In the figure below, regular hexagon ABCDEF is inscribed in circle O; OA = 12. Find each of the following quantities.

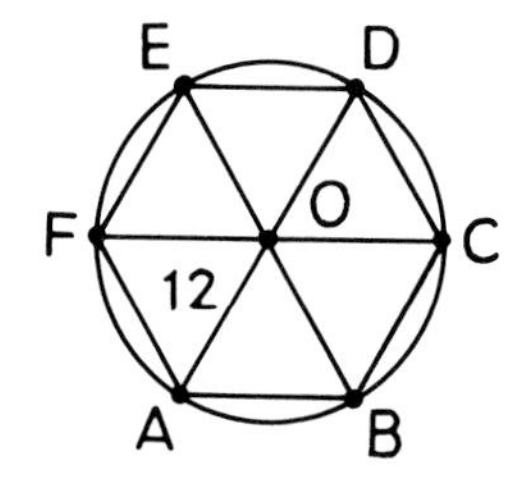

61. The measure of $\overparen{AB}$ in degrees.

62. The exact length of $\overparen{AB}$.

63. The perimeter of ABCDEF.

64. The area of ABCDEF.

Two similar cones have altitudes of 5 and 10 units respectively.

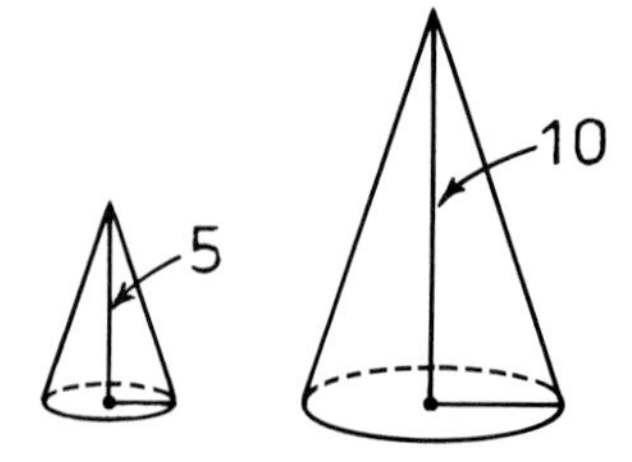

65. Find the surface area of the larger cone given that the surface area of the smaller cone is 36 square units.

66. Find the volume of the larger cone given that the volume of the smaller cone is 15 cubic units.

As the number of sides of a regular polygon inscribed in a circle increases, what does each of the following quantities get closer and closer to?

67. The area of the polygon.

68. The perimeter of the polygon.

69. The apothem of the polygon.

In the figure below, $\overline{CA}$ and $\overline{CE}$ are secant segments to circle O; CB = 6, CD = 7, and DE = 5.

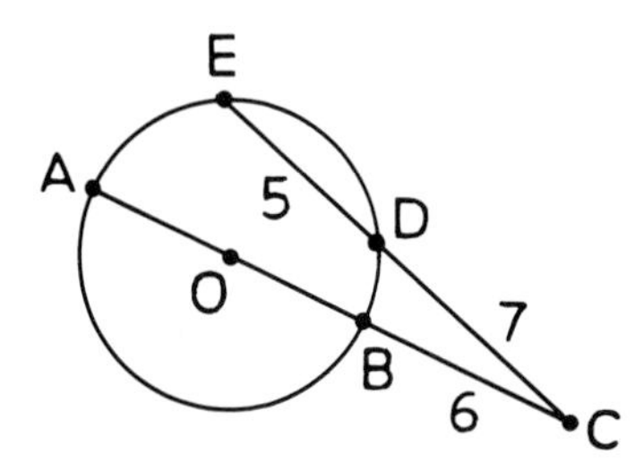

70. Find CA.

71. Find OB.

Find the exact area of the shaded region in each of these figures.

72.

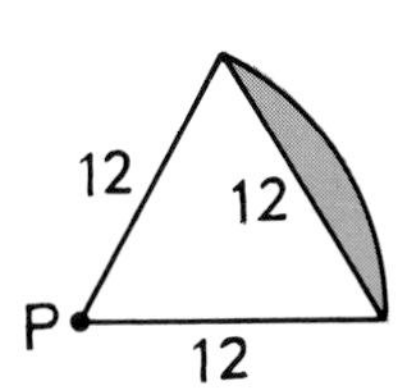

The arc has its center at point P.

73.

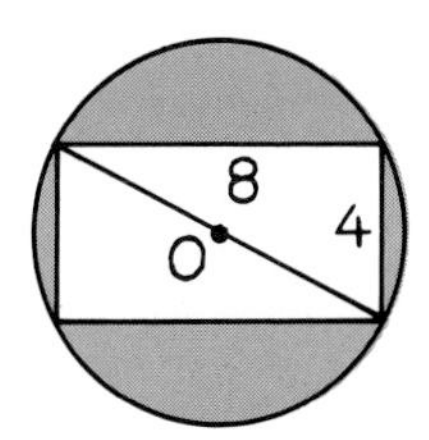

A rectangle is inscribed in circle O.

The figures below represent a right cylinder and a hemisphere.

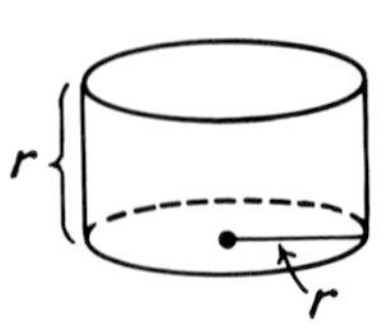

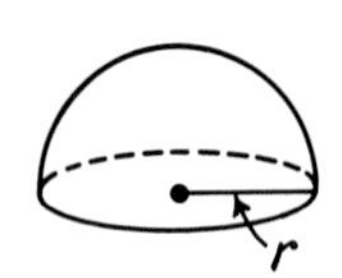

Comparing the second solid with the first,

74. find the ratio of the areas of their curved surfaces.

75. find the ratio of their volumes.

Make the following constructions as accurately as you can.

76. Draw a circle with a radius 2 centimeters long and mark a point P 5 centimeters from its center. Construct the tangents from P to the circle.

77. Construct a triangle having sides 5, 6, and 8 centimeters long. Then inscribe a circle in the triangle.

78.

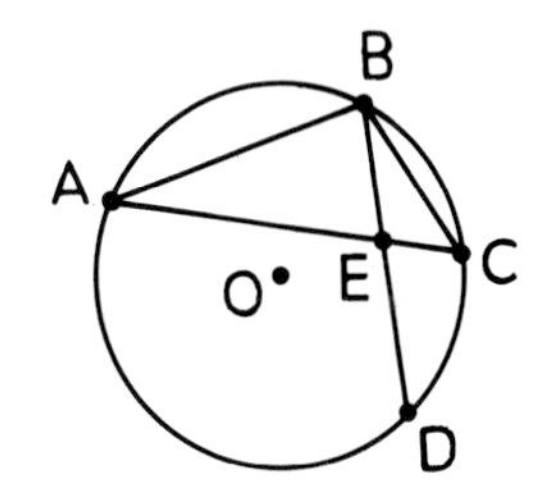

Given: Circle O with chords $\overline{AC}$ and $\overline{BD}$ and $m\overset{\frown}{BC} = m\overset{\frown}{CD}$.

Prove: $\triangle ABC \sim \triangle BEC$.

79.

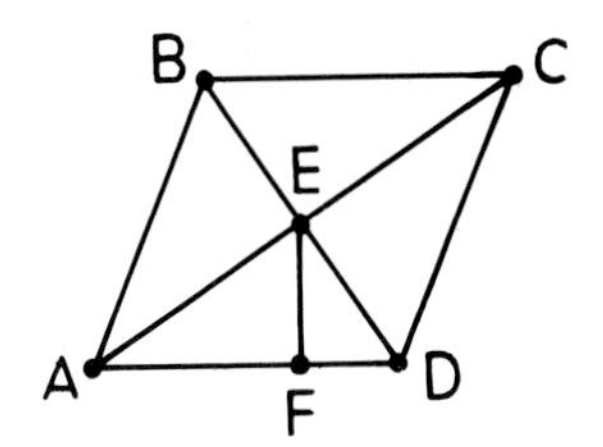

Given: Quadrilateral ABCD is a rhombus with diagonals $\overline{AC}$ and $\overline{BD}$; $\overline{EF} \perp \overline{AD}$.

Prove: $\frac{AB}{AE} = \frac{AE}{AF}$.

80.

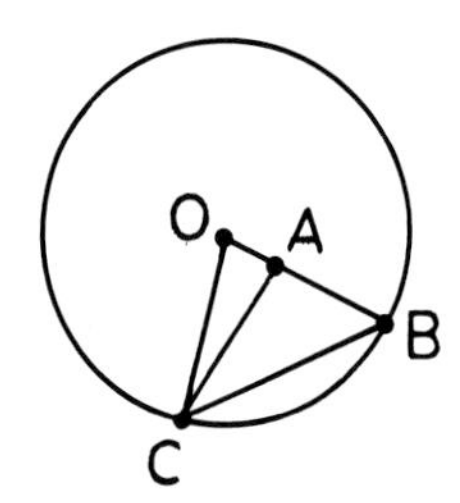

Given: $\overline{OB}$ and $\overline{OC}$ are radii of circle O.

Prove: AC > AB.

Glossary

Acute angle: An angle whose measure is less than 90°. 91

Acute triangle: A triangle all of whose angles are acute. 121

Altitude of a prism: A line segment that connects the planes of the bases of a prism and that is perpendicular to both of them. 551

Altitude of a pyramid: The perpendicular line segment joining the vertex of the pyramid to the plane of its base. 556

Altitude of a quadrilateral that has parallel sides: A perpendicular line segment that joins a point on one of the parallel sides to the line that contains the other side. 324

Altitude of a triangle: A perpendicular line segment from a vertex of the triangle to the line of the opposite side. 318

Angle: A pair of rays that have the same endpoint. 85

Apothem of a regular polygon: A perpendicular line segment from the center of the polygon to one of its sides. 500

Area of a circle: The limit of the areas of the inscribed regular polygons. 522

Betweenness of Points: Suppose that points A, B, and C are collinear with coordinates a, b, and c, respectively. Point B

A B C
a b c

is between points A and C (written A-B-C) iff either $a < b < c$ or $a > b > c$. 65

Betweenness of Rays: Suppose that $\overrightarrow{OA}$, $\overrightarrow{OB}$, and $\overrightarrow{OC}$ are in a half-rotation with coordinates a, b, and c, respectively.

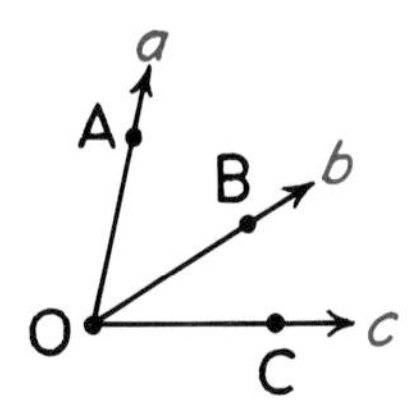

$\overrightarrow{OB}$ is between $\overrightarrow{OA}$ and $\overrightarrow{OC}$ (written $\overrightarrow{OA}$-$\overrightarrow{OB}$-$\overrightarrow{OC}$) iff either $a < b < c$ or $a > b > c$. 94

Biperpendicular quadrilateral: A quadrilateral that has a pair of sides both of which are perpendicular to a third side. 586

Bisector of an angle: A ray that is between the sides of the angle and that divides it into two equal angles. 95

Center of a regular polygon: The center of the circumscribed circle of the polygon. 500

Central angle of a circle: An angle whose vertex is the center of the circle. 500

Central angle of a regular polygon: An angle formed by radii drawn to two consecutive vertices of the polygon. 500

Centroid of a triangle: The point in which the medians of the triangle are concurrent. 484

Cevian of a triangle: A line segment that joins a vertex of the triangle to a point on the opposite side. 477

Chord of a circle: A line segment that joins two points of the circle. 421

Circle: The set of all points in a plane that are at a given distance from a given point in the plane. 420

Circumcenter of a polygon: The center of the circle circumscribed about the polygon. 463

Circumference of a circle: The limit of the perimeters of the inscribed regular polygons. 520

Circumscribed circle about a polygon: The circle that contains all of the vertices of the polygon. 463

Collinear points: Points that lie on the same line. 51

Complementary angles: Two angles the sum of whose measures is 90°. 100

Concave polygon: A polygon that is not convex. 75

Concentric circles: Circles that lie in the same plane and have the same center. 421

Concurrent lines: Lines that contain the same point. 463

Concyclic points: Points that lie on the same circle. 462

Cone: Suppose that A is a plane, R is a circular region in plane A, and P is a point

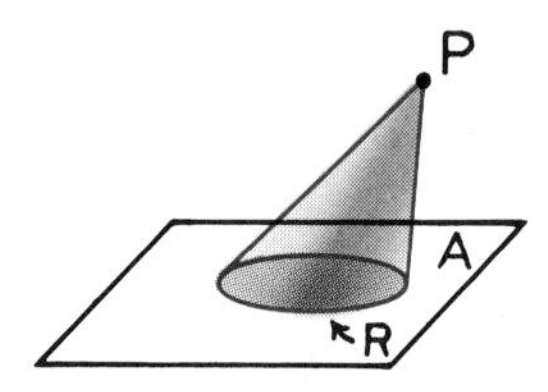

not in plane A. The set of all segments that join P to a point of region R form a cone. 560

Congruent figures: Two figures for which there is an isometry such that one figure is the image of the other. 291

Congruent polygons: Two polygons for which there is a correspondence between their vertices such that all of their corresponding sides and angles are equal. 126

Contrapositive of a conditional statement: The statement formed by interchanging the hypothesis and conclusion of the statement and denying both. 15

Converse of a conditional statement: The statement formed by interchanging the hypothesis and conclusion of the statement. 15

Convex polygon: A polygon such that, for each line that contains a side of the polygon, the rest of the polygon lies on one side of the line. 75

Coplanar points: Points that lie in the same plane. 51

Corollary: A theorem that can be easily proved as a consequence of another theorem. 141

Cosine of an acute angle of a right triangle: The ratio of the length of the adjacent leg to the length of the hypotenuse. 410

Cross section of a geometric solid: The intersection of a plane and the solid. 551

Cube: A rectangular solid all of whose dimensions are equal. 542

Cyclic polygon: A polygon for which there exists a circle that contains all of its vertices. 463

Cylinder: Suppose that A and B are two parallel planes, R is a circular region in one plane, and ℓ is a line that intersects

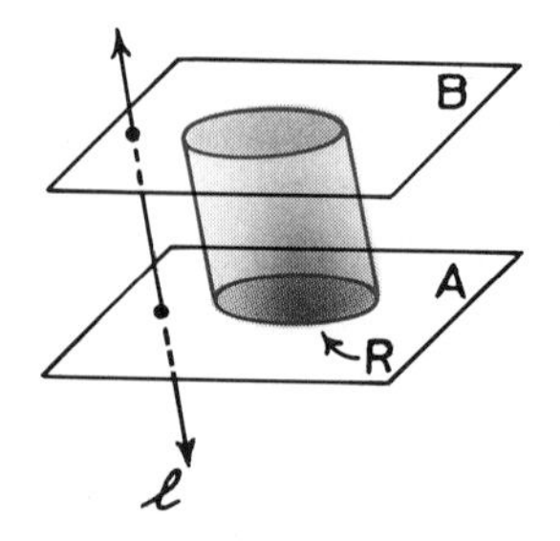

both planes but not R. The set of all segments parallel to line ℓ that join a point of region R to a point of the other plane form a cylinder. 561

Decagon: A polygon that has 10 sides. 75

Degree measure of a major arc: 360° minus the measure of the corresponding minor arc. 432

Degree measure of a minor arc: The measure of its central angle. 432

Degree measure of a semicircle: 180°. 432

Diagonal of a polygon: A line segment that joins any two nonconsecutive vertices of the polygon. 75

Diameter of a circle: A chord that contains the center of the circle. 421

Distance between two parallel lines: The length of any perpendicular segment joining one line to the other. 215

Distance from a point to a line: The length of the perpendicular segment from the point to the line. 215

Dodecagon: A polygon that has 12 sides. 75

Equiangular polygon: A polygon all of whose angles are equal. 121

Equilateral polygon: A polygon all of whose sides are equal. 121

Exterior angle of a polygon: An angle that forms a linear pair with one of the angles of the polygon. 173

Geometric mean: The number b is the geometric mean between the numbers a and c iff a, b, and c are positive and $\frac{a}{b} = \frac{b}{c}$. 349

Great circle of a sphere: A set of points that is the intersection of the sphere and a plane containing its center. 580

Heptagon: A polygon that has 7 sides. 75

Hexagon: A polygon that has 6 sides. 75

Hypotenuse of a right triangle: The side opposite the right angle of the triangle. 121

Incenter of a polygon: The center of the inscribed circle of the polygon. 472

Inscribed angle of a circle: An angle whose vertex is on the circle and each of whose sides intersects the circle in another point. 437

Inscribed circle in a polygon: A circle for which each side of the polygon is tangent to the circle. 472

Inverse of a conditional statement: The statement formed by denying both the hypothesis and conclusion of the statement. 16

Isometry: A transformation that preserves distance. 275

Isosceles trapezoid: A trapezoid whose legs are equal. 256

Isosceles triangle: A triangle that has at least two equal sides. 121

Lateral area of a prism: The sum of the areas of the lateral faces of the prism. 547

Length of a line segment: The distance between the endpoints of the line segment. 69

Line segment: The set of two points and all the points between them. 69

Linear pair: Two angles that have a common side and whose other sides are opposite rays. 86

Median of a triangle: A cevian that joins a vertex of the triangle to the midpoint of the opposite side. 483

Midpoint of a line segment: A point between the endpoints of the line segment that divides it into two equal segments. 69

Midsegment of a triangle: A line segment that joins the midpoints of two sides of the triangle. 260

Nonagon: A polygon that has 9 sides. 75

Noncollinear points: Points that do not lie on the same line. 51

Oblique line and plane (or two planes): A line and plane (or two planes) that are neither parallel nor perpendicular. 537

Obtuse angle: An angle whose measure is more than 90° but less than 180°. 91

Obtuse triangle: A triangle that has an obtuse angle. 121

Octagon: A polygon that has 8 sides. 75

Opposite rays: Rays $\overrightarrow{AB}$ and $\overrightarrow{AC}$ are opposite rays iff B-A-C. 85

Orthocenter of a triangle: The point in which the lines containing the altitudes of the triangles are concurrent. 484

Parallel line and plane: A line and plane that do not intersect. 536

Parallel lines: Lines that lie in the same plane and do not intersect. 111

Parallel planes: Planes that do not intersect. 537

Parallelogram: A quadrilateral in which both pairs of opposite sides are parallel. 237

Pentagon: A polygon that has 5 sides. 75

Perimeter of a polygon: The sum of the lengths of the sides of a polygon. 75

Perpendicular bisector of a line segment: The line that is perpendicular to the line segment and that divides it into two equal parts. 157

Perpendicular line and plane: A line and plane that intersect such that the line is perpendicular to every line in the plane that passes through the point of intersection. 537

Perpendicular lines: Lines that form a right angle. 111

Perpendicular planes: Planes such that one plane contains a line that is perpendicular to the other plane. 537

Polar points: The points of intersection of a line through the center of the sphere with the sphere. 581

Polygon: Let $P_1, P_2, \ldots, P_n$ be a set of at least three points in a plane such that

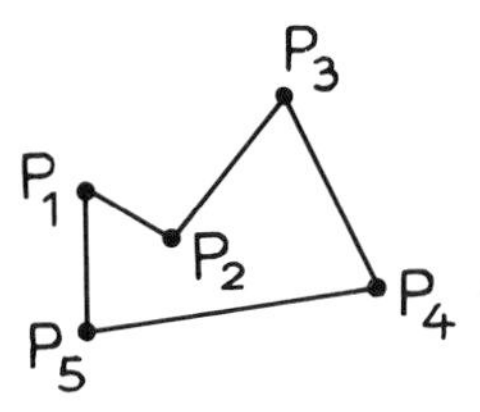

no three consecutive points are collinear. If the segments $\overline{P_1P_2}, \overline{P_2P_3}, \ldots, \overline{P_nP_1}$ intersect only at their endpoints, they form a polygon. 74

Polygonal region: The union of a finite number of nonoverlapping triangular regions in a plane. 309

Polyhedron: A solid bounded by parts of intersecting planes. 541

Postulate: A statement that is assumed to be true without proof. 40

Prism: Suppose that A and B are two parallel planes, R is a polygonal region in

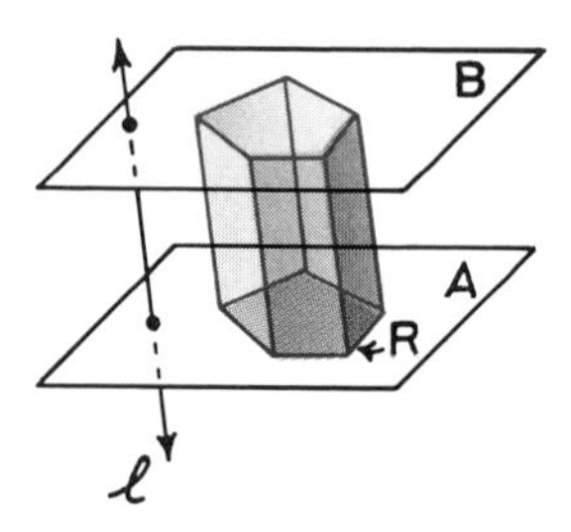

one plane, and ℓ is a line that intersects both planes but not R. The set of all segments parallel to line ℓ that join a point of region R to a point of the other plane form a prism. 546

Proportion: An equality between two ratios. 348

Pyramid: Suppose that A is a plane, R is a polygonal region in plane A, and P is a

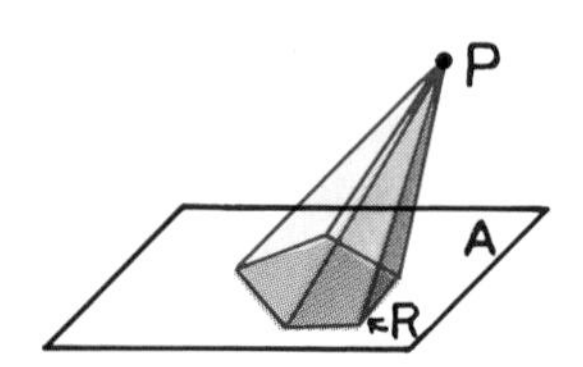

point not in plane A. The set of all segments that join P to a point of region R form a pyramid. 555

Pythagorean triple: A set of three integers that can be the lengths of the sides of a right triangle. 397

Quadrilateral: A polygon that has 4 sides. 75

Radius of a circle: A line segment that joins the center of the circle to a point on the circle. 421

Radius of a regular polygon: A line segment that joins the center of the polygon to a vertex of the polygon. 500

Ratio: The ratio of the numbers a to b is the number $\frac{a}{b}$. 348

Ray: A ray $\overrightarrow{AB}$ is the set of points A, B, and all points X such that either A-X-B or A-B-X. 85

Rectangle: A quadrilateral all of whose angles are equal. 237

Rectangular solid: A polyhedron that has six rectangular faces. 542

Reflection of a point through a line: The reflection of point P through line ℓ is

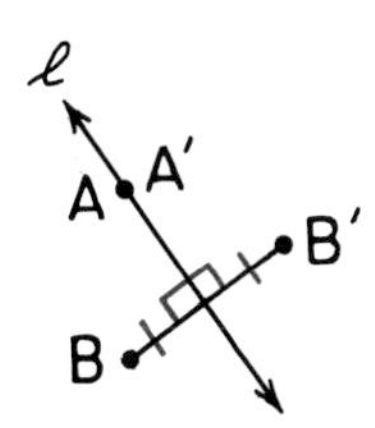

point P itself if P is on ℓ or the point P′ such that ℓ is the perpendicular bisector of $\overline{PP'}$ if P is not on ℓ. 271

Reflection symmetry: A figure has reflection symmetry with respect to a line iff it coincides with its reflection image through the line. 295

Regular polygon: A convex polygon that is both equilateral and equiangular. 498

Regular pyramid: A pyramid whose base is a regular polygon and whose lateral edges are equal. 556

Rhombus: A quadrilateral all of whose sides are equal. 237

Right angle: An angle whose measure is 90°. 91

Right triangle: A triangle that has a right angle. 121

Rotation: A transformation that is the composite of two successive reflections through intersecting lines. 287

Rotation symmetry: A figure has rotation symmetry with respect to a point iff it coincides with its rotation image about the point. 296

Saccheri quadrilateral: A biperpendicular quadrilateral whose legs are equal. 586

Scalene triangle: A triangle that has no equal sides. 121

Secant: A line that intersects a circle in two points. 442

Secant angle: An angle whose sides are contained in two secants of a circle so that each side intersects the circle in at least one point other than the angle's vertex. 442

Secant segment: A segment that intersects a circle in two points, exactly one of which is an endpoint of the segment. 453

Sector of a circle: A region bounded by an arc of the circle and the two radii to the endpoints of the arc. 526

Semiperimeter of a triangle: The number that is half the perimeter of the triangle. 335

Similar polygons: Two polygons for which there is a correspondence between their vertices such that the corresponding sides of the polygons are proportional and the corresponding angles are equal. 360

Sine of an acute angle of a right triangle: The ratio of the length of the opposite leg to the length of the hypotenuse. 409

Slope: The slope m of a nonvertical line that contains the points $P_1(x_1,y_1)$ and $P_2(x_2,y_2)$ is $m = \frac{y_2 - y_1}{x_2 - x_1}$. 616

Sphere: The set of all points in space that are at a given distance from a given point. 565

Square: A quadrilateral that is both equilateral and equiangular. 237

Straight angle: An angle whose measure is 180°. 91

Supplementary angles: Two angles the sum of whose measures is 180°. 100

Tangent: A line in the plane of a circle that intersects the circle in exactly one point. 426

Tangent of an acute angle of a right triangle: The ratio of the length of the opposite leg to the length of the adjacent leg. 405

Tangent segment: Any segment of a tangent line to a circle that has the point of tangency as one of its endpoints. 447

Theorem: A statement that is proved by reasoning deductively from already accepted statements. 33

Transformation: A one-to-one correspondence between two sets of points. 275

Translation: A transformation that is the composite of two successive reflections through parallel lines. 282

Transversal: A line that intersects two or more lines that lie in the same plane in different points. 192

Trapezoid: A quadrilateral that has exactly one pair of parallel sides. 237

Triangle: A polygon that has three sides. 75

Triangular region: The union of a triangle and its interior. 308

Vertical angles: Two angles such that the sides of one angle are opposite rays to the sides of the other angle. 86

Postulates and Theorems

Chapter 2

Postulate 1 If there are two points, then there is exactly one line that contains them. 51

Postulate 2 If there is a line, then there are at least two points on the line. 51

Postulate 3 If there are three noncollinear points, then there is exactly one plane that contains them. 51

Postulate 4 If two points lie in a plane, then the line that contains them lies in the plane. 52

Postulate 5 *The Ruler Postulate* The points on a line can be numbered so that
a) to every point there corresponds exactly one real number called its coordinate,
b) to every real number there corresponds exactly one point,
c) to every pair of points there corresponds exactly one real number called the distance between the points,
d) and the distance between two points is the absolute value of the difference between their coordinates. 57

Theorem 1 *The Betweenness of Points Theorem* If A-B-C, then AB + BC = AC. 65

Postulate 6 *The Midpoint Postulate* A line segment has exactly one midpoint. 70

Theorem 2 The midpoint of a line segment divides it into segments half as long as the line segment. 70

Chapter 3

Postulate 7 *The Protractor Postulate* The rays in a half-rotation can be numbered so that
a) to every ray there corresponds exactly one real number called its coordinate,
b) to every real number from 0 to 180 inclusive there corresponds exactly one ray,
c) to every pair of rays there corresponds exactly one real number called the measure of the angle that they determine,
d) and the measure of an angle is the absolute value of the difference between the coordinates of its rays. 90

Theorem 3 *The Betweenness of Rays Theorem* If $\overrightarrow{OA}$-$\overrightarrow{OB}$-$\overrightarrow{OC}$, then $\angle AOB + \angle BOC = \angle AOC$. 94

Theorem 4 *The Angle Bisector Theorem* A ray that bisects an angle divides it into angles half as large as the angle. 95

Theorem 5 Complements of the same angle (or equal angles) are equal. 100

Theorem 6 Supplements of the same angle (or equal angles) are equal. 101

Theorem 7 If two angles are a linear pair, then they are supplementary. 106

Theorem 8 If the two angles in a linear pair are equal, then each is a right angle. 106

Theorem 9 If two angles are vertical angles, then they are equal. 106

Theorem 10 If two lines are perpendicular, they form four right angles. 111

Theorem 11 Any two right angles are equal. 111

Chapter 4

Postulate 8 *The S.A.S. Congruence Postulate* If two sides and the included angle of one triangle are equal to two sides and the included angle of another triangle, the triangles are congruent. 126

Postulate 9 *The A.S.A. Congruence Postulate* If two angles and the included side of one triangle are equal to two angles and the included side of another triangle, the triangles are congruent. 127

Theorem 12 Two triangles congruent to a third triangle are congruent to each other. 137

Theorem 13 If two sides of a triangle are equal, the angles opposite them are equal. 140

Corollary If a triangle is equilateral, it is also equiangular. 141

Theorem 14 If two angles of a triangle are equal, the sides opposite them are equal. 141

Corollary If a triangle is equiangular, it is also equilateral. 141

Theorem 15 *The S.S.S. Congruence Theorem* If the three sides of one triangle are equal to the three sides of another triangle, the triangles are congruent. 146

Theorem 16 In a plane, two points each equidistant from the endpoints of a line segment determine the perpendicular bisector of the line segment. 157

Chapter 5

Theorem 17 *The Exterior Angle Theorem* An exterior angle of a triangle is greater than either remote interior angle. 174

Theorem 18 If two sides of a triangle are unequal, the angles opposite them are unequal and the larger angle is opposite the longer side. 179

Theorem 19 If two angles of a triangle are unequal, the sides opposite them are unequal and the longer side is opposite the larger angle. 179

Theorem 20 *The Triangle Inequality Theorem* The sum of the lengths of any two sides of a triangle is greater than the length of the third side. 183

Chapter 6

Theorem 21 If two lines form equal corresponding angles with a transversal, then the lines are parallel. 193

Corollary 1 If two lines form equal alternate interior angles with a transversal, then the lines are parallel. 193

Corollary 2 If two lines form supplementary interior angles on the same side of a transversal, then the lines are parallel. 193

Corollary 3 In a plane, two lines perpendicular to a third line are parallel to each other. 193

Theorem 22 In a plane, through a point on a line there is exactly one line perpendicular to the line. 197

Theorem 23 Through a point not on a line, there is exactly one line perpendicular to the line. 198

Postulate 10 *The Parallel Postulate* Through a point not on a line, there is exactly one line parallel to the line. 204

Theorem 24 In a plane, two lines parallel to a third line are parallel to each other. 204

Theorem 25 If two parallel lines are cut by a transversal, then the corresponding angles are equal. 209

Corollary 1 If two parallel lines are cut by a transversal, then the alternate interior angles are equal. 209

Corollary 2 If two parallel lines are cut by a transversal, then the interior angles on the same side of the transversal are supplementary. 210

Corollary 3 In a plane, if a line is perpendicular to one of two parallel lines, it is also perpendicular to the other. 210

Theorem 26 The perpendicular segment from a point to a line is the shortest segment joining them. 215

Theorem 27 If two lines are parallel, every perpendicular segment joining one line to the other has the same length. 215

Theorem 28 The sum of the measures of the angles of a triangle is 180°. 220

Corollary 1 If two angles of one triangle are equal to two angles of another triangle, then the third pair of angles are equal. 220

Corollary 2 The acute angles of a right triangle are complementary. 220

Corollary 3 Each angle of an equilateral triangle has a measure of 60°. 220

Theorem 29 An exterior angle of a triangle is equal in measure to the sum of the measures of the remote interior angles. 221

Theorem 30 *The A.A.S. Congruence Theorem* If two angles and the side opposite one of them in one triangle are equal to the corresponding parts of another triangle, then the triangles are congruent. 226

Theorem 31 *The H.L. Congruence Theorem* If the hypotenuse and a leg of one right triangle are equal to the corresponding parts of another right triangle, then the triangles are congruent. 226

Chapter 7

Theorem 32 The sum of the measures of the angles of a quadrilateral is 360°. 237

Corollary Each angle of a rectangle is a right angle. 238

Theorem 33 The opposite sides of a parallelogram are equal. 241

Theorem 34 The opposite angles of a parallelogram are equal and its consecutive angles are supplementary. 242

Theorem 35 The diagonals of a parallelogram bisect each other. 242

Theorem 36 A quadrilateral is a parallelogram if both pairs of opposite sides are equal. 247

Theorem 37 A quadrilateral is a parallelogram if two opposite sides are both equal and parallel. 247

Theorem 38 A quadrilateral is a parallelogram if both pairs of opposite angles are equal. 248

Theorem 39 A quadrilateral is a parallelogram if its diagonals bisect each other. 248

Theorem 40 All rectangles are parallelograms. 251

Theorem 41 The diagonals of a rectangle are equal. 252

Theorem 42 All rhombuses are parallelograms. 252

Theorem 43 The diagonals of a rhombus are perpendicular. 252

Theorem 44 The base angles of an isosceles trapezoid are equal. 257

Theorem 45 The diagonals of an isosceles trapezoid are equal. 257

Theorem 46 *The Midsegment Theorem* A midsegment of a triangle is parallel to the third side and half as long. 260

Chapter 8

Transformation Theorem 1 Reflection of a pair of points through a line preserves distance. 271

Transformation Theorem 2 An isometry preserves collinearity and betweenness. 276

Transformation Theorem 3 An isometry preserves angle measure. 277

Transformation Theorem 4 A triangle and its image under an isometry are congruent. 277

Chapter 9

Postulate 11 *The Area Postulate* To every polygonal region there corresponds a positive number called its area such that
a) triangular regions bounded by congruent triangles have equal areas,
b) and polygonal regions consisting of two or more nonoverlapping regions are equal in area to the sum of their areas. 309

Postulate 12 The area of a rectangle is the product of the lengths of its base and its altitude. 313

Corollary The area of a square is the square of the length of its side. 313

Theorem 47 The area of a right triangle is half the product of the lengths of its legs. 317

Theorem 48 The area of a triangle is half the product of the lengths of any base and corresponding altitude. 318

Corollary Triangles with equal bases and equal altitudes have equal areas. 319

Theorem 49 The area of a rhombus is half the product of the lengths of its diagonals. 319

Theorem 50 The area of a parallelogram is the product of the lengths of any base and corresponding altitude. 324

Theorem 51 The area of a trapezoid is half the product of the length of its altitude and the sum of the lengths of its bases. 324

Theorem 52 *The Pythagorean Theorem* In a right triangle, the square of the hypotenuse is equal to the sum of the squares of the legs. 329

Theorem 53 If the square of one side of a triangle is equal to the sum of the squares of the other two sides, then the triangle is a right triangle. 330

Theorem 54 *Heron's Theorem* The area of a triangle with sides of lengths a, b, and c and semiperimeter s is $\sqrt{s(s-a)(s-b)(s-c)}$. 336

Corollary The area of an equilateral triangle with sides of length a is $\frac{a^2}{4}\sqrt{3}$. 338

Chapter 10

Theorem 55 *The Side-Splitter Theorem* If a line parallel to one side of a triangle intersects the other two sides in different points, it divides the sides in the same ratio. 354

Corollary If a line parallel to one side of a triangle intersects the other two sides in different points, it cuts off segments proportional to the sides. 356

Theorem 56 *The A.A. Similarity Theorem* If two angles of one triangle are equal to two angles of another triangle, then the triangles are similar. 364

Corollary Two triangles similar to a third triangle are similar to each other. 365

Theorem 57 *The S.A.S. Similarity Theorem* If an angle of one triangle is equal to an angle of another triangle and the sides including these angles are proportional, then the triangles are similar. 369

Theorem 58 Corresponding altitudes of similar triangles have the same ratio as the corresponding sides. 370

Theorem 59 *The Angle Bisector Theorem* An angle bisector in a triangle divides the opposite side into segments that have the same ratio as the other two sides. 376

Theorem 60 The ratio of the perimeters of two similar polygons is equal to the ratio of the corresponding sides. 381

Theorem 61 The ratio of the areas of two similar polygons is equal to the square of the ratio of the corresponding sides. 381

Chapter 11

Theorem 62 The altitude to the hypotenuse of a right triangle forms two triangles that are similar to each other and to the original triangle. 391

Corollary 1 The altitude to the hypotenuse of a right triangle is the geometric mean between the segments into which it divides the hypotenuse. 391

Corollary 2 Each leg of a right triangle is the geometric mean between the hypotenuse and its projection on the hypotenuse. 391

The Pythagorean Theorem In a right triangle, the square of the hypotenuse is equal to the sum of the squares of the legs. 396

Theorem 63 *The Isosceles Right Triangle Theorem* In an isosceles right triangle, the hypotenuse is $\sqrt{2}$ times the length of one leg. 400

Corollary Each diagonal of a square is $\sqrt{2}$ times the length of one side. 400

Theorem 64 *The 30°-60° Right Triangle Theorem* In a 30°-60° right triangle, the hypotenuse is twice the length of the shorter leg and the longer leg is $\sqrt{3}$ times the length of the shorter leg. 400

Corollary An altitude of an equilateral triangle is $\frac{\sqrt{3}}{2}$ times the length of one side. 401

Chapter 12

Corollary to the definition of a circle All radii of a circle are equal. 421

Theorem 65 If a line through the center of a circle is perpendicular to a chord, it also bisects the chord. 422

Theorem 66 If a line through the center of a circle bisects a chord that is not a diameter, it is also perpendicular to the chord. 422

Theorem 67 The perpendicular bisector of a chord (in the plane of a circle) passes through the center of the circle. 422

Theorem 68 If a line is tangent to a circle, it is perpendicular to the radius drawn to the point of contact. 427

Theorem 69 If a line is perpendicular to a radius at its outer endpoint, then it is tangent to the circle. 427

Postulate 13 *The Arc Addition Postulate* If C is on $\overset{\frown}{AB}$, then $m\overset{\frown}{AC} + m\overset{\frown}{CB} = m\overset{\frown}{ACB}$. 432

Theorem 70 If two chords of a circle are equal, their minor arcs are equal. 433

Theorem 71 If two minor arcs of a circle are equal, their chords are equal. 433

Theorem 72 An inscribed angle is equal in measure to half its intercepted arc. 437

Corollary 1 Inscribed angles that intercept the same arc or equal arcs are equal. 438

Corollary 2 An angle inscribed in a semicircle is a right angle. 438

Theorem 73 A secant angle whose vertex is inside a circle is equal in measure to half the sum of the arcs intercepted by it and its vertical angle. 442

Theorem 74 A secant angle whose vertex is outside a circle is equal in measure to half the positive difference of its intercepted arcs. 443

Theorem 75 The tangent segments to a circle from an external point are equal. 447

Theorem 76 *The Intersecting Chords Theorem* If two chords intersect in a circle, the product of the lengths of the segments of one chord is equal to the product of the lengths of the segments of the other. 453

Theorem 77 *The Secant Segments Theorem* If two secant segments are drawn to a circle from an external point, the product of the lengths of one secant segment and its external part is equal to the product of the lengths of the other secant segment and its external part. 454

Chapter 13

Theorem 78 Every triangle is cyclic. 463

Corollary The perpendicular bisectors of the sides of a triangle are concurrent. 463

Theorem 79 A quadrilateral is cyclic iff a pair of its opposite angles are supplementary. 467

Theorem 80 The angle bisectors of a triangle are concurrent. 473

Corollary Every triangle has an incircle. 473

Theorem 81 *Ceva's Theorem* Three cevians $\overline{AY}$, $\overline{BZ}$, and $\overline{CX}$ of $\triangle ABC$ are concurrent iff $\frac{AX}{XB} \cdot \frac{BY}{YC} \cdot \frac{CZ}{ZA} = 1.$ 478

Theorem 82 The medians of a triangle are concurrent. 484

Theorem 83 The lines containing the altitudes of a triangle are concurrent. 484

Chapter 14

Theorem 84 Every regular polygon is cyclic. 499

Theorem 85 The perimeter of a regular polygon having n sides is $2Nr$, in which r is the length of its radius and $N = n \sin \frac{180}{n}$. 505

Theorem 86 The area of a regular polygon having n sides is Mr^2, in which r is the length of its radius and $M = n \sin \frac{180}{n} \cos \frac{180}{n}$. 511

Theorem 87 The circumference of a circle is $2\pi r$, in which r is the length of its radius. 521

Theorem 88 The area of a circle is πr^2, in which r is the length of its radius. 522

Theorem 89 The area of a sector whose arc has a measure of $m°$ is $\frac{m}{360}\pi r^2$, in which r is the radius of the circle. 527

Theorem 90 The length of an arc whose measure is $m°$ is $\frac{m}{360}2\pi r$, in which r is the radius of the circle. 527

Chapter 15

Theorem 91 The length of a diagonal of a rectangular solid is $\sqrt{l^2 + w^2 + h^2}$, in which l, w, and h are its dimensions. 542

Corollary The length of a diagonal of a cube is $e\sqrt{3}$, in which e is the length of one of its edges. 542

Postulate 14 *Cavalieri's Principle* Consider two geometric solids and a plane. If every plane parallel to this plane that intersects one of the solids also intersects the other so that the resulting cross sections have the same area, then the two solids have the same volume. 552

Postulate 15 The volume of any prism is Bh, in which B is the area of one of its bases and h is the length of its altitude. 552

Corollary 1 The volume of a rectangular solid is lwh, in which l, w, and h are its length, width, and height. 552

Corollary 2 The volume of a cube is e^3, in which e is the length of one of its edges. 552

Theorem 92 The volume of any pyramid is $\frac{1}{3}Bh$, in which B is the area of its base and h is the length of its altitude. 556

Theorem 93 The volume of a cylinder is $\pi r^2 h$, in which r is the radius of its bases and h is the length of its altitude. 561

Theorem 94 The volume of a cone is $\frac{1}{3}\pi r^2 h$, in which r is the radius of its base and h is the length of its altitude. 562

Theorem 95 The volume of a sphere is $\frac{4}{3}\pi r^3$, in which r is its radius. 566

Theorem 96 The surface area of a sphere is $4\pi r^2$, in which r is its radius. 568

Theorem 97 The ratio of the surface areas of two similar solids is equal to the square of the ratio of any pair of corresponding segments. 572

Theorem 98 The ratio of the volumes of two similar solids is equal to the cube of the ratio of any pair of corresponding segments. 572

Chapter 16

Theorem 99 The summit angles of a Saccheri quadrilateral are equal. 586

Theorem 100 The line segment joining the midpoints of the base and summit of a Saccheri quadrilateral is perpendicular to both of them. 587

Theorem 101 If the legs of a biperpendicular quadrilateral are unequal, then the summit angles are unequal and the larger angle is opposite the longer leg. 587

Theorem 102 If the summit angles of a biperpendicular quadrilateral are unequal, then the legs are unequal and the longer leg is opposite the larger angle. 587

The Lobachevskian Postulate The summit angles of a Saccheri quadrilateral are acute. 592

Lobachevskian Theorem 1 In Lobachevskian geometry, the summit of a Saccheri quadrilateral is longer than its base. 592

Lobachevskian Theorem 2 In Lobachevskian geometry, a midsegment of a triangle is less than half as long as the third side. 592

Lobachevskian Theorem 3 In Lobachevskian geometry, the sum of the measures of the angles of a triangle is less than 180°. 597

Corollary In Lobachevskian geometry, the sum of the measures of the angles of a quadrilateral is less than 360°. 597

Lobachevskian Theorem 4 In Lobachevskian geometry, if two triangles are similar, they must also be congruent. 597

Chapter 17

Theorem 103 *The Distance Formula* The distance between the points $P_1(x_1,y_1)$ and $P_2(x_2,y_2)$ is $\sqrt{(x_2 - x_1)^2 + (y_2 - y_1)^2}$. 612

Theorem 104 Two nonvertical lines are parallel iff their slopes are equal. 620

Theorem 105 Two nonvertical lines are perpendicular iff the product of their slopes is -1. 620

Theorem 106 *The Midpoint Formula* The midpoint of the line segment whose endpoints are $P_1(x_1,y_1)$ and $P_2(x_2,y_2)$ has the coordinates $\left(\frac{x_1 + x_2}{2}, \frac{y_1 + y_2}{2}\right)$. 625

Answers to Selected Exercises

Chapter 1, Lesson 1 (pages 7–8)

2. Not certain. (Mother may have sent boxes of candy to other people as well.) 10. False. 18. Not certain (although the statement does seem quite probable). 24. False. (We do *not* know this.) 28. Not certain.

Chapter 1, Lesson 2 (pages 11–12)

7. It is snowing. 10. If a year is a leap year, then it has 366 days. 14. If money grew on trees, then Smokey the Bear wouldn't have to do commercials for a living. 26. The second. 28. Two. 30. a and d.

Chapter 1, Lesson 3 (pages 16–18)

4. a and d. 5. They are equivalent.
10. If your temperature is not more than 102°, then you do not have a fever. 11. No.
22. If you are more than six feet tall, you are not a U.S. astronaut. 23. Yes. 29. Original statement. 30. Contrapositive. 37. If the vampires are out, the moon is full.

Chapter 1, Lesson 4 (pages 20–22)

6. Not a good definition because, if it is a holiday, it is not necessarily New Year's Day.
9. A good definition, because frozen carbon dioxide is dry ice. 15. No. 16. No.
17. Yes. 20. b and c.

Chapter 1, Lesson 5 (pages 25–26)

2. Yes. 3. No.

6. $a \rightarrow b$
   b
   Therefore, a.

11. $a \rightarrow b$
    b
    Therefore, a.
    Invalid—the converse error.

18. $a \rightarrow b$
    not a
    Therefore, not b.
    Invalid—the inverse error.

20. No deduction possible. 24. If someone is a night owl, then he hoots it up. Fred is not a night owl.

Chapter 1, Lesson 6 (pages 30–31)

3. The hypothesis. 6. No. 10. Yes.
17. No. 21. If you know what's what, then you're sure of where's where. 25. If you eat in an expensive restaurant, it helps to have a mint.

Chapter 1, Lesson 7 (pages 33–35)

3. Yes. 6. If you don't realize what time it is, you will be late for school.
12. If someone is a baby, he is illogical.
    If someone is illogical, he is despised.
    If someone is despised, he cannot manage crocodiles.

Chapter 1, Lesson 8 (pages 37–39)

1. Suppose it would not speak foul language.
5. Suppose that *the player using X did not go first.*
   If *the player using X did not go first,* the player using O went first.
   If the player using O went first, then it is O's turn to play now.
   If *it is O's turn to play now,* then X will lose the game.
   If X loses the game, then X is stupid.
   This contradicts the fact that *neither player is stupid.*
   Therefore, what we supposed is false and the player using X went first.
8. No.

Chapter 1, Lesson 9 (pages 41–42)

8. "Star" and "tail." 12. Statements 1 and 3.
17. To have a tourist visa, you must have a passport. 19. Suppose Colonel Mustard did not do it. 22. If a person is a felon, he has committed a serious crime. If a person has committed a serious crime, he is a felon.

Chapter 1, Review (pages 44–46)

4. Statement c. 5. Its inverse. 8. Where there's fire, there's smoke. 13. No conclusion. 17. The postulates. 19. Deductive reasoning. 25. The meaning of *same noise*.

Chapter 1, Algebra Review (page 48)

1. 8. 2. 12. 3. -20. 4. -5. 5. -10. 6. 24. 7. 3. 8. -10. 9. -63. 10. -16. 11. -0.25. 12. 27. 13. 0. 14. 15. 15. 5. 16. 18. 17. 8. 18. $2x + 3y$. 19. $x^2 + y^3$. 20. $7x$. 21. $6x^2$. 22. $4x^2$. 23. x^8. 24. $8x^3$. 25. $9x^6$. 26. $11 + x$. 27. $28x$. 28. $28 + 4x$. 29. $3x$. 30. $-40x^2$. 31. 5. 32. $5x - x^2$. 33. $9x + 2y$. 34. $3x$. 35. x^3. 36. $x^5 - x^4$. 37. $7x + 70x^2$. 38. $17x + 1$. 39. $9 + x^4 + x^5$. 40. $20x^9$.

Chapter 2, Lesson 1 (pages 52–54)

2. At least two points. 11. True. 18. C. 24. A line contains at least two points. 25. Points are noncollinear iff there is no line that contains all of them. 29. Suppose that the lines intersect in more than one point. 30. If there are two points, then there is exactly one line that contains them. 33. Figure 1 shows *lines ℓ and m intersecting in point A.* 35. Points A, B, and C are *noncollinear.* 39. If it does, then points A and B lie in plane P, but line ℓ *does not.*

Chapter 2, Lesson 2 (pages 58–59)

5. 21. 20. Two. 22. One. 24. *n*. 33. 3.

Chapter 2, Lesson 3 (pages 62–63)

7. Reflexive. 8. Addition. 13. $y = x + 7$. 15. $|x| = \sqrt{y}$. 20. $12 + 2y = 5y$. 24. Fact 1 and the symmetric property. 30. Facts 1 and 3 and the substitution property. 35. Facts 1 and 2 and the transitive property.

Chapter 2, Lesson 4 (pages 65–67)

2. A. 5. YA + AP = YP. 7. $4 < 8 < 9$. 13. May be true. 15. Must be true. 19. No. 20. CA + AK = *CK*. 26. R-B-K.

Chapter 2, Lesson 5 (pages 71–73)

9. Infinitely many. 16. AM = $\frac{1}{2}$AY. 22. LS = 2SA. 27. IN + NA = IA, not $\overline{IN} + \overline{NA} = \overline{IA}$. 30. $\frac{1}{2}(c - a)$. 32. $m + r$ or $m - r$.

Chapter 2, Lesson 6 (pages 76–78)

5. Convex. 10. Its segments do not intersect only at their endpoints. 18. $2a + 2b$. 24. Three. 25. Four.

Chapter 2, Review (pages 79–81)

3. Two. 6. They are coplanar. 10. BE + EL = BL. 14. 13. 17. Three noncollinear points determine a plane. 21. -6. 28. Facts 1 and 2 and the substitution property. 31. L-T-U.

Chapter 2, Algebra Review (page 82)

1. $13x$. 2. $9x$. 3. $10x^2$. 4. $-8x$. 5. $48x$. 6. $11x + 2$. 7. $5x - 6$. 8. $10x - 5$. 9. $5x + 15$. 10. $44 - 11x$. 11. $18x + 6$. 12. $40 - 56x$. 13. $x^2 + x$. 14. $60 - 12x$. 15. $9x^2 - 18x$. 16. $20x + 6x^2$. 17. 10. 18. 8. 19. 3. 20. 6. 21. 13. 22. 9. 23. -7. 24. -5. 25. 60. 26. 0. 27. -12. 28. 37. 29. -2. 30. 4.5. 31. 12.

Chapter 3, Lesson 1 (pages 86–88)

2. $\overrightarrow{EK}$ and $\overrightarrow{EG}$. 7. Yes. 9. ∡BOL (or ∡LOB). 18. Six. 26. Eight. 27. Incorrect. $\overrightarrow{TS}$ and $\overrightarrow{TH}$ are opposite rays.

Chapter 3, Lesson 2 (pages 91–92)

1. The coordinates of the rays. 6. The measure of angle ABC. 14. 100°. 28. 33°20′. 31. 75.

Chapter 3, Lesson 3 (pages 95–97)

3. $\overrightarrow{OR}$-$\overrightarrow{OS}$-$\overrightarrow{OE}$ because $155 > 80 > 35$. 5. ∠ROS + ∠SOE = ∠ROE. 8. No. 14. Fact 3 and the subtraction property. 25. $x > 90$. 27. 14.

Chapter 3, Lesson 4 (pages 102–103)

2. ∠A = ∠B. 13. The sum of their measures is 180°. 16. They are equal. 18. No. 22. 49°. 24. 30.

Chapter 3, Lesson 5 (pages 106–108)
7. No. 13. Yes. 14. They are opposite rays. 17. They are a linear pair and they are supplementary. 21. 45°. 23. -16.

Chapter 3, Lesson 6 (pages 111–114)
6. Convex. 12. False. 15. $\measuredangle AOC$, $\measuredangle AOD$, $\measuredangle BOC$, and $\measuredangle BOD$ are right angles. 18. $\overline{GD}$, $\overline{DH}$, $\overline{OL}$, and $\overline{LS}$. 24. Ollie.

Chapter 3, Review (pages 116–117)
1. $\angle A + \angle B = 180°$. 3. $\overrightarrow{DF}$-$\overrightarrow{DE}$-$\overrightarrow{DG}$, $\angle FDE = \angle EDG$. 6. They are supplementary. 10. They form four right angles. 15. Five. 18. Must be true. 21. False. 25. -3.

Chapter 3, Algebra Review (page 118)
1. (8, 8). 2. (7, 14). 3. (11, 9). 4. (3, -4). 5. (-1, 3). 6. (0, -1). 7. (5, 14). 8. (14, 5). 9. (3.5, 10.5). 10. (18, 9). 11. (45, 9). 12. (-2, 1). 13. (-19, -2). 14. (6, -7). 15. (5, 12). 16. (0, -5).

Chapter 4, Lesson 1 (pages 121–123)
7. $\overline{FE}$ and $\overline{FB}$. 9. $\measuredangle B$. 11. Scalene. 15. Yes. 18. $\triangle MRA$ and $\triangle MRC$. 23. $\triangle ALP$. 26. Six. 27. Two.

Chapter 4, Lesson 2 (pages 127–129)
2. Z. 3. HAZE $\leftrightarrow$ TLNU. 5. $\angle P = \angle T$. 11. $AS = EW$. 14. $\measuredangle T$. 17. S.A.S. 20. No conclusion possible. 26. Six. 33. Reflexive. 35. The midpoint of a line segment divides it into two equal segments.

Chapter 4, Lesson 3 (pages 132–134)
1. S.A.S. 3. C.A.C. 6. 9. 12. Must be true. 15. May be true. 20. Must be true.

Chapter 4, Lesson 4 (pages 137–139)
3. Theorem. Its converse is false. 8. $KA = AK$. 10. $\angle OKA = \angle EAK$. 12. $\measuredangle RAI$ and $\measuredangle IAN$ are a linear pair. 15. $RA = AN$. 21. If two triangles are congruent, their corresponding parts are equal. 23. If a ray bisects an angle, it divides it into two equal angles. 26. If the two angles in a linear pair are equal, each is a right angle.

Chapter 4, Lesson 5 (pages 142–144)
2. A triangle is equilateral iff *all of its sides are equal.* 6. If a triangle is equilateral, *it is also equiangular.* 8. True. 12. $\measuredangle PEI$. 14. $\overline{IE}$. 18. No. 24. May be true.

Chapter 4, Lesson 6 (pages 147–149)
3. No. 11. Yes; S.S.S. theorem. 15. Must be true. 19. Appears to be false.

Chapter 4, Lesson 7 (pages 154–156)
3. Its midpoint. 6. O. 9. The arc numbered 3. 13. The arc numbered 1.

Chapter 4, Lesson 8 (pages 159–161)
6. They appear to be collinear. 8. $AR = AK$. 10. They are equidistant from the endpoints of $\overline{KT}$. 17. If a ray divides an angle into two equal angles, it bisects it. 27. They seem to be parallel and to have equal lengths.

Chapter 4, Review (pages 163–165)
6. $\triangle NLE$. 10. VOLGA $\leftrightarrow$ *RHINE.* 13. No. 16. Yes. 20. The ray that bisects the angle.

Chapter 4, Algebra Review (page 166)
1. (51, -16). 2. (25.5, 17.5). 3. (-2, 3). 4. (3, -11). 5. (18.5, 0.9). 6. (-1, 7). 7. (-1, -10). 8. (-4, -5). 9. (11, -4). 10. (7, 1). 11. (7, 3). 12. (9, 2). 13. (4, 0). 14. (-1, 5). 15. (8, -2). 16. (0, -8). 17. (5, -3). 18. (5, -1).

Chapter 5, Lesson 1 (pages 170–172)
2. $OA < AD$. 9. $\angle ULB > \angle UEB > \angle UVB$. 11. Whole greater than its part. 16. Facts 2 and 3 and the transitive property. 20. Fact 1 and the multiplication property. 27. The largest angle is opposite the longest side. 33. $7 > -4$.

Chapter 5, Lesson 2 (pages 175–177)
2. $\measuredangle O$ and $\measuredangle S$. 4. They are a linear pair and they are supplementary. 6. No. 15. True. 16. $\measuredangle 4$; $\angle 4 > \angle 2$ and $\angle 4 > \angle 6$. 19. 36.

Chapter 5, Lesson 3 (pages 180–182)
8. $\angle E > \angle T$. 11. $\measuredangle L$. 20. Yes. 23. Yes. 26. Yes; $DA > SD$. 27. No.

Chapter 5, Lesson 4 **(pages 184–186)**
4. $JE + JO > OE$. 7. Not possible; $2 + 5 < 8$. 14. $ED + EA + EV > 25$. 19. No.

Chapter 5, Review **(pages 187–188)**
1. It seems to be longer. 4. True. 12. Fact 3 and the "whole greater than its part" property. 13. 62°. 15. $\overline{TR}$. 17. May be true. 19. Must be true.

Chapter 5, Algebra Review
(pages 189–190)
1. $6x + 1$. 2. $2x - 4y$. 3. $4x^2 + 8$. 4. $3x^2 + 6x - 5$. 5. $7x + 4$. 6. $6x$. 7. $-x^2 - 11x$. 8. $2x^2 + 10x - 1$. 9. $x^2 + 14x + 24$. 10. $6x^2 - 7x - 20$. 11. $100x^2 - 20x + 1$. 12. $16x^2 - 9$. 13. $3(2x - 7)$. 14. $x(x + 10)$. 15. $x^2(3x - 2)$. 16. $(x + 7)(x - 7)$. 17. $(x + 4)^2$. 18. $(x + 3)(x + 13)$. 19. $(x - 6)(x + 7)$. 20. $(2x + 1)(x + 7)$. 21. $(3x - 5)(x + 2)$. 22. $6x(x + 1)(x - 1)$. 23. $x(x + 2y)$. 24. $(3x + y)(3x - y)$. 25. $(x + 5y)(x + y)$.

Chapter 6, Lesson 1 **(pages 194–196)**
3. $\measuredangle 6$. 5. $\measuredangle 3$ and $\measuredangle 5$. 8. c and k. 11. y and a. 16. $\overline{AB}$ and $\overline{WL}$ (or $\overline{EL}$). 22. $f \parallel e$. 24. No lines.

Chapter 6, Lesson 2 **(pages 200–202)**
4. It suggests that there are infinitely many. 7. False. 9. True. 14. Must be true. 16. May be true. 23. $\angle O = 50°$, $\angle PDN = 130°$.

Chapter 6, Lesson 3 **(pages 205–206)**
4. A definition. 11. No. 12. Two points determine a line. 15. The Protractor Postulate. 20. $\overline{UH} \parallel \overline{OS}$. 24. In a plane, two lines perpendicular to a third line are parallel to each other.

Chapter 6, Lesson 4 **(pages 210–212)**
3. Alternate interior angles. 7. $\measuredangle 2$ and $\measuredangle 4$. 9. $\measuredangle N$. 11. 12. 16. -16. 22. $\measuredangle ETS$.

Chapter 6, Lesson 5 **(pages 216–217)**
2. $\overline{ON}$. 6. $\overline{BR}$. 9. In a plane, two lines perpendicular to a third line are parallel to each other. 12. The perpendicular segment from a point to a line is the shortest segment joining them. 14. 1.5 cm.

Chapter 6, Lesson 6 **(pages 221–224)**
3. Obtuse. 7. Equilateral. 17. 120°. 19. 80°. 22. 13. 27. -4.

Chapter 6, Lesson 7 **(pages 227–229)**
2. A.A.S. 6. $\angle S = \angle O$. 9. The triangles are congruent by A.A.S. 13. No. 20. 4.

Chapter 6, Review **(pages 231–233)**
2. False. 6. False. 11. F appears to be the midpoint of $\overline{BC}$. 13. $\overline{AB} \perp \overline{BC}$. 18. 65°. 22. 64°. 26. Must be true. 29. May be true.

Chapter 6, Algebra Review **(page 234)**
1. $2\sqrt{7}$. 2. $7\sqrt{2}$. 3. $6\sqrt{10}$. 4. 47. 5. $5\sqrt{x}$. 6. x^8. 7. $5\sqrt{3}$. 8. $3\sqrt{5}$. 9. $5\sqrt{7}$. 10. $5\sqrt{21}$. 11. $2\sqrt{15}$. 12. $6\sqrt{5}$. 13. 14. 14. 10. 15. 5. 16. $5\sqrt{5}$. 17. $7\sqrt{6}$. 18. 150. 19. 11. 20. $2\sqrt{13}$. 21. $24\sqrt{10}$. 22. 63. 23. $3\sqrt{5}$. 24. $5\sqrt{2}$. 25. 10. 26. 23. 27. $9 + \sqrt{3}$. 28. $4\sqrt{3} + 1$. 29. $2\sqrt{x} + 2\sqrt{y}$. 30. $x + 2\sqrt{xy} + y$.

Chapter 7, Lesson 1 **(pages 238–240)**
2. Rectangle. 12. Parallelograms, rhombuses, rectangles, and squares. 14. Rectangles and squares. 18. 116°. 21. 120°. 26. A.S.A. 29. A trapezoid.

Chapter 7, Lesson 2 **(pages 242–244)**
5. $\angle B = \angle F$, $BC = FG$, $\angle C = \angle G$, $CD = GH$, and $\angle D = \angle H$. 9. No. 12. 4. 16. 35. 21. -9.

Chapter 7, Lesson 3 **(pages 248–250)**
7. Yes. A quadrilateral is a parallelogram if both pairs of opposite sides are equal. 8. No. 13. $\angle BCI = \angle BUI$. 16. They are equal (or, they are right angles). 20. True. 25. True.

Chapter 7, Lesson 4 **(pages 253–254)**
1. Parallel. 7. Isosceles. 10. The diagonals of a rhombus are perpendicular. 15. A parallelogram. 17. No. 19. It is equilateral. 22. 68.

Chapter 7, Lesson 5 **(pages 257–259)**
3. No. 9. $\overline{NI}$ and $\overline{EC}$. 15. Must be true. 19. False. 23. True. 24. 18.

Chapter 7, Lesson 6 (pages 261–263)

1. $\overline{AL}$. 3. Three. 5. A parallelogram. 10. 50. 15. Equiangular (and equilateral). 18. 8. 23. -6. 30. A midsegment of a triangle is parallel to the third side. 33. Substitution.

Chapter 7, Review (pages 265–266)

1. Parallelograms. 5. A quadrilateral is a parallelogram if two opposite sides are both equal and parallel. 9. False. 10. True. 17. A trapezoid (because $\overline{AT} \parallel \overline{LC}$). 20. -7. 28. $\angle TFL = \frac{1}{2}\angle NFL$, $\angle FLT = \frac{1}{2}\angle FLI$. 30. It is a right angle.

Chapter 7, Algebra Review (pages 267–268)

1. 0.75. 2. 0.05. 3. 1.8. 4. $0.58\overline{3}$. 5. $\frac{2}{9}$. 6. $\frac{x^2}{3}$. 7. $\frac{4}{7}$. 8. $\frac{1}{x+5}$. 9. $\frac{7}{6}$. 10. $\frac{x}{20}$. 11. $\frac{x+10}{2x}$. 12. $\frac{4x-3}{x^3}$. 13. $\frac{2}{3}$. 14. $\frac{x^5}{6}$. 15. $\frac{5}{7}$. 16. $\frac{x+2}{2}$. 17. $\frac{3x}{4}$. 18. $\frac{1}{5}$. 19. $\frac{x+3}{3x}$. 20. $\frac{50}{x^2}$. 21. $\frac{y+x}{xy}$. 22. $\frac{2x-6}{2x-3}$. 23. $\frac{x-y}{y}$. 24. $\frac{1}{x}$.

Chapter 8, Lesson 1 (pages 272–274)

3. R. 5. The reflection of U through $\overleftrightarrow{ML}$. 10. They appear to intersect on line ℓ. 15. It appears to bisect ∡NAN′ and ∡G′AG. 18. ℓ_2. 23. They appear to be collinear. 26. Line ℓ is the perpendicular bisector of $\overline{AA'}$ and $\overline{BB'}$.

Chapter 8, Lesson 2 (pages 277–280)

2. $\overline{OE}$. 5. ∡R. 6. No. 9. Yes. 17. No.

Chapter 8, Lesson 3 (pages 282–284)

1. A translation. 2. A reflection is an isometry, and a triangle and its image under an isometry are congruent. 8. △GHI. 12. △JKL. 18. They seem to be parallel. 20. They seem to be collinear. 23. They seem to be perpendicular to ℓ_1 and ℓ_2. 26. 5.0 cm. 30. $x + y$.

Chapter 8, Lesson 4 (pages 288–290)

1. A reflection through ℓ_1. 2. A translation. 6. $\overline{GH}$. 9. $\overline{GH}$. 11. Their center. 13. △ABC. 17. Collinearity and betweenness (also distance). 19. The vertices of the triangles lie on them. 22. 50°. 25. Its magnitude.

Chapter 8, Lesson 5 (pages 293–294)

2. A reflection. 4. Yes. 7. ABCDE ↔ GFJIH. 10. Two points determine a line. 14. A translation. 16. The angles are congruent because there is an isometry such that one angle is the image of the other. 23. △DEF is the reflection image of △ABC through ℓ. 26. EFGH is a rotation image of ABCD.

Chapter 8, Lesson 6 (pages 296–298)

2. Two. 11. H, I, M, O, T, U, V, W, X, and Y. 14. Reflection symmetry (one line). 21. 180.

Chapter 8, Review (pages 299–301)

5. True. 6. False. 8. False. 10. No. 17. Translations and reflections. 20. No. 26. They are equidistant from point P. 31. Yes. Two figures are congruent if there is an isometry such that one figure is the image of the other. 36. 72° (approximately).

Chapter 9, Lesson 1 (pages 309–311)

11. No. 15. False. 16. Must be true. 24. $\alpha\triangle SHK = \frac{1}{4}\alpha SHEI$. 29. 6.

Chapter 9, Lesson 2 (pages 313–315)

1. Perimeter, 20; area, 24. 6. Perimeter, $2\sqrt{3} + 2\sqrt{6}$; area, $3\sqrt{2}$. 8. 258 feet. 10. 3780 square feet. 12. 1296. 16. $5\sqrt{2}$ inches. 18. 49.7025 square inches. 22. 24 square feet. 24. 128 square feet. 29. 33. 32. 26.

Chapter 9, Lesson 3 (pages 320–322)

1. Perimeter, 36; area, 54. 5. Perimeter, $10 + 5\sqrt{2}$; area, 12.5. 7. Perimeter, $3b$; area, $\frac{1}{2}ab$. 16. No. 19. 840. 21. 30.

23. 234. 26. 69.5. 29. If two lines are parallel, every perpendicular segment joining one line to the other has the same length.
30. $\frac{1}{2}ah$. 35. 20.

Chapter 9, Lesson 4 (pages 325–327)
6. Perimeter, 40; area, 77. 9. Perimeter, 54; area, 168. 15. Perimeter, $2a + b$; area, $\frac{1}{2}ac$. 18. $(c + 5)^2 = c^2 + 10c + 25$.
21. $(a + 4b)^2 = a^2 + 8ab + 16b^2$. 25. $20x$.
28. 40. 31. 94.

Chapter 9, Lesson 5 (pages 330–334)
4. $WM^2 = WA^2 + AM^2$. 6. Yes.
7. $5^2 + 12^2 = 13^2$. 10. 64, 36, and 100; $x = 10$. 14. $2\sqrt{10}$. 18. 196. 21. 306.
22. 60. 27. $\alpha ABDE = \frac{1}{2}(a + b)^2$.
30. $b - a$.

Chapter 9, Lesson 6 (pages 338–340)
1. Two. 4. 84. 7. $20\sqrt{11}$. 12. 6.9.
13. $\frac{25}{4}\sqrt{3}$. 17. Such a triangle cannot exist because $4 + 6 = 10$. 21. $36\sqrt{3}$.
24. Perimeter, 20; area, $12\sqrt{3}$. 27. Perimeter, 56; area, 168. 29. $100 - 25\sqrt{3}$.

Chapter 9, Review (pages 343–345)
1. 16. 6. 625. 13. Yes. 14. 40.
17. $16\sqrt{3}$. 22. 40. 26. 44. 29. 30.
32. 63.

Chapter 9, Algebra Review (page 346)
1. 500. 2. 0.2. 3. 2.5. 4. 10 and -10.
5. -3. 6. 42. 7. 15. 8. 20. 9. 41.
10. 3.5. 11. 0. 12. 8. 13. 5. 14. 5.
15. -36. 16. $-ab$. 17. $\frac{a^2}{b}$. 18. a.
19. $\frac{a + b}{c}$. 20. $\frac{a - 1}{b}$.

Chapter 10, Lesson 1 (pages 350–352)
4. 2.7. 6. 6. 8. 3.14. 11. 9.
17. $\frac{3}{1.25} = 2.4$. 19. $\frac{x}{12} = \frac{y}{5}$. 22. Divide by ab. 25. $x = 12$, $y = 125$.

Chapter 10, Lesson 2 (pages 356–358)
3. The midpoint of a line segment divides it into two equal segments. 5. $\frac{SA}{AH} = \frac{TY}{YH}$.
7. $\frac{YT}{HT} = \frac{AS}{HS}$. 11. a, c, d, and e. 13. 2.4.
18. 16. 21. 6.

Chapter 10, Lesson 3 (pages 361–362)
3. ABCDE $\leftrightarrow$ HIJFG. 5. $\frac{AB}{HI} = \frac{DE}{FG}$.
12. $\frac{8}{y} = \frac{6}{x} = \frac{12}{8}$. 15. Division.
20. Supplements of equal angles are equal.
24. Division. 28. No.

Chapter 10, Lesson 4 (pages 366–368)
4. 345 ft. 8. 18 ft. 14. Must be true.
17. May be true. 21. 9. 23. 36.
26. $\frac{2}{3}$.

Chapter 10, Lesson 5 (pages 371–373)
5. Yes. 13. 12. 16. 8. 18. 54.
20. $\frac{3}{5}$.

Chapter 10, Lesson 6 (pages 377–378)
5. $\frac{BE}{EL} = \frac{WO}{OL}$. 7. $\frac{OL}{WL} = \frac{EL}{BL}$. 9. 6.
11. 5.6. 13. 6.75. 16. It is a right triangle.
18. AD = 5 cm and DB = 3 cm. 22. 40.
24. No.

Chapter 10, Lesson 7 (pages 382–384)
1. $\frac{4}{5}$. 5. $\frac{16}{25}$. 13. 4. 14. $\frac{1}{3}$. 16. 18.
18. 35. 20. $\frac{16}{9}$. 23. 115. 24. 675.
27. $\sqrt{2}$. 28. $3\sqrt{2}$. 30. $\frac{1}{2}\sqrt{11}$. 31. $\frac{3}{2}$.

Chapter 10, Review (pages 386–387)
8. 6.4. 9. 5.6. 12. 12.8. 14. Yes (by A.A.). 15. No. 16. $\frac{3}{4}$. 19. $\frac{4}{3}$.
22. No. 24. No. 27. 126. 29. 36.
30. 56.

Chapter 10, Algebra Review (page 388)
1. 4 and -7. 2. 1 and $-\frac{1}{5}$. 3. $4 + \sqrt{11}$

and $4 - \sqrt{11}$. 4. 7 and -7. 5. $-2 + 2\sqrt{6}$ and $-2 - 2\sqrt{6}$. 6. $\frac{1}{3}$. 7. $\frac{4}{3}$ and -3. 8. 2 and 5. 9. -8 and -1. 10. $\frac{1 + \sqrt{26}}{5}$ and $\frac{1 - \sqrt{26}}{5}$.

Chapter 11, Lesson 1 (pages 392–393)
3. $\overline{RN}$ and $\overline{NI}$. 5. $\overline{RN}$. 6. $\overline{RI}$ and $\overline{RN}$. 9. $\triangle WIN \sim \triangle WDI$. 12. $\frac{WN}{WI} = \frac{WI}{WD}$. 17. 4. 19. 6. 22. 10. 24. 4. 26. 1.8 cm.

Chapter 11, Lesson 2 (pages 396–398)
1. $3\sqrt{2}$. 5. $3\sqrt{3}$. 9. $10\sqrt{10}$. 19. $7^2 + 24^2 = 49 + 576 = 625 = 25^2$. 23. 10, 24, 26 and 15, 36, 39. 27. 10. 29. 8. 32. $3\sqrt{5}$.

Chapter 11, Lesson 3 (pages 401–403)
1. Divide c by $\sqrt{2}$. 6. $8\sqrt{2}$. 8. 4. 10. $x = 12$, $y = 6\sqrt{3}$. 13. $x = 5$, $y = 10$. 14. $x = 10\sqrt{3}$, $y = 20\sqrt{3}$. 18. $10\sqrt{2}$. 20. $5\sqrt{3}$. 32. $h = 8\sqrt{3}$; area, $200\sqrt{3}$. 34. h = 9; area, 103.5. 36. $OT = 3\sqrt{2}$, $DT = 3\sqrt{3} + 3$.

Chapter 11, Lesson 4 (pages 406–408)
3. $\frac{OW}{TW}$. 7. $\tan A = \frac{5}{12}$, $\tan B = \frac{12}{5}$. 14. $\frac{\sqrt{3}}{3}$. 18. 15. 21. 27.5. 25. 20°. 29. 32 m.

Chapter 11, Lesson 5 (pages 412–413)
1. $\sin A = \frac{GM}{AM}$. 2. $\cos B = \frac{BL}{BE}$. 8. $\frac{\sqrt{2}}{2}$. 14. $\frac{\sqrt{3}}{2}$. 17. 49°. 20. 20°. 23. 5.3. 25. 58°. 29. 81 ft.

Chapter 11, Review (pages 415–418)
2. 23°. 3. $3\sqrt{2}$. 5. $6\sqrt{3}$. 8. $2\sqrt{30}$. 11. $\tan 1 = \frac{CN}{RN}$. 15. $\sin 2 = \frac{NA}{RA}$. 20. 76°. 21. 6.3. 26. 37°. 27. $12 + 4\sqrt{3}$. 30. 70. 31. 139 ft.

Chapter 12, Lesson 1 (pages 422–425)
2. Two. 7. They seem to have the same center. 21. 10. 25. $2\sqrt{5}$. 29. 20.

Chapter 12, Lesson 2 (pages 428–429)
1. $\overline{BD} \perp \overline{ND}$. 5. 9. 20. 30°. 29. 16. 32. 10.

Chapter 12, Lesson 3 (pages 433–435)
3. $\widehat{PSE}$. 14. 260°. 16. 6. 25. 17.3. 27. 60°.

Chapter 12, Lesson 4 (pages 438–440)
4. $\widehat{IDT}$. 6. 70°. 15. No. 17. 184°. 21. No. 23. 74°. 27. No. 28. $(180 - x)°$. 30. $\left(\frac{1}{2}x\right)°$.

Chapter 12, Lesson 5 (pages 443–445)
13. 75° 16. 30°. 19. 65°. 25. $\angle BEL = m\widehat{BL}$, $\angle ZEA = m\widehat{ZA}$. 30. $\angle BEL = \frac{1}{2}(m\widehat{BL} + m\widehat{ZA})$.

Chapter 12, Lesson 6 (pages 448–450)
4. 6. 5. $2\sqrt{3}$. 8. $5\sqrt{6}$. 13. It was bisected. 21. The radius. 29. $\tan Y = \frac{EN}{YE} = \frac{EN}{1} = EN$.

Chapter 12, Lesson 7 (pages 454–456)
5. $\angle OSL = \frac{1}{2}(m\widehat{OL} + m\widehat{CT})$. 7. $BA \cdot BE = BR \cdot BS$. 9. EA = SR. 10. 10. 15. $2\sqrt{6}$. 16. 10. 22. 9. 24. 62°. 26. 118°. 29. 27. 31. 6.

Chapter 12, Review (pages 458–460)
14. False. 17. False. 20. 16.5. 22. 100°. 24. 3. 29. 16. 31. 36.

Chapter 13, Lesson 1 (pages 464–466)
1. They are concyclic. 4. They are equidistant from point O. 8. The perpendicular bisectors of the sides of a triangle are concurrent. 13. At the midpoint of its hypotenuse. 16. $4\sqrt{3}$. 19. 10. 27. 5. 28. 53°.

Chapter 13, Lesson 2 (pages 468–471)
2. They are supplementary. 3. 95°. 8. Yes. 14. 50°. 16. 120°. 26. 12.

Chapter 13, Lesson 3 (pages 474–476)
9. It is equidistant from them. 22. 20. 24. $\sqrt{2}$. 26. 30°-60° right triangles. 27. $3x$. 28. $2x\sqrt{3}$. 30. $3x^2\sqrt{3}$.

Chapter 13, Lesson 4 (pages 480–482)
4. $\overline{NY}$ and $\overline{KW}$. 7. $\frac{PI}{IN} \cdot \frac{NE}{EO} \cdot \frac{OH}{HP} = 1$.
10. AX = 4 cm, YB = 3 cm, CZ = 2 cm, and ZB = 3 cm. 11. $\frac{1}{2}$. 16. Not necessarily.
18. $\frac{1}{2}$. 20. 12. 23. SL = 9 and LU = 21.
26. ON = 24 and SK = 24. 27. 30.
30. 20.

Chapter 13, Lesson 5 (pages 485–487)
1. C (the angle bisectors). 9. F (the circumcenter). 16. It is the vertex of the right angle. 19. They are equidistant from them.
26. They seem to divide it into three equal segments. 32. A parallelogram.

Chapter 13, Lesson 6 (pages 491–492)
7. a) $\overline{BA}$ was bisected.
b) A semicircle was drawn with the midpoint of $\overline{BA}$ as its center and $\frac{1}{2}$BA as its radius.
c) An angle inscribed in a semicircle is a right angle.

Chapter 13, Review (pages 494–496)
5. The circumcenter. 6. The midsegments.
9. The altitudes. 10. The orthocenter.
12. They seem to be concyclic. 15. 3.2.
18. That these lines bisect its angles.
22. False. 23. Must be true.

Chapter 14, Lesson 1 (pages 500–502)
1. True. 3. False. 8. False. 17. Yes.
18. It decreases. 20. 36°. 23. They are equal to it. 25. They are perpendicular.
29. 8.1.

Chapter 14, Lesson 2 (pages 506–508)
2. S.S.S. 5. It bisects it. 12. 42.
14. An isosceles right triangle. 17. 40.
19. A 30°-60° right triangle. 23. $18\sqrt{3}$.
24. 31.176. 27. $\sin \frac{180}{n}$ decreases.
31. 37. 35. c.

Chapter 14, Lesson 3 (pages 511–513)
2. An altitude. 5. αVAMPI = 5α△VRI.
6. $\frac{1}{2}$VI · RE. 10. $3\sqrt{2}$. 12. $12\sqrt{3}$.
15. 2. 19. $\sin \frac{180}{n} \cos \frac{180}{n}$ decreases.
23. 77. 27. b. 28. 41.6. 30. 29.9.
31. 83.1 33. 27.7.

Chapter 14, Lesson 4 (pages 517–519)
2. $\frac{1}{n}$. 4. 0. 6. $3n$. 10. Zero, point, followed by n 3's. 11. They get larger.
15. $\frac{12}{5} = 2.4$. 19. 3. 20. $\frac{4}{5} = 0.8$.
24. $-\frac{199{,}999}{1{,}000{,}000} = -0.199999$. 26. 3.
32. 3.14. 35. 1.

Chapter 14, Lesson 5 (pages 523–525)
1. The circumference. 3. $c = \pi$; $a = \frac{1}{4}\pi$.
5. 30 cubits. 8. 3. 16. $\frac{3}{4}$. 18. 3.1429.
20. Six. 21. 16. 23. 3.5. 25. 20π.
27. $196 - 49\pi$. 29. $25\pi - 48$. 32. 254.
34. Approximately 584,000,000 miles.

Chapter 14, Lesson 6 (pages 527–529)
1. 12π. 3. π. 7. 3π. 10. 5π.
12. 10π. 13. $\widehat{NI}$ and $\widehat{EL}$. 17. 9 in^2.
19. $8\pi + 30$. 21. $16\pi + 120$. 24. 16π.
27. $48\pi + 32$. 29. $16\sqrt{3} - 8\pi$. 30. 2π.
32. $72 - 18\pi$. 35. $8\pi - 16$.

Chapter 14, Review (pages 531–534)
2. False. 3. False. 5. True. 10. 70°.
12. $\frac{11}{5} = 2.2$. 16. The triangle, 52; the square, 57; the pentagon, 59. 17. The triangle, 130; the square, 200; the pentagon, 238.
22. It decreases. 25. 13.2 m. 26. 13.9 m^2.
30. $32\pi - 32$. 32. $72 - 18\pi$.
35. $36\sqrt{3} - 12\pi$.

Chapter 15, Lesson 1 (pages 538–539)
1. Parallel. 4. Perpendicular. 10. No.
11. Yes. 13. Yes. 21. True. 26. True.

Chapter 15, Lesson 2 (pages 543–545)
3. Twelve. 4. C. 6. $\overline{AD}$, $\overline{DH}$, $\overline{BC}$, and $\overline{CG}$.
11. False. 15. 90°. 16. No. 23. 13.
25. $4\sqrt{5}$. 27. 7. 30. 11.
35. $48\sqrt{3} \approx 83$.

Chapter 15, Lesson 3 (pages 547–549)
4. No. 8. Yes; triangles. 12. A pentagonal prism. 13. 10. 15. 7. 17. 9. 19. A

right square prism. 24. $4e^2$. 27. 162. 30. 96. 35. $\frac{3e^2}{2}\sqrt{3}$.

Chapter 15, Lesson 4 (pages 552–554)
2. 27. 7. 108. 10. $45\sqrt{3}$. 13. 1300 cm^2. 15. 24. 17. It is multiplied by 4. 20. It is multiplied by 8. 23. 84. 27. 572 ft^2. 28. 1040 ft^2. 32. 925 ft^2. 34. 7875 ft^3. 36. 10 ft^3.

Chapter 15, Lesson 5 (pages 557–559)
1. F. 6. Isosceles triangles. 10. Eight. 12. 128. 15. $24\sqrt{3}$. 19. 8. 22. No. 26. 36 cubic units. 28. 17. 29. 544. 32. They appear to be similar. 34. 10.8.

Chapter 15, Lesson 6 (pages 562–564)
3. Its axis. 6. 20π. 8. 21π. 11. $72\pi\sqrt{3}$. 14. $\frac{1}{3}\pi a^2 b$. 15. $\frac{2}{3}\pi a^2 b$. 18. It is doubled. 21. $2\pi rh$. 23. $6\pi a^2 b$. 26. 8π. 28. 103 in^3. 29. 6. 32. $\pi r\ell$.

Chapter 15, Lesson 7 (pages 568–570)
2. Area, 324π; volume, 972π. 7. It is multiplied by 4. 9. $\frac{2}{3}\pi r^3$. 18. 201,000,000 square miles. 20. $\frac{2}{3}\pi a^3$. 25. $\frac{4}{3}\pi r^3$. 26. πe^2. 28. $\frac{\pi}{6}$. 32. Approximately 14.1 cubic feet. 33. Approximately 733 pounds.

Chapter 15, Lesson 8 (pages 573–575)
1. 9. 2. 27. 5. $\frac{3}{4}$. 9. $6a^2$ and $6b^2$. 15. $\frac{5}{2}$. 23. 500. 26. 320. 32. 9¢. 35. Approximately 173 pounds.

Chapter 15, Review (pages 577–578)
1. True. 5. False. 7. False. 11. $90\sqrt{3}$. 13. k. 15. k^3. 19. A, 104; B, 94. 22. The surface area of the Great Pyramid is 52,900 times that of the small pyramid. 23. 138,000 square meters. 27. $\frac{2}{3}$. 32. $v = \frac{1}{6}\pi d^3$.

Chapter 16, Lesson 1 (pages 582–584)
8. No. 10. Yes. 14. Yes. 18. Yes. 21. Yes. 23. No. 26. No. 34. It is more than 180°.

Chapter 16, Lesson 2 (pages 587–589)
5. The summit. 7. The legs of a Saccheri quadrilateral are equal. 10. S.S.S. 13. The summit angles of a Saccheri quadrilateral are equal. 18. The "three possibilities" property. 20. A Saccheri quadrilateral.

Chapter 16, Lesson 3 (pages 593–595)
1. In Lobachevskian geometry, the summit angles of a Saccheri quadrilateral are acute. 3. Substitution. 13. It is a biperpendicular quadrilateral with equal legs. 14. In Lobachevskian geometry, the summit of a Saccheri quadrilateral is longer than its base. 19. It is less than 15. 21. Must be true. 23. False. 26. May be true. 31. Biperpendicular quadrilaterals.

Chapter 16, Lesson 4 (pages 598–600)
2. ∡F and ∡H. 3. $\overline{FB}$ and $\overline{HC}$. 11. Addition. 20. No. 24. $\angle 1 + \angle 2 = 180°$. 31. It is acute. 36. True. 39. No.

Chapter 16, Review (pages 602–604)
3. True. 8. No (because it is not equilateral). 12. Saccheri quadrilaterals. 13. Acute. 17. Multiplication. 22. The Pythagorean Theorem is not true in Lobachevskian geometry. 24. Two points determine a line. 27. The sum of the measures of the angles of a triangle is less than 180°.

Chapter 17, Lesson 1 (pages 607–609)
3. The third. 6. (2, 6). 9. (-4, 0). 14. The y-coordinate is twice the x-coordinate. 15. C. 19. (17, 0). 22. (0, -10). 29. $2\sqrt{10}$. 30. In the first quadrant. 35. A and D.

Chapter 17, Lesson 2 (page 613)
1. 10. 4. $\sqrt{53}$. 12. No. (AC = $\sqrt{41}$.) 16. Yes. 20. $18\sqrt{5}$. 21. 90. 28. 40. 29. $\sqrt{a^2 + b^2}$. 32. a.

Chapter 17, Lesson 3 (pages 617–618)
5. $\frac{2}{3}$. 6. $x_1 = -3$ and $y_1 = 4$. 8. -5. 10. -1. 11. 2. 13. $\frac{4}{5}$. 17. OB. 19. The slope of $\overleftrightarrow{OA}$. 23. $\overline{BC}$ and $\overline{AC}$. 30. $m\overline{AC} = \frac{1}{3}$ and $m\overline{BD} = -3$. 32. $\frac{b}{a}$. 35. -1.

Chapter 17, Lesson 4 (pages 621–622)
1. $\overline{AD}$ and $\overline{BC}$. 6. They are perpendicular. 8. A rectangle. 9. -4. 11. $\ell_1 \perp \ell_3$. 13. $\frac{3}{4}$. 16. It is undefined. 22. No. 26. Yes. 31. 7. 32. -27.

Chapter 17, Lesson 5 (pages 625–626)
4. (2, 6). 6. (-2, 3.5). 8. (6, -10). 12. (3, 1). 14. They bisect each other. 18. (1, 8). 20. -2. 24. (1, -2). 25. $2\sqrt{5}$. 27. (-3, 0). 30. (4, 3). 31. $\frac{3}{2}$. 32. $-\frac{2}{3}$.

Chapter 17, Lesson 6 (pages 630–632)
1. b. 2. $\frac{b}{c}$. 4. $c - a$. 6. $(0, a)$. 9. $m\overline{AB} = 0$ and $m\overline{DC} = 0$. 11. AB = a and DC = a. 18. $\frac{1}{2}\sqrt{a^2 + b^2}$. 20. $\sqrt{(a + b)^2 + c^2}$. 23. Subtraction. 26. Point D is on the y-axis.

Chapter 17, Review (pages 633–634)
1. B. 4. $\overline{AB}$. 7. 20. 9. $-\frac{4}{3}$. 14. BC = 5 and DA = 5. 18. $\frac{3}{4}$. 24. It is a square. 25. $8\sqrt{5}$. 26. 20. 29. $\frac{1}{2}\sqrt{b^2 + c^2}$. 31. $\left(\frac{d}{2}, \frac{e}{2}\right)$.

Index